P9-DFN-049

# ANNUAL REVIEW OF ASTRONOMY AND ASTROPHYSICS

# ANNUAL REVIEW OF ASTRONOMY AND ASTROPHYSICS

VOLUME 35, 1997

GEOFFREY BURBIDGE, *Editor*
University of California, San Diego

ALLAN SANDAGE, *Associate Editor*
Observatories of the Carnegie Institution of Washington

FRANK H. SHU, *Associate Editor*
University of California, Berkeley

http://annurev.org science@annurev.org 415-493-4400

ANNUAL REVIEWS INC. 4139 EL CAMINO WAY P.O. BOX 10139 PALO ALTO, CALIFORNIA 94303-0139

ANNUAL REVIEWS INC.
Palo Alto, California, USA



*International Standard Serial Number: 0066-4146*
*International Standard Book Number: 0-8243-0935-9*
*Library of Congress Catalog Card Number: 63-8846*



♾ The paper used in this publication meets the minimum requirements of American National Standard for Information Sciences—Permanence of Paper for Printed Library Materials. ANSI Z39.48-1992.



TYPESET BY TECHBOOKS, FAIRFAX, VA
PRINTED AND BOUND IN THE UNITED STATES OF AMERICA

# PREFACE

This volume was planned at a meeting held on May 13, 1995, in Victoria, British Columbia. Those who attended the meeting included Geoffrey Burbidge (Editor), Allan Sandage (Associate Editor), Anne Cowley, John Leibacher, Anneila Sargent, Tom Soifer, and Scott Tremaine (Editorial Committee Members); guests David Hartwick, Paul Hodge, Frank Shu, and George Wallerstein; and Production Editor David Couzens.

In the preface to Volume 34, I pointed out that 25 articles were scheduled for this volume. In fact, 17 are contained here. For the next volume (Volume 36, 1998), 35 articles are scheduled.

In the latter part of 1996, Frank Shu became Associate Editor, succeeding David Layzer, who had been Associate Editor since the first volume appeared in 1963. I would like to thank David for working for so long and so effectively to help maintain the high standards that are a hallmark of this series.

Late in 1996, Naomi Lubick took over as Production Editor. She has done an excellent job on this volume.

GEOFFREY BURBIDGE
EDITOR
March 1997

*Annual Review of Astronomy and Astrophysics*
*Volume 35 (1997)*

# CONTENTS



*(continued)*

INDEXES

## SOME RELATED ARTICLES IN OTHER *ANNUAL REVIEWS*

From the *Annual Review of Earth and Planetary Sciences*

Volume 24, (1996)

*Probing Planetary Atmospheres with Stellar Occulations*, J. L. Elliot and C. B. Olkin

Volume 25, (1997)

*Beta-Pictoris: An Early Solar System?*, P. Artymowicz

*Determination of the Composition and State of Icy Surfaces in the Outer Solar System*, R. H. Brown and D. P. Cruikshank

ANNUAL REVIEWS INC. is a nonprofit scientific publisher established to promote the advancement of the sciences. Beginning in 1932 with the *Annual Review of Biochemistry*, the Company has pursued as its principal function the publication of high-quality, reasonably priced *Annual Review* volumes. The volumes are organized by Editors and Editorial Committees who invite qualified authors to contribute critical articles reviewing significant developments within each major discipline. The Editor-in-Chief invites those interested in serving as future Editorial Committee members to communicate directly with him. Annual Reviews Inc. is administered by a Board of Directors, whose members serve without compensation.



ANNUAL REVIEWS OF

Anthropology
Astronomy and Astrophysics
Biochemistry
Biophysics and Biomolecular Structure
Cell and Developmental Biology
Computer Science
Earth and Planetary Sciences
Ecology and Systematics
Energy and the Environment
Entomology
Fluid Mechanics
Genetics
Immunology
Materials Science
Medicine
Microbiology
Neuroscience
Nuclear and Particle Science
Nutrition
Pharmacology and Toxicology
Physical Chemistry
Physiology
Phytopathology
Plant Physiology and Plant Molecular Biology
Psychology
Public Health
Sociology

SPECIAL PUBLICATIONS

Excitement and Fascination of Science, Vols. 1, 2, 3, and 4
Intelligence and Affectivity, by Jean Piaget

For the convenience of readers, a detachable order form/envelope is bound into the back of this volume.

Charles H. Townes

*Annu. Rev. Astron. Astrophys. 1997. 35:xiii–xliv*


# A PHYSICIST COURTS ASTRONOMY

*Charles H. Townes*
Department of Physics, University of California, Berkeley, California 94720

## *Introduction*

My approach to science may be compared with that of an exploring naturalist. I am curious, I like to explore new things, and I like to push limits. This attitude developed rather early, when I was a boy living on a small farm on the outskirts of the city of Greenville, South Carolina. Although my father was a lawyer, he owned this and a couple of larger farms more distant from the city where my brother and I could also explore the streams and the woods. One of these farms was where my father himself had lived as a boy, on a large farm owned by his own lawyer father, and it had interesting woods, streams, and swamps. I believe my parents made special efforts to see that my older brother and I, with whom I shared many activities, had opportunities to tinker with equipment, to construct things, to engage in various hobbies, and to populate our house and yard with a variety of insects, reptiles, mammals, and other creatures that might seem a nuisance in some households. I also looked at the sky and probably immediately recognized more constellations then than I do now. And I liked to figure things out.

Astronomy has always had a strong appeal for me. However, it was not taught in the mid-1930s at Furman University where I took my undergraduate degree. In addition, I found my first physics course, which I took as a sophomore, especially appealing because it offered the possibility of thinking things through rigorously and quantitatively in a way that seemed to allow one to be rather sure of being either right or wrong. As an undergraduate at Furman University, I also vividly remember reading of Jansky's discovery of radio waves coming from outer space. It made a lasting impression—a striking observation that was said not to be understood. Another lasting impression was my first encounter with relativity. My junior year physics class had not proceeded far enough to cover material on special relativity in our text. So I studied that part in the summer. I still can visualize the mossy rock over a stream near my grandmother's mountain vacation cottage where I sat and puzzled through that text, marveling at the profound conclusions that could be drawn from Einstein's simple and beautiful assumptions.



0066-4146/97/0915-xiii$08.00

After an MA at Duke, I headed for the California Institute of Technology (Caltech) in the fall of 1936 and for a PhD in physics. It was my understanding that Caltech was the best institution in the country at that time for physics. So with $500 to keep me going, I took a bus to Pasadena and entered graduate school there. At that time, scholarships and assistantships were not abundant, and since Furman University was not well known in the sciences, there was no chance of my getting financial aid before I arrived. But fortunately, by the middle of the year I was given a teaching assistantship and could continue.

Caltech was of course a renowned place for both astronomy and physics. My teachers included Fritz Zwicky, Richard Tolman, and IS Bowen, the latter famous for identifying the "nebulium" lines. I enjoyed them all. Zwicky was self-centered and difficult, but fascinating and very general in his thinking. Bowen was a more meticulous and conservative scientist, who seemed always modestly right. He was also very kind to me, inviting me for Thanksgiving dinner and on excursions and hikes with Bengt Edlèn, who was visiting from Sweden. Bowen had invited him to Caltech because of his recent apparent identification of coronal lines as due to highly ionized iron. This was striking but initially unbelievable to Bowen because it seemed to require temperatures at the Sun on the order of a million degrees, which obviously could not be consistent with then-current thought.

Although my own thesis research was under WR Smythe on the spins of nuclei, while I was a graduate student I went to Zwicky, telling him that I was rather interested in astronomy and would be happy if he could give me some interesting project to do on the side. He welcomed me and immediately assigned the job of peering through a microscope at photographic plates to compare ones taken at different times and thereby find new supernovae for him. I did that for about two hours and then told Zwicky that I did not want to spend much more time on that routine. This ended my first venture into astronomy, although in my spare time I made trips up to Mount Wilson, slept overnight in the woods there, and watched the telescope operators. I also tried to engage Hubble in conversation from time to time when he would appear at the Athenaeum, Caltech's faculty club where teaching assistants were privileged to stay.

Perhaps my thesis research was already somewhat indicative of my own tastes and approach to science. At that time, nuclear physics was the hot subject. Before that, spectroscopy had been an important field and, although by the late 1930s it was not a very hot subject, some spectroscopy was still being done. Modern solid state physics was just being born. I did not feel like doing standard nuclear physics; certainly it was of interest, but everybody else was doing it and why should I just join a gang? So instead of working with the usual Van de Graaf machines, cyclotrons, or cloud chambers, each of which were big things of that day, I separated isotopes and tried to measure the nuclear spins of some of the rare isotopes of common elements such as carbon, oxygen, and nitrogen, using spectroscopic techniques. As an explorer of nature, I have very commonly shied away from fields that are well recognized and popular. If many other people are going in that direction, why does there need to be another one such as myself? Instead, I tend to look for those areas I feel may be of importance but which are being neglected. They seem to me more inviting for exploration.

Willie Fowler was at that time a somewhat senior postdoc and was right in the middle of the active program at Caltech, which was trying to understand the light nuclei. Oppenheimer spent spring semesters at Caltech, bringing down his very loyal students with him from the University of California at Berkeley (UC Berkeley). This allowed me to listen to Oppenheimer's superb lectures and get acquainted with people like Phil Morrison, Hartland Snyder, Bob Christy, and George Volkoff. We all saw a lot of each other; George Volkoff and I went on a number of excursions together.

Oppenheimer put together many elaborate theories of the structure of light nuclei. He himself later said none of them worked, which was true. However, the theses of Hartland Snyder and of George Volkoff, which he suggested and supervised, represent landmark work on black holes and neutron stars, respectively, and of course Phil Morrison is noted among other things for his later suggestion that we might receive microwave communications from extraterrestrial beings.

## *Bell Laboratories and Microwaves*

My preference has always been for academic work, but academic jobs were very scarce when I finished my PhD in 1939, and I was fortunate to obtain a good position at the Bell Telephone Laboratories. Initially, my boss at Bell Labs, Harvey Fletcher, gave me a great deal of freedom in doing physics research. However, about nine months before World War II began for the United States, I was suddenly assigned to help invent and design radar bombing and navigation systems, with Dean Woolridge as my immediate supervisor. This experience at Bell Telephone Laboratories had an enormous impact on my career. Of immense importance is that I learned electronics, with which few physicists were very familiar before World War II, and I learned microwave technology. It was thus natural that, while in Florida to test our radar bombing and navigation systems, and on days when the airplanes could not fly, my mind turned to Jansky's discovery. I wanted to get going on some basic scientific research when possible. Here was something that I understood had not been explained and that seemed to me quite striking. In my spare time I played around with trying to find what might possibly be producing this radio radiation.

Fairly quickly I decided that electronic collisions in ionized gas would be a reasonable source for Jansky's radio radiation and began to work out how much radiation should be expected. I formulated an answer, calculating the radiation characteristics and integrating over a thermal distribution so that the radiation intensity and frequency dependence could be expressed with a likely distribution of electron velocities. At Bell Labs at that time, any likely publication was first written up more or less as a memo and circulated particularly to the patent department so that any possible bearing it might have on useful patents could be covered before the material was made public. Of course I sent around the memorandum on my calculations to my friends in physics research at Bell Labs to also get their comments. One of these friends, who had done his thesis in X rays, came around to my desk and asked if I realized how similar this was to a calculation made sometime before by the Dutch theorist Kramers, who produced a theory for production of X rays by high speed electrons passing by nuclei (Kramers 1923). I looked up Kramers's paper and found that indeed the principles and even the resulting expressions for radiation were essentially similar. Kramers had, of course, not applied such theory to the particulars of radio radiation nor integrated over a thermal distribution of electrons, so I felt I still had something to say that might be useful. I had also been talking with microwave personnel at Bell Labs, including Jansky, who was then at the Holmdel Laboratory, and his boss Harold Friis. Jansky was very friendly and glad to tell me about his own work. He had by then stopped measuring radio waves coming in from outer space primarily, it seemed to me, because he felt there was not much more that he could do that would be useful. However, he did have some additional measurements that had not been published because he felt that they were not especially significant from a scientific point of view. But they seemed to me significant in light of my own interpretation and, as a result, he was glad to turn them over to me for whatever use in publication I might wish.

I also talked with various astronomers about the possible source of Jansky's radiation and found my way to Don Menzel at Harvard. He pointed out that there had already been theories more or less along the line of my approach and gave me references to them (Reber 1940, Henyey & Keenan 1940). None of these made any connection with Kramers's earlier work. In addition, they were not very complete and some involved basic errors, so I decided to proceed and publish a paper giving the theory with a thermal distribution of electrons and interpreting the radiation detected at the Earth as well as I could (Townes 1946a). A problem was that Jansky's measurements at longer wavelengths indicated a temperature up to about one million degrees in the ionized gas that was producing the radiation. While no one seemed to think the temperature could be that high, I was reminded of Edlèn's discovery of the high temperatures of the corona that did not make sense initially and decided to note that there might in fact be temperatures as high as $10^{6\circ}$C. Of course, I had missed synchrotron radiation as a possible cause of this particular part of the radiation. The synchrotron mechanism was suggested later by Ginzburg and by Sklovskii (cf Ginzburg 1990). It did indeed require electrons of very high energy as the radio waves indicated, but not a gas in equilibrium at such high temperatures.

During the latter part of World War II, those who made decisions about radar systems chose, as a system we should design and build, a radar at a wavelength of 1.25 cm. This was shorter than the previous 3-cm wavelength with which we had worked and hence would obtain higher directivity with the antenna sizes possible in airplanes. But I was familiar with a memorandum written by JH Van Vleck (1942) about possible absorption of this range of frequencies by water vapor, and I studied the situation rather carefully to try to decide whether the radar would in fact work. It seemed to me that it would not. Unfortunately, I found that I was too young to persuade those in power that they were wrong. The answer given me was not based on scientific logic, but simply, well, this decision has been made and we must go ahead. In fact, the 1.25-cm radar was to be used over the Pacific, which was the primary theater of war at that time, and water vapor was dense enough there that the 1.25-cm radar proved to be of no military value. Its parts were junked, which was a great boon to my own immediate postwar research because I was able to use these very cheap and plentiful parts.

I was eager to return to physics after World War II and thought at some length about just what I should do. My two possible choices included microwave spectra of molecules, where I realized that very high spectral resolution was possible at low gas pressures partly because of my study of the water vapor spectrum and the theory of collisional broadening by Van Vleck & Weisskopf (1945), of which I had a preprint. Normal spectroscopists had never seen lines as narrow as theory predicted for the microwaves, and there was considerable skepticism. Jim Fisk, head of the physics department at Bell Labs, particularly asked the theorist Arnold Nordsieck to check out what I was saying to see whether it was really right. He fortunately readily agreed with me. The other research possibility I saw, besides developing high resolution microwave spectroscopy, was that I might go into radio astronomy. Bell Labs seemed an excellent base for such astronomical measurements, and it was a new, still unexplored field that attracted me. I hence made a point of visiting IS Bowen, my old teacher, who by then was Director of Mt. Palomar and Mt. Wilson Observatories, to ask his views as to what were the most important things to try to do if I were to take up measurements of these radio waves coming in from outer space. He was not encouraging, commenting that the long radio waves would give poor directivity and probably no useful information about astronomy. This was a completely logical answer, in a sense, but limited. I realized that Bowen was probably overly conservative, but since I myself did not have many ideas about what was best to measure in radio astronomy and I knew that microwave spectroscopy gave me a clear-cut opening for interesting research, I decided to pursue that field in the postwar years.

### *Peripheral Work in Astronomy—From East Coast to West Coast*

The first high resolution spectroscopy of molecules, that of ammonia (Townes 1946b), immediately revealed many new phenomena, and the field caught on quickly, particularly at those places that possessed discarded 1.25-cm radar equipment such as RCA, General Electric, Westinghouse, and Bell Labs, where I was busy. Its success resulted a little later in an invitation to me to go to Columbia University. At Columbia there was good equipment available and, more importantly, a number of other physicists interested in the general field of radio spectroscopy. With this and my long-standing attraction to the academic world, I accepted. Microwave spectroscopy is perhaps the one field I have felt that I should follow until it became really mature from the point of view of physical principles. Furthermore, I had many good students and colleagues who were interested in the subject. Columbia was a hotbed of outstanding fundamental science at that time, with emphasis on new varieties of spectroscopy and on high energy physics. I have counted up 18 Nobel Laureates in physics who overlapped my eight years at Columbia but did not have such prizes when I first knew them as students, postdocs, or faculty members.

Columbia was indeed a rich place for physics, and I found microwave spectroscopy very interesting and rewarding. But in 1955, after I had earned a sabbatical and had finished a book with Art Schawlow covering microwave spectroscopy (Townes & Schawlow 1955), I felt that the field of microwave spectroscopy was pretty well explored from the point of view of physicists and henceforth would be primarily a subject for chemists. It was time, I thought, for me to look in other directions and decide where to explore next. In 15 months on sabbatical (two summers plus the usual 9 months), I spent considerable time in England, France, and Japan, visiting laboratories and talking to colleagues to become more widely acquainted with other fields. By then, radio astronomy had become a well-established research field, but it was still of interest to me because I thought I saw new things that needed to be done. Before I left on sabbatical, I had played with the possibility of detecting particular atoms in space and molecules of various types. Discussions of these possibilities with one of my students, Alan Barrett, led him to decide he wanted to go into radio astronomy and look for astronomical sources of OH after he finished his degree. Some of my other students and Art Schawlow, who was then a postdoc, had obtained the OH spectrum in the microwave region in our laboratory at Columbia, so we had a reasonably good value for the OH frequency (Dousmanis et al 1955). Alan gave it a good try by taking a postdoctoral position with the radio astronomy group at the Naval Research Laboratory (NRL). His efforts to find cosmic OH there failed (Barrett & Lilley 1957), but fortunately he was a persistent researcher and this paid off later. My students Gerald Ehrenstein and Mirek Stevenson soon made a point of measuring precisely the Λ splitting frequency of the lowest OH rotational line (Ehrenstein et al 1959). With this information, Barrett was to try again with more sensitivity and succeed (Weinreb et al 1963).

During my sabbatical in Europe, I talked with many astronomers as well as physicists and expressed some of my interest in molecules and radio astronomy. As a result, Hendrik van der Hulst invited me to attend an IAU meeting in England to give a talk about the possibility of radio detection of molecules in space (Townes 1957). It was 1955 and I was still in Europe on sabbatical, so I was delighted to attend the meeting and discuss such possibilities. I had picked out in particular OH, water, ammonia, CO, and about a dozen other compounds along with possible atomic fine structure lines for which searches might be conducted. Van der Hulst was pleased, but the subject seemed to die at that point, except for Alan Barrett's continued interest.

Before I left Columbia for a sabbatical, Jim Gordon, Herb Zeiger, and I had made the first masers using beams of ammonia (Gordon et al 1954). During my sabbatical I did some maser experiments and had further ideas about the field, partly generated by my interactions in Paris and Tokyo. And when I returned, I decided that my primary push should be on masers at that time, while also finishing off some special projects in microwave spectroscopy. I wanted to make a maser amplifier specifically to use in radio astronomy and thereby take advantage of the very large potential increase in sensitivity. I was also very interested in masers as frequency standards. This perhaps helped fulfill my instinct for pushing on limits. I have always been interested in extending precision measurements to their limits and thus testing theory as rigorously as is practical.

While in Tokyo, Shimoda, Takahasi, and I had written a paper giving a fairly detailed and complete theory of the noise properties of maser amplifiers and explaining why they could be a couple of orders of magnitude more sensitive than existing electronic amplifiers (Shimoda et al 1957). After returning to the states, I was approached by Walter Higa, from the Jet Propulsion Laboratory, who had worked with me briefly on masers at Columbia. He knew of our work and asked whether I thought parametric amplifiers might in principle achieve the same kind of sensitivity. That had not previously occurred to me, and as we discussed it I woke up to the fact that he was quite correct—parametric amplifiers do offer somewhat comparable sensitivity. They were not in use at that time, but by now they have fulfilled much of the role of sensitive amplifiers for which I had envisaged maser use, including prominent use in radio astronomy. Masers can still achieve the very best sensitivity, although the improvement they offer over parametric amplifiers is, for many applications, unimportant. At the time, parametric amplifiers were not yet developed, and I was rather committed to developing maser amplifiers and trying to do radio astronomy with them. With two excellent students, Lee Alsop and Joe Giordmaine, we made a good ruby amplifier. With the help of Connie Mayer, who was at the NRL, we put it on the NRL's 50-foot antenna dish and made the first astronomical measurements with the maser's excellent sensitivity (Giordmaine et al 1959).

Immediately following our initial work at the NRL, another one of my students, Arno Penzias, began developing a maser amplifier to do what I felt might be a worthwhile enterprise—to look with the utmost sensitivity for the 21-cm line of hydrogen in intergalactic space. There had been some speculation about the possible abundance of hydrogen in interstellar space and its velocity distribution. Detection of hydrogen in the intergalactic medium would clearly be difficult, but we felt it was worth a try. Penzias developed an excellent maser and looked carefully, but there was no signal. His subsequent experience at Bell Labs, and his detection of the "Big Bang" radiation with Bob Wilson (Penzias & Wilson 1965), was considerably more rewarding than this particular thesis. Although a maser amplifier was not crucial to this discovery, I was pleased that they had used one.

A major hiatus appeared in my career at that point. Sputnik had gone up in 1957, and the missile crisis was upon us. There were very few well-known scientists in Washington, and the administration was both eager to take appropriate steps for national defense and to obtain expertise in the latest technologies. I was approached to go to Washington as Vice President and Director of Research for the Institute of Defense Analyses (IDA), a newly formed think tank with trustees who were largely university presidents, including the president of my own Columbia University. At that time, IDA was a key institution in advising the government on technology and technological policy. I agreed to go for two years, and during that time I could only do research on a part-time basis. Art Schawlow and I had written our paper (Schawlow & Townes 1958) on how the maser idea could be extended into wavelengths as short as the visible, which stirred many groups to try to make such a system, later to be called a laser, actually work. This was also one of the rather hot topics for discussion in new technology.

Although I continued to publish a few papers with my students and colleagues from Columbia, the only scientific paper I published as a result of work in Washington was in fact something that might be called an astronomical use of the laser. I was struck by the suggestion of Cocconi & Morrison (1959) that interstellar communication could be carried out in the microwave region and that perhaps frequencies near the then-famous hydrogen line would be the natural place for such communications. My reaction was that with lasers at shorter wavelengths it was possible to obtain much higher directivity, so that even though the quanta were larger, the energy costs could be lower by using lasers rather than radio waves. RN Schwartz, one of the younger members of the IDA, was interested and was coauthor with me of a paper showing somewhat quantitatively why a very advanced civilization, or one that had taken a slightly different technological path than did our own, might in fact prefer to communicate with lasers in the infrared region (Schwartz & Townes 1961). So few people in the field paid much attention that about 20 years later I wrote another, still more detailed discussion of the pros and cons of infrared versus microwave Search for Extra-Terrestrial Intelligence (SETI) (Townes 1983).

I had agreed to go to Washington for only two years, but then I was caught up in an administrative job as Provost at Massachussetts Institute of Technology (MIT) for the next five years. Although I was able to do some research in the field of nonlinear optics using the new and exciting laser possibilities with a couple of outstanding students, Ray Chiao and Elsa Garmire, and with further help from more senior personnel, including Boris Stoicheff, there was no time to branch into other new fields. I did have the pleasure of presiding at a news conference while at MIT when Alan Barrett announced discovery of the OH molecule in interstellar space in 1963. He had stuck to his search for it and, with the help of several associates, including the outstanding engineer Sandy Weinreb, had succeeded (Weinreb et al 1963).

Although I found administrative work moderately interesting and could tolerate it, research seemed clearly more fun to me. Hence, after a new president was chosen for MIT, I decided I had taken my turn at administration and would resign to go seriously into astronomy. For preparation, I took a year off, which was spent half at Harvard trying to soak up astronomy in the Department of Astronomy there and about half at MIT finishing up work on nonlinear optics. Leo Goldberg was then Chair of the Astronomy Department at Harvard. I had known him for a long time and saw a good deal of him during that year. I attended seminars and some classes, including rather regular attendance of a course on planets given by the then-young assistant professor Carl Sagan.

There were many attractive places to which I might make a move in order to do astronomy, but UC Berkeley had good radio and optical observatories, and for my taste, UC Berkeley seemed the best. The UC Berkeley campus was just then in the middle of student disturbances, but I felt that students of physics or astrophysics had to work seriously enough that political affairs would not capture a disastrous fraction of their attention, and this proved to be the case.

## *The Search for Microwave Spectra of Interstellar Molecules*

My plan upon arrival in Berkeley in the fall of 1967 was to continue some nonlinear optics, an active field in which I was already very much engaged, and at the same time move into astrophysics. Raymond Chiao, my student at MIT in nonlinear optics, had been appointed assistant professor at UC Berkeley and was an important part of the nonlinear optics program. David Rank came as a young postdoc and spectroscopist from Michigan. Some students moved out to UC Berkeley from MIT at the same time as I, and they, with other UC Berkeley students, quickly made a busy research group.

The fields of astronomy that appealed to me immediately at that time were, first, to take a serious look for additional molecules in space. No one, so far as I could discover, had followed up on my earlier suggestions or Barrett's discovery of OH and tried to look for other molecules. OH was of course very similar in type to the three molecules previously found in interstellar space by optical spectroscopy, CN, CH, and $CH^+$—a diatomic free radical. The presence of other types of molecules seemed discounted by most astronomers, but I felt they were worth a try. My second goal was to try to develop infrared astronomy, particularly in regions beyond the near infrared. Pierre Connes had done beautiful work on planets in the near infrared with Fourier transform interferometry. The Caltech group—Bob Leighton and Gerry Neugebauer—and Frank Low at Arizona had been doing interesting continuum work in the infrared. Otherwise, the field seemed to me quite neglected in view of the potent spectroscopy, the increasing sensitivity obtainable in the infrared, and the wealth of possible astronomical results. UC Berkeley already had very active work in the microwave range. Professors Harold Weaver and Jack Welch were very helpful in getting me started in that area. I also attended seminars and some classes in the astronomy department. George Field was another very stimulating colleague from whom I learned a great deal during this early period.

The basic reason that astronomers had not bothered to look for molecules in interstellar space even after Barrett discovered OH seemed to be the firm general conviction that the density of gas in interstellar clouds was quite low. This had been concluded from measurements of the 21-cm hydrogen line, which indicated densities in the range of a few atoms per cubic centimeter. Not only did the material seem to be atomic (and ultraviolet radiation was likely to disassociate any occasional molecules that might be in interstellar space), the density was so low that even if molecules existed, excited energy levels of molecules would not be much populated by collisional excitation. However, I had encountered a few papers indicating that in dust clouds no atomic hydrogen was found at all, and these papers even speculated that the material might possibly be molecular, which would result in no detectable atomic hydrogen. There was really no conclusive evidence that clouds denser than those detected by atomic hydrogen did not exist: The evidence was simply that low density clouds did exist. And in addition, the dust clouds where no gas seemed to be detected was where it seemed to me logical that there could be substantial molecular gas. But many people predicted a fruitless search, and George Field also advised strongly against looking for ammonia because he saw no known mechanism to form ammonia in interstellar space (Field & Chaisson 1985).

In the late fall of 1967, a graduate student, Albert Cheung, agreed to do his PhD thesis on a search for ammonia. I also enlisted the help of David Rank, who was working with me at the time on nonlinear optics, and that of Professor Jack Welch, who was then directing the University of California radio telescope in Northern California. Jack Welch also highly recommended engineer Doug Thornton to assist us in the engineering aspects of putting a system on the University's radio antenna. At that time, very little astronomy was done at wavelengths as short as a centimeter, so we needed to construct an amplifier for 1.25-cm wavelengths near the important ammonia inversion spectrum. I thought we should first look for inversion in the lowest possible rotational state of ammonia, the $J = 1, K = 1$ level, because somewhat higher densities were required to excite higher rotational levels and low densities were thought to be a problem. The amplifier was constructed and hitched up to other equipment on the 20-foot antenna at Hat Creek in accordance with instructions from Thornton and Welch. We initiated the search for the ammonia inversion line at 23.6 GHz in the direction of Sagittarius B2, which was one of the several sources Harold Weaver thought might be most promising. There indeed was the line! We soon tuned our system to the inversion line of ammonia's $J = 2, K = 2$ rotational state, and there it was also (Cheung et al 1968).

Although finding a stabile molecule in interstellar space was striking, I believe the most important immediate change in astronomical understanding came as the result of the cloud densities deduced from such findings. The disequilibrium between excitation of the two ammonia states detected and the isotropic microwave radiation immediately showed that the density had to be $>10^3$ cm$^{-3}$ for excitation to occur via collisions. Also, hydrogen clearly had to be in a molecular state. A number of colleagues believed that such densities could not really be correct and that some other mechanism must be at work. However, no other ones could be seriously proposed. Further molecular work of course showed that such densities, and densities even a few orders of magnitude higher, are not uncommon in interstellar clouds and that many complex molecules are commonplace in them.

The amplifier we had built for finding ammonia lines did not give us ready access to many other lines of molecules that I thought might be present. Other amplifiers at different frequencies would have to be built, for example for CO, which fell at millimeter wavelengths. However, the water line at 1.35-cm wavelength, well known from laboratory spectroscopy, was nearby and could be looked for. It seemed improbable that this line would appear, as it represented a rather high rotational state and would need quite high density to be excited adequately. On the other hand, by then I was ready for any additional new surprises and felt that it was so easy to look for the water line that we should do it. Jack Welch also wanted to try for the water line. So, soon after the ammonia lines were seen, we looked. And the water line also showed up (Cheung et al 1969)!

We later learned that Louis Snyder and David Buhl, both of whom were young postdocs at the National Radio Astronomy Observatory (NRAO), had applied for an opportunity to look for water vapor there but had been turned down. Of course, shortly after our discovery, they and others were given the opportunity to search actively for molecular lines, and a large number of other lines and molecules began to turn up.

We were initially puzzled why in fact the water line would be present and with such intensity. But our success at UC Berkeley meant that I could reasonably ask for time on still larger telescopes, and having already used the 50-foot telescope at the NRL for our maser amplifier experiments, I naturally contacted Connie Mayer there to ask if we might use that telescope in order to get somewhat better angular resolution. I felt the water line might well represent maser action and wanted to understand its angular distribution and effective temperature. In some sources, the intensity was indeed spectacular. During the Christmas season of 1968, I was giving a party for the rest of my research group while Al Cheung was at Hat Creek observing. During the party he called up, exclaiming, "It must be raining in Orion." The water lines there were indeed intense.

The OH line had been proven already to be a maser by work at MIT using interferometry for high angular resolution (Moran et al 1967). Al Cheung, Dave Rank, and I took our amplifier to the NRL, which had an antenna larger than the one at Hat Creek, and with Connie Mayer and Steve Knowles, we quickly found that the water line did indeed show effective radiation temperatures at least as high as 50,000 K and that the intensity of some sources varied with time. It was clearly maser-type emission (Knowles et al 1969).

After discovery of ammonia and water lines, further searches for molecular lines began to spring up in many centers for radio astronomy. Snyder and Buhl had been quickly given opportunities to use the NRAO facilities. Their earliest efforts were made in conjunction with Ben Zuckerman and Patrick Palmer, young scientists at Harvard, and were rewarded with discovery of formaldehyde (Snyder et al 1969). Along with our own observations at UC Berkeley, we were also busy developing maser amplifiers for further molecular work. The number of molecules detected and the variety of masers discovered grew rapidly, with many of the further discoveries by Snyder and Buhl at NRAO, Zuckerman and Palmer at Harvard, Barrett at MIT, Penzias and Wilson at Bell Labs, Thaddeus at Goddard and later Harvard, and their associates. Wilson, Jefferts, and Penzias built a millimeter wave receiver and made the first detection of CO (Wilson et al 1970), a particularly important molecular tracer for interstellar clouds because of its abundance and stability. Some of these observers were newcomers to the field, but others, like Barrett, Penzias, and Thaddeus, had been with me at Columbia and were already very familiar with microwave spectroscopy. In addition to Al Cheung, molecular astronomy was the center of thesis work for a number of excellent UC Berkeley students who continue work in radio astronomy, including Neal Evans, Paul Goldsmith, Andy Harris, Demetrios Matsakis, and Dick Plambeck.

At present, 111 species of molecules have been detected in interstellar space, 7 of them with as many as 9 atoms. A number of the molecules had not previously been known on earth. Patrick Thaddeus has been involved in discovering 20 of the present 111. So far, there are also about 110 different masing lines, according to a list recently assembled by Carl Menten, and to which a large number of observers have contributed. Masers have been found under a wide variety of conditions, including the recent comet Shoemaker-Levy impact on Jupiter, which set off a water maser (Cosmovici et al 1996). In retrospect, it is obvious that conditions in interstellar space and surrounding energetic astronomical sources are just what is needed for masers, and we should have expected them at a much earlier time. I sometimes point out that, if the scientific community had been more active in radio astronomy, masers might have been discovered in astronomical objects much earlier and we would not have needed to invent them on Earth.

Observation of microwave molecular lines quickly made it clear that there are many clouds, that they are commonly several orders of magnitude more dense than had been previously expected, and that many complex molecules are formed in interstellar space and in circumstellar gas. It was initially puzzling how there could be so many large clouds of density $10^4$ molecules/cm$^3$ or higher, since the gravitational collapse time for such clouds is relatively short. This made many astronomers continue to doubt the densities that were determined from molecular spectra. For some time there was also debate about whether the clouds were quite clumpy or of rather uniform density. However, Doppler shifts evident in molecular measurements soon provided good evidence for considerable turbulence, shock waves, and presumably magnetic fields. A beautiful example of recent measurements is the map of the Orion molecular cloud in the $NH_3$ [1,1] line made by Wiseman & Ho (1996) with a spatial resolution of 0.02 pc and a velocity resolution of 0.3 km/s. It shows very complex structure of gas streamers, gas clumps, and complex kinematics. The turbulence, shock waves, and magnetic fields are often strong enough to clearly dominate thermal velocities or even gravitational fields. These plus stellar formation now make a complex but believable explanation for the number of large dense clouds in galaxies and notable structure on almost all scales. Molecular lines have provided important measurements for many parameters in these clouds—velocities, densities, temperatures, shocks, and isotopic abundances—as well as information on molecular formation, a field still poorly understood partly because past cloud histories can be only surmised, not measured.

I believe some of my friends have been puzzled over why I have left the field of microwave observations of molecular species in interstellar clouds, since it continues to yield many valuable results. This comes from my instincts about exploration. If there are many skillful people and many powerful instruments working in the field, is it more important to join them or to move to more neglected fields? Logically or not, I usually choose the latter.

### *The Infrared*

I was surprised recently to discover in my laboratory notebook an entry of April, 1968, labeled a "List of Present Problems." There I listed research projects envisioned at that time, about seven months after I came to UC Berkeley. Many of them had the name of a student or one of my postdoctoral associates associated with the project who was either working on it or I thought was a good prospect. In addition to several "problems" in nonlinear optics, the list was the following:

- $NH_3$
- OH sources
  - [infrared] IR radiation
  - mechanisms
  - simulation
- Up conversion
- IR spectra
- Airplane experiments
- IR of QSOs:
  - size
  - spectrum
  - polarization
  - time variation
- IR interferometry
- IR of Galactic Center
- Millimeter-wave maser

IR "sky noise" and compensation (not a thesis)

Michelson-Morley experiment

Photon sputtering

Some of these—work on QSOs, photon sputtering of dust particles, and the Michelson-Morley experiment, which was supposed to make a major advance in precision—were dreams that never materialized. So-called IR sky noise, which turned out to be mostly due to variation in radiation from telescope structures, continues to bother IR astronomers but, as my list already indicated, was hardly a fit thesis problem. But the other problems listed and their ramifications have led me into what has become almost three decades of fascinating investigations.

Work on $NH_3$ was already well under way at that time and was to lead to the molecular radio astronomy that kept me busy for some years thereafter, as is discussed briefly above. We were also puzzling over the excitation of OH masers and considering IR spectroscopy on them to clarify conditions in their sources. We made a little progress later, on collisional inversion of CH and possibly OH masers (Bertojo et al 1976), extending what seemed to me a convincing explanation (Townes & Cheung 1969) of the striking observation that $H_2CO$ actually had an excitation temperature lower than the all-pervasive 3° background radiation (Palmer et al 1969). (The $H_2CO$ situation could properly be called a dasar, for darkness amplification by stimulated absorption of radiation.) $H_2O$ masers seemed more puzzling than most of the others largely because of their intensities, and they clearly required extreme conditions such as shock waves.

The up-conversion idea mentioned in my 1968 notes was to use nonlinear optics to up-convert weak IR radiation to visible light by mixing it with a strong optical laser beam, thus making a corresponding optical image of an astronomical object that could be easily recorded using already available high resolution optical imaging such as photography. This was a project successfully carried out in the theses of two students, Howard Smith and Robert Boyd, and we did do a little astronomy-related work with it (Boyd & Townes 1977, 1978). However, the rapid advance in technology of sensitive multipixel IR detectors, which have now become remarkably effective, meant that up conversion was never particularly productive or useful for astronomy.

Shortly after I made this notebook list of problems, IR spatial interferometry was initiated at UC Berkeley. It was to occupy me off and on for some time and is presently my primary experimental effort. Such interferometry is discussed at more length later.

As I began intensive research in astronomy at UC Berkeley in 1967–1968, no IR spectral lines of wavelengths beyond the near IR had been detected from outside the Solar System. This seemed to me a wide-open and enticing field. Pierre Connes' beautiful work in the near IR on planets and some of the brighter stars (cf Connes 1970) had convinced many who were interested in IR spectroscopy that Fourier-transform systems, which Connes had used, provided the appropriate technique for most high spectral resolution work. However, such systems have an advantage only if the primary noise comes from the detector rather than from an external source. Another technique attracting some attention was the use of low resolution filter wheels, but they provided no promise of high spectral resolution. I myself was convinced that Fabry-Perot etalons were most appropriate for relatively inexpensive high spectral resolution and good sensitivity, and they had few proponents at that time. Hence we began working in this direction to measure astronomical spectra in the atmospheric windows near 10- and 20-micron wavelengths, as well as in the far IR (50–200 microns).

Frank Low at the University of Arizona had been and was at that time very productive in continuum IR observations of astronomical objects, and I learned much from him. Ed Ney at the University of Minnesota was also very active in IR continuum observations. Low was manufacturing bolometers and the appropriate dewars for detection as a by-product of his pioneering IR work, so preliminary detecting equipment was available. And he was even pushing into measurements at far IR wavelengths. We hoped to extend his far IR work to detection of spectral lines with Fabry-Perot spectrometers, and soon began trying out relatively simple systems, first with a 12-inch telescope in a small jet aircraft flying out of NASA's base at (Moffett Field) Ames just south of Berkeley, and later with a 36-inch telescope in NASA's C141 plane flying from the same base. The latter plane was named the Kuiper Astronomical Observatory (KAO) for Gerard Kuiper, another pioneer in the field of observations from aircraft, and I was to have a long-term relationship with the KAO.

Fabry-Perot work in the 10-micron region was initiated as a thesis for my first UC Berkeley student, Jim Holtz, and we set our sights on three fine structure lines of the ground states of NeII, AIII, and SIV. These were at 12.8, 9.0, and 10.5 microns, respectively, all transmitted adequately through the atmosphere. They were also from common atoms and required ionizing energies of 21.6, 27.6, and 34.8 eV, respectively. They thus provided a range of nicely spaced energies as possible tests of ionization conditions. Holtz was later joined by postdoc Dave Rank and student Tom Geballe. We worked intensively towards getting a sensitive high resolution spectrometer, designing a Fabry-Perot system with NaCl etalons and a gallium-doped germanium bolometer detector. The system was first tried out at UC Berkeley's Leuschner Observatory nearby, then at the Lick Observatory on Mount Hamilton. By 1970 we had detected our first line: SIV emission from NGC7027 (Rank et al 1970). This was by then the second mid-IR line detected from a source outside the Solar System, not the first. Fred Gillett and Wayne Stein, former students of Frank Low and

Ed Ney, respectively, had already beaten us in finding the first mid-IR line from outside the solar system (Gillett & Stein 1967). They reported detection of the NeII line in IC418, using a filter wheel. The resolution and frequency precision of our Fabry-Perot system was, of course, much higher than that of the filter wheel, which had a resolution of about 15 $cm^{-1}$ compared to our 0.3 $cm^{-1}$. But we were still not resolving the SIV line width, and so we pushed on for better performance. Subsequent student theses were to extend these results considerably to wavelengths ranging from a few to 30 microns. Good photoconducting detectors became available, and we used double Fabry-Perots for higher resolution. Many more sources were examined, one of the most important of which is the galactic center, which is separately discussed.

The last word in spectral resolution came from a different development—heterodyne spectroscopy in the mid-IR region. This was initially developed for stellar interferometry, which is described later. Two students, Mike Johnson and Al Betz, began heterodyne detection in the 10-$\mu$m region from scratch, building a $CO_2$ laser as a local oscillator and making detectors from a chip of germanium. This provided a bandwidth, for both sidebands, of 1.5 GHz. Arbitrarily high resolution could be obtained with standard radio-type filtering, and in principle arbitrarily high sensitivity as the bandwidth was decreased. However, for normal astronomy there is no point in decreasing the bandwidth much below the Doppler width of lines because the signal then goes down faster than the noise. The net result is that, at the minimum pass band needed for astronomy, heterodyne detection at 10-$\mu$m wavelength is close to the sensitivity of present direct photodetectors, but it is much less sensitive for broader bandwidths. The ease of obtaining very high resolution, however, paid off for Johnson and Betz. With Ed Sutton and Bob McLaren, they were able to measure $CO_2$ lines on Mars and Venus in some detail with resolving powers greater than $10^8$ (Betz et al 1976). This provided a precision in wind velocities measured on Venus of about 2 m/s! By also measuring the atmosphere's kinetic gas temperatures from $CO_2$ line shapes, they showed that some line intensities were an order of magnitude above what could be expected at thermal equilibrium (Johnson et al 1976). Laser action due to excitation by solar radiation was suggested by these results, the first natural laser. This conclusion was later clearly shown by Mike Mumma and his group at Goddard Space Flight Center (Mumma et al 1981); they were also active in heterodyne detection. Heterodyne techniques have produced excellent high spectral resolution results, and Betz has continued such work in both the far and mid-IR regions.

## *Exploring the Galactic Center*

The initial thesis work of Jim Holtz on mid-IR Fabry-Perot spectroscopy was followed by the theses of Tom Geballe, Eric Wollman, Larry Greenberg, John Lacy, Gene Serabyn, and Sara Beck. We studied a variety of objects, using a range of IR wavelengths from 5- to 20-$\mu$m windows; the 20-$\mu$m region is still relatively little used by astronomers. Perhaps the most interesting problem—which was begun by Wollman (Wollman et al 1976, 1977) at the Cerro Tololo Observatory in Chile and continued with collaboration and extensive follow-up at Carnegie's Las Campanas Observatory by Geballe, Lacy, Serabyn, and by postdocs Dave Rank and Fred Baas—was an examination of ion densities and velocities in the galactic center. Although IR of the galactic center was on my list of priorities in 1968, it was not until 1975 that we were ready with well-tested equipment to make visits to Chile and obtain good quality measurements on spectral lines near the galactic center. The fine structure transition of the NeII ground state and of other common ions were to be our entries into the study of this region. In the meantime, Aitken, Jones, and Penman had beaten us to the first NeII observation in the galactic center (Aitken et al 1974). Their low spectral resolution instrument could obtain no significant velocity information, but it did tell us that the intensity of NeII was plenty high for our instrument to easily do high spectral and angular resolution work. And with the Fabry-Perot, Wollman and coworkers (Wollman et al 1976) immediately found high NeII velocities that indicated a mass of about $4 \times 10^6$ $M_\odot$ within a radius of about 0.8 pc around the center, a discovery that was to generate much further work and some controversy.

John Lacy's thesis provided detailed measurements of NeII velocities (and a search for AIII and SIV) at many individual spots around the galactic center, gave details of the density and temperature of ionized gas there, and showed a wide range of velocities that increase in magnitude as the measurements approach the center (Lacy et al 1979, 1980, 1982, Townes et al 1983). This provided a reasonably strong indication of a black hole of a few million solar masses. However, the situation was soon further clarified by microwave interferometric measurements. Ekers, Van Gorkom, Schwarz, & Goss mapped the ionized cloud structure around the galactic center and interpreted it as spiral infall of materials (Ekers et al 1983). Lo & Claussen (1983) mapped out the ionized clouds and deduced a more or less elliptical structure that rotated about the center, as well as the presence of some streamers. Mapping point by point with the 10-$\mu$m spectrometer to obtain these broad scale patterns had not been practical for us, and the detailed radio maps gave important new insight. The next step was clearly to make systematic spectral measurements around the apparent ellipse and along the streamer that extended from the edge of the ellipse in towards the center, which became known as the northern arm. Eugene Serabyn's thesis did an excellent job on these structures. He was able to show that the ellipse represented rather well a ring of ionized gas moving in an approximately circular path about the center and the northern arm an infalling streamer moving along an elliptical orbit into a region very close to the center.

The dynamics of both were well explained by a point mass in the location of Sagittarius A* of magnitude of about $3 \times 10^6\ M_\odot$ (Serabyn & Lacy 1985). The relatively low gas densities in much of the central region and some variations in velocity along the circumference of circulating material also indicated some quite energetic explosive event within the last 100,000 years.

Far IR measurements in NASA's KAO are discussed later, but it should be noted here that they have also contributed importantly to understanding of the galactic center. These far IR measurements had limited angular resolution ($\sim$30 arcsec) due to diffraction of the KAO's 36-inch telescope, which has not allowed detailed information very close to the center. But they have provided valuable understanding of the surrounding regions, such as transitions in the state of gas from ionic to atomic and then molecular with increasing distance from the center. They have also shed light on the densities, temperatures, excitation mechanisms, and shock waves within a radius of a few hundred parsecs from the center (Storey et al 1981, Watson et al 1984, Genzel et al 1989, Harris et al 1994, Timmerman et al 1996). With an instrument newly constructed at Cornell, Stacey and collaborators (Latvakoski et al 1996) have recently obtained a particularly clear map of the ellipse around the galactic center, and with good angular resolution, using the KAO at 31.5- and 37.7-$\mu$m wavelength.

I was rather convinced by our various measurements that there was a highly concentrated mass at the galactic center of a few million solar masses, too concentrated to be explained by any current theory excepting a black hole. However, completely firm identification is difficult, and a number of scientists have expressed strong doubts. Stellar spectroscopists who followed our work with measurement of velocities of CO in the atmospheres of late-type stars near the galactic center initially believed that they found a substantial difference between what was indicated by stellar velocities and what we had found from the gas velocities; namely, they believed the region was much more like a stellar cluster. In addition there was a problem, which still exists today, of why the total radiation from the galactic center is as small as it is if in fact there is a large black hole in the midst of the many gas and ionized clouds that were found there. These clouds and their velocities were obviously capable of feeding in enough mass to provide a great deal more radiant energy.

In time, the stellar spectroscopic groups of Sellgren et al and Haller et al did intensive work on CO velocities associated with stars in the galactic center, and both groups found that these velocities increased towards the galactic center as if accelerated by a central mass close to the value indicated by the gas velocities (Sellgren et al 1990, Haller et al 1996). Doubts about the significance of the stellar measurements and particularly of the gas velocities still remained. Some suggested that the ionized gas velocities might be primarily generated by magnetic fields or winds from the central region, including winds from very hot stars that had by then been found close to the center. There seemed to be no good way of making a quantitative explanation on this basis, however. Also, limits set on magnitudes of the magnetic field from a few measurements available indicated that they were much too small to control the kinetic energy for the densities of gas that were there. It would be rather difficult for these fields and certainly for winds from the center to provide the particular geometry and high velocities of circulation around the center that were observed. Evidence for a large central mass from either the gas measurements or the stellar measurements can be characterized by a signal-to-noise ratio of perhaps 5 or 6. Genzel and his associates have recently made still another examination of stellar velocities, somewhat more extensive than previous ones. These provide a signal-to-noise ratio of about 6–8 for evidence of a velocity gradient characteristic of a central mass of a few million solar masses (Genzel et al 1996). This signal-to-noise ratio may not be completely overwhelming, but the combination of measurements would ordinarily be quite convincing. What has maintained doubts about reality of the black hole is primarily the paucity of radiant energy, because for some time theoretical discussions inevitably predicted much larger amounts of radiation than are observed.

There is good evidence for large black holes in a number of external galaxies, and in fact a good fraction of these have relatively little radiant energy, some even less than that from our own galactic center. However, in those cases we do not know enough about the material immediately surrounding the central black hole and hence cannot be as certain that material should frequently be falling in as we can for the center of our own galaxy. Fortunately, the center of our galaxy provides an excellent laboratory, close enough for us to measure many of its parameters in some detail and thus to test the black hole phenomena involved and theoretical ideas about them.

Quite recently, Eckhart & Genzel (1996) of the Max Planck Institute in Garching have mapped the IR stars in the galactic center very precisely, using multiple very short exposures of a near IR camera on a large European telescope in Chile. Over a two-year period, they have been able to detect motions of a number of stars and to show that these motions are more rapid close to the galactic center than they are further away. With these measured velocities added to others previously measured spectroscopically, they obtain a curve of velocity variation with radius from the galactic center down to distances as small as 1/10 parsec. The results agree very well with the previous expectation of a central mass of a few million solar masses, and such agreement makes the experimental case for a central black hole still much stronger (Eckhart & Genzel 1996).

In spite of present difficulty with understanding the low radiation from the presumed black hole in the galactic center, I believe the evidence for a highly concentrated mass is overwhelming. Furthermore, gravitational attraction and kinetic energy relations are enormously simpler than are the dynamics of infall of a plasma into a black hole and subsequent radiation, and hence we might trust our understanding of them more. Infall is faced with considerable uncertainty in the dynamic conditions and instabilities. The presumption is that further theoretical work and developments will help us understand why the radiation is in fact so low. There is perhaps even some possibility that the concentrated mass is in some form other than a black hole, but there is no theoretical model for such a case.

A suggested explanation for the low radiation is that material does fall into the black hole from time to time, generating much energy and perhaps producing the recent energetic events indicated by the gas distribution and velocities. Such events could blow away material from immediately around the black hole and make large energy emissions episodic, with a period of low infall at present. However, measurements with the Russian spacecraft GRANAT show a lack of scattered X rays from distances of many hundreds of light-years from the galactic center, indicating that there has been no very violent eruption of energy from the center for many hundreds, or perhaps a few thousands, of years (Sunyaev et al 1993).

### *Far Infrared and the Kuiper Astronomical Observatory*

Far IR astronomy was in an early stage of development in the early 1970s, when we began work in this field at UC Berkeley. Our work progressed in a sequence of steps but always towards high sensitivity and high spectral resolution with Fabry-Perot interferometers.

We were fortunate to begin almost immediately with what seemed like an interesting and relatively simple measurement, even though a new spectrometer needed to be put together for it. Martin Harwit and his associates at Cornell had just published observations from high altitude rocket observations indicating a remarkably intense isotropic flux in the 0.4–1.3-mm wavelength range (Shivanandan et al 1968, Houck & Harwit 1969). The flux was about 25 times more than expected from a 2.7-K blackbody field. It appeared strong enough and at wavelengths where the atmosphere was transparent enough that we should be able to observe the radiation from a high altitude location on Earth. Mike Werner, a newly arrived postdoc at UC Berkeley, and John Mather, a graduate student interested in our research, were willing to try to measure this exciting but very puzzling radiation. Could it even be some intense spectral line? I asked Paul Richards, a fellow professor experienced with bolometers and far IR, for help to speed up the work, and fortunately he was also interested. These three put together both a tunable and a fixed Fabry-Perot interferometer with nickel mesh reflectors, an indium-antimonide bolometer detector, a chopper, and a focuser of 8-cm aperture (Mather et al 1971). The system was set up at a 12,500-ft altitude on White Mountain in eastern California, and spectra were taken in the 0.7–1.7-mm wavelength range with a resolving power of about 100. The radiation apparently detected by rocket flights didn't seem to be there! Rocket measurements are of course difficult, and this was not the only time that rocket measurements were to give misleading results in measuring the isotropic background radiation. The work clearly interested Paul Richards, who then moved into rocket measurements of the background radiation. John Mather became Paul's student, and eventually was to lead a spectacularly successful experiment, with the COBE satellite, to measure the background radiation and apparently really get it right.

The experiment on White Mountain helped us get started, and my own research group, including Mike Werner, continued development of systems to measure astronomical spectra in the far IR from airplanes and at wavelengths somewhat shorter than those of this initial ground-based experiment. Our first operating system used a bolometer detector and reflectors made of metal mesh structures deposited on quartz. At the time, unsupported mesh seemed insufficiently stable to withstand aircraft vibrations. Mike Werner and Bob McLaren, along with a student, Don Brandshaft, flew the system in NASA's Lear jet and were able to measure radiation of the Orion region between 60 and 100 $\mu$m, but with a resolution of only 4 $\mu$m (Brandshaft et al 1975). By 1976, NASA's much larger C141 plane was available with its larger telescope, 36 inches in diameter. With it, we could measure an ammonia rotational line in the Jovian atmosphere at 85-$\mu$m wavelength with a resolution of about 1 $\mu$m (Greenberg et al 1977).

The next major steps, carried out by student Dan Watson and postdoc John Storey, involved Fabry-Perot reflectors made of thin metal mesh tightly stretched on a circular frame. There were two Fabry-Perots in series, one fixed and one tunable. In addition, Kandiah Shivanandan of the NRL lent us a gallium-doped germanium detector. By 1978, this system had the sensitivity and resolution required to give a resolving power of 1000 and good measurements of the fine structure lines of OIII at 88 $\mu$m and OI at 63 $\mu$m (Storey et al 1979). The first far IR lines from outside the Solar System had already been detected by a group under Martin Harwit at Cornell (Ward et al 1975) and a French-European Space Agency group including Baluteau and Moorwood (Baluteau et al 1976). Harwit's group had used gratings to detect the OIII line; the French-European Space Agency group had used a Michelson interferometer and achieved very high resolution. These were historical firsts; the Cornell group obtained a clear-cut detection as early as 1975, but with a resolution of only 1.3 $\mu$m. The European group had excellent resolution, near 0.02 $\mu$m, but their spectral line detection was a bit marginal. We worked away at Fabry-Perot systems, believing them to be the most powerful simple systems for detecting and mapping lines. However, grating systems are certainly competitive; an excellent recent grating system has been used by Ed Erickson of NASA Ames and others (Erickson et al 1995).

We were fortunate that Eugene Haller, a solid-state physicist at UC Berkeley, was making and doing research on doped germanium detectors. He provided us with gallium-doped germanium detectors, and by 1980, we could use one of Haller's antimony-doped germanium detectors, which sensitively detected photons of wavelength longer than the 120-$\mu$m limit of gallium-doped germanium. This allowed detection of rotational lines of CO in the Orion nebula (Watson et al 1980). These and OH lines (Storey et al 1981) gave good evidence of shocks in Orion, and they allowed determination of gas densities and temperatures.

During the following several years, the double Fabry-Perot system, using antimony- or gallium-doped germanium detectors and flying at about 41,000-ft altitudes in NASA's KAO, brought in much valuable information about atoms, ions, and molecules in our own and other galaxies. Under favorable conditions and at longer wavelengths, it achieved a resolving power of 30,000 and a sensitivity of $2 \times 10^{-15}$ watts/Hz$^{0.5}$.

Reinhard Genzel had come to UC Berkeley in 1982 on a Miller postdoctoral fellowship. He worked with the far IR group and helped very much in extending the work begun by postdocs Storey and Crawford and graduate student Dan Watson. He was also soon appointed to the academic staff.

By the mid-1980s we took another step in development by putting three detectors in a row for more rapid mapping. These detectors could also be mechanically squeezed to extend their sensitivity somewhat towards wavelengths longer than their normal cutoff. Gordon Stacey, who did his thesis on far IR astronomy at Cornell with Martin Harwit, had come as a postdoc and was an important player in putting this system into operation.

Reinhard Genzel went back to Germany as a director of the Max Planck Institute at Garching in late 1986, but we continued to work closely together. The UC Berkeley and Garching groups jointly constructed the next far IR spectrometer, which first flew in the KAO in 1989. Its improvements included a choice of either two or three Fabry-Perots in series to allow either very high or medium-high spectral resolution and a 5-by-5 assembly of 25 detectors to provide rapid mapping along with good sensitivity and spectral resolution. Each detector had its own optics cone to put all the IR radiation in a given angular resolution element on its particular detector. This new system, dubbed FIFI (Far-IR Imaging Fabry-Perot Interferometer) (Poglitsch et al 1991), has been used in the KAO by the UC Berkeley–Garching groups and by guest observers to map many spectral lines in regions of our galaxy and in external galaxies. One of Genzel's students, Norbert Geis, came to work with the UC

Berkeley group for some years on our joint projects with FIFI. Resolution of the 36-inch Kuiper telescope in the far IR is limited by diffraction to 30–50 arcsec. However, this is enough to significantly resolve features of a number of nearby galaxies.

NASA's grounding of the KAO in late 1995, in order to save funds to build a still better system, ended my own far IR observations. However, I am happy that this next step to a new system called SOFIA (Stratospheric Observatory Far Infrared Astronomy) is being taken. Among other advantages, it will carry a 2.4-m telescope with diffraction-limited angular resolution 2.8 times higher than that of the KAO and with sensitivity on small objects about an order of magnitude better. Genzel and his group at the Max Planck Institute have begun planning the next instrumental improvements towards still better spectral line measurement and mapping, and they will be one of the many groups that will keep SOFIA quite busy.

### *Spatial Interferometry in the Infrared*

The development of lasers has brought optical technology conceptually much closer to that of radio or microwaves. It also makes relatively easy the measurement or control of distances to almost arbitrarily good precision (e.g. less than the diameter of an atomic nucleus) with laser distance interferometers. These developments and the tremendous importance of high angular resolution to understanding distant astronomical objects led me into work on spatial interferometry, a difficult but potentially very rewarding field. Another attraction was that IR detector developments had very much increased the practicality of such interferometry at IR wavelengths.

I believe that as of the late 1960s, IR stellar interferometry had never before been attempted. Since Michelson's first measurement of a stellar diameter at visible wavelengths in 1921 at Mount Wilson (Michelson & Pease 1921), there had been a number of stellar interferometric projects to measure stellar diameters at visible wavelengths, but the technique was clearly difficult and no such project had led to a long-term series of good measurements. Perhaps one exception was the intensity correlation interferometer of Hanbury Brown in Australia, which was useful for a selection of bright hot stars (Hanbury Brown 1968), but work on this interferometer has also to come to a close. On the other hand, multiple telescope interferometry at microwave frequencies had become tremendously successful and even routine. Part of this relative success at radio frequencies is due to the longer wavelength, which makes interferometry less sensitive to pathlength fluctuations, in particular atmospheric "seeing." Part is because of the nature of radio sources. There are intense extended radio sources due to high temperature ionized material and also extremely high temperature "point-source" maser and continuum sources of microwaves. Masers are fairly common and give strong signals even when they are quite small. No lasers of comparable intensity have so far been found at wavelengths below 100 $\mu$m. In spite of such disadvantages for the shorter wavelengths, successful spatial interferometry at visible or IR wavelengths clearly seems to give extremely important information.

After getting well settled at UC Berkeley and with burgeoning IR technology in view, I felt it was time to develop IR interferometry. One of the new developments was that of semiconductor photodetectors in the mid-IR that are very fast, allowing heterodyne detection bandwidths at least as large as a gigahertz when an IR signal is mixed with a suitable IR laser output. I was fortunate that an outstanding student, Michael Johnson, who arrived at UC Berkeley from MIT at the same time I did, was interested in giving IR stellar interferometry a try. He was joined before long by another student, Albert Betz, an exceptionally able experimentalist. They built $CO_2$ lasers for local oscillators following a then-new design of Freed (1968). They doped germanium crystals with copper, and with these put together two heterodyne detection systems giving two detected sidebands, each about $10^9$-Hz wide and with detection quantum efficiency of about 5%. This was remarkable performance at the time, but clearly we looked forward to future improvement. The detectors needed to be kept at liquid helium temperatures, and the two $CO_2$ laser local oscillators, operated with $^{13}CO_2$ gas to avoid the atmospheric absorption $^{12}CO_2$ frequencies would suffer, had to be locked together in phase. This then produced IF signals that could be sent along coaxial cables and beat together to produce fringe signals in accordance with techniques very familiar to radio interferometry, though of somewhat exceptionally large bandwidths ($\sim$2 GHz).

The McMath Solar Telescope at Kitt Peak is a 90-inch heliostat mounted 30 m above ground level, and on either side of it is a smaller 36-inch heliostat. The latter two heliostats are separated by 5.5 m on an approximately E-W baseline and, because they are smaller than the large one, are not frequently used. This was just what we needed, and fortunately the Kitt Peak National Laboratory let us use these two telescopes over a considerable period of time and were very helpful with general accommodations and access to supplies and workshops. Extended and hard work brought in the first good fringe signal from an astronomical object, the bright edge of Mercury, in 1974 (Johnson et al 1974).

Johnson and Betz were soon joined by Bob McLaren, John Storey, and Ed Sutton, who helped use heterodyne detection for ultra-high spectral resolution, as discussed above, and also carried out further interferometry. The next major step was Ed Sutton's thesis project, interferometric examination of some well-known stars.

David Spears of Lincoln Laboratory headed a group working on semiconducting detectors, and about that time his work fortunately provided a major advance in fast detectors for the 10-micron region (Spears 1977). These were HgCdTe photodiodes, which needed only liquid nitrogen temperatures rather than those of liquid helium, had bandwidths of about 1.5 GHz, and had quantum efficiencies of about 0.2. This was, overall, a gain in sensitivity over our own homemade detectors of a factor of $\sim$5 but still left considerable room for further hoped-for improvements. With these and other refinements of our system at the McMath solar telescopes, Sutton was able to resolve the dust shells around half a dozen well-known stars (Sutton et al 1977, 1978).

Our fixed E-W baseline of 5.5 m gave only a rather limited total range of effective baselines over which measurements could be made. Hence, although dust shells were partially resolved in $\alpha$ Ori, $\alpha$ Sco, R Leo, o Ceti, IRC+10216, and VY CMa, their dust shells could only be modeled with simple assumptions as to their form, such as a Gaussian pattern. Later measurements were to show much more detail. In the case of Mira, we were also able to see changes of visibility with the stellar luminosity phase. Our prototype system on the McMath solar telescopes allowed us on occasion to obtain good fringe phase measurements and hence relative stellar positions, which gave encouragement regarding the practicality of astrometry (Sutton et al 1982).

To develop IR interferometry further, we clearly needed a wider variety of baselines, along with substantially larger and possibly movable telescopes. Construction of a suitable telescope system required at least a few million dollars, and my applications for this to several government agencies were characteristically turned down. However, finally, in 1983, ARPA and the Office of Naval Research jointly funded the construction of two movable telescopes for interferometry.

Because the telescope diameter over which atmospheric seeing allows diffraction-limited imaging is proportional to $\lambda^{6/5}$ according to the Kolmogorov approximation, where $\lambda$ is the wavelength, the usual 10-cm diameter characteristic of diffraction-limited seeing under good conditions for visible wavelengths increases to 3.8 m for 10-$\mu$m wavelengths. We wanted a telescope somewhat smaller than this limit, but otherwise as large as could be used within a convenient movable housing such as a trailer. The so-called Pfund design seemed best adapted for this. It involves a flat coelostat mirror on an altazimuth mount illuminating a coaxial parabola. An 80-inch flat and a 65-inch parabola seemed the appropriate compromise to be fitted inside a standard trailer, which could not only be moved in position easily on a given site but could be readily moved on the road or shipped—possibly eventually to Chile. From my earlier observations at Las Campanas in northern Chile I had been impressed by atmospheric conditions there. I believe it is the best of any site presently used for astronomy—and atmospheric pathlength stability or good seeing is critically important to high quality interferometry. We initially set up, however, at Mt.

Wilson Observatory, which appears to have the best seeing of any established observatory in the continental United States. Concrete slabs were laid down for parking the telescopes to allow baselines varying from 4–35 m. The 4-m baseline was to provide an overlap with and approximately the same resolution as the filled disk of the largest optical telescopes.

A number of scientific colleagues—Ed Sutton and Bernard Sadoulet during the early stages, Bill Danchi and Manfred Bester on a more continuous basis, some very good specially-hired engineers, and Walter Fitelson (the technician-engineer who has participated in most of my enterprises at UC Berkeley)—helped design and get the new interferometer going over the next few years. Its use has been partially supported by the Office of Naval Research and partially by the National Science Foundation. The instrument, christened the Infrared Spatial Interferometer (ISI), has been described in various stages and has been continuously upgraded (Townes et al 1986, Bester et al 1992b). By 1995, it had been successful enough to convince the National Science Foundation to fund the construction of a third telescope to allow visibility measurements to be made at more baselines faster and also to obtain phase closure. Baselines are also to be extended in length to about 75 m and enriched as to orientation.

While Danchi and Bester have been the backbone of this work, we have enjoyed the help of Cuno Degiacomi and Lincoln Greenhill, now at the University of Köln and Harvard University, respectively, and collaboration with Robert Treuhaft of the Jet Propulsion Lab and Bruno Lopez and Djamel Mékarnia of the University of Nice.

New instrumentation capable of making measurements never available before can always be expected to reveal new phenomena, either unimagined before or only covered up by simplifying approximations that are not good enough for the new information revealed. That is of course the point of instruments with improved performance. What follows is some of the new information that interferometry in the mid-IR has turned up so far.

An encounter subsidiary to the study of astronomical objects but inherently linked to stellar interferometry is with fluctuations in relative pathlengths through the atmosphere for two paths separated by the interferometric baseline. Any such variation naturally becomes less troublesome to interferometry as the wavelength increases; radio waves represent an extreme in this direction. The Kolmogorov-Taylor approximation for atmospheric turbulence predicts an improvement proportional to $\lambda^{6/5}$ rather than simply the wavelength $\lambda$. Our measurements indicate an improvement with wavelength even somewhat faster than $\lambda^{6/5}$ (Bester et al 1992a). In addition, under good seeing conditions, there is an "outer scale" or limit to the magnitude of fluctuations so that they frequently do not increase as baselines are lengthened beyond 10–20 m (Bester et al 1992b). These effects all tend to make fluctuations still less troublesome with increasing wavelength than simple approximations have predicted. Of course, ideally one would like to have no relative pathlength variations, such as may only be achieved outside the Earth's atmosphere. But we have encountered some remarkable occasions on Mt. Wilson when over the better part of an hour, relative pathlengths 4 m apart have been essentially fixed to within a small fraction of our 11-$\mu$m wavelength. On other occasions there can be changes of more than a wavelength occurring within $10^{-2}$ s.

It has been well known that pathlength fluctuations near the ground are likely to be greater than those some distance above the ground. The ISI has naturally led to a quantitative measure of this phenomenon on Mt. Wilson. Not only is turbulence larger near the ground than high above it under usual conditions, but the larger pathlength fluctuations a few meters above ground are strongly correlated with those as much as 40 m higher, at least under good seeing conditions on Mt. Wilson (Treuhaft et al 1995). As a result, local measurements near ground level can frequently be used to calculate and compensate for more than half the total relative pathlength fluctuations.

As for measurements on stars themselves with the ISI, some of the most common and striking phenomena found are the episodic emission of dust-forming gas by a large fraction of the stars we have measured. All of these were previously known to be surrounded by dust. Most of such stars are variables, and dust is produced by some of them on essentially every cycle of luminosity variation, as might be expected. However, about half of such stars emit shells of dust separated in time by 10–200 years, much longer than the typical 1–2 year period of their luminosity cycle (Danchi et al 1994, Monnier et al 1997). This unexpected distribution of emitted material also has a strong effect on the occurrence of masers. Betelgeuse has a shell of dust around it that was probably emitted about 1836, when Herschel (1840) alerted the astronomical community to an unusual change in its luminosity (Danchi et al 1994). Whether this particular dust shell was emitted then, or in the early 1940s when other, somewhat smaller changes occurred, will be better determined after the distance to Betelgeuse measured with the Hipparcos satellite has been reported. We have found that the star also made a rather different and smaller emission in 1994, so small that it will probably fade from view within a few years (Bester et al 1996).

The ISI has measured the diameters of $\alpha$ Ori and $\alpha$ Sco at 11-$\mu$m wavelength and found them about 10% larger than indicated by the best measurements at visible wavelengths. This is in good agreement with theoretical estimates of limb darkening. The apparent changes in size of $\alpha$ Ori indicated by earlier interferometric measurements in the visible region seem likely to be due primarily to hot spots on its surface. With the coming new three-telescope system and longer baselines, we hope to measure the sizes of a number of stars and their deviation from sphericity. The approximately linear dependence on temperature of mid-IR intensity of stellar surfaces as opposed to exponential dependence at visible wavelengths makes these longer wavelengths ideal for good size and ellipticity measurements.

Not only do many stars emit material episodically, some of those that appear to emit on each luminosity cycle may behave somewhat differently in succeeding cycles. Mira shows remarkable variations in the characteristics of dust immediately surrounding it at similar phases of successive cycles (Lopez et al 1997). Our measurements on some dust-surrounded stars vary so strangely that they must be watched in still more detail before their patterns of variation can be understood.

Heterodyne detection of radiation imposes fundamental quantum fluctuations, or noise, which are substantially larger than those of direct photon counting, and heterodyne detectors imply the use of relatively narrow band detection ($<1\ \mathrm{cm}^{-1}$) (Townes 1984). These characteristics are of little consequence in the radio or microwave region, where quanta have very small energies, but become more troublesome at shorter wavelengths and make heterodyne detection relatively insensitive compared to direct detection if a broad band of wavelengths is appropriate. I do not expect heterodyne detection to be very useful in the near IR for these reasons. However, there are advantages of heterodyne detection in terms of quality of visibility measurements, partly because the only radiation power detected is the fraction that is spatially coherent. Such detection is also very convenient and does not lose signal-to-noise ratio when multiple telescopes are used. These factors made it competitive with direct detection techniques; I expect it to give excellent results on a large number of stars that are bright in the mid-IR. In addition, heterodyne interferometry is probably unique in its ability to examine narrow spectral lines in dust shells around stars. This should allow determination of the distribution of molecules in various energy states within the gas surrounding stars that lose mass and, hence, should allow an understanding of their formation and excitation.

Very useful interferometry has been carried out with so-called speckle interferometry (Labeyrie 1970), as well as by masking apertures on large telescopes (McCarthy et al 1977, Tuthill et al 1995). These methods are productive, but generally limited to baselines smaller than the diameter of the largest optical telescopes. Stellar interferometry with two or more separate telescopes is a demanding and technically difficult field. But my assessment is that present technology can now make it quite rewarding, assuming very substantial and clever efforts. Other groups of astronomers seem to agree and there are many now active in the field. Much of the recent work has been in the visible region, an example of which is the very successful interferometry done on Mt. Wilson under the direction of Mike Shao of the Jet Propulsion Laboratory and Ken Johnston of the Naval Observatory. There are a number of other excellent and active groups. There is also increasing interest in the IR. The near IR will probably be particularly rewarding for the study of stellar surfaces. The mid-IR is especially needed for the study of surrounding dust or for the penetration of dust to examine stars deeply embedded in dust. I look forward to getting at objects in the galactic center, for example, with a mid-IR interferometer. The next decade should reveal much more about the real place and contributions of interferometry to the realm of astronomical studies.

### *Summarizing Comments*

Experimental astronomy is in some ways very similar to the exploration of woods, fields, and streams I enjoyed as a youngster. It is observational; we observe and examine what is there, and we try to understand it. But we cannot normally make tests by tweaking the object studied or changing its conditions to see what happens, as we do in many physics experiments. Astronomy does, however, have a happy commonality with all physical sciences in that it is subject to quantitative, rigorous theoretical logic and tests. And good instrumentation technology is critically important to the sharpness and variety of our observations. During the last half century, we have experienced a remarkable period for astronomy in this respect. There has been an extension of important observations to wavelengths in the radio region, IR, and X rays. Major telescope improvements have come, along with observations from outside the Earth's atmosphere thanks to the NASA space program. The latter has also provided contact with or very close views of much of our Solar System. New kinds of objects have been found—neutron stars, black holes, quasars, the isotropic radiation, and a steadily enriched variety of galaxies. We can look forward to more—what will there be? All of our observational techniques and the theoretical power for handling complex problems that computers provide will surely continue to improve. So also, I believe, will the resolution and detail of imaging—through space work, adaptive optics, and further interferometric developments. But what surprising discoveries will be made can only be known in retrospect, after the interesting decades that are to come.

The rapid development of science depends very much on interactions between scientists. I myself am very pleased to have had an opportunity to be rather fully engaged with research in physics and astrophysics and with their communities. I feel especially privileged to have interacted and worked with many talented individuals. It is these individuals and the community of researchers as well as the fascinating explorations involved that have made such a career especially rewarding.

*Literature Cited*


Aitken DK, Jones B, Penman JM. 1974. *MNRAS* 169:35
Baluteau JP, Bussoletti E, Anderegg M, Moorwood AFM, Coran N. 1976. *Ap. J.* 210:L45
Barrett AH, Lilley AE. 1957. *Astron. J.* 62:5
Bertojo M, Cheung AC, Townes CH. 1976. *Ap. J.* 208:914
Bester M, Danchi WC, Degiacomi CG, Greenhill LJ, Townes CH. 1992a. *Ap. J.* 392:357
Bester M, Danchi WC, Hale D, Townes CH, Degiacomi CG, et al. 1996. *Ap. J.* 243:336
Bester M, Degiacomi CG, Danchi WC, Greenhill LJ, Townes CH. 1992b. In *Robotic Telescopes in the 1990s*, ed. AV Filippenko, *Astron. Soc. Pac. Conf. Ser.* 34:213
Betz AL, Johnson MA, McLaren RA, Sutton EC. 1976. *Ap. J.* 208:L141
Boyd RW, Townes CH. 1977. *Ap. J. Lett.* 31:440
Boyd RW, Townes CH. 1978. In *Proc. 4th Rochester Conf., Coherence and Quantum Optics IV*, ed. L Mandel, E Wolf. New York: Plenum. 333 pp.
Brandshaft D, McLaren RA, Werner MW. 1975. *Ap. J.* 199:L115
Cheung AC, Rank DM, Townes CH, Thornton DD, Welch WJ. 1968. *Phys. Rev.* L21:1701
Cheung AC, Rank DM, Townes CH, Thornton DD, Welch WJ. 1969. *Science* 163:1055
Cocconi G, Morrison P. 1959. *Nature* 184:844
Connes P. 1970. *Annu. Rev. Astron. Astrophys.* 8:209
Cosmovici CB, Montebugnoli S, Orfei A, Pagrebenko S, Colom P. 1996. *Planet. Space Sci.* 44:735
Danchi WC, Bester M, Degiacomi CG Greenhill LJ, Townes CH. 1994. *A. J.* 107:1469
Dousmanis GC, Sanders TM Jr, Townes CH. 1955. *Phys. Rev.* 100:1735
Eckhart A, Genzel R. 1996. *MNRAS* 284:576
Ehrenstein G, Townes CH, Stevenson MJ. 1959. *Phys. Rev. Lett.* 3:40
Ekers RD, van Gorkom JH, Schwarz UJ, Goss WM. 1983. *Astron. Astrophys.* 122:143
Erickson EF, Haas MR, Colgan SWJ, Simpson JP, Rubin RH. 1995. *Astron. Soc. Pac. Conf. Ser.* 73:523
Field GB, Chaisson EJ. 1985. *The Invisible Universe*. Boston: Birkhäuser
Freed C. 1968. *IEEE J. Quant. Electron.* 4:404
Genzel R, Thatte N, Krabbe A, Kroker H, Tacconi-Garman LE. 1996. *Ap. J.* 474:153
Genzel R, Watson DM, Townes CH, Dinerstein HL, Hollenbach D, et al. 1984. *Ap. J.* 276:551
Gillett FC, Stein WA. 1967. *Ap. J. Lett.* 149:L97
Ginzburg VL. 1990. *Annu. Rev. Astron. Astrophys.* 28:1
Giordmaine JA, Alsop LE, Mayer CH, Townes CH. 1959. *Proc. Inst. Radio Eng.* 47:1062
Gordon JP, Zeiger HJ, Townes CH. 1954. *Phys. Rev.* 95:282L
Greenberg L, McLaren R, Stoller L, Townes CH. 1977. In *Symp. Recent Results in Infrared Astrophysics, NASA Tech. Mem. NASA TMX-73*, ed. P Dyal, pp. 9–13
Haller JW, Rieke MJ, Rieke GH, Tamblyn P, Close L, et al. 1996. *Ap. J.* 456:194
Hanbury Brown R. 1978. *Annu. Rev. Astron. Astrophys.* 6:13
Harris AI, Krenz T, Genzel R, Krabbe A, Lutz D, et al. 1994. In *The Nuclei of Normal Galaxies: Lessons from the Galactic Center*, ed. R Genzel, AI Harris, p. 223. Dordrecht: Kluwer
Henyey LG, Keenan PC. 1940. *Ap. J.* 91:625
Herschel J. 1840. *MNRAS* 5:11
Houck JR, Harwit MO. 1969. *Ap. J.* 157:L45
Johnson MA, Betz AL, McLaren RA, Sutton EC, Townes CH. 1976. *Ap. J.* 208:L145
Johnson MA, Betz AL, Townes CH. 1974. *Phys. Rev. Lett.* 33:1617
Knowles SH, Mayer CH, Cheung AC, Rank DM, Townes CH. 1969. *Science* 163:1055
Kramers HA. 1923. *Philos. Mag.* 46:836
Labeyrie A. 1970. *Astron. Astrophys.* 6:85
Lacy JH, Baas F, Townes CH, Geballe TR. 1979. *Ap. J. Lett.* 227:L17
Lacy JH, Townes CH, Geballe TR, Hollenbach DJ. 1980. *Ap. J.* 241:132
Lacy JH, Townes CH, Hollenbach DJ. 1982. *Ap. J.* 262:120
Latvakoski HM, Stacey GJ, Hayward TL, Gull GE. 1996. In *4th ESO/CTIO Workshop: The Galactic Center*, ed. R Gredel, *Astron. Soc. Pac. Conf. Ser.* 102:106
Lo KY, Claussen MJ. 1983. *Nature* 306:647
Lopez B, Danchi WC, Bester M, Hale DDS, Lipman EA, et al. 1997. *Ap. J.* In press
Mather JC, Werner MW, Richards P. 1971. *Ap. J.* 170:L59
McCarthy DW, Low FJ, Howell R. 1977. *Ap. J. Lett.* 214:85
Michelson AA, Pease FG. 1921. *Ap. J.* 53:249
Monnier JD, Bester M, Danchi WC, Johnson M, Lipman E, et al. 1997. *Ap. J.* In press
Moran JM, Barrett AM, Rogert AEE, Burke BF, et al. 1967. *Ap. J.* 148:L69
Mumma MJ, Buhl D, Chin G, Deming D, Espenak F, et al. 1981. *Science* 212:45
Palmer P, Zuckerman B, Buhl D, Snyder LE. 1969. *Ap. J.* L156:147
Penzias AA, Wilson RW. 1965. *Ap. J.* 142:419
Poglitsch A, Beeman JW, Geis N, Genzel R, Haggerty M, et al. 1991. *Int. J. Infrared Millimeter Waves* 12:859
Rank DM, Holtz JZ, Geballe TR, Townes CH. 1970. *Ap. J.* 161:L185
Reber G. 1940. *Inst. Radio Eng.* 28:68

Schawlow AL, Townes CH. 1958. *Phys. Rev.* 112:1940
Schwartz RN, Townes CH. 1961. *Nature* 190: 205
Sellgren K, McGinn MT, Becklin EE, Hall DNNB. 1990. *Ap. J.* 359:112
Serabyn E, Lacy JH. 1985. *Ap. J.* 293:445
Shimoda K, Takahasi H, Townes CH. 1957. *J. Phys. Soc. Jpn.* 12:686
Shivanandan K, Houck JR, Harwit MO. 1968. *Phys. Rev. Lett.* 21:1460
Snyder LE, Buhl D, Zuckerman B, Palmer P. 1969. *Phys. Rev.* L22:679
Spears DL. 1977. *Infrared Phys.* 17:5
Storey JWV, Watson DM, Townes CH. 1979. *Ap. J.* 233:109
Storey JWV, Watson DM, Townes CH. 1981. *Ap. J.* 244:L27
Sunyaev RA, Markevich M, Pavlinsky J. 1993. *Ap. J.* 407:606
Sutton EC, Storey JWV, Betz AL, Townes CH, Spears DL. 1977. *Ap. J.* 217:L97
Sutton EC, Storey JWV, Townes CH, Spears DL. 1978. *Ap. J. Lett.* 224:123
Sutton EC, Subramanian S, Townes CH. 1982. *Astron. Astrophys.* 110:324
Timmerman R, Genzel R, Poglitsch A, Lutz D, Madden S, et al. 1996. *Ap. J.* 466:242
Townes CH. 1946a. *Ap. J.* 105:235
Townes CH. 1946b. *Phys. Rev.* 70:665
Townes CH. 1957. In *IAU Symp. No. 4*, ed. HS van de Hulst, p. 92. Cambridge: Cambridge Univ. Press
Townes CH. 1983. *Proc. Natl. Acad. Sci. USA* 80:1147
Townes CH. 1984. *J. Astrophys. Astron.* 5:111
Townes CH, Cheung AC. 1969. *Ap. J.* 157: L103
Townes CH, Danchi WC, Sadoulet B, Sutton EC. 1986. In *Advanced Technology Optical Telescope III, Proc. SPIE*, ed. LD Barr, 628:281. Bellingham, WA: Soc. Photo-Opt. Eng.
Townes CH, Schawlow AL. 1955. *Microwave Spectroscopy.* New York: McGraw Hill
Townes CH, Lacy JH, Geballe TR, Hollenbach DJ. 1983. *Nature* 301:661
Treuhaft RN, Lowe S, Bester M, Danchi WC, Townes CH. 1995. *Ap. J.* 453:522
Tuthill PG, Haniff CA, Baldwin JE. 1995. *MNRAS.* 277:1541
Van Vleck JH. 1942. *Report 43-2 of the M.I.T. Radiation Laboratory.* 32 pp.
Van Vleck JH, Weisskopf VF. 1945. *Rev. Mod. Phys.* 17:227
Ward DB, Dennison B, Gull G, Harwit M. 1975. *Ap. J.* 202:L31
Watson DM, Genzel R, Townes CH, Werner MW, Storey JWV. 1984. *Ap. J.* 279:L1
Watson DM, Storey JWV, Townes CH, Haller EE, Hansen WL. 1980. *Ap. J.* 239:L129
Weinreb S, Barrett AH, Meeks ML, Henry JC. 1963. *Nature* 200:829
Wilson RW, Jefferts KB, Penzias AA. 1970. *Ap. J.* 161:L43
Wiseman JJ, Ho PTP. 1996. *Nature* 382:139
Wollman ER, Geballe TR, Lacy JH, Townes CH, Rank DM. 1976. *Ap. J.* 205:L5
Wollman ER, Geballe TR, Lacy JH, Townes CH, Rank DM. 1977. *Ap. J.* 218:L103

*Annu. Rev. Astron. Astrophys. 1997. 35:1–32*


# ETA CARINAE AND ITS ENVIRONMENT

*Kris Davidson and Roberta M. Humphreys*
Astronomy Department, University of Minnesota, Minneapolis, Minnesota 55455;
e-mail: kd@ea.spa.umn.edu

KEY WORDS: massive stars, variable stars, LBVs or luminous blue variables

ABSTRACT

Eta Carinae (Eta) is one of the most remarkable of all well-studied stars and perhaps the most poorly understood. Observations with the Hubble Space Telescope and other modern instruments have solved a few of the mysteries concerning this object while opening a comparable number of new ones. In this review we first recount some essential background information concerning Eta, then we sketch most of the observational developments of the past few years, related to the star itself and to its ejecta. Throughout, we propose a series of specific unsolved observational and theoretical problems that seem especially interesting or important at this time.

## 1. INTRODUCTION

> A strange field of speculation is opened by this phenomenon... here we have a star fitfully variable to an astonishing extent, and whose fluctuations are spread over centuries, apparently with no settled period and no regularity of progression. What origin can we ascribe to these sudden flashes and relapses? What conclusions are we to draw as to the comfort and habitability of a system depending for its supply of light and heat on so uncertain a source?
>
> —JFW Herschel (1847)

Eta Carinae ($\eta$ Car) is arguably the most remarkable stellar object that is close enough to be observed in great detail. It is so special that enthusiasts often refer simply and familiarly to "Eta"—a usage we would deplore for any other star with a Bayer designation. Eta's fame and uniqueness rest on a combination of three different but related lists of attributes, which we merely sketch here at the outset of this review, postponing references until later.



0066-4146/97/0915-0001$08.00

First, a list of superlatives. The most luminous evolved star that can be studied closely, Eta is also the survivor of the greatest well-documented non-terminal stellar explosion. Its mass probably exceeds 100 $\mathcal{M}_\odot$, which would make it rare even if it were a more stable star near the main sequence. For two decades of the nineteenth century, Eta was visually one of the dozen brightest stars in the sky, even at a distance of more than 2 kpc; the total luminous energy radiated during the whole outburst was comparable to that of a supernova event. In recent years, its luminosity has been several million times that of the Sun. Today it is the brightest extra–solar-system infrared source in the sky at $\lambda \sim 20\ \mu$m, and it has the brightest known stellar wind at millimeter or centimeter wavelengths. Eta has produced one of the most elegant of all known bipolar nebulae.

Second, as John Herschel (1847) wrote with only mild exaggeration, "There is perhaps no other sidereal object which unites more points of interest than this." Stellar physics?—The interplay of radiation pressure and opacity (maybe rotation too) causes Eta to be irregularly, chaotically unstable, with occasional titanic eruptions. Stellar evolution?—This star appears to be an extreme example showing how instability limits evolution at the top of the Hertzsprung-Russell (H-R) diagram. Star formation?—Eta was formed in a rich association that includes several other very massive (and therefore very unusual) stars. Gas dynamics?—Bipolar flows created by the aforesaid instability are beautifully symmetric, but at the same time they have conspicuous localized asymmetric features that no one predicted. Microphysics and radiative transfer?—Certain bright emission lines in the ejecta are excited by fluorescent processes that are uniquely effective around Eta. Dust formation? Chemical enrichment of the surrounding medium?—Eta is relevant to these and to other topics as well, in nonroutine largely unexplored ways.

Third, this object is very poorly understood. We do not know why or how Eta's giant eruptions occurred, and its energy budget has never been properly audited. Although we do know that Eta is evolved, we are not sure whether it is typical of the most massive stars; it might be a rapid rotator or a close binary system. Some authors suggest that the star exhibits surprising periodicities. Some of the most obvious characteristics of its bipolar ejecta are difficult to explain, as are the brightnesses and ratios of some of its most conspicuous emission lines. As new instruments have settled particular observational questions about Eta, they have opened new questions to replace the old ones—a classic tendency shared with other unusually interesting astrophysical topics.

We recommend Eta as a subject for all sorts of theoretical and observational research projects. Therefore, throughout the review we mark some of the most crucial or interesting questions with $\{Q\}$ labels, e.g. Herschel's classic mystery that still deserves top billing:

{$Q1$} What origin can we ascribe to these sudden flashes and relapses?

Eta and its vicinity were the topics of a meeting held in 1993, whose proceedings contain many useful papers (see Niemela et al 1995). For a review of the class of star to which Eta most likely belongs, see Humphreys & Davidson (1994).

## 2. AN UPPER-CLASS NEIGHBORHOOD

From our vantage point, the constellation Carina marks the most spectacular direction in the Milky Way. There our line of sight passes along the Carina spiral arm, which can be traced to distances of more than 10 kpc at visual wavelengths (Humphreys 1976) and as far as 23 kpc with CO observations (Grabelsky et al 1988). This major Galactic feature is defined by large numbers of massive stars and prominent star-formation regions, such as the Carina Nebula (NGC 3372) and its surrounding association Car OB1 ($\ell \approx 284.2°$ to $288.3°$, $b \approx -2.2°$ to $+0.9°$, $D \approx 2.5$ kpc). With a more impressive concentration of very luminous stars than any other place within a few kiloparsecs of the Sun, the Carina Nebula is one of the best available sites for studying the formation and evolution of the most massive stars. Its two central clusters, Trumpler 14 and Trumpler 16 (Tr 14 and Tr 16), are particularly unusual. When we refer here to Tr 16, we also include a set of stars called Collinder 228 (Cr 228). Together these are now considered to be a single cluster, with an illusory separation between Tr 16 and Cr 228 caused by a superimposed dust lane (see Walborn 1995). A census of hot stars in Tr 14 and Tr 16, based on the work of Walborn (1982a,b), Morrell et al (1988), Levato & Malaroda (1981, 1982), and Massey & Johnson (1993), includes 36 stars earlier than spectral type B0 and 3 WN-A stars. These provide enough ultraviolet (UV) photons to ionize the large nebula and to heat the dust seen in infrared observations (Harvey et al 1979; Ghosh et al 1988). Far more remarkably, Tr 14 and Tr 16 contain 6 of the 17 known O3-type stars in our Galaxy; because O3 stars are at the hot end of the normal spectral sequence, they are very massive and quite rare. Knowing this, perhaps one is not very surprised to learn that Eta is a member of Tr 16.

The larger association Car OB1 and the outer clusters IC 2581, NGC 3293, and NGC 3324 contain numerous OB stars and evolved supergiants, including many M supergiants, whose presence implies an age of about 10 million years for the oldest massive stars in this stellar complex. Published distance moduli for the outer clusters range from 11.9–12.5 mag, and the distance to the association is usually quoted as $\sim$2.5 kpc (see Humphreys 1978). The clusters Bochum 10 and 11 are apparently close to the Carina Nebula and contain several OB stars; Fitzgerald & Mehta (1987) derived ages of 7 and 3 million years and distance moduli of 12.2 and 12.7 mag for Bo 10 and Bo 11, respectively. Tr 15 is usually considered to be behind the H II region (Walborn 1995).

Distances to Tr 14 and Tr 16 have been made more uncertain by anomalous extinction within or near the Carina Nebula. Feinstein et al (1973), Herbst (1976), and Forte (1978) reported abnormally high ratios of extinction to reddening there, i.e. the wavelength dependence of the obscuration is relatively weaker than it is in the normal interstellar medium. In a multiwavelength study, Thé et al (1980) found that the wavelength dependence of the extinction varies from star to star in the vicinity of the Carina Nebula. The required average correction for interstellar extinction, which is usually deduced from the reddening of stars with known intrinsic colors, has therefore been controversial for Tr 14 and Tr 16. Independent infrared (IR) studies by Smith (1987) and Tapia et al (1988) confirmed the anomalous and position-dependent behavior of extinction for stars embedded in the nebula, whereas stars at the outer edge of the nebula seem to obey a normal interstellar extinction law. According to Thé & Graafland (1995), extinction estimates for stars in the clusters Tr 14, Tr 15, and Tr 16 are not correlated with spectral type or luminosity class. The extinction/reddening ratio $R = A_V/E_{B-V}$ ranges from a "normal" 3 to as high as 6 for a few stars. Such variations are attributed to highly variable distributions of dust grain sizes: Larger grains tend to give larger values of $R$. (Unusually large average grain sizes may occur in the ejecta of Eta; see Sections 4.1 and 5.2 below.)

Thé & Graafland's (1995) results for Tr 14 and Tr 16 show that $R$ is in the range 3–5 for the O-type stars on which the estimated distances to these clusters are based. The average value appears to be near 4. Using assumed absolute magnitudes for 19 O-type stars, Walborn (1995) estimates that the distance to Tr 16 is $D \approx 0.8^{(R-4)} \times 2250$ pc, with a formal error of only a few percent. Tr 14 is about 30% farther for a given value of $R$ (which, however, can differ between the two clusters, so they may be at about the same distance).

In this review, we adopt $D \approx 2300 \pm 200$ pc for Eta, corresponding to $R_{av} \approx 3.9$ for Tr 16. Allen & Hillier (1993) used a completely different method based on expansion of Eta's ejecta to find a similar result, ~2200 pc. (Most papers on Eta in the past 15 years assumed larger distances, usually 2500 or 2800 pc.) We also suppose that the visual-wavelength interstellar extinction toward Eta is $A_V \approx 1.7$ mag, not including circumstellar extinction (Feinstein et al 1973; Thé et al 1980; Feinstein 1982; Walborn 1995, and other references cited therein). With this distance and extinction, the difference between apparent and absolute visual magnitudes is $m_V - M_V \approx 13.5$, a number that will be useful in Section 3 below.

Tr 14 and Tr 16 are clearly very young clusters, judging from their stellar content. Whiteoak's (1994) radio continuum survey of the Carina Nebula revealed no nonthermal sources, i.e. no supernova remnants. The lifetime of a very massive star like Eta is about 3 million years; because Eta seems to have evolved well away from the main sequence (Section 3 below), whereas three WN stars may also be associated with Tr 16, the most likely age of Tr 16 is also on the order of 3 million years (cf Feinstein 1982, 1995; Massey & Johnson 1993). Tr 14 may be younger, perhaps only 1 million years old (Walborn 1995; Penny et al 1993).

Thus, in the entire association and region around Eta, we find a record of massive star formation over several million years. Is star formation still occurring there? Infrared and molecular surveys have revealed few obvious embedded sources, including only one embedded IRAS source near Tr 16. However, recent deep JHK imaging by Megeath et al (1996) shows a number of very reddened sources dominated by one strong one, the first clear evidence of continuing star formation in the Carina Nebula. Because the CO survey by Grabelsky et al (1988) shows that the Carina Nebula is part of a giant molecular cloud 130-pc across with a mass of about $10^{5.7} \mathcal{M}_{\odot}$, we expect that massive stars will continue to form there.

The O3 stars in Tr 14 and Tr 16, along with $\eta$ Car (Eta), remind us of another important question, which also pertains to other places such as 30 Dor and NGC 3603 and which has been asked for a long time (e.g. by Humphreys & Davidson 1984):

{*Q*2} Where a freakishly massive star with $\mathcal{M} > 50\ \mathcal{M}_{\odot}$ is found, there are usually others nearby. Why?

## 3. THE PRE-1990 HISTORY OF ETA AND ITS BASIC PHYSICAL STATE

Many of the phenomenological points in this section were mentioned in three earlier discussions (Davidson 1987a, 1989; Humphreys & Davidson 1994), but they are essential background for later sections of this review.

We note the visual-photometric record first, then its modern interpretation. Eta's photometric history was reviewed by, for example, Innes (1903), Gratton (1963), Walborn & Liller (1977), van Genderen & Thé (1984), Viotti (1995), and Feast et al (1994). Apart from the shadowy possibility of ancient observations (Davidson 1987a, Humphreys & Davidson 1994), our knowledge of the object begins around the year 1600. From that time until the 1830s, it was usually reported as a second-magnitude but sometimes as a fourth-magnitude star. In those days it was called $\eta$ Argus or $\eta$ Navis or (by some loyal Englishmen) $\eta$ Roburis. It may have become unusually active around 1820 or 1830; then, beginning in 1837, Eta brightened more than usual and fluctuated between first and zero magnitude for almost 20 years—an event that we call the Great

Eruption. In 1843, it was briefly the second brightest extra–solar-system object in the sky, with $m_V \approx -1$ mag. After 1856 it faded, apparently stabilizing at seventh or eighth magnitude around 1870. Except for a Lesser Eruption between 1887 and 1895, Eta has appeared more stable since 1870 than it was in the centuries preceding the Great Eruption. Davidson (1989), Maeder (1992), and Humphreys & Davidson (1994) have commented that this behavior pattern is reminiscent of a geophysical geyser or a volcano: Irregular activity precedes a big eruption, followed by relative quiescence—possibly as a recurrent sequence of events. After 1900, a very pretty expanding circumstellar ejecta-nebula became visible (Section 4 below). Since 1940, Eta has gradually brightened, and it is lately of sixth-going-on-fifth magnitude (van Genderen et al 1993, 1994).

Twentieth-century astronomers realized that Eta is more than 1-kpc away, which makes the Great Eruption an extremely luminous event (Thackeray 1956a). They speculated that it might be an unusual nova, a slow supernova, or a massive protostar, or some sort of nonthermal object. Payne-Gaposchkin (1957) and Burbidge (1962) came closest to the truth when they called it an unstable yellow supergiant. Then Westphal & Neugebauer (1969) made the crucial discovery that most of Eta's luminosity now emerges at infrared wavelengths beyond 10 $\mu$m, in an amount comparable to what was seen at visual wavelengths during the Great Eruption. At about the same time, Pagel (1969a,b) estimated the reddening of some emission lines, which indicated considerable extinction by circumstellar dust. This combination of circumstellar visual extinction with powerful infrared emission clarified the situation: The central object's present luminosity is far greater than had been recognized before 1969 and is mainly at UV wavelengths; most of this energy flux is absorbed by dust that has formed in the ejecta from the Great Eruption; and the dust reradiates it as thermal IR emission. (Most of the visible-wavelength and UV light that does escape is first scattered by that same dust; see Section 4.1 below.) In Pagel's analysis, the central spectrum seemed likely to be nonstellar, but Davidson (1971) noted that a very–low-gravity supergiant star hotter than 24,000 K would be satisfactory and may well be unstable in the observed way. This completed the essential first stage toward our modern view of Eta.

With a few numbers, we can now interpret the visual-photometric history described above. First, the luminosity: Integrating the data summarized in Figure 3 of Cox et al (1995), the observed energy flux at wavelengths longer than 1 $\mu$m is $(2.7 \pm 0.4) \times 10^{-5}$ erg cm$^{-2}$ s$^{-1}$, a little higher than previous estimates (cf Davidson et al 1986). This represents thermal emission by dust; the correction for visual and UV light that escapes absorption by dust is probably between 5 and 30%, and we assume 10% (the details are too lengthy to discuss here). The total luminosity ($\mathcal{L}$) is thus found to be $1.9 \times 10^{40}$ erg s$^{-1} \approx 10^{6.7}\, \mathcal{L}_{\odot}$ if the distance is 2300 pc. The uncertainty may be about $\pm 20\%$.

In order to avoid violating the Eddington limit at the estimated luminosity, the present-day stellar mass must exceed 90 $\mathcal{M}_\odot$. The luminosity is appropriate for an evolved star whose initial mass was roughly 160 $\mathcal{M}_\odot$ (see e.g. Meynet et al 1994); because stars like this lose much of their mass during evolution, $\sim$120 $\mathcal{M}_\odot$ is a plausible guess for its present-day mass.

Following the examples of other very massive unstable evolved stars (see references cited in Humphreys & Davidson 1994), let us conjecture that pre-1830 variations represented changes in bolometric correction rather than luminosity. In other words, suppose that the luminosity remained nearly constant while fluctuations in radius and apparent temperature caused the spectral distribution to shift back and forth between visual and UV wavelengths. The luminosity quoted above corresponds to absolute bolometric magnitude $M_{bol} \approx -12.0$. If the circumstellar extinction was small before the Great Eruption, then, with $D \approx 2300$ pc and interstellar extinction $A_V \approx 1.7$ as mentioned in Section 2 above, the apparent magnitude was related to the absolute magnitude by

$$m_V \approx 13.5 + M_V \approx 13.5 + M_{bol} - BC \approx 1.5 - BC,$$

where $BC$ denotes the bolometric correction, $M_{bol} - M_V$, which depends mainly on the photospheric temperature. Sometimes Eta was seen as a fourth-magnitude star in the centuries before 1830; then $BC \approx -2.5$ mag, appropriate for a hot supergiant which radiates mainly in the UV range. The star seems to be in a similar state today (Davidson et al 1995), but now circumstellar dust further reduces the visual brightness. The 8-mag decline in apparent visual brightness following the Great Eruption was partly due to rapid formation of dust in the ejecta, and the twentieth-century brightening has largely been due to decreasing absorption by dust as the ejecta expand.

What about those pre-1837 occasions when Eta was seen at second magnitude rather than fourth? Then $BC$ was close to zero, which indicates an apparent temperature between 5000 and 10,000 K, so that most of the radiation emerged at visual wavelengths. Either the stellar radius had expanded by a factor of 10, or, more likely, the apparent photosphere was then located in a dense, opaque stellar wind far outside the normal stellar surface. For reasons explained by Davidson (1987b), the apparent temperature of a sufficiently dense stellar wind or explosion tends to be in the range 6500–9000 K, insensitive to details of the mass flow. (This is not to say that the star's radius remained constant; the stellar radius is essentially invisible under these conditions and may be ill-defined anyway.) In summary, the sparse historical record suggests that Eta oscillated fitfully between two modes before 1830, "hot quiescent" and "rapid mass loss." Other interpretations are possible but less straightforward.

The luminosity did increase in the Great Eruption of 1837–1857. During very rapid mass loss, the bolometric correction of a star (or rather its opaque wind) tends to be close to zero, as noted above, but it cannot be very positive. Therefore, when Herschel noticed that Eta had brightened to $m_V \approx 0.0$ in December 1837, the absolute bolometric magnitude must have been close to $M_{bol} \approx -13.5$. The luminosity had risen to $\sim 10^{7.3}\ \mathcal{L}_\odot$, exceeding the Eddington limit by a factor on the order of 4. The "photosphere" must have been almost as big as the orbit of Saturn. What made the situation truly remarkable was that it persisted so long; the fundamental dynamical timescale of this star or its envelope can be 20 days or 20 weeks, but it cannot be 20 years, the duration of the giant outburst. The total energy radiated in the whole event was on the order of $10^{49.5}$ ergs. The ejected mass was at least 1 $\mathcal{M}_\odot$ and may have been several times that, with ejection speeds of a few hundred kilometers per second (Section 4.1 below) and a kinetic energy on the order of $10^{48.8}$ ergs.

Since the late 1970s, a likely evolutionary context for Eta has been recognized: It may be an extreme member of a class of very massive stars called LBVs (Luminous Blue Variables, perhaps not an adequately descriptive term), which we have reviewed at length elsewhere (Humphreys & Davidson 1979, 1984, 1994). According to an empirical scenario based on the observed distribution of stars in the upper H-R diagram, a star with mass $>60\ \mathcal{M}_\odot$ evolves in roughly the following manner. After leaving the main sequence, it expands at roughly constant luminosity, moving directly to the right in the H-R diagram. It is potentially unstable because its $\mathcal{L}/\mathcal{M}$ ratio is of the same order of magnitude as the Eddington limit. When it reaches some critical radius or surface temperature, the star suddenly loses mass in one or more LBV eruptions, whose causes are not yet understood. Such drastic mass loss causes the star to shrink and move back to the left in the H-R diagram. As the rapid mass loss causes the $\mathcal{L}/\mathcal{M}$ ratio to increase, making the star fundamentally less stable, the instability limit follows the star toward the left. At this point the star is an LBV. Eventually it will probably become a Wolf-Rayet star, but the instability prevents it from ever experiencing the pleasures of red supergianthood. (Authors of modern evolutionary tracks for very massive stars sometimes forget to emphasize that rapid LBV-style mass loss is an empirical, not a theoretically understood, feature of their models.) In most ways Eta fits the qualitative description of an LBV: It is in the expected part of the H-R diagram, it is indeed evolved (the surface material has been processed by nuclear reactions in the core of the star; see Davidson et al 1982, 1986), and its pre-1830 events looked like LBV eruptions. The Great Eruption was more dramatic than any other well-studied LBV event, but that is probably acceptable because Eta is more luminous than the other known Galactic LBVs. Today the consensus view is that Eta is probably an extreme LBV; if it is something else, then it is even more mysterious. Thus we add another question to our special list:

{$Q3$} Is Eta really an LBV?

The most promising hypothetical mechanisms to explain LBV eruptions involve high opacities and radiation pressure in the outer stellar layers and seem quite applicable to Eta. One class of instability occurs in layers whose temperatures are on the order of $10^{5.5}$ K, where high iron opacities cause dynamical trouble in the presence of strong radiation pressure (Stothers & Chin 1993, 1994; Glatzel & Kiriakidis 1993; Cox et al 1993; Soukup et al 1994). Instabilities of this type are so complex that it is difficult to be sure that competing models are fundamentally different. Another possible type of instability involves a "modified Eddington limit" that is conjectured to occur in or near the photosphere. This idea is essentially that radiation pressure becomes more dangerous when the temperature decreases below 30,000 K because the opacity tends to increase as concentrations of $H^0$, $Fe^{++}$, and $Fe^+$ become perceptible. (The classical Eddington limit includes only scattering by free electrons and is almost independent of temperature and density.) The modified Eddington limit has never been proven—in fact we know of no serious attempt to demonstrate it by a careful stability analysis—even though it seems intuitively plausible and was proposed long ago (see Davidson 1971, 1987a, 1989; Humphreys & Davidson 1984; Appenzeller 1986, 1989; Lamers & Fitzpatrick 1988; Pauldrach & Puls 1990; Gustafsson & Plez 1992; Lamers & Noordhoek 1993). Section 5 of Humphreys & Davidson (1994) should also be consulted for cautionary remarks about this possible surface instability. Our fourth crucial question is

{$Q4$} does the modified Eddington limit produce an instability of the desired sporadic type?

Whether it does or not, the other, deeper type of instability mentioned above must also be reckoned with.

In general, one or more of the proposed LBV instability mechanisms may explain Eta's Great Eruption and earlier eruptions, but other possibilities are conceivable and we are disappointed that so little theoretical work has been done on this problem. Herschel's original question that we quoted in Section 1 ({$Q1$}, "What origin can we ascribe to the sudden flashes and relapses?") brings to mind supplementary problems and subquestions:

{$Q1a$} Was the Great Eruption caused by the same mechanism as the smaller outbursts that preceded it?

{$Q1b$} Why did the brightness fluctuate drastically and quickly during the Great Eruption? (See historical references cited at the beginning of this section.)

{*Q*1*c*} Why did that giant event end? (A simple pertinent stellar-structure calculation was recommended in Davidson 1987a and on p. 105 of Davidson 1989.)

{*Q*1*d*} Considering that Eta has generally fluctuated less since 1870 than it did before 1830, so that the Great Eruption seems to have stabilized it for the time being, why did an isolated 1-mag Lesser Eruption occur around 1890?

Modern discoveries related in the following sections have not answered these questions, but some of them probably convey significant hints.

## 4. THE HOMUNCULUS NEBULA

Some of the most familiar images made with the Hubble Space Telescope (HST) have been those of Eta's bipolar ejecta (Figure 1) (cf pictures in Humphreys & Davidson 1994; Morse 1996a,b; Currie et al 1996). As we shall explain below, the two polar lobes were recognized in the pre-HST era and qualitatively are not very surprising. The ragged equatorial midplane debris-disk, however, first became obvious in HST data and puzzles us in several ways.

The low-resolution visual appearance of the circumstellar nebula 50 years ago led Gaviola (1950) to call it the Homunculus, a name that has stuck regardless of

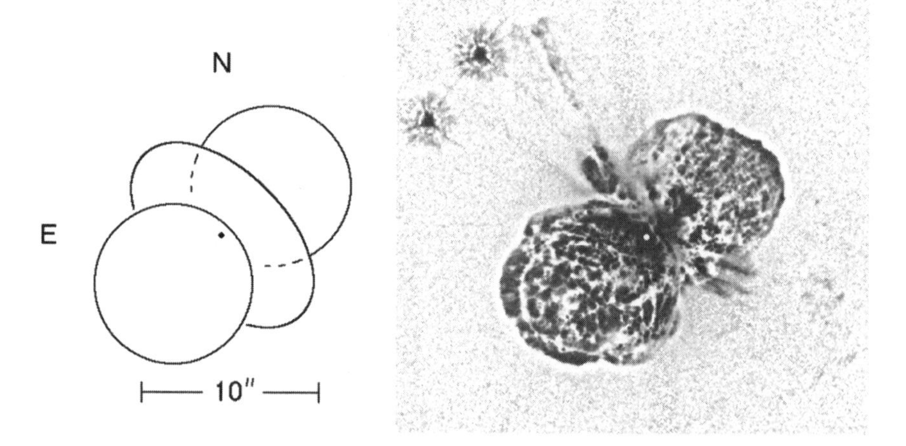


*Figure 1* The "Homunculus Nebula" of ejecta around Eta. The picture on the right is an unsharp-masked image based on several violet-wavelength images obtained with the HST/WFPC2(PC), combined in such a way as to nearly eliminate diffraction spikes. A very wide range of surface brightnesses is depicted here. The long jet-like feature to the upper left is a real structure, and the two large spots near it are stars.

its connotations. The nebula's expansion has been observed over many years, leaving little doubt that the polar lobes were ejected in the Great Eruption, probably in the early 1840s (Thackeray 1949; Gaviola 1950; Ringuelet 1958; Walborn et al 1978; Currie et al 1996). The outward velocity at the pole or apex of each lobe is close to 650 km/s, and the tilt angle between the polar axis and our line of sight is $57 \pm 10°$, based on radial velocities, proper motions, and the projected shape of the equatorial debris disk (Thackeray 1961a; Walborn et al 1978; Meaburn et al 1987, 1993b; Hillier & Allen 1992; Duschl et al 1995). Polarimetry and spectroscopy show that the visual-wavelength Homunculus is mainly a reflection nebula, wherein dust grains scatter light that came from the central object, although some intrinsic emission lines are also present (Thackeray 1956b, 1961a; Visvanathan 1967; Warren-Smith et al 1979; Meaburn et al 1987, 1993b; Hillier & Allen 1992). Most of Eta's luminosity emerges from the Homunculus as thermal IR emission from the dust grains, which are heated by the star's UV light to temperatures ranging from 200–1000 K (Westphal & Neugebauer 1969; Gehrz et al 1973; Harvey et al 1978; Hyland et al 1979; Hackwell et al 1986; Russell et al 1987; Meaburn et al 1993b; Smith et al 1995; Aitken et al 1995; Rigaut & Gehring 1995). As noted in Section 3 above, the total luminosity is probably about $10^{6.7} \mathcal{L}_{\odot} \approx 10^{40.3}$ erg s$^{-1}$.

The word "bipolar" was often applied to the Homunculus around 1980, even though images did not show much detail at that time (Warren-Smith et al 1979; Hyland et al 1979; Mitchell et al 1983). The IR data obtained by Hackwell et al (1986) clearly showed a pair of hollow lobes, and Meaburn et al (1987) and Allen (1989) accurately described them in light of other information. The first published HST image was initially misinterpreted, largely because its central region was very overexposed (Hester et al 1991); but the two lobes and equatorial material were visible there. The two-lobe and midplane morphology was obvious in other early HST images (Ebbets et al 1993a; Davidson 1993). Later and better images are shown, for example, by Morse (1996a,b) and Currie et al (1996), and in Figure 1. Ironically, the first good discussion of the newly recognized equatorial–ejecta disk did not involve HST data; Duschl et al (1995) belatedly described its appearance in an image that had been obtained in 1985 by a special ground-based technique. Duschl and coworkers' paper was important because the equatorial structure presently seems harder to understand than the polar lobes. Here we discuss the lobes first, then the equatorial material. There is much to be said about the bipolar lobes, but in our opinion the most interesting developments are recounted in Section 4.2 and later parts of this review.

## 4.1 *The Bipolar Lobes*

The two lobes appear nearly circular, and each has a diameter of about $8.5''$ $\approx 0.1$ pc $\approx 10^{17.5}$ cm. As a first approximation they are hollow, because they show limb-brightening in infrared images and some emission lines have radial velocity components that correspond to both the near and far sides (Hackwell et al 1986; Allen 1989; Hillier & Allen 1992; Meaburn et al 1993b; Smith et al 1995; Aitken et al 1995). Particularly interesting velocity data were shown by Hillier & Allen (1992); their Figure 12 and Figure 13 give a good impression of how complicated the situation is. According to Smith et al (1995), some lower-density material in the interior cavities is also detectable—this may be dust in the continuous stellar wind or slower-moving condensations ejected in the Great Eruption, or both. The lobes have variously been interpreted as separated ellipsoids connected by conical flows, as intersecting ellipsoids, or as polar caps on a single spheroidal surface (e.g. see Hillier & Allen 1992, Meaburn et al 1993b), but most of the radial velocity data and projected shapes in the images are consistent with a model in which two nearly spherical lobes are tangent to each other at the star's location ("osculating spheres"; Aitken et al 1995).

High gas densities in the equatorial plane tend to form ejecta into bipolar lobes (cf Frank et al 1995; Königl 1982). Such a morphology was once presumed to be evidence for a close binary star system (Warren-Smith et al 1979; Mitchell et al 1983), but a more recent consensus view is that moderate (not necessarily fast) rotation of a single star can create an equatorial density enhancement of the required type (e.g. Lamers & Pauldrach 1991; Bjorkman & Cassinelli 1993; Owocki et al 1994). This seems especially likely for a very luminous star, whose stability may be reduced near the equator. If lower surface temperatures make the star less stable, then it is possible for the equatorial effect to increase even though the star is expanding. Internal redistribution of angular momentum may also play a role. Thus a single star may produce a bipolar nebula after all, perhaps in more than one way. The binary- vs single-star question is still quite undecided for Eta (see Section 5.3 below). In either case, quantitative details of the lobes should convey information about the Great Eruption.

Most of the visible-wavelength luminosity escaping from the Homunculus is light from the central star, scattered by dust grains in the lobes, as noted above. The present-day apparent magnitude of the entire object is $m_V \approx 5.7$ (van Genderen & Thé 1984; van Genderen et al 1993, 1994; Whitelock et al 1994), far brighter than light coming directly from the central star ($m_V \approx 8.4$; Davidson et al 1995). If the star is hot, say 25,000 or 30,000 K, and has the luminosity quoted above, then its apparent brightness would be roughly $m_V \approx 4$ if there were no circumstellar dust. (All of these quoted magnitudes do include interstellar extinction.) Therefore, the fraction of visual-wavelength light that escapes after scattering in the Homunculus is on the order of 20% (5.7 vs 4 mag). This rather crude result is consistent with analyses based on emission-line ratios (e.g. Pagel 1969a; Viotti 1969; Davidson 1971; Leibowitz 1977; Hillier & Allen 1992), and it suggests that the average extinction thickness of each lobe is roughly $A_V \sim 1$ or 2 mag (see Figure 2 of Davidson & Ruiz 1975, even though the geometry is oversimplified there). With a "normal" gas-to-dust ratio, a simple thin shell of the observed size produces that amount of extinction if its mass is of the order of 1 $\mathcal{M}_\odot$, i.e. a total of $\sim$2 $\mathcal{M}_\odot$ for the two lobes. Typical gas densities in the lobe shells are probably $10^5$–$10^6$ atoms cm$^{-3}$.

The lobes have a mottled appearance in the HST images (Figure 1) and are also clumpy in IR images (Hackwell et al 1986; Allen 1989; Smith et al 1995). The clumps are much larger than their internal speed of sound multiplied by the expansion age. Presumably they were formed by gas-dynamic instabilities, perhaps involving the presence of a continuous stellar wind inside each massive ejecta shell (cf instabilities shown by García-Segura et al 1996). Because the individual blobs are well resolved in the best HST images, one can measure characteristic size scales on the order of $10^{16}$ or $10^{16.5}$ cm for the instability or instabilities, on the leading edge of the southeast (nearer) lobe and also on the sides of both lobes. Thus we propose a theoretical question:

{*Q*5} What physical information can be deduced from these observable size scales?

Perhaps, for instance, the stellar wind speed and density inside each lobe can be constrained in this way. A radiative transfer model of scattering among the clumps, and information about the geometrical thickness of each lobe, would also be useful.

Most of Eta's thermal infrared emission is produced in the lobe-shells, predominantly by dust grains with temperatures between 200 and 400 K (Harvey et al 1978; Hackwell et al 1986; Russell et al 1987). Grain temperatures are correlated with distance from the central object. Some hotter grains are present near the star, but significant absorption begins roughly $10^{16.7}$ cm from the star where grain temperatures are around 400 K. Models of the dust-grain parameters are of interest partly because they help to indicate the total mass of the ejecta; moreover, the grain formation has occurred under unusual conditions, in gas where most of the CNO is nitrogen rather than oxygen and carbon (see Section 4.3 below). Unfortunately, each analysis of dust in the Homunculus so far has used only a small subset of the observational information now available.

In the 1970s and 1980s, some authors attempted to explain the observed infrared continuum and the profile of the silicate emission feature near $\lambda \sim 10$ $\mu$m. The problem was not fully solved, though, and there have not been many recent analyses in the same vein. Most pre-1990 models assumed spherical or circular rather than bipolar distributions of dust. Mitchell & Robinson (1978) briefly reviewed earlier work and presented models with several different grain materials. The 10-$\mu$m feature tends to be broader at larger distances from the star, indicating that basic parameters of the grains are not uniform throughout the

Homunculus. Mitchell & Robinson (1978) proposed that the mixture of grain compositions varies with location. Mitchell & Robinson (1986) and Robinson et al (1987) later preferred a model wherein the average grain size increases with distance from the star. They assumed a mixture of two types of grains, silicates and a spectrally "featureless" component such as iron. Large grains with radii $\sim$1 $\mu$m were needed in the outermost regions. Hackwell et al (1986), Russell et al (1987), and Allen (1989) found that the true dust distribution is bipolar but "clumpy." Excellent maps of the 12.5- and 17-$\mu$m emission have been provided by Aitken et al (1995) and Smith et al (1995).

An alternative approach to the dust problem concentrates on the observed polarization of the reflected visual-wavelength light. In particular, the apparent polarized fraction is $\sim$25% at the southeast (nearer) end, where typical scattering angles are around 70°, and $\sim$37% at the northwest end, where scattering angles are 110° or so. Warren-Smith et al (1979) and Carty et al (1979) concluded that grains with radii on the order of 0.15 $\mu$m account for most of the visual-wavelength scattering. Because this result applies to the same locations where most of the UV absorption is ascribed to $\sim$1-$\mu$m grains in some models of grain heating, there may be a quantitative conflict here even if the grain-size distribution is broad. Meaburn et al (1993b) have improved the reflection data by observing polarization as a function of radial velocity in the H$\beta$ emission line at several locations in the Homunculus; this allowed them to discriminate between different scattering angles corresponding to the near and far sides of each lobe. Like earlier authors, Meaburn et al (1993b) found grain sizes around 0.15 $\mu$m necessary to account for the visual-wavelength polarization.

If grain parameters are known, the total mass of emitting dust can be estimated from its infrared luminosity. On this basis, total dust masses of 0.01–0.03 $\mathcal{M}_\odot$ have been quoted for the Homunculus (e.g. Mitchell & Robinson 1978; Hackwell et al 1986; Cox et al 1995), which correspond to a total ejected mass of 1–3 $\mathcal{M}_\odot$ if we assume a conventional gas/dust ratio. Robinson et al (1987) commented that an estimate of this type varies by a factor of perhaps two across a range of 0.1–1 $\mu$m in assumed grain size. Moreover, the gas/dust ratio may not be "conventional" (recall that the material is oxygen- and carbon-poor; Davidson et al 1986). In summary, the mass of the Homunculus based on infrared emission is consistent with that based on visual-wavelength scattering noted above—a reasonable "best bet" is 3 $\mathcal{M}_\odot$ altogether—but both estimates are quite imprecise. One may question whether it is logically consistent to measure both the total luminosity and the total dust mass from the infrared emission; for instance, if dust is thick enough to intercept most of the star's luminosity, then we should not be surprised if excess dust in some directions is shielded, unheated, and thus not included in the mass estimate.

Obviously a new analysis of the situation is needed to help answer the following questions:

{$Q6$} What is the total mass of the emitting dust?

{$Q7$} What is the latitudinal distribution of the dust in the Homunculus, and thus of the ejecta?

{$Q8$} What is the gas/dust mass ratio?

This last quantity is needed for an estimate of the total ejected mass and energy. A satisfying model would account for the position-dependent IR spectrum, the visual-wavelength polarization, and the rather "gray" (wavelength-insensitive) extinction curve for the central star at blue and UV wavelengths (Section 5.2 below). Regarding latitudinal dependences, models described by Allen (1989) suggest that most of the mass ejected in the Great Eruption was within 20° of the equatorial plane; IR images (Hackwell et al 1986; Aitken et al 1995) and the HST images give us the same impression. Such an equatorial concentration (not the same thing as a disk) would be relevant to the problem of Eta's basic instability mechanism.

Dust may be expected to condense in the present-day stellar wind at distances on the order of $10^{16}$ cm $\sim 0.2''$ from the star, where grain temperatures are around 1000 K (typical formulae for $T_{dust}$ are given, for example, by Smith et al 1995 and Davidson et al 1997). Andriesse et al (1978) and Men'shchikov et al (1989) attempted to model the grain formation process, but they invoked unrealistically high mass-loss rates for the star, between 0.01 and 0.1 $\mathcal{M}_\odot$ year$^{-1}$. Assuming that the mass flow covers a large solid angle, rates higher than $\sim$0.003 $\mathcal{M}_\odot$ year$^{-1}$ would cause a significant fraction of the star's luminosity to be absorbed by hot grains at small distances, which would produce more short-wavelength IR radiation than is observed (Davidson et al 1986). Modern estimates of Eta's current mass-loss rate, based on gas rather than on dust, are indeed much less than 0.01 $\mathcal{M}_\odot$ year$^{-1}$ (Section 5.2 below). In principle, a large mass flow might be concealed in the equatorial plane, but there is no evidence that this is the case, and the models by Andriesse and coworkers and Men'shchikov and coworkers were not of such a type.

The question of dust formation during the Great Eruption leads to an interesting ambiguity: We do not really know when that event ended. Eta began to fade rapidly at visual wavelengths after the year 1856. That was also about the time when ejecta with typical speeds of $\sim$500 km s$^{-1}$ reached the dust-formation distance, say $10^{16.4}$ cm with the high luminosity then prevalent. Therefore the visual fading may have been caused by dust formation and circumstellar extinction rather than cessation of the outburst; the eruption may then have continued for an indefinite number of years after 1856. On the other hand, perhaps the luminosity did begin to decrease before much dust had formed, in which case the diminished heating would have triggered rapid grain formation. We do not know how to decide between these possibilities, and the truth may be a combination of the two.

Spectroscopy of emission lines produced by gas in the lobes (i.e. not the reflected light) is too complex a topic to discuss here, especially since almost no theoretical work has been done yet concerning the heating and excitation. The Homunculus UV-to-near-IR spectroscopy has been described by Davidson et al (1986), Baratta et al (1995a,b), Dufour (1989), Hillier & Allen (1992), and Hamann et al (1994); most other reported spectroscopy of Eta has included the star and the bright atypical ejecta near it (Section 5 below), which makes the Homunculus contribution difficult to separate. Spectroscopic data on condensations in the Homunculus have also been obtained with the HST but have not yet been published (RJ Dufour and coworkers, in preparation; and K Davidson and coworkers, in preparation). Theoretical analyses would be useful.

A final interesting detail about the polar lobes is that Briggs & Aitken (1985) found the IR emission to be slightly polarized. This is a much smaller effect than the polarization of shorter-wavelength reflected light, and it indicates that the nonspherical grains tend to be locally aligned. A magnetic field imbedded in the gas is probably responsible. Aitken et al (1995) show maps of the IR polarization that give an impression of magnetic field lines compressed near the surface of each bipolar lobe. It is not yet clear whether a surprisingly strong magnetic field is required.

## 4.2 *Surprising Equatorial Structure*

The ragged midplane ejecta disk seen in Figure 1 was not clearly recognized in the pre-HST era. It cannot be seen in IR maps like those of Hackwell et al (1986), Allen (1989), and Smith et al (1995). Warren-Smith et al (1979) did notice evidence in their polarization data for an equatorial disk or toroid. A few other authors interpreted a bright spot east-northeast of the star as one side of a disk or toroid (e.g. Hyland et al 1979; Rigaut & Gehring 1995), but that may have been a false clue, because HST images show a dense region in the northern "shoulder" of the southeast lobe at that location (cf Falcke et al 1996). The large-scale radial streaks in the equatorial debris are particularly interesting, as we discuss below.

At present, little is known about the illumination or excitation of the midplane structure and therefore about its density and mass. Our recent 2400- to 5000-Å HST spectroscopic data show that intrinsic emission lines (i.e. not reflected light) are more conspicuous in some of the equatorial condensations than they are in the bipolar lobes. On the other hand, radial velocities shown in Figure 12 of Hillier & Allen (1992) indicate that He I $\lambda7065$ in the equatorial "fan" structure about $3''$ northwest of the star is mainly reflected light. Part of the northwest lobe about $5''$ north of the star appears to be partially obscured in

HST images, most likely by dust in an intervening part of the equatorial debris-disk. But if the equatorial material contains enough dust to cause reflection and extinction, why doesn't it produce noticeable amounts of thermal IR emission? A tentative conjecture is that equatorial dust close to the star is thick enough to shield the visible outer structure from UV light but that the red-wavelength light can get through. Because most of the original luminosity is UV, the large-scale equatorial dust is therefore not heated very strongly and may be cooler than 200 K. Because it is cool and also thinner than the dust in nearby parts of the polar lobes, the equatorial dust may thus be faint at $\lambda \sim 10$ or 20 $\mu$m. However, this is only a working hypothesis and a quantitative model is needed. Fortunately, HST spectroscopy of equatorial condensations is expected to become available in the near future.

{*Q*9} What is the mass of the equatorial ejecta?

At this time we can only venture a plausible guess. If dust in the plane has an average optical thickness $\tau \sim 0.5$ at visual wavelengths (based on the reflected He I $\lambda$7065 and obscuration arguments noted above), and if we use a conventional gas/dust ratio, then the indicated total mass is on the order of 0.1 or 0.2 $\mathcal{M}_{\odot}$. This may be a serious underestimate, because it is easy for additional mass to be inconspicuous, and material within $10^{16.8}$ cm of the star is not really included in our "guesstimate."

{*Q*10} When was the equatorial material ejected?

The answer is not as obvious as one might expect. Outward motions of several condensations in the Homunculus were observed during the years 1914–1945 (see Gaviola 1950; Ringuelet 1958). Those representing edges of the polar lobes were found to have extrapolated origins around 1840, during the Great Eruption. However, the estimated proper motions of two condensations—one northwest of the star, the other northeast—implied an origin around 1890, and those bright spots can be identified with features now seen in the equatorial material. Moreover, the proper motions and radial velocities of slow-moving equatorial blobs within $0.3''$ of the star, described in Section 5.1 below, indicate that they most likely originated several decades after the Great Eruption (Weigelt et al 1996; Davidson et al 1997). It is tempting to ascribe the entire equatorial structure to the Lesser Eruption seen around 1890, an unexpected development if true! Measurements of HST images reported by Currie et al (1996) appear at first sight to contradict this idea. However, those authors were seeking to determine the time of origin of the lobes (their result was about 1841), rather than looking for anomalous motions in the equatorial region, a task for which their HST data were somewhat unsuitable. Part of their analysis required the assumption that proper motions between 3″ and 7″ from the star obey one simple linear relation. Hence the age of the equatorial structure remains an unknown, with obvious relevance to theories of the star's instability. We hope that new HST data (radial velocities as well as images) will settle it in the near future.

{*Q*11} How were the radial streaks formed?

The existence of the large-scale equatorial disk does not seem very surprising, but maybe its detailed structure is. One can easily imagine that either a binary system or a rotating star with Eta's parameters can form an equatorial disk at radii of the order of 1 AU or $10^{13}$ cm (see e.g. Bjorkman & Cassinelli 1993; Owocki et al 1994), and that entrainment by surrounding outflow or some other process might then lead to a much larger disk with a size scale of $\sim 10^{17}$ cm, as observed. One might naively expect the large-scale structure to be either rather uniform or randomly mottled. In fact, the observed morphology consists largely of long radial "streaks," "spokes," "rays," or "fans" (Figure 1). Inspection of the images suggests that these are real alignments of ejecta, not merely illusions caused by shadowing. So far the main discussions of the radial streaks have been a few remarks by Duschl et al (1995) and Davidson et al (1997).

We doubt that these features can result directly from asymmetric eruptions. As noted by Davidson et al (1997), in such a model, each streak must be ejected within a time of 30 days or less, to avoid azimuthal smearing by stellar rotation in a single-star scenario or by orbital motion in a binary model. The local rate of kinetic energy production is then implausibly high.

A more appealing, fairly obvious idea is that the streaks may radiate outward like comet tails from compact concentrations of gas fairly close to the star. In this scenario, the streaks are material ablated from the hypothetical gas concentrations by fast-moving ejecta or conceivably by radiation pressure acting on dust grains. The slow-moving compact ejecta discussed in Section 5.1 below may be the gas concentrations needed in a model of this type. As usual, the devil is in the details; for example, the objects just mentioned seem insufficiently massive. Quantitative gas-dynamic models are needed. We note that most models for the radial streaks may entail serious departures from azimuthal symmetry in at least some of the ejection events.

We regard the radial streaks as warning arrows, pointing inward toward some extraordinary phenomenon near the central star.

## 4.3 *Unexplained Features Outside the Homunculus*

Thackeray (1949) reported the presence of outer nebulosities 10–25″ from the star ($10^{17.5}$–$10^{18}$ cm, or 0.1–0.3 pc), and Walborn (1976) gave them names that are often used today, such as the S (South) condensation and W (West) arc. (Some of these labels are misleading; for instance, the S condensation is roughly west of the central star, not south.) Their proper motions confirm that they are ejecta from the star, as noted below. Peculiar rope-like structures also exist outside the polar lobes but have not been studied much; for pictures of the outer structure, see Walborn et al (1978), Meaburn et al (1993a), and Figure 2 of Currie et al (1996).

The S, NS, and NN condensations are fast-moving outlying features related to radial streaks in the equatorial structure. They were ejected either soon after the Great Eruption or else a few decades later, depending on whether they decelerated (Walborn & Blanco 1988; Ebbets et al 1993b). HST images show that the S condensation is several condensations, not one. (They are faintly visible near the right edge of Figure 1.) The gas there has a strong intrinsic emission-line spectrum, which can be used to show that it was once processed in the CNO cycle in the stellar core; the He/H and N/H abundance ratios are somewhat larger than normal, while carbon and oxygen are less abundant than nitrogen (Davidson et al 1982, 1986; Dufour 1989). Allen et al (1985) also found a helium overabundance in the near-IR spectrum of the star, but results based on the low-density essentially nebular S condensation are less subject to geometrical and radiative-transfer complications. The processed composition of the ejecta is critical in one or two ways: It shows that the star is an evolved object, and it may cause the grain formation process to be abnormal.

The NN condensation is the outer end of a linear jetlike feature of striking appearance. (This is the jet in the upper left part of Figure 1.) Based on spectroscopic data, Meaburn et al (1993a) proposed that this structure has been formed by a series of fast-moving condensations ("bullets"). (How such bullets can be formed is not obvious, but the near-polar rope-like structures mentioned above may require an equally surprising explanation.) A feature whose shape reminds many astronomers of a bow shock is associated with the NN jet; Meaburn et al (1993a) argue that this is not really a bow shock but may represent gas "squirted" sideways in shocks. This idea reminds us that Dufour (1989, 1994) found velocities exceeding 2000 km $s^{-1}$ associated with the S condensation, much faster than expected, and that we have sometimes wondered whether "squirting" in a shock surface is a possible explanation. Altogether, these condensations offer several entertaining gas-dynamical puzzles.

The other outlying nebulosities are probably not equatorial and may have been ejected in outbursts that occurred before the Great Eruption. Walborn et al (1978) and Walborn & Blanco (1988) measured proper motions and found that some condensations must either have been ejected centuries before that event, or else they have been severely decelerated within the past century. Either possibility is quite remarkable! In the former case, we should like to know whether major outbursts occur every few hundred years; in the latter, it is not obvious why some gas was shot out at much higher speeds than were the bipolar lobes. Shocks decelerating the outlying ejecta may be responsible for the observed soft X-rays in that region (Chlebowski et al 1984; Corcoran et al 1995). Thus, another worthwhile question for research is as follows:

{$Q12$} Were the outer condensations ejected in the Great Eruption, or much earlier?

Continued proper motion measurements would be useful, but a new approach would be to inquire whether the suspected deceleration is physically reasonable. It should be possible to obtain good new ground-based spectroscopy of the condensations, then to use such data to estimate gas densities and masses, and finally to combine these results with shock-excitation models and X-ray data in order to assess the likely deceleration.

## 5. THE STAR AND DENSE MATERIAL NEAR IT

At normal ground-based spatial resolutions of $0.5''$ or worse, Eta—the central star—is part of a bright "core" at the center of the Homunculus Nebula. At visual wavelengths, this core is a seventh-or eighth-magnitude object with a tremendously complex emission-line spectrum, including very bright hydrogen features and hundreds of Fe II, [Fe II], and other lines (Figure 2). Our line of sight to the core passes through an edge of the southeast Homunculus lobe where local extinction is patchy (Figure 1); for reasons explained below, we think that the star is obscured more than some slow-moving ejecta near it. Consequently the apparent spectrum of the core is not dominated by the star as one might expect, but rather is a roughly equal mix of stellar and compact-nebulous components. In the past few years speckle techniques and HST data have made it possible to separate the components. Results for compact ejecta in the core have turned out to be suggestive in ways that no one anticipated. We shall discuss all the components of the core region in this section even though the ejecta are related to the topics of Section 4 above; our reason is that the core has been *observationally* somewhat distinct from the larger-scale Homunculus. High–spatial-resolution observations of the star and of ejecta in the core have usually been reported together.

There have been many observations of the total spectrum of the core region. Visual and near-IR wavelengths have been observed from the ground, and UV data were obtained with the International Ultraviolet Explorer (IUE) space telescope in the 1980s. The total resulting data set is very inhomogeneous in coverage and resolution (both spatial and spectral), and the spectrum may have varied among the almost random times of observation. The spectrum of the core in wavelength intervals from UV to near IR has been described by Gaviola

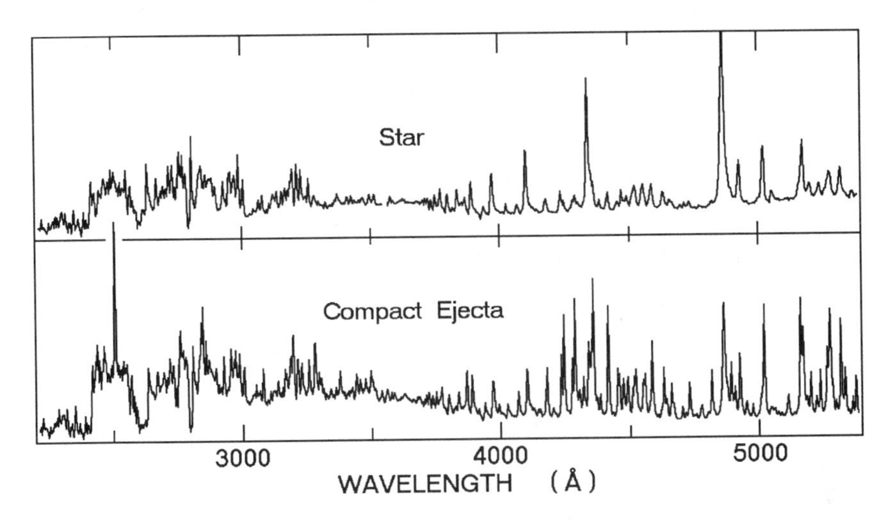


*Figure 2* Spectra of Eta obtained with the HST/FOS (Davidson et al 1995). The upper tracing shows $F_\lambda$ for the star itself, with only broad permitted lines (this scale is insufficient to show their widths). The lower tracing represents the three brightest slow-moving ejecta mentioned in Section 5.1, only about 0.2″ from the star, with many narrow permitted and forbidden lines; the Fe II and [Fe II] forest continues to much longer wavelengths beyond those shown here. The vertical scales are approximately the same for both tracings.

(1953), Thackeray (1961b, 1962, 1967), Aller (1966), Aller & Dunham (1966), Rodgers & Searle (1967), Aitken et al (1977), Johansson (1977), Walborn & Liller (1977), Zanella et al (1984), Allen et al (1985), Davidson et al (1986), McGregor et al (1988), Viotti et al (1989), Hillier & Allen (1992), Hamann et al (1994), and Johansson et al (1995). In general, the spectrum is too complex to discuss here, and several types of theoretical models are needed.

## 5.1 *Peculiar Compact Ejecta Near the Star*

Weigelt & Ebersberger (1986) and Hofmann & Weigelt (1988) used speckle techniques to detect three remarkably compact objects between 0.1″ and 0.3″ northwest of the star. Initially, these objects seemed theoretically implausible whether they were regarded as companion stars or as ejected blobs (Davidson & Humphreys 1986). In the 1990s, we have found that they are ejecta moving at low speeds on the order of 50 km $s^{-1}$. Their outward proper motions have been observed, they produce forbidden emission lines, and numerous other similar but fainter objects have been detected (Weigelt et al 1995, 1996; Davidson et al 1995). Their distances from the star are on the order of $10^{16}$ cm, a few light-days.

An initial reluctance in the 1980s to interpret these objects as ejecta was based on their surprising brightness relative to the star. One of them appears about 1/10 as bright as the star at far-red wavelengths and more than 1/2 as bright at UV wavelengths, but it is too small to intercept more than 1/300 of the energy flowing out from the star. This paradox can be explained by supposing that our view of the star is more heavily obscured by dust in the core region and in the southeast Homunculus lobe, so that the apparent ejecta/star brightness ratio is misleadingly high (Humphreys & Davidson 1994; Weigelt et al 1995; Davidson et al 1995). Line ratios reflected in the bipolar lobes confirm that forbidden lines from the low-speed ejecta would be much less conspicuous relative to the stellar spectrum if we could view the core region from a direction nearer the polar axis (Hillier & Allen 1992).

The mass of [Fe II]–emitting gas in one of the brightest blobs is most likely on the order of 0.002 $\mathcal{M}_{\odot}$, and the corresponding total within $10^{16.3}$ cm of the star may be 10 times larger (Hamann et al 1994; Davidson et al 1995). Additional cool, dense, nonemitting gas may also be present. The size of each blob is most likely on the order of $10^{15}$ cm, with internal gas densities between $10^7$ and $10^{10}$ atoms cm$^{-3}$. Because the internal velocity dispersions are on the order of 10 km s$^{-1}$, these objects produce narrow emission lines. They also reflect the broad-line stellar spectrum; evidently some dust is present.

Radial velocities measured with the HST/Goddard High-Resolution Spectograph (GHRS) show that the slow-moving blobs are near the equatorial plane (Davidson et al 1997). Their outward speeds relative to the star are about 50 km s$^{-1}$, and the brightest ones are moving obliquely toward us. (Incidentally, these observations required resolutions of both $0.1''$ and 0.1 Å. So far as we know, this may have been the first occasion when such high spatial and spectral resolutions were needed together and were attained.) Among the questions opened by this intriguing result, perhaps the most important, are as follows:

{$Q$13} Why did the star eject such slow-moving material in its equatorial zone?

{$Q$14} Why was the eruption azimuthally asymmetric; why were such massive condensations produced?

Intuitively, one tends to associate low ejection speeds with a low escape velocity. Maybe the rotation speed is very significant in this case, or perhaps locally cooler temperatures near the star's equator changed the opacity enough to greatly reduce the effective gravity (true gravity minus the effect of radiation pressure) there. Either way, the eruption process appears to have been surprisingly inhomogeneous. The slow-moving equatorial ejecta may be evidence against a close-binary scenario, because orbital velocities of hundreds of kilometers per second are needed in such a model, and one might therefore expect characteristic speeds in the equatorial plane to be high, not low. But this is only an impression; quantitative models are needed! We believe that the slow-moving ejecta are crucial for understanding the stellar eruption mechanism.

Another unexpected result concerns the ejection epoch of the slow-moving ejecta. Proper motions indicate an age of roughly 100 years rather than 150 years (Weigelt et al 1996), and so do the radial velocity data (Davidson et al 1997). Either of these estimates has low weight by itself, but they agree and they are essentially independent of each other. An origin during the Great Eruption seems excluded for these objects at about a 95% level of confidence. Pending further results, one naturally suspects that the slow compact ejecta were produced in the Lesser Eruption around 1890. Question $\{Q9\}$ in Section 4.2, concerning the age of the larger-scale equatorial ejecta, also includes these smaller, denser objects.

$\{Q15\}$ Are the large-scale radial alignments in the equatorial midplane related to these dense slow-moving condensations?

Perhaps each radial streak is like a comet tail of gas streaming outward from one of the compact objects, a qualitative idea noted in Section 4.2. The number of compact blobs is comparable to the number of radial streaks (Weigelt et al 1996). The three brightest blobs are roughly aligned with the "fan" structure (sometimes called a "paddle") about 3″ northwest of the star, and polarization data may show a continuous connection (Falcke et al 1996). On close examination, the two most obvious blobs mark the sides of the "fan," not its center. A model connecting the radial streaks to the compact blobs has at least one quantitative difficulty: The total mass of the blobs based on [Fe II] lines, $\sim$0.02 $\mathcal{M}_\odot$, is a factor of 10 less than our estimate of the large-scale equatorial mass that was hypothetically ablated from the blobs in this model (question $\{Q9\}$ in Section 4.2 above). Both mass estimates are very crude, though.

Strange details have been found in the emission-line spectra of the slow-moving ejecta. Some of their narrow emission features require unusual fluorescent excitation mechanisms (Viotti et al 1989; Johansson & Hamann 1993; Hamann et al 1994; Johansson et al 1995). One set of four narrow Fe II lines in the HST/GHRS data is particularly mysterious (Johansson et al 1996; Davidson et al 1997). These lines have wavelengths near 2507 Å (they appear as one bright feature in Figure 2) and are thought to be radiatively excited, pumped by hydrogen Ly $\alpha$ or perhaps N I emission. They occur weakly in a few other objects, but Eta and the possibly similar star AE Andromedae in M31 are the only places where they are known to be intense (cf Viotti et al 1989; Humphreys & Davidson 1994; Davidson et al 1995). The large total brightness of the $\sim$2507 Å lines in the slow-moving gas near Eta is not understood, but we are most puzzled by their intensity ratios. These four lines represent transitions between two upper and two lower levels of $Fe^+$, and their intensity ratios are known from laboratory measurements and confirmed in the spectra of a few astronomical objects where the lines are weak features. In Eta's slow-moving ejecta, however, the brighter pair of lines (transitions to one of the lower levels) are far more intense than the other pair. One lower level is favored much more than it should be, seemingly in defiance of the atomic data. This anomaly is so difficult to explain that Johansson et al (1996) have proposed it as evidence for a natural UV laser, a recourse that is amazing if true.

{*Q*16} Why are some of these radiatively excited Fe II lines uniquely bright near Eta, and why do they have such peculiar intensity ratios?

## 5.2 *The Star*

The HST's spatial resolution has allowed us to observe the stellar spectrum of Eta separated from its nearby ejecta (Davidson et al 1995, 1997). The characteristic temperature and mass-loss rate appear to be in the previously expected ranges, and there is evidence that the stellar wind is not spherically symmetric.

For reasons explained in Section 2.4 of Humphreys & Davidson (1994), effective temperature $T_{eff}$ is a rather unsatisfactory parameter for a star with a very dense wind; a characteristic temperature $T_c$ defined by the continuum energy distribution is preferable. With reasonable corrections for extinction, the slope of Eta's visual-wavelength continuum clearly indicates that most of the luminosity is in the UV range. The total luminosity (Section 3 above) and the visual-wavelength brightness are mutually consistent if $T_c \sim 30{,}000$ K. This seems consistent with the very bright hydrogen emission lines, whereas Hillier & Allen (1992) have noted that helium lines would be brighter than observed if the temperature were much higher than this. Old-fashioned Zanstra estimates that led to the same result (Davidson 1971; Aitken et al 1977; Allen et al 1985) may have a rough indirect validity even though these researchers erroneously assumed that the hydrogen lines come from a compact nebula (Davidson et al 1995). The stellar wind absorbs most of the Lyman continuum photons ($\lambda < 912$ Å), and the luminosity emerging from the wind should be mainly at wavelengths between 1000 and 2000 Å. The photospheric radius is on the order of $10^{12.8}$ cm $\approx 100\ \mathcal{R}_\odot \approx 0.4$ AU, though the star itself may be somewhat smaller if its wind is opaque.

Unfortunately, the extinction by circumstellar dust (including the Homunculus) is poorly known for $\lambda < 4000$ Å. The visual-wavelength circumstellar obscuration is at least 4 mag, since the star's apparent visual magnitude is $m_V \approx 8.4$ in the HST/Faint Object Spectograph (FOS) data [see values of $F_\lambda$ quoted by Davidson et al (1995)], whereas it would be around 4 if there were no circumstellar extinction (Sections 3 and 4 above). With a normal wavelength dependence of the extinction, the apparent flux shown in Figure 2 would decline very noticeably toward shorter wavelengths. Because this does not occur, we conclude that the extinction is not as sensitive to wavelength as it is in the normal interstellar medium; this is reminiscent of the unusual extinction/reddening ratio for other stars around Eta noted in Section 2 above. The most likely explanation involves large average grain sizes in the Homunculus, **which are** consistent with the IR thermal emission models cited in Section 3.1 but not necessarily with the reflection-polarization models. Anyway, it is difficult to correct the spectrum of the star for extinction. Moreover, wavelengths between 2000 and 2400 Å are effectively blocked by large numbers of Fe II absorption lines in the stellar wind or the Homunculus, or both.

With such intense emission lines, Eta must have an unusually dense wind and a high mass-loss rate. White et al (1994) estimated a rate of $10^{-3.5}$ $\mathcal{M}_{\odot}$ year$^{-1}$ from radio observations, but their 3.5-cm data refer to a large region more than $10^{16}$ cm across because optical depths are high at such wavelengths. Since we know that the geometrical situation is complex at that size scale (Section 5.1 above), the result is quantitatively doubtful. Cox et al (1995) used millimeter waves to observe a region about $10^{15}$ cm across, and their average result was $\sim 10^{-2.6}$ $\mathcal{M}_{\odot}$ year$^{-1}$. They noticed fluctuations that were probably real. Davidson et al (1995) used the brightness, width, and wings of the H$\beta$ emission line in the HST/FOS data to find a mass-loss rate of $\sim 10^{-3}$ $\mathcal{M}_{\odot}$ year$^{-1}$. In principle this was a more fundamental approach than the millimeter-wave observations because it referred to the stellar wind proper rather than a large surrounding region; however, uncertainties concerning the extinction correction were severe, and a simplified analysis was used. The millimeter-wave and H$\beta$ results are consistent enough to be very useful: Evidently Eta's present-day mass-loss rate is impressively high, probably exceeding $10^{-3.3}$ $\mathcal{M}_{\odot}$ year$^{-1}$, but it is not $10^{-2}$ $\mathcal{M}_{\odot}$ year$^{-1}$ or more as was suggested occasionally in the past (e.g. Andriesse et al 1978; Men'shchikov et al 1989) (cf remarks in Section 4.1 above). As White et al (1994) noted, the kinetic energy flux in the wind is smaller than the radiation flux by a factor on the order of 200. (A conceivable loophole in our argument is that a large, slow mass flow in the equatorial plane might be hard to detect. There is no reason to suspect that this is happening, however.)

Various hints suggest that Eta's wind is nonspherical, which, perhaps, is not very surprising given the morphology of the Homunculus. The simplified H$\beta$ profile analysis mentioned above (Davidson et al 1995) required the emission line photons to come from a wider range of wind velocities than one would normally see in a spherical configuration. Viotti et al (1989) made a similar remark based on asymmetric line profiles in their UV observations of the core region (cf Damineli et al 1993b). Hillier & Allen (1992) noted that the equivalent width of He I $\lambda$7065 differs between the core region and the spectrum reflected by the Homunculus, which possibly suggests that the stellar spectrum depends on latitudinal viewing angle. X-Ray observations are highly relevant; Chlebowski (1984) and Corcoran et al (1995) report hard thermal X rays in the core region that are most likely caused by wind speeds on the order of 2000 km $s^{-1}$. Because Eta's typical wind speed is much slower than that, we suspect that the speed varies with location above the stellar surface—perhaps qualitatively like a B[e] star, with a fast polar wind and a slow equatorial wind. Lamers & Pauldrach (1991) described a similar type of model that may be applicable to Eta, whereas polarization measurements of massive stars suggest that nonspherical objects are not uncommon (Schulte-Ladbeck et al 1993).

Hence our next question:

{*Q*16} What is the structure of the outer layers of Eta?

We hope that improved HST spectroscopy of the star will soon become available, and then models will be needed to assess temperatures, possible latitude dependences, and the mass-loss rate. A realistic model should not be spherical, nor should it assume a conventional wind velocity law $v(r)$.

## 5.3 *Fluctuations and Periodicities*

Although it is steadier than it was before 1830, Eta is continually fluctuating in brightness. There is also a long-term trend; the total observed brightness is gradually increasing at visual and near-IR wavelengths (van Genderen et al 1993, 1994; Whitelock et al 1994; Hamann et al 1994) and fading by perhaps 2% per year at $\lambda \sim 10\ \mu$m (Russell et al 1987; Smith et al 1995). These effects are largely due to expansion of the dusty Homunculus, but the star is not necessarily constant. A difficult long-term observational question is:

{*Q*17} What is the rate of change of Eta's total luminosity?

The star must have been far from thermal equilibrium after its Great Eruption and is probably readjusting on a thermal time scale for the outer layers, perhaps a few hundred years long. de Groot & Lamers (1992) and Lamers & de Groot (1992) have found such a trend for P Cygni, although they did not interpret it in quite this way.

Occasionally some of the narrow emission lines in Eta's core region almost disappear for durations of a few weeks (Ruiz et al 1984; Zanella et al 1984; Bandiera er al 1989; Bidelman et al 1993; Damineli et al 1993a, 1995). These narrow lines come from the slow-moving compact ejecta (Davidson et al 1997), so each event of this type probably represents a sensitive response of the excitation and ionization in the ejecta to a less obvious change in the star's energy distribution, as conjectured by Zanella et al (1984). (For instance, maybe $T_c$ becomes slightly cooler, and this substantially affects the supply of ionizing photons.) Radio and hard–X-ray outbursts reported by Duncan et al (1995) and Corcoran et al (1995) almost surely involve the same phenomenon, and a general synthesis is needed:

{$Q18$} How are the visual-wavelength spectroscopic events, the radio event, and the hard thermal X-ray flux related to changes in the star?

With regard to the X rays, perhaps there has been an increase in the fraction of the star's surface where the wind speed is very high (see Davidson et al 1997). Above some locations on the star's surface, a high-speed wind might arise where lower speeds formerly prevailed; then the present wind moving outward into the slower-moving material ejected earlier can produce hot X rays. High-speed–wind locations may have higher than average local photospheric temperatures, so the total UV flux is also affected. This is merely a qualitative speculation, and a quantitative model may not be so easy. One reason why the X rays are difficult to interpret is that their total luminosity is dwarfed by the other energy flows in the system; only a miniscule fraction of the wind's kinetic energy is converted to X rays.

Searches for periodic behavior in Eta have a long history; for examples from two different centuries, see Wolf (1863) and Feinstein & Marraco (1974). Recent attempts have led to one very interesting possible result, plus another that is a major surprise. The first has been reported by van Genderen et al (1994, 1995) and Sterken et al (1996), who find photometric evidence for a period between 50 and 60 days. They interpret this as evidence for a close binary companion, but the phenomenon is not as regular as one might expect in that case, and the suggested period may be reasonable for oscillations or rotation of a single star. Because decreased stability tends to lengthen a star's fundamental pulsation period, Eta's fundamental period may be weeks rather than days. A characteristic dynamic time in the envelope might be as long as 60 days; calculations are needed.

A much longer and more surprising period has recently emerged; in this case the periodic events are not at all subtle. An obvious fluctuation can be seen in the near-IR photometry of Whitelock et al (1994), where three peaks are separated by intervals of 5 or 6 years. Damineli (1996) has noticed the following points: (*a*) those peaks coincide with the spectroscopic events discussed above; (*b*) earlier spectroscopic events can be used to extend the time baseline and give a consistent period; and (*c*) some brightness peaks seen more than 150 years ago are consistent with the same period! With such a long baseline, Damineli (1996) estimates the period to be 5.52 years. Even if we distrust the nineteenth-century part of this story, it appears that all of the spectroscopic events observed since 1948 have been accurately consistent with a recurrence period of 5.5 years. (This was not noticed earlier because there are gaps in the record; several events probably occurred without being observed.) Damineli's hypothesis makes a very definite, easily testable prediction: There should be another spectroscopic event around the end of 1997. If it does transpire, this event will presumably be studied far more intensively than the previous ones.

A period of 5.5 years does not have an obvious theoretical interpretation for Eta! It is far longer than any fundamental dynamic time scales for this star, longer than the thermal time scale for the most unstable outer layers, and shorter than the thermal time scale for the entire star. In a single-star model, it may represent a thermal time scale for some particular set of outer layers. In a binary-star model, a period of 5.5 years would correspond to an orbital radius of about 15 AU, too large for strong interactions between the components unless the orbital eccentricity is 0.8 or more.

{$Q$19} If the 5.5 year period is real, what does it represent?

This brings us to the final, most inevitable question on our list:

{$Q$20} Is Eta a close interacting binary?

Many authors have found a binary scenario appealing from various points of view (e.g. Bath 1979; Warren-Smith et al 1979; Mitchell et al 1983; Baratta & Viotti 1989; Gallagher 1989; Viotti et al 1989; Men'shchikov et al 1989; van Genderen et al 1993). On the other hand there is no clear need to invoke a binary model; Eta's luminosity and surface conditions seem consistent with instability anyway (Section 3 above), whereas bipolar structure may be due to moderate rotation according to current views. If this object is a close binary star, it will not behave in the same way as more familiar massive binary systems. For instance, as noted in Section 5.5 of Humphreys & Davidson (1994), mass exchange cannot occur through the normal Lagrange point; instead, a primary component this close to the Eddington limit will probably mimic a rapid rotator, losing mass on the side facing away from its companion. Attempts to decide whether or not Eta is binary deserve high priority. We suspect that very careful radial velocity monitoring is the most promising approach.

When we undertook to write this article, we did not forsee that it would be a game of 20 Questions. Indeed, perhaps it is not; many other problems in addition to those that we have proposed here are also worth investigation. But the 20 that we have emphasized are a powerful subset, and most of them are intrinsically interesting.

ACKNOWLEDGMENTS

Our recent work on Eta has been supported by grants GO-6041 and GO-6501 from the Space Telescope Science Institute, which is operated by the Association of Universities for Research in Astronomy, Inc, under NASA contract NAS5-26555. It is a pleasure to thank our many coinvestigators in these projects, especially Dennis Ebbets, Sveneric Johansson, and Jon Morse, for a continuing series of puzzles and surprises regarding this topic, and, occasionally, solutions to some of the older conundrums. Joe Cassinelli gave us the idea of proposing so many specific research questions.



*Literature Cited*

Aitken DK, Jones B, Bregman JD, Lester DF, Rank DM. 1977. *Ap. J.* 217:103–7
Aitken DK, Smith CH, Moore TJT, Roche PF. 1995. *MNRAS* 273:359–66
Allen DA. 1989. *MNRAS* 241:195–207
Allen DA, Hillier DJ. 1993. *Proc. Astron. Soc. Aust.* 10:338–41
Allen DA, Jones TJ, Hyland AR. 1985. *Ap. J.* 291:280–90
Aller LH. 1966. *Proc. Natl. Acad. Sci. USA* 55:671–76
Aller LH, Dunham T. 1966. *Ap. J.* 146:126–41
Andriesse CD, Donn BD, Viotti R. 1978. *MNRAS* 185:771–88
Appenzeller I. 1986. In *Luminous Stars and Associations in Galaxies, IAU Symposium 116,* ed. H Lamers, C de Loore, pp. 139–43. Dordrecht: Reidel
Appenzeller I. 1989. In *Physics of Luminous Blue Variables, IAU Colloq. 113,* ed. K Davidson, AFJ Moffat, H Lamers, pp. 279–80. Dordrecht: Kluwer
Bandiera R, Focardi P, Altamore A, Rossi C, Stahl O. 1989. In *Physics of Luminous Blue Variables, IAU Colloq. 113,* ed. K Davidson, AFJ Moffat, H Lamers, pp. 279–80. Dordrecht: Kluwer
Baratta GB, Viotti R. 1989. In *Physics of Luminous Blue Variables, IAU Colloq. 113,* ed. K Davidson, AFJ Moffat, H Lamers, pp. 277–78. Dordrecht: Kluwer
Baratta GB, Cassatella A, Viotti R. 1995a. *Astron. Astrophys. Suppl.* 113:1–19
Baratta GB, Cassatella A, Viotti R. 1995b. *Rev. Mex. Astron. Astrofis. Ser. Conf.* 2:41–45
Bath GT. 1979. *Nature* 282:274–76
Bidelman WP, Galen TA, Wallerstein G. 1993. *Publ. Astron. Soc. Pac.* 105:785–86
Bjorkman JE, Cassinelli JP. 1993. *Ap. J.* 409:429–49
Briggs GP, Aitken DK. 1985. *Proc. Astron. Soc. Aust.* 6:145–47
Burbidge GR. 1962. *Ap. J.* 136:304–7
Carty TF, Perkins HG, Warren-Smith RF, Scarrott SM. 1979. *MNRAS* 189:299–304
Chlebowski T, Seward FD, Swank J, Szymkowiak A. 1984. *Ap. J.* 281:665–72
Corcoran MF, Rawley GL, Swank JH, Petre R. 1995. *Ap. J.* 445:L121–24
Cox AN, Morgan SM, Soukup MS, Guzik JA. 1993. *Bull. Am. Astron. Soc.* 25:1441
Cox P, Mezger PG, Sievers A, Najarro F, Bronfman L, et al. 1995. *Astron. Astrophys.* 297:168–74
Currie DG, Dowling DM, Shaya EJ, Hester JJ, Scowen P, et al. 1996. *Astron. J.* 112:1115–27
Damineli A. 1996. *Ap. J.* 460:L49–52
Damineli A, Cassatella A, Viotti R, Baratta GB, Carranza GJ, Villada M. 1995. *Rev. Mex. Astron. Astrofis. Ser. Conf.* 2:41–45
Damineli A, Viotti R, Cassatella A, Baratta GB. 1993a. *Space Sci. Revs.* 66:211–14
Damineli Neto A, Viotti R, Baratta GB, de Araujo FX. 1993b. *Astron. Astrophys.* 268:183–86
Davidson K. 1971. *MNRAS* 154:415–27
Davidson K. 1987a. In *Instabilities of Luminous Early Type Stars,* ed. H Lamers, C de Loore, pp. 127–41. Dordrecht: Reidel
Davidson K. 1987b. *Ap. J.* 317:760–64
Davidson K. 1989. In *Physics of Luminous Blue Variables, IAU Colloq. 113,* ed. K Davidson, AFJ Moffat, H Lamers, pp. 101–8. Dordrecht: Kluwer
Davidson K. 1993. In *Massive Stars: Their*

*Lives in the Interstellar Medium, ASP Conference Ser. 35,* ed. JP Cassinelli, EB Churchwell, pp. 483–88. San Francisco: Astron. Soc. Pac.

Davidson K, Dufour RJ, Walborn NR, Gull TR. 1986. *Ap. J.* 305:867–79

Davidson K, Ebbets D, Johansson S, Jorse JA, Hamann FW, et al. 1997. *Astron. J.* In press

Davidson K, Ebbets D, Weigelt G, Humphreys RM, Hajian AR, et al. 1995. *Astron. J.* 109:1784–96

Davidson K, Humphreys RM. 1986. *Astron. Astrophys.* 164:L7–9

Davidson K, Ruiz MT. 1975. *Ap. J.* 202:421–24

Davidson K, Walborn NR, Gull TR. 1982. *Ap. J.* 254:L47–51

de Groot M, Lamers HJGLM. 1992. *Nature* 355:422–23

Dufour RJ. 1989. *Rev. Mex. Astron. Astrofis.* 18:87–98

Dufour RJ. 1994. *Rev. Mex. Astron. Astrofis.* 29:213

Duncan RA, White SM, Lim J, Nelson GJ, Drake SA, Kundu MR. 1995. *Ap. J.* 441:L73–76

Duschl WJ, Hofmann K-H, Rigaut F, Weigelt G. 1995. *Rev. Mex. Astron. Astrofis. Ser. Conf.* 2:17–22

Ebbets D, Garner H, White R, Davidson K, Malumuth E, Walborn N. 1993a. In *Circumstellar Media in the Late Stages of Stellar Evolution, 34th Herstmonceux Conf.*, ed. RES Clegg, R Stevens, WPS Meikle, pp. 95–97. Cambridge: Cambridge Univ. Press

Ebbets D, Malumuth E, Davidson K, White R, Walborn NR. 1993b. In *Massive Stars: Their Lives in the Interstellar Medium, ASP Conf. Ser. 35,* ed. JP Cassinelli, EB Churchwell, pp. 263–65. San Francisco: Astron. Soc. Pac.

Falcke H, Davidson K, Hofmann K-H, Weigelt G. 1996. *Astron. Astrophys.* 306:L17–20

Feast MW, Whitelock PA, Warner B. 1994. *Astron. Astrophys.* 285:199–200

Feinstein A. 1982. *Astron. J.* 87:1012–21

Feinstein A. 1995. *Rev. Mex. Astron. Astrofis. Ser. Conf.* 2:57–67

Feinstein A, Marraco HG, Muzzio JC. 1973. *Astron. Astrophys. Suppl.* 12:331–50

Feinstein A, Marraco HG. 1974. *Astron. Astrophys.* 30:271–73

Fitzgerald PM, Mehta S. 1987. *MNRAS* 228:545–55

Forte JC. 1978. *Astron. J.* 87:1199–205

Frank A, Balick B, Davidson K. 1995. *Ap. J.* 441:L77–80

Gallagher JS. 1989. In *Physics of Luminous Blue Variables, IAU Colloq. 113,* ed. K Davidson, AFJ Moffat, H Lamers, pp. 185–94. Dordrecht: Kluwer

García-Segura G, Mac Low M-M, Langer N. 1996. *Astron. Astrophys.* 305:229–44

Gaviola E. 1950. *Ap. J.* 111:408–13

Gaviola E. 1953. *Ap. J.* 118:234–51

Gehrz RD, Ney EP, Becklin EE, Neugebauer G. 1973. *Astrophys. Lett.* 13:89–93

Ghosh SK, Iyengar KVK, Rengarajan TN, Tandon SN, Verma RP, Daniel RR. 1988. *Ap. J.* 330:928–36

Glatzel W, Kiriakidis M. 1993. *MNRAS* 263:375–84

Grabelsky DA, Cohen RS, Bronfman L, Thaddeus P. 1988. *Ap. J.* 331:181–96

Gratton L. 1963. In *Star Evolution* , ed. L Gratton, pp. 297–311. New York: Academic

Gustafsson B, Plez B. 1992. In *Instabilities in Evolved Supergiants* , ed. C de Jager, H Nieuwenhuijzen, pp. 86–90. Amsterdam: North-Holland

Hackwell JA, Gehrz RD, Grasdalen GL. 1986. *Ap. J.* 311:380–99

Hamann F, DePoy DL, Johansson S, Elias J. 1994. *Ap. J.* 422:626–41

Harvey PM, Hoffmann WF, Campbell MF. 1978. *Astron. Astrophys.* 70:165–68

Harvey PM, Hoffmann WF, Campbell MF. 1979. *Ap. J.* 227:114–20

Herbst W. 1976. *Ap. J.* 208:923–31

Herschel JFW. 1847. *Results of Astronomical Observations Made During the Years 1834–1888 at the Cape of Good Hope,* pp. 32–37. London: Smith Elder

Hester JJ, Light RM, Westphal JA, Currie DG, Groth EJ, et al. 1991. *Astron. J.* 102:654–57

Hillier DJ, Allen DA. 1992. *Astron. Astrophys.* 262:153–70

Hofmann K-H, Weigelt G. 1988. *Astron. Astrophys.* 203:L21–22

Humphreys RM. 1976. *Publ. Astron. Soc. Pac.* 88:647–55

Humphreys RM. 1978. *Ap. J. Suppl.* 38:309–50

Humphreys RM, Davidson K. 1979. *Ap. J.* 232:409–20

Humphreys RM, Davidson K. 1984. *Science* 223:243–49

Humphreys RM, Davidson K. 1994. *Publ. Astron. Soc. Pac.* 106:1025–51 (Note journal cover picture of the Homunculus)

Hyland AR, Robinson G, Mitchell RM, Thomas JA, Becklin EE. 1979. *Ap. J.* 233:145–53

Innes RTA. 1903. *Cape Ann.* 9:75B–78B

Johansson S. 1977. *MNRAS* 178:17P–20P

Johansson S, Davidson K, Ebbets D, Weigelt G, Balick B, et al. 1996. In *Science with the Hubble Space Telescope, STSCI/ST-ECF Workshop* , ed. P Benvenuti, FD Macchetto, EJ Schreier, pp. 361–62. Baltimore: Space Telescope Sci. Inst.

Johansson S, Hamman FW. 1993. *Phys. Scr.* T47:157–64

Johansson S, Wallerstein G, Gilroy KK, Joueizadeh A. 1995. *Astron. Astrophys.* 300:521–24

Königl A. 1982. *Ap. J.* 261:115–34

Lamers HJGLM, de Groot. 1992. *Astron. Astrophys.* 257:153–62

Lamers HJGLM, Fitzpatrick E. 1988. *Ap. J.* 324:279–85

Lamers HJGLM, Noordhoek R. 1993. In *Massive Stars: Their Lives in the Interstellar Medium, ASP Conf. Ser. 35,* ed. JP Cassinelli, EB Churchwell, pp. 517–21. San Francisco: Astron. Soc. Pac.

Lamers HJGLM, Pauldrach AWA. 1991. *Astron. Astrophys.* 244:L5–8

Leibowitz EM. 1977. *MNRAS* 178:271–77

Levato H, Malaroda S. 1981. *Publ. Astron. Soc. Pac.* 93:714–18

Levato H, Malaroda S. 1982. *Publ. Astron. Soc. Pac.* 94:807–10

Maeder A. 1992. In *Instabilities in Evolved Supergiants,* ed. C de Jager, H Nieuwenhuijzen, pp. 138–45. Amsterdam: North-Holland

Massey P, Johnson J. 1993. *Ap. J* 105:980–1001

McGregor PJ, Hyland AR, Hillier DJ. 1988. *Astrophys. J.* 324:1071–98

Meaburn J, Gehring G, Walsh JR, Palmer JW, López JA, et al. 1993a. *Astron. Astrophys.* 276:L21–24

Meaburn J, Walsh JR, Wolstencroft RD. 1993b. *Astron. Astrophys.* 268:283–93

Meaburn J, Wolstencroft RD, Walsh JR. 1987. *Astron. Astrophys.* 181:333–42

Megeath ST, Cox P, Bronfman L, Roelfsema PR. 1996. *Astron. Astrophys.* 305:296–307

Men'shchikov AB, Tutukov AV, Shustov BM. 1989. *Sov. Astron.* 33:416–20 (*Astron. Zh.* 66:801–8. In Russ.)

Meynet G, Maeder A, Schaller G, Schaerer D, Charbonnel C. 1994. *Astron. Astrophys. Suppl.* 103:97–105

Mitchell RM, Robinson G. 1978. *Ap. J.* 220:841–52

Mitchell RM, Robinson G. 1986. *MNRAS* 222:347–55

Mitchell RM, Robinson G, Hyland AR, Jones TJ. 1983. *Ap. J.* 271:133–42

Morrell N, Garcia B, Levato H. 1988. *PASP* 100:1431–35

Morse J. 1996a. *Astronomy* 24(9):25

Morse J. 1996b. *Sky Telesc.* 92(4):13

Niemela V, Morrell N, Feinstein A, eds. 1995. *Rev. Mex. Astron. Astrofis., Ser. Conf.* 2. Mexico DF: Consejo Nacl. Ciencia y Technol.

Owocki SP, Cranmer SR, Blondin JM. 1994. *Ap. J.* 424:887–904

Pagel BEJ. 1969a. *Nature* 221:325–27

Pagel BEJ. 1969b. *Astrophys. Lett.* 4:221–24

Pauldrach AWA, Puls J. 1990. *Astron. Astrophys.* 237:409–24

Payne-Gaposchkin C. 1957. *The Galactic Novae.* Amsterdam: North-Holland (New York: Dover, 1962)

Penny LR, Gies DR, Hartkopf WI, Mason BD, Turner NH. 1993. *Publ. Astron. Soc. Pac.* 105:588–94

Rigaut F, Gehring G. 1995. *Rev. Mex. Astron. Astrofis. Ser. Conf.* 2:27–35

Ringuelet AE. 1958. *Z. f. Astrophys.* 46S:276–78

Robinson G, Mitchell RM, Aitken DK, Briggs GP, Roche PF. 1987. *MNRAS* 227:535–42

Rodgers AW, Searle L. 1967. *MNRAS* 135:99–119

Ruiz MT, Melnick J, Ortiz P. 1984. *Ap. J.* 285:L19–21

Russell RW, Lynch DK, Hackwell JA, Rudy RJ, Rossano GS, Castelaz MW. 1987. *Ap. J.* 321:937–42

Schulte-Ladbeck RE, Clayton GC, Leitherer C, Drissen L, Robert C, et al. 1993. *Space Sci. Rev.* 66:193–98

Smith CH, Aitken DK, Moore TJT, Roche PF, Puetter RC, Piña RK. 1995. *MNRAS* 273:354–58

Smith RG. 1987. *MNRAS* 227:943–65

Soukup MS, Cox AN, Guzik JA, Morgan SM. 1994. *Bull. Am. Astron. Soc.* 26:907

Sterken C, de Groot MJH, van Genderen AM. 1996. *Astron. Astrophys. Suppl.* 116:9–14

Stothers R, Chin CW. 1993. *Ap. J.* 408:L85–88

Stothers R, Chin CW. 1994. *Ap. J.* 426:L43–46

Tapia M, Roth M, Marraco H, Ruiz MT. 1988. *MNRAS* 232:661–81

Thackeray AD. 1949. *Observatory* 69:31–33

Thackeray AD. 1956a. *Observatory* 76:103–5

Thackeray AD. 1956b. *Observatory* 76:154–55

Thackeray AD. 1961a. *Observatory* 81:99–104

Thackeray AD. 1961b. *Observatory* 81:102–4

Thackeray AD. 1962. *MNRAS* 124:251–62

Thackeray AD. 1967. *MNRAS* 135:51–81

Thé PS, Graafland F. 1995. *Rev. Mex. Astron. Astrofis. Ser. Conf.* 2:75–82

Thé PS, Takker R, Tjin A, Djie HRE. 1980. *Astron. Astrophys.* 89:209–13

van Genderen AM, de Groot MJH, Thé PS. 1993. *Space Sci. Rev.* 66:219–223

van Genderen AM, de Groot MJH, Thé PS. 1994. *Astron. Astrophys.* 283:89–110

van Genderen AM, Sterken C, de Groot M, Stahl O, Andersen J, et al. 1995. *Astron. Astrophys.* 304:415–30

van Genderen AM, Thé PS. 1984. *Space Sci. Rev.* 39:317–73

Viotti R. 1969. *Astrophys. Space Sci.* 5:323–32

Viotti R. 1995. *Rev. Mex. Astron. Astrofis. Ser. Conf.* 2:1–10

Viotti R, Rossi L, Cassatella A, Altmore A, Baratta GB. 1989. *Ap. J. Suppl.* 71:983–1009

Visvanathan N. 1967. *MNRAS* 135:275–86

Walborn NR. 1976. *Ap. J.* 204:L17–19

Walborn NR. 1982a. *Ap. J.* 254:L15–17
Walborn NR. 1982b. *Astron. J.* 87:1300–3
Walborn NR. 1995. *Rev. Mex. Astron. Astrofis. Ser. Conf.* 2:51–55
Walborn NR, Blanco B. 1988. *Publ. Astron. Soc. Pac.* 100:797–800
Walborn NR, Blanco B, Thackeray AD. 1978. *Ap. J.* 219:498–503
Walborn NR, Liller M. 1977. *Ap. J.* 211:181–86
Warren-Smith RF, Scarrott SM, Murdin P, Bingham RG. 1979. *MNRAS* 187:761–68
Weigelt G, Albrecht R, Barbieri C, Blades JC, Boksenberg A, et al. 1995. *Rev. Mex. Astron. Astrofis. Ser. Conf.* 2:11–16
Weigelt G, Davidson K, Hofmann K-H. 1996. *Astron. Astrophys.* In press
Weigelt G, Ebersberger J. 1986. *Astron. Astrophys.* 163:L5–L6
Westphal JA, Neugebauer G. 1969. *Ap. J.* 156:L45–48
White SM, Duncan RA, Lim J, Nelson GJ, Drake SA, Kundu MR. 1994. *Ap. J.* 429:380–84
Whitelock PA, Feast MW, Koen C, Roberts G, Carter BS. 1994. *MNRAS* 270:364–72
Whiteoak JB. 1994. *Ap. J.* 429:225–32
Wolf R. 1863. *MNRAS* 23:208–10
Zanella R, Wolf B, Stahl O. 1984. *Astron. Astrophys.* 137:79–84

*Annu. Rev. Astron. Astrophys. 1997. 35:33–67*

# THE SUN'S VARIABLE RADIATION AND ITS RELEVANCE FOR EARTH[1]

*Judith Lean*
EO Hulburt Center for Space Research, Naval Research Laboratory, Washington, DC 20375; e-mail: lean@demeter.nrl.navy.mil

KEY WORDS: solar variability, climate change, ozone, space weather, upper atmosphere, ionosphere

ABSTRACT

To what extent are changes in the Earth's global environment linked with fluctuations in its primary energy source, the radiation from a variable star, the Sun? A firm scientific basis for policy making with regard to anthropogenic greenhouse warming of climate and chlorofluorocarbon depletion of ozone requires a reliable answer to this question. Reduction of the vulnerability of spacecraft operations and communications to space weather necessitates knowledge of solar induced variability in Earth's upper atmosphere. Toward these goals, solar radiation monitoring and studies of solar variability mechanisms facilitate an understanding of the sources and amplitudes of the Sun's changing radiation. Interdisciplinary studies that link these changes with a wide array of terrestrial phenomena over the longer time scales of global change and the shorter time scales of space weather address the relevance of solar radiation variability for Earth. However, although numerous associations are apparent between solar and terrestrial fluctuations, full comprehension of the physical mechanisms responsible for the many facets of radiative Sun-Earth coupling remains to be accomplished.

## 1. INTRODUCTION

Solar radiation is an unfailing source of energy for the Earth. Without visible and infrared (IR) radiation from the Sun, Earth's surface temperature would be too cold to support life. Nor would there be energy to fuel photosynthesis or power the circulations of the lower atmosphere and oceans that profoundly influence living organisms. Lacking solar ultraviolet (UV) radiative inputs, Earth's middle atmosphere would be devoid of ozone and its upper atmosphere cold and unionized. Living things would be exposed to damaging high-energy solar photons. Society would lack the many benefits of Earth-orbiting spacecraft and global communication. For these reasons monitoring, understanding, and predicting the Sun's radiation variability and its terrestrial impacts is a component of the US Global Change Research Program (Committee on Earth Sciences 1989), the National Space Weather Program (1995), NASA's Mission to Planet Earth (Hartmann et al 1993), Space Environment Effects (NASA 1994) and Sun-Earth Connections Programs, and one of five key current topics in space physics research (National Research Council 1995).

Emergent from the Sun's atmosphere at wavelengths spanning the entire electromagnetic spectrum from X rays to radio waves, solar radiation achieves maximum flux levels at the visible wavelengths that are essential for photosynthesis and human processes. After nearly five billion years of evolution, the Sun has been such a reliable, steadily emitting star that its total radiative energy at the Earth is known historically as the solar "constant," a quantity whose value at the mean distance of the Earth from the Sun is measured to be $S = 1366$ W/m$^2$ (with an uncertainty of $\pm 3$ W/m$^2$).

Although more than a century of ground-based monitoring failed to determine whether the Sun's total radiative energy was truly constant (Hoyt 1979), fluctuations in other solar phenomena—especially the changing patterns of features on the Sun's surface—have long intrigued both solar and terrestrial observers. Dark sunspots appear to move across the face of the Sun as it rotates once every 27 days. Sunspots exhibit, as well, a pronounced quasiregular 11-year cycle in their number (discovered by Schwabe in 1843) that has been particularly strong in recent decades. Purported correlations of solar variability and climate have been reported for more than a century (Meadows 1975, Eddy 1976), then abandoned under statistical scrutiny and for lack of adequate physical mechanisms (Pittock 1978), only to reemerge (Reid 1991, Friis-Christensen & Lassen 1991) for renewed scrutiny (Kelly & Wigley 1992, Schlesinger & Ramankutty 1992, Reid 1995).

Discovery of the ozone layer in the 1920s, and recognition of the critical role of the Sun's UV radiation in its formation, initiated curiosity about the likely impact of solar variability on Earth's middle atmosphere (from 15–50 km above the surface). But attempts to identify the causes of long-term ozone and middle atmosphere changes evident in ground-based monitoring were inconclusive (Angell & Korshover 1973). Apparent correlations of ozone, zonal winds, and temperature with solar variability were detected (Paetzold 1973, Nastrom & Belmont 1978, Quiroz 1979), but the amplitudes of UV radiation variability were too poorly known to facilitate definitive simulations (Penner &

Chang 1979). And, like weather and climate near the surface, Earth's middle atmosphere experiences large seasonal and biannual variability that obscures decadal trends.

Robust associations did, however, emerge between solar variability and the behavior of neutral and ion densities in the Earth's upper atmosphere (more than 100 km above the surface). Radio wave propagation via the ionosphere (Ellison 1969) and the orbits of artificial satellites (Jacchia 1963) exhibited strong associations with the Sun's 11-year cycle. Rocket-based solar observations made above the Earth's atmosphere detected significant levels of extreme ultraviolet (EUV) radiation, and variations of these emissions (Hinteregger 1970) offered a physical mechanism for the observed relationships (Nicolet & Swider 1963, Roble 1976, Roble & Emery 1983).

Solar monitoring during the past few decades now affords a broader perspective of solar variability and its mechanisms. The familiar 11-year sunspot cycle is but one manifestation of our intrinsically inconstant Sun. Essentially all solar phenomena exhibit 11-year cycles, including radiative output (White 1977), particle and plasma emissions, interior oscillations, and perhaps even fundamental processes in the Sun's nuclear burning core (Cox et al 1991, Sonett et al 1991). Solar variability is also evident on other time scales. Oscillations below the Sun's visible surface and flaring of its outer atmosphere modulate the Sun's radiation on time scales of minutes to hours (Hudson 1987, Donnelly 1976). Proxies of solar activity archived by $^{10}$Be and $^{14}$C cosmogenic isotopes in ice cores and tree rings, respectively, indicate variability modes with periods near 87, 206, 512, and 2400 years (Stuiver & Braziunas 1993, Beer et al 1988).

Since the first curiosity-driven discoveries of solar terrestrial linkages, a more urgent imperative has emerged to quantify the relevance for Earth of the Sun's variable radiation (National Research Council 1994). Changes in our terrestrial environment can have extensive consequences for habitability, biodiversity, and technological infrastructures. Requisite for policy making to regulate greenhouse gas and chlorofluorocarbon (CFC) emissions is a firm assessment of the extent to which observed climate and ozone changes are of anthropogenic versus natural origin. Efficient utilization of the space environment requires operational specification of how changing neutral and ion densities impact the performances, lifetimes, and reentries of Earth orbiting spacecraft and the efficiency of radio frequency communications, with adequate predictive capability to support national needs.

## 2. SPECTRUM OF SOLAR RADIATION

A nuclear burning core generates the Sun's energy. Radiative processes transport this energy outward from 0.3 to 0.7 solar radii, and convective motions in the outer third of the Sun's interior deliver it to the visible solar surface. Above the Sun's surface—defined as that layer of the Sun from which radiation at 500 nm emerges at unit optical depth—a strongly ionized plasma extends for thousands of kilometers into space. This plasma comprises the solar atmosphere, the source of the Sun's radiation (for details see Foukal 1990, Cox et al 1991).

Temperature and composition at the Sun's surface and within the solar atmosphere (shown in Figure 1*a*) dictate the spectral structure of emitted radiation. In a solid angle subtended at the Earth at a distance of 1 AU, the Sun's radiant energy produces an irradiance spectrum (shown in Figure 1*b*) that has maximum levels near 500 nm and decreases to more than six orders of magnitude less in the X-ray and radio spectral regions. Some 48% of the Sun's total (spectrally integrated) irradiance of 1366 $W/m^2$ is at visible and near-IR

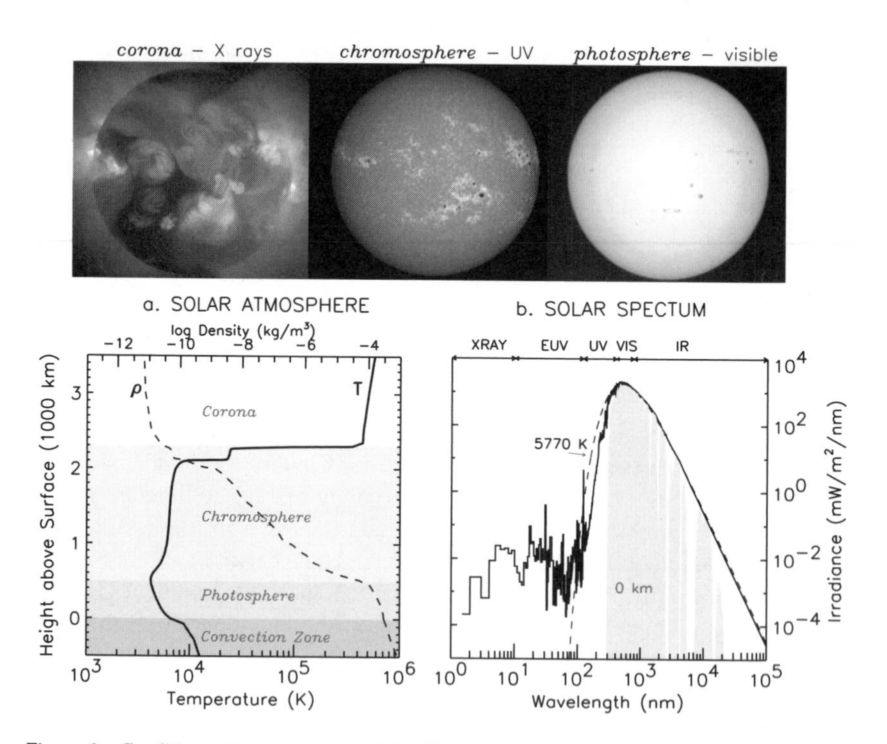


*Figure 1* Conditions of temperature and density at the Sun's surface and in its atmosphere (*a*) determine the solar irradiance spectrum (*b*). Magnetic fields distributed throughout the solar atmosphere alter the mean temperature (T) and density ($\rho$) profiles in (*a*), generating inhomogeneities in the distribution of radiation emitted from the solar disk, shown on January 10, 1992, for X-ray emission from the corona (*left image*), Ca II K emission from the chromosphere (*middle image*, made by the Big Bear Solar Observatory) and visible light from the photosphere (*right image*). Evolution of these surface inhomogeneities in turn modulates the irradiance spectrum.

wavelengths from 400–800 nm and emerges from the lowest layer of the solar atmosphere—the photosphere (Figure 1*a*)—with a spectral shape similar to the curve of black-body emission at 5770 K (Figure 1*b*).

Numerous spectral features punctuate the smooth continuum of photospheric radiation. Absorption and emission processes of ionized constituents in the Sun's atmosphere produce spectral features with widths from a few tenths to more than one angstrom. The least absorption by the solar atmosphere of the underlying photospheric continuum emission occurs in the vicinity of 1.6 microns ($\mu$) (1600 nm), and this radiation thus emerges from the deepest observable layer of the Sun. The spectrum at both longer and shorter wavelengths originates at greater heights because the opacity of the solar atmosphere is larger than at 1.6 $\mu$ (Vernazza et al 1976). In the UV spectrum (120–400 nm), absorption by many species—Al, Mg, Ca, O, Fe, H—depletes the underlying continuum radiation so much that the equivalent blackbody temperature of radiation near 160 nm is reduced from 5770 K to 4500 K. This temperature occurs at about 500 km above the Sun's visible surface and defines a minimum in the solar atmosphere temperature that delineates the upper boundary of the photosphere (Figure 1*a*). The spectrum from 75 $\mu$ to 300 $\mu$ also originates in the temperature minimum region as a result of increasing photospheric absorption at longer IR wavelengths.

Temperatures in the solar atmosphere rise slowly for another 1500 km above the temperature minimum (throughout the chromosphere), then increase dramatically in the transition region and corona. These hotter and higher atmospheric layers emit radiation shorter than 160 nm and longer than 300 $\mu$, primarily by processes other than thermal radiation. Local conditions of density and temperature produce complex EUV (10–120 nm) and X-ray (0.1–10 nm) spectra that are dominated by line emission from multiply ionized species with fluxes well in excess of the photospheric black-body curve (Mariska 1992) (Figure 1).

## 3. SOLAR RADIATION VARIABILITY

Extensive evidence identifies the Sun as a variable star. Diverse solar parameters such as those in Figure 2 record this variability over wide ranging spectral and temporal scales. Like the sunspot number (Figure 2*e*), the total radiative output (Figure 2*c*) and the entire solar spectrum exhibit pronounced quasi-11-year and 27-day cycles, illustrated in Figure 2 for X rays (Figure 2*a*), UV radiation (Figure 2*b*), and radio waves (Figure 2*d*). Linking all aspects of solar radiation variability, as reviewed previously by Newkirk (1983) and Hudson (1988), is a common source, the magnetic activity of the Sun.

### 3.1 *Solar Activity*

Magnetic fields weave through the outer third of the Sun's interior—the convection zone (Figure 1*a*)—and penetrate the solar surface, extending outward

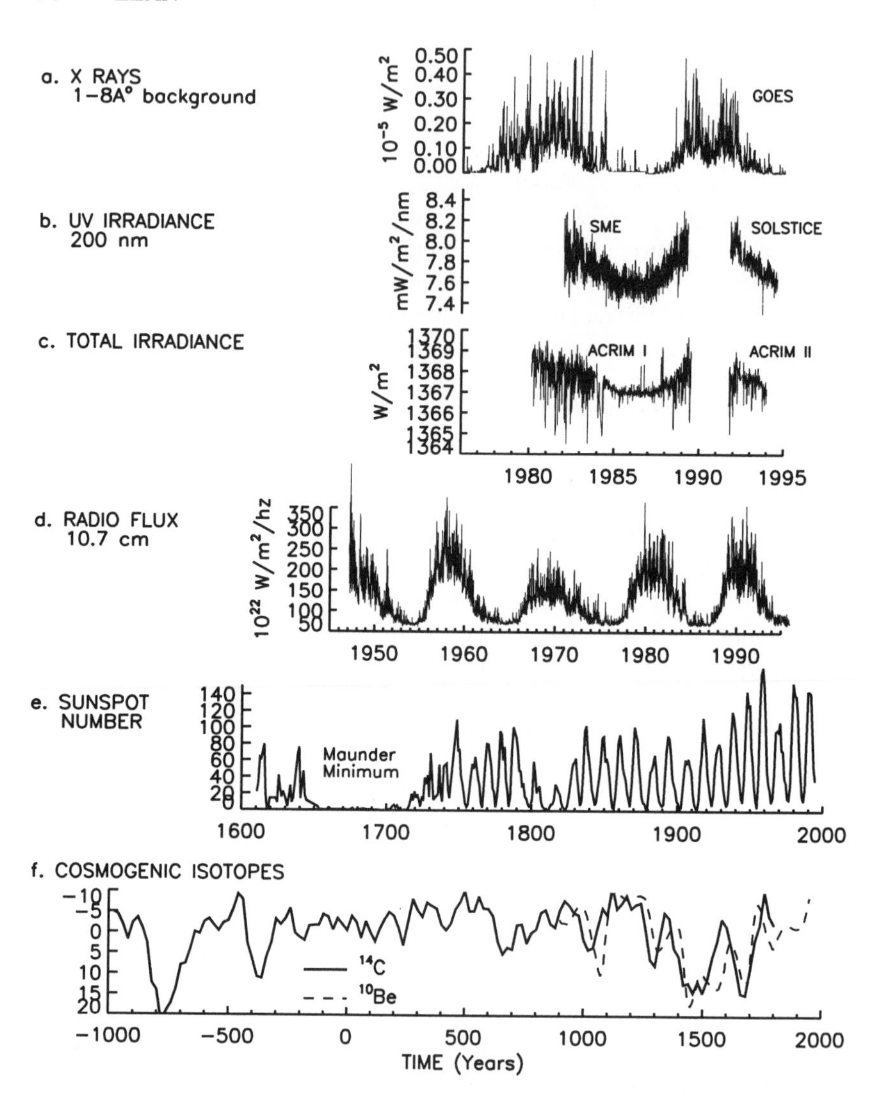


*Figure 2* Variations occur in the Sun's radiation at wavelengths throughout the spectrum, seen in (*a*) the Geostationary Operational Environmental Satellite (GOES) 1–8 Å background flux, (*b*) the UV 200-nm irradiance measured by the Solar Mesosphere Explorer (SME) (Rottman 1988) and the Solar Stellar Irradiance Comparison Experiment (SOLSTICE) (Woods et al 1996), (*c*) the total irradiance measured by the Active Cavity Radiometer Irradiance Monitor (ACRIM) instruments on the Solar Maximum Mission (SMM) and the Upper Atmosphere Research Satellite (UARS) (Willson & Hudson 1991, Willson 1994), and (*d*) the 10.7-cm radio flux. These variations occur in concert with solar activity, whose long-term changes produce the variable sunspot record (Hoyt et al 1994) in (*e*) and affect as well the $^{14}C$ and $^{10}Be$ cosmogenic isotopes (*f*) in tree rings and ice cores, respectively (McHargue & Damon 1991, Stuiver & Braziunas 1993).

through the solar atmosphere. In the lower solar atmosphere, these field lines delineate patterns of granule and supergranule cells, defining the boundaries of the turbulent convective motions responsible for their transport to the surface. Higher in the solar atmosphere, the field lines fan out and either loop back to the surface to form large-scale emission structures, reconnect in situ, or continue outward into the solar wind and heliosphere. Magnetic fields nearer the Sun's equator rotate faster than those at higher heliocentric latitudes. This differential solar rotation stretches and shears the field lines, organizing patterns of magnetism that repeat with 11-year cycles. The combination of convective motions and differential rotation is thought to constitute a dynamo that drives solar activity (Nesme-Ribes et al 1996) and generates variability in a plethora of solar phenomena including radiative output. Because the polarity of the magnetic flux in each solar hemisphere reverses in successive 11-year cycles, the fundamental cycle length is actually 22 years.

In different solar atmosphere regimes, magnetic fields perpetrate distinctive features, which are displayed for the photosphere, chromosphere, and corona in Figure 1. In the photosphere (*far right image*), clumps of very strong magnetic fields—thousands of gauss—form sunspots that are darker and cooler than the surrounding photosphere because the magnetic fields somehow inhibit the upward flow of energy from the convection zone to the surface. Sunspots are compact in shape because gas pressure in the photosphere is large enough to balance the magnetic field strength. Less compact aggregates of field lines form faculae, which are slightly brighter than the surrounding photosphere and barely detectable in visible light images of the solar disk except near the limb (e.g. Foukal 1990).

Hundreds of kilometers above the Sun's surface the primary evidence of magnetic fields are bright regions called plages, which are evident in the image of the Sun's chromosphere in Figure 1 (*middle image*) made in the core of the singly ionized line at 393.4 nm—the Ca K Fraunhofer line. Plages correlate spatially with photospheric faculae and are sites of significant UV and EUV emission from the upper photosphere, chromosphere, and transition region (Dupree et al 1973, Cook et al 1980).

Magnetic fields in the solar corona are manifest in large-scale loops that confine extended volumes of bright emission, as seen in the soft X-ray (*far left*) image in Figure 1 made by the Soft X-ray Telescope (SXT) on the Yohkoh spacecraft (Tsuneta et al 1991, Acton 1996, Hara 1996). At coronal heights, the reduced pressure of the ionized solar atmosphere relative to magnetic pressure allows magnetic field lines to expand into large-scale complexes that can extend over much of the solar disk during times of high solar activity. Hot coronal loops generally overlay sites of the sunspots and plages (Noyes et al 1985), a correspondence evident in the images in Figure 1.

Also present on the Sun is a population of smaller-scale magnetic features that comprise newly emerging magnetic flux and remnants of decayed active regions (Harvey et al 1975, Tang et al 1984). Subsurface convective motions are thought to sweep these smaller magnetic features to the boundaries of granule and supergranule cells, where they trace out a network of bright emission over the entire surface of the Sun. This network disappears in radiation emitted at temperatures greater than $10^6$ K (Reeves 1976), presumably subsumed in coronal loops of enhanced emission that dominate on large spatial scales. But in the chromosphere and photosphere, network emission is evident even during minima of 11-year activity cycles, possibly as a result of accumulated residual magnetism from past cycles, and is a source of variability in the contemporary Sun even when it is quietest (Gurman et al 1996).

## 3.2 *Sources of Solar Radiation Variability*

Magnetic phenomena are sources of significant solar radiation variability. Waxing and waning magnetic activity throughout the solar cycle produces changing sunspots, faculae, plages, and network that modify the Sun's net radiative output by altering temperature and density in the otherwise homogeneous solar atmosphere. Magnetic regions occur frequently when solar activity is high, typified by the January 1992 images in Figure 3, but they are sparse or absent during low activity epochs, such as in the February 11, 1994, image.

Solar activity does not influence the Sun's spectrum uniformly (Lean 1991). Figure 3 illustrates the changes in X-ray, UV, and total radiation concurrent with the evolution of solar magnetism from high activity in 1992 to lower levels in 1994. Dark sunspots and bright faculae act in opposition to modulate total irradiance and the spectrum at near-UV, visible, and IR wavelengths; the sunspot influence increases relative to the influence of faculae with increasing wavelength. In comparison, plage emission alone controls photospheric and chromospheric radiation variability in the UV and EUV spectrum at wavelengths less than about 300 nm. For example, while total irradiance in Figure 3*c* decreased 0.09% from January 16 to 31, 1992—because sunspot darkening exceeded facular brightening—the 200-nm irradiance actually increased by 5.5% (Figure 3*b*) in response to UV facular brightening alone. Coronal radiation (Figure 3*a*) generally tracks UV radiation because plasma emission is enhanced in magnetic field loops overlying large active regions that include both photospheric spots and chromospheric plages, but variations are significantly larger (145% from January 16 to 31, 1992).

Magnetic sources are apparently responsible for most, if not all, of the Sun's radiation fluctuations during the 27-day and 11-year cycles. Numerical simulations that combine the emission depletion in sunspots with enhancements in plages, faculae, and network replicate rotational modulation recorded in broad

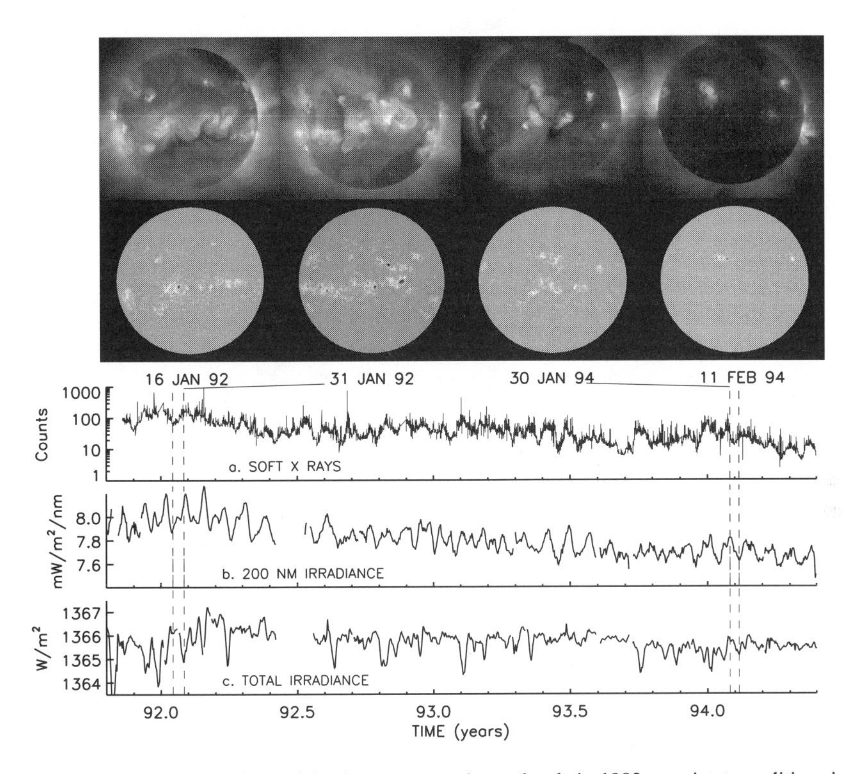


*Figure 3* Evolution of solar activity from near maximum levels in 1992 to quieter conditions in 1994 alters the occurrence of bright and dark magnetic features on the Sun's disk, seen in coronal soft X-rays measured by the Soft X-ray Telescope on Yohkoh (*upper images*) and in chromospheric Ca II K images measured at the Big Bear Solar Observatory (*lower images*). The disk-integrated soft X-ray fluxes measured by Yohkoh in (*a*), the UV 200-nm irradiance measured by SOLSTICE in (*b*), and total irradiance measured by ACRIM on UARS in (*c*) track the overall decline in solar activity and also the Sun's 27-day rotation, which alters the distribution of active regions projected toward Earth.

UV and total irradiance bands shown in Figure 4 (Foukal & Lean 1990, Lean et al 1997a). Parameterizations that incorporate enhanced Ca K emission from full disk images (e.g. Figure 3) as a surrogate for radiation brightness successfully reproduce observed total, 200-nm, and C IV transition region irradiances during both the 27-day rotational modulation and the recent 11-year cycle decline (Lean et al 1997b, Warren et al 1996).

However, the brevity of solar monitoring—less than two 11-year cycles—combined with some sensitivity drifts in the monitors precludes unambiguous identification of all potential sources of radiation variability. The apparent

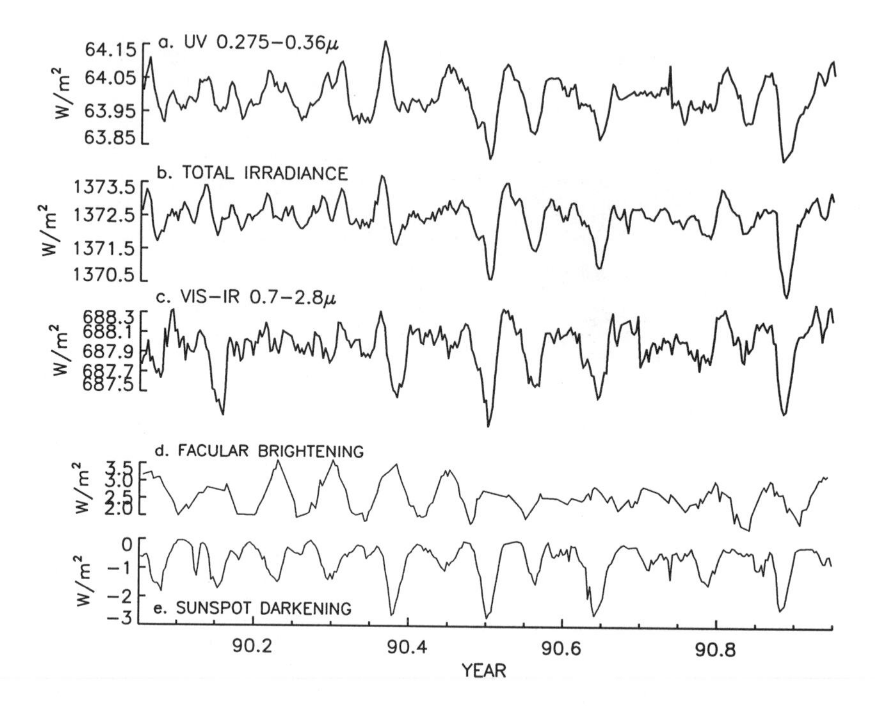


*Figure 4* Variations occur in the Sun's radiation spectrum not only at the shorter UV and X-ray wavelengths but also in (*a*) the near UV and (*c*) visible to near IR radiation, as well as in (*b*) total radiation. Filter radiometers on the Nimbus 7 spacecraft (Kyle et al 1993) detected solar rotation modulation of these broad wavelength bands. Comparisons with proxies for facular brightening (*d*) and sunspot darkening (*e*) illustrate that the variations occur primarily because of the changing impacts of these magnetic features, whose opposing influences depend on wavelength.

inability of global brightness proxies to account fully for increases evident in some total irradiance measurements near activity maxima motivates speculation of a global brightness component—such as a few-degree change in solar surface temperature—in addition to magnetic sources (Willson & Hudson 1991, Kuhn & Libbrecht 1991, Fröhlich 1994). Nor are the variations in some strong emission lines—notably H I 121.6 nm—necessarily connected to magnetic sources by simple linear parameterizations of brightness proxies (Hoegy et al 1993, Woods & Rottman 1996).

Although solar activity itself exhibits multidecadal and centennial fluctuations, whether sources of solar radiation variability exist over these time scales is unknown. A lack of sunspots in the seventeenth century Maunder Minimum (Figure 2*e*) that was coincident with high cosmogenic isotope concentrations (Figure 2*f*), decreased solar rotation, and increased solar diameter points to anomalously low solar activity relative to contemporary levels (Eddy 1976, Eddy et al 1976, Gilliland 1981, Nesme-Ribes et al 1993, Fiala et al 1994). That Sun-like stars without apparent activity cycles, possibly analogous to the Maunder Minimum Sun, are less bright than their cycling counterparts (Baliunas & Jastrow 1990) suggests that the Sun's radiation was also depleted during this period (White et al 1992, Lean et al 1992). Proposed mechanisms include reduction of the Sun's chromospheric network and basal emission from supergranule cells (White et al 1992) and reduced convective flow strengths that bring magnetic fields to the Sun's surface (Hoyt & Schatten 1993, Nesme-Ribes et al 1993). More generally, comparisons of solar and stellar radiation suggest that the Sun is potentially capable of a wider range of variability than that which we are witnessing in the contemporary era (Lockwood et al 1992, Haisch & Schmitt 1996, Ayres 1997), but sources of this prospective variability, though speculated, are undetermined (Newkirk 1983).

## 3.3 *Amplitudes of Solar Radiation Variability*

Measuring the variability of the Sun's radiation spectrum with sufficient reliability to define Sun-Earth coupling processes is a challenging task for contemporary space-based solar monitoring. With this goal, solar irradiance monitors have flown on various spacecraft since the 1970s (Lean 1991), producing, for example, the total, 200-nm, and X-ray irradiance data in Figure 2.

The variability best quantified observationally is that of the Sun's total (spectrally integrated) radiative output. Overlapping cross-calibrated measurements made by active cavity radiometers since November 1978 compose a total irradiance database with sufficient long-term precision to identify an 11-year total irradiance cycle of about 0.1% amplitude during solar cycles 21 and 22, in phase with solar activity (Willson & Hudson 1991, Willson 1994, Lee III et al 1995). The total irradiance cycle may be as much as 0.15% (Hoyt et al 1992, Fröhlich 1994) and may precede the phase of solar activity (Schatten 1988) if trends evident in some portions of the database arise from real solar variability rather than drifts in radiometer sensitivities.

Irradiance variability amplitudes in most parts of the solar spectrum must be deduced from intermittent measurements made with inadequately calibrated instruments that lack in-flight sensitivity monitoring (Lean 1991, Tobiska 1993); these data exist only at wavelengths less than 400 nm and are shown in Figure 5. Most reliable is the 8 $\pm$ 1% irradiance cycle at 200 nm, for which two flight-calibrated instruments on the Upper Atmosphere Research Satellite (UARS) in cycle 22 (Woods et al 1996) agree with earlier Solar Mesosphere Explorer (SME) observations in cycle 21 (Rottman 1988) and with empirical proxy parameterizations (Cebula et al 1992, Lean et al 1997a). At wavelengths longer than 300 nm, the $\pm$1% long-term precision of UARS solar UV monitoring is

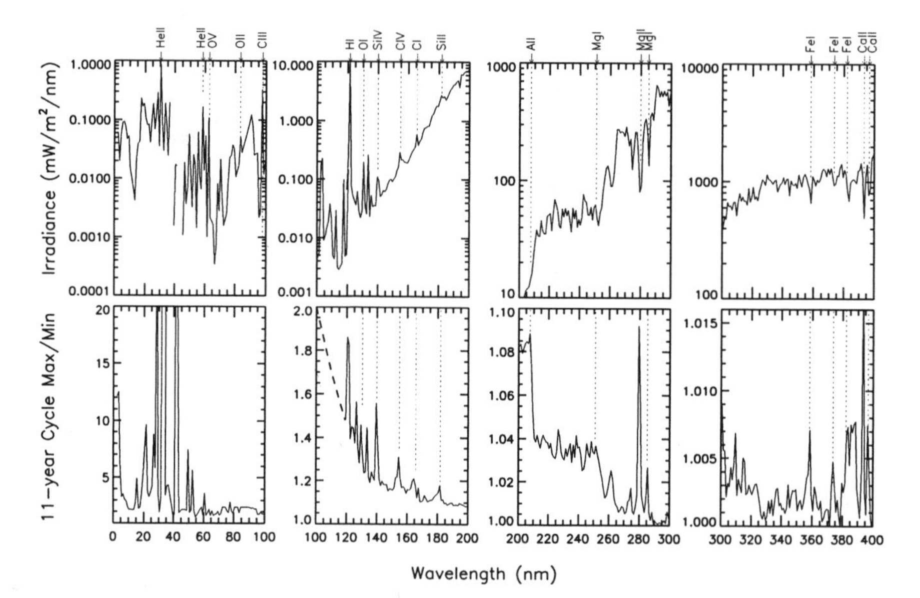


*Figure 5* Shown in the upper panel is the Sun's spectral irradiance at wavelengths from 0 to 400 nm and in the lower upper panel its 11-year cycle variability. EUV spectra are obtained from the Hinteregger et al (1981) empirical model, which is based on Atmospheric Explorer-E data. The 100- to 400-nm spectra were measured by SOLSTICE on UARS (Woods et al 1996). At wavelengths longward of 200 nm, the variations are estimates from a proxy variability model that scales the irradiance modulation detected during solar rotation (e.g. Figure 4) to the longer time scale of the 11-year cycle (Lean et al 1997a).

insufficient to detect spectral irradiance variations that are estimated to be less than 1% (Figure 5). In the vicinity of H I 121.5 nm, irradiance changes of 80–100% measured by the UARS instruments exceed proxy model estimates of a 50–60% reduction during the descending phase of solar cycle 22 (Chandra et al 1995) and of the 60% variability measured by SME during cycle 21. The EUV spectral irradiance variability amplitudes, deduced primarily from four years of Atmospheric Explorer-E measurements (Hinteregger et al 1981), exceed factors of two (Figure 5) but are uncertain by 50–100%. In lieu of adequate observational data, empirical variability models estimate irradiances from parameterizations of various solar activity proxies either by fitting the extant database (Hinteregger et al 1981, Tobiska 1993) or by scaling the better-defined rotational modulation to the longer 11-year cycle (Cebula et al 1992, Lean et al 1997a). Poor observational constraints continue to limit the reliability of these models (Lean 1990).

Solar radiation variability amplitudes remain poorly characterized on the very short time scales (minutes to hours) of solar flares, except in certain X-ray bands measured by GOESS and Yohkoh. Eruptive events in the Sun's atmosphere can cause significant amplitude variations in EUV radiation (White 1977) and possibly in UV emissions as well (Brekke et al 1996), but the time resolution of solar radiation monitoring in general has been insufficient to properly quantify these events.

On multidecadal and centennial time scales, amplitudes of solar irradiance variability are as yet unmeasured. Diagnosed associations between sources of contemporary irradiance variability and appropriate solar activity proxies that extend over longer time spans permit the irradiance reconstructions shown in Figure 6. In addition to the 11-year cycle, a longer-term variability component is postulated to account for a 0.24% irradiance reduction in the Maunder Minimum based on interpretation of Ca II emission from the Sun and Sun-like stars (Lean et al 1995). Other reconstructions yield long-term variability amplitudes that range from 0.1 to 0.6% (Hoyt & Schatten 1993, Nesme-Ribes et al 1993, Jirikowic & Damon 1994, Zhang et al 1994).

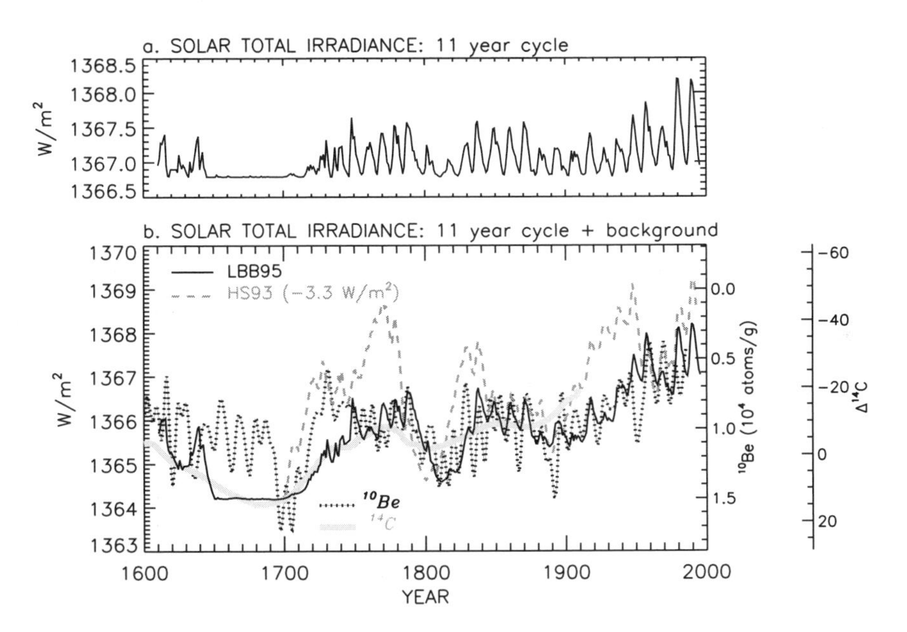


*Figure 6* The reconstruction of solar total irradiance shown in (*a*) is of the 11-year cycle alone (Foukal & Lean 1990), whereas the dark solid line in (*b*) combines the 11-year activity cycle and a longer term component based on the average amplitude of each sunspot cycle (Lean et al 1995). This latter irradiance reconstruction is compared with $^{10}$Be (*small squares*) and $^{14}$C (*thick gray line*) cosmogenic isotope records (Beer et al 1988, Stuiver & Braziunas 1993) and with the Hoyt & Schatten (1993) irradiance reconstruction (*gray dashed line*) in which longer term changes are based on the length of the 11-year solar activity cycle (rather than average amplitude).

## 4. EARTH'S ABSORPTION OF SOLAR RADIATION

Earth's atmosphere, a blanket of gases extending hundreds of kilometers into space, is relatively transparent to visible and near-IR radiation, allowing some 99% of the Sun's energy—the spectrum longward of 310 nm—to penetrate to below 15 km, into the troposphere where most of the Earth's atmosphere resides, weather and climate occur, and the biosphere exists (Figure 7). The Earth reflects about 30% of this incident solar energy back into space and receives from the remaining 70% a globally averaged energy input of 239 $W/m^2$ ($1366 \times 0.7/4$ $W/m^2$ ). Heated to 255 K by the Sun, the Earth radiates its own energy, which gases in its lower atmosphere—$CO_2$, $H_2O$, $O_3$, $N_2O$, and CFCs—absorb, trapping additional energy in the troposphere. This greenhouse effect provides further surface warming to 288 K (Peixoto & Oort 1992, Hartmann 1994).

The UV radiation from the Sun also provides significant energy to the terrestrial system, although not in the form of direct surface heating. Energy deposition from radiation at wavelengths less than 310 nm, in the UV, EUV, and X-ray portions of the spectrum, takes place at altitudes shown in Figure 7*a*

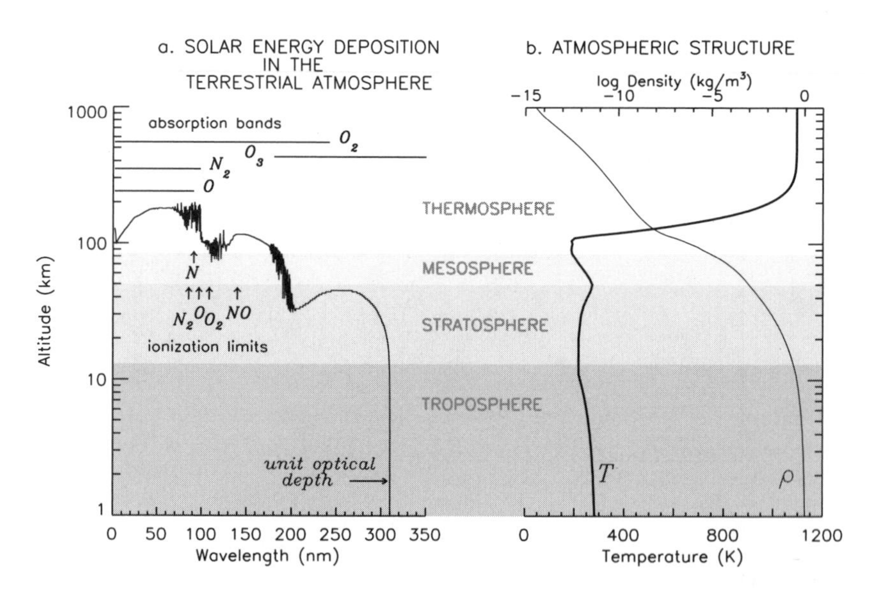


*Figure 7* Different regions of the solar spectrum penetrate to different altitudes in the Earth's atmosphere because atmospheric species absorb radiation at different wavelengths. Shown in (*a*) is the altitude at which the Earth's atmosphere attenuates incident radiation from an overhead Sun by a factor of 1/*e* (Meier 1991). Deposited solar energy determines the atmosphere's temperature profile (T) in (*b*), on which atmospheric regions are based.

for the Sun directly overhead, which are determined by the concentrations and absorption cross sections of the primary atmospheric gases ($O_2$, $N_2$, O) and $O_3$ (a minor species) (Meier 1991). Although a mere 1% of the Sun's total radiative output, this radiation controls the overall thermal profile of the atmosphere above about 15 km and much of the chemical, dynamical, and radiative processes occurring there (Banks & Kockarts 1973, Brasseur & Solomon 1984). Ozone absorption of 200- to 300-nm radiation provides sufficient energy to heat the stratosphere from 216 K at altitudes near 15 km to 270 K at 50 km. Molecular oxygen absorption of radiation at wavelengths shorter than 242 nm is a prime energy source for the mesosphere and lower thermosphere (altitudes 75–200 km) and creates oxygen atoms that initiate ozone formation. All three of the upper atmosphere's primary species absorb radiation at wavelengths shorter than 100 nm, providing energy that can heat the thermosphere to more than 1000 K at altitudes above 200 km (Figure 7*b*).

Solar photons at wavelengths less than 102.7 nm have sufficient energy to remove electrons from $O_2$, $N_2$, and O (Figure 7*a*). Longer wavelength radiation (primarily H I 121 nm) ionizes NO, and chemical reactions among ions, electrons, and neutrals of major and minor species create additional forms of ionization. As a result, a weakly ionized plasma called the ionosphere is embedded in the neutral upper atmosphere from about 50 to 1000 km (Rishbeth & Garriot 1969).

Both neutral and ionized constituents can transport solar energy from one part of the atmosphere to another, coupling different atmospheric regions with each other and with the Earth's surface. Geographical and altitudinal differences in solar heating drive atmospheric motions that transport heat and chemical species in the lower atmosphere from the tropics to the poles, in Hadley circulation cells. Flows at higher altitudes rise during the summer hemisphere and sink in the winter hemisphere. Turbulent eddy motions and vertical processes such as gravity and planetary waves intermingle with and perturb these meridional motions.

## 5. SOLAR RADIATIVE FORCING OF GLOBAL CHANGE

Changes in the Earth's global environment—whether caused by the Sun's variable radiation or by other natural or human-made processes—can have ecological, social, and economic consequences (Committee on Earth Sciences 1989, Committee on Environment and Natural Resources 1996). Of paramount concern in the present era are changes arising from anthropogenic influences. Warming of a few degrees Celsius is forecast for Earth's surface temperature in the twenty-first century from increasing concentrations of $CO_2$ and other greenhouse gases in the atmosphere. Depletion of atmospheric ozone by CFCs injected into the stratosphere during the past few decades threatens increased surface fluxes of biologically damaging UV radiation [National Research Council 1989, Intergovernmental Panel on Climate Change (IPCC) 1992, 1995].

## 5.1 *Climate*

Earth's surface temperature, shown in Figure 8*a*, increased 0.7°C over the past 350 years, and 0.5°C since the beginning of the twentieth century. Present levels of annual $CO_2$ concentrations, shown in Figure 8*b*, are 350 ppm, 27% higher than the 275 ppm preindustrial level (Boden et al 1994). Concentrations of other polyatomic gases—$CH_4$, $N_2O$, and CFCs—have also increased, contributing to a 2.4 $W/m^2$ radiative climate forcing since 1850 (Hansen et al 1993) by trapping IR energy radiated by the Earth to space. This greenhouse gas forcing is identified as the most likely cause of climate change in the past century (IPCC 1995).

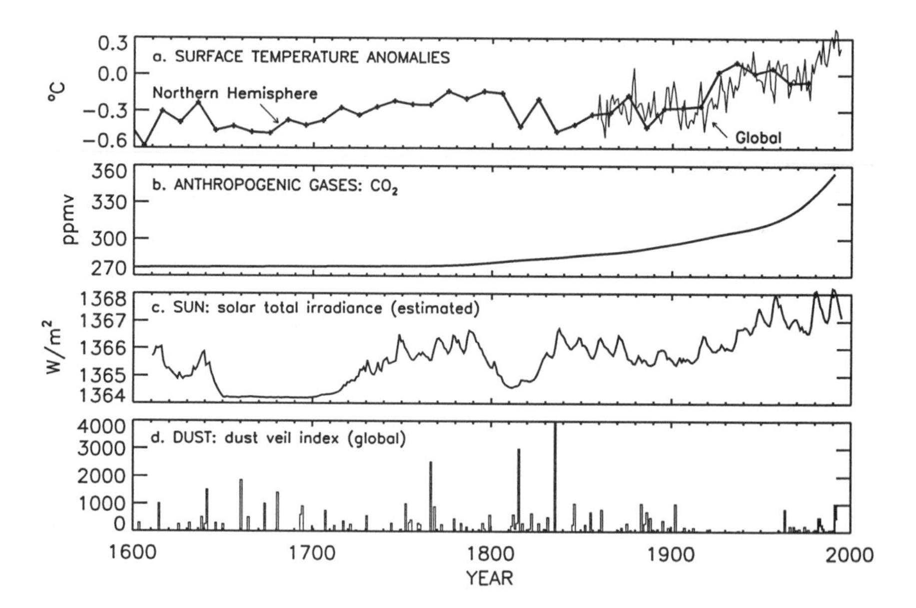


*Figure 8* Compared in (*a*) are the Bradley & Jones (1993) reconstructed record of decadal NH surface temperature since 1600 (*solid line with plus signs*) and the IPCC (1992) global instrumental record since 1850 (*thin line*). Both natural and anthropogenic influences may have contributed to the observed surface warming since 1850. Shown are (*b*) annual averages of the concentration of $CO_2$ (Boden et al 1994), (*c*) estimated solar total irradiance (Lean et al 1995), and (*d*) volcanic aerosol loading according to the global dust veil index (Lamb 1977, Robuck & Free 1995).

Climate adjusts to radiative forcing $\Delta F$ by seeking a new thermal equilibrium. Providing the forcing persists long enough for complete adjustment, the accompanying surface temperature change is $\Delta T = \kappa \Delta F$, where $\kappa$ is the climate system sensitivity. This sensitivity is the net of the various amplification or attenuation processes by which the climate system responds to changes in the energy it receives, depending on the spatial distribution, altitude, and time history of the specific forcing. With $\kappa$ in the range 0.3–1°C per W/m$^2$ (Wigley & Raper 1990, IPCC 1995), the greenhouse gas forcing of 2.4 W/m$^2$ since 1850 should have increased surface temperature in the range of 0.7–2.4°C.

The discrepancy between the observed 0.5°C warming since 1850 and the estimated 0.7–2.4°C greenhouse gas warming cautions that climate change forcings and responses in the industrial era are not yet fully understood. Unlike greenhouse gas forcing, Earth's surface temperature did not increase steadily (Figure 8*a*)—statistical analyses of surface temperatures since 1850 reveal significant decadal and interdecadal variability on global and regional spatial scales (Allen & Smith 1994, Mann & Park 1994, Lau & Weng 1995). Moreover, preindustrial surface temperatures apparently varied independently of greenhouse gas concentrations, which commenced their increase in the nineteenth century.

Other anthropogenic and natural factors likely influenced Earth's climate in the recent past. Figure 9 compares amplitudes of their radiative forcings since 1850, which are known only with very low confidence. Enhanced solar radiative output (Figure 8*c*) may have contributed to surface warming as solar activity increased from the anomalously low activity of the seventeenth century Maunder Minimum. Over the past century, increased industrial aerosol concentrations in the troposphere have potentially cooled Earth's surface by scattering and deflecting incoming solar radiation and by affecting cloud formation (Penner et al 1994, Schwartz & Andreae 1996). Volcanic aerosols in the stratosphere likewise reduce incoming solar radiative energy inputs to the Earth (Robuck & Free 1995). That the atmosphere has been relatively free of volcanic aerosols for most of the twentieth century (Figure 8*d*) implies warmer conditions than in the previous century. Some surface cooling is expected in the past few decades from the depletion of stratospheric ozone by CFCs because ozone is a greenhouse gas (Schwarzkopf & Ramaswamy 1993). Modified patterns of vegetation coverage through human activity have altered surface reflectivity (Hannah et al 1994)—albedo—and the Earth's ability to absorb incoming solar radiation.

In the preindustrial decades from 1610 to 1800, surface temperature anomalies apparently tracked solar activity, increasing 0.26°C as irradiance increased 2 W/m$^2$ (a 0.35 W/m$^2$ climate forcing), based on the reconstructed solar irradiance and temperatures shown in Figure 10. Extending this relationship to

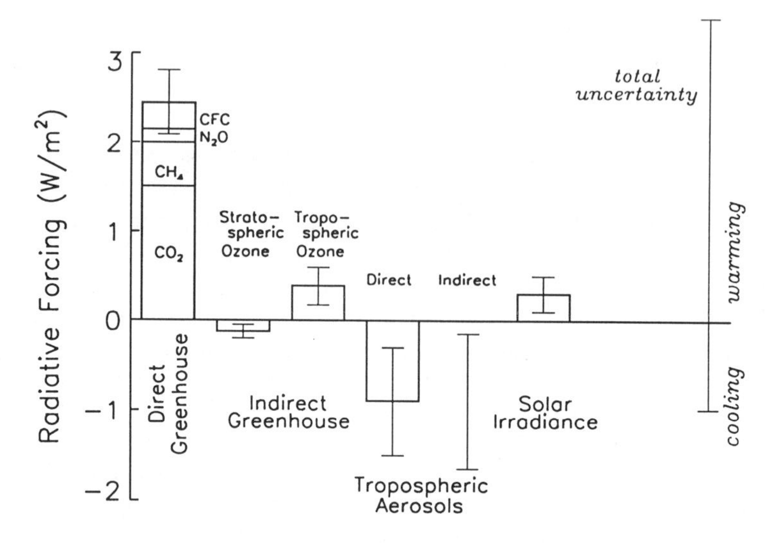


*Figure 9* Amplitudes of natural and anthropogenic climate forcings from 1850 to 1990 are shown from IPCC (1995). Each individual forcing is expected to impact the climate system in different ways depending on its latitude, altitude, and history. However, climate change assessments lack the complexity to account for the myriad pathways of the climate system response, and global scale studies often assume a common climate sensitivity to the different forcings.

the present yields a solar induced surface temperature increase of about 0.25°C since 1850, or about half the observed warming. Other empirical studies similarly infer that one third to one half of centennial scale climate change in the recent Holocene may be solar related (Reid 1991, Schlesinger & Ramankutty 1992, Crowly & Kim 1996). Adding further support, a simulation with the Goddard Institute for Space Studies (GISS) general circulation climate model estimates a northern-hemisphere equilibrium surface temperature change of 0.49°C for a 0.25% irradiance reduction (Rind & Overpeck 1993), which is in surprisingly good agreement with the 0.45°C increase since the Maunder Minimum determined empirically from the data in Figure 10. But whether solar total irradiance actually did increase 0.25% over the past 350 years is uncertain, and attributing as much as half of the surface warming since 1850 to solar forcing requires substantial offsetting of greenhouse warming by other mechanisms—aerosol cooling and albedo changes, for example. For these reasons, the reality of solar variability influences on centennial climate change deduced from circumstantial evidence such as the empirical relationship in Figure 10 is controversial.

Solar variability may influence climate on decadal as well as centennial time scales. Cycles of 11 and 22 years exist in a wide variety of contemporary

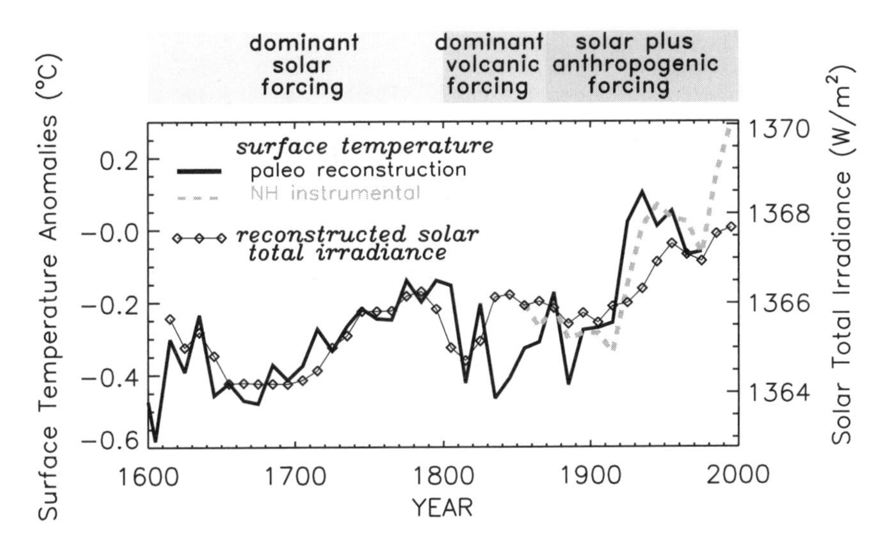


*Figure 10* Compared are decadally averaged values of solar total irradiance reconstructed by Lean et al (1995) and NH summer temperature anomalies from 1610 to the present. The dark solid line is the Bradley & Jones (1993) NH summer surface temperature reconstruction from paleoclimate data (primarily tree rings), scaled to match the NH instrumental data (IPCC 1992) (*gray dashed line*) during the overlap period. (Updated from Lean et al 1995.)

climate records, including temperatures of land and ocean surfaces and of the troposphere, US drought, rainfall, forest fires, and cyclones. These cycles also occur in longer-term climate proxy records, including $\Delta^{18}$O concentrations in ice cores and tropical corals (see Lean & Rind 1997 for references). Since the 1950s, a decadal component in global sea surface temperature, shown in Figure 11, has varied in phase with the 11-year solar activity cycle (White et al 1997), increasing roughly 0.1°C (peak to peak) as solar irradiance increased 1.3 W/m$^2$ during the two most recent solar cycles. Wavelet transform analysis of the northern-hemisphere temperature record (land plus ocean) likewise detects an 11-year cycle of amplitude ~0.1°C, which is in phase with solar activity in recent cycles (Lau & Weng 1995).

But like other climate cycles, including the 3- to 7-year El Niño Southern Oscillation band, the 11- and 22-year cycles are capricious (Burroughs 1992). They are neither present in all climate proxy records nor are they always present in any one record or always in phase with the solar activity cycle. Instead of solar variability, an internal oscillation of the ocean-atmosphere system is frequently proposed as the origin of observed decadal climate change (Metha & Delworth 1995). Climate simulations with general circulation models replicate such an oscillation in the absence of any external forcing (James & James 1989),

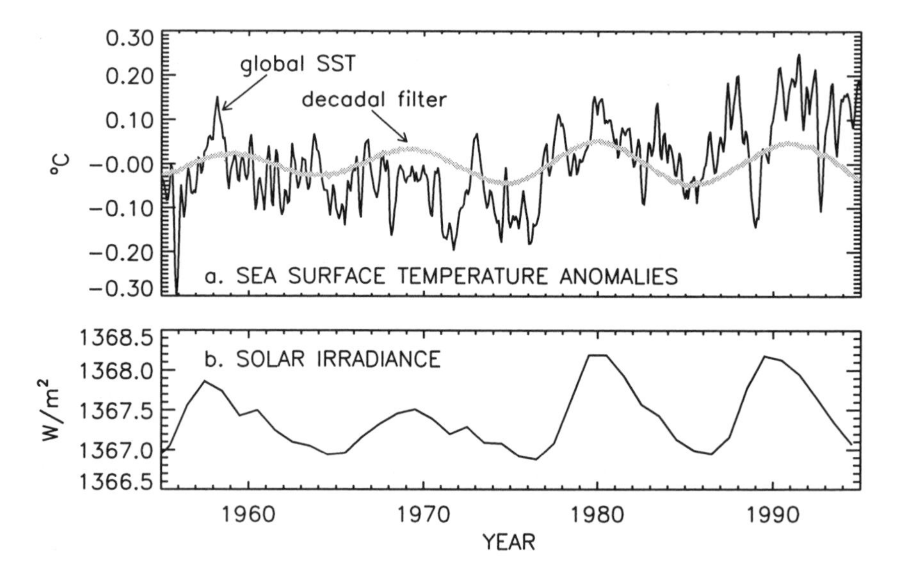


*Figure 11* White et al (1997) identified in globally averaged sea surface temperature anomalies compiled from bathythermographs (BT), significant annual and interannual variability including a decadal component shown in (*a*) that tracks solar irradiance reconstructed by Lean et al (1995), shown in (*b*).

although not on the global scale evident in the connections in Figure 11. However, the indication by the GISS general circulation climate model simulation of regional surface temperature inhomogeneities for a 0.25% irradiance reduction may provide an explanation for the non-stationarity of solar-induced decadal cycles—dynamical patterns set up by differential heating of the land and ocean induce regional effects, evolution of which could cause apparent ambiguities in site-specific climate proxy records.

Current climate change theory assumes that the thermal inertia of the oceans prevents the climate system from reaching an equilibrium in response to solar irradiance variations over the 11-year cycle, realizing surface temperature changes of only 0.02–0.03°C (Wigley & Raper 1990, North & Kim 1995), which are factors of three to five smaller than the apparent relationship between decadal solar and global ocean surface temperature fluctuations in Figure 11. There are some indications, however, that the sensitivity of the real climate system to small solar radiative perturbations may be greater and more complex than is presently assumed. Using a recently developed scheme for detecting subtle variability modes in climate parameters, Stevens & North (1996) attributed somewhat larger signals of 0.05°C to 11-year solar forcing in the instrumental surface temperature record since 1850. On much longer time scales, general circulation climate model simulations of Ice Age occurrences forced by Earth's orbital motions about the Sun (the Milankovitch effect) (Hays et al 1976) cannot account properly for the unexpectedly prominent 100,000-year periodicity in the paleoclimate record, which is associated with the eccentricity of the Earth's orbit (Rind et al 1989). This changing Sun-Earth distance modulates the solar radiation incident on the Earth, as does solar activity.

## 5.2 *Ozone*

Ozone is Earth's biological UV shield (de Gruijl 1995), as well as a greenhouse gas (National Research Council 1989). Although present in the atmosphere only in trace amounts—about 40 parts per billion volume near the Earth's surface—ozone molecules exert a strong control on, and are in turn affected by, the physical state of the middle atmosphere (Brasseur & Solomon 1984).

Fundamental to the existence of ozone is the interaction of solar UV radiation with atmospheric oxygen compounds. Solar UV radiation at wavelengths less than 242 nm creates oxygen atoms by photodissociating oxygen molecules. Atomic and molecular oxygen combine to produce a layer of triatomic oxygen—ozone—with peak concentrations of 10 ppm at altitudes near 25 km. By photodissociating other atmospheric constituents, solar radiation also creates varieties of hydrogen, nitrogen, oxygen, and chlorine radicals that destroy ozone by catalytic chemical cycles. In particular, solar UV radiation near 200 nm liberates Cl atoms from anthropogenic CFCs that drift up to the middle atmosphere from the Earth's surface (McElroy & Salawitch 1989). And solar radiation itself destroys ozone, which dissociates upon absorption of UV radiation at wavelengths from 200 to 300 nm. This absorption prevents UV photons from reaching the biosphere and supplies the middle atmosphere with energy to drive large-scale dynamical motions that transport atmospheric species to different latitudes and altitudes.

Globally averaged total column ozone concentrations above the Earth's surface, shown in Figure 12, exhibit a long-term downward trend during the past 15 years that is anticipated to enhance fluxes of UVB radiation (290–320 nm) at the Earth's surface (Madronich & deGruijl 1993, Herman et al 1996). This trend is attributed to ozone destruction by CFCs (World Meteorological Organization 1991) and is estimated to be occurring at a rate of $0.27 \pm 0.14\%$ per year (Stolarski et al 1991, Hood & McCormack 1992). Also, industrially produced $CO_2$ may be cooling the stratosphere (Rind et al 1990, Ramaswamy et al 1996) and impacting ozone through the strong temperature dependence of ozone's reactions with other atmospheric species.

Superimposed on the overall downward ozone trend in Figure 12 are significant ozone fluctuations related to seasonal cycles, the quasibiennial oscillation (known as the QBO) of wind direction in the tropical lower stratosphere,

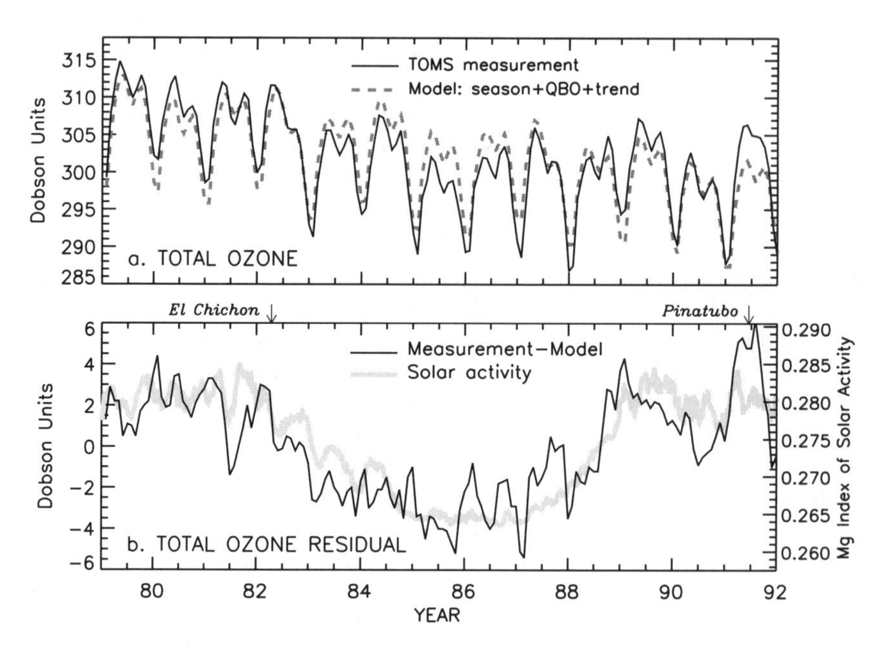


*Figure 12* Shown are the results of an analysis of the variability modes evident in global column ozone since 1978, by Hood & McCormack (1992). The Total Ozone Mapping Spectrometer ozone measurements in (*a*) (*solid line*) are compared with a fitted statistical model of seasonal, linear trend, and quasibiennial oscillation terms (*gray dashed line*). The residual of the measured minus model ozone is shown in (*b*) to track the Mg index of solar activity (provided by L Puga, NOAA).

volcanic eruptions, and the 11-year solar activity cycle (Hood & McCormack 1992, Reinsel et al 1994, Randel & Cobb 1994, Solomon et al 1996). These naturally occurring influences combine to almost obscure ozone's long-term depletion since 1979, and they add considerable uncertainty to the deduction of its true magnitude. For example, the increase in solar activity from minimum levels in 1986 to high levels near the cycle 22 maximum in 1990 produced UV radiation enhancements that increased total ozone in the range 1.5 to 1.8% (Wuebbles et al 1991, Hood & McCormack 1992), more than countering the 1.35% anthropogenic depletion over the same five year period.

The impacts of variable solar radiation and other natural and anthropogenic influences on ozone depend on geographical location and altitude. Long-term CFC-induced ozone depletion is pronounced at higher latitudes and at altitudes below the peak ozone concentrations—in the lower stratosphere (Randel & Cobb 1994, Reinsel et al 1994), whereas ozone responds to varying solar radiation mainly in regions where it is photochemically controlled—over the subtropics in the upper stratosphere. Ozone also responds to the apparent overall response of the middle atmosphere to solar activity through temperature-dependent chemistry and by dynamical transport. Oscillations with periods of 10 to 12 years in phase with solar activity are evident in middle atmosphere temperatures, pressures, and winds. These oscillations vary in strength with geographical locations and altitude and are enhanced at some locations by the phase of the quasibiennial oscillation (Labitzke & van Loon 1992, 1993, Hood et al 1993). However, the sparseness and brevity of comprehensive, global middle atmosphere data prompts questions about the robustness of these relationships (Salby & Shea 1991) and the reality of solar-induced decadal scale variability in the middle atmosphere is still somewhat controversial, although less so than in the lower atmosphere.

Middle atmosphere theory can account for only some of the empirical connections evident among solar variability, ozone and the middle atmosphere. Coupled two-dimensional chemical-dynamical models do predict that enhanced UV radiation near maxima of the 11-year solar cycle increases ozone in the upper stratosphere, but by about a factor of two less than the 5–7% observed enhancement (Chandra & McPeters 1994). Agreement between models and measurements is better for the shorter 27-day solar rotational modulation (Fleming et al 1995). But these models fail to account for emerging empirical evidence of solar-induced fluctuations in the lower stratosphere (Reinsel et al 1994, Hood 1997), which are thought to arise from dynamical amplification of the initial solar radiative forcing of ozone (Hood et al 1993). Nor do they simulate adequately the coupling of solar-induced variability in the lower thermosphere with the middle atmosphere. When transported to the stratosphere in the polar night, enhanced levels of NO produced in the lower thermosphere by solar EUV radiation and X rays during high solar activity are calculated to destroy sufficient ozone such that total column concentrations vary out of phase with solar activity (Brasseur 1993), in contrast to observations (Figure 12).

Whether caused by anthropogenic or natural influences, changes in middle atmosphere ozone and heating have the potential to impact climate in the troposphere through radiative and dynamic coupling processes. Key to these putative impacts are changes in ozone's vertical concentration profile. Depending on altitude, such changes can affect absorption of both solar UV and terrestrial IR energy, producing either a net cooling or warming of the Earth's surface (Lacis et al 1990). In addition, model simulations of solar-induced wind and temperature patterns in the stratosphere produce thermal gradients between the troposphere and stratosphere that alter the strength of tropospheric Hadley cell circulation and the generation and propagation of planetary waves around the globe, which affects weather and climate (Haigh 1994, 1996, Rind & Balachandran 1995).

## 6. SPACE WEATHER IMPACTS OF SOLAR RADIATION

Solar-extreme and far-UV radiation (at wavelengths less than 170 nm) is the upper atmosphere's primary energy input and creates, as well, its embedded ionosphere (Banks & Kockarts 1973). Unlike the subpercent fluctuations typical of the visible and IR radiative energy inputs to the lower atmosphere, solar EUV radiation exhibits substantial variability, of factors of two or more (Figure 5), that alters significantly the thermodynamic, chemical, and radiative state—the "weather"—of the thermosphere (Roble & Emery 1983) and ionosphere (Ratcliffe 1972, Jursa 1985). During the Sun's 11-year activity cycle, upper atmosphere temperatures (shown in Figure 13) fluctuate by factors of two, and neutral and electron densities by factors of ten. These solar-induced changes exceed by two orders of magnitude the suspected decadal

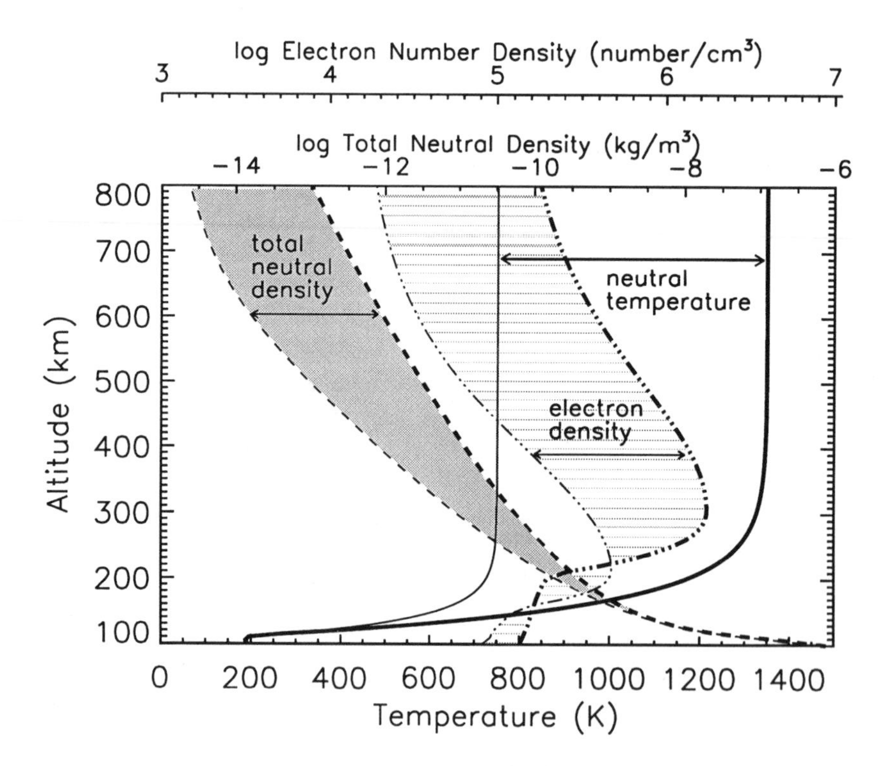


*Figure 13* Increases in extreme and far UV radiation from low ($F_{10.7} = 70$) to high ($F_{10.7} = 230$) solar activity heats the entire upper atmosphere and increases total neutral and electron densities by more than an order of magnitude. The Mass Spectrometer and Incoherent Scatter (MSIS) and International Reference Ionosphere (IRI) models were used to calculate these profiles.

anthropogenic impact of greenhouse gas cooling on either upper atmosphere temperatures or densities (Rishbeth & Roble 1992, Bremer 1992). As a fundamental driver of space weather at altitudes between 50 and 1000 km, the Sun's variable EUV radiation is of socioeconomic relevance (National Space Weather Program 1995)—its variability can affect the operation and performance of complex space-borne systems and networks utilized with increasing reliance for national and societal needs of communication, navigation, surveillance, and commerce (Goodman & Aarons 1990, Gorney 1990, Lanzerotti 1994).

## 6.1 *Spacecraft*

Including spacecraft and debris, more than 8000 catalogued objects now orbit the Earth in both circular and elliptical orbits at altitudes ranging from a few hundred to many thousands of kilometers (Committee on Transportation Research and Development 1995). Depending on its velocity and ballistic coefficient, an Earth-orbiting object experiences deceleration due to friction with the upper atmosphere that limits its lifetime (King-Hele 1987). Unboosted spacecraft in circular orbits at altitudes less than 500 km eventually reenter the Earth's atmosphere, usually burning up and generating additional debris, as was the case for the Solar Maximum Mission (Covault 1989), but sometimes impacting the Earth's surface, as did SkyLab in 1979.

"Drag" on an object at a given altitude depends on the density of the upper atmosphere, which, as seen in Figure 13, is not constant. Responding to elevated heating by enhanced EUV and UV radiation during times of high solar activity, the atmosphere expands outward from the Earth, bringing higher densities to a given altitude. Density fluctuations also occur with somewhat smaller amplitudes on time scales of days and months (White et al 1994), primarily in response to the Sun's 27-day rotational modulation of EUV radiation. As an example, atmospheric drag on the Solar Maximum Mission spacecraft in the vicinity of 400 km during a period of moderate solar activity, shown in Figure 14 just prior to reentry, caused orbital decay rates as high as 0.5 km/day near radiation peaks of the 27-day cycle.

Efficient planning, management, and operation of space-based resources require the ability to specify and forecast the orbits of both spacecraft and debris (for collision avoidance) (Space Environment Services Center 1982). On multiyear time scales, this capability benefits planning for mission resources such as operations, battery storage, and fuel loadings (NASA 1994), and it is needed for coordinating the construction and maintenance of the International Space Station (ISS). At a planned orbital altitude of 400 km, ISS is scheduled to commence operation at near record high levels of activity predicted for solar cycle 23 maximum, and it will require periodic reboosting throughout its multidecadal

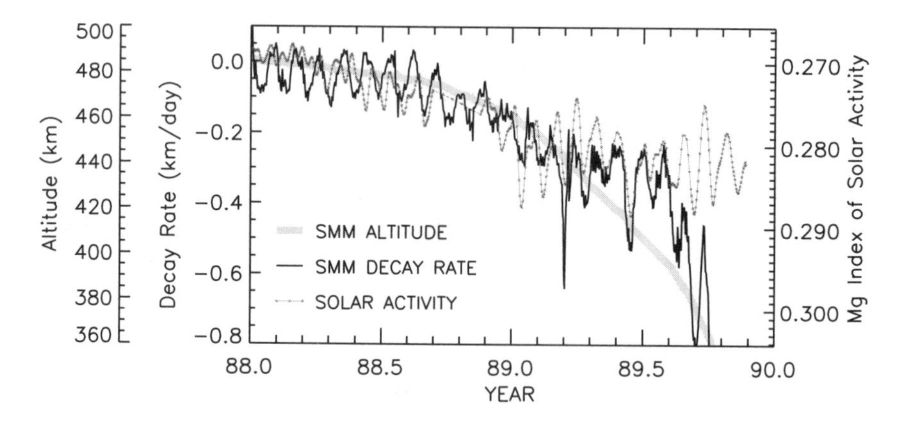


*Figure 14* Superimposed on the overall orbital decay of the Solar Maximum Mission (SMM) spacecraft at altitudes from 490 to 360 km (*thick gray line*) are notable fluctuations in decay rate (*thin dark line*) associated with modulation of the Sun's EUV radiation by solar rotation, indicated by the Mg index proxy (*gray stars*). (D Messina provided the SMM altitude data, and L Puga the Mg indices.)

operational lifetime. During the previous solar cycle 22, inadequacies of the current state of the art for predicting solar activity and the response of the upper atmosphere to attendant EUV radiation fluctuations caused alarm that launch of the Hubble Space Telescope into low Earth orbit at that time would severely curtail its lifetime (Withbroe 1990). Orbital anomalies and premature vehicle reentry arising from uncertainties in upper atmosphere densities are also of economic concern for commercial ventures, which propose to utilize arrays of small, low Earth orbit spacecraft for global communications networks (e.g. IRIDIUM).

When differences between predicted and observed positions of space objects from one orbit to the next exceed specified tolerances, near real-time decisions must be made by the US Air Force and Naval Space Commands (which have responsibility for operational tracking of objects in Earth orbit) to attribute the anomaly to either unexpected atmospheric drag or object misidentification. Minimizing uncertainties in orbital tracking due to atmospheric drag can help reduce the occurrence of such anomalies and also improve the accuracies of alerts and warnings of potential collisions between spacecraft, such as that witnessed recently between two orbiting objects (David 1996). Collisions with orbital debris is a serious issue for the ISS because maneuvering to avoid infrastructure damage will perturb the microgravity and aspect environments for ISS experiments.

Incorporating into operational orbital tracking tools state-of-the-art upper atmosphere density models promises to reduce uncertainties in the locations of spacecraft and debris. In lieu of routine solar EUV irradiance measurements for input, present semi-empirical upper atmosphere models such as those of Jacchia (1977) and MSIS (Mass Spectrometer and Incoherent Scatter) (Hedin 1991) utilize the 10.7-cm radio flux as a proxy. Since four of the five strongest solar emission lines that heat the thermosphere (Roble 1987) emerge from the Sun's chromosphere, adopting a chromospheric proxy for solar EUV radiation variability rather than the primarily coronal 10.7-cm radio flux may help improve these models. One candidate is the Mg index proxy (L Puga, private communication) used in Figure 14.

## 6.2 *Communication*

Electrons in the Earth's ionosphere form conducting layers that inhibit the dissipation to space of radio waves transmitted upward from the surface at certain frequencies, reflecting them instead. This process enables communications from one site to another around the globe, first demonstrated by Marconi in 1901, as well as from spacecraft to the ground and between spacecraft (Goodman 1992).

Different ionospheric layers play different roles in either facilitating or impeding radio communications over a range of frequencies (Jursa 1985). Electron concentrations peak at a few hundred kilometers above the Earth's surface (Figure 13). This uppermost ionospheric layer, called the F region, reflects high frequencies (3–30 MHz) that have a range of 100 to 10 m and are utilized for broadcasting and over-the-horizon radar surveillance. Electrons at lower altitudes—95 to 140 km, in the E region—reflect medium frequencies (300–3000 kHz) with wavelengths 1 to 0.1 km. Although absorbed by the underlying D region during the daytime, medium frequency waves can propagate considerable distances during the night. Electrons at heights less than 90 km ($<75$ km during the daytime) reflect low and very low frequencies (3–300 kHz), which facilitate communication over distances ranging from several hundred to many thousands of kilometers that are used primarily by the military for navigation and surveillance. Electron and ion densities that are present over a very large range of altitudes—above 50 km—affect extremely low frequencies, which can propagate on a global scale.

Management of the frequency spectrum to ensure uninterrupted communication among different sites on the globe requires operational specification of the ability of the ionosphere to reflect or absorb radio waves at different frequencies. This depends on the total number and height distribution of electrons, which depend, in turn, on the field of solar ionizing radiation. Diurnal, seasonal, and solar activity cycles are thus critical factors that determine the state of the ionosphere and electron concentrations at a given altitude and geographical location (Rishbeth & Garriott 1969).

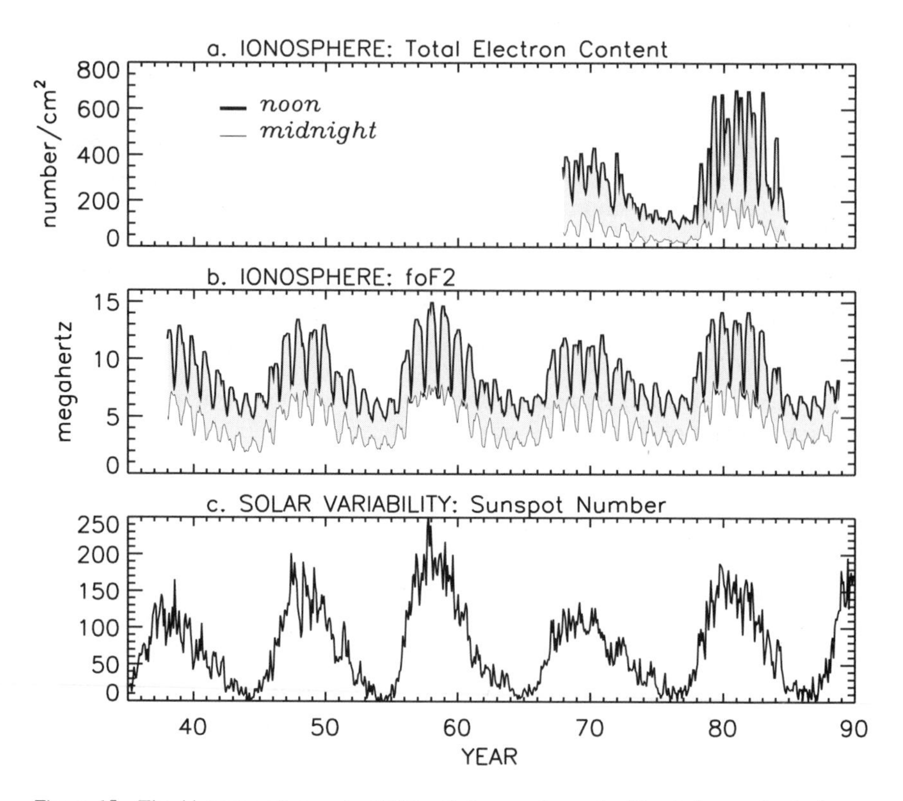


*Figure 15* The 11-year cycle in solar EUV radiation produces significant changes in the ionosphere, including in (*a*) the total number of electrons and (*b*) the critical frequency, foF2, above which radio waves are lost because the ionosphere is unable to reflect them to Earth (Davies & Conkright 1990). (Data provided by R Conkright of NOAA's WDC.)

Changing levels of solar activity continuously alter EUV and X-ray ionization rates, and hence electron concentrations, in the atmosphere on time scales from minutes to the 11-year cycle (Balan et al 1994). For example, the limiting or critical frequency, foF2, above which a radio wave is no longer reflected by an ionized atmospheric layer, depends on the square root of the electron density in the layer, and this in turn varies with the intensity of the ionizing radiation. Monthly averaged noontime foF2 values are seen in Figure 15 to vary from 6 to 12 MHz as total electron column increases by factors of more than five from low to high solar activity levels, as indicated by the sunspot number (Davies & Conkright 1990).

Less regular, sudden ionospheric disturbances (SIDs) occur when solar flare enhancements of X-ray and short-wavelength EUV emission cause rapid increases in D-region electron concentrations, thickening this lowest ionospheric region and lowering the height of maximum electron density on the sunlit side of the Earth. Major operational impacts of SIDs last from minutes to a few hours. These include fadeouts and frequency increases of high frequency waves, which experience increased absorption as they pass through the D region and encounter increased ionization in the E and F regions. Phase anomalies, sudden signal enhancements, and increased background noise occur in communications at very low frequencies in response to the lowering and strengthening of the D region, which affect the accuracies of navigation systems that use these frequencies (LORAN and OMEGA).

Total electron content is also of great importance to precision satellite-based positioning systems such as Global Positioning Systems (GPS), Very Long Baseline Interferometry (VLBI), and Satellite Laser Ranging (SLR) because the ionosphere is a dispersive medium with respect to the radio signals that they employ (Vallance Jones 1993). Since the time delay in the information carried by a wave depends on its frequency, the use of dual frequencies at fairly widely spaced bands reduces errors by eliminating ionospheric refraction. But much navigation and positioning still relies on a single frequency for which specification is needed of the variable total electron content and its rate of change along a path from receiver to transmitter—quantities that are strongly affected by variable solar EUV radiation. Like the upper atmosphere neutral density models, empirical and operational ionospheric models utilize either the 10.7-cm flux or the sunspot number as proxies for the geophysically relevant but observationally unavailable EUV irradiances.

## 7. SUMMARY

Numerous associations are evident between solar variability and terrestrial parameters that range from the Earth's surface to hundreds of kilometers above it, on time scales from days to centuries. Global solar and terrestrial parameters derived from space-based and ground-based monitoring, primarily in the past few decades, are facilitating renewed efforts, begun over a century ago, to authenticate apparent Sun-Earth connections by deducing their physical mechanisms.

Decadal cycles in phase with the Sun's activity are evident in temperatures at the Earth's surface and throughout the atmosphere, from the troposphere to the thermosphere. Also apparent is an association of surface temperature with overall solar activity during the past few centuries. But whether the Sun's variable radiation is responsible for these connections is in many instances ambiguous. Confidence in this hypothesis increases with height above the Earth's surface, as the variability of the solar radiative energy deposited there increases and the density of the atmosphere decreases. Least certain is the extent to which tenths percent changes in visible and IR radiation modify global surface temperature and climate, in competition with impacts of other natural and anthropogenic influences. Higher, in the middle atmosphere, there is now cautious confidence that 11-year UV radiation cycles with amplitudes of a few percent do indeed drive changes in temperature, ozone, and winds—associations that were originally debunked in the 1970s as "autosuggestions." In fact, global column ozone concentrations vary during the 11-year cycle by amounts comparable to anthropogenic CFC depletion over the same period. The control of upper atmosphere temperature, density, and ionization by solar EUV radiation is undisputed. But for lack of global upper atmosphere data and coincident EUV irradiance monitoring, the mechanistic details are insufficiently developed for assessment and forecasting needed in spacecraft and communication management, which continues to rely on radio flux or sunspot number proxies.

During the past few decades, minima to maxima amplitudes of solar activity cycles have been near the largest in the 400-year record, and the overall activity level of the Sun is at a historically high level. Historically high levels of radiation are assumed, too, based on the dependence of the Sun's radiative output on the level of solar activity. Although prediction of the course of future solar activity with any certainty is not yet possible, the Sun's radiation will undoubtedly continue to vary, potentially impacting the Earth's surface temperature, ozone concentrations, and upper atmosphere composition in the future. With greenhouse gas concentrations forecast to be double preindustrial levels sometime in the twenty-first century, rigorous attribution of all climate forcings and effects will be needed on regional and global scales to secure reliable early warning of the anticipated climatic impacts. Equally challenging will be evaluation of CFC restrictions implemented in the late 1980s to arrest further ozone depletion. Future monitoring of the expected recovery of ozone to pre-anthropogenic concentration levels must account for natural variability induced by the Sun's variable UV radiation. And in the International Space Station era, the ever-growing amount of orbital space debris will increase the urgency to specify solar radiation impacts on upper atmosphere neutral and ion densities to secure efficient utilization and protection of space-based national and commercial assets.

Critical for these tasks is knowledge of solar radiation variability during the 11-year cycle and over the longer term. To secure this knowledge, continuous, reliable space-based monitoring of the Sun's total and spectral irradiance for the indefinite future is necessary, with instruments that have notably improved accuracy and long-term precision relative to the present state of the art. Complementary coordinated studies of the Sun's global radiation variability mechanisms—both magnetic and nonmagnetic, over a wide range of times scales—are needed to identify the sources of variability and to secure physically justifiable relationships with more widely available solar activity proxies. These studies must develop an understanding of connections between global solar radiative output and continually erupting, evolving, and decaying magnetic fields in the Sun's atmosphere. A lack of such global connections will obfuscate efforts to better quantify the terrestrial relevance of the Sun's variable radiation.

Equally critical are investigations and model simulations of the Earth's response at its surface and throughout its atmosphere to changes in incoming radiation from the Sun over many time scales. Present inability to adequately quantify all climate and ozone forcings adds ambiguities to assessments of the causes of recent global change and cautions against neglect of natural and anthropogenic contributions other than by increasing concentrations of greenhouse gases and CFCs. Improved knowledge of how the terrestrial system itself has varied in the recent past and how forcings in one regime couple to another will help in this effort.

Both the Sun and the Earth exhibit complex, multifaceted, nonlinear, nonstationary variabilities. Neither entities are yet understood sufficiently well to define the causes of their individual variabilities or the impacts of the one on the other. Difficulties inherent in this endeavor should not be underestimated: Caution is required before dismissing empirical associations for lack of understanding within the paradigm of incomplete models, and focused interdisciplinary studies should lead to future progress. Over a decade ago, Evans (1982) characterized the topic of the Sun's relevance for Earth at that time as "intriguing, tantalizing, and far from settled." Although still not settled, the advent of global solar and terrestrial data sets and advances in interpretive and theoretical techniques are enticing new and fruitful explorations of the intrigue and speculation with the added imperative of the recognized societal relevance of Sun-Earth links.

ACKNOWLEDGMENTS

Discussions with many solar and terrestrial physicists over the past decade, and data kindly provided by many more, have benefited this review considerably. Particularly valued are ongoing collaborations with Peter Foukal, David Rind, Dick White, Andy Skumanich, Bill Livingston, Gary Rottman, Tom Woods, Claus Fröhlich, John Cook, Warren White, Ray Bradley, and Juerg Beer. The National Oceanic and Atmospheric Administration World Data Center and its staff, especially Helen Coffey, Dan Wilkinson, and Ray Conkright, have provided continued support for solar terrestrial data requests: Ray Conkright provided the ionospheric data in Figure 15. Loren Acton and Keith Strong have helped in discussing Yohkoh SXT data. Bill Marquette and Anders Johannesson supplied Ca K images, Dick Willson the total irradiance data, Minze Stuiver the $^{14}C$ data in Figure 2, Warren White the SST data in Figure 11, Larry Puga the Mg index data in Figures 12 and Figures 14, and Lon Hood the ozone data in Figure 12. Harry Warren helped with accessing Ca K and GOES data. Lee Kyle, John Hickey, and Doug Hoyt answered many questions and provided the Nimbus 7 solar filter data. Owen Kelley ran the IRI model to generate the ionospheric data used in Figure 13. The NRL's Space Science Division provides a stimulating environment for investigations of this topic in ongoing interactions among members of its Solar Terrestrial Relationships and Upper Atmospheric Physics Branches—Bob Meier, Mike Picone, John Mariska, and Dave Siskind. In addition to numerous discussions, some members of these branches read parts of the text. NASA's Space Physics Supporting Research and Technology and Mission to Planet Earth UARS Guest Investigator programs jointly funded this review.



*Literature Cited*


Acton LW. 1996. In *Proc. 9th Coolstar Workshop,* 3–6 October 1995. Florence, Italy

Allen MR, Smith LA. 1994. *Geophys. Res. Lett.* 21:883–86

Angell JK, Korshover J. 1973. *Mon. Weather Rev.* 101:426–43

Ayres TR. 1997. *J. Geophys. Res.* 102:1641–51

Balan N, Bailey GJ, Jenkins B, Rao PB, Moffett RJ. 1994. *J. Geophys. Res.* 99:2243–53

Baliunas S, Jastrow R. 1990. *Nature* 348:520–23

Banks PM, Kockarts G. 1973. *Aeronomy.* New York & London: Academic. Part A, 430 pp. Part B, 355 pp.

Beer J, Siegenthaler U, Bonani G, Finkel RC, Oeschger H, et al. 1988. *Nature* 331:675–79

Boden TA, Kaiser DP, Sepanski RJ, Stoss FW, eds. 1994. *Trends'93: A Compendium of Data on Global Climate Change.* Carbon Dioxide Information Analysis Center. Publ. No. ORNL/CDIAC-65, Oak Ridge Natl. Lab., Oak Ridge, Tenn.

Bradley RS, Jones PD. 1993. *Holocene* 3(4): 367–76

Brasseur G. 1993. *J. Geophys. Res.* 98:23079–90

Brasseur G, Solomon S. 1984. *Aeronomy of the Middle Atmosphere.* Dordrecht: Reidel. 441 pp.

Brekke P, Rottman GJ, Fontenla J, Judge PG. 1996. *Astrophys. J.* 468:418–32

Bremer J. 1992. *J. Atmos. Terrestr. Phys.* 54: 1505–11

Burroughs WJ. 1992. *Weather Cycles Real or Imaginary?* Cambridge: Cambridge Univ. Press. 207 pp.

Cebula RP, DeLand MT, Schlesinger BM. 1992. *J. Geophys. Res.* 97:11613–20

Chandra S, Lean JL, White OR, Prinz DK, Rottman GJ, Brueckner GE. 1995. *Geophys. Res. Lett.* 22:2481–84

Chandra S, McPeters RD. 1994. *J. Geophys. Res.* 99:20665–71

Committee on Earth Sciences. 1989. *Our Changing Planet: The FY 1990 Research Plan.* US Global Change Res. Prog. Washington, DC: Fed. Coord. Council Sci. Eng. Technol.

Committee on Environment and Natural Resources. 1996. *Our Changing Planet: The FY 1996 U.S. Global Change Research Plan.* Washington, DC: Natl. Sci. Technol. Council

Committee on Transportation Research and Development. 1995. *Interagency Report on Orbital Debris.* Washington, DC: Off. Sci. Technol. Policy

Cook JW, Brueckner GE, VanHoosier ME. 1980. *J. Geophys. Res.* 85:2257–68

Covault C. 1989. *Aviation Week Space Technol.* Oct. 16:23–24

Cox AN, Livingston WC, Matthews MS, eds.

1991. *Solar Interior and Atmosphere.* Tucson: Univ. Arizona Press. 1403 pp.
Crowley TJ, Kim K-Y. 1996. *Geophys. Res. Lett.* 23:359–62
David L. 1996. *Space News* Aug. 26–Sept. 1:4, 19
Davies K, Conkright R. 1990. In *The Effect of the Ionosphere on Radiowave Signals and System Performance,* ed. JM Goodman, pp. 1–11. Ionospheric Effects Symp., Govt. Print. Off. 1990 0-278-020
de Gruijl FR. 1995. *Consequences* 1:13–21
Donnelly RF. 1976. *J. Geophys. Res.* 81:4745–53
Dupree AK, Huber MCE, Noyes RW, Parkinson WH, Reeves EM, Withbroe GL. 1973. *Astrophys. J.* 182:321–33
Eddy JA. 1976. *Science* 192:1189–202
Eddy JA, Gilman PA, Trotter DE. 1976. *Solar Phys.* 46:3–14
Ellison MA. 1969. *The Sun and Its Influence.* New York: Elsevier. 240 pp.
Evans JV. 1982. *Science* 216:467–74
Fiala AD, Dunham DW, Sofia S. 1994. *Solar Phys.* 152:97–104
Fleming EL, Chandra S, Jackman CH, Considine DB, Douglass AR. 1995. *J. Atmos. Terr. Phys.* 57:333–65
Foukal P. 1990. *Solar Astrophysics.* New York: Wiley. 475 pp.
Foukal P, Lean J. 1990. *Science* 247:556–58
Friis-Christensen E, Lassen K. 1991. *Science* 254:698–700
Fröhlich C. 1994. In *The Sun as a Variable Star,* IAU Colloquium 143, ed. JM Pap, C Fröhlich, HS Hudson, SK Solanki, pp. 28–36. Cambridge: Cambridge Univ. Press. 355 pp.
Gilliland R. 1981. *Astrophys. J.* 248:1144–55
Goodman JM. 1992. *HF Communications Science and Technology.* New York: Van Nostrand Reinhold. 631 pp.
Goodman JM, Aarons J. 1990. *Proc. IEEE* 78:512–27
Gorney DJ. 1990. *Rev. Geophys.* 28:315–36
Gurman JB, Delaboudinière JP, Artzner G, Garbiel A, Maucherat A, et al. 1996. *Bull. Am. Astron. Soc.* 28:880 (Abstr.)
Haigh JD. 1994. *Nature* 370:544–46
Haigh JD. 1996. *Science* 272:961–84
Haisch B, Schmitt JHMM. 1996. *Publ. Astron. Soc. Pac.* 108:113–29
Hannah L, Lohse D, Hutchinson C, Carr JL, Lankerani A. 1994. *Ambio* 23:246–50
Hansen J, Lacis A, Ruedy R, Sato M, Wilson H. 1993. *Natl. Geogr. Res. Explor.* 9:142–58
Hara H. 1996. *Structures and heating mechanisms of the solar corona.* PhD thesis. Natl. Astronom. Obs., Japan. 164 pp.
Hartmann DL. 1994. *Global Physical Climatology.* San Diego: Academic. 411 pp.
Hartmann DL, Barkstrom BR, Crommelynck D, Foukal P, Hansen JE, et al. 1993. *Earth Obs.* 5:23–27
Harvey KL, Harvey JW, Martin SF. 1975. *Solar Phys.* 40:87–102
Hays JD, Imbrie J, Shakleton NJ. 1976. *Science* 194:1121–32
Hedin A. 1991. *J. Geophys. Res.* 96:1159–72
Herman JR, Bhartia PK, Ziemke J, Ahmad Z, Larko D. 1996. *Geophys. Res. Lett.* 23:2117–20
Hinteregger HE. 1970. *Ann. Géophys.* 26:547–54
Hinteregger HE, Fukui K, Gilson BG. 1981. *Geophys. Res. Lett.* 8:1147–50
Hoegy WR, Pesnell WD, Woods TN, Rottman GJ. 1993. *Geophys. Res. Lett.* 20:1335–38
Hood LL. 1997. *J. Geophys. Res.* 102:1355–70
Hood LL, Jirikowic JL, McCormack JP. 1993. *J. Atmos. Sci.* 50:3941–58
Hood LL, McCormack JP. 1992. *Geophys. Res. Lett.* 19:2309–12
Hoyt DV. 1979. *Rev. Geophys. Space Phys.* 17: 427–58
Hoyt DV, Kyle HL, Hickey JR, Maschhoff RH. 1992. *J. Geophys. Res.* 97:51–63
Hoyt DV, Schatten KH. 1993. *J. Geophys. Res.* 98:18895–906
Hoyt DV, Schatten KH, Nesmes-Ribes E. 1994. *Geophys. Res. Lett.* 21:2067–70
Hudson HS. 1987. *Rev. Geophys.* 25:651–62
Hudson HS. 1988. *Annu. Rev. Astron. Astrophys.* 26:473–507
Intergovernmental Panel on Climate Change (IPCC). 1992. *Climate Change 1992. Suppl. Rep IPCC Sci. Assess.,* ed. JT Houghton, BA Callander, SK Varney. Cambridge, MA: Cambridge Univ. Press
Intergovernmental Panel on Climate Change (IPCC). 1995. *Climate Change 1994, Radiative Forcing of Climate Change and an Evaluation of the IPCC 1992 Emission Scenarios,* ed. JT Houghton, LG Meira Filho, J Bruce, H Lee, BA Callander, E Haites, N Harris, K Maskell. Cambridge, MA: Cambridge Univ. Press
Jacchia LG. 1963. *Rev. Mod. Phys.* 35:973–91
Jacchia LG. 1977. *Thermospheric temperature, density, and composition models.* Smithsonian Astrophysical Observatory Special Report 375, Cambridge, Mass.
James IN, James PM. 1989. *Nature* 342:53–55
Jirikowic JL, Damon PE. 1994. In *The Medieval Warm Period,* ed. MK Hughes, HF Diaz, pp. 309–16. Dordrecht: Kluwer Academic. 342 pp.
Jursa AS, ed. 1985. *Handbook of Geophysics and the Space Environment.* Air Force Geophys. Lab., Air Force Syst. Command.
Kelly PM, Wigley TML. 1992. *Nature* 360: 328–30

King-Hele D. 1987. *Satellite Orbits in an Atmosphere: Theories and Applications*. Glasgow, London: Blackie
Kuhn JR, Libbrecht KG. 1991. *Astrophys. J.* 381:L35–37
Kyle HL, Hoyt DV, Hickey JR, Maschhoff RH, Vallette BJ. 1993. *NASA Ref. Publ. 1316*
Labitzke K, van Loon H. 1992. *Geophys. Res. Lett.* 19:401–3
Labitzke K, van Loon H. 1993. *Ann. Geophys.* 11:1084–94
Lacis AA, Wuebbles DJ, Logan JA. 1990. *J. Geophys. Res.* 95:9971–81
Lamb HH. 1977. *Clim. Monit.* 6:57–67
Lanzerotti LJ. 1994. In *Solar Terrestrial Energy Program*, ed. DN Baker, VO Papitashvili, MJ Teague, pp. 547–55. COSPAR Colloquia Series, 5. Oxford, UK: Pergamon
Lau K-M, Weng H. 1995. *Bull. Am. Meteorol. Soc.* 76:2391–402
Lean J. 1990. *J. Geophys. Res.* 95:11933–44
Lean J. 1991. *Rev. Geophys.* 29:505–35
Lean J, Beer J, Bradley R. 1995. *Geophys. Res. Lett.* 22:3195–98
Lean J, Rind D. 1997. *J. Climate.* In press
Lean J, Skumanich A, White OR. 1992. *Geophys. Res. Lett.* 19:1591–94
Lean JL, Cook J, Marquette W, Johannesson A, Willson RC. 1997b. *Astrophys. J.* In press
Lean JL, Rottman GJ, Kyle HL, Woods TN, Hickey JR, Puga LC. 1997a. *J. Geophys. Res.* In press
Lee III RB, Gibson MA, Wilson RS, Thomas S. 1995. *J. Geophys. Res.* 100:1667–75
Lockwood GW, Skiff BA, Baliunas SL, Radick RR. 1992. *Nature* 360:653–55
Madronich S, de Gruijl FR. 1993. *Nature* 366: 23
Mann ME, Park J. 1994. *J. Geophys. Res.* 99: 25819–33
Mariska JT. 1992. *The Solar Transition Region*, ed. RF Carswell, DNC Lin, JE Pringle, Cambridge Astrophys. Ser. 23. Cambridge, MA: Cambridge Univ. Press
McElroy MB, Salawitch RJ. 1989. *Science* 243:763–70
McHargue LR, Damon PE. 1991. *Rev. Geophys.* 29:141–58
Meadows AJ. 1975. *Nature* 256:95–97
Meier RR. 1991. *Space Sci. Rev.* 58:1–185
Metha VM, Delworth T. 1995. *J. Clim.* 8:172–90
NASA. 1994. *The Natural Space Environment: Effects on Spacecraft.* NASA Ref. Publ. 1350, Marshall Space Flight Center, Alabama
Nastrom GD, Belmont AD. 1978. *Geophys. Res. Lett.* 5:665–68
National Research Council. 1989. *Ozone Depletion, Greenhouse Gases, and Climate Change.* Washington, DC: Natl. Acad. 122 pp.
National Research Council. 1994. *Solar Influences on Global Change.* Washington, DC: Natl. Acad. 163 pp.
National Research Council. 1995. *A Science Strategy for Space Physics,* Space Studies Board. Washington, DC: Natl. Acad. 81 pp.
National Space Weather Program. 1995. *The Strategic Plan FCM-P30-1995.* Washington, DC/Silver Springs, MD: Off. Fed. Coord. Meteorol. Serv. Supp. Res. 18 pp.
Nesme-Ribes E, Baliunas SL, Sokoloff D. 1996. *Sci. Am.* 275(August):46–52
Nesme-Ribes E, Ferreira EN, Sadourny R, Le Truet H, Li ZX. 1993. *J. Geophys. Res.* 98: 18923–35
Newkirk G Jr. 1983. *Annu. Rev. Astron. Astrophys.* 21:429–67
Nicolet M, Swider W Jr. 1963. *Planet. Space Sci.* 11:1459–82
North GR, Kim K-Y. 1995. *J. Clim.* 8:409–17
Noyes RW, Raymond JC, Doyle JG, Kingston AE. 1985. *Astrophys. J.* 297:805–15
Paetzold HK. 1973. *Pure Appl. Geophys.* 106–108:1308–11
Peixoto JP, Oort AH. 1992. *Physics of Climate.* New York: Am. Inst. Phys. 520 pp.
Penner JE, Chang JS. 1979. *J. Geophys. Res.* 85:5523–28
Penner JE, Charlson RJ, Hales JM, Laulainen NS, Leifer R, et al. 1994. *Bull. Am. Meteorol. Soc.* 75:375–400
Pittock AB. 1978. *Rev. Geophys. Space Phys.* 16:400–20
Quiroz RS. 1979. *J. Geophys. Res.* 84:2415–20
Ramaswamy V, Schwarzkopf MD, Randel WJ. 1996. *Nature* 382:616–18
Randel WJ, Cobb JB. 1994. *J. Geophys. Res.* 99:5433–47
Ratcliffe JA. 1972. *An Introduction to the Ionosphere and Magnetosphere.* Cambridge: Cambridge Univ. Press. 256 pp.
Reeves EM. 1976. *Solar Phys.* 46:53–72
Reid G. 1991. *J. Geophys. Res.* 96:2835–44
Reid G. 1995. *Rev. Geophys. Suppl.* July:535–38
Reinsel GC, Tam W-K, Ying LH. 1994. *Geophys. Res. Lett.* 21:1007–10
Rind D, Balachandran NK. 1995. *J. Clim.* 8:2080–95
Rind D, Overpeck J. 1993. *Quat. Sci. Rev.* 12:357–74
Rind D, Peteet D, Kukla G. 1989. *J. Geophys. Res.* 94:12851–71
Rind D, Suozzo R, Balachandran NK, Prather MJ. 1990. *J. Atmos. Sci.* 47:475–94
Rishbeth H, Garriott OK. 1969. *Introduction to Ionospheric Physics.* New York; London: Academic. 331 pp.
Rishbeth H, Roble RG. 1992. *Planet. Space Sci.* 40:1011–26

Roble RG. 1976. *J. Geophys. Res.* 81:265–68
Roble RG. 1987. In *Solar Radiative Output Variations,* Workshop Proc., Nov. 9–11, Boulder, ed. P Foukal, pp. 1–25. Cambridge, MA: Cambridge Res. Instrum.
Roble RG, Emery BA. 1983. *Planet. Space Sci.* 31:597–614
Robuck A, Free MP. 1995. *J. Geophys. Res.* 100:11549–67
Rottman GJ. 1988. *Adv. Space Res.* 8(7):53–66
Salby ML, Shea DJ. 1991. *J. Geophys. Res.* 96:22579–95
Schatten KH. 1988. *Geophys. Res. Lett.* 15: 121–24
Schlesinger ME, Ramankutty N. 1992. *Nature* 360:330–33
Schwartz SE, Andreae MO. 1996. *Science* 272: 1121–22
Schwarzkopf MD, Ramaswamy V. 1993. *Geophys. Res. Lett.* 20:205–8
Solomon S, Portmann RW, Garcia RR, Thomason LW, Poole LR, McCormick MP. 1996. *J. Geophys. Res.* 101:6713–27
Sonett CP, Giampapa MS, Matthews MS, eds. 1991. *The Sun in Time.* Tucson: Univ. Ariz. Press. 990 pp.
Space Environment Services Center. 1982. *Proc. of a Workshop on Satellite Drag.* March 18–19. Space Environ. Lab., Boulder, CO.
Stevens MJ, North GR. 1996. *J. Atmos. Sci.* 53:2594–608
Stolarski RS, Bloomfield P, McPeters RD, Herman JR. 1991. *Geophys. Res. Lett.* 18:1015–18
Stuiver M, Braziunas TF. 1993. *Holocene* 3(4): 289–305
Tang F, Howard R, Adkins JM. 1984. *Solar Phys.* 91:75–86
Tobiska K. 1993. *J. Geophys. Res.* 98:18879–93
Tsuneta S, Acton L, Bruner M, Lemen J, Brown W, et al. 1991. *Solar Phys.* 136:37–67
Vallance Jones A, ed. 1993. *Environmental Effects on Spacecraft Positioning and Trajectories. Geophys. Monogr. 73.* Washington, DC: Am. Geophys. Union. 173 pp.
Vernazza JE, Avrett EH, Loeser R. 1976. *Astrophys. J. Suppl. Ser.* 30:1–60
Warren HP, Mariska JT, Lean J, Marquette W, Johannesson A. 1996. *Geophys. Res. Lett.* 23:2207–10
White OR, ed. 1977. *The Solar Output and Its Variation.* Boulder: Colo. Assoc. Univ. Press. 526 pp.
White OR, Rottman GJ, Woods TN, Knapp BG, Keil SL, et al. 1994. *J. Geophys. Res.* 99:369–72
White OR, Skumanich A, Lean J, Livingston WC, Keil SL. 1992. *Publ. Astron. Soc. Pac.* 104:1139–43
White WB, Lean J, Cayan D, Dettinger M. 1997. *J. Geophys. Res.* 102:3255–66
Wigley TM, Raper SCB. 1990. *Geophys. Res. Lett.* 17:2169–72
Willson RC. 1994. In *The Sun as a Variable Star,* ed. JM Pap, C Fröhlich, HS Hudson, SK Solanki, IAU Colloq. 143, pp. 4–10. Cambridge: Cambridge Univ. Press
Willson RC, Hudson HS. 1991. *Nature* 351:42–44
Withbroe GL. 1990. *Adv. Astronaut. Sci.* 71: 727–43
Woods TN, Prinz DK, Rottman GJ, London J, Crane PC, et al. 1996. *J. Geophys. Res.* 101:9541–69
Woods TN, Rottman G. 1996. *J. Geophys. Res.* Submitted
World Meteorological Organization. 1991. *Scientific Assessment of Ozone Depletion. Global Ozone Research and Monitoring Project—Report No. 25.* Geneva, Switzerland: World Meteorol. Org.
Wuebbles DJ, Kinnison DE, Grant KE, Lean J. 1991. *J. Geomag. Geoelectr. Supp.* 43:709–18
Zhang Q, Soon WH, Baliunas SL, Lockwood GW, Skiff BA, Radick RR. 1994. *Astrophys. J.* 427:L111–14

*Annu. Rev. Astron. Astrophys. 1997. 35:69–100*


# LUMINOUS SUPERSOFT X-RAY SOURCES

*P. Kahabka and E. P. J. van den Heuvel*
Anton Pannekoek Astronomical Institute and Center for High Energy Astrophysics, University of Amsterdam, Kruislaan 403, 1098 SJ Amsterdam, The Netherlands; e-mail: ptk@astro.uva.nl and edvdh@astro.uva.nl

ABSTRACT

Luminous supersoft X-ray sources were discovered with the Einstein observatory and have been established as an important new class of X-ray binaries on the basis of observations with the Roentgen Satellite (ROSAT). They have extremely soft spectra (equivalent blackbody temperatures of ~15–80 eV) and are highly luminous (bolometric luminosities of $10^{36}$–$10^{38}$ erg $s^{-1}$). Correcting for the heavy extinction of soft X rays by interstellar neutral hydrogen, their numbers in the disks of ordinary spiral galaxies like our own and M31 are estimated to be of the order of $10^3$. Their observed characteristics are consistent with those of white dwarfs, which are steadily or cyclically burning hydrogen-rich matter accreted onto the surface at a rate of order $10^{-7}$ $M_{\odot}$ $year^{-1}$. The required high accretion rates can be supplied by mass transfer on a thermal time scale ($10^6$–$10^7$ years) from close companion stars that are more massive than the white dwarf accretor, typically 1.3–2.5 $M_{\odot}$. Steady burning can also occur in a post-nova stage, but for shorter time scales, and it has been observed in a few classical novae and symbiotic novae. A few supersoft sources have been found to be recurrent transients. They are possibly connected with very massive white dwarfs accreting at high rates. Luminous supersoft sources may make a considerable contribution to the Type Ia supernova rate in spiral and irregular galaxies.

## 1. INTRODUCTION

The luminous supersoft X-ray sources (SSS) were recognized as an important new class of intrinsically bright X-ray sources by Trümper et al (1991) (see also Greiner et al 1991). A careful analysis of the X-ray spectral properties of the first few sources of this type found in the Large Magellanic Cloud (LMC) showed



0066-4146/97/0915-0069$08.00

that while their X-ray luminosities are of the order of the Eddington limit (about $10^{38}$ erg s$^{-1}$), their X-ray spectra are extremely soft, peaking at energies in the range 15–80 eV, on average of order 30–40 eV. This corresponds to a blackbody temperature typically of ~300,000–500,000 K, some two orders of magnitude lower than for that of the classical X-ray binaries which contain neutron stars or black holes. As these peak energies of around 40 eV are below the lower energy limit of the ROSAT detectors, the extrapolations of the observed tails of the energy distributions towards lower energies are, of course, rather uncertain, especially since at these energies extinction by interstellar neutral hydrogen plays a large role. Nevertheless, taking realistic error margins in these extrapolations into account, the total X-ray luminosities of these LMC sources always come out close to the Eddington limit (Greiner et al 1991, van Teeseling et al 1996a). Therefore, there can be no doubt that they are intrinsically highly luminous. The discovery of a large ionization region with a diameter of several tens of parsecs around the supersoft LMC source CAL 83 (Remillard et al 1995) is a further confirmation of their very high soft X-ray luminosities.

So far, some 35 luminous SSS have been found with ROSAT, 16 in the Andromeda Nebula (M31), 11 in the two Magellanic Clouds, 7 in our own Galaxy and 1 in the Local Group galaxy NGC 55 (see Table 1). In view of the very large interstellar extinction of soft X rays, the sources in the other galaxies can only be observed when they are near the outer edge of the interstellar neutral hydrogen layer, at our side of that galaxy. Taking this into account, the total numbers of sources in M31 and in our Galaxy are estimated to be some two orders of magnitude larger than the observed numbers, and in the LMC some 20 times larger (cf Rappaport et al 1994b). The SSS therefore constitute a major new population of highly luminous X-ray sources in spiral galaxies like our own.

In this article, we review the observed characteristics of these luminous SSS ($L_x = 10^{36}$–$10^{38}$ erg s$^{-1}$) and the theoretical models that have been put forward for them. [Other types of less luminous supersoft sources have also been discovered with ROSAT in our Galaxy, in our local neighborhood, for example PG 1159 stars, certain CV systems, and some nuclei of planetary nebulae (see Greiner 1996b). These are not the subject of this review. The same holds for active galactic nuclei (AGNs) with supersoft spectra.] The now generally accepted model for the luminous SSS is that, with few exceptions, these sources are accreting white dwarfs (WDs) in binaries, which are burning hydrogen in their envelopes in a steady or intermittend way (van den Heuvel et al 1992). In Section 2, we review how the luminous supersoft sources were first recognized as a separate class of X-ray sources and how the accreting binary WD model with surface nuclear burning has arisen. In Section 3, we review the observations in terms of source samples, both in X rays and at other wavelengths, in external galaxies as well as our own Galaxy. Section 5 is devoted to physical models of accreting WDs with nuclear burning and Section 6 to the evolutionary history of the different types of luminous SSS. We also review in Section 6 the results of population synthesis calculations, and we compare these results with the observations. Finally, in Section 7 we discuss the possible relation between SSS and Type Ia supernovae (SN Ia). Important recent publications on the subject of SSS can be found in the proceedings of a recent workshop on this subject (Greiner 1996a).

## 2. THE RECOGNITION OF THE LUMINOUS SSS AS A NEW CLASS AND THE WD MODEL WITH SURFACE NUCLEAR BURNING

### 2.1 *Discovery*

Four of the luminous SSS, CAL 83 and CAL 87 in the LMC and 1E 0035.4-7230 and 1 E0056.8-7154 in the Small Magellanic Cloud (SMC), had already been discovered with the Einstein satellite around 1980 by Long et al (1981) and Seward & Mitchell (1981), respectively. It was recognized that the sources have unusually soft spectra, but since the Einstein satellite had a different energy range and a lower spectral resolution than ROSAT, they were not recognized as a separate class. Since black hole sources often also have quite soft spectra, the first idea was that CAL 83 and CAL 87 are black holes in binary systems (Smale et al 1988, Cowley et al 1990, Cowley 1992, Wang et al 1991, Wang & Wu 1992). What is called soft in connection with the black hole sources is, however, still a fairly hard type of X ray (i.e. with a large flux in the 0.1- to 1-keV range); in addition, black hole sources always have an important hard component (Tanaka & Lewin 1995), which is completely lacking in the SSS. It was only thanks to ROSAT that the SSS could be clearly distinguished from the classical strong point X-ray sources (accreting neutron stars and black holes in binaries) and from the X-ray emission of coronae of nearby solar type stars, on the basis of their observed ROSAT X-ray spectra, as depicted in Figure 1. The classical strong point X-ray sources typically have blackbody temperatures on the order of several times $10^7$ K, and therefore their main emission is observed in the hardest ROSAT energy band, with hardly any emission in the softest ROSAT bands. Stellar coronae (solar-type stars) typically have temperatures of about 1–3 $\times$ $10^6$ K, which causes them to produce, in addition to a strong signal in the lowest energy bands of ROSAT, a significant signal also in the harder ROSAT band. Figure 1 shows that the SSS can be immediately recognized as a distinct class with extremely soft spectra. The peak of their energy distributions is actually lower than the lower cutoff of the softest ROSAT band.

**Table 1** Summary of all known luminous supersoft X-ray sources[a]

| Name | Count rate (cts/s) | T (eV) | $L_{bol}$ (erg/s) | Type | Period | Ref.[b] |
|---|---|---|---|---|---|---|
| LMC | | | | | | |
| RX J0439.8-6809 | 1.35 | 21–27 (wd) | $0.6–1.5 \times 10^{37}$ | CV | 3.37 h | 1–4 |
| RX J0513.9-6951 | <0.06–2.0 | 34–54 (wd) | $1.2–4.8 \times 10^{37}$ | CBSS | 18.3 h | 2, 5–10 |
| RX J0527.8-6954 | 0.004–0.25 | 27–68 (wd) | $0.038–3.0 \times 10^{37}$ | CBSS? | | 2, 7, 11–15 |
| RX J0537.6-7034 | 0.02 | 18–30 (bb) | $0.6–2 \times 10^{37}$ | | | 16–17 |
| CAL 83 | 0.98 | 34–54 (wd) | $0.38–4.8 \times 10^{37}$ | CBSS | 1.04 day | 12, 18–19 |
| CAL 87 | 0.09 | 68–86 (wd) | $1.2–9.5 \times 10^{37}$ | CBSS | 10.6 h | 18–22 |
| RX J0550.0-7151 | <0.02–0.9 | 25–40 (bb) | | | | 2, 7 |
| SMC | | | | | | |
| 1E0035.4-7230 | 0.33 | 34–54 (wd) | $0.38–1.2 \times 10^{37}$ | CV | 4.1 h | 23–25 |
| RX J0048.4-7332 | 0.19 | 25–45 (wd) | $0.48–1.2 \times 10^{37}$ | Sy-N | | 22, 26–29 |
| RX J0058.6-7146 | <0.001–0.7 | 15–70 (bb) | $2 \times 10^{36}$ | | | 22 |
| 1E0056.8-7154 | 0.29 | 27–43 (wd) | $1.5–3.8 \times 10^{37}$ | PN | | 30 |
| Milky Way | | | | | | |
| RX J0019.8+2156 | 2.0 | 21–27 (wd) | $3–9 \times 10^{36}$ | CBSS | 15.8 h | 43–45 |
| RX J0925.7-4758 | 1.0 | 70–100 (wd) | $3–7 \times 10^{35}$ | CBSS | 3.8 day | 40–42 |
| Nova 1983 Mus | 0.1 | 25–35 (bb) | $1–2 \times 10^{38}$ | CV-N | 85 min | 31, 32–36 |
| 1E 1339.8+2837 | 0.01–1.1 | 20–45 (bb) | $0.12–10 \times 10^{35}$ | | | 46–47 |
| AG Dra | 1.0 | 10–15 (bb) | $1.4 \times 10^{36}$ | Sy | 554 day | 49–50 |
| RR Tel | 0.18 | 14 (wd) | $1.3 \times 10^{37}$ | Sy-N | 387 day | 29, 48 |
| Nova V1974 Cygni | 0.03–76 | | $2 \times 10^{38}$ | CV-N | 1.95 h | 38, 39 |
| M31 | | | | | | |
| a. RX J0037.4+4015 | $0.3 \times 10^{-3}$ | 43 (bb) | | | | 51 |
| b. RX J0038.5+4014 | $0.8 \times 10^{-3}$ | 45 (bb) | | | | 51 |
| c. RX J0038.6+4020 | $1.7 \times 10^{-3}$ | 43 (bb) | | | | 51 |

| | | | | | | |
|---|---|---|---|---|---|---|
| d. RX J0039.6+4054 | $0.4 \times 10^{-3}$ | 45 (bb) | | | | 51 |
| e. RX J0040.4+4009 | $0.8 \times 10^{-3}$ | 42 (bb) | | | | 51 |
| f. RX J0040.7+4015 | $1.3 \times 10^{-3}$ | 42 (bb) | | | | 51 |
| g. RX J0041.5+4040 | $0.3 \times 10^{-3}$ | 40 (bb) | | | | 51 |
| h. RX J0041.8+4059 | $0.5 \times 10^{-3}$ | 43 (bb) | | | | 51 |
| i. RX J0042.4+4044 | $1.7 \times 10^{-3}$ | 43 (bb) | | | | 51 |
| j. RX J0043.5+4207 | $2.2 \times 10^{-3}$ | 45 (bb) | | | | 51 |
| k. RX J0044.0+4118 | $2.5 \times 10^{-3}$ | 42 (bb) | | | | 51 |
| RX J0045.4+4154 | $<10^{-5}$–0.03 | 70–90 (wd) | | | | 52 |
| l. RX J0045.5+4206 | $3.1 \times 10^{-3}$ | 20–48 (bb) | | | | 53 |
| m. RX J0046.2+4144 | $2.1 \times 10^{-3}$ | 38 (bb) | | | | 51 |
| n. RX J0046.2+4138 | $1.1 \times 10^{-3}$ | 40 (bb) | | | | 51 |
| o. RX J0047.6+4205 | $1.0 \times 10^{-3}$ | 39 (bb) | | | | 51 |
| NGC 55 | | | | | | |
| RX J0016.0-3914 | $4.5 \times 10^{-3}$ | | | | | 54 |

[a]Excluding PG 1159-type stars and supersoft AGNs. Given for each source are the name (column 1); the ROSAT PSPC count rate (column 2); the best-fit X-ray temperature (bb = blackbody, wd = white dwarf) (column 3); the bolometric luminosity (column 4); the type of the system (CBSS = close binary supersoft source, PN = planetary nebula, Sy = symbiotic system, N = Nova, CV = cataclysmic variable) (column 5); the binary period (column 6); references (column 7) (from Greiner 1996a,b). Columns 3 and 4 in part from Kahabka & Trümper 1996; P Kahabka, in preparation.

[b]References: (1) Greiner et al (1994); (2) Cowley et al (1996); (3) van Teeseling et al (1996b); (4) Schmidtke & Cowley (1996); (5) Schaeidt et al (1993); (6) Pakull et al (1993); (7) Cowley et al (1993); (8) Crampton et al (1996); (9) Hutchings et al (1995); (10) Schaeidt (1996); (11) Trümper et al (1991); (12) Greiner et al (1991); (13) Orio & Ögelman (1993); (14) Hasinger (1994); (15) Greiner et al (1996b); (16) Orio & Ögelman (1993); (17) Orio et al (1996); (18) Long et al (1981); (19) Smale et al (1988); (20) Pakull et al (1988); (21) Schmidtke et al (1993); (22) Kahabka et al (1994); (23) Wang & Wu (1992); (24) Orio et al (1994); (25) Schmidtke et al (1996); (26) Vogel & Morgan (1994); (27) Morgan (1992); (28) Jordan et al (1996); (29) Mürset et al (1996); (30) Heise et al (1994); (31) Krautter & Williams (1989); (32) Diaz & Steiner (1989); (33) Diaz & Steiner (1994); (34) Diaz et al (1995); (35) Ögelman et al (1993); (36) Shanley et al (1995); (37) Krautter et al (1996); (38) Shore et al (1993); (39) Shore et al (1994); (40) Motch et al (1994); (41) Motch (1996); (42) Ebisawa et al (1996); (43) Beuermann et al (1995); (44) Greiner & Wenzel (1995); (45) Gänsicke et al (1996) (46) Verbunt et al (1993); (47) Hertz et al (1993); (48) Jordan et al (1994); (49) Viotti et al (1996); (50) Greiner et al (1996d); (51) Greiner et al (1996b); (52) White et al (1994); (53) R Supper, in preparation; (54) Singh et al (1995).

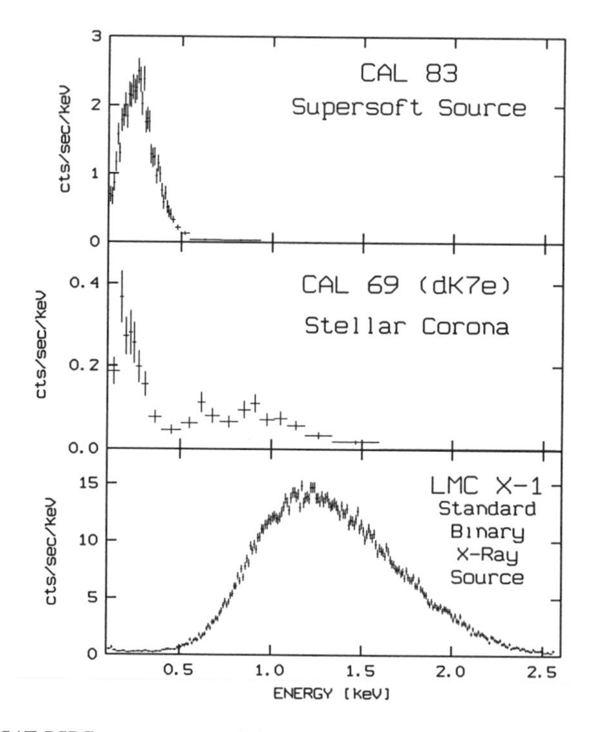


*Figure 1* ROSAT PSPC count spectra of three objects in the Large Magellanic Cloud (LMC) field: the SSS CAL 83, the dK7e foreground star CAL 69, and the black hole candidate LMC X-1 (similar to Figure 2 of Trümper et al 1991).

The effects of extinction by interstellar neutral hydrogen on their ROSAT spectra is schematically illustrated in Figure 2. For large column densities, the soft part of the spectrum is completely blocked by the interstellar medium (ISM), and only the "hard" tail of the spectrum is observed. The figure illustrates the expected ROSAT response to a 300,000-K blackbody spectrum, for a hydrogen column density of $2.10^{20}$ cm$^{-2}$ (SA Rappaport, private communication). As the column density towards a source is not known beforehand, the analysis of the observed ROSAT spectra yields combinations of $L_X$ and $T_{eff}$ (assuming a blackbody) that still depend on the assumed column density.

## 2.2 *The Binary Nature of CAL 83 and CAL 87*

CAL 83 was optically identified with a star of magnitude V = 16.8 by Smale et al (1988) and CAL 87 with a star of magnitude about V = 20 by Pakull et al (1988). The optical studies of CAL 83 by Smale et al (1988) revealed regular photometric variations with a period of 1.04 days. The light curve

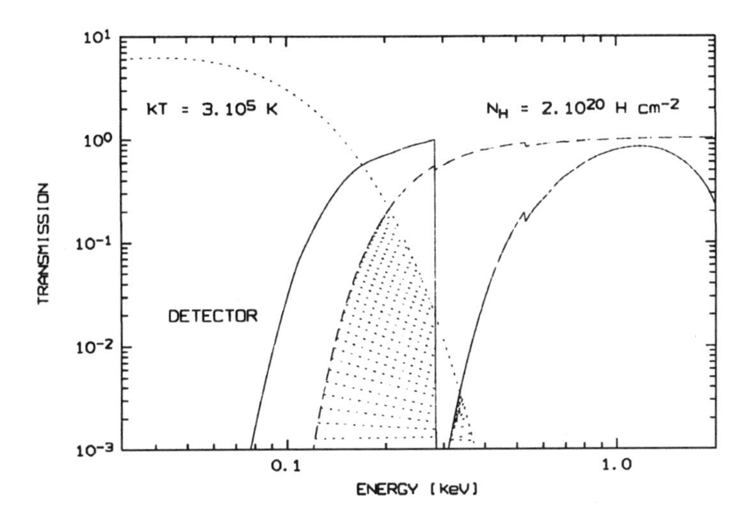


*Figure 2* ROSAT PSPC efficiency (*solid curve*), transmission of ISM for hydrogen column of $2.10^{20}$ H atoms $cm^{-2}$ (*dashed line*), distribution of a $3.10^5$-K blackbody spectrum (*dotted line*) and folded (observed) distribution (*hatch marks*) (SA Rappaport, private communication).

is approximately sinusoidal with an amplitude of about 0.1 magnitude (mag). Similarly, CAL 87 was found to show regular photometric variations with a period of 10.6 h (Cowley et al 1990). The light curve resembles that of an eclipsing binary with a deep primary minimum (about 1.3 mag) and a small and variable secondary minimum (cf Schmidtke et al 1993). Figure 3 depicts the average light curves of CAL 83 and CAL 87, together with a proposed schematic explanation of the shapes in terms of a model with an optically bright accretion disk and a companion that is strongly heated on one side (see Figure 3 caption).

As for their optical spectra, both sources—as well as other LMC sources—show strong emission of He II$\lambda$4686 as well as the Balmer lines of hydrogen, but the CIII/NIII-complex near $\lambda$4640 (Figure 4) usually seen in low-mass X-ray binaries is absent in CAL 87 and only weakly present in CAL 83. [The similarity of the optical spectra with those of low-mass X-ray binaries led at first to the suggestion that SSS might be neutron stars (cf Greiner et al 1991, Kylafis & Xilouris 1993).] In CAL 87, the strengths of the emission lines are not strongly dependent on orbital phase (Cowley et al 1990), which indicates that they must arise in a volume that is large in comparison to the eclipsing donor star: They probably arise in a corona around the binary system or in a stellar wind emanating from the disk, the X-ray source, or the heated side of the donor star (van den Heuvel et al 1992). The X rays in CAL 87 do show an eclipse

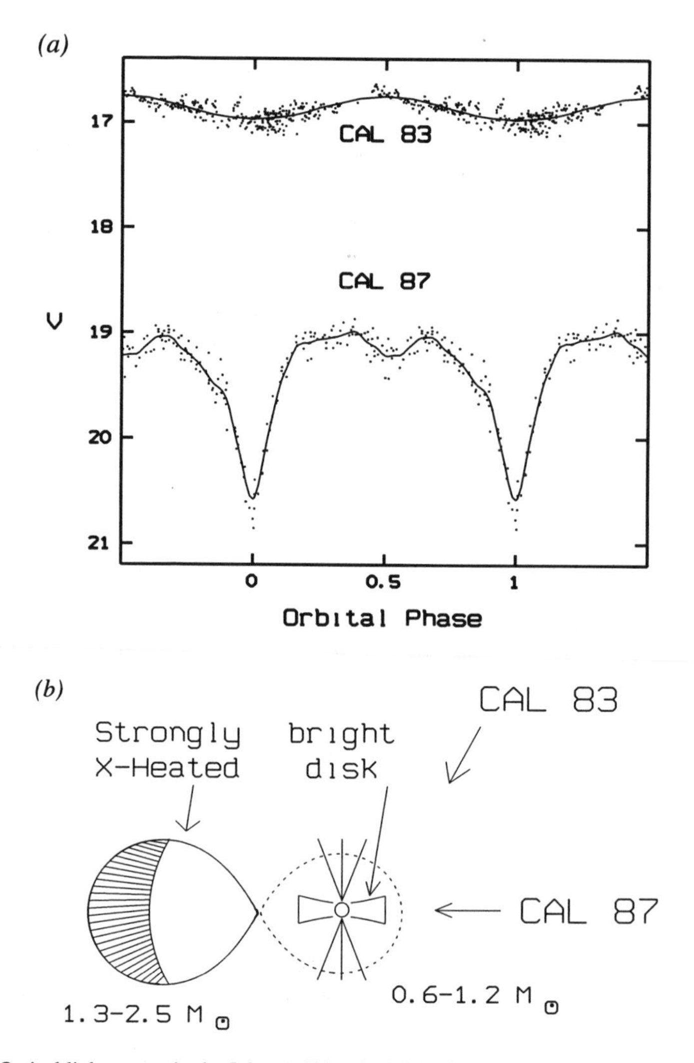


*Figure 3* (*a*) Optical light curves in the Johnson V-band of CAL 83 and CAL 87 plotted on the same scale for comparison. The *solid curves* give the mean light curves. The upper light curve is adapted from Smale et al (1988), the lower light curve from Schmidtke et al (1993). (*b*) Schematic model for explaining the optical light curves of CAL 83 and 87: The main light sources in the systems are the very bright accretion disk and the X-ray heated side of the donor star. In CAL 87 the accretion disk is regularly eclipsed; CAL 83 is seen at low inclination, such that only the heating effect is observed [after van den Heuvel et al (1992); for a refined model, see S Schandl et al (1996); see also section on The "Standard" CBSS].

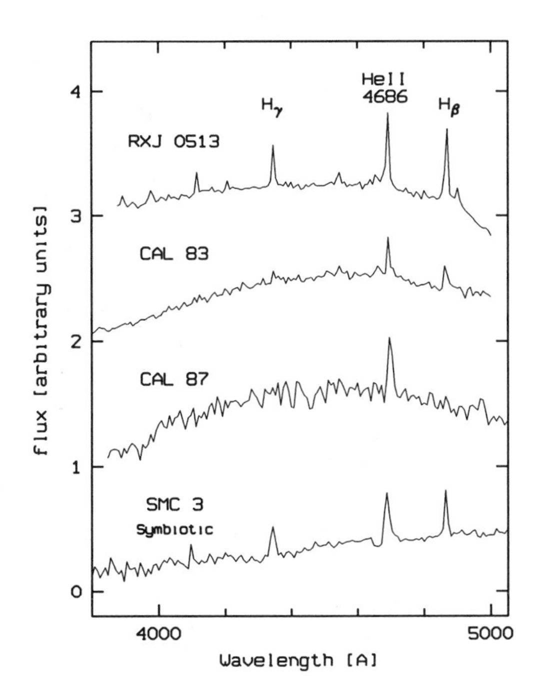


*Figure 4* Optical spectra of RX J0513.9-6951 and CAL 83 (from Cowley et al 1993), CAL 87 (from Pakull et al 1988), and SMC 3 (from Morgan 1992).

(Kahabka et al 1994), although not as deep as the optical photometric eclipse. They therefore arise in a smaller volume than the emission lines. Recently, from HST observations, Deutsch et al (1996) discovered that CAL 87 has two very close optical companions (distances 0.88″ and 0.66″) that contribute significant light at the time of minimum light. These are a mid-G type subgiant and a late A main-sequence star. Outside eclipse, they together contribute 20–25% of the light. Therefore, the eclipse of the system is in fact deeper than depicted in Figure 3, and also, the close optical companions, with a combined spectrum of type F, dominate the optical spectrum of the system observed at minimum light, as they provide some 70% of the light at this orbital phase. Hence, unfortunately, this spectrum does not provide accurate information about the intrinsic spectral type (unheated) that one expects to be observed at the dark side of the star.

## 2.3 *Why Nuclear Burning on the Surface of a WD?*

The luminosity and effective temperature of a star allow one to derive its radius by using Stefan-Boltzman's law:

$$L = 4\pi R^2 \sigma T^4, \tag{1}$$

**Table 2** Energy gain from accretion onto a 1-$M_\odot$ black hole, neutron star, and white dwarf, compared with energy gain by nuclear burning of hydrogen

| | Energy release | |
|---|---|---|
| Compact object | Accretion | Nuclear burning |
| Black hole | (0.1–0.42) $mc^2$ | — |
| Neutron star | 0.15 $mc^2$ | 0.007 $mc^2$ |
| White dwarf | 0.00025 $mc^2$ | 0.007 $mc^2$ |

which yields

$$R = 9 \times 10^8 (L_{37.5})^{1/2} (T_e/40\text{ eV})^{-2}\text{ cm}, \qquad (2)$$

where $L_{37.5}$ is the X-ray luminosity in units of $10^{37.5}$ erg/s, and $T_e$ is the effective temperature in electron volts.

Equation 2 shows that for values characteristic for the luminous SSS, namely $L_{37.5} = 1$ and $T_e = 40$ eV, the emitting object has a radius of about 9000 km, i.e. similar to that of a WD. For this reason, Grindlay suggested (JE Grindlay, personal communication) that, in analogy with the accreting neutron stars and black holes in the classical X-ray binaries, the X rays are generated here by accretion of matter onto a WD. In this case, in order to generate $10^{37.5}$ erg/s, as Table 2 shows, a solar mass WD with a radius of 6000 km should accrete some $2.10^{-6}$ $M_\odot$ year$^{-1}$, a huge amount!

A simple calculation shows, however, that with such an accretion rate, the inflowing matter would have an optical thickness large enough to block the soft X rays from coming out.

There is, however, a great difference between the energy generation by accretion onto neutron stars and black holes on the one hand and WDs on the other, as Table 2 shows, namely the following: The energy release by accretion of hydrogen onto neutron stars and black holes per unit mass is some 20 times more than the energy release by nuclear fusion of hydrogen in the same amount of mass. Therefore, in the case of neutron stars and black holes, nuclear fusion makes a negligible contribution to the energy generation. On the other hand, in the case of a WD, the energy release by accretion is some 30 times smaller than the energy release by nuclear fusion in the same amount of matter. van den Heuvel et al (1992) realized that if the accreted matter on the surface of a WD begins to burn steadily, an accretion rate is necessary that is some 30 times lower than the rate required for energy generation purely by accretion. The required rate then is only of order $10^{-7}$ $M_\odot$ year$^{-1}$, and in this case the soft X rays can be shown to be able to escape.

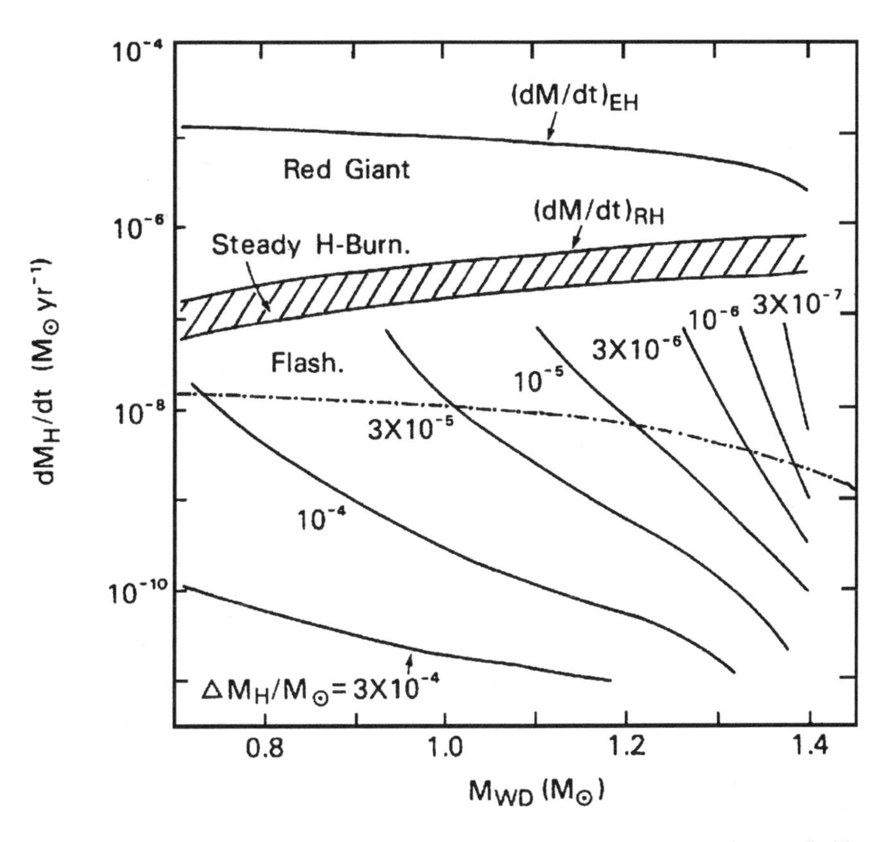


*Figure 5* Regimes of steady nuclear burning, weak flashes (cyclic burning), and strong flashes (novae) in the $\dot{M}$–$M_{WD}$ plane (cf Fujimoto 1982a,b, Nomoto 1982, DiStefano & Rappaport 1995). The $\Delta M_H$ values indicate envelope masses (for a given accretion rate) at which burning is ignited. Below the *dash-dot line*, flashes produce nova explosions.

The condition for various types of nuclear burning on the surface of a WD had already been studied about a decade earlier by various authors, e.g. Paczyński & Zytkow (1978), Prialnik et al (1978), Sion et al (1979), Sienkiewicz (1980), Nomoto (1982), Fujimoto (1982a,b), Iben (1982), and more recently by Prialnik & Kovetz (1995). These authors found that there are, broadly speaking, three possible regimes of nuclear burning in the surface layers of a WD, which depend on the accretion rate and the WD mass as follows (Figure 5) (after Nomoto 1982; SA Rappaport, private communication).

1. In a narrow range of accretion rates, roughly 1–4 $\times$ $10^{-7}$ $M_\odot$/year for a solar mass WD, the accreted hydrogen burns steadily, without much radius

expansion of the WD (Nomoto et al 1979). This is the hatch-marked region in Figure 5.

2. For accretion rates below the range for steady burning, the accreted matter burns in flashes, which for decreasing accretion rates have longer intervals between them and become more and more violent. In typical novae (cataclysmic variables, CVs), the accretion rates in general range between $10^{-9}$ and $10^{-8}$ solar masses per year (see, for example, Livio 1994).

3. For very high accretion rates, above the range for steady burning without radius expansion, the radius expands to red giant dimensions, and the matter keeps burning steadily in a thin shell around the WD. This is, in fact, the standard model for a red giant of mass lower than about 8 solar masses (cf Kippenhahn & Weigert 1994).

Recent more-refined calculations have shown (Hachisu et al 1996) that in the regime of very high accretion rates, instead of forming a red giant envelope, the burning of the accreted matter may cause the formation of a very strong stellar wind, opaque to soft X rays, which flows out at high speeds from the WD (velocity typically on the order of the escape velocity from the WD surface, i.e. about 5000 km/s). In any case, for accretion rates above the hatch-marked region in Figure 5, one does not expect the X rays to escape. We return to this in more detail in Sections 3.4 (Symbiotic Systems; Long-Term Variations) and 5.

From the above, it should be clear that in binaries with donor stars that are able to provide accretion rates in the region hatch-marked in Figure 5, a WD will become a steady, very luminous supersoft X-ray source. We now consider which companions will be able to provide just this required rate.

## 2.4 *Which Companions Are Able to Provide the Right Accretion Rate for Producing a Luminous SSS?*

NEAR MAIN-SEQUENCE COMPANIONS MORE MASSIVE THAN THE WHITE DWARFS The simplest type of binary configuration in which accretion rates as high as 1–4 $\times$ $10^{-7}$ $M_{\odot}$/year can be achieved is that in which the companion has a mass larger than that of the WD and has a radiative envelope. A possible companion is a star with a mass larger than about 1.3 $M_{\odot}$, which is still fairly close to the main sequence, i.e. a slightly evolved star of spectral type earlier than about F5. In such a system, once the companion overflows its Roche lobe (Roche lobe overflow, RLOF), mass transfer will take place on a thermal time scale of the donor star. This is due to the fact that mass transfer from the more massive to the less massive component of a binary causes the orbit to shrink, whereas the thermal equilibrium radius of a star that has an evolved (i.e. He-rich) core does not decrease when mass is taken away from its envelope. As a result, the mass transfer and associated shrinking of the orbit will cause the donor to become thermally unstable: It will start transferring mass on its thermal timescale until it has become the less massive of the two, and the orbit can expand when further mass transfer occurs (cf Paczyński 1971, van den Heuvel 1994). The thermal timescale of a star in or close to the main sequence is roughly given by

$$t_{th} = 3 \times 10^7 \Big/ \left(\frac{M}{M_\odot}\right)^{-2} \text{ years.} \tag{3}$$

As $\dot{M} \simeq M/t_{th}$, the result is roughly

$$\dot{M}_{th} \simeq 3 \times 10^7 \left(\frac{M}{M_\odot}\right)^3 M_\odot/\text{year.} \tag{4}$$

Thus for stars with $M \gtrsim 1.3\ M_\odot$, one has $\dot{M} \gtrsim 10^{-7}\ M_\odot/\text{year}$.

Typically, near–main-sequence companions in the mass range 1.3–2.5 $M_\odot$ can thus provide $\dot{M}$ in the range 1–4 $\times\ 10^{-7}\ M_\odot/\text{year}$ (van den Heuvel et al 1992). It is striking that this is indeed the estimated mass range of the companions of CAL 83 and CAL 87, as derived from their orbital periods and absolute luminosities. At present, five of the luminous SSS have been identified with binaries with periods in the range 10 h to a few days, which fits this model (see Table 1). We call these systems close binary supersoft sources (CBSS).

SYMBIOTIC SYSTEMS Symbiotic systems are wide binaries consisting of a red giant star and a WD (cf Kenyon 1986, Friedjung 1988, Mikolajewska & Kenyon 1992, Viotti et al 1996). This is the second obvious class of binary systems in which high mass-transfer rates towards WDs can be achieved such that nuclear burning in the surface layers can be ignited. Sion & Starrfield (1994) were the first to realize this (see also Iben & Tutukov 1994, Yungelson et al 1996b).

Among the symbiotics, one expects two different types of companions that provide different modes of mass transfer, as follows:

1. An approximately one–solar-mass red giant, less massive than the WD, which is overflowing its Roche lobe in a binary with an orbital period longer than about 125 days. Here, mass transfer will be driven by the nuclear evolution of the red giant, which has a degenerate He core. The expected accretion rate in this case (Verbunt 1989, Verbunt & van den Heuvel 1995) is roughly given by

$$\dot{M} \approx (P_o/12.5d)10^{-8}\ M_\odot/\text{year}, \tag{5}$$

where $P_o$ is the orbital period at the onset of mass transfer. So, for orbital periods longer than about 125 days, $\dot{M}$ will be sufficiently large to ensure that steady nuclear burning on the WD can take place. Also, the companion will transfer its envelope on a time scale on the order of a few million years.

2. Wide symbiotic systems in which the red giant is on the asymptotic giant branch (AGB) and is not filling its Roche lobe, but has a very strong stellar wind. The mass-loss rates in AGB winds are on the order of $10^{-5}$ $M_{\odot}$/year, and the wind velocities are low, about 30 km/s or less. This can yield an accretion rate on the order of $10^{-7}$ $M_{\odot}$/year on a WD companion, which is sufficient to power a SSS. In this case, the SSS phase probably lasts at most only a few hundred thousand years.

Both types of symbiotic systems have been found among the SSS (see Table 1 and Section 3.4).

# 3. THE OBSERVED SAMPLE

## 3.1 *The ROSAT All Sky Survey*

Hasinger (1994) presented the first complete list of SSS candidates, selected out of the 50,000 sources of the ROSAT All Sky Survey (RASS), later refined by Kahbaka & Trümper (1996). Some of the sources in this list were not luminous SSS, i.e. they are AGNs or PG 1159 stars. These were later rejected. The latest "clean" version of the catalogue of SSS with $L_X > 10^{36}$ ergs/s has been presented by Greiner (1996b). These are the 35 sources listed in Table 1. Of these, 7 are in the LMC, 4 in the SMC, 15 steady plus 1 transient (White et al 1994) in M31 (Greiner et al 1996d), 1 in the nearby spiral galaxy NGC 55, and 7 in our own galaxy. Figure 6 shows the positional distribution of the 16 sources in M31, after Greiner et al (1996d). Motch et al (1994) concluded that the finding of only two galactic SSS with binary periods similar to CAL 87 and CAL 83 is consistent with a distribution that is more concentrated towards the plane of the Milky Way than the distribution of the low-mass X-ray binaries.

One observes from Table 1 that apart from the five systems with detected orbital periods close to one day (like CAL 83 and CAL 87) and the symbiotic systems, there is a third class of objects, a subgroup of the CVs with very short orbital periods: Some novae appear as luminous SSS for a limited period, up to ten years, after the nova outburst. A search for soft X rays from a sample of recent novae showed this phenomenon to be quite rare: 2 of a sample of 26 recently exploded novae show it (Ögelman & Orio 1995). The first system of this type detected is GQ Mus (Nova Muscae 1983), which was discovered by Ögelman et al (1987, 1993) to be a luminous SSS; since its discovery it has "turned off." The second galactic example is Nova Cyg 1992 (V1974 Cyg). The systems RX J0439.8-6809 and 1E 0035.4-7230 in the LMC and SMC,

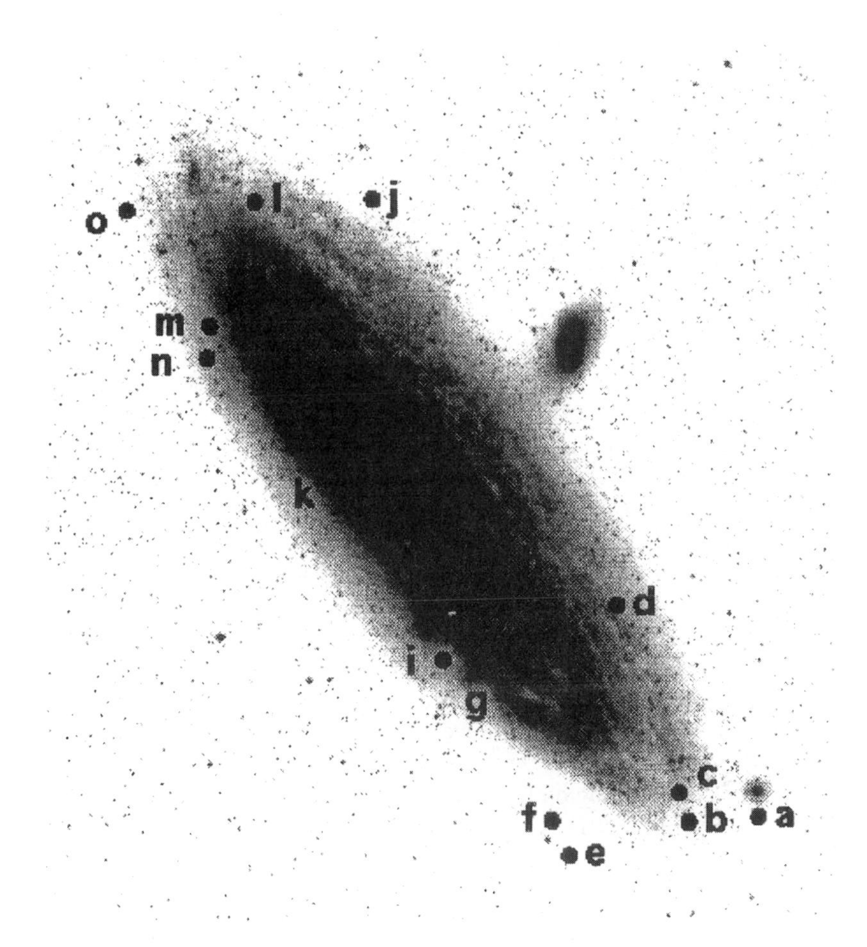


*Figure 6* The location of the 16 SSS in M31 plotted over an optical image of this galaxy. Letters (*a* to *o*) denote M31 X-ray sources from Table 1. The source RX J0045.4+4154 (White et al 1994) is shown as a square (from Greiner et al 1996d).

respectively, may well be of the same type. Their orbital periods are in the range of those of CVs. The only difference with the two galactic SSS novae is that the latter two have orbital periods below the period gap of CVs, whereas the Magellanic Cloud systems both have periods above this gap. It is as yet not clear whether this difference is of any physical significance (P Kahabka & E Ergma, in preparation). The reason that nova systems may in some cases for some time appear as luminous SSS is that not all the accreted matter is ejected during the outburst. The possible reasons for this are discussed by MacDonald (1996).

## 3.2 *Classification*

A first classification scheme of SSS has been introduced by DiStefano & Nelson (1996): CVs, CBSSs, wide binary SSS (WBSSs), and symbiotics. Another, more refined scheme has been proposed by Kahabka (1996b). In this review, for simplicity, we use the scheme of DiStefano & Nelson (1996).

## 3.3 *Luminosities and Temperatures of SSS Derived Using WD Atmosphere Models*

In the foregoing sections, for the sake of argument the sources were assumed to have a blackbody spectrum. For hot WDs, this is a poor approximation, as was shown by Heise et al (1994) (see also van Teeseling et al 1996a). The flux distributions of local thermodynamic equilibrium (LTE) WD atmosphere models, calculated by the above authors (as a function of log $T_e$ and log g), show absorption edges due to various heavier elements (from carbon onward). These edges have recently been observed with the Japanese X-ray observatory ASCA in RX J0925.7-4758 and CAL 87 (Ebisawa et al 1996; K Ebisawa, K Mukai, F Nagase, et al, in preparation).

Extrapolating from the flux observed in the softest ROSAT band, which is in the tail of the flux distribution of the WD, one obtains with LTE WD atmosphere models a considerably lower bolometric luminosity than if one uses the blackbody approximation. The resulting bolometric X-ray luminosities are not larger than the Eddington limit, and they fit well to the models for stable nuclear burning on the WD surface (van Teeseling et al 1996a).

## 3.4 *Optical Characteristics and Variability of Sources*

For a review of the optical observations of LMC sources, see Cowley et al (1996). For a discussion of the ultraviolet (UV) spectra of RX J0019.8+2156, CAL 83 and RX J0513.9-6951 we refer to Gänsicke et al (1996) and for a discussion of the UV spectrum of CAL 87 to Hutchings et al (1995). Optical identifications with blue emission line objects have been achieved for most SSS in the Magellanic Clouds and for nearly all galactic sources. We now briefly describe the results of these optical studies. We restrict ourselves to the Close Binary SSS (CBSS) and the Symbiotic systems.

THE "STANDARD" CBSS *CAL 83 and 87* The optical light curves and spectra of these sources were already briefly discussed in Section 2.2, where it was shown that the light variations of these binaries can be understood in terms of a model in which the two dominant light sources in the systems are a very bright accretion disk and the X-ray heated side of the donor star (see Figure 3; cf van den Heuvel et al 1992). This model appears to be the "standard" model for understanding the optical light curves of the five presently known CBSS discussed in this section. In CAL 87, our line of sight practically coincides with the orbital plane such that regular deep eclipses of the bright disk occur. The secondary minimum occurs when the disk eclipses a part of the strongly heated side of the donor star.

The CAL 83 system is seen at much lower inclination, such that no eclipses of the disk occur. The sinusoidal optical modulation is simply due to the one-sided heating of the donor star.

Although this model qualitatively explains the observed light curve of CAL 87, quantitatively it needs refinement in order to explain (*a*) the width and asymmetry of the primary eclipse and (*b*) the depth of the secondary eclipse. Refined models were therefore constructed by Schandl et al (1996) and Naber (1996). Schandl et al (1996) showed that in order to explain the asymmetry and width of the primary eclipse, the disk must have a rim of considerable height, highest on the trailing side (the place where the accretion stream impacts on the disk). Both Schandl et al (1996) and Naber (1996) showed that in order to explain the depth of the secondary eclipse, there must be a disk corona or wind from the disk of considerable optical depth, which blocks a sizeable part of the light from the heated side of the donor. The variability of the depth of the secondary eclipse can then be explained by variations in the density in the corona or wind.

*The variable source RX J0513.9-6951* This LMC source was discovered by Schaeidt et al (1993); it closely resembles CAL 83 in its X-ray, UV, and optical luminosities and spectra (Pakull et al 1993). However, while CAL 83 is a practically permanent X-ray source (only in April 1996 was it briefly absent; cf Kahabka 1996c, Kahabka et al 1996), RX J0513.9-6951 shows long "off" states, lasting up to many months, with irregularly spaced "on" intervals in between (see Figure 7).

Radial velocity measuremants by Crampton et al (1996) and photometry on 6 nights by Motch & Pakull (1996) suggested a period close to 0.76 days, confirmed by MACHO-photometry over more than 400 nights by Southwell et al (1996a), which yields $P = 0.^{d}76278$. The blue light curve derived from the latter observations is nearly sinusoidal with an amplitude of only 0.0213 mag. The same interpretation as the light curve of CAL 83 is consistent with a one-sided heated donor star seen at very low orbital inclination.

The spectrum of this source, depicted in Figure 8, is most remarkable in that the bright emission lines ($He\ II\lambda4686$ and $H_{\beta}$) show blue- and red-shifted

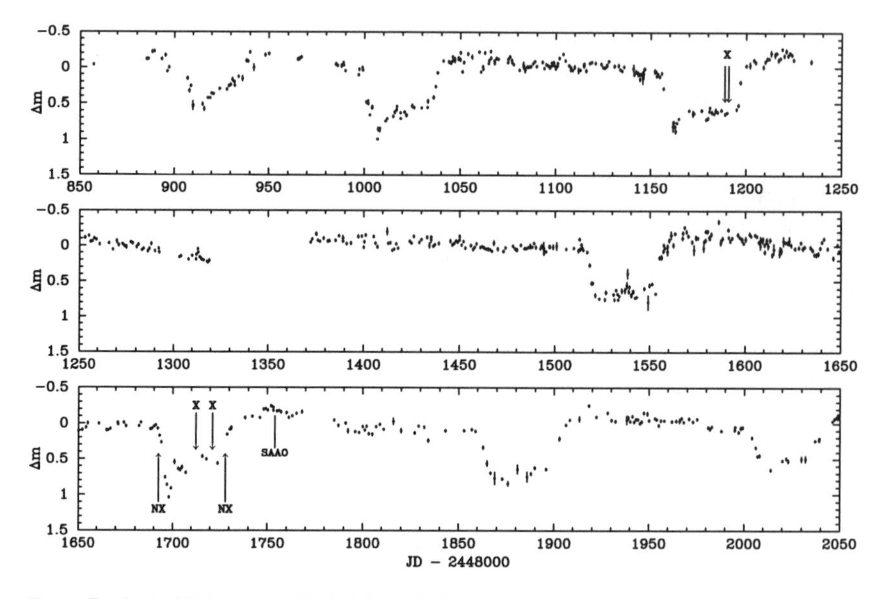


*Figure 7* Optical light curve of RX J0513.9-6951 from August 22, 1992, to November 27, 1995, obtained with the MACHO project. Downward and upward *vertical arrows* indicate times at which the system was known to be on (X) or off (NX) in X rays, respectively (from Southwell et al 1996b).

satellites, corresponding to Doppler velocities of 3800 km/s (Pakull et al 1993, Cowley et al 1996). As in SS 433, this strongly suggests a bipolar outflow pattern, with velocities somewhat larger than this value (we see only the projection on the line of sight). In such outflows, the outflow velocities are expected to be of the same order as the escape velocity from the stellar surface (Casinelli 1979, Castor et al 1975). As the observed velocities are of the order of the escape velocity from the surface of the WD, this provides strong evidence that the central object here is indeed a WD, as realized by Southwell et al (1996b). In view of the low inclination of the system, we are looking almost perpendicular to the orbital plane and the accretion disk. The observations are therefore consistent with the jets oriented perpendicularly to the disk, as in the usual configuration of jets from accretion disks (e.g. see Shu et al 1988, Appl & Camenzind 1992, Ostriker & Shu 1995).

*The galactic CBSS sources RX J0019.8+2156 and RX J0925.7-4758* The first-mentioned source, RX J0019.8+2156, was the first luminous SSS found in our Galaxy (Reinsch et al 1993). It has a relatively bright optical counterpart of 12.5 mag. This star was studied photometrically and spectroscopically by Beuermann et al (1995), who derived a photometric and spectroscopic period

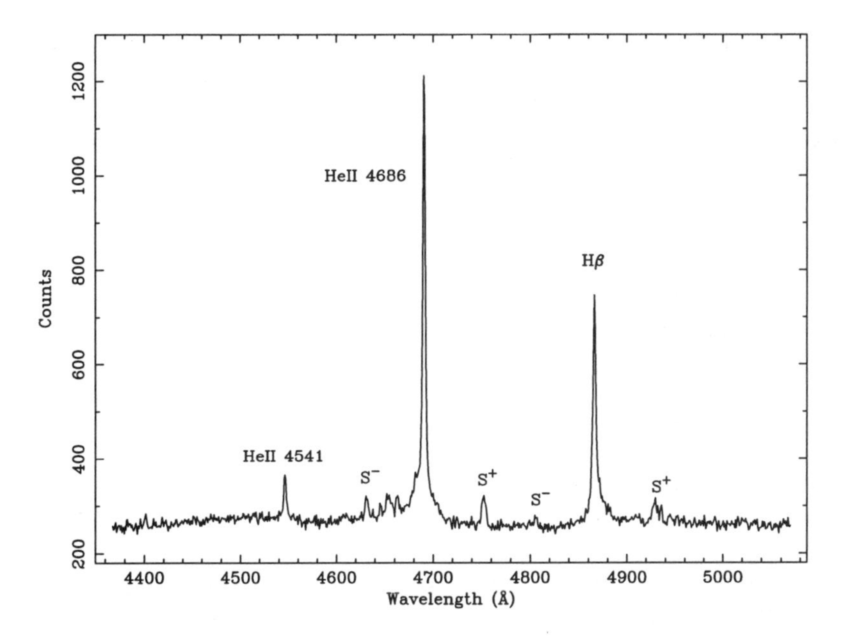


*Figure 8* Average blue spectrum of RX J0513.9-6951. The principal He II and H emission features are marked, along with their associated Doppler-shifted components (from Southwell et al 1996b).

of about 0.66 days. Greiner & Wenzel (1995), from observations taken in the interval 1955–1993, found a period of 0.6604565( ± 15) days, while Will & Barwig (1996) derived for the interval September 1992 to October 1995 a period of 0.6604721 ( ± 72) days. The light curve derived by Will & Barwig (1996) resembles that of CAL 87 in that it has a fairly deep primary minimum (about 0.3 mag) as well as a secondary minimum. The smaller depth of the primary minimum than in CAL 87 suggests that the orbit is seen at lower inclination than in this source.

The second important galactic SSS is RX J0925.7-4758, discovered by Motch et al (1994). It was identified by them with a highly reddened star of V = 17 mag [E(B-V) = 2.1 mag]. It shows the usual emission spectrum of SSS, including CIII/NIII$\lambda$4640–60. V-band photometry suggests an orbital period of $3.^d79 \pm 0.^d24$, in agreement with a spectroscopic period of $3.^d78$ (cf Motch et al 1994, Motch 1996).

SYMBIOTIC SYSTEMS Two of the three symbiotic SSS (Table 1) are symbiotic novae: RR Tel and SMC 3 ( = RX J0048.4-7332) (cf Mürset et al 1996). These systems show nova-like optical outbursts. In 1944, RR Tel brightened by ~7 mag to reach $M_V = -5$ mag, returning to its original state in subsequent years.

The cool component of RR Tel is a low-mass red giant that normally shows Mira-type light variations (Mayall 1949). SMC 3 has been in outburst since 1981 (Morgan 1992). The ROSAT PSPC count rate of this source is higher than that of RR Tel, even though its distance is 20 times larger (Kahabka et al 1994). Its minimum effective temperature (260,000 K) is about 2 times that of RR Tel, and its X-ray luminosity (12000 $L_{\odot}$) some 4 times larger than that of RR Tel. For details, see Mürset et al (1996). The orbital periods of symbiotic novae are on the order of a year, e.g. in RR Tel: 387 days. Therefore, it is likely that in some phase of the Mira pulsations, the cool component may overflow its Roche lobe, giving rise to the outburst. Alternatively, it has been suggested that the outburst is generated by a hydrogen flash in matter that was accreted by the WD from the wind of the giant during the quiet phase between outbursts. In view of the relatively short orbital periods for symbiotic stars, the RLOF explanation seems the more likely one to us.

The third symbiotic SSS, AG Dra, is not known to be a symbiotic nova. Like all symbiotics, it shows brightness increases by several magnitudes ("outbursts") at irregularly spaced times (cf Viotti et al 1996). During its most recent outburst (1994/1995), its soft X-ray flux greatly decreased and returned to its original value in the decline of the burst (Greiner et al 1996a). Greiner et al (1996a) suggest that this anticorrelation of optical and X-ray brightness is due to a temporary expansion of the photosphere of the burning compact star, owing to a temporarily increased state of mass transfer.

LONG-TERM VARIATIONS Apart from the symbiotic systems, several of the sources show interesting long-term variations. We briefly describe here some well-studied cases.

*RX J0513.9-6951 ("Schaeidt's Source")* As mentioned above, RX J0513.9-6951 shows bipolar outflows with V = 3800 km/s and exhibits long X-ray off phases. Thanks to its location in a MACHO field (Alcock et al 1996), Southwell et al (1996a,b) could derive a very accurate long-term optical light curve, which is shown in Figure 7. During the flat optical maxima, the X-ray source is off, while during the optical minima (depth in B about 1 mag) the X-ray source is on (see also Reinsch et al 1996). This behavior strongly suggests that the optically bright phases are phases of enhanced mass transfer, such that the photosphere of the burning compact star has expanded by a factor of two or more, causing the effective temperature to drop and the soft X-ray flux to be converted into an enhanced optical (and UV) flux. [This explanation is similar to Greiner et al's (1996a) explanation for the anticorrelation between X-ray and optical brightness in AG Dra.]

This indicates that the mass-transfer rate $\dot{M}$ in this system is usually higher than the maximum $\dot{M}$ for a steady burning without radius expansion, which is indicated in Figure 5.

This is confirmed by the outflows from the system in the form of jets: As mentioned in Section 2.3, one expects from the work of Hachisu et al (1996) that if $\dot{M}$ is above the upper boundary for a steady burning SSS in Figure 5, a strong stellar wind develops. In the presence of a disk, such a wind will be focused to form bipolar jets perpendicular to the disk, as discussed above.

*Recurrent shell flashes* The relatively bright ($V \sim 12$ mag) optical counterpart of RX J0019.8+2156, studied on photographic plates taken over the past 100 years, appears to exhibit abrupt rises in optical magnitude by ~0.5 mag, followed by a decay to its initial magnitude in ~20–30 years (Greiner & Wenzel 1995).

Meyer & Meyer-Hofmeister (1996) discuss several possible explanations. They favor an explanation in terms of a WD accreting at a moderate rate ($\sim 10^{-7}$ $M_{\odot}$year$^{-1}$) undergoing hydrogen shell flashes (cf Kahabka 1995, 1996b). Such behavior in terms of recurrent shell flashes was predicted, for example, by Iben (1982) and Fujimoto (1982a,b) for accretion rates somewhat below the lower boundary for stable nuclear burning in Figure 5. The recurrence time depends, among other factors, on the WD mass and the accretion rate. A semianalytical model for such flashes was worked out by Kahabka (1995, 1996b). In this model the WD envelope mass can be related to the recurrence time scale of the source: Given the X-ray on time and recurrence time, the envelope mass $\Delta$M can be calculated, which, in turn (with Figure 5), yields the allowed range of WD masses. Applying this, one finds that the 40-year recurrence time scale implies a WD mass of $\sim 1.0$–$1.1 M_{\odot}$.

*RX J0527.8-6954* This LMC source has an X-ray spectrum very similar to that of CAL 83 (Trümper et al 1991). Since its discovery with ROSAT in 1990, the X-ray brightness has steadily declined by more than a factor of 20 in 1800 days (Greiner et al 1996b,c). These authors suggest cooling of a massive WD following a H-shell flash on its surface as a likely explanation for the decline. The WD may remain in the stable burning range, but a decrease of a factor of 2 in its bolometric luminosity will then move its $T_{eff}$ so far down that in the ROSAT channels its flux drops by a factor of 20. Assuming a time of 5 years for returning to minimum, and a recurrence time of ~10 years, one derives with the above-mentioned semiempirical model of Kababka (1995) a WD mass $M_{Wd} \approx 1.2$–$1.35$ $M_{\odot}$ and $\dot{M}_a \approx 2$–$7 \times 10^{-7}$ $M_{\odot}$ year$^{-1}$. The high WD mass is also compatible with the relatively high value of $T_{eff} \approx 5$–$6 \times 10^5$ $K$ of the source (Greiner et al 1996b,c).

## 4. IONIZATION NEBULAE

The SSS are expected to ionize the surrounding interstellar medium out to distances of several parsecs due to the large ionizing flux (a few times $10^{37}$ erg s$^{-1}$)

and the long on times (a few times $10^6$ year). Therefore a search for nebulae of this type (H II regions) surrounding all known SSS has been initiated (Rappaport et al 1994a, Remillard et al 1995). Surprisingly, only for one source, i.e. CAL 83, has a large ionization nebula been detected. This has been explained by the high density of the surrounding local interstellar medium. The time variable, i.e. recurrent or cyclic illumination (excitation) of the interstellar medium, may make such nebulae less luminous (cf E Chiang & SA Rappaport, in preparation), but it is more likely that the absence of detectable ionization nebulae around other sources is due to a much lower local interstellar density for the other SSS (cf Remillard et al 1995).

## 5. PHYSICAL PROCESSES

### 5.1 *Types of Nuclear Burning as a Function of $\dot{M}$ and $M_{WD}$*

As mentioned in Section 2.3, steady burning of hydrogen (by the CNO cycle) without radius expansion occurs only for a limited range of accretion rates close to $10^{-7}$ $M_{\odot}$ year$^{-1}$, which depends on the WD mass, as depicted in Figure 5. The precise way in which the burning takes place depends, in addition, on the thermal history of the WD (cf Iben 1982, Fujimoto 1982a,b, MacDonald 1983, Prialnik & Kovetz 1995). For simplicity we do not consider this extra complication here.

The dependence of the type of burning on $\dot{M}$ and $M_{WD}$ was calculated by the above-mentioned authors and by Iben et al (1992), Kato & Hachisu (1994), and Sion & Starrfield (1994).

For accretion rates below the lower bound of $\dot{M}$ for steady burning, the accreted matter will burn in flashes. The strength of the flashes depends on the degree of the degeneracy of the matter in the burning shell, which in turn depends on the accretion rate: The lower the accretion rate, the higher the degree of degeneracy of the matter at the moment of ignition, and the stronger the flash. The flashes reach the strengths of nova explosions for $\dot{M} \lesssim 10^{-8}$ $M_{\odot}$ year$^{-1}$ at low WD masses to $\lesssim 10^{-9}$ $M_{\odot}$ year$^{-1}$ for high WD masses, as depicted in Figure 5 (cf Starrfield et al 1974, 1985, Prialnik et al 1978, Prialnik 1986, José et al 1993). In this figure, the *dashed lines* are isolines of the critical envelope mass $\Delta M$ at which burning is ignited, and a flash occurs.

### 5.2 *Inferred WD Masses*

Mass determinations on the basis of radial velocity variations of the spectral lines (mostly emission lines are observed) are not expected to be reliable, as one does not know where in the systems these lines are formed.

To determine masses, one can use the core-mass luminosity relation for cold WDs accreting hydrogen (cf Iben & Tutukov 1996):

$$L/L_{\odot} \approx 4.6 \times 10^4 \, (M_{core}/M_{\odot} - 0.26). \tag{6}$$

Alternatively (as mentioned in Long-Term Variations above), for sources with recurrent X-ray flashes, one can derive limits on $M_{WD}$ from the recurrence time of flashes. As mentioned, all these sources appear to harbor quite massive WDs, $M_{WD} \gtrsim 1.1\ M_{\odot}$. The same holds for the recurrent source in M31 (White et al 1994, Kahabka 1995, Kahabka 1996b). For the steady sources, the $T_{eff}$ and $L$ combination yield a value for the radius and for the mass (cf van Teeseling et al 1996a). For CAL 87, one finds that $M_{WD} > 1.2\ M_{\odot}$. These high-mass values do not necessarily imply that one is dealing with O-Ne-Mg WDs: The WDs may well have started out as CO dwarfs of moderate mass, which accreted a He layer of considerable mass due to steady burning. Whether or not this He layer will explode and lead to a double detonation SN Ia is not yet clear (cf Nomoto & Yamaoka 1992, Kato 1996). We return to this problem in Section 7. The reason the WD masses of the SSS in the LMC and M31 tend to be high has been explained by Rappaport et al (1994b): This is due to a selection effect, as the SSS with higher-mass WDs tend to be hotter and more luminous than those with low-mass WDs and therefore are more easily detected (see also DiStefano 1996).

# 6. EVOLUTION AND POPULATION SYNTHESIS

## 6.1 *Origin of the CBSS*

The origin of the standard CBSS with orbital periods on the order of one day and donor masses in the range 1.3–2.5 $M_{\odot}$ is very similar to that of CVs, to which they are closely related. Both types of systems are close binaries consisting of a WD and a star of relatively low mass ($\leq$2.5 $M_{\odot}$). The only difference between them is that in the CVs the donor star is less massive than the WD, whereas in the CBSS the donor is more massive than the WD.

The formation history of both types of systems is therefore expected to be basically the same: They formed through common envelope (CE) evolution out of initially very wide binary systems that consisted of a red giant with a degenerate CO (or O-Ne-Mg) core, together with an unevolved main-sequence star of lower mass ($\leq$2.5 $M_{\odot}$), as depicted in the top frame of Figure 9.

When the envelope of the red giant (which at this stage was on the AGB) overflowed its Roche lobe, it engulfed the lower-mass companion. A system resulted that consisted of the latter star and the core of the giant, orbiting each other inside this CE. The large frictional drag on the orbital motion made the companion and the core rapidly spiral inwards, while the frictional heat production caused the envelope to be expelled in the process (cf Meyer & Meyer-Hofmeister 1979, Taam et al 1978, Taam 1996). The time scale of the expulsion is on the order of $10^3$ years (Taam 1996). This resulted in the presently observed very close systems.

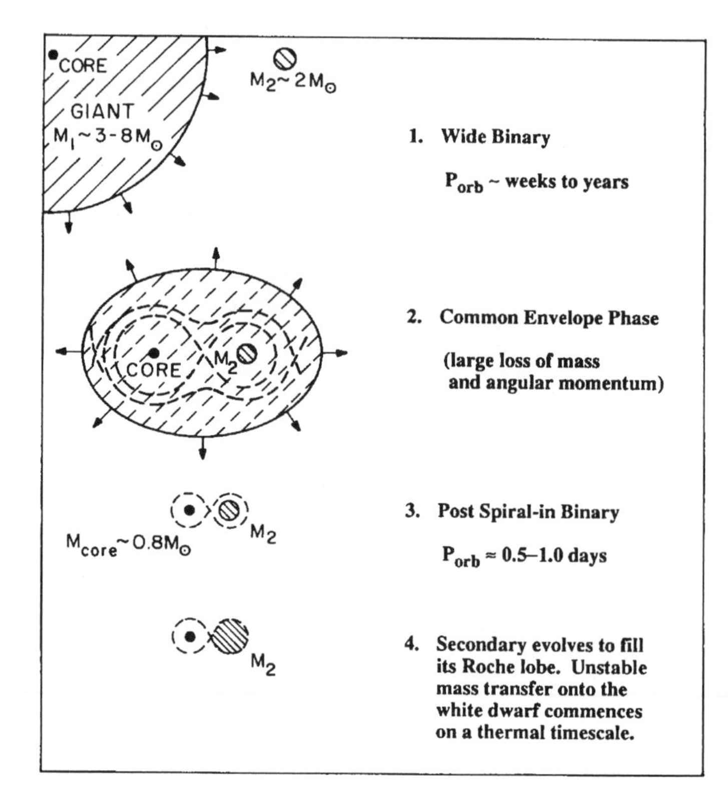


*Figure 9* Evolutionary scenario for the formation of a close binary supersoft X-ray source (cf text; from Rappaport & DiStefano 1996).

At a much later stage, when the companion evolves away from the main sequence and overflows its Roche lobe, the system will become a CBSS or a CV. In particular, CVs will be produced if the companion is of low mass ($\lesssim M_\odot$) and the orbital period after "spiral-in" is sufficiently short ($\lesssim$12 h) that it decays within a Hubble time, owing to the emission of gravitational waves or to angular momentum losses by magnetic breaking (cf Verbunt 1989, Verbunt & van den Heuvel 1995). The mass-transfer rates driven by these processes are typically $10^{-10}$–$10^{-9}$ $M_\odot$ year$^{-1}$, as is indeed observed in CVs.

## 6.2 *The Origin of Symbiotic Systems*

Here the initial system was so wide that the two stars evolved in large part independently of each other. The more massive star evolved into a red giant and finally ejected its envelope without much interaction with the at that time still unevolved partner. This resulted in a wide system consisting of a WD and a main-sequence star, resembling the systems of Sirius AB and Procyon AB. At a much later stage, the companion evolved into a red giant and the system became a symbiotic star. For details we refer the reader to Yungelson et al (1995).

### 6.3 *Results of Population Synthesis Studies*

Pioneering numerical population synthesis calculations for the formation of the CBSS were carried out by Rappaport et al (1994b, hereafter RDS). A review and critical discussion is given by DiStefano & Nelson (1996). For a review of population synthesis techniques with binary evolution, see De Kool (1996).

In population synthesis calculations, one starts out with a population of unevolved binary systems, composed such that it mimics the real population of unevolved binaries in the Galaxy. For composing this input population, one uses the best available observational data (on masses, mass ratios, orbital periods) of unevolved binaries. In addition, one needs a stellar evolution code that is able to incorporate all possible types of evolution of binaries, including CE evolution.

By constructing a population of a large number of binaries (typically $10^7$) and evolving them, and assuming a steady state of star formation in the Galaxy (calibrated with the presently observed rate), one typically obtains a realistic population that contains all types of products of binary evolution, including the CBSS- and CV-type systems. The calculations yield the expected distributions of orbital periods, donor masses, WD masses, etc. Figure 10 shows an example of these predicted distributions obtained by RDS (1994). These authors obtained a steady-state population of CBSS in the Galaxy on the order of $10^3$. Depending on the adopted input distribution of orbital parameters—some of which are not well known—and input parameters in the evolutionary code, these predicted numbers can still range from 100–1500 in the Galaxy, as shown in Table 3, where the predicted numbers for M31, LMC, and SMC also are shown. These estimates agree with the early analytic estimates ($\sim$360) of van den Heuvel et al (1992). The galactic population inferred from the observations, after taking into account the effects of strong interstellar extinction, is between 400 and 3000 (Rappaport & DiStefano 1994b), roughly consistent with the theoretical predictions.

A completely independent population synthesis calculation by Yungelson et al (1996b), using a different code, and including CV-type and symbiotic SSS, yields the numbers given in the lower part of Table 3. Yungelson et al (1996a,b) argue that symbiotics will be extremely hard to observe because their soft X rays will practically always be completely absorbed ("shielded") by the winds of their red giant companions. Within the limits for the uncertainties of the input distributions and evolutionary parameters (such as the efficiency

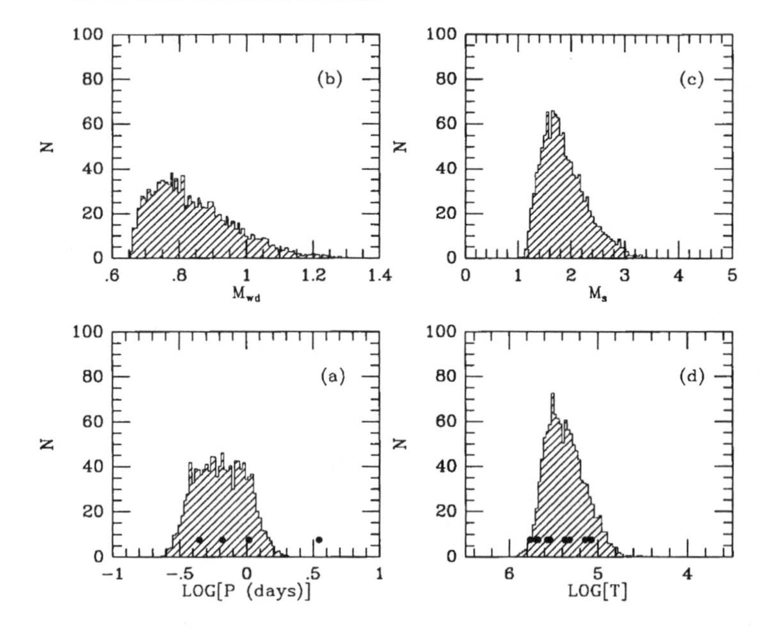


*Figure 10* Expected distributions of white dwarf (WD) ($M_{WD}$) and donor ($M_s$) mass, orbital period $P_{orb}$, and effective temperature $T_{eff}$ as derived in the population synthesis study of Rappaport & DiStefano (1996), together with observed quantities (*black dots*).

parameter $\alpha$ of CE evolution), the numbers of CBSS calculated by the above two groups are in agreement with each other, as well as with the latest results of DiStefano & Nelson (1996), in which mass and angular momentum losses from the systems were taken into account. The latter yields a galactic number of CBSS between 1000 and 4000.

## 7. SN IA PROGENITORS

Supernovae of Type Ia (SN Ia) are most likely to be associated with the thermonuclear explosions of mass-accreting CO WDs, a now widely accepted proposition (Arnett 1969; for reviews see Branch et al 1995, Livio 1994, 1996). The source of accretion can be stellar wind from the donor, RLOF, or the result of the coalescence of two WDs. The exploding WD is usually assumed to be near the Chandrasekhar limit ($\sim$1.4 $M_\odot$), but also the possibility of explosions at sub-Chandrasekhar mass has recently been put forward: Off-center detonations of He on CO WDs in the mass range 0.6–0.9 $M_\odot$, after accreting 0.15–0.20 $M_\odot$ of He (produced by H-shell burning) can trigger explosions that in many respects are similar to SN Ia (Livne 1990, Limongi & Tornambé 1991,

**Table 3** Numbers of supersoft X-ray sources (SSS) in different galaxies predicted from population synthesis calculations, compared with numbers from observations[a]

| Galaxy | Population synthesis | Inferred from observations | |
|---|---|---|---|
| M31 | 400–6000 | 800–5000 | |
| Milky Way | 100–1500 | 400–3000 | |
| LMC | 20–300 | 13–60 | |
| SMC | 5–60 | 9–40 | |
| **Type** | **Cataclysmic variables** | **Subgiant systems** | **Symbiotic systems** |
| Permanent SSS | 130 | 400 | 450 |
| Recurrent SSS | 350 | 70 | 45 |

[a]Upper part: numbers of SSS predicted from population synthesis by Rappaport et al (1994b) and inferred from observations by DiStefano & Rappaport (1994); lower part: numbers of SSS for the Milky Way calculated by Yungelson et al (1996b). LMC, Large Magellanic Clouds; SMC, Small Magellanic Clouds.

Woosley & Weaver 1994). This is the so-called double detonation model (see also Ruiz-Lapuente et al 1995, Canal et al 1996).

The most attractive candidate for near-Chandrasekhar mass SN Ia is the merger of a close pair of CO WDs due to orbital decay by losses of gravitational radiation (cf Webbink 1984, Iben & Tutukov 1984), which works in any type of galaxy, i.e. also in elliptical galaxies, which only contain very old stars. Unfortunately, so far no such close massive binary WDs have been found observationally. Alternative candidates put forward are symbiotic stars (Munari & Renzini 1992). These do occur in old stellar systems. Yungelson et al (1995) showed that the WDs in these systems accrete too little mass to allow them to reach the Chandrasekhar limit. They showed, however, that if sub-Chandrasekhar double detonation models (see above) indeed produce SN Ia, then symbiotics might produce at most one third of the SN Ia. With the discovery of the SSS, a new class of potential SN Ia progenitors has been found, as was suggested by RDS and subsequently by many other authors (see Branch et al 1995, Livio 1996, Kato 1996, DiStefano 1996). Here the CBSS are the most interesting subclass, since in the CVs, the accreted mass is violently ejected in nova explosions (cf Livio 1994), and symbiotics yield a too-low rate (see above).

In practically all cases, the CBSS will accrete sufficient mass to at least lead to a sub-Chandrasekhar mass double detonation. With the number of $\sim$1–4 $\times$ $10^3$

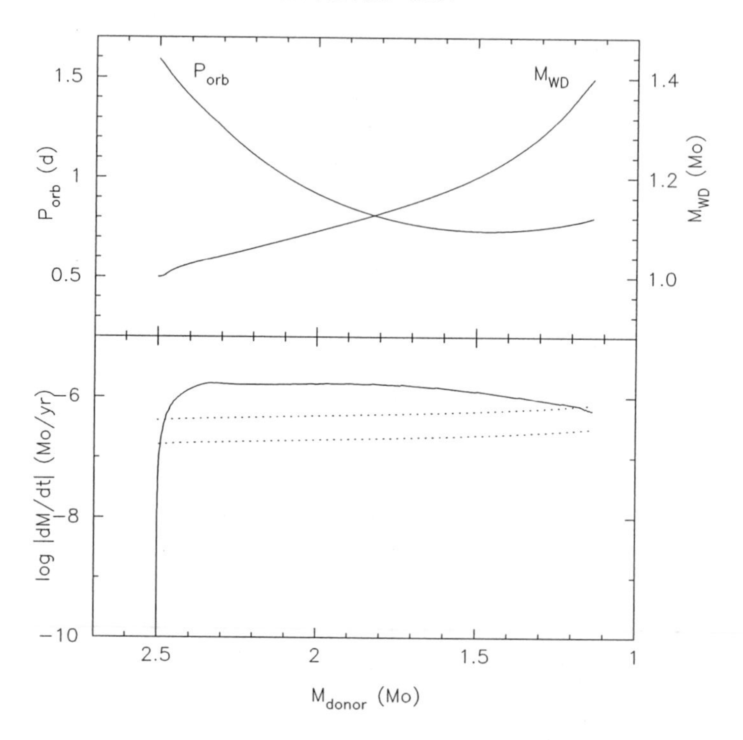


*Figure 11* Evolution of orbital period, white dwarf (WD) mass, and $\dot{M}$ of a CBSS with an initial donor mass of 2.5 $M_\odot$ and an initial WD mass of 1 $M_\odot$, in which the mass-transfer rate is mostly $\sim 10^{-6}$ $M_\odot$year$^{-1}$ (from X Li, in preparation). *Dashed lines*: boundaries of the stable burning region without radius expansion.

CBSS in the Galaxy (DiStefano & Nelson 1996), which roughly increase in mass by $\sim 0.2$ $M_\odot$ in $\sim 10^6$ years ($\dot{M} \sim 2.10^{-7}$ $M_\odot$ year$^{-1}$), this may lead to a double detonation rate of order 1–2 $\times$ $10^{-3}$ year$^{-1}$, i.e. almost as high as SN Ia rate in spiral galaxies like our own (3 $\times$ $10^{-3}$ year$^{-1}$, cf Branch et al 1995). It should be pointed out here that since the donor stars in CBSS have masses $>1.2$ $M_\odot$, they live no longer than $\sim 5 \times 10^9$ years. Therefore CBSS are not good SN Ia progenitor candidates in elliptical galaxies. On the other hand, in spiral and irregular galaxies, they might make a considerable contribution to the SN Ia rate. If SN Ia require growth of the WD to the Chandrasekhar mass, the event rate produced by CBSS will be lower but perhaps not by a large factor. This is because recent full evolutionary calculations (X Li, in preparation) of systems with donors in the mass range 1.5–3.0 $M_\odot$ showed that He detonation may be prevented in many cases, since $\dot{M}$ is sufficiently high for the He to burn only in weak flashes (cf Kato 1996). Mass and angular momentum losses by wind were included in the calculations for evolutionary stages where $\dot{M}$ is too large for stable burning SSS. Figure 11 shows, as an example, the evolution of a CBSS with an initial donor mass of 2.5 $M_\odot$ and an initial WD mass of 1 $M_\odot$, in which the mass-transfer rate is mostly $\sim 10^{-6}$ $M_\odot$ year$^{-1}$. It is clear that the SN Ia problem has not yet been solved, but, as Branch et al (1995) concluded, in view of their calculated numbers and birth rates, merging double degenerates and close binary SSS are now probably the two most promising progenitor candidates for SN Ia.

ACKNOWLEDGMENTS

We acknowledge discussions with many colleagues and friends contributing to the new field of SSS. We especially thank the participants of several conferences and workshops who provided us with new and exciting results. We thank K Beuermann, Anne Cowley, Rosario DiStefano, Ken Ebisawa, Ene Ergma, Jochen Greiner, John Heise, I Iben, Stephan Jordan, Joachim Krautter, F Meyer, X Li, Mario Livio, Joanna Mikolajewska, Christian Motch, Ken-ito Nomoto, Hakki Ögelman, Manfred Pakull, D Prialnik, S Rappaport, Klaus Reinsch, Paul Schmidtke, Karen Southwell, J Trümper, Jan van Paradijs, Andre van Teeseling, and Lev Yungelson.



*Literature Cited*

Alcock C, Allsman RA, Alves D, Axelrod TS, Bennett DP et al. 1996. *MNRAS* 280:L49–L53
Appl S, Camenzind M. 1992. *Astron. Astrophys.* 256:354–70
Arnett WD. 1969. *Astrophys. Space Sci.* 5:180–212
Beuermann K, Reinsch K, Barwig H, Burwitz V, de Martino D, et al. 1995. *Astron. Astrophys.* 294:L1–L4
Branch D, Livio M, Yungelson LR, Boffi FR, Baron E. 1995. *PASP* 107(717):1019–29
Canal R, Ruiz-Lapuente P, Burkert A. 1996. *Ap. J.* 456:L101–5
Casinelli JP. 1979. *Annu. Rev. Astron. Astrophys.* 17:275–308
Castor JI, Abbott DC, Klein RI. 1975. *Ap. J.* 195:157–74
Cowley AP. 1992. *Annu. Rev. Astron. Astrophys.* 30:287–310
Cowley AP, Schmidtke PC, Crampton D, Hutchings JB. 1990. *Ap. J.* 350:288–94
Cowley AP, Schmidtke PC, Crampton D, Hutchings JB. 1996. See van Paradijs et al 1996, pp. 439–44
Cowley AP, Schmidtke PC, Hutchings JB, Crampton D, McGrath TK. 1993. *Ap. J.* 418:L63–L66
Crampton D, Hutchings JB, Cowley AP, Schmidtke PC, McGrath TK, et al. 1996. *Ap. J.* 456:320–28
De Kool M. 1996. In *Evolutionary Processes in Binary Stars*, ed. RAMJ Wyers, MB Davies, CA Tout, pp. 365–84. Dordrecht: Kluwer. 411 pp.
Deutsch EW, Margon B, Wachter S, Anderson SF. 1996. *Ap. J.* 471:979–86
Diaz MP, Steiner JE. 1989. *Ap. J.* 339:L41–L43
Diaz MP, Steiner JE. 1994. *Ap. J.* 425:252–63

Diaz MP, Williams RE, Phillips MM, Hamut M. 1995. *MNRAS* 277:959–64
DiStefano R. 1996. See Greiner 1996a, pp. 193–204
DiStefano R, Nelson LA. 1996. See Greiner 1996a, p. 3–14
DiStefano R, Rappaport S. 1994. *Ap. J.* 437:733–41
DiStefano R, Rappaport S. 1995. In *17th Texas Symposium on Relativistic Astrophysics and Cosmology,* ed. H Böhringer, GE Morfill, JE Trûmper, pp. 328–31. New York: NY Acad. Sci. 328 pp.
Ebisawa K, Asai K, Mukai K, Smale A, Dotani T, et al. 1996. See Greiner 1996a, pp. 91–98
Friedjung M. 1988. In *The Symbiotic Phenomenon,* ed. J Mikolajewska, M Friedjung, SJ Kenyon, R Viotti, *IAU Colloq.* 103:199–204
Fujimoto MY. 1982a. *Ap. J.* 257:752–66
Fujimoto MY. 1982b. *Ap. J.* 257:767–79
Gänsicke BT, Beuermann K, de Martino D. 1996. See Greiner 1996a, pp. 107–14
Greiner J, ed. 1996a. *Workshop on Supersoft Sources. Lect. Notes Phys. 472.* Berlin: Springer-Verlag. 350 pp.
Greiner J. 1996b. See Greiner 1996a, pp. 299–337
Greiner J, Bickert K, Luthardt R, Viotti R, Altamore A, González-Riestra R. 1996a. See Greiner 1996a, pp. 267–78
Greiner J, Hasinger G, Kahabka P. 1991. *Astron. Astrophys.* 246:L17–L20
Greiner J, Hasinger G, Thomas H-C. 1994. *Astron. Astrophys.* 281:L61–L64
Greiner J, Schwarz R, Hasinger G, Orio M. 1996b. *Astron. Astrophys.* 312:88–92
Greiner J, Schwarz R, Hasinger G, Orio M. 1996c. See Greiner 1996a, pp. 145–52
Greiner J, Supper R, Magnier EA. 1996d. See Greiner 1996a, pp. 75–82
Greiner J, Wenzel W. 1995. *Astron. Astrophys.* 294:L5–L8
Hachisu I, Kato M, Nomoto K. 1996. *Ap. J.* 470:L97–L100
Hasinger G. 1994. *Rev. Mod. Astron.* 7:129–49
Heise J, van Teeseling A, Kahabka P. 1994. *Astron. Astrophys.* 288:L45–L48
Hertz P, Grindlay JE, Bailyn CD. 1993. *Ap. J.* 410:L87–L90
Hutchings JB, Cowley AP, Schmidtke PC, Crampton D. 1995. *Astron. J.* 110:2394–99
Iben I Jr. 1982. *Ap. J.* 259:244–66
Iben I Jr, Fujimoto MY, MacDonald J. 1992. *Ap. J.* 388:521–40
Iben I Jr, Tutukov AV. 1984. *Ap. J. Suppl.* 54:335–72
Iben I Jr, Tutukov AV. 1994. *Ap. J.* 431:264–72
Iben I Jr, Tutukov AV. 1996. *Ap. J. Suppl.* 105:145–80
Jordan S, Mürset U, Werner K. 1994. *Astron. Astrophys.* 283:475–82
Jordan S, Schmutz W, Wolff B, Werner K, Mürset U. 1996. *Astron. Astrophys.* 312:897–904
José J, Hernanz M, Isern J. 1993. *Astron. Astrophys.* 269:291–300
Kahabka P. 1995. *Astron. Astrophys.* 304:227–34
Kahabka P. 1996a. *Astron. Astrophys.* 306:795–802
Kahabka P. 1996b. See Greiner 1996a, pp. 215–22
Kahabka P. 1996c. *IAU Circ. No. 6432*
Kahabka P, Haberl F, Parmar AN. 1996. *IAU Circ. No. 6467*
Kahabka P, Pietsch W, Hasinger G. 1994. *Astron. Astrophys.* 288:538–50
Kahabka P, Trümper J. 1996. See van Paradijs et al 1996, pp. 425–38
Kato M. 1996. See Greiner 1996a, pp. 15–32
Kato M, Hachisu I. 1994. *Ap. J.* 437:802–26
Kenyon SJ. 1986. *The Symbiotic Stars.* Cambridge: Cambridge Univ. Press. 288 pp.
Kippenhahn R, Weigert A, eds. 1994. *Stellar Structure and Evolution.* Berlin: Springer-Verlag. 468 pp. 3rd ed.
Krautter J, Ögelman H, Starrfield S, Wichmann R, Pfefferman E. 1996. *Ap. J.* 456:788–97
Krautter J, Williams RE. 1989. *Ap. J.* 341:968–73
Kylafis ND, Xilouris EM. 1993. *Astron. Astrophys.* 278:L43–L46
Limongi M, Tornambé A. 1991. *Ap. J.* 371:317–31
Livio M. 1994. In *Interacting Binaries,* ed. H Nussbaumev, A Orr, pp. 135–262. Berlin: Springer-Verlag
Livio M. 1995. In *Millisecond Pulsars: A Decade of Surprise,* ed. A Fruchter, M Tavani, D Backer, *ASP Conf. Ser.* 105 pp.
Livio M. 1996. see Greiner, 1996a, pp. 183–91
Livne E. 1990. *Ap. J.* 354:L53–L55
Long KS, Helfand DJ, Grabelsky DA. 1981. *Ap. J.* 248:925–44
MacDonald J. 1983. *Ap. J.* 267:732–46
MacDonald J. 1996. In *Cataclysmic Variables and Related Objects,* ed. A Evans, JH Wood, pp. 281–87. Dordrecht: Kluwer
Mayall MW. 1949. *Harv. Bull.* 919:15–17
Meyer F, Meyer-Hofmeister E. 1979. *Astron. Astrophys.* 78:167–76
Meyer F, Meyer-Hofmeister E. 1996. See Greiner 1996a, pp. 153–58
Mikolajewska J, Kenyon SJ. 1992. *MNRAS* 256:177–85
Morgan DH. 1992. *MNRAS* 258:639–46
Motch C. 1996. See Greiner 1996a, pp. 83–90
Motch C, Hasinger G, Pietsch W. 1994. *Astron. Astrophys.* 284:827–38

Motch C, Pakull MW. 1996. See Greiner 1996a, pp. 127–30

Munari U, Renzini A. 1992. *Ap. J.* 397:L87–L90

Mürset U, Jordan S, Wolff B. 1996. See Greiner 1996a, pp. 251–58

Naber R. 1996. *Modelling the optical light curve of CAL 87.* Masters thesis. Univ. Amsterdam. 52 pp.

Nomoto K. 1982 *Ap. J.* 253:798–810

Nomoto K, Nariai K, Sugimoto D. 1979. *PASJ* 31:287–98

Nomoto K, Yamaoka H. 1992. In *X-Ray Binaries and Recycled Pulsars,* ed. EPJ van den Heuvel, SA Rappaport, pp. 189–205. Dordrecht: Kluwer

Ögelman H, Krautter J, Beuermann K. 1987. *Astron. Astrophys.* 177:110–16

Ögelman H, Orio M. 1995. In *Cataclysmic Variables,* ed. A Bianchini, M Della Valle, M Orio, pp. 11–19. Dordrecht: Kluwer

Ögelman H, Orio M, Krautter J, Starrfield S. 1993. *Nature* 361:331–33

Orio M, Della Valle M, Massone G, Ögelman H. 1994. *Astron. Astrophys.* 289:L11–L14

Orio M, Della Valle M, Massone G, Ögelman H. 1996. In *Cataclysmic Variables and Related Objects,* ed. A Evans, JH Wood, p. 429. Dordrecht: Kluwer

Orio M, Ögelman H. 1993. *Ap. J.* 273:L56–L58

Ostriker EC, Shu FH. 1995. *Ap. J.* 447:813–18

Paczyński B. 1971. *Annu. Rev. Astron. Astrophys.* 9:183–208

Paczyński B, Zytkow AN. 1978. *Ap. J.* 222: 604–11

Pakull MW, Beuermann K, van der Klis M, van Paradijs J. 1988. *Astron. Astrophys.* 203:L27–L30

Pakull MW, Motch C, Bianchi L, Thomas HC, Guibert J, et al. 1993. *Astron. Astrophys.* 278:L39–L42

Prialnik D. 1986. *Ap. J.* 310:222–37

Prialnik D, Kovetz A. 1995. *Ap. J.* 445:789–810

Prialnik D, Shara MM, Shaviv G. 1978. *Astron. Astrophys.* 62:339–48

Rappaport S, Chiang E, Kallman T, Malina R. 1994a. *Ap. J.* 431:237–46

Rappaport S, DiStefano R. 1996. See van Paradijs et al 1996, pp. 415–24

Rappaport S, DiStefano R, Smith JD. 1994b. *Ap. J.* 426:692–703

Reinsch K, Beuermann K, Thomas H-C. 1993. *Astron. Gesell Abstr. Ser.* 9:41

Reinsch K, van Teeseling A, Beuermann K, Abbott TMC. 1996. *Astron. Astrophys.* 309:L11–L14

Remillard RA, Rappaport S, Macri LM. 1995. *Ap. J.* 439:646–51

Ruiz-Lapuente P, Burkert A, Canal R. 1995. *Ap. J.* 447:L69–L72

Schaeidt S, Hasinger G, Trümper J. 1993. *Astron. Astrophys.* 270:L9–L12

Schaeidt SG. 1996. See Greiner 1996a, pp. 159–64

Schandl S, Meyer-Hofmeister E, Meyer F. 1996. See Greiner 1996a, pp. 53–64

Schmidtke PC, Cowley AP. 1996. *Astron. J.* 112:167–70

Schmidtke PC, Cowley AP, McGrath TK. 1996. *Astron. J.* 111:788–93

Schmidtke PC, McGrath TK, Cowley AP, Frattare LM. 1993. *PASP* 105:863–66

Seward FD, Mitchell M. 1981. *Ap. J.* 243:736–43

Shanley L, Ögelman H, Gallagher JS, Orio M, Krautter J. 1995. *Ap. J.* 438:L95–L98

Shore SN, Sonneborn G, Starrfield S, Gonzalez-Riestra R. 1993. *Astron. J.* 106:2408–28

Shore SN, Sonneborn G, Starrfield S, Gonzalez-Riestra R, Polidan RS. 1994. *Ap. J.* 421:344–49

Shu FH, Lizano S, Ruden SP, Najita J. 1988. *Ap. J.* 328:L19–L23

Sienkiewicz R. 1980. *Astron. Astrophys.* 85: 295–301

Singh KP, Barrett P, White NE, Giommi P, Angelini L. 1995. *Ap. J.* 455:456–67

Sion EM, Acierno MJ, Tomczyk K. 1979. *Ap. J.* 230:832–38

Sion EM, Starrfield SG. 1994. *Ap. J.* 421:261–68

Smale AP, Corbet RHD, Charles PA, et al. 1988. *MNRAS* 233:51–63

Southwell KA, Livio M, Charles PA, Sutherland W, Alcock C, et al. 1996a. See Greiner 1996a, pp. 165–72

Southwell KA, Livio M, Charles PA, O'Donoghue D, Sutherland WJ. 1996b. *Ap. J.* 470:1065

Starrfield S, Sparks WM, Truran JW. 1974. *Ap. J.* 192:647–55

Starrfield S, Sparks WM, Truran JW. 1985. *Ap. J.* 291:136–46

Taam RE. 1996. See van Paradijs et al 1996, pp. 3–15

Taam RE, Bodenheimer P, Ostriker JP. 1978. *Ap. J.* 222:269–80

Tanaka Y, Lewin WHG. 1995. In *X-Ray Binaries,* ed. WHG Lewin, J van Paradijs, EPJ van den Heuvel, pp. 126–74. Cambridge: Cambridge Univ. Press

Trümper J, Hasinger G, Aschenbach B, Bräuninger H, Briel EG, et al. 1991. *Nature* 349:579–83

van den Heuvel EPJ. 1994. In *Interacting Binaries,* ed. H Nussbaumer, A Orr, pp. 263–474. Berlin: Springer-Verlag

van den Heuvel EPJ, Bhattacharya D, Nomoto K, Rappaport SA. 1992. *Astron. Astrophys.* 262:97–105

van Paradijs J, van den Heuvel EPJ, Kuulkers E, eds. 1996. *Compact Stars in Binaries. IAU Symp. No. 165*. Dordrecht: Kluwer. 546 pp.
van Teeseling A, Heise J, Kahabka P. 1996a. See van Paradijs et al 1996, pp. 445–450
van Teeseling A, Reinsch K, Beuermann K. 1996b. *Astron. Astrophys.* L49–L52
van Teeseling A, Reinsch K, Beuermann K, et al. 1996c. See Greiner 1996a, pp. 115–22
Verbunt F. 1989. In *Neutron Stars and Their Birth Events*, ed. W Kundt, pp. 179–218. Dordrecht: Kluwer
Verbunt F, Hasinger G, Johnston HM, Bunk W. 1993. *Adv. Space Res.* 131:151–60
Verbunt F, van den Heuvel EPJ. 1995. In *X-Ray Binaries,* ed. WHG Lewin, J van Paradijs, EPJ van den Heuvel, pp. 457–94. Cambridge: Cambridge Univ. Press
Viotti R, González-Riestra R, Montagni F, Mattei J, Maesano M. 1996. See Greiner 1996a, pp. 259–66
Vogel M, Morgan DH. 1994. *Astron. Astrophys.* 288:842–48
Wang Q. 1991. *MNRAS* 252:P47–P49
Wang Q, Hamilton T, Helfand DJ, Wu X. 1991. *Ap. J.* 374:475–95
Wang Q, Wu X. 1992. *Ap. J. Suppl.* 78:391–401
Webbink RF. 1984. *Ap. J.* 277:355–60
White NE, Giommi P, Heise J, Angelini L, Fantasia S. 1994. *Ap. J.* 445:L125–28
Will T, Barwig H. 1996. See Greiner 1996a, pp. 99–106
Woosley SE, Weaver TA. 1994. *Ap. J.* 423:371–79
Yungelson L, Livio M, Tutukov A, Kenyon SJ. 1995. *Ap. J.* 447:656–79
Yungelson L, Tutukov A, Fedorova A, et al. 1996a. In *Cataclysmic Variables and Related Objects,* ed. A Evens, JH Wood, pp. 417–20. Dordrecht: Kluwer
Yungelson L, Livio M, Truran JW, Tutukov A, Fedorova A. 1996b. *Ap. J.* 466:890–910

*Annu. Rev. Astron. Astrophys. 1997. 35:101–36*


# OBSERVATIONAL SELECTION BIAS AFFECTING THE DETERMINATION OF THE EXTRAGALACTIC DISTANCE SCALE

*P. Teerikorpi*
Tuorla Observatory, Turku University, FIN-21500 Piikkiö, Finland;
e-mail: pekkatee@sara.cc.utu.fi

KEY WORDS: galaxies, Tully-Fisher relation, Malmquist bias, distance scale, cosmology

ABSTRACT

The influence of Malmquist bias on the studies of extragalactic distances is reviewed, with brief glimpses of the history from Kapteyn to Scott. Special attention is paid to two kinds of biases, for which the names Malmquist biases of the first and second kind are proposed. The essence of these biases and the situations where they occur are discussed.

The bias of the first kind is related to the classical Malmquist bias (involving the "volume effect"), while the bias of the second kind appears when standard candles are observed at different (true) distances, whereby magnitude limit cuts away a part of the luminosity function. In particular, study of the latter bias in distance indicators such as Tully Fisher, available for large fundamental samples of galaxies, allows construction of an unbiased absolute distance scale in the local galaxy universe where approximate kinematic relative distances can be derived. Such investigations, using the method of normalized distances or of the Spaenhauer diagram, support the linearity of the Hubble law and make it possible to derive an unbiased value of the Hubble constant.

## 1. INTRODUCTORY REMARKS

A red line in the history of astronomy is the extension of distance measurements into deep space as a result of efforts to take advantage of what Nature offers astronomers. Ptolemy (ca 150) stated that "none of the stars has a noticeable



0066-4146/97/0915-0101$08.00

parallax (which is the only phenomenon from which distances can be derived)," which illustrates the situation of when we see a class of celestial bodies and also have in mind a sound method but cannot apply it because of insufficient observational means. Of course, only after Copernicus was there strong theoretical pressure to search for stellar parallaxes as a crucial cosmological test and as a method of distance determination. However, long before this triangulation method was successfully applied by Bessel, Henderson, and Struve in the 1830s (see Hoskin 1982), the photometric method, based on the inverse square law of light flux from a point source introduced by Kepler, had been recognized as a possible way of getting information on stellar distances. In 1668, James Gregory applied it to the distance of Sirius, using the Sun as the calibrator star. In Gregory's method of "standard candles," one had to assume that stars are other suns, identical to our Sun, and are observed through the transparent Euclidean space where Kepler's inverse square law is valid.

## 1.1 *Stars and Galaxies are Gathered from the Sky and not from Space*

The above notes illustrate how determination of distances is intimately related to our general astronomical knowledge and assumptions, and also to our abilities to measure the directions and fluxes of weak photon streams. Though knowledge and observational methods, and hence construction of the distance ladder, are steadily advancing, there are fundamental difficulties that will always haunt the measurers of the universe.

The astronomer makes observations from a restricted vantage point in time and space. In fact, he or she does not observe celestial bodies in space, but in the sky (as traces on photographic plates or CCD images). Fortunately, very luminous objects exist that may be detected even from large distances. This diversity in the cosmic zoo, which allows one to reach large distances and hence makes cosmology possible, also involves problematic aspects.

First, from large distances only highly luminous objects are detectable, and the photons usually do not carry information on how much the objects differ from their average properties: There are no genuine Gregory's standard candles. Second, objects in the sky that are apparently similar, i.e. have the same distance modulus, actually have a complicated distribution of true distances. Distances larger than those suggested by the distance modulus are favored because of the volume ($r^2 dr$) effect. Third, at large distances there is much more space than within small distances from our position. Because very luminous objects are rare, these are not found in our vicinity. We can see objects that as a class might be useful indicators of large distances but perhaps cannot make the crucial step of calibration, which requires a distance ladder to reach the nearest of such objects (Sandage 1972). As an extreme example, even if one could find standard candles among luminous quasars (Teerikorpi 1981), there is no known method to derive their distances independently of redshift.

When one uses "standard candles" or "standard rods," calibrated on a distance ladder, systematic errors creep into the distance estimates. When related to the above problems, they are often collectively called Malmquist-like biases.

According to Lundmark's (1946) definition, a distance indicator is a certain group of astronomical objects that has the same physical properties in different galaxies. The principle of uniformity of natural law is assumed to be valid when one jumps from one galaxy to another. Slightly expanding Lundmark's definition, one may say that a distance indicator is a method where a galaxy is placed in three-dimensional space so that its observed properties agree with what we know about galaxies, their constituents, and the propagation of light. An ideal distance indicator would restrict the galaxy's position somewhere in a narrow range around the true distance if that indicator has an intrinsic dispersion in its properties, and a set of such indicators should lead to a consistent picture of where the galaxies are. In practice, even a "good" indicator is affected by sources of systematic error due to the intrinsic dispersion.

## 1.2 *Distances and Hubble Law*

The Hubble law between distance and cosmological redshift is a blessing for the cosmographer. A great motivation for investigation of the distance scale, it is also helpful for tackling the problems mentioned above.

Systematic errors in obtained distances are often recognized as a deviation from the linear Hubble law, and the reality and speed of galaxy streams, for example, closely depend on how well distances to galaxies are known. When one speaks about the choice of one or another distance scale, this is intimately connected with the Hubble constant $H_0$. In the Friedmann model, $H_0$ together with $q_0$ allows one to extend the distance scale to high cosmological redshifts where classical distance indicators are lost. However, this extension is not the topic of my review [for discussion of such questions, see e.g. Rowan-Robinson (1985)].

To discuss biases in extragalactic distances, one might like to know what "distance" represents. As McVittie (1974) says, distance is a degree of remoteness; in some sense or another, faint galaxies must be remote. Only a cosmological model gives the exact recipe for calculating from the observed properties of an object its distance (which may be of different kinds). Because our basic data consist of directions and fluxes of radiation, we are usually concerned with luminosity or diameter distances. An error, say, in the luminosity distance in a transparent space means that if one puts a genuine standard candle beside the distance indicator in question, its photometric distance modulus is not the same.

Among the variety of distance concepts, one would like to think that there exists a fundamental one, corresponding to meter sticks put one after another from the Sun to the center of a galaxy. For instance, in the Friedmann universe, the theoretical and not directly measurable "momentary" proper distance is often in the background of our minds as the basic one (Harrison 1993), and the luminosity and other types of distances are the necessary workhorses. This review refers to the local galaxy universe where the different distance types are practically the same; in any case, the tiny differences between them are overwhelmed by selection biases and other sources of error. Another allowance from Nature is that in this distance range, evolutionary effects can be forgotten: In an evolving universe, distance indicators often require that the look-back time should be much less than the global evolutionary time scale.

## 1.3 *Aim of the Present Review*

In the recent literature, comprehensive and technical discussions of various aspects of the Malmquist bias exist (Willick 1994, Strauss & Willick 1995, Hendry & Simmons 1995). The reviews by Sandage (1988a, 1995) put the subject in the general context of observational cosmology. I have in mind a wide audience, from the general astronomer to the quantum cosmologist, who may need an introduction to the Malmquist biases. This problem is so characteristic of the special nature of astronomy that it should be included in introductory astronomy text books. In the otherwise versatile work on the cosmological distance ladder by Rowan-Robinson (1985), the problem of Malmquist bias receives only a passing note, and the extensive review article on distance indicators by Jacoby et al (1992) is also relatively silent in this respect.

The concepts of the different kinds of Malmquist biases are rather simple, though it is possible to dress the discussions in mathematics "complicated enough to be but dimly understood except, perhaps, by their authors" as Sandage (1988a) noted. On the other hand, one sometimes sees references to "the well-known Malmquist bias," which is usually a sign of an insufficient treatment of this problem. The present review is written in the spirit of a useful middle path.

I do not discuss in detail different distance indicators but concentrate on the central issues of the biases. Also, this is not a discussion of the Hubble constant, though it occasionally appears. I also leave aside the question of local calibration. Of course, the biases discussed here are not the only problems of the distance scale. For instance, the supposed distance indicators may not at all fulfill Lundmark's definition, an example of which is Hubble's brightest stars in distant galaxies that were actually HII regions (Sandage 1958).

I take examples mostly from the Tully-Fisher (TF) indicator, which is the most widely applied method in the local universe, with samples of several thousands of galaxies and the calibration of which can now be based on an increasing number of Cepheid distances. Understanding the biases in such large and fundamental samples will always be the test bench on which to build the distance scale and where differing alternative scales must be ultimately compared.

The theory of how to deal with the selection effects affecting distances has gradually evolved to such a level that one may already speak about an emerging general theory of distance indicators. Relevant results are often scattered in articles concerned with a variety of topics, and though I have tried to find a representative reference list, I apologize if some interesting aspects go unnoticed. Finally, I should admit that my several years of pleasant collaboration with the Meudon and Lyon groups must somewhat "bias" this review.

## 2. SOME HISTORY FROM KAPTEYN TO SCOTT

Of course, this is not a proper place to write a history of selection biases, or how they have been invented and reinvented, considered, or neglected in astronomical works of the present century. However, it seems helpful to introduce the reader to the current discussion of this subject by picking from the past a few important fragments.

### 2.1 *Kapteyn's Problem I and Problem II*

In a paper on the parallaxes of helium stars "together with considerations on the parallax of stars in general," Kapteyn (1914) discussed the problem of how to derive the distance to a stellar cluster, presuming that the absolute magnitudes of the stars are normally distributed around a mean value. He came upon this question after noting that for faint stars the progress of getting kinematical parallaxes is slow and "can extend our knowledge to but a small fraction of the whole universe." A lot of magnitude data exist for faint stars, but how to put them to use? He formulated Problem I as follows:

> Of a group of early B stars, all at practically the same distance from the sun, we have given the average apparent magnitude $\langle m \rangle$ of all the members brighter than $m_0$. What is the parallax $\pi$ of the group?

Changing a little notation and terminology, Kapteyn's answer to this question may be written as an integral equation, where the unknown distance modulus $\mu$ appears:

$$\langle \mathrm{m} \rangle = \frac{\int_{-\infty}^{m_l} \mathrm{m} \exp\left[-(\mathrm{m}-\mu-\mathrm{M_o})^2/2\sigma^2\right] dm}{\int_{-\infty}^{m_l} \exp\left[-(\mathrm{m}-\mu-\mathrm{M_o})^2/2\sigma^2\right] dm}. \tag{1}$$

Because values for the parameters $M_o$ and $\sigma$ of the gaussian luminosity function are known, one may solve the distance modulus $\mu$. Note that the integration over apparent magnitudes is made from $-\infty$ to $m_l$, the limiting magnitude. Kapteyn calculated a table for practical use of his equation, so that from the observed value of $\langle m \rangle$, one gets the distance modulus $\mu$. If one simply uses the mean absolute magnitude $M_0$ and calculates the distance modulus as $\langle m \rangle - M_0$, a too-short distance is obtained. Though Kapteyn did not discuss explicitly this bias, his method was clearly concerned with the Malmquist bias of the second kind, a typical problem in photometric distance determinations (see Section 3).

Kapteyn recognized that the situation is different if stars are scattered at varying distances, leading to his Problem II:

> Of a group of early B stars, ranging over a wide interval of distance, given the average apparent magnitude of all the stars brighter than $m_0$ we require the average parallax of the group.

In this scenario, if one now uses Kapteyn's table mentioned above, taking the $\mu$ corresponding to $\langle m \rangle$, one generally obtains an incorrect average distance modulus $\langle \mu \rangle$. Also, as in Problem I, one cannot take $\langle \mu \rangle = \langle m \rangle - M_0$, either. This Kapteyn's problem (for which he did not offer a complete solution) is related to what is called the classical Malmquist bias.

In Problem I, one has information on relative distances; in this particular case the distances are equal. The necessity of having relative distances indicates that the solution to Problem I is not applicable to a "group" of one star. In Problem II there is no a priori information on relative distances. On the other hand, in order to calculate a mean distance modulus, one needs such information as must be extracted from the only data available, i.e. from the distribution of apparent magnitudes. Again, a sample of a single star with $m = m_0$ cannot be a basis for solving Problem II (as an answer to the question "what is the most probable distance of this star?"), unless one makes some assumption on how the magnitudes of the other stars are distributed.

## 2.2 *The Classical Malmquist Bias*

In his work, "A study of the stars of spectral type A," Malmquist (1920) investigated how to derive the luminosity function of stars from their proper motions, provided that it is gaussian and one knows the distribution of apparent magnitudes up to some limiting magnitude. This led Malmquist to investigate the question of what is the average value (and other moments) of the quantity $R$, or the reduced distance, as earlier introduced by Charlier:

$$R = 10^{-0.2M}. \tag{2}$$

Malmquist made three assumptions: 1. There is no absorption in space. 2. The frequency function of the absolute magnitudes is gaussian ($M_0$, $\sigma$). 3. This function is the same at all distances. The third assumption is the principle of uniformity as implied in Lundmark's (1946) definition of distance indicators.

Using the fundamental equation of stellar statistics, Malmquist derived $\langle R^n \rangle$ and showed that it may be expressed in terms of the luminosity function constants $M_o$ and $\sigma$ and the distribution $a$(m) of apparent magnitudes, connected with the stellar space density law. Especially interesting for Malmquist was the case $n = -1$, or the mean value of the "reduced parallax," that appears in the analysis of proper motions. However, for distance determination, the case $n = 1$ is relevant because it allows one to calculate, from the mean value of the reduced distance, the average value of the distance $\langle r \rangle$ for the stars that have their apparent magnitude in the range $m \pm 1/2\, dm$ or their distance modulus $\mu = m - M_o$ in the range $\mu \pm 1/2\, d\mu$. The result is, written here directly in terms of the distance modulus distribution $N(\mu)$ instead of $a$(m),

$$\langle r \rangle_\mu = r(\mu) \exp(0.5b^2\sigma^2) N(\mu + b\sigma^2)/N(\mu), \tag{3}$$

where $b = 0.2 \cdot \ln 10$. This equation is encountered in connection with the general Malmquist correction in Section 6. Naturally, in Malmquist's paper one also finds his formula for the mean value of M for a given apparent magnitude m:

$$\langle M \rangle_m = M_o - \sigma^2 d[\ln a(m)]/dm. \tag{4}$$

The term including the distribution of apparent magnitudes (or distance moduli in Equation 3) reduces to a simple form when one assumes that the space density distribution $d(r) \propto r^{-\alpha}$:

$$\langle M \rangle_m = M_o - (3 - \alpha) \cdot 0.461\, \sigma^2. \tag{5}$$

With $\alpha = 0$, one finally obtains the celebrated Malmquist's formula valid for a uniform space distribution:

$$\langle M \rangle_m = M_o - 1.382\, \sigma^2. \tag{6}$$

Hubble (1936) used Malmquist's formula (Equation 6) when, from the brightest stars of field galaxies (and from the magnitudes of those galaxies), he derived the value of the Hubble constant. Hubble derived from a local calibrator sample the average (volume-limited) absolute photographic magnitude and its dispersion for the brightest stars. As "the large-scale distribution of nebulae and, consequently, of brightest stars is approximately uniform," he derived the expected value for the mean absolute magnitude of the brightest stars $\langle M_s \rangle$ for a fixed apparent magnitude. His field galaxies were selected, Hubble maintained, on the basis of the apparent magnitudes of the brightest stars, which justified the calculation and use of $\langle M_s \rangle$. In the end, he compared the mean apparent magnitudes of the brightest stars in the sample galaxies with $\langle M_s \rangle$, calculated the average distance $\langle r \rangle$, and derived the value of the Hubble constant (526 km/s/Mpc). Hence, it is important to recognize that this old value of $H_0$, canonical for some time, already includes an attempt to correct for the Malmquist bias. Also, it illustrates the role of assumptions in this type of correction (what is the space density law, what is the mode of selection of the sample, etc).

## 2.3 *The Scott Effect*

Forty years ago, Scott (1957) published an important paper on "The Brightest Galaxy in a Cluster as a Distance Indicator." Looking back to this study, one may find it contained many basic points that were later discussed in connection with the Malmquist bias of the second kind. Her concern was how the availability of a distant cluster of galaxies influences the use of its brightest galaxy as a standard candle. Availability means that (*a*) at least *n* cluster members must be brighter than the limiting magnitude $m_1$ of the plate, and (*b*) the apparent magnitude of the brightest galaxy must be brighter than another limit $m_2$, which is needed for measurements of magnitude and redshift. Let us pick up a few conclusions of Scott (1957, p. 249):

- (A)t any given distance, a cluster with many members is more likely to be available to the observer than a cluster containing fewer galaxies, or

- if a very distant cluster is available to the observer, then this cluster must be unusual, and

- the brightest galaxies actually observed in very distant clusters must have a tendency to possess brighter absolute magnitudes than the average brightest galaxies in the nearer clusters, hence,

- for distant clusters, the simple use of the brightest galaxy in the cluster as a distance indicator leads to an underestimate of the distance.

Scott used both numerical simulations and an analytical model to show the size of systematic errors that the condition of availability causes to the derived distance at different true distances, and she concluded that the selection effect is bound to influence seriously the Hubble m-z diagram constructed for brightest cluster galaxies. She (1957, p. 264) also concluded that

- an interpretation of the deviations from linearity in the magnitude redshift relation that occur near the threshold of available instruments cannot be made with confidence without appropriate allowance for selection bias.

She (1957, p. 264) also pondered about how to make a difference between a real and selection-induced deviation from linearity in the Hubble law:

- (T)he question as to whether or not the apparent deviation from linearity is an effect of selection [or a real effect] may perhaps be solved by the accumulation of further data using instruments corresponding to larger values of both $m_1$ and $m_2$.

Of the above two extracts, Hubble's application is clearly concerned with the classical Malmquist bias (or Kapteyn's Problem II), whereas the discussion by Scott has a relation to Kapteyn's Problem I, though the effect in galaxy clusters is more complicated. In fact, there is no mention of the classical Malmquist bias in Scott's paper and really no need for it.

## 3. TWO KINDS OF MALMQUIST BIAS

In recent years, Kapteyn's Problems I and II have been discussed (by authors unaware of that early work) in connection with extragalactic distances from methods like Tully-Fisher (TF) and Faber-Jackson, for determination of the Hubble constant and peculiar velocity maps. Both of these problems have been sometimes referred to as Malmquist bias. I have collected in Table 1 examples of the used nomenclature.

### 3.1 *Biases of the First and Second Kind*

In this review, I call these two problems Malmquist bias of the first and second kinds: The first kind is the general Malmquist bias, directly connected with the classical treatment by Malmquist (1920). One might briefly define these two aspects as follows:

- Malmquist bias of the first kind is the systematic average error in the distance modulus $\mu$ for a class of galaxies with "derived" $\mu = \mu_{\text{der}} =$ constant.

- Malmquist bias of the second kind is the systematic error in the average derived $\langle \mu_{\text{der}} \rangle$ for the class of galaxies with "true" $\mu = \mu_{\text{true}} =$ constant.

In discussions of the first bias, one is interested in the distribution of true distance moduli for the constant derived modulus, whereas the second bias is concerned with the distribution of derived distance moduli at a constant true

**Table 1** Various names used for the two kinds of biases

| Author | Nomenclature | |
|---|---|---|
| Kapteyn (1914) | Problem II | Problem I |
| Landy & Szalay (1992) | General Malmquist bias | — |
| Han (1992) | Geometry bias | Selection effect |
| Teerikorpi (1990, 1993) | "V against r" | "r against V" |
| Sandage (1994a) | Classical | Distance-dependent |
| Hendry & Simmons (1994) | Bayesian | Frequentist |
| Willick (1994) | Inferred-distance problem | Calibration problem |
| Strauss & Willick (1995) | Malmquist bias | Selection bias |
| Teerikorpi (1995) | M-bias of the first kind | M-bias of the second kind |

distance (e.g. at a constant redshift, if the Hubble law is valid). These two biases have appeared under different names. Teerikorpi (1990, 1993) separated the study of the Hubble diagram into "distance against velocity" and "velocity against distance." Hendry & Simmons (1994) spoke about "Bayesian" and "frequentist" approaches in terms of mathematical statistics, whereas Willick (1994) emphasizes "inferred-distance problem" and "calibration problem." In his bias properties series, Sandage (e.g. 1994a) also made a clear difference between the classical Malmquist bias and the distance-dependent effect, and he showed in an illuminating manner the connection between the two.

More recently, Strauss & Willick (1995) have used the terms Malmquist bias and selection bias, whereby they emphasize by "selection" the availability of galaxies restricted by some limit (flux, magnitude, angular diameter). On the other hand, one may say that the first kind of Malmquist bias is also caused by a selection effect dictated by our fixed position in the universe and the distribution of galaxies around us, and in the special case of the inverse relation (Section 5), it also depends on the selection function.

The first kind of bias is closely related to the classical Malmquist (1920) bias; hence the name is well suited. However, Lyden-Bell (1992) prefers to speak about Eddington-Malmquist bias: Malmquist (1920) acknowledged that Eddington had given Equation 6, which corresponds to the special case of constant space density, before he had. As for the second bias, the situation is not so clear. However, in Malmquist (1922), in his Section II.4, one finds formally the equations needed for the calculation of the second bias, when the luminosity function has a gaussian distribution in M. Malmquist does not seem to have this application in mind, but in view of the widespread habit of speaking about the Malmquist bias (or Malmquist effect) in this connection, it might be appropriate to name the distance-dependent effect after him as well [another possibility would be to speak about Behr's (1951) effect, cf Section 3.3].

## 3.2 *Situations Where the First or General Malmquist Bias Appears*

Let us consider a class of galaxies having a gaussian luminosity function $G(M_p, \sigma)$ acting in the role of a standard candle. Such a class may be defined, e.g. via a fixed value of the TF parameter $p = \log V_{max}$, and it is assumed that the calibration has been made, i.e. the volume-limited value of $M_p$ is known. A sample of these galaxies with magnitudes measured up to some limit gives us a collection of derived distance moduli $\mu = m - M_p$. If in addition the redshifts are available, one may construct the $\mu$-log V Hubble diagram, and try to determine the Hubble constant H from the following expected relation:

$$\log V = \log H - 0.2\mu + \text{const.} \quad (7)$$

If the error in the distance modulus has a gaussian distribution $G(0, \sigma)$, one can take simple averages $\langle \log V \rangle$ and $\langle \mu \rangle$, and solve for $\langle \log H \rangle$. In reality, there is a systematic error in each subaverage $\langle \mu_i \rangle$ ($\mu_i$ = constant), and this systematic error generally depends on the value of $\mu_i$ if the space distribution of galaxies is not uniform. The error $\Delta\mu(\mu_i)$ is given by Malmquist's formula (Equation 3). One point should be especially emphasized: The ratio $N(\mu + b\sigma^2)/N(\mu)$ refers to the distance moduli in the ideal case when the sample is not influenced by a magnitude limit, i.e. when $N(\mu)$ contains information on the real space distribution of the galaxies.

One implication of the general Malmquist bias is that the Hubble diagram does not necessarily show a linear Hubble law, because the bias depends on the real distribution of galaxies and is different for different (derived) distances. Also, in different directions of the sky, one may derive by the mentioned method different values of the Hubble constant, possibly interpreted, but incorrectly, to be caused by our own peculiar motion or by large scale streams of galaxies. Such explanations have been applied to the Rubin-Ford effect (Sandage & Tammann 1975a,b, Fall & Jones 1976, James et al 1991), the supergalactic anisotropy (Teerikorpi & Jaakkola 1977), and the Great Attractor velocity field (Landy & Szalay 1992).

The general Malmquist correction has in recent years been discussed particularly in connection with the attempts to derive maps of peculiar velocities or galaxy streams. The radial component of a peculiar velocity for a galaxy with true distance r and observed correct radial velocity $V_{obs}$ is as follows:

$$V_{pec} = V_{obs} - H_o \cdot r. \tag{8}$$

Let us look at galaxies in some direction, with their derived distances within a narrow range $r_{der} \pm 1/2dr$. Perhaps these galaxies are also in the real space so close by that they have a common peculiar velocity (bulk flow or stream), though different cosmological components. Then $\langle V_{obs} \rangle - H_o \cdot \langle r_{der} \rangle$ does not give the correct peculiar velocity, unless a proper correction for $\langle r_{der} \rangle$ is made, i.e. the Malmquist correction of the first kind, which gives the correct average distance for the galaxies in the $r_{der} \pm 1/2dr$ range.

Lynden-Bell et al (1988) used such a correction in their Great Attractor paper, though they assumed a uniform distribution of galaxies. This means that for their diameter (vs velocity dispersion) indicator for elliptical galaxies, the bias formula is analogous to the classical Malmquist formula:

$$\langle r \rangle = r_{est}\left(1 + 3.5\sigma_D^2\right). \tag{9}$$

### 3.3 *Situations Where the Second Malmquist Bias Is Important*

A simple way to see the second kind of Malmquist bias in action is to take the sample above, calculate log H for each galaxy, and plot log H against redshift.

As demonstrated in several studies during the last twenty or so years, log H stays first roughly constant and then starts to increase. So clear and dramatic is this phenomenon, and also so much expected from simple reasoning as due to bias, that it is appropriate to repeat the words of Tammann et al (1980): "If an author finds $H_o$ to increase with distance he proves in the first place only one thing, i.e., he has neglected the Malmquist effect! This suspicion remains until he has proved the contrary." To this statement, one should add that here "distance" means the true distance or at least a true relative distance (e.g. redshift in the case of the Hubble law). If "distance" is the measured or inferred distance, $H_o$ does not necessarily change with the distance, though it may have a wrong average value—we have come back to the first kind of bias. This also serves as a warning that simple comparison of the "linearity" of photometric distances from two methods may hide a common bias, as seen in the comparisons of two or more distances by separate indicators discussed by de Vaucouleurs (1979). Each of the indicators suffers the same type of bias properties.

It seems that Behr (1951) was the first to point out, after comparison of the width of the Local Group luminosity function to that of the field galaxies, that application of the standard candle method may lead to systematically short distances at large true distances. As we discussed above, this idea was transformed into a quantitative model by Scott (1957) for the selection of brightest cluster galaxies. It was then reinvented by Sandage & Tammann (1975a) and Teerikorpi (1975a,b), in connection with concrete field samples of spiral galaxies with van den Bergh's (1960a,b) luminosity classes. The basic reasoning is illustrated by the formula that connects the derived distance modulus to apparent magnitude, the assumed standard absolute magnitude $M_p$, and to the magnitude limit $m_{lim}$:

$$\mu_{der} = m - M_p \leq m_{lim} - M_p. \tag{10}$$

Clearly, there is a maximum derived distance $\mu_{max} = m_{lim} - M_p$. However, because the standard candle actually has a dispersion $\sigma$ in absolute magnitude, some galaxies can be seen from true distances beyond $\mu_{max}$ that necessarily become underestimated.

Under the assumption of a gaussian luminosity function $G(M_p, \sigma)$ and a sharp magnitude limit $m_{lim}$, it is a straightforward task to calculate the magnitude of the second Malmquist bias $\langle\mu\rangle - \mu_{true}$ at each true distance modulus (Teerikorpi 1975b; Sandage 1994a), or how much $\langle \log H \rangle$ will increase due to the magnitude limit cutting galaxies away from the fainter wing of the luminosity function. Such an analytical calculation needs the correct distance scale and the relevant dispersion $\sigma$ as input, which restricts its application in practice. However, it gives the general behavior of the bias, which is quite similar to the observed behavior of $\langle \log H \rangle$ vs kinematic distance, and forms the basis for recognition and correction methods that are independent of $H_o$ (Section 4).

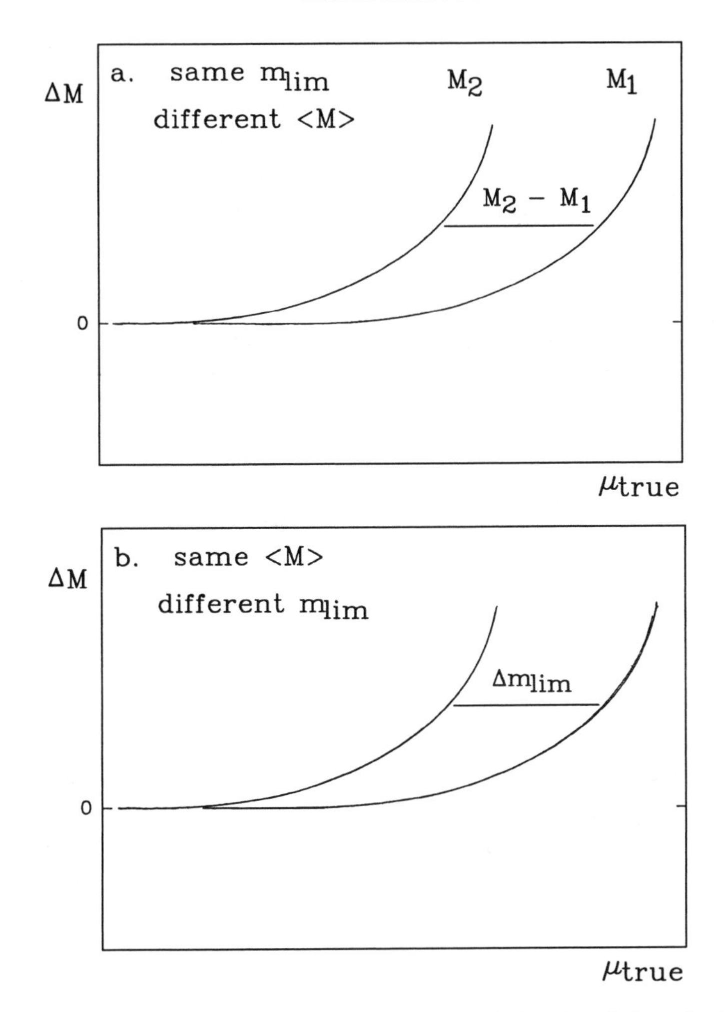


*Figure 1* Illustrations of how the bias ΔM in the average magnitudes of standard candles depends on true distance modulus $\mu_{true}$: a. When the limiting magnitude $m_{lim}$ is fixed, the bias curves are simply shifted by $M_2 - M_1$. b. Increasing $m_{lim}$ shifts the bias curve of a single standard candle class by $\Delta m_{lim}$ in $\mu$. At small true distances, one sees the unbiased plateau.

A fundamental property of the theoretical bias curves is that the curves for standard candles with different means, $M_p$, will show a shift along the axis of true distance (redshift) (see Figure 1*a*). This, in the first place, led to the recognition of the bias by Teerikorpi (1975a), where different van den Bergh's luminosity classes were inspected, and by analogy, to the proposal that de Vaucouleurs's parabolic velocity–distance relation (de Vaucouleurs 1972) was caused by a related effect.

An earlier example of the second Malmquist bias at work, is found in Hawkins (1962), where the composite Hubble diagram for field galaxies (using the magnitudes as distance indicators) was interpreted as supporting the quadratic law $z = k \cdot r^2$. This was suggested to be expected from the gravitational redshift in a static uniform universe, contrary to the linear law expected for the expansion redshift in a homogeneous and isotropic universe, described by the Robertson-Walker line element (Robertson 1955). If written formally as a velocity law, the quadratic law assumes the form $\log H = 0.5 \log V + \text{const}$, which roughly describes the run of the data points between the unbiased region ($\log H = 0 \cdot \log V + \text{const}$) and the strongly biased one ($\log H = 1.0 \cdot \log V + \text{const}$). One sometimes sees, though not quite appropriately, "Lundmark's law" mentioned in connection with the quadratic law. True, Lundmark (1924, 1925) produced the first diagram where a dependence between redshift and distance was discernable, and he suggested the representation $z = a + b \cdot r - c \cdot r^2$. In Lundmark's formal solution, the negative quadratic term does not seem to be related to any selection or other real effect—the scatter from galaxy diameters as distance indicator was large, and he might have easily suggested a linear law. Perhaps he was motivated by the interesting prediction that redshifts have a maximum value (3000 km/s), which was soon to be contradicted by Humason's measurements.

A special case of the second kind of Malmquist bias is related to Kapteyn's Problem I and sometimes called cluster population incompleteness bias. Here the galaxies are at the same true distance, and the bias in the derived distance modulus is caused and calculated similarly as above. This bias makes the distances to clusters of galaxies, calculated by the TF method, progressively too short, and as it does for field galaxies, making log H increase with true distance. Assertions in literature (e.g. Aaronson et al 1980, de Vaucouleurs & Corwin 1985, Bothun et al 1992) have stated that there is no Malmquist bias in clusters (because the galaxies in clusters have no volume effect like the field galaxies), implying that one may utilize the TF method without bias. However, there were indications in the 1980s that the clusters, in comparison with Malmquist corrected field galaxies, give $H_0$ values that are too large, which led to recognition of the cluster incompleteness bias by Teerikorpi (1987), Bottinelli et al (1987), and Kraan-Korteweg et al (1988).

## 4. ATTEMPTS TO OVERCOME THE SECOND KIND OF BIAS

The solution, Equation 1, proposed by Kapteyn for his Problem I, gives the distance to a cluster or to any subsample of galaxies known, e.g. from the

Hubble law or other velocity field model, to be at the same distance. However, this presupposes that (*a*) there is a sharp magnitude limit, with data complete up to $m_{lim}$, (*b*) the standard candle has a gaussian distribution $G(M_p, \sigma)$, (*c*) the mean $M_p$ is known from local considerations (calibration), and (*d*) the dispersion $\sigma$ is known. These are rather strict conditions. Fortunately, there are situations and aims that do not necessarily require a complete knowledge of all these factors. For instance, study of the linearity of the Hubble law does not need an absolute calibration of the standard candle, and knowledge of $\sigma$ is not necessary in all methods for deriving the Hubble constant.

## 4.1 *The Bias-Free Redshift Range*

Sandage & Tammann (1975a) introduced the concept of the bias-free distance (redshift) range in their Hubble parameter H-vs-log $V_o$ diagram for luminosity classified spiral galaxies, which showed an increase of H with redshift. They explained this increase as caused by the truncation effect of the limiting magnitude, which makes the derived distances too small.

One expects an unbiased region at small true distances (redshifts) because the sample can be distance-limited rather than flux-limited, hence no part of the faint end of the luminosity function is truncated at appropriately small distances. It is only here that the Hubble constant can be derived without correction for bias. Such a region of about constant H was clearly visible in the H-vs-log $V_o$ diagram of Sandage & Tammann (1975). As a first approximation, they cut away galaxies more distant than a fixed $V_o$ ($\approx$2000 km/s), independently of the morphological luminosity class. Actually, each luminosity class has its own limiting distance, as Sandage & Tammann recognized. Using one fixed $V_o$, one (*a*) loses high-$V_o$ data, part of it possibly unbiased, and (*b*) allows a remaining bias due to intrinsically faint luminosity classes with their proper limit $\langle V \rangle_o$. This was inspected by Teerikorpi (1976), where from the ST data (and their luminosity class calibration) the low value of $H_o = 41$ was derived, in comparison with $H_o = 57$ by Sandage & Tammann (1975). I give this reference because it was a step toward the method of normalized distances, later applied to samples of galaxies with TF measurements. The similarity with $H_o \approx 43$ by Sandage (1993) is probably not just a coincidence; both determinations relied on M101. An up-to-date discussion of the bias in the luminosity class method was given by Sandage (1996a).

## 4.2 *The Method of Normalized Distances for Field Galaxies*

Assume that we have two standard candle (galaxy) classes, each having gaussian luminosity functions $G(M_1, \sigma)$ and $G(M_2, \sigma)$, hence simply shifted in M by $\Delta M = M_1 - M_2$. If both are sampled up to a sharp magnitude limit $m_{lim}$, it is easy to see that in the bias vs true distance modulus $\mu$ diagram, the bias of the second kind suffered by these two candles is depicted by curves of the same form but separated horizontally by constant $\Delta\mu = -\Delta M$ (see Figure 1*a*). The curve of the brighter candle achieves only at larger distances the bias suffered by the fainter candle already at smaller distances. In this way, simultaneous inspection of two or more standard-candle classes gives a new dimension to the problem of how to recognize a bias. Figure 1*b* shows another important property of the bias behavior. If one keeps the standard candle the same but increases the limiting magnitude by $\Delta m_{lim}$, the bias curve shifts to larger distances by $\Delta\mu = \Delta m_{lim}$. This is the basis of what Sandage (1988b) calls the "adding of a fainter sample" test.

van den Bergh's morphological luminosity classes clearly showed this effect (Teerikorpi 1975a,b), and even gave evidence of the "plateau" discussed below. In the beginning of 1980s, extensive studies started to appear where the relation (Gouguenheim 1969, Bottinelli et al 1971, Tully & Fisher 1977) between the magnitude (both B and infrared) and maximum rotational velocity of spiral galaxies $V_{max}$ was used as a distance indicator. In the following, the direct regression (M against fixed log $V_{max} \equiv p$) form of this TF relation is written as

$$M = a \cdot p + b. \tag{11}$$

There was at that time some uncertainty about which slope to use in distance determinations to individual galaxies—direct, inverse, or something between? Then Bottinelli et al (1986) argued that in order to control the Malmquist bias of the second kind, it is best to use the direct slope, so that the regression line is derived as M against the fixed observed value of $p$, without attempting to correct the slope for the observational error in $p$. The observed value of $p$ devides the sample into separate standard candles analogous to Malmquist's star classes. A similar conclusion was achieved by Lynden-Bell et al (1988) in connection with the first kind of bias.

The importance of the direct relation is the fact that this particular slope allows one to generalize the mentioned example of two standard candles to a continuum of $p$ values and in this way to recognize and investigate how the Malmquist bias of the second kind influences the distance determinations [that such a bias must exist also in the TF method was suspected by Sandage & Tammann (1984) and introduced on a theoretical basis in Teerikorpi (1984)]. If one inspects the whole sample (all $p$ values clumped together) in the diagnostic log H vs $d_{kin}$ diagram (cf Figure 1), the bias may not be very conspicuous. On the other hand, if one divides the sample into narrow ranges of $p$, each will contain a small number of galaxies, which makes it difficult to see the behavior of the bias for each separate standard candle within $p \pm 1/2dp$. For these reasons, it is helpful to introduce so-called normalized distance $d_n$ (Teerikorpi 1984, Bottinelli et al 1986), which transforms the distance axis in the log H vs $d_{kin}$ diagram so that the separate $p$ classes are shifted one over the other and the

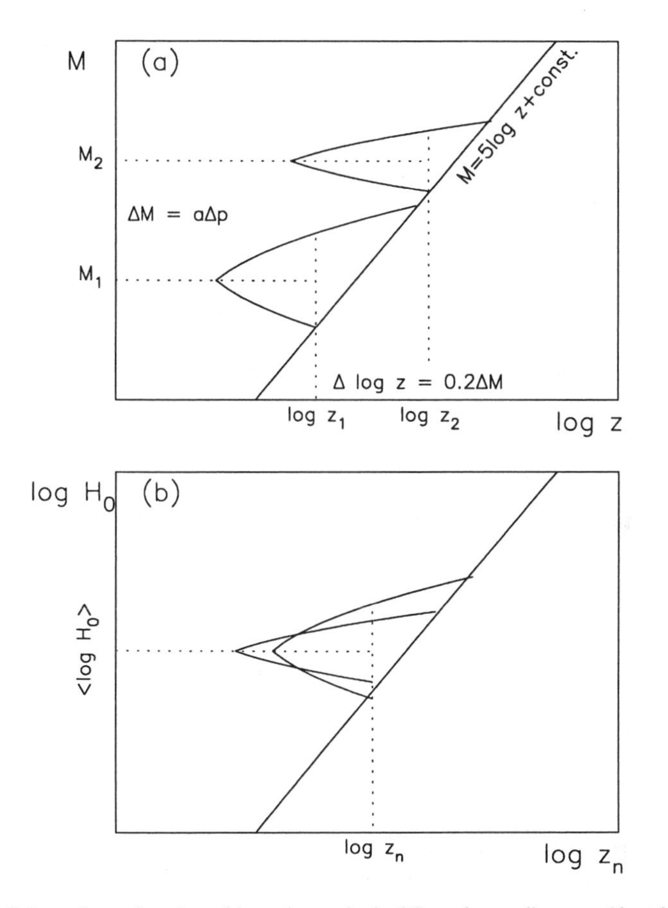


*Figure 2* Schematic explanation of how the method of Spaenhauer diagrams (the triple-entry correction or TEC) and the method of normalized distances (MND) are connected. Upper panel shows two Spaenhaeur (M vs log *z*) diagrams corresponding to TF parameters $p_1$ and $p_2$. The inclined line is the magnitude limit. When the data are normalized to the log H vs log$z_n$ diagram, using the TF slope *a*, the separate Spaenhauer diagrams "glide" one over the other and form an unbiased plateau that, among other things, can be used for determination of the Hubble constant.

bias behavior is seen in its purity (see Figure 2 in this regard):

$$\log d_n = \log d_{kin} + 0.2(a \cdot p + b - \text{const}). \tag{12}$$

In fact, one might also term this transformed distance as the effective one. This method of normalized distances (MND) uses as its starting point and test bench an approximative kinematical (relative) distance scale ($d_{kin}$, e.g. as provided by the Hubble law or Virgo-centric models) used with observed redshifts. The method usually investigates the bias as seen in the Hubble parameter log H, calculated from the (direct) TF distance for each galaxy using the (corrected) radial velocity. If the Hubble law is valid, i.e. there exists a Hubble constant $H_0$, then one expects an unbiased plateau at small normalized distances, a horizontal part from which the value of $H_0$ may be estimated from the plot of the apparent $H_0$ vs $d_n$. Bottinelli et al (1986) applied the method to a sample of 395 galaxies having B magnitudes and the TF parameter $p$ and could identify clearly the plateau and determine $H_0$ from it. Certain subtle points of the method were discussed by Bottinelli et al (1988a), who also presented answers to the criticism from de Vaucouleurs & Peters (1986) and Giraud (1986). A somewhat developed version of it has recently been applied to the KLUN (kinematics of the local universe) sample constructed on the basis of the Lyon-Meudon extragalactic database and containing 5171 galaxies with isophotal diameters $D_{25}$ (Theureau et al 1997).

## 4.3 *Spaenhauer Diagrams and the Triple-Entry Correction by Sandage*

The triple-entry correction (TEC) of Sandage (1994a,b) is an approach and method for bias recognition, derivation of unbiased TF relations, and calculation of unbiased Hubble constant, which in some respects differs from the method of normalized distances though is based on similar basic reasoning. The theory of the TEC method is given in a clear manner in the two articles by Sandage (1994a,b), and it is applied to the Mathewson et al (1992a,b) sample of 1355 galaxies in Federspiel et al (1994). The method, originally suggested by Sandage (1988b,c), is based on Spaenhauer (1978) log $V_0$-M diagrams, which reveal how the average absolute magnitude M of a standard candle changes with increasing kinematical distance $V_0$ when the magnitude limit cuts away a progressively larger part of the luminosity function.

Because these two approaches are the most developed ways of deriving the direct TF relation and the value of $H_0$ from large field spiral samples, it is important to see clearly how they are connected. This is best done with the aid of the absolute magnitude M-vs-log $V_0$ diagram (Figure 2, *upper panel*), similar to Figure 3 by Sandage (1994b), though with only two inserted Spaenhauer patterns corresponding to TF parameters $p_1$ and $p_2$. The inclined straight line corresponds to the magnitude limit, cutting away everything to the right of the line. The tips, i.e. the apex $\langle M \rangle$, of the Spaenhauer patterns give the average TF magnitudes $M_1$ and $M_2$ for $p_1$ and $p_2$, respectively.

Of course, the data used in the MND can be given exactly the same Spaenhauer representation as in Figure 2. Then what happens when the data are tranformed into the form used in the MND? First, calculate the Hubble parameter log H for each galaxy, using an approximate TF relation. This shifts the tips of the Spaenhauer patterns to about the same vertical level in the log H-vs-log $V_0$

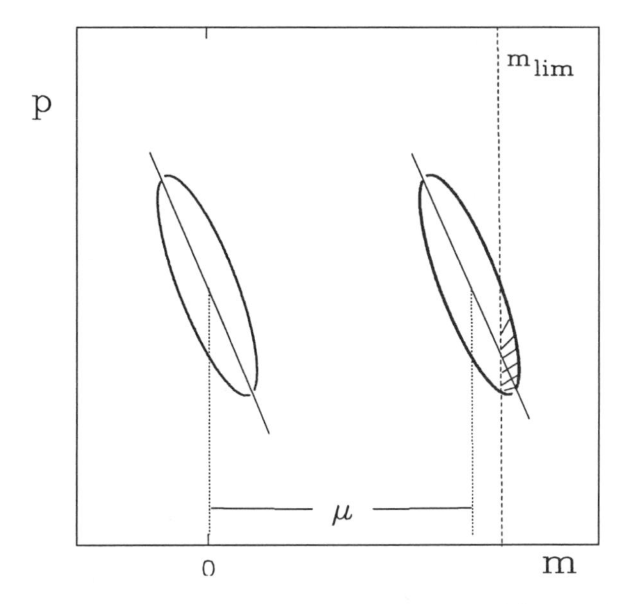


*Figure 3* Schematic explanation of how the inverse TF relation ($p$ vs M) may under ideal conditions overcome the Malmquist bias of the second kind. The nearby calibrator sample is made to glide over the distant sample so that the regression lines overlap. The condition for this is given by Equation 14, with $\mu = \mu_{est}$.

diagram. Second, shift the kinematical distances (log $V_o$) of the $p_2$ class by the factor $0.2(M_2 - M_1) = 0.2a(p_2 - p_1)$. This normalization puts the bias curve for $p_2$ on top of the bias curve for $p_1$, and the Spaenhauer patterns lie one over the other (Figure 2, *lower panel*). The individual unbiased parts of the patterns thus amalgamate together to form the common unbiased plateau. Refering to Figure 2, the MND means letting the upper Spaenhauer pattern glide down along the limiting magnitude line until it settles over the lower pattern.

In both methods, the TF relation $M = a \cdot p + b$ is derived from the unbiased data: in the TEC using the undistorted part of the Spaenhauer pattern and in the MND using the unbiased plateau in an iterative manner. A difference is that in the MND, the dispersion $\sigma$ is generally assumed to be a constant for all $p$, whereas the TEC allows different $\sigma$s for different $p$ classes. The same basic idea lies behind the two, with TEC leaving the $p$ classes separated where their individual behavior is better inspected, whereas MND unites them into one ensemble, so that the collective and common bias behavior is more easily seen. With the large KLUN sample now available, it is also possible with MND to conveniently cut the sample into many subsamples according to type, inclination, etc, in order to investigate in detail their influence. In this manner, the inclination correction and the type dependence have been studied by Bottinelli et al (1995) and Theureau et al (1996), respectively. An indication of the close relationship between the MND and TEC are the similar values of $H_0$ that were derived from field samples by Theureau et al (1997) and Sandage (1994b), of $55 \pm 5$ and $48 \pm 5$ km/s/Mpc.

Finally, it should be noted that MND does not depend on any assumption on the space density distribution, as sometimes has been suspected. An advantage of both MND and TEC is that they are relatively empirical in essence: A minimum of assumptions is needed. More analytical methods, though needed in some situations, rely on ideal mathematical functions and assumed behavior of the selection function (Staveley-Smith & Davies 1989; Willick 1994) that often are not met.

Sandage (1994a) subtitled his paper "The Hubble Constant Does Not Increase Outward." He emphasized that inspection of the Spaenhauer diagrams of $M_{cal}$ vs $\log v$ for samples with different limiting magnitudes and different TF parameter $p$ values allows one to exclude any such significant systematic deviation from the linear Hubble law as repeatedly proposed by Segal (cf Segal & Nicoll 1996) with the quadratic redshift-distance law of his chronogeometric cosmology. Especially, addition of information in the form of several standard candles breaks the vicious circle, which allows one to interpret the data either in terms of the linear Hubble law plus selection bias or the quadratic law plus little selection bias (meaning very small dispersion of the luminosity function). The same thing can be said of the method of normalized distances: One would not expect the constant plateau for H, built by galaxies of widely different $p$ values and widely different redshifts, if $H_0$ actually does not exist. The method is also the same as adding a fainter sample of the same type of indicator and testing if the bias properties of the extended sample moves toward fainter magnitudes by the difference in the magnitude limits of the two lists (Sandage 1988b) (see Figure 1*b*).

## 4.4 *The Cluster Incompleteness Bias*

When one constructs the log H-vs-p diagram for members of a galaxy cluster using the direct TF relation, one generally observes log H increasing towards smaller values of the TF parameter $p$, i.e. towards the approaching magnitude limit. In one distant cluster, with a small number of measured galaxies, this trend may be difficult to recognize, and it is especially hard to see the expected unbiased plateau at large values of $p$, which gives the correct distance (and log H). Using data from several clusters and normalizing, instead of distance, the parameter $p$ to take into account the vicinity of the magnitude limit, the combined log H-vs-$p_{norm}$ diagram reveals better the bias behavior. This was shown by Bottinelli et al (1987) using 11 clusters (B-magnitude TF relation), and furthermore by Bottinelli et al (1988b) for clusters with infrared data. In both cases, the distance scale and the Hubble constant were derived in good agreement with the method of normalized distances for field galaxies. Kraan-Korteweg et al (1988) and Fouqué et al (1990) analyzed the distance of the Virgo cluster and clearly also gathered evidence of the influence of the cluster incompleteness bias.

A graphic way of determining the distance to a cluster (in fact, solving Kapteyn's formula of Problem I for an ensemble of clusters and deriving the value of $H_0$) was used in Bottinelli et al (1988b). Willick (1994) has described an iterative approach.

Recently, Sandage et al (1995) gave a very illuminating exposition of the cluster incompleteness bias, using field galaxies in redshift bins to imitate clusters at different distances. Again, using the principle that adding a fainter sample shows the presence of bias, if it exists, from their Spaenhauer approach, these authors show the presence and behavior of the bias in the imitated clusters. An important empirical result concerns how deep in magnitudes, beyond the brightest galaxy, one must penetrate a cluster so as to avoid the bias. This depends on the dispersion $\sigma$ of the TF relation and on the value of the TF parameter $p$. For small values of $p$ (slowest rotators, faintest galaxies), the required magnitude range may be as wide as 8 mag. It is clear that for distant clusters such deep samples are beyond reach, and one must always be cautious of the incompleteness bias because it always leads to too large a value of $H_0$ if uncorrected.

Another important point emphasized by Sandage et al (1995) is the artificial decrease in the apparent scatter of the TF relation, if derived from magnitude-limited cluster samples. This was also pointed out by Willick (1994), and it was implicit in the conclusion by Bottinelli et al (1988a) that this bias changes the slope of the TF relation. Sandage et al (1995) calculate the dependence of the apparent $\sigma$ on the magnitude range $\Delta m$ reached for a cluster. For instance, if the true dispersion is 0.6 mag, penetration by $\Delta m = 3$ mag gives $\sigma \approx 0.4$ mag, and $\Delta m = 6$ mag is needed to reveal the true $\sigma$. They conclude that the sometimes-mentioned small dispersions for the TF relations below $\sigma \approx 0.4$ mag are probably caused in this way (for other viewpoints, see Willick et al 1996, Bernstein et al 1994).

The problem of artificially decreased scatter is a dangerous one because it may lead to the conclusion that the selection bias, which is generally dependent on $\sigma^2$, is insignificant. In this manner, the bias itself produces an argument against its presence.

## 4.5 *Normalization: Other Applications*

Normalization is not restricted to kinematic distances of field galaxies or to TF parameters of cluster members. It is a useful approach to try when a range of values exists for some parameter on which the bias depends. The aim is to reveal the bias from the observed trend and to recognize the unbiased plateau. Teerikorpi (1986) inspected how selection affects the inner rings in galaxies as a distance indicator by constructing a normalized parameter from the quantity *k*, which indicates the dependence of the inner ring size on de Vaucouleurs's morphological galaxy class, as calibrated by Buta & de Vaucouleurs (1983). The analysis dropped Buta & de Vaucouleurs's $H_0 = 93 \pm 4$ to $75 \pm 3$ [or to 58 for the Sandage & Tammann (1975a) primary calibration], which is close to what happens with the TF indicator. There are two factors at work: 1. A relation between ring size and galaxy luminosity makes the former larger when the latter is larger. 2. At large distances, small rings cannot be measured. These are akin to a problem inherent in HII region size as a distance indicator, which is also clearly seen by normalization (Teerikorpi 1985). The possibilities of the inner ring method of Buta & de Vaucouleurs (1983) do not yet seem to be fully exploited.

## 5. THE INVERSE TULLY-FISHER RELATION AND THE SECOND KIND OF BIAS

de Vaucouleurs (1983) differentiated between what he called the Malmquist effect (the progressive truncation of the luminosity function at increasing distances in a magnitude-limited sample) and the Malmquist bias in the distances derived from such a sample. One may intuitively think that if there is a way of classifying galaxies into absolute magnitude bins, for example, by using de Vaucouleurs's luminosity index or the TF relation $M = ap + b$, the Malmquist effect, as defined above, will certainly cut away fainter galaxies from the sample, but then the parameter $p$ "glides" simultaneously. de Vaucouleurs argues that this compensates for the systematic distance dependent effect. However, the theory of the Malmquist bias of the second kind in direct TF distance modulus shows that such a compensation is not complete: Average $p$ glides to larger values, but still, no matter what the value of $p$ is, the corresponding distribution of true M is cut at a common $M_{lim}$ that depends only on the distance. One cannot escape this fact, which means that in the observed sample, the distance indicator relation $\langle M \rangle = ap + b$ is necessarily distorted and causes the second kind of Malmquist bias. However, at each distance the bias is smaller by the factor $\sigma^2/(\sigma^2 + \sigma_M^2)$, as compared with the simple truncation effect of the luminosity function with dispersion $\sigma_M$.

### 5.1 *The Ideal Case of the Inverse Relation*

In the ideal case, the TF parameter $p$ is not restricted by any such observational limit as $M_{lim}$. Hence, at any distance, the distribution of observed $p$ corresponding to a fixed M, and especially its average $\langle p \rangle_M$, is the same. Schechter (1980) thus realized that the inverse relation

$$p = a'\mathrm{M} + b' \tag{13}$$

has the useful property that it may be derived in an unbiased manner from magnitude-limited samples, if there is no selection according to $p$. He used this relation in a study of the local extragalactic velocity field, which requires that kinematic distances minimize the $p$ residuals (see also Aaronson et al 1982).

In what manner could one use the inverse relation as a concrete distance indicator? Assume that there is a cluster of galaxies at true distance modulus $\mu$. Derive the distance modulus for each galaxy $i$ that has $p_i$ measured, using the inverse relation as a "predictor" of M: $\mu_i = \mathrm{m} - (1/a')(p - b')$. Teerikorpi (1984) showed that the distance estimate $\langle\mu_i\rangle$ is unbiased, under the condition that there is no observational restriction to $p$. This result was supported by numerical simulations in Tully (1988).

Our ordinary way of thinking about distance indicators is closely linked to the direct relation: Measure $p$, determine from the relation what is the expected $\langle\mathrm{M}\rangle$, and calculate $\mu = \mathrm{m} - \langle\mathrm{M}\rangle$ for this one object. The use of the inverse relation is at first intuitively repugnant because one tends to look at the predictor of M, $(1/a')(p - b')$, similarly as one looks at the direct relation. The direct distance moduli are "individuals," whereas the inverse relation is a kind of collective distance indicator: Measure the average $p$ for the sample and calculate from $\langle\mathrm{m}\rangle$ and $\langle\mathrm{p}\rangle$ the distance modulus. Restriction to one galaxy, which is so natural with the direct relation, means restricting the value of $p$ to the one observed, which is not allowed with the inverse relation.

From a m $-$ $p$ diagram (Figure 3) showing a "calibrator" (nearby) cluster and a more distant cluster, one can easily explain the secret of the inverse relation. Let us put the calibrator sample at 10 pc, so that m $=$ M. The cluster to be measured is at the unknown distance modulus $\mu$ and is cut by the magnitude limit $\mathrm{m}_{\mathrm{lim}}$. Glide the calibrator cluster along the m axis by the amount of $\mu$. Then the inverse regression lines are superimposed. This means that the observed average of $p$ at m is $\langle p\rangle_{\mathrm{m}}$ for the second cluster, which is the same as for the calibrator cluster at $\mathrm{M} = \mathrm{m} - \mu$. From this, it follows that $\langle\mu\rangle_{\mathrm{m}} = \mathrm{m} - (\langle p\rangle_{\mathrm{m}} - b')/a'$ and, by averaging over all m, that

$$\mu_{est} = \langle\mathrm{m}\rangle - (\langle p\rangle - b')/a'. \tag{14}$$

The ($p$, M) data form a scattered bivariate distribution, and without further knowledge of the reason for the scatter, one has the freedom, within the limits of what is the application and what is known about the selection of $p$ and M parameters, to use either the direct or the inverse relation. Even if the scatter is not due to errors in $p$ or natural processes that shift $p$ at constant M, Figure 3 shows that one may use the inverse relation if the bivariate distributions of the calibrator and distant samples are the same. On the other hand, even if there is error in *p*, one may choose to use the direct relation if the application requires it (Bottinelli et al 1986; Lynden-Bell et al 1988). Naturally, another problem and source of biases is that the bivariate distribution may not fulfill the conditions of gaussianity, which are required in the derivation of the regression lines (Bicknell 1992; Ekholm & Teerikorpi 1997), or the calibrator and distant samples have, for example, different measurement accuracy in magnitude (Teerikorpi 1990; Fouqué et al 1990).

Finally, the inverse relation does not require that its calibrators form a volume-limited sample, which is necessary for the correct calibration of the direct relation. This is also illustrated by Figure 3 because the regression line of the calibrator sample is not changed if a portion $m >$ some $m_{lim}$ is cut away from it.

## 5.2 *Problems in the Use of the Inverse Relation*

In principle, the inverse TF relation seems to be a good solution to the bias of the second kind, though the scatter in the average distance modulus is larger than for the direct relation in the unbiased plateau ($1/\sigma^2_{\mathrm{inv}} = 1/\sigma^2_{\mathrm{dir}} - 1/\sigma^2_{\mathrm{M}}$, where $\sigma_{\mathrm{M}}$ is the dispersion of the general luminosity function). In practice, there are a few more serious problems.

It is essential that there should be no selection according to *p*, say, working against distant, very broad HI-line galaxies. Also noteworthy is that the calibrator slope for the inverse relation, derived from bright nearby galaxies, is not necessarily the correct slope for distant galaxies. This was shown in a concrete manner in the study of the Virgo cluster by Fouqué et al (1990). If the magnitude or diameter measurements are less accurate for the distant sample than for the calibrators, then the correct slope differs from the calibrator slope. If one ignores this problem, the inverse relation will give distances that are too small or a value of $H_0$ that is too high. Theoretically, Teerikorpi (1990) concluded that a solution is to use the slope obtained for the distant sample. However, this requires that the general luminosity function is symmetric around a mean value $M_0$.

The correct slope for the inverse relation is especially important because the aim is to extend measurements at once to large distances, i.e. to extreme values of m and *p*. A small error in the slope causes large errors at large distances. Of course, it is also important to have the direct relation correct, but in any case, caution is required with regard to the expected Malmquist bias of the second kind, whereas the very absence of the bias is motivation to use the inverse relation.

Hendry & Simmons (1994) made numerical experiments in order to see what is needed of the calibrators in order to produce the correct inverse slope. Adjusting their experiments to the Mathewson et al (1992a,b) data, they concluded that if the number of calibrators is less than 40, the uncertainty $\sigma_{slope} > |direct$

*slope* − *inverse slope*|. In other words, with a small calibrator sample, we may think that we use the inverse relation, whereas, in fact, we actually have determined and therefore use the direct one.

Recently, Hendry & Simmons (1994, 1995) formulated the inverse TF distance estimator within the framework and language of mathematical statistics, which confirmed the earlier conclusions on its unbiased nature as regards the Malmquist bias of the second kind. Further discussion on the statistical properties of the inverse TF relation as a distance indicator may be found in Triay et al (1994, 1996) [see also Appendix of Sandage et al (1995) for illuminating notes].

## 6. THE BIAS OF THE FIRST KIND IN DIRECT AND INVERSE TF DISTANCE MODULI

In their useful study, Landy & Szalay (1992) initiated the recent discussions on the general or "inhomogeneous" Malmquist bias and its correction, i.e. how to deal with the bias of the first kind in the case of a general space density distribution. True, they encountered some problems in the practical application of the formula derived in their paper. In fact, the basic problem was that they assumed in the beginning a distance indicator that has a zero bias of the second kind (which is unbiased at all distances) and then derived an expression for the bias of the first kind.

In Teerikorpi (1993) the problem of the general correction was discussed with explicit reference to direct and inverse TF relations, which served to clarify some of the points raised by Landy & Szalay (1992). It was noted that Malmquist's formula (Equation 3) was a general one, and in modern terms is best interpreted as applicable to the direct TF relation, for a constant value of $p$ (Malmquist's "star class") and requires that the limiting magnitude $(m_l) = \infty$. In that case, the distribution of distance moduli $N(\mu)$ refers to all moduli that could be observed without any cutoff in the magnitudes.

Interestingly, Feast (1972) had already given a formula quite similar to Equation 3, but now with a quite different meaning for the distribution of $N(\mu)$: In his formulation, $N(\mu)$ must be regarded as the distribution of the derived distance moduli of the galaxies in the observed sample. Inspection of Feast's (1972) derivation reveals the implicit assumption that the bias of the second kind is zero, and the end result was the same as that of Landy & Szalay (1992). As the inverse TF relation distance moduli have the second bias = 0, one can conclude that Feast's and Landy & Szalay's variant of Malmquist's formula applies to the inverse TF distance moduli.

Landy & Szalay's (1992) paper gave rise to a burst of independent discussions. Similar conclusions as in Teerikorpi's (1993) paper on the general correction and the inverse TF relation were given by Feast himself (1994) and Hendry & Simmons (1994) (see also Hudson 1994 and Strauss & Willick 1995).

To reiterate, Malmquist's formula (Equation 3), with $N(\mu)$ refering to all distance moduli ($m_{lim} = \infty$) in the considered sky direction, applies to the direct TF moduli. In the case of the inverse TF relation, $N(\mu)$ is the observed distribution. Hence, for making corrections of the first kind for the direct distance moduli, one needs information on the true space density distribution of galaxies, and the selection function (magnitude limit) does not enter the problem. Corrections for the inverse moduli depend directly on the distribution of apparent magnitudes (distance moduli) in the sample and, hence, on the selection function. In this sense, the biases of the first and second kind for the direct and inverse distance moduli have curious complementary properties.

One might be content with the above conclusions and try to use the inverse relation in situations where the bias of first kind appears: The correction does not need the knowledge of the true space distribution. In practice, the high inhomogeneity of the local galaxy universe requires corrections for differing directions, which divides galaxy samples into small subsamples where the detailed behavior $N(\mu)$ is difficult to derive.

The general Malmquist bias for the inverse distance moduli, even in the case of a homogeneous space distribution, can be quite complicated. Because the observed $N(\mu)$ usually first increases, reaches a maximum, and then decreases to zero, the bias is first negative (too small distances), then goes through zero, and at large derived distances is positive (i.e. the distances are too long). Something like this is also expected simply because the average bias of the whole sample should be zero.

An important special case of the first Malmquist bias of the direct distance moduli is that of a homogeneous space distribution, which shifts standard candles in a Hubble diagram by a constant amount in magnitude and leaves the expected slope of 0.2 intact. In this manner, Soneira (1979) argued for the local linearity of the Hubble law, in contrast to a nonlinear redshift-distance relationship espoused by Segal, where he argues against debilitating bias effects.

It has been suggested that the distribution of galaxies at least up to some finite distance range could be fractal in nature, with fractal dimension $D \approx 2$ (e.g. Di Nella et al 1996). In that case one might be willing to use, instead of the classical Malmquist correction (for $D = 3$ or homogeneity), Equation 5, which corresponds to the average density law proportional to $r^{D-3}$. However, though the classical formula applies to every direction in a homogeneous universe, the deformed ($D = 2$) formula is hardly useful in any single direction of a fractal universe because of the strong inhomogeneities. Note however that in this case the argument by Soneira (1979) would be equally valid.

## 7. MALMQUIST BIAS AND THE GREAT ATTRACTOR

A good example of the difficulties in establishing the distance scale within the local space is provided by the attempts to determine whether there is a differential infall velocity field around the Great Attractor at about $V_{cosm} \approx 4500$ km/s (Lynden-Bell et al 1988), or alternatively, a bulk flow of galaxies in this same direction of the Hydra-Centaurus supercluster, as earlier proposed by Tammann & Sandage (1985). In principle, with good distances to galaxies in this direction, extending beyond the putative Great Attractor, one would be able to construct a Hubble diagram, velocity vs distance, which in the former case should show both the foreground and backside infall deviations from the linear Hubble law. Such deviations have been detected in the direction of the Virgo cluster (Tully & Shaya 1984; Teerikorpi et al 1992). When aiming at the Great Attractor or the Hydra-Centaurus complex, one must look four to five times farther, hampered by the additional complication of the Zone of Avoidance at low galactic latitudes.

Though different studies agree that there is a flow towards the Great Attractor region, the question of the backside infall remains controversial, after some evidence supporting its detection (Dressler & Faber 1990a,b). Landy & Szalay (1992) suggested that the Malmquist bias (of the first kind) could cause an apparent backside infall signal behind a concentration of galaxies: In the velocity-vs-distance diagram, the distances in front of the concentration, where the density increases, will come out smaller, whereas those calculated behind the concentration, where the density decreases, will be larger on the average than predicted by the classical Malmquist formula.

Mathewson et al (1992a) concluded on the basis of a sample of 1332 southern spirals with TF I-mag distances (Mathewson et al 1992b) that the Hubble diagram does not show any backside infall into the Great Attractor. They thus questioned its very existence. This and Landy & Szalay's (1992) suggestion led Ekholm & Teerikorpi (1994) to investigate the outlook of the velocity-distance and distance-velocity diagrams in the presence of a mass concentration, utilizing a synthetic spherically symmetric supercluster and making comparisons with the data of Mathewson et al (1992a,b). The background method was given by a discussion of four alternative approaches (direct and inverse TF, velocity vs distance, and distance vs velocity) (Teerikorpi 1993). We concluded there that the most promising approach for the detection of the background infall is the analysis of the data in the sense "distance vs velocity." The other way via the classical Hubble diagram needs uncertain Malmquist corrections of the first kind. On the other hand, a *p*-class analysis (related to MND and TEC) showed that the available sample is not deep enough for the background infall to be detectable using the direct TF relation (Ekholm & Teerikorpi 1994). However, the inverse TF relation is capable of revealing the backside infall, if it exists. For this, it is crucial to know the relevant inverse slope and to resolve the other problems hampering the inverse relation (see Section 5.2).

Federspiel et al (1994) made an extensive analysis of the Mathewson et al (1992a,b) sample using the Spaenhauer diagrams with their "triple entry corrections" (TEC). They identified clearly the Malmquist bias of the second kind and concluded that the dispersion in the I-mag TF relation is relatively large ($\sigma = 0.4$–$0.7$ mag), depending on the TF parameter $p$. After making first order bias corrections, based on an underlying Hubble law, they concluded that there is no detectable backside infall, though there is in the foreground a bulk flow of about 500 km/s, apparently dying out before the putative Great Attractor is reached. A question is whether the first order correction, based on the assumed Hubble law, could lead to a null result or a vicious circle. A demonstration against such a vicious circle was made by Ekholm (1996) using the synthetic supercluster. The Hubble law and the TEC formalism for "fast rotators" should reveal the backside infall, if it exists.

Hudson (1994) calculated inhomogeneous Malmquist corrections to the direct $D_n$-$\sigma$ distance moduli in the Mark II sample of E and SO galaxies: He used the density field that he had derived from another, larger sample of galaxies with redshifts known. As a result, Hudson concluded that the apparent backside infall, visible in the Hubble diagram when the homogeneous Malmquist correction was made, was significantly reduced after the new corrections. This is in agreement with the conclusion by da Costa et al (1996).

A special problem is posed by galactic extinction. For instance, the center of the putative Great Attractor, at $l \approx 307°$, $b \approx 9°$, is situated behind the Zone of Avoidance, and the Local Group apex of Rubin et al's (1976) motion observations, at $l \approx 163°$, $b \approx -11°$, was also at low latitudes. Whatever the role of extinction is in those particular cases, in general its influence is subtle and needs special attention. One is not only concerned with good estimates of extinction in the directions of sample galaxies. The variable galactic extinction also creeps in indirect ways into the analyses of galaxy samples. An extinction term must be added into the formula of normalized distances in MND, because galaxies behind an enhanced extinction look fainter in the sky; hence, their "effective" limiting magnitude is brighter (the normalized distance is shorter). In the methods where one uses the limiting magnitude to calculate the importance of the bias of the second kind, one should take into account the extra factor due to extinction (also the inclination effect of internal extinction; Bottinelli et al 1995). Change in the effective limiting magnitude means that other things being equal, the galaxies in the direction of the enhanced extinction are influenced by the second bias already at smaller true distances than are their counterparts in more transparent parts of the sky. This effect is not allowed for just by making normal corrections to the photometric quantities in the TF relation.

Influence of the extinction is also felt when one uses infrared magnitudes where the individual extinction corrections are small. Such samples are usually based on B-mag or diameter-limited samples that have been influenced by variable B extinction, which changes the effective limiting magnitude in different directions of the sky. Depending on the correlation between B and the magnitude in question, this influence of extinction propagates as a kind of Gould's effect (Section 8.3), into the bias properties of the latter.

# 8. SOME RECENT DEVELOPMENTS

In this section I gather together some of the most recent results and ideas concerning the Malmquist biases of the first and second kind.

## 8.1 *Sosies Galaxies and Partial Incompleteness*

Witasse & Paturel (1997) have discussed the effect of partial incompleteness on the derivation of the Hubble constant, when Kapteyn's Equation 1 is used to make the bias (of the second kind) correction at each redshift. They inspected the completeness of their sample of sosies galaxies ["look-alike"; the distance indicator introduced by Paturel (1984)] and concluded that there is partial incompleteness starting at $B = 12.0$ mag, relative to the assumed $10^{0.6m}$ law. (Another possibility, which Witasse & Paturel (1997) mention for the first time in this kind of work, is that the apparent incompleteness may actually reflect a fractal distribution of galaxies.) Inserting the empirical selection probability into Kapteyn's formula, they could reduce the Hubble constant by about 15%, as compared with the assumption of a complete sample and obtained $H_o \approx 60$ from 181 sosies galaxies. In comparison, Sandage (1996b) applied Paturel's look-alike idea to the sosies galaxies of M31 and M101 and derived by the Spaenhauer method $H_o \approx 50$.

## 8.2 *About the Unbiased Plateau in the MND*

In the method of normalized distances (MND), the basic concept is the unbiased plateau, which corresponds to the separate unbiased parts of the Spaenhauer diagram in TEC. After its first utilization by Bottinelli et al (1986), several new developments have matured the understanding of the unbiased plateau. Bottinelli et al (1995) recognized that in the Equation 12 of MND, one should add terms describing how the effective magnitude limit changes due to internal extinction (inclination effect) and galactic extinction; it is also possible that the limiting magnitude depends on the TF parameter $p$ (Bottinelli et al 1988a). One must also include the type dependence in the method (Theureau et al 1997).

Numerical simulations by Ekholm (1996) supported the reliability of MND. An interesting result was that in certain kinds of studies (e.g. for determination of the slope of the TF relation), it is admissible to use galaxies somewhat beyond the unbiased plateau, which increases the sample. In fact, it is a handicap in MND, as well as in TEC, that the number of "useful" galaxies remains small in the unbiased regions, e.g. when one determines the value of $H_0$. One remedy is to use increasingly large samples. For instance, Bottinelli et al (1986) used the total number of galaxies of 395, and the size of the adopted plateau was 41. Theureau et al (1997) used a KLUN sample (with diameters) that was, after necessary restrictions, 4164, and the adopted plateau contained 478 galaxies. It seems that generally the visible empirical plateau contains about 10% of the total sample. Theureau et al (1997) confirmed this by an analytical calculation, where the cumulative error of $\langle \log H \rangle$ was seen to reach about 1% when the fraction of the sample is 0.1.

Because of the small number of plateau galaxies, one would be willing to use for the determination of $H_0$ the inverse relation, where, in principle, one could use the whole sample. However, the various problems with the inverse relation (Section 5) must be solved before it can be safely used as an independent distance indicator.

## 8.3 *Gould's Effect*

Gould (1993) pointed out a complication present when a sample of galaxies to be used, for example, for infrared I-mag TF relation, is constructed from a sample originally based on selection criteria other than those of I mag, e.g. apparent diameter. The Malmquist bias of the first kind in the distance moduli from the I-mag TF relation does not now generally depend on the squared dispersion $\sigma_I^2 = \langle \epsilon_I^2 \rangle$ of the I-mag TF relation nor on the squared dispersion $\sigma_D^2 = \langle \epsilon_D^2 \rangle$ of the diameter relation, but on the covariance $\langle \epsilon_I \epsilon_D \rangle$ between the corresponding logarithmic distance errors $\epsilon$. An interesting extreme case is when this covariance is zero, i.e. the deflections about the two TF relations are independent. Then there should be no Malmquist bias in the distance moduli from the I-mag TF relation. This is easy to understand: Though the original D-limited sample was selected "from the sky," the second set of I-mag measurements produces symmetrical residuals around the TF relation because of the assumed independence on D residuals and because in this case there is no I-mag limit (cf also Section 3.2 in Landy & Szalay 1992).

In practice, it may be difficult to find such pairs of observables that correlate with a common distance-independent parameter (e.g. TF parameter $p$), but have independent deflections (larger-than-average galaxies tend to be also more luminous than average). Also, it should be noted that the above argument is valid only if one could measure I for all the galaxies first taken from the

D-limited sample, i.e. if the I limit was really $=\infty$. In fact, though to measure the relevant covariance is one approach, Gould (1993) also sees the described problem as supporting the use of the "good old" B-band TF-relation. [For further discussions of Gould's effect, see Willick (1994) and Strauss & Willick (1995).]

In the KLUN project (e.g. Paturel et al 1994), the sample of 5174 spirals has been selected on the basis of apparent size $D_{25}$ (in B), and in the analysis of the diameter TF distance moduli, Gould's effect should not appear. However, in such cases a somewhat related problem is that the measured diameters contain measurement error, and when one constructs a diameter-limited sample from measured galaxies, there is a Malmquist effect due to the dispersion $\sigma_\epsilon$ in the measurement error: The sample contains an excess of overestimated apparent diameters. Ekholm & Teerikorpi (1997) pointed out that this may have a significant influence on the results, especially on those from the inverse TF relation where all galaxies (in view of the method's supposedly unbiased nature) are used. Assume now that the apparent sizes of such a diameter-limited sample are once more measured. Because the first and second measurement errors are independent, their influence vanishes from Gould's covariance, and now the second sample has correct measured apparent sizes, on the average. In practice, such a remeasurement of large samples is out of the question, and one has to be aware of the problem.

## 8.4 *Simulation Approach*

As noted above, Scott (1957) was a pioneer in making numerical simulations, at that time using tables of random numbers in the study of how selection effects influence the distribution function of standard candles and hence the inferred distances. More recently, computers have often been used in this manner, either to show the existence of or to illuminate some effect, for testing a correction method, or for using a Monte Carlo procedure as an integral part of the method. An advantage of such experiments, found in many of the mentioned references, is that one can construct a synthetic galaxy universe where the true distances and values of other relevant parameters are known and where imposition of the known selection effects simulates what the astronomer sees "in the sky."

Simulations have been used to show what happens in the classical Hubble diagram when the line of sight traverses a concentration of galaxies (e.g. Landy & Szalay 1992; Ekholm & Teerikorpi 1994; Section 6.5.2 in Strauss & Willick 1995). However, the most extensive applications have concerned the methods and Malmquist corrections needed when one tries to map the peculiar velocity field, especially using the POTENT algorithm (Dekel et al 1993). Indeed, Dekel et al (1993) made the classical (homogeneous) Malmquist correction to distances inferred from the direct relation, whereas Newsam et al (1995) suggest the use of the inverse relation in their iterative variant of POTENT. Strauss & Willick (1995) describe their "Method II$^+$," which incorporates random peculiar velocities in a maximum likelihood method, applicable either to the direct or, preferably, to the inverse distance indicator. Nusser & Davis (1995) use the inverse relation in their method for deriving a smoothed estimate of the peculiar velocity field, and they support the method using simulations on a synthetic data set. Freudling et al (1995) gives several examples of how the peculiar velocity field is deformed because of unattended bias, including Gould's effect.

Clearly, theoretical understanding of how the Malmquist biases affect studies of the peculiar velocities is rapidly advancing. On the balance, it is worthwhile to be cautious of how successfully one can apply the actual inverse relation, which is favored in such theoretical studies, to the real data.

## 9. CONCLUDING REMARKS

This review concentrates on the basic behavior of the selection biases affecting classical photometric (or diametric) distance indicators, which will always have a central role when large numbers of galaxies are analyzed (morphological luminosity classification, TF, and Faber-Jackson relations). Large samples have the special advantage that one may investigate in detail their properties, completeness, and composition and recognize the relevant selection effects, and hence put the distance indicator on a safe basis.

Along with such major indicators, which are practical for large fundamental samples, there are more specialized methods that complement them, especially within the Local Supercluster, and may provide more accurate distances for a number of individual galaxies. If the distance scale that a "new generation" indicator erects significantly deviates from that obtained from the classical, carefully scrutinized methods, then the first task is to ask whether the new methods have some source of bias. As historically evidenced, distance indicators need sufficient time to mature, and in the long run all the methods, when better understood, should result in consistent distances. Such problems have been suspected in recent years, e.g. in the methods of planetary nebulae luminosity function (PNLF) and of surface brightness fluctuations, which have given systematically smaller distances than Cepheids in galaxies where comparison has been possible (Gouguenheim et al 1996; Tammann et al 1996). Generally, selection effects tend to bias true distances to incorrect smaller values. This rule seems to apply also to the method of PNLF when one attempts to extend it to galaxies in or beyond the Virgo cluster, though it may work very well for close galaxies (Bottinelli et al 1991).

The Hubble Space Telescope has given remarkable building blocks for the distance ladder, especially by the measurement of many new Cepheid distances for galaxies that can be used for calibrating secondary distance indicators (e.g.

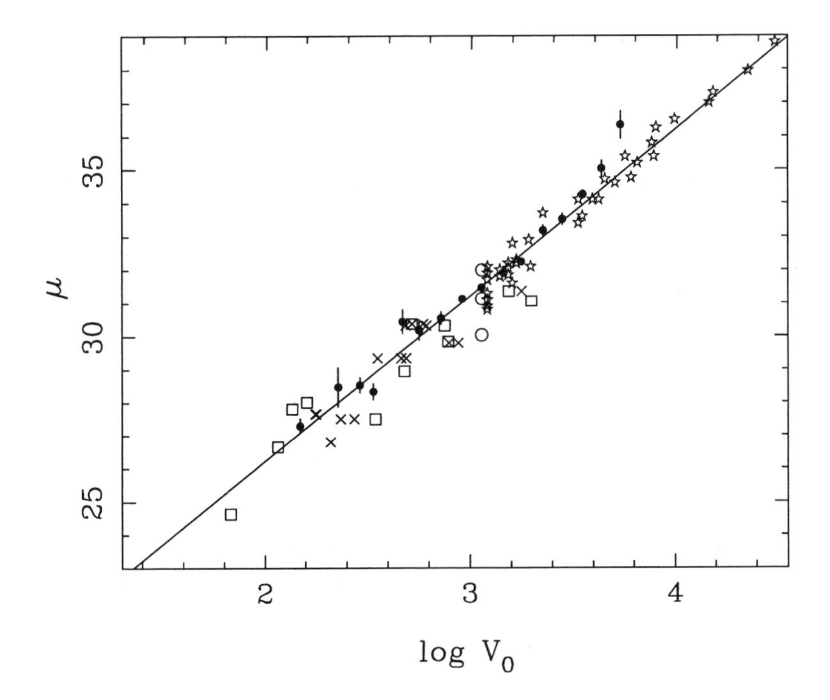


*Figure 4* Three distance indicators and the Hubble law are illustrated: SN Ia (*stars*), the B-magnitude TF distance moduli $\mu$ from the unbiased plateau of the normalized distance method (*dots*: averages of several galaxies), and Cepheids used for calibrating the supernovae method and the TF relation (*crosses*: individual Cepheid-galaxies, *squares*: groups containing a Cepheid-galaxy). Supernovae data are from Tammann et al (1996) and TF data from Theureau et al (1997). Radial velocity $V_o$ is corrected for the Virgo-centric velocity field. The Hubble line corresponds to $H_o = 56$ km/s/Mpc.

Freedman 1996; Tammann et al 1996). This allows different distance indicators to be based on roughly similar sets of calibrating galaxies and, without intermediaries other than the Hubble law, allows comparison of the resulting distance scales. Differences should then directly reflect problems in one or both of the methods. Presently, there is one pair of distance indicators that can be usefully compared in this way: supernovae of type Ia (SN Ia) (Tammann et al 1996) and the TF relation for $D_{25}$ and B, as applied by Theureau et al (1997) on the KLUN-sample. Figure 4 shows the data, together with single Cepheid galaxies, on the $\mu$-vs-log $V_o$ diagram. The KLUN galaxies have been taken from the unbiased plateau of MND; each symbol refers generally to an average of a few tens of galaxies. The straight line corresponds to the Hubble law with $H_o = 56$ km/s/Mpc, which describes the distribution of individual Cepheid galaxies, groups containing a Cepheid galaxy, TF galaxies, and the SN Ia galaxies in the range of $25 < \mu < 39$. Rather than claiming that this diagram gives any final distance scale, I show it as an encouraging sign of how different distance indicators begin to give consistent results when special attention is paid to the problem of selection bias. It should be noted that SN Ia, with their small dispersion $\sigma$, high luminosity, and detection requirements that are not solely dictated by the maximum luminosity magnitude, are expected to be little affected by the Malmquist bias of the second kind in the distance range shown.

It has been said that "cosmology is a study of selection effects" (Y Baryshev, private communication). This view of what cosmologists are doing may seem a little too dull; nevertheless, there is a deep truth in it, too. It should be an intimate part of the methodology to try to recognize different kinds of selection effects when one attempts to build an unbiased picture of the universe (leaving aside the profound question whether such an enterprise is ultimately at all possible). Often in the forefront of scientific discoveries, one rather likes to ignore selection effects because in order to positively identify these, one should have enough collected data, and this rarely happens in the avant-garde phase. Also, selection effects sometimes produce apparent phenomena that, if true, would certainly be more interesting than the mechanisms of selection and bias that naturally attract less attention.

Another point worthy of emphasis is that the selection biases do not vanish anywhere, even though the astronomical data are accumulated beyond old magnitude limits. Problems are shifted towards larger distances and fainter magnitudes, and new generations of astronomers have to learn how the biases reappear.

## ACKNOWLEDGMENTS

I am very much indebted to my French colleagues and friends, L Bottinelli, L Gouguenheim, G Paturel, and G Theureau. Collaboration with them always has been pleasant and we have learned much about the Malmquist bias by inspecting together their large and wonderful galaxy samples. I thank my young Finnish collaborators, T Ekholm and M Hanski, for help with the figures included in this review. I am very grateful to A Sandage for several useful suggestions. My teacher in extragalactic astronomy was Toivo Jaakkola (1941–1995). I wish to dedicate this article to his memory.



*Literature Cited*

Aaronson M, Huchra J, Mould J, Schechter PL, Tully RB. 1982. *Ap. J.* 258:64–76
Aaronson M, Mould J, Huchra J, Sullivan WT III, Schommer RA, Bothun GD. 1980. *Ap. J.* 239:12–37
Behr A. 1951. *Astron. Nach.* 279:97–104
Bernstein GM, Guhathakurta P, Raychaudhury S, Giovanelli R, Haynes MP, et al. 1994. *Astron. J.* 107:1962–76
Bicknell GV. 1992. *Ap. J.* 399:1–9

Bottinelli L, Chamaraux P, Gerard E, Gouguenheim L, Heidmann J, et al. 1971. *Astron. Astrophys.* 12:264–70
Bottinelli L, Fouqué P, Gouguenheim L, Paturel G, Teerikorpi P. 1987. *Astron. Astrophys.* 181:1–13
Bottinelli L, Gouguenheim L, Paturel G, Teerikorpi P. 1986. *Astron. Astrophys.* 156:157–71
Bottinelli L, Gouguenheim L, Paturel G, Teerikorpi P. 1988a. *Ap. J.* 328:4–22
Bottinelli L, Gouguenheim L, Paturel G, Teerikorpi P. 1991. *Astron. Astrophys.* 252:550–56
Bottinelli L, Gouguenheim L, Paturel G, Teerikorpi P. 1995. *Astron. Astrophys.* 296:64–72
Bottinelli L, Gouguenheim L, Teerikorpi P. 1988b. *Astron. Astrophys.* 196:17–25
Buta R, de Vaucouleurs G. 1983. *Ap. J. Suppl.* 51:149–70
da Costa LN, Freudling W, Wegner Y, Giovanelli R, Haynes MP, Salzer JJ. 1996. *Ap. J.* 468:L5–8
Dekel A, Bertschinger E, Yahil A, Strauss M, Davis M, Huchra J. 1993. *Ap. J.* 412:1–21
de Vaucouleurs G. 1972. In *External Galaxies and Quasi Stellar Objects*, ed. DS Evans, pp. 353–58. Dordrecht: Reidel
de Vaucouleurs G. 1979. *Ap. J.* 227:729–55
de Vaucouleurs G. 1983. *MNRAS* 202:367–78
de Vaucouleurs G, Peters WL. 1986. *Ap. J.* 303:19–24
Di Nella H, Montuori M, Paturel G, Pietronero L, Sylos Labini F. 1996. *Astron. Astrophys.* 308:L33–36
Dressler A, Faber SM. 1990a. *Ap. J.* 354:13–17
Dressler A, Faber SM. 1990b. *Ap. J.* 354:L45–48
Ekholm T. 1996. *Astron. Astrophys.* 308:7–16
Ekholm T, Teerikorpi P. 1994. *Astron. Astrophys.* 284:369–85
Ekholm T, Teerikorpi P. 1997. *Astron. Astrophys.* In press
Fall SM, Jones BGT. 1976. *Nature* 262:457–60
Feast MW. 1972. *Vistas Astron.* 13:207–21
Feast MW. 1994. *MNRAS* 266:255–62
Federspiel M, Sandage A, Tammann GA. 1994. *Ap. J.* 430:29–52
Fouqué P, Bottinelli L, Gouguenheim L, Paturel G. 1990. *Ap. J.* 349:1–21
Freedman WL. 1996. In *Proceedings of Science with the Hubble Space Telescope-II*, ed. P Benvenuti, FD Macchetto, EJ Schreier, p. 3–8. Baltimore, MD: Space Telescope Sci. Inst.
Freudling W, da Costa LN, Wegner G, Giovanelli R, Haynes MP, Salzer JJ. 1995. *Astron. J.* 110:920–33
Giraud E. 1986. *Ap. J.* 301:7–16
Gouguenheim L. 1969. *Astron. Astrophys.* 3:281–307
Gouguenheim L, Bottinelli L, Theureau G, Paturel G, Teerikorpi P. 1996. *Rev. Mod. Astron.* 9:127–37
Gould A. 1993. *Ap. J.* 412:L55–58
Han M. 1992. *Ap. J.* 395:75–90
Harrison E. 1993. *Ap. J.* 403:28–31
Hawkins GS. 1962. *Nature* 225:563–64
Hendry MA, Simmons JFL. 1994. *Ap. J.* 435:515–27
Hendry MA, Simmons JFL. 1995. *Vistas Astron.* 39:297–314
Hoskin M. 1982. *Stellar Astronomy.* Cambridge: Sci. Hist. 195 pp.
Hubble E. 1936. *Ap. J.* 84:158–79
Hudson MJ. 1994. *MNRAS* 266:468–74
Jacoby GH, Branch D, Ciardullo R, Davies RL, Harris WE, et al. 1992. *Publ. Astron. Soc. Pac.* 104:599–662
James PA, Joseph RD, Collins CA. 1991. *MNRAS* 248:444–50
Kapteyn JC. 1914. *Ap. J.* 15:339–413
Kraan-Korteweg RC, Cameron LM, Tammann GA. 1988. *Ap. J.* 331:620–40
Landy SD, Szalay A. 1992. *Ap. J.* 391:494–501
Lundmark K. 1924. *MNRAS* 84:747–57
Lundmark K. 1925. *MNRAS* 85:865–94
Lundmark K. 1946. *Lund Medd. Ser. I* 163:17–27
Lynden-Bell D. 1992. In *Statistical Challenges in Modern Astronomy*, ed. ED Feigelson, GJ Babu, pp. 201–16. New York: Springer-Verlag
Lynden-Bell D, Faber SM, Burstein D, Davies RL, Dressler A, et al. 1988. *Ap. J.* 326:19–49
Malmquist KG. 1920. *Lund Medd. Ser. II* 22:1–39
Malmquist KG. 1922. *Lund Medd. Ser. I* 100:1–52
Mathewson DS, Ford VL, Buchhorn M. 1992a. *Ap. J.* 389:L5–L8
Mathewson DS, Ford VL, Buchhorn M. 1992b. *Ap. J. Suppl.* 81:413–659
McVittie GC. 1974. *Q.J.R. Astron. Soc.* 15: 246–63
Newsam A, Simmons JFL, Hendry MA. 1995. *Astron. Astrophys.* 294:627–38
Nusser A, Davis M. 1995. *MNRAS* 276:1391–401
Paturel G. 1984. *Ap. J.* 282:382–86
Paturel G, Bottinelli L, Di Nella H, Fouqué P, Gouguenheim L, Teerikorpi P. 1994. *Astron. Astrophys.* 289:711–14
Ptolemy. ca 150. *Almagest.* (1984. Transl. GJ Toomer). London: Duckworth. 419 pp.
Robertson HP. 1955. *Publ. Astron. Soc. Pac.* 67:82–98
Rowan-Robinson M. 1985. *The Cosmological Distance Ladder.* New York: Freeman

Rubin VC, Thonnard N, Ford WK Jr, Roberts MS. 1976. *Astron. J.* 81:719–37
Sandage A. 1958. *Ap. J.* 127:513–26
Sandage A. 1972. *QJRAS* 13:282–96
Sandage A. 1988a. *Annu. Rev. Astron. Astrophys.* 26:561–630
Sandage A. 1988b. *Ap. J.* 331:583–604
Sandage A. 1988c. *Ap. J.* 331:605–19
Sandage A. 1993. *Ap. J.* 402:3–14
Sandage A. 1994a. *Ap. J.* 430:1–12
Sandage A. 1994b. *Ap. J.* 430:13–28
Sandage A. 1995. In *The Deep Universe*, ed. B Binggeli, R Buser, pp. 1–232. Berlin: Springer-Verlag
Sandage A. 1996a. *Astron. J.* 111:1–17
Sandage A. 1996b. *Astron. J.* 111:18–28
Sandage A, Tammann GA. 1975a. *Ap. J.* 196:313–28
Sandage A, Tammann GA. 1975b. *Ap. J.* 197:265–80 (Appendix)
Sandage A, Tammann GA. 1984. *Nature* 307:326–29
Sandage A, Tammann GA, Federspiel M. 1995. *Ap. J.* 452:1–15
Schechter PL. 1980. *Astron. J.* 85:801–11
Scott EL. 1957. *Astron. J.* 62:248–65
Segal IE, Nicoll JF. 1996. *Ap. J.* 465:578–94
Soneira RM. 1979. *Ap. J.* 230:L63–65
Spaenhauer AM. 1978. *Astron. Astrophys.* 65:313–21
Staveley-Smith L, Davies RD. 1989. *MNRAS* 241:787–826
Strauss MA, Willick JA. 1995. *Phys. Rep.* 261:271–431
Tammann GA, Labhardt L, Federspiel M, Sandage A, Saha A, et al. 1996. In *Proceedings of Science with the Hubble Space Telescope-II*, ed. P Benvenuti, FD Macchetto, EJ Schreier, p. 9–19. Baltimore, MD: Space Telescope Sci. Inst.
Tamman GA, Sandage A. 1985. *Ap. J.* 294:81–95
Tammann GA, Sandage A, Yahil A. 1980. In *Physical Cosmology*, ed. R Balian, J Adouze, DN Schramm, pp. 53–125. Amsterdam: North-Holland
Teerikorpi P. 1975a. *Observatory* 95:105–7
Teerikorpi P. 1975b. *Astron. Astrophys.* 45:117–24
Teerikorpi P. 1976. *Astron. Astrophys.* 50:455–58
Teerikorpi P. 1981. *Astron. Astrophys.* 98:309–15
Teerikorpi P. 1984. *Astron. Astrophys.* 141:407–10
Teerikorpi P. 1985. *Astron. Astrophys.* 143:469–74
Teerikorpi P. 1986. *Vestnik Leningrad State Univ.* 2:121–23
Teerikorpi P. 1987. *Astron. Astrophys.* 173:39–42
Teerikorpi P. 1990. *Astron. Astrophys.* 234:1–4
Teerikorpi P. 1993. *Astron. Astrophys.* 280:443–50
Teerikorpi P. 1995. *Astrophys. Lett. Comm.* 31:263–68
Teerikorpi P, Bottinelli L, Gouguenheim L, Paturel G. 1992. *Astron. Astrophys.* 260:17–32
Teerikorpi P, Jaakkola T. 1977. *Astron. Astrophys.* 59:L33–36
Theureau G, Hanski M, Ekholm T, Bottinelli L, Gouguenheim L, et al. 1997. *Astron. Astrophys.* In press
Theureau G, Hanski M, Teerikorpi P, Bottinelli L, Ekholm T, et al. 1996. *Astron. Astrophys.* In press
Triay R, Lachièze-Rey M, Rauzy S. 1994. *Astron. Astrophys.* 289:19–34
Triay R, Rauzy S, Lachièze-Rey M. 1996. *Astron. Astrophys.* 309:1–8
Tully RB. 1988. *Nature* 334:209–12
Tully RB, Fisher JR. 1977. *Astron. Astrophys.* 54:661–73
Tully RB, Shaya EJ. 1984. *Ap. J.* 281:31–55
van den Bergh S. 1960a. *Ap. J.* 131:215–23
van den Bergh S. 1960b. *Ap. J.* 131:558–73
Willick J. 1994. *Ap. J. Suppl.* 92:1–31
Willick J, Courteau S, Faber SM, Burstein D, Dekel A, Kolatt T. 1996. *Ap. J.* 457:460–89
Witasse O, Paturel G. 1997. *Astron. Astrophys.* In press

*Annu. Rev. Astron. Astrophys. 1997. 35:137–77*


# MODEL ATMOSPHERES OF VERY LOW MASS STARS AND BROWN DWARFS

*France Allard,*[1] *Peter H. Hauschildt,*[2] *David R. Alexander,*[1] *and Sumner Starrfield*[3]

[1]Department of Physics, Wichita State University, Wichita, Kansas 67260-0032; e-mail: allard@eureka.physics.twsu.edu; dra@twsuvm.uc.twsu.edu

[2]Department of Physics and Astronomy, University of Georgia, Athens, Georgia 30602-2451; e-mail: yeti@hal.physast.uga.edu

[3]Department of Physics and Astronomy, Arizona State University, Tempe, Arizona 85287-1504; e-mail: sumner.starrfield@asu.edu

KEY WORDS: stellar atmospheres, stellar fundamental parameters, low mass stars, brown dwarfs

ABSTRACT

As progressively cooler stellar and substellar objects are discovered, the presence first of molecules and then of condensed particulates greatly complicates the understanding of their physical properties. Accurate model atmospheres that include these processes are the key to establishing their atmospheric parameters. They play a crucial role in determining structural characteristics by setting the surface conditions of model interiors and providing transformations to the various observational planes. They can reveal the spectroscopic properties of brown dwarfs and help establish their detectability. In this paper, we review the current state-of-the-art theory and modeling of the atmospheres of very low mass stars, including the coolest known M dwarfs, M subdwarfs, and brown dwarfs, i.e. $T_{eff} \leq 4{,}000$ K and $-4.0 \leq [M/H] \leq +0.0$. We discuss ongoing efforts to incorporate molecular and grain opacities in cool stellar spectra, as well as the latest progress in (*a*) deriving the effective temperature scale of M dwarfs, (*b*) reproducing the lower main sequences of metal-poor subdwarfs in the halo and globular clusters, and (*c*) results of the models related to the search for brown dwarfs.



0066-4146/97/0915-0137$08.00

# 1. INTRODUCTION

The crop of extremely cool stars and substellar objects has been meager until very recently, when marked improvements in detection ability have finally started to yield a rich harvest. Stars with masses as low as $0.1M_{\odot}$ and white dwarfs as cool as 5000 K (Paresce et al 1995, Richer et al 1995, Cool et al 1996, Renzini et al 1996) have been resolved in nearby globular clusters using the Wide Field/Planetary Camera 2 on board the Hubble Space Telescope (HST). Charge-coupled device (CCD) astrometry has unveiled uncharted M dwarfs in the immediate vicinity of our Solar System (Henry et al 1995, Stone et al 1996, Tinney 1996). Growing numbers of halo carbon dwarfs have been discovered by deep multicolor CCD surveys (Heber et al 1993, Warren et al 1993, Deutsch 1994, Liebert et al 1994). Young deuterium-burning brown dwarfs have been identified in nearby open clusters (Rebolo et al 1995, Basri et al 1996, Rebolo et al 1996) and in the solar neighborhood (Thackrah et al 1997) using the Keck telescope. Cryogenic coronographic imaging and astrometric surveys are revealing cool, evolved brown dwarfs hiding in the solar neighborhood (Nakajima et al 1995, Oppenheimer et al 1995, Mazeh et al 1996, Williams et al 1997), and high signal/noise radial velocity and astrometric surveys are now sensitive to massive planets around nearby Sun-like stars (Mayor & Queloz 1995, Butler & Marcy 1996, Marcy & Butler 1996a,b, Butler et al 1997, Marcy et al 1997). These exciting developments, and those of the MACHO, EROS, and OGLE microlensing surveys (Aubourg 1995, Alcock et al 1996), will soon enable a reconstruction of the faint end of the galactic initial mass function and the determination of the baryonic fraction of the missing mass (Chabrier et al 1996b, Flynn et al 1996, Graff & Freese 1996). Very low mass (VLM) stars and brown dwarfs are probably the most numerous objects in the galaxy (Gould et al 1996, Méra et al 1996a). Nevertheless, until recently, little about their atmospheres, evolution, or spectral characteristics was clearly understood. The presence of both a wide variety of molecular absorbers (each with hundreds of thousands to millions of spectral lines) and numerous condensates greatly complicates accurate modeling of these cool stellar atmospheres. The extension of the convection zone to the outermost photospheric layers means that evolutionary models depend critically on accurate handling of the surface boundary. Recently, theoretical calculations of important molecular absorbers such as $H_2O$, TiO, CN, and CO, as well as grain opacities, have made it possible to generate greatly improved models of cool stellar and brown dwarf atmospheres, their high resolution spectra, and their evolution. In view of the latest progress both on the observational and theoretical fronts, it seems timely to summarize here our present understanding of the atmospheres and spectroscopic properties of cool VLM stars and substellar brown dwarfs. Previous reviews of these topics can be found in Liebert & Probst (1987), Stevenson (1991), Bessell & Stringfellow (1993), Burrows & Liebert (1993), and Gustafsson & Jørgensen (1994).

## 2. GENERAL SPECTROSCOPIC PROPERTIES OF COOL DWARFS

A VLM star generally refers to a main sequence star with a spectral type ranging from mid K to late M and a mass from about 0.6 $M_\odot$ to the hydrogen-burning minimum mass (0.075–0.085 $M_\odot$, depending on metallicity). Such stars span a wide range of populations, from the youngest metal-rich M dwarfs in open clusters such as the Pleiades (Simons & Becklin 1992, Hambly et al 1993, Williams et al 1996, Zapatero Osorio et al 1997), the Hyades (Leggett & Hawkins 1989, Reid 1993, Bryja et al 1994, Leggett et al 1994), $\rho$ Ophiuchus (Cameron et al 1993), $\alpha$ Persei (Zapatero Osorio et al 1996), and the galactic disk (Gliese & Jahreiss 1991), to the several billion–year-old metal-poor subdwarfs of the galactic halo (Green et al 1991, Monet et al 1992, Green & Margon 1994) and globular clusters (Paresce et al 1995, Richer et al 1995, Cool et al 1996, Renzini et al 1996). Even within the solar neighborhood, M dwarfs do not form a homogeneous sample of unique age and metallicity, but rather they span up to $\approx 10^3$ years in age and $\pm 0.5$–1.0 dex around the solar value in metallicity (Burrows & Liebert 1993).

In VLM star and brown dwarf atmospheres, most of the hydrogen is locked in $H_2$ and most of the carbon in CO, with excess oxygen bound in molecules such as TiO, VO, and $H_2O$. The energy distribution of a typical late-type M dwarf is entirely governed by the absorption of TiO and VO in the optical, and $H_2O$ in the infrared, leaving no window of true continuum in the emergent spectrum. With their extremely low intrinsic faintness ($10^{-2}$–$10^{-5}$ $L_\odot$), in particular in the $V$ bandpass, the painstaking spectral classification of the nearby stars aiming toward a complete census of the luminosity function is still in progress (Reid 1994, Kirkpatrick & Beichman 1995, Liebert et al 1995, Reid et al 1995a). Fortunately, the groundwork necessary to construct an effective temperature scale at the lower end of the main sequence has already been laid by Boeshaar (1976) in the visual (0.44–0.68 $\mu$m), with (*a*) the first identification of CaOH bands at 0.54–0.556 $\mu$m in dwarfs later than about M3.5 (these bands are excellent temperature indicators and good discriminants between field M dwarfs and background red giant stars); (*b*) the first report of a saturation of the visual TiO band strengths in M dwarfs later than M5; and (*c*) the introduction of the VO to TiO band strength index now being used to classify M dwarfs and substellar candidates later than M5 (Henry et al 1994, Kirkpatrick et al 1995, Martín et al 1996). Boeshaar's classifications soon were extended beyond even the limits of the classical Morgan & Keenan spectral sequence, i.e. to types M9.5–>M10, by Kirkpatrick et al (1995) in the optical to near-infrared regime (0.65–1.5 $\mu$m) and by Davidge & Boeshaar (1993), Jones et al (1994), and Leggett et al (1996) in the near infrared (1.1–2.5 $\mu$m).

Figure 1 summarizes a typical near-infrared spectral sequence of M dwarfs to brown dwarfs. The near-infrared water vapor bands become slowly stronger with the spectral type of M dwarfs. The CO overtones near 2.3 $\mu$m (and 4.5 $\mu$m,

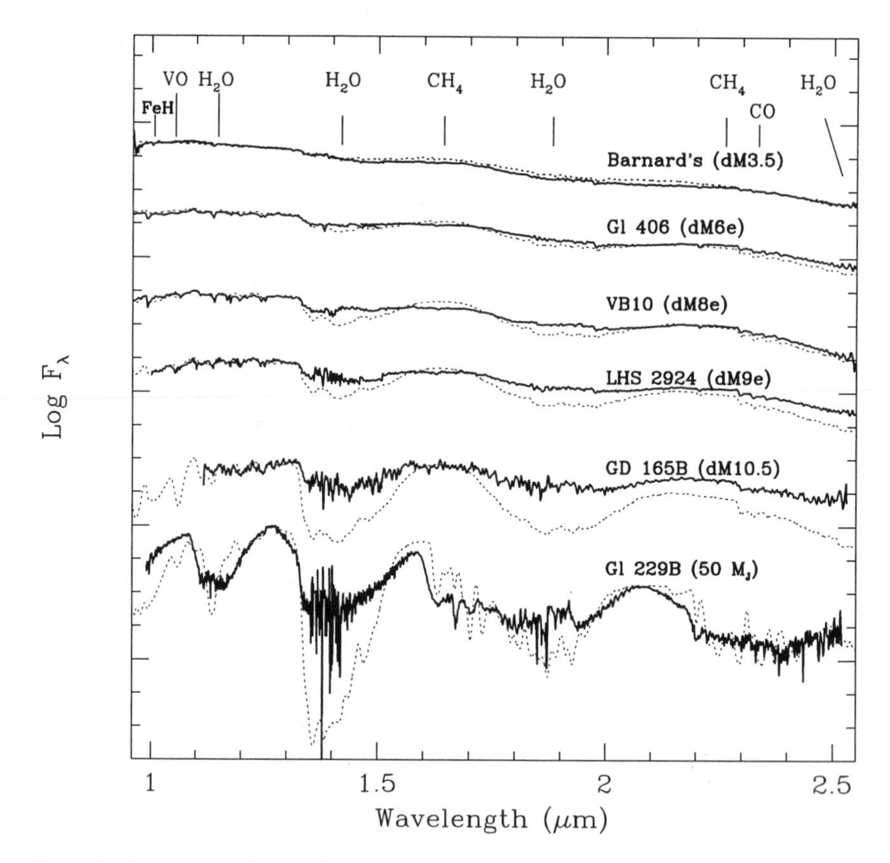


*Figure 1* A near-infrared spectral sequence of M dwarfs to brown dwarfs. The observed spectral distributions (*full lines*) were obtained at UKIRT for the M dwarfs by Jones et al (1994), and for the brown dwarf Gl229B by Geballe et al (1996). A comparison to OS models with, from top to bottom, $T_{eff}$ = 3400, 3000, 2700, 2600, 2000, and 1000 K (F Allard & PH Hauschildt, in preparation) (*dotted lines*) reveals a growing overestimation of water vapor band strengths with decreasing mass. The peculiar optical spectrum of GD 165B forces an arbitrary choice of the model parameters (here set to those of a star at the hydrogen-burning limit) for this object.

not shown in Figure 1) are still apparent, although much weaker than in late-type giant stars due to the stronger $H_2O$ "continuum" in the dwarfs. At the hydrogen-burning limit, i.e. at a spectral type of M10.5 and a $T_{eff}$ of about 2000 K (Baraffe et al 1995), the peculiar spectral distribution of GD 165B (Zuckerman & Becklin 1992, Kirkpatrick et al 1993a) suggests that all signs of the TiO bands disappear from the optical spectral distribution, leaving only atomic lines and perhaps VO bands (Davis 1994, Kirkpatrick et al 1995), CaH, CaOH, and/or FeH bands. As the effective temperature drops into the brown dwarf regime, methane ($CH_4$) features begin to appear (Tsuji et al 1995, Allard et al 1996, Marley et al 1996), and corundum ($Al_2O_3$), perovskite ($CaTiO_3$), iron, enstatite ($MgSiO_3$), and forsterite ($Mg_2SiO_4$) clouds may form, enhancing the carbon:oxygen abundance ratio and profoundly modifying the thermal structure and opacity of the photosphere (Sharp & Huebner 1990, Fegley & Lodders 1996).

The chemistry of cool dwarf atmospheres is, therefore, a complex nonlinear problem requiring a detailed knowledge of the concentration of atoms and molecules, which prevents a straightforward derivation of quantities such as excitation temperatures and metallicities from line ratios, as is possible for hotter stars. The most reliable way to estimate effective temperatures and metallicities of VLM stars and to identify substellar brown dwarfs is by a direct comparison of observed and model spectra.

## 3. A BRIEF HISTORY OF THE MODELS

Advances in atmospheric modeling of cool stars have been slowed by the twin bottlenecks of (*a*) incomplete molecular opacity data bases and (*b*) the inability to handle convection rigorously. Once these problems are addressed reasonably well, we still face other challenges: incorporating the effects of photospheric grain formation, chromospheres, magnetic fields, departures from local thermodynamic equilibrium, spatial variations in atmospheric structure due to starspots, cloud formation, and eventually weather patterns. Model atmospheres incorporating such processes have only become possible within the past two decades with the work of Mould (1975, 1976), Allard (1990), Kui (1991), Brett & Plez (1993), Allard & Hauschildt (1995b), Brett (1995a,b), and Tsuji et al (1996a) for M dwarfs; Saumon et al (1994) for zero-metallicity subdwarfs; and Tsuji et al (1995, 1996b), Allard et al (1996), and Marley et al (1996) for substellar brown dwarfs.

Mullan & Dermot (1987) have reviewed early efforts in modeling M dwarf atmospheres. Mould (1975, 1976) was the first to produce an extensive grid of convective M dwarf model atmospheres between 4750 and 3000 K. The models effectively combined the ATLAS code (Kurucz 1970), TiO band model opacities and chemical equilibrium by Tsuji (1966, 1973), $H_2O$ opacities by

Auman (1967), and a mixing-length treatment of convection (Bøhm-Vitense 1958, Kippenhan 1962). Mould also incorporated atomic line blanketing in the form of an Opacity Distribution Function (ODF; see Kurucz 1970, Mihalas 1978). However, the coarseness of his opacity grid kept him from adequately reproducing the observed spectral characteristics of the coolest M dwarfs.

It took another 15 years before model calculations finally broke the "3000-K barrier" in $T_{eff}$, with the work of Allard (1990) and Kui (1991). Both adapted their model codes from that of Wehrse (1972), who had treated the more extreme atmospheric conditions of cool white dwarfs ($T_{eff} \simeq 7000$ K). Both authors also handled molecular opacity using band models and straight mean (SM) techniques that made it possible to include—beyond the dominant TiO and $H_2O$ opacities—a number of important molecular bands such as those of the hydrides (CaH, MgH, SiH, OH, CH), which are important in low-metallicity subdwarfs, as well as the red and infrared bands of VO (Keenan & Schroeder 1952) and CO, respectively, which act as sensitive temperature indicators (Henry et al 1994, Kirkpatrick et al 1995, Martín et al 1996). From the Allard (1990) grid, Kirkpatrick et al (1993b) derived a revised temperature sequence for M dwarfs that casts new light on traditional results based on blackbody methods. This new sequence yielded values of $T_{eff}$ as much as 500-K higher at a given luminosity and shifted the positions of the late-type dwarfs in the HR diagram from cooling tracks to the blue side of theoretical lower main sequences (D'Antona & Mazzitelli 1985, Burrows et al 1989, 1993). This made it more likely that field late-type M dwarfs were hydrogen-burning stars rather than young, contracting, substellar brown dwarfs. Subsequent improvements to these models—such as the introduction of (*a*) laboratory oscillator strengths for the TiO bands (Davis et al 1986) instead of the smaller (by a factor of 2–3) empirically derived astrophysical values of Brett (1989, 1990) and (*b*) the FeH Wing-Ford bands near 0.98 $\mu$m (Phillips et al 1987)—allowed Allard (1994) to resolve most of the remaining discrepancies in the optical model spectra that had been pointed out by Kirkpatrick et al (1993b), Gustafsson & Jørgensen (1994), and Jones et al (1994).

Despite the initial successes, comparison with observed near-infrared spectra uncovered another problem: The models failed to match the infrared spectrum governed by the water vapor opacity profile (Allard & Hauschildt 1995b, Bessell 1995, Tinney et al 1995). This situation is illustrated in Figure 1, which shows that the water bands are clearly too strong in the metal-rich models. The peak of the energy distribution of M dwarfs is located in the near infrared, at around 1 $\mu$m. For brown dwarfs, most of the emitted flux emerges between 1 and 10 $\mu$m. One difficulty in determining the quality of model spectra is due to telluric absorption in the Earth's atmosphere. Telluric water bands filter the light of these faint objects over most of the infrared range. While some near-infrared spectra from about 0.9–2.5 $\mu$m can be obtained from ground-based facilities (e.g. the UKIRT spectra of Figure 1), these are unreliable in intervals where the water bands are strongest. A proper calibration of the measured fluxes becomes even more delicate for faint brown dwarfs in close binary systems (for an illustration of the uncertainties in calibrating the "K" band fluxes in the spectrum of Gl 229B, see e.g. Oppenheimer et al 1995, Matthews et al 1995, Geballe et al 1996). Beyond 2.5 $\mu$m, the Earth's atmosphere is nearly opaque and red dwarfs must be observed with infrared space-based facilities such as the HST, NICMOS, ISO, and the planned SIRTF, NGST, and DARWIN missions. But while there remain uncertainties in the absolute calibration of ground-based spectrophotometry of faint M dwarfs, these cannot completely account for the observed flux discrepancy in the infrared spectra of M dwarfs. For example, Figure 1 indicates that the predicted $H_2O$ bands grow in strength more rapidly with decreasing $T_{eff}$ than those of observed M dwarfs (Kirkpatrick et al 1995). This comparison supports the conclusion that there are shortcomings in the models. One of those shortcomings is clearly the treatment of opacity in very cool atmospheres.

## 4. MOLECULAR OPACITIES

Most of the molecules that play an important role in cool star atmospheres have been known since the early 1930s from the work of Russell (1934) and later De Jager & Neven (1957). Some of the most extensive studies of cool stellar atmosphere chemistry are by Vardya (1966), Morris & Wyller (1967), Tsuji (1973), and Gurvich (1981), who published equilibrium constants for an extensive list of diatomic and polyatomic species. More recent studies such as those of Sauval & Tatum (1984), Rossi et al (1985), Irwin (1987, 1988), Cherchneff & Barker (1992), Neale & Tennyson (1995), and Sharp & Huebner (1990) provide partition functions for most molecules directly, which allows for more flexible atmospheric calculations.

In the absence of detailed lists of transitions, or sometimes to cope with restricted computational facilities, atmospheric modelers often resort to band models or to average opacities such as the Just Overlapping Line Approximation (JOLA), SM, or ODF techniques, which approximate (by a continuum distribution) the absorption within a band or a predefined wavelength bin (Kurucz 1970, Mihalas 1978, Tsuji 1994). While computationally economical, they make the assumption that the rotational fine structure is smeared out; i.e. the lines overlap without being saturated. Such conditions are never truly met even for the strongest bands of TiO and $H_2O$ in the densest of the VLM stellar atmospheres, and these methods tend to overestimate the resulting molecular blanketing by trapping photons that would have otherwise escaped from between the lines.

A far more accurate account of molecular and atomic opacities in model atmospheres is achieved by applying an Opacity Sampling (OS) treatment of transitions lists on a prespecified fine grid of wavelengths (Peytremann 1974, Sneden et al 1976). This can be done either dynamically within the atmospheric calculations (Kurucz 1992b, Hauschildt et al 1992, Allard & Hauschildt 1995b) or be pretabulated as a function of pressure, temperature, isotopic ratios, and wavelengths (Plez et al 1992, Brett 1995a, Kipper et al 1996). While the advantage of a dynamical approach lies in the flexibility of handling depth-dependent mechanisms such as pressure broadening, departures from local thermodynamic equilibrium, microturbulence, and abundance variations, the more efficient pretabulation of the OS opacities gives the modeler freedom to incorporate his or her choice of complete line lists.

In an attempt to address the too-strong infrared water band problem in M dwarf models, Brett & Plez (1993), Allard et al (1994, 1996), Brett (1995a,b), and F Allard & PH Hauschildt (in preparation) used the OS treatment of molecular opacities to compute a new generation of M dwarf model atmospheres, which brought important breakthroughs in the understanding of M dwarf atmospheres. In the next sections, we summarize the most significant improvements in the treatment of opacities due to TiO and $H_2O$.

## 4.1 *Optical Bands*

The strengths of TiO bands define the optical (0.4–1.2 $\mu$m) spectral distribution of late K to M stars. Together with the VO bands and a few other optical spectral features, they constitute the primary $T_{eff}$ indicators in very cool stars. There currently exist three TiO line lists generated (*a*) from first principles by Collins & Faÿ (1974) and more recently extended for isotopic species and the $\epsilon$ system by Jørgensen (1994), (*b*) empirically from molecular levels assigned in laboratory experiments by Kurucz (1993), and (*c*) by Plez et al (1992). While substantial errors in the Kurucz (1993) line list have been acknowledged by the author, the Plez et al (1992) and Jørgensen (1994) line lists lead to great improvements upon previous models based on SM treatment of opacities (Mould 1975, Kui 1991, Allard & Hauschildt 1995b) in the modeling of M dwarfs. Each TiO line list applied in an OS treatment of the opacities leads to better agreement with the observed optical absolute magnitudes of M dwarfs (see e.g. Brett 1995a,b, Chabrier et al 1996a). The new models also show excellent agreement with the measured parameters of the only two known M dwarfs in eclipsing binaries (see Section 9 below; also see Bessell 1991, 1995, Chabrier & Baraffe 1995).

Unfortunately, the TiO line lists of Plez et al (1992) and Jørgensen (1994) give poor line positions and relative band strengths that prevent accurate high-resolution spectral syntheses of M dwarfs (Piskunov et al 1996, Schweitzer et al 1996). They also fail to reproduce the optical *R-I* colors of late-type dwarfs (F Allard & PH Hauschildt, in preparation), which may reflect either some remaining inaccuracies in the current estimates of the oscillator strengths (Davis et al 1986, Doverstal & Weijnitz 1992, Hedgecock et al 1995) or an incomplete account of VO or other opacity in the "R" bandpass. Indeed, despite the existence of a few spectroscopic studies of VO systems (Davis 1994, Merer et al 1987, Bauschlicher & Langhoff 1986), no list of transitions and oscillator strengths adequate for stellar atmosphere modeling is yet available for this important molecule. The Berkeley program has generated extensive line lists for FeH (Phillips et al 1987, Phillips & Davis 1993; see also Balfour & Klynning 1994), which, however, lack matching oscillator strengths. Moreover, the complexity of the FeH molecule has prevented theoretical models (Langhoff & Bauschlicher 1990, 1991) from reproducing the observed spectrum of FeH (Langhoff & Bauschlicher 1994). A similar situation also prevails for the electronic systems of CaOH (Bernath & Brazier 1985, Ziurys et al 1992, 1996), despite their importance as one of the strongest visual bands in the spectra of M-type dwarfs. Modelers have resorted to band models for most of these molecular systems (Brett 1989, 1990, Brett & Plez 1993, Allard & Hauschildt 1995b), which overestimate the resulting opacity and compromise both high-resolution spectral analysis and the determination of accurate atmospheric parameters. Fortunately, a new ab initio calculation of TiO is currently under way (SR Langhoff & CW Bauschlicher, Jr, in preparation), which should soon enable improved modeling of some aspects of cool M dwarfs.

### 4.2 *$H_2O$ Bands*

In view of the initial success obtained with an OS treatment of the TiO opacities for the optical spectral distribution of M dwarfs, Alexander et al (1989) and later Plez et al (1992) developed an OS table of randomly distributed $H_2O$ lines derived from line strength and line spacing data measured in the laboratory (Ludwig 1971). Brett & Plez (1993) and Brett (1995a,b) then used the Plez et al (1992) table in their models of M dwarfs, but this treatment still failed to reproduce the infrared spectra and colors of M dwarfs (Bessell & Stringfellow 1993, Bessell 1995).

In retrospect, this result was to be expected because $H_2O$ lines overlap more than those of TiO, so SM treatment of opacities is more appropriate for $H_2O$. Schryber et al (1995) therefore argued, based on results of their ab initio calculations for $H_2O$, that the $H_2O$ laboratory cross sections obtained by Ludwig (1971), used by both groups in the form of either SM (Allard & Hauschildt 1995b) or OS (Plez et al 1992, Brett & Plez 1993, Brett 1995a,b) may be intrinsically overestimated when applied to gas hotter than about 1500 K. Theoretical lists of transitions that include "steam" or "hot" band transitions—based on molecular levels assigned in laboratory experiments (semiempirical; e.g.

Kurucz 1992a) or on a molecular model from first principles (ab initio; e.g. Miller et al 1994)—are of far greater relevance for atmospheric calculations and are essential for an adequate account of molecular opacities in cool star and brown dwarf atmospheres. Over the past decade, efforts have converged in the development of improved theoretical opacity data for molecules of astrophysical interest with the creation of the Kurucz (1992a) and SCAN (Jørgensen 1992) data bases, and with the fruitful work of the University of the College of London (Miller et al 1994) and NASA Ames (Langhoff & Bauschlicher 1994) centers of quantum chemistry calculations.

Theoretical line lists for hot $H_2O$ from three independent sources have recently been released by Jørgensen et al (1994), Miller et al (1994), and Partridge & Schwenke (1997). The Miller et al (1994) list (6.2 million lines) uses a laboratory potential surface (Jensen 1989), while the Partridge & Schwenke (1997) list (300 million lines) uses a purely theoretical potential but the same computational approach. Both preliminary lists were computed up to $J$ values of about 30; i.e. they do not include all the necessary hot or steam bands. The Jørgensen et al (1994) list (20 million transitions) on the other hand, while also based on the Jensen (1989) potential energy functions, was computed with the goal of completeness for the atmospheres of cool giants with some compromise on the treatment of the molecular binding. For example, they use a rigid rotator approximation with an a posteriori correction to the Hamiltonian. The three data sets lead to very different opacity profiles, with the Jørgensen et al and Partridge & Schwenke lists reproducing the results obtained previously with the Ludwig opacities. Only the Miller et al line list led to an improved fit of the infrared spectral distribution of M dwarfs (Allard et al 1994, Jones et al 1995, 1996, Leggett et al 1996), as well as to an excellent agreement of early-type M dwarfs with a whole new generation of evolutionary models that include improved non-gray surface boundary conditions (Baraffe et al 1995, 1997, Chabrier et al 1996a, Leggett et al 1996). However, none of the current $H_2O$ line lists can explain the apparent saturation of the water vapor bands observed in the latest-type M dwarfs and illustrated in Figure 1. The cause of those discrepancies may therefore lie elsewhere, as is discussed in Section 5 below. A more accurate knowledge of the water vapor opacity profile is clearly needed and is now being addressed by the work of Viti et al (1995) and Partridge & Schwenke (1997).

## 5. GRAINS

The current generation of M dwarf model atmospheres (Brett 1995b, Allard et al 1996) does not include the condensation of molecules to grains. Condensation clearly must be included in the calculations as indicated by the work of Sharp & Huebner (1990), who report the abundance of condensates as a function of the gas conditions. If $ZrO_2$—one of the first condensates to appear at gas temperatures $\approx$2000 K—is not an important species in M dwarf atmospheres, the condensation of corundum at 1800 K and iron, VO, and enstatite at 1600 K most certainly affects the spectral distribution of late M dwarfs and brown dwarfs because of the large extinction of solid particles. The importance of condensation in the atmospheres of late-type M dwarfs and brown dwarfs has been confirmed by Tsuji et al (1996a,b) and Fegley & Lodders (1996), who find large concentrations of such condensates in their model atmospheres.

The impact of condensation on the spectral distribution and atmosphere of a cool dwarf is to gradually deplete the gas phase abundance of titanium, iron, vanadium, and oxygen. If we ignore for the moment the opacity of the grains, the result is a more transparent optical spectral distribution because the TiO-, VO-, FeH-, and metal-line opacities decline with decreasing effective temperature of the star. This should be reflected by an observed saturation of these molecular bands in the latest-type M dwarfs and brown dwarfs, a behavior that is presently difficult to ascertain without accurate model atmospheres that incorporate the effects of condensation. Perhaps a confirmation can be found in the peculiar optical spectrum of the coolest known M dwarf, GD 165B, mentioned in Section 2 above. However, the true nature of GD 165B's atmosphere is uncertain because this object, the companion of an old pulsating DA white dwarf within an orbital distance of $\approx$128 AU (Becklin & Zuckerman 1988, Bergeron & McGraw 1990, Zuckerman & Becklin 1992, Bergeron et al 1993, Kirkpatrick et al 1993a), may be more metal-poor and/or more carbon-rich than other nearby stars as a result of the white dwarf's prior evolution.

Tsuji et al (1996a) were the first to calculate model atmospheres for M dwarfs and brown dwarfs including not only grain formation but also grain opacities, the so-called dusty models. Their results showed that including corundum, iron, and enstatite opacities, while assuming arbitrarily spherical grains with sizes set to 0.1 $\mu$m, could heat the photospheric layers and change the overall structure of the atmosphere. The resulting dusty spectral distributions of late-type M dwarfs were redder with weaker molecular spectral features than models without grain opacities, and they were shown to reproduce the infrared broadband fluxes of the latest-type M dwarfs, including GD 165B. If confirmed, this greenhouse effect, caused by the presence of photospheric grains, may help explain the observed saturation of the near-infrared water vapor bands discussed in Section 4 and illustrated in Figure 1, as well , as well as perhaps the *R-I* colors (see Section 4.1) of late-type dwarfs, which the grainless models of Allard & Hauschildt (1995b) and Brett (1995a,b) fail to reproduce. The calculations presented by Tsuji et al (1996a), however, are coarse, and a better treatment of both the molecular and grain opacities, as well as the formal inclusion of dust scattering in the solution of the radiative transfer equation, can be achieved.

Early attempts to compute the opacity of grains were made by Cameron & Pine (1973) and Alexander (1975). More detailed calculations including the effects of chemical equilibrium calculations and grain-size distributions were reported by Alexander et al (1983) and Pollack et al (1985). Alexander & Ferguson (1994a,b) have described the computation of the opacity of grains with the inclusion of equilibrium condensation abundances, the effects of the distribution of grain sizes, and the effect of grain shape through the continuous distribution of the ellipsoid model of Bohren & Huffman (1983). These calculations include the absorption and scattering due to magnesium silicates, iron, carbon, and silicon carbide grains for a wide range of chemical compositions down to temperatures of 700 K. The direct inclusion of the equilibrium calculations of Sharp & Huebner (1990) in the future will allow for more detailed treatment of the effects of trace condensates, lower temperature opacity sources, and the effects of different elemental abundances. The inclusion of high-temperature condensates such as $Al_2O_3$ and $CaTiO_3$ may have significant effects on the opacity in cool star atmospheres, even though their abundance is quite small because of the high absorption and scattering efficiency of grains. For lower temperatures, the optical effects of species such as FeS, $Fe_3O_4$, and $H_2O$ need to be included. Pollack et al (1994) have produced opacities for water, ammonia, methane, and other low-temperature condensates. They assume complete condensation of all condensible species and extend the temperature range down to 300 K. These opacities offer an excellent basis for future brown dwarf and Jovian-type planet atmosphere calculations. However, the extinction caused by grains in a stellar atmosphere depends critically on the rate of grain formation and the resulting size distribution.

Moreover, constraints imposed by the lack of detection of cloud layers in Jupiter by the Galileo atmospheric probe (Isbell & Morse 1996, Keane et al 1996), and of any trace of scattering by grains in the evolved brown dwarf Gl 299B (Allard et al 1996, Tsuji et al 1996b), may imply an inhomogeneous vertical and/or horizontal distribution of the grains, such as scarce cloud distribution, gravitational settling, and sedimentation and rains of condensates in substellar dwarf atmospheres. While grains are likely to be destroyed by the radiative and convective heat in the inner layers of the atmosphere, the main effects of the sedimentation and rains of condensates should be a radial abundance gradient (Muchmore 1987, Guillot et al 1994) and a gradual depletion of the upper photosphere from its condensible elements over time.

The effect of grain formation and of its opacity on the atmospheric structure of M dwarf atmospheres will, therefore, not be fully understood until grain formation and time-dependent grain growth calculations incorporating the effects of sedimentation, diffusion, coagulation, and coalescence are included. Gail & Sedlmayr (1988) and Dominik et al (1989) (see references therein) have developed a formalism to account for the phenomena in the outflows from cool giants (Beck et al 1992), supergiants (Seab & Snow 1989), and nova atmospheres (Beck et al 1995). Grain growth models have also been developed for the atmospheres of cool carbon-rich white dwarf (Zubko 1987) and Jovian planet (Rossow 1978, Dobrijevic et al 1992) atmospheres. However, as yet, no results have been obtained for oxygen-rich dwarf atmospheres.

## 6. LINE ABSORPTION

The contribution of atomic and ionic line transitions to photospheric opacities is relatively less important for M dwarfs than for cool giants and hotter stars. This result arises not only from the fact that molecular absorption bands dominate opacity, but also because the lower photospheric temperatures cause the number densities ($N_i$) of atoms in higher excitation and ionization levels, such as those of the hydrogen Balmer series, to be quenched ($N_i \propto e^{-\chi_i/kT}$, where $T$ is the gas temperature and $\chi_i$ is the excitation potential or ionization energy relative to the ground state). Moreover, the "locking" of elements into molecular compounds and further condensation of such elements to grains also reduces the available abundances of atomic species, as is the case for hydrogen, which is about 70–85% $H_2$ in the photospheres of M dwarfs.

As a result, only the strongest resonance and subordinate lines, with the low excitation energy of mostly alkali and earth-alkali elements, prevail in the spectra of M dwarfs. Those lines can be very broad owing to van der Waals (vdW) pressure broadening, and they often contrast greatly with the narrow emission and weak absorption lines that originate in the chromospheric layers of active M stars (such as the Balmer series and the Ca H and K lines). Only a few of the atomic lines that are created in the photospheric layers can be detected within the haze of molecular lines and provide diagnostics of the photospheric parameters. Examples include the Na I–D lines at λ5889,5896 Å, as well as other Na I resonance transitions at λ8183,8195 Å and λ10746,10749,10835 Å and those of K I at λ6911,6939, λ7665,7699, λ9950,9954, and λ10480,10482,10487 Å. Lines of Rb I at λ7950 Å and Ba I at λ7911,7913 Å are also particularly strong (relative to the local continuum) in late-type M dwarfs and brown dwarf candidates.

Despite the relative scarcity of directly observable atomic lines in their spectra, an accurate modeling of M dwarf atmospheres nevertheless requires the use of a complete atomic line list that includes lines of ionized elements for a complete account of the opacity in the hotter layers (typically about 8000 K in M dwarfs) of the inner atmosphere. A failure to do so may result in atmospheric structures that are too cool globally, as the efficient convection zone assures the transfer of inner atmospheric heat to the outer photospheric layers. The most complete list of atomic transitions currently available is that by Kurucz (1994) and its revisions. Several other line lists, such as those generated from first principle model atoms of the Opacity Project (Seaton 1992, Seaton et al 1992) or semiempirically using atomic levels assigned in laboratory experiments (see Verner et al 1996 and revisions), are also available but are still too incomplete for the purpose of model atmosphere calculations.

## 6.1 *Line Broadening Mechanisms*

The high densities prevailing in VLM star atmospheres cause strong spectral lines to be significantly broadened. Because the gas temperatures are not high enough to sustain a significant amount of ionization, the electron and proton densities are much smaller than the densities of the most important neutral and molecular species. Consequently, the contribution of Stark broadening to the total damping constant is very small, even in stars with very low metallicities. The total thermal plus microturbulent line widths are always much smaller than the line width owing to vdW broadening:

$$\gamma_{\mathrm{vdW}} = 17 \cdot C_6^{2/5} v^{3/5} N_p, \tag{1}$$

which describes the interaction between two different, unpolarized neutral particles within the impact or static approximation, with $\gamma_{\mathrm{vdW}}$ the full-width half-maximum damping constant of the resulting Lorentz profile, $v$ the relative velocity between perturber and absorber, and $N_p$ the number density of perturbers. While the interaction constant $C_6$ can be determined exactly for both the ground and excited states of a perturbed atom when the perturber is atomic hydrogen (Michelis 1976), no exact method has yet been developed for the case of collisions with the much slower molecular hydrogen perturbers that dominate the atmospheres of VLM stars, brown dwarfs, and Jovian-type planets (Guillot et al 1994). In those cases, the collisions are not instantaneous and the profiles not strictly Lorentzian (Kunde et al 1982, Goody & Yung 1989), but in the absence of accurate alternatives, modelers often resort to using the hydrogenic approximation formulated by Unsöld (1955) for collisions with neutral hydrogen with some ad hoc modifications:

$$C_6 = 1.01 \times 10^{-32}(Z+1)^2 \times \left[\frac{E_{\mathrm{H}}^2}{(E-E_l)^2} - \frac{E_{\mathrm{H}}^2}{(E-E_u)^2}\right] \mathrm{cm}^6\,\mathrm{s}^{-1}, \tag{2}$$

where $Z$ is the charge of the absorber, $E$ the ionization energy (e.g $E_{\mathrm{H}} = 13.6$ eV), and $E_l$ and $E_u$ the lower and upper level excitation energies of the absorber. Investigations by Weidemann (1955), for instance, showed that the values as calculated above are in good agreement with observed line widths for alkali metals but not for other elements, such as iron (Kusch 1958). This has led to the introduction of correction factors to the "classical" formula, which can range from 10 $C_6$ in the Sun (Takeda et al 1996) to $10^{1.8}$ $C_6$ in white dwarfs for non–alkali-like species (Wehrse & Liebert 1980). No corrections are required for alkali elements. The Unsöld (1955) approximation, combined with this correction factor for non-alkali elements and with an explicit account of the different polarizabilities of each perturber ($\alpha_p/\alpha_{\rm H}$ $C_6$, where the subscript $p$ refers to the perturber), leads to improved profiles that appear to describe well the atomic lines observed in late-type M dwarfs (Schweitzer et al 1996). The most abundant perturbers in M dwarfs and their polarizabilities are given by Weast (1988) and Schweitzer et al (1996).

While the situation is poor for atomic line broadening, it is even worse for molecular lines, for which only a few sources and techniques exist (Lazarev & Pnomarev 1992, Kurucz 1993, Guillot et al 1994). Fortunately, individual molecular lines are usually not saturated, so that broadening is less important for them than for strong atomic lines. Moreover, molecular lines often overlap so strongly that their wings are completely masked (Schweitzer et al 1996), and only the Gaussian line cores of the strongest molecular transitions are observed. The atmospheres of VLM stars and brown dwarfs are therefore only weakly sensitive to the adopted value of the vdW damping constant in the bands of several of the most important molecular absorbers (e.g TiO and $H_2O$). This may, however, not be the case for some hydride bands and for the infrared CO overtones that show larger typical line spacings (Kui 1991, Davis 1994, Tsuji 1994).

## 7. CONVECTION

The thin radiative skin above the convective region in an M dwarf determines the surface boundary conditions for the entire temperature structure of the fully convective photosphere and interior. This radiative zone is often limited to the outermost optically thin regions of the photosphere in early-type M dwarfs (Allard 1990, Kui 1991, Burrows et al 1993, Allard & Hauschildt 1995b): i.e. to optical depths below about $10^{-3}$. Figure 2 illustrates how the outer atmosphere of a typical M dwarf is affected by the atomic and molecular opacities and convection. At such low optical depths ($\tau = 10^{-3}$ corresponds to $\log P_{gas} \approx 3.8$ in this model), the structure of the atmosphere is sensitive to the strong opacities of TiO and $H_2O$. Early-type M dwarf atmospheres, spectra, colors, and even their evolution are, therefore, very dependent upon elemental abundances and the treatments of molecular opacities and possibly convection (see Section 9 below; see Baraffe et al 1995 for an illustration of these effects). Early-type M dwarfs should serve as excellent stellar laboratories in which to study convection.

The standard mixing length theory (Bøhm-Vitense 1958, Kippenhan 1962, Mihalas 1978) used to model convective energy transport in stars is only a crude

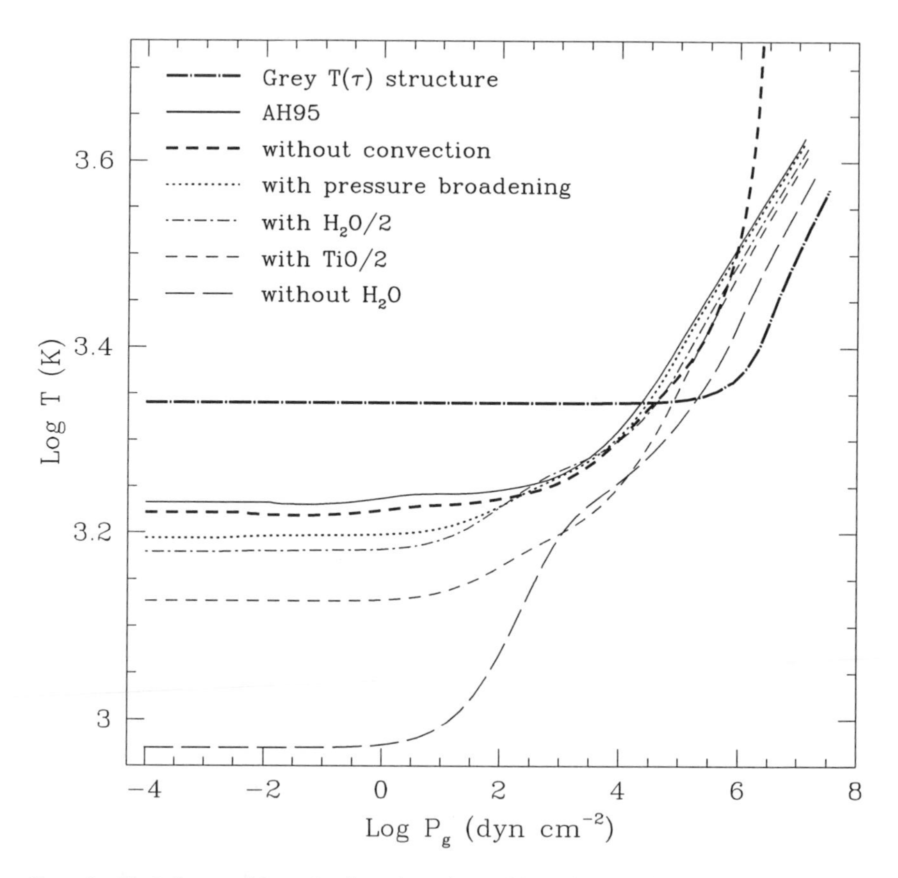


*Figure 2* The influence of the molecular and atomic opacities and convection upon the atmospheric structure of a typical model atmosphere; here the $T_{eff} = 2800$ K, $\log g = 5.0$, and solar metallicity model of Allard & Hauschildt (1995b). A corresponding gray structure without convection (*bold dot-dashed*) is also shown for comparison. While the complete neglect of $H_2O$ opacities causes a dramatic cooling (by CO) of the atmosphere (*long-dashed curve*), uncertainties by a factor of two in the $H_2O$ opacity cross sections cause only negligible changes in the atmospheric structure (*thin dot-dashed* relative to *dotted curve*). A similar drop in the opacity cross section of TiO, however (*thin short-dashed* relative to *dotted curve*), causes a much more significant cooling of the atmosphere.

approximation. While nonlocal treatments of convection exist (for review, see Chan et al 1991, Gustafsson & Jørgensen 1994, Grossman 1996, Kim et al 1996) that may be better suited to the optically thin medium of cool stellar atmospheres, they are very computationally prohibitive and have not been applied to models of M dwarfs. Fortunately or sadly, depending on your point of view, the large opacities in M dwarfs mean convection is nearly adiabatic for values of the mixing length ($\ell$) comparable to the atmospheric pressure scale height. The atmospheres and synthetic spectra of M dwarfs therefore show very little sensitivity to changes in $\ell$ over the range typical of solar-type atmospheres; i.e. $\ell/H_P = 1.2$ to 2.2 (Brett 1995a, Baraffe et al 1997). Moreover, models indicate that the convection zone gradually retreats with decreasing mass in late-type M dwarfs because of their decreasing luminosity and with decreasing metallicity due to decreasing photospheric opacities (Allard 1990). In cool brown dwarf models such as those computed for Gl 229B, for example, the convection zone reaches no higher than optical depths of about unity (although models show signs of a second, separate convective layer closer to the surface in such cool objects). The spectroscopic and photometric properties of late-type M dwarfs, metal-poor subdwarfs, and possibly brown dwarfs are therefore relatively insensitive to the details of convection (see e.g. Brett 1995a). This means that standard mixing length approximations are probably suitable for these stars (which is good news for brown dwarf modelers), but that these same stars are not very good laboratories to study convection (which is bad news for convection modelers).

## 8. STELLAR ACTIVITY

The "solar dynamo" model, also known as the alpha-omega dynamo, which operates at the radiative convective boundary layer in the solar interior, predicts a correlation of activity with rotation that is observed in solar-type stars (Noyes et al 1984, Marilli et al 1986, Rutten 1986). When coupled with the decrease of the rotation rate as the star ages (owing to loss of angular momentum during its lifetime), a rotation-activity-age correlation is expected and has also been observed (Wilson & Skumanich 1964). The fully convective lowest mass stars, however, like pre–main-sequence solar-type stars, are known to be very active even if no spot-cycle variability can yet be confirmed in any of them. Several flare periodically [see e.g. Linsky et al (1995) for a report of recent flare outbursts in VB10], and a large fraction of them (up to 60% in M5 dwarfs) show chromospheric $H_\alpha$ emission. The dynamo generation of their fields must therefore occur from a different, or at least modified, mechanism. And indeed, the surface activity in the M dwarfs has been observed to exhibit general characteristics that contrast with those of solar-type stars: 1. The incidence of chromospheric and coronal activity in M dwarfs grows with decreasing stellar mass (Joy & Abt 1974, Giampapa & Liebert 1986, Reid et al 1995a,b, Hawley et al 1996). 2. The coronal and chromospheric luminosity and the luminosity that occurs in flares all decrease with stellar mass (Fleming 1988, Petterson 1989); however, the fraction of the luminosity that appears in these magnetic indicators relative to the total stellar luminosity ($L_{H_\alpha}/L_{bol}$, $L_X/L_{bol}$, hereafter "activity level") remains nearly constant (Fleming et al 1995, Reid et al 1995a, Mullan & Fleming 1996). 3. While, as in solar-type stars, M dwarfs show little or no activity when the rotation rate reaches below a certain threshold (~5 km/s; cf Marcy & Chen 1992), stars with rotation rates above this threshold show weak or no correlation between rotation and the activity level for both chromospheric and coronal emissions (Stauffer & Hartmann 1986, Rutten et al 1989, Hawley et al 1996). 4. Coronal and chromospheric activity show a correlation with scale height from the galactic plane, metallicity, and probably age of the star (Fleming et al 1995, Reid et al 1995a, Hawley et al 1996).

M dwarfs are therefore more active than solar-type stars. This clearly indicates that some change in the magnetic field generation and/or the interaction of the field with the stellar atmosphere has occurred. Two possible ideas have been suggested to explain the differing M dwarf behavior: 1. Noyes et al (1984) and later Peterson (1989) pointed out that the volume of the convection zone is an important parameter in the generation of the magnetic field, and this volume begins to decrease in proportion to the mass once the stars are mostly convective (in M dwarfs). The field strength may become saturated in the lowest mass stars, and hence no strong rotation-activity connection would be expected (see e.g. Rosner et al 1985, Stauffer et al 1991). Moreover, the relative neutrality of the gas in M dwarfs compared with hotter stars may also alter the behavior of magnetic field lines (P Ulmschneider, private communication). Direct measurements of the magnetic field strength and its stellar surface coverage have been obtained, using Zeeman splitting of highly magneto-sensitive lines, by Robinson (1980), Gray (1984), Saar (1988), Mathys & Solanki (1989), and Basri & Marcy (1994) for a number of late-type stars, which confirm the presence of strong magnetic fields in M dwarfs. Saar (1994) and Johns-Krull & Valenti (1996), for example, report magnetic field strengths for the dMe dwarfs AD Leo, EV Lac, and AU Mic of 4.0–4.3 kG with covering factors between 55 and 85%. 2. On the other hand, the propagation of acoustic shocks (which originate at the convective-radiative surface boundary), and the resulting acoustic heating of the chromosphere, should become most efficient in the strongly convective M dwarf photospheres and may therefore play an important role in the energy budget of their atmospheres and coronae (Schrijver 1987, Mathioudakis & Doyle 1992, Mullan & Cheng 1993). Since the extension of the convection zone depends sensitively on the atmospheric parameters (see Section 7 above), a correlation of the chromospheric activity with metallicity and age in M dwarfs would therefore be likely. The first steps in integrating detailed photosphere models with acoustically heated M dwarf chromospheres were taken by Buchholz (1995) and Mullan & Cheng (1993, 1994). Mullan & Cheng (1994) report a more effective penetration of acoustic waves into the coronae of M dwarfs compared with the case of more massive solar-type stars. They find that acoustic heating can maintain a corona with a temperature on the order of 0.7–1 $\times$ $10^6$ K and a surface X-ray flux as large as $10^5$ ergs cm$^{-2}$ s$^{-1}$, and they suggest that relatively inactive M dwarfs that display X-ray emissions below this limit may be candidates for acoustically maintained coronae.

However, these ideas still leave some unanswered questions. How can we explain, for example, stars of the same age or the same mass and rotation rates above the threshold but with widely different activity levels? In an attempt to explain the dilemma raised by the two $\geq$M9.5-type field dwarfs PC 0025+0447 (Graham et al 1992) and BRI 0021 $-$ 012 (Tinney 1993) that have widely different activity levels, Basri & Marcy (1995) suggested that there could also be a threshold temperature below which acoustic and Alfvén waves become inefficient and stellar activity subsides. Below this temperature threshold, the heating of the chromosphere and corona and the wind generation would be sufficiently prohibited to prevent the formation of chromospheric emission lines and to slow the rotation of the star. However, this interpretation still leaves the case of hotter stars unaddressed. An example is the triple VLM star system LHS 1070 (Leinert et al 1994; F Allard et al, in preparation), in which the two faint companions of the same age and metallicity are also similar both in mass ($\approx$0.085 $M_{\odot}$) and $T_{\rm eff}$ (2600–2700 K), but where only the faintest of the three components is inactive. Several other systems have been observed where only the more massive primary star is active (Hawley et al 1996). While most cases can be explained if the primaries are in fact unresolved close-binary stars in which enhanced rotation (and activity) is maintained by tidal interactions, this does not seem to explain the lack of activity in the close-binary star LHS 1070C.

Modeling an M dwarf chromosphere is a complex problem owing to the complexity of the radiative transfer calculations in such a cool, dense environment. Cram & Mullan (1985) modeled an M dwarf chromosphere using hydrogen Balmer emission line observations. Giampapa et al (1982) used Ca II observations to model M dwarf chromospheres, with limited success. Houdebine & Panagi (1990) investigated the effects of changing the model hydrogen atom used in the calculations, but they did not fit their models to data. Hawley & Fisher (1994) have developed chromospheric flare models, which incorporate a full nonlocal thermodynamic equilibrium (NLTE) treatment of the statistical equilibrium and radiative transfer in the important optically thick chromospheric lines and a helium ionization equilibrium computed self-consistently with the downward X-ray flux from the corona. Yet they were unsuccessful at fitting both the Ca II and hydrogen Balmer lines in their quiescent and flare observations. More recently, Mauas & Falchi (1996) were also unable to match both the observed hydrogen Balmer line strengths in their quiescent model of a well-observed active M dwarf. To date, no model has successfully predicted all the major chromospheric lines observed in an active M dwarf atmosphere.

The success of chromospheric modeling may be limited by these workers' assumptions of monotonically rising atmospheres or even two-component models. Hawley et al (1996) found that active M dwarfs in the field show systematic spectral differences relative to nonactive stars, which may lead to a better understanding of the atmospheric heating mechanisms: Early-type dMe appear (*a*) brighter by $\approx$0.5 mag from *V-K*, (*b*) $\leq$0.1-mag redder in (V-I) and (V-K), and (*c*) to show systematic differences in the relative strengths of some near-infrared TiO subbands compared with those of nonactive dM stars of the same spectral type. These effects may in part be expected if dMe stars are systematically younger with larger radii. However, they are only observed in the most massive M dwarfs. On the other hand, while direct effects of magnetic fields on the structure of the atmosphere (e.g. through the magnetic pressure term) and the Zeeman splitting of atomic lines are negligible for all but the outermost photosphere, the impinging radiation upon the upper photosphere by a magnetically or acoustically heated chromosphere can be important. The radiation temperature of the chromosphere is generally much higher than that of the photosphere, and even a relatively small irradiation of the photosphere (0.1% of the total flux of the star) by the chromosphere can introduce important NLTE effects that may change the temperature structure of the outermost layers. Since the outermost thin radiative skin of an M dwarf regulates the entire structure of the convective photosphere and interior (see Section 7 above), these effects may couple back to the dynamo-generated magnetic and acoustic heating. Systematic development of (magneto) hydrodynamical studies of chromospheric activity must be tied to realistic photosphere models to truly understand cool dwarfs.

## 9. THE $T_{eff}$ SCALE OF M DWARFS

While bolometric luminosities can be derived from a careful integration of the observed stellar radiation for single stars within the accuracy of their known parallaxes (Tinney et al 1993, 1995), and stellar masses can be derived for close binaries down to the hydrogen-burning limit (Henry & McCarthy 1993), the calibration of the observed magnitudes and spectral types as a function of the physical atmospheric parameters of the stars still remains difficult. The determinations of M dwarf effective temperatures have been refined considerably since the work of Veeder (1974), Peterson (1980), and Read & Gilmore (1984), who fit blackbody curves through broadband colors and points of assumed observed continuum. But even the current empirical methods (Berriman et al 1992, Jones et al 1994) still assume that nearly pure thermal radiation escapes from dM atmospheres at some wavelength(s). Such an assumption is

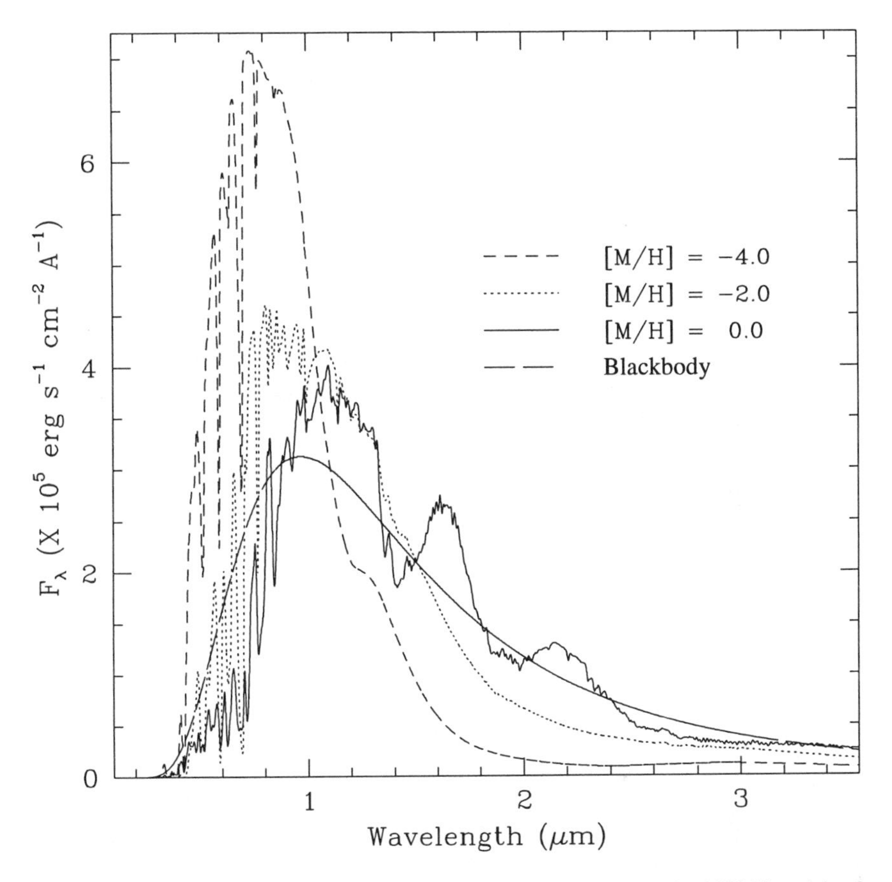


*Figure 3* Spectral distributions of emerging fluxes at the stellar surface for 3000 K models of metallicities corresponding roughly to the solar neighborhood ([M/H] = 0.0), halo ([M/H] = −2.0), and Population III ([M/H] = −4.0) stars. A blackbody of the same effective temperature (*smooth curve*) is shown for comparison.

secure only for optically thick layers of a nonconvective atmosphere, but models strongly suggest that M dwarf atmospheres are convective out to optical depths as low as $\tau \sim 10^{-3}$ (see Section 7 above). The hazards of any type of Planck flux fitting to an M dwarf spectrum are apparent from Figure 3 and have been emphasized by Allard & Hauschildt (1995a), who showed how strong molecular absorption and flux redistribution obliterate all evidence of the original continuum shape.

Fortunately, two double-line spectroscopic and eclipsing M dwarf binary systems can offer some guidance in the subsolar mass regime: CM Draconis and

YY Geminorum. Lacy (1977) and later Habets & Heintze (1981) determined the $T_{eff}$ of M dwarfs in these systems based on the observed masses and radii. Figure 4 compares the latest OS models of Brett (1995b), F Allard & PH Hauschildt (in preparation), and Kurucz 1992b to these fundamental stellar calibrators. The $T_{eff}$ scales derived from the spectral synthesis of individual stars of Kirkpatrick et al (1993b) [using the SM models of Allard (1990)] and Leggett et al (1996)

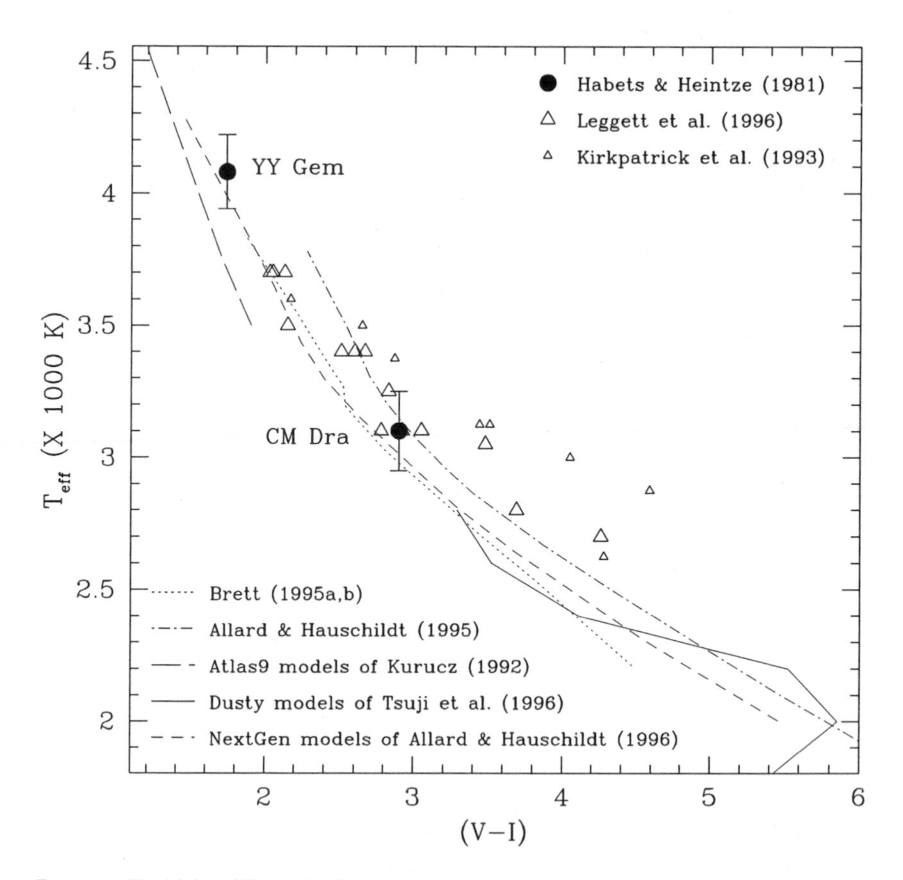


*Figure 4* The M dwarf $T_{eff}$ scale. Current model-dependent effective temperature scales for cool stars down to the hydrogen-burning limit. *Open triangles* feature results from spectral synthesis of selected stars from the works of Kirkpatrick et al (1993b) and Leggett et al 1996 as indicated. The new generation of OS models by Brett (1995b) and F Allard & PH Hauschildt (in preparation), as interpolated onto theoretical isochrones by Chabrier et al (1996a), reproduce closely the independently determined positions of M dwarfs in the eclipsing binaries CM Dra and YY Gem (Habets & Heintze 1981). Much uncertainty remains, however, in the lowermost portion of the main sequence where effects of grain formation can become important.

[using OS synthetic spectra drawn from the Allard & Hauschildt (1995b) model structures] are also shown, which illustrate the tendency of theoretical $T_{eff}$ to become cooler with developments in the treatment of opacities in the models.

As an inspection of Figure 2 reveals, M dwarf photospheric structures are much more sensitive to TiO opacities than to the current uncertainties in the $H_2O$ opacity profiles discussed in Section 4. The result of the use of an OS treatment of the main molecular opacities, in particular for TiO, appears therefore to be a breakthrough in the agreement of modeled $T_{eff}$ scales with observations of early-type M dwarfs. Note that the OS models of Kurucz (1992b), on the other hand, suffer from an inaccurate TiO absorption profile and a complete lack of $H_2O$ opacities, and the models are therefore clearly inadequate in the regime of VLM stars (i.e. below $T_{eff} \approx 4500$ K and $M \approx 0.8\ M_{\odot}$), where molecular opacities dominate the stellar spectra and atmospheric structures.

However, the situation still remains uncertain at $T_{eff} \leq 3500$ K, where the SM models of Allard & Hauschildt (1995b) seem to yield a better agreement, both with the current $T_{eff}$ estimate for CM Draconis and with the *VRI* colors of the Leggett (1992) disk stars sample. This could be understood in terms of an incompleteness of current optical opacities (see Section 4.1 for details). The SM technique used for the treatment of TiO opacities in the models of Allard & Hauschildt (1995b) compensates for missing optical opacities and may represent a better intermediate solution in this regime. Kinematics indicate that CM Draconis is most likely an old disk system and may be slightly metal-depleted compared with young-disk main-sequence stars (Rucinski 1978, Saumon et al 1995a, Chabrier & Baraffe 1995). Recently, new determinations of the masses and radii of CM Draconis have been obtained by Metcalfe et al (1996), and a revision of the effective temperatures and metallicity of the system is under way (Viti et al 1996), which may help improve the temperature estimate for CM Draconis and shed some light on this region of the lower main sequence.

The models of F Allard & PH Hauschildt (in preparation) and Brett (1995a,b) presently neglect the effects of both condensation and grain opacities that may affect stars cooler than 2700 K. The preliminary dusty models of Tsuji et al (1996a) (also shown in Figure 4) indicate that condensation effects may drive the $T_{eff}$ scale of these M dwarfs to cooler values at a given color. Tsuji et al's models also indicate that when grain opacities begin to alter the thermal equilibrium of the atmosphere, the lowest-mass stars become gradually hotter and redder, mimicking more closely the behavior of blackbodies.

Much uncertainty remains, therefore, at the lowermost portion of the main sequence. The effects of grain formation and more complete opacities of TiO promise a better understanding of the stars and brown dwarfs in the vicinity of the hydrogen-burning limit [the location of which is roughly indicated in Figure 4 by the termination point of the Allard & Hauschildt model sequence], but they still remain to be ascertained. These questions are currently being addressed by Tsuji et al (1996b) and F Allard & PH Hauschildt (in preparation).

# 10. THE DETECTABILITY OF BROWN DWARFS

Brown dwarfs are substellar objects not massive enough to sustain stable nuclear fusion and therefore cannot successfully stabilize on the hydrogen-burning main sequence (Burrows et al 1993, Saumon et al 1995a). While this boundary between stars and brown dwarfs is fairly well defined, the recently reported massive extrasolar planets, some of which have orbital eccentricities more reminiscent of stellar binary systems (e.g. $e = 0.4$ for 70 Vir; see Marcy & Butler 1996b), blur the distinction between planets and brown dwarfs. The minimum mass for deuterium burning ($\approx$13 $M_J$) may represent a more physically meaningful criterion to distinguish Jovian planets from classical brown dwarfs (Saumon et al 1996).

To develop search strategies for substellar objects, we propose three categories for objects in the brown dwarf regime: (*a*) transitional objects: massive brown dwarfs ($M \geq 0.06\ M_\odot$) younger than about $10^9$ years that are burning deuterium or even hydrogen temporarily; (*b*) lithium brown dwarfs: low-mass brown dwarfs ($M \leq 0.06\ M_\odot$) younger than about $10^7$ years that are still burning deuterium but have not depleted their initial reservoir of lithium and beryllium; and (*c*) evolved brown dwarfs: brown dwarfs older than $10^7$–$10^9$ years that have exhausted their nuclear fuel and have cooled and faded below the parameters of the coolest main-sequence stars [i.e. $T_{eff} = 2000$ K and $L = 10^{-4}\ L_\odot$ (Burrows et al 1993, Baraffe et al 1995)]. Many brown dwarfs of all types are expected to be in the solar neighborhood and nearby clusters, so it was puzzling that concerted searches did not reveal any (Martín et al 1994, Marcy et al 1994, Henry 1996).

## 10.1 *Transitional Objects*

If it is massive enough, a brown dwarf can initiate thermonuclear fusion early in its evolution, before fading to invisibility as a component of the dark matter (Burrows et al 1993). Such young objects can approach atmospheric conditions similar to those of M dwarf stars ($T_{eff} \leq 3000$ K) and can easily be confused with them. An extrapolation of the observed mass–spectral type relation of red dwarfs in short-period binary systems seems to support the existence of a large population of unrecognized brown dwarfs among red dwarfs with spectral types later than M7 (Henry & McCarthy 1993, Henry et al 1994, Kirkpatrick et al 1994, 1995, Kirkpatrick & McCarthy 1994). However, further extensions of the Henry & McCarthy (1993) mass-spectral type relation and new evolution models (Baraffe & Chabrier 1996) reveal a sharp drop in the mass-luminosity and mass-color relations from about 0.1 $M_\odot$ to the hydrogen-burning limit, which rules out this naive extrapolation. The possibility that such objects may masquerade as known field M dwarfs must still be confirmed.

Transitional objects in the field with unknown age and mass should betray their youth with the following signatures that distinguish them from more massive M dwarfs: (*a*) a lower surface gravity ($\log g < 5.0$), since the brown dwarf is still early on its evolutionary track, and (*b*) a higher rotation rate accompanied by more chromospheric and coronal activity. Unfortunately, both these criteria are difficult to apply in practice: Metal-rich model spectra for the latest-type dwarfs show very little sensitivity to changes in gravity (Allard & Hauschildt 1995a). Moreover, the wings of resonance lines that are sensitive to the pressure stratification of the photosphere are also sensitive to slightly increased $T_{eff}$ and/or metallicity that can compensate for a lower gravity (Schweitzer et al 1996). As for rotation, the relation between age, rotation, and activity in late-type dwarfs is very poorly understood (see Section 8 above) and is not a reliable discriminant of transitional objects either. This situation is best illustrated by the striking contrast between the candidates PC 0025+0447 (classified dM9.5; Graham et al 1992), which shows one of the strongest $H_\alpha$ emission lines of the lower main sequence, and the brown dwarf candidate BRI 0021−012 (classified <dM9.5; Tinney 1993), which rotates 20 times faster than other field nonemission M stars (Basri & Marcy 1995). Faced with these ambiguities, nobody has yet been able to confirm any transitional object in the field. Young stellar clusters, however, have yielded more positive results (see below).

## 10.2 *Lithium Brown Dwarfs*

The key to recognizing such substellar objects is in their spectra. Since an M dwarf is completely convective, the entire star is mixed and the surface abundances reflect the elemental abundances in the core. If the central temperatures are low enough or the star is young enough, then nuclei should survive in the atmosphere that would otherwise be completely destroyed in hotter, more massive stars. This is the case for $^7$Li nuclei, which are destroyed by proton captures at relatively low temperatures of a few ($\sim$2) million degrees in the interior. Therefore, if you can detect and measure the strength of $^7$Li lines in a stellar spectrum and you are armed with accurate atmospheric and evolution models, you should be able to estimate the mass and age of the star (Rebolo et al 1992, Maggazzú et al 1993).

To date, the detection of Li I lines is the most decisive spectral indicator of substellarity for young brown dwarfs with masses below about 0.06 $M_\odot$. This test is best used for nearby young clusters where the age is reasonably well known and the 6707-Å region can be studied at relatively high dispersion to obtain a precise abundance of lithium. If the cluster is old enough that only the very lowest mass stars should retain any surface lithium, as is the case for the Pleiades, then this is a very powerful test of substellarity. The fact that early searches for Li I in Pleiades brown dwarf candidates yielded no detections prompted Pavlenko et al (1995) and later Allard & Hauschildt (1995a,b) to investigate a number of concurrent physical processes that could prevent the Li I 6708 Å doublet from being seen in those objects. They explored possible molecular bonding involving lithium and departures from LTE in the Li I lines, all with negative results: The Li I lines should be observable in young brown dwarfs. Indeed, recent spectroscopic observations of Teide 1, Calar 3, and PPl 15 (Rebolo et al 1995, Zapatero Osorio et al 1997, Stauffer et al 1994) show that they are cluster members, have low luminosities, and have retained lithium at their surface (Basri et al 1996, Rebolo et al 1996). Teide 1 and Calar 3 are most likely brown dwarfs with estimated masses of $55 \pm 15$ $M_J$ (Zapatero Osorio et al 1997). One more candidate, the first found in the field, has also been confirmed to have Li in its atmosphere (Thackrah et al 1997). Several more Pleiades candidates may soon be confirmed (or rejected) by the Li test (Martín et al 1996).

Accurate estimation of brown dwarf masses depends critically upon an understanding of the depletion of lithium at the surface as a function of mass (Nelson et al 1993). This in turn requires evolutionary calculations for low-mass stars that include the most modern equations of state and non-gray atmospheres for their outer boundary conditions, as have been performed by Baraffe et al (1995, 1997), Chabrier & Baraffe (1995), Chabrier et al (1996a), Baraffe & Chabrier (1996), and Allard et al (1996). The evolutionary calculations are largely sensitive to the treatment of the equations of state, which requires the inclusion of the thermodynamic properties of a strong-coupled Coulomb plasma, electron degeneracy, and pressure ionization (Dorman et al 1989, Nelson et al 1993, Burrows et al 1993). Non-gray atmospheres were found to cause the interiors to become systematically cooler (and the hydrogen-burning minimum mass smaller) than when using gray models such as those of Burrows et al (1993) for a given mass. This is because energy is transported in the interiors of these objects by convection, and their structure can be approximated by that of a polytrope of index, $n = 1.5$ [see Burrows & Liebert (1993) for a discussion of the polytropic nature of low mass stars and brown dwarfs]. The polytropic interior characteristics are only specified once the atmospheric structure is determined, and improvements in the atmospheric properties, therefore, directly impact the determinations of the ages, masses, and location in the Hertzprung-Russell (HR) diagram of these stars. A proper treatment of the atmosphere is therefore essential to the predicted mass–Li abundance relation as a function of the age of brown dwarfs (Chabrier et al 1996a).

EVOLVED BROWN DWARFS A major breakthrough in finding the missing link between Jovian planets and low mass stars was the discovery of the first evolved field brown dwarf Gliese (Gl) 229B by Oppenheimer et al (1995) and Nakajima et al (1995). Unfortunately, Gl 229B—and other candidates like the astrometrically discovered HD 114762 (Mazeh et al 1996, Williams et al 1997), the ZZ Ceti companion GD 165B (Becklin & Zuckerman 1988, Zuckerman & Becklin 1992), and the massive extrasolar giant planet candidate around 70 Vir (Marcy & Butler 1996b)—are all binary companions that are too faint and too close to their stellar primaries to be tested for an optical lithium signature. Fortunately, their low temperatures offer an advantage, since unique changes in molecular chemistry that occur across the temperature transition from the coolest M dwarfs ($\approx$2000 K) to the Jovian planets ($\leq$150 K) result in distinctive spectral signatures.

There currently exist three sets of model atmospheres and synthetic spectra for the $T_{eff}$ regime of cool brown dwarfs: (*a*) The Tsuji et al (1996a,b) models cover the range from 4000 K (although only from 2700 K with grains) to 1000 K; (*b*) the Allard et al (1996) models reach down (from 10,000 K) to 800 K; and (*c*) the Marley et al (1996) models cover the range from 1000 K to the temperature of Jupiter, i.e. 150 K. Tsuji et al (1995) were the first to introduce detailed models and synthetic spectra for cool brown dwarfs. While unsuitable for high-resolution spectral synthesis, their models (based on a band model treatment of the molecular opacities) predict the growing intensity of infrared $CH_4$ bands with $T_{eff}$ cooler than about 1800 K. This signature was later identified in the near-infrared spectral distribution of the brown dwarf Gl 229B (Matthews et al 1995, Geballe et al 1996) and helped confirm the substellar nature of the brown dwarf. The study of Gl 229B also led to the important realization that, although grain formation must occur in such cool photospheres and may indeed affect the spectroscopic properties of late M dwarfs (Tsuji et al 1996a), none of the predicted greenhouse effects of grain opacities seem to be present in Gl 229B Allard et al (1996) and Tsuji et al (1996b). These authors showed that grainless models provide a better description of the spectral distribution of the brown dwarf when a homogeneous atmosphere is assumed (see Section 5 above for details; Tsuji et al 1996b). The grainless brown dwarf models of Tsuji et al (1995), however, are based on band-model opacities that tend to overestimate the strength of molecular features and cannot be used reliably to derive accurate atmospheric parameters and absolute fluxes for evolved brown dwarfs.

The brown dwarf model atmospheres of Allard et al (1996), on the other hand, are based on a direct OS treatment of an ab initio line list for $H_2O$ (Miller et al 1994) and lists of transitions observed in planetary atmospheres [e.g. the RADEN, GEISA, HITRAN, and ATMOS projects by Farmer & Norton (1989), Husson et al (1992), Rothman et al (1992), Farmer & Norton (1989),

Kuznetsova et al (1993), respectively] and that reproduce more accurately the spectroscopic properties of evolved brown dwarfs. This can be appreciated in Figure 1, where the most recent observations of the brown dwarf Gl 229B by Geballe et al (1996) are compared to the best fitting model of Allard et al (1996). Their models, combined with the latest brown dwarf evolution models, led to a $T_{eff}$ of 900–1000 K and a mass of about 30–50 $M_J$ for Gl 229B. The lack of condensation in their calculations, however, prevented them from completely covering the range of relevant brown dwarfs parameters; i.e. brown dwarfs cooler than about 800 K, in which water vapor begins to condense out of the gas phase, leaving profoundly transformed photospheres and synthetic spectra. This limitation was avoided by Marley et al (1996) by simply neglecting elements that are expected to be condensed at the effective temperatures of Jovian planets and by including only the $H_2$, He, $CH_4$, $NH_3$, $H_2O$, and $H_2S$ species in their chemical equilibrium calculations. This allowed them to compute models that cover the regime of brown dwarfs and extrasolar planets cooler than 1000 K. Despite their approximations in the treatment of molecular opacities (K-coefficient technique) and convection (adiabatic mixing only), their model successfully reproduced most of the observed spectral characteristics of the brown dwarf Gl 229B.

The models of Allard et al (1996) and Marley et al (1996) are compared in Figure 5 to summarize the predicted absolute fluxes that brown dwarfs would have at a distance of 50 parsec (pc). Both the model spectra and blackbody distributions of the same $T_{eff}$ are shown, which indicate the range of possible spectroscopic characteristics of brown dwarfs (from a grainless to a fully dusty atmosphere). As can be seen in that figure, brown dwarfs are most readily detected at 4.5 $\mu$m: the peak of their spectral energy distribution. At 5 $\mu$m, the hotter (younger or more massive) brown dwarfs and stars show strong CO bands that cause their flux to drop by nearly 0.5 dex relative to that at 4.5 $\mu$m. Between 4.5 and 10 $\mu$m, opacities of $CH_4$ (and $H_2O$ in the hot brown dwarfs) cause the flux to drop by 0.5–1.0 dex. Searches in the 4.5–5 $\mu$m region and redwards of 10 $\mu$m should therefore offer the best possibilities for finding and resolving brown dwarfs in binaries. The detection limits of current and planned ground-based and space-based telescopes (from Saumon et al 1996) are also indicated in Figure 5, which show that brown dwarfs (dusty or not) within 50 pc would be easily detected by Space Infrared Telescope Facility in the 4.5–5.0 $\mu$m region. The drop in sensitivity of the various instruments redward of 10 $\mu$m implies, however, that only brown dwarfs as hot as or hotter than Gl 299B have a good chance of detection at that distance.

The general spectral distributions of brown dwarfs hotter than about 800 K are relatively well reproduced by current models. In contrast, the incompleteness of lists of observed molecular transitions for several important molecular absorbers

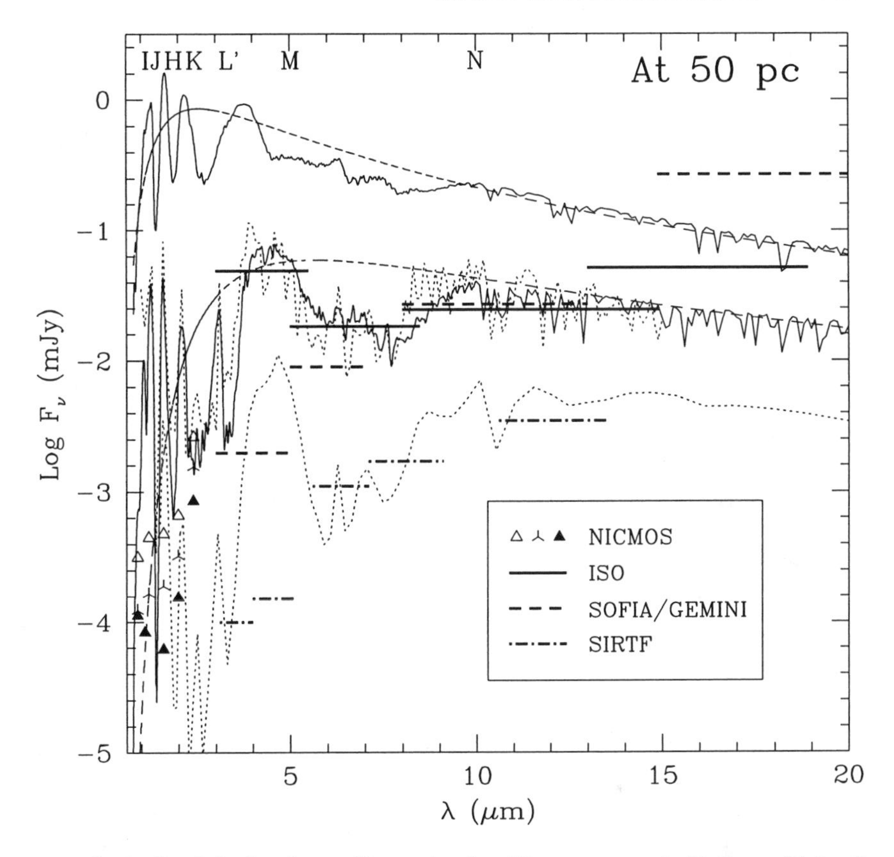


*Figure 5* Predicted absolute fluxes of brown dwarfs at 50 pc as compared with the sensitivity of ground- and space-based platforms that will be or are currently applied to the search for brown dwarfs and extrasolar planets. The latter are values reported for the 5$\sigma$ detection of a point source in 1 h of integration, except for the three NICMOS cameras where the integration is limited to 40 min (see Saumon et al 1996 for details). Models of both Allard et al (1996) (*solid line*) and Marley et al (1996) (*dotted line*) are shown, which simulate (*a*) a brown dwarf near the hydrogen-burning limit (*top-most spectrum*: $T_{eff}$ = 2000 K), (*b*) an evolved brown dwarf similar to Gl 229B (*central spectra*: $T_{eff}$ = 900 K and 960 K), and (*c*) a brown dwarf closer to the deuterium-burning limit (*lower-most spectrum*: $T_{eff}$ = 500 K). The corresponding blackbody (*dashed line*) are also shown for comparison.

may introduce uncertainty in the predicted absolute fluxes and colors of cooler brown dwarfs where water is no longer the dominant infrared opacity. Opacities for $CH_4$, for example, are clearly incomplete because only the strongest lines of $CH_4$ are available from the Gestion et Etude des Informations Spectroscopiques Atmospheriques and High Resolution Transmision data bases, while none are available blueward of 1.6 $\mu$m where systems of $CH_4$ are known to cause strong absorption features in the spectra of planets [e.g. in Jupiter and Titan; see Mickelson & Larson 1992, Bernath et al 1995, Baines et al 1993, Strong et al 1993, Larson & Mickelson 1996). The complexity of the $CH_4$ compound has so far prevented the accurate modeling of the methane spectrum beyond 2000 $cm^{-1}$ (see e.g. Tyuterev et al 1994), where a significant fraction of the flux emerges from a brown dwarf or Jovian planet.

## 11. THE METAL-DEFICIENT VLM STARS

Accurate knowledge of the compositions of metal-poor VLM subdwarf M (hereafter sdM) stars and their positions in the Hertzprung-Russell diagram is essential to a full understanding of the chemical history of our Galaxy. The reconstruction of the initial mass function in the old disk, halo, and globular clusters—founded on an accurate mass-luminosity relation that is a sensitive function of the stellar chemical composition (Chabrier et al 1996a, D'Antona & Mazzitelli 1996, von Hippel et al 1996)—also depends upon it.

Unfortunately the metallicity, atmospheric parameters, and mass-luminosity relation of sdM stars have long remained uncertain owing to the lack of VLM stellar model atmospheres and the resulting lack of bolometric corrections and synthetic photometry to transform theoretical evolution models into various empirical planes such as color-magnitude diagrams (see e.g. Greenstein 1989 for details).

This situation began to change in the early 1990s with the Allard (1990) grid of VLM models that explored a wide range of parameter space with metallicities ranging from solar values to as low as 1/10,000th solar (i.e. a logarithmic ratio of metal to hydrogen abundances of $[M/H] = -4.0$). This encompassed all relevant age/metallicity populations of cool dwarfs, including disk, halo, and even unobserved Population III VLM subdwarfs. These models were later updated by Allard & Hauschildt (1995b), for an OS treatment of more complete atomic opacities, and most recently by F Allard & PH Hauschildt (in preparation), for an OS treatment of the molecular opacities as well. Figure 3 presents three typical model-flux distributions obtained by F Allard & PH Hauschildt (in preparation) for 3000-K dwarfs: one for metal-rich conditions of young disk dwarfs, the other two for their higher-gravity halo and Population III subdwarf counterparts. The collision-induced absorption (CIA) of $H_2$-$H_2$, $H_2$-H, and $H_2$-He (Borysow et al 1989, Borysow & Frommhold 1989, 1990, Zheng & Borysow 1994) defines the near-infrared continuum in these low-metallicity subdwarfs. Centered at 2 $\mu$m, $H_2$ CIA depresses the infrared continuum such that most of the flux emerges only in bluer passbands. Molecular absorption bands of hydrides such as MgH and CaH are the predominant features in the optical region when double metals such as TiO and VO are depleted (Mould &

Wyckoff 1978, Boeshaar 1976, Bessell 1982). The optical spectrum of an sdM is therefore much more transparent than that of a metal-rich M dwarf, where the continuum is defined by $H^-$ opacity. The result is a spectral distribution that becomes bluer with decreasing stellar metallicity.

Figure 6 illustrates the behavior of metal-poor VLM stars in the $M_V$–($V$-$I$) color-magnitude diagram. As can be seen, the work of Monet et al (1992) and Dahn et al (1995) have revealed a number of old disk and halo subdwarfs,

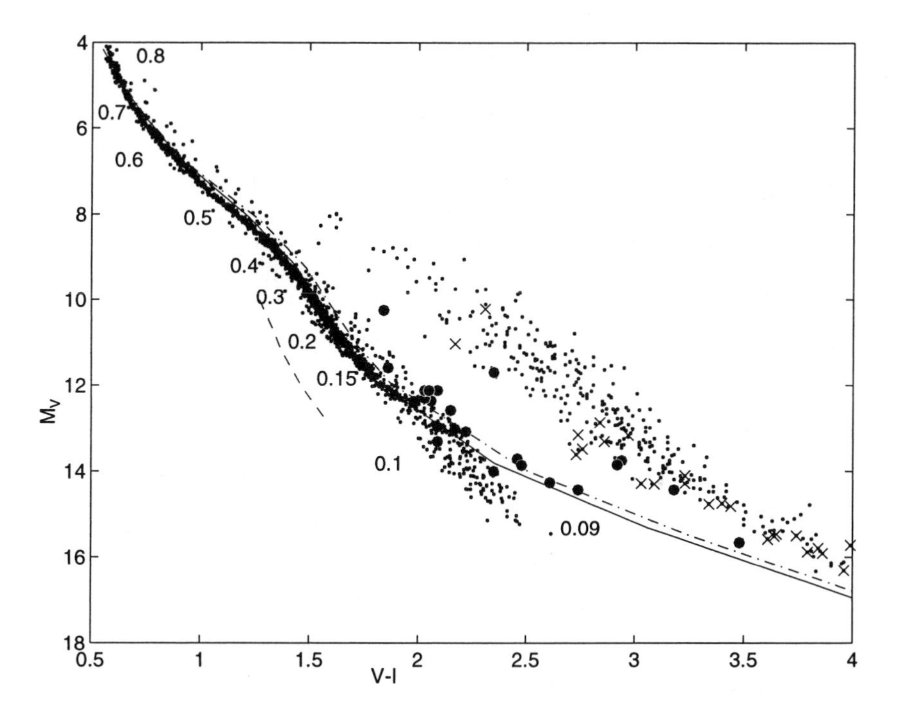


*Figure 6* Color-magnitude diagram of the lower main sequence for disk to globular clusters stars. The excellent agreement obtained by Baraffe et al (1997) to the [Fe/H] $= -1.9$ globular cluster NGC6397 of Cool et al (1996) (*small dots* to the blue of the diagram) illustrates the recent success of synthetic photometry and evolution models to reproduce metal-poor VLM stars. Two model sequences by Baraffe et al (1997) are shown for [M/H] $= -1.5 \approx$ [Fe/H] $= -1.9$ (*full line*) and for [M/H] $= -1.3$ i.e. $\approx$ [Fe/H] $= -1.0$ (*dot-dashed line*). Points along the theoretical sequences are labeled with the stellar mass, in units of the solar mass. The latter isochrone gives a correct description of the most metal-poor halo subdwarfs of Monet et al (1992) (*full circles*). The disk stars of Monet et al (1992) (*crosses*) and Dahn et al (1995) (*small dots*) to the red and the zero-metallicity models of Saumon et al (1994) (*dashed line*) to the blue display the possible range of so-called subluminosity for metal-poor VLM subdwarfs.

some of which are nearly two magnitudes bluer than previously known field sdM stars. D'Antona (1995) suggested that the extreme subdwarfs of the Monet et al (1992) sample belong to the galactic halo. But despite the availability of VLM model atmospheres and synthetic spectra, it has still proven difficult to untangle the effects of reduced metallicity from those of increased gravity or reduced effective temperature, which all affect the pressure structure of the photosphere in similar ways (Allard 1990, Kui 1991, Allard & Hauschildt 1995b). This difficulty, combined with the lack of accurate line lists for molecular hydride absorption bands, has hampered efforts to obtain the atmospheric parameters of observed M subdwarfs (sdM) via spectral synthesis. Thus far, only a few stars have been analyzed in any detail (cf Dahn et al 1995).

The M subdwarfs in binary and multiple systems, for which the metallicity of the hotter primary is known, and clusters with VLM stars of the same age and metallicity offer more promise to test Population II evolutionary and atmosphere models. Two early-type M subdwarfs in binary systems have been recently reported by Martín et al (1995): the faint, wide proper-motion companion to G116-009 and the fourth and faintest ($M_V = 12.2$) companion in the $[M/H] = -1.7 \pm 0.4$ multiple system G176-46 (see also Laird et al 1988, Latham et al 1992). A spectral synthesis of these M subdwarfs with the most recent OS model atmospheres should soon permit calibration of the $T_{eff}$ scale of similar halo subdwarfs (F Allard & PH Hauschildt, in preparation).

One of the unique contributions of the HST to the understanding of the lower main-sequence star formation in the early Universe and the contribution of low-mass stars to the universal dark matter has been the recent detection of large numbers of VLM stars and white dwarfs in the nearest globular clusters (Paresce et al 1995, Richer et al 1995, Cool et al 1996, Renzini et al 1996). Figure 6 depicts the globular cluster NGC6397 as observed with the Wide Field/Planetary Camera 2 by Cool et al (1996). This cluster displays a genuine Fe underabundance of $[M/H] = -1.9$ (i.e. $1/80$ times solar) and must have formed some $15 \times 10^{-3}$ years ago. Its main sequence is narrow and well defined, without the age, activity, and metallicity dispersion characteristic of samples of field stars. Fits to that main sequence below about 0.5 $M_\odot$ do not suffer from the same high sensitivity to assumed values of the helium abundance fraction and mixing length as do fits to more massive parts of the main sequence (cf Baraffe et al 1997). Two characteristic kinks also typical of the mass-luminosity relation shape the lower portion of the main sequence. The first kink at $\approx$0.4 $M_\odot$ is caused by $H_2$ dissociation (Copenland et al 1970). The second, below $\approx$0.1 $M_\odot$, is close to the HST detection limit for several clusters and was first revealed in the Population II models of D'Antona (1987, 1990, 1995). The first kink is due to the onset of electron degeneracy in the interior, which is also the cause for the sharp drop in the mass-luminosity relation below this threshold (Chabrier & Baraffe 1997). Its exact location depends upon the pressure structure and hence the opacities of the atmospheres (see also Section 7). The shape of the lower main sequence of globular clusters represents a nearly parameter-free test bed for stellar evolution and atmospheric physics in a metallicity regime where models are less prone to the incompleteness of $H_2O$ and TiO opacities.

D'Antona & Mazzitelli (1985) and D'Antona (1987) laid the ground work for the theoretical modeling of VLM stars down to the hydrogen-burning limit, and their models long provided among the best descriptions of the lower main sequence of VLM stars. More recently, D'Antona & Mazzitelli (1994, 1996) have computed a series of pre–main-sequence tracks for Population II stars with masses ranging from 0.015–2.5 $M_{\odot}$ using both standard mixing-length theory and the Canuto & Mazzitelli (1991, 1992) theory of turbulent convection. However, a critical element to the handling of convection that describes the stellar structures in these fully convective objects comes (*a*) from the equations of state, i.e. the adiabatic gradient (Saumon et al 1995b), and (*b*) from the treatment of the surface boundary, i.e. the atmosphere as we pointed out in Sections 7 and 10. Even with updated versions of the Magni & Mazzitelli (1979) equations of state, the models of D'Antona & Mazzitelli (1996) are still systematically too hot by $\approx$200 K across the VLM range and could not reproduce the observed lower main sequences of globular clusters.

Recently, Alexander et al (1996) and Baraffe et al (1997) computed evolution models based on the equation of state of Saumon et al (1995b) and found a remarkable agreement of their models with the observed main sequence of the cluster NGC6397 all the way down to the detection limit, corresponding to masses of $\approx$0.13 $M_{\odot}$. This success is illustrated in Figure 6, where the models of Baraffe et al (1997) are plotted over the photometry. (Note that the dispersion in the diagram at magnitudes fainter than $\approx$14.5 mag is likely due to foreground stars.) The two groups, however, disagree on the metallicity of NGC6397: Alexander et al (1996) use bolometric corrections and synthetic colors of Allard & Hauschildt (1995b) and find [M/H] = −2.0. Baraffe et al (1997) use the more recent, less blanketed (F Allard & PH Hauschildt, in preparation) models both as surface boundary conditions and for color transformations, and they find agreement for [M/H] = −1.5, a value that is more consistent with a history of oxygen and other enrichment in old stellar populations (Ryan et al 1991 and references therein). On the other hand, the halo subdwarfs of Monet et al (1992) (*full circles* in Figure 6) show a wider dispersion in metallicity. Baraffe et al (1995) and Alexander et al (1996) derived isochrones for the most extreme subdwarfs of the Monet et al (1992) sample and obtained a metallicity of [M/H] = −1.5, while the analysis of Baraffe et al (1997) (shown in Figure 6) led to consistant values of [M/H] = −1.3 to −1.5 for the same subdwarfs.

Clearly, these results illustrate the progress brought about by more accurate stellar equations of states, model atmospheres, and synthetic photometry to the understanding of the lower main sequence of halo and globular cluster stars.

These successes increase our confidence that the present Population II stellar models are now sufficiently accurate to derive reliable mass-luminosity relations and the stellar mass functions that rely upon them. The mass function in the stellar halo has been derived from the Dahn et al (1995) luminosity function by Méra et al (1996b,c) and Chabrier et al (1996b), using the theoretical mass-luminosity relations drawn from the Baraffe et al (1995, 1997) evolution models. These authors found a halo mass function which is rising all the way down to the hydrogen-burning limit, suggesting a large population of substellar objects in the halo. But to determine exactly the amount of dark mass in the form of metal-poor substellar brown dwarfs, we need to extrapolate the mass function into the substellar domain. Indeed, despite all the advances in cool star models and detection techniques, we still have identified only a handful of halo M subdwarfs and none below the hydrogen-burning limit. A hint of the nature of the missing mass in the halo is nevertheless already provided by the frequency of events reported by the EROS and MACHO microlensing surveys: The average time of recorded events indicates the existence of a rich population of halo objects with masses of at least 0.3–0.5 $M_\odot$ (Aubourg 1995, Alcock et al 1996). Metal-poor brown dwarfs seem therefore eliminated as a strong contender for the missing mass (measured to be $\approx 10^{12}$ $M_\odot$, i.e. 10 times the visible mass) in the halo. Rather, the average masses of the microlenses suggest at least two possibilities, either main sequence VLM subdwarf stars or white dwarf stellar remnants. However, although the results of astrometric surveys like 2MASS, DENIS, and more accurate parallaxes for nearby stars from, for example, the USNO, CCD, and HIPPARCOS surveys may give us larger halo samples in the near future, HST pencil surveys and the Dahn et al (1995) luminosity function of the halo seem for now to eliminate the possibility of a large population of VLM stars in the halo (Bahcall et al 1994, Elson et al 1996, Flynn et al 1996, Gould et al 1996, Graff & Freese 1996).

## 12. CONCLUSIONS

Realistic model atmospheres of VLM stars and brown dwarfs are essential if we are ever to fully understand the population of the lower main sequence, the mass-luminosity relation, and the initial mass function and its dependance on the chemical history of the Galaxy. The last few years have brought significant improvements in the models, as well as the first convincing detections of substellar objects that can be used to test the models. This progress on both theoretical and observational fronts has led to several noteworthy advances in our knowledge of the lower main sequence, including the following:

1. The spectroscopic properties of brown dwarfs and their luminosities as a function of mass and age can now be reliably predicted. We confidently expect that most brown dwarfs within 50 pc of the Sun are within detection range of either existing instruments or facilities planned for the next decade. This bodes well for studies of the low-mass end of the initial mass function for the solar neighborhood.

2. The mass-luminosity relation for cool field stars of solar metallicity is now reproducible by models, and the underlying physics is well understood (Chabrier et al 1996a). The sharp drop in the relation at masses below 0.1 $M_\odot$ due to the onset of electron degeneracy in the interior provides a natural explanation for the corresponding drop in the observed luminosity function as one approaches the hydrogen-burning limit. These models and the results of microlensing observations all point to a rising-disk mass function all the way down to the hydrogen-burning limit and suggest a substantial number of brown dwarfs in the galactic disk (Han & Gould 1996, Méra et al 1996a,c).

3. It is clear that non-gray model atmospheres must be invoked to handle correctly the evolution and internal structure of M dwarfs, brown dwarfs, and metal-poor VLM stars. The first major victory in attacking these problems has occured not in the field, but in metal-poor globular clusters. The lower main sequences of such clusters can now be fitted by interior models using the latest non-gray surface boundary and color transformations. In light of this success, we can confidently assert that the Monet et al (1992) subdwarfs are low-luminosity members of the galactic halo with metallicities of [M/H] $\simeq -1.5 \pm 0.2$ (i.e. [Fe/H] $\simeq -1.9 \pm 0.2$) and that, if current stellar luminosity functions for the halo are accurate, such main-sequence subdwarfs cannot make up a significant fraction of the halo missing mass.

The progress outlined above and the good agreement among brown dwarf model spectra generated by various independent model codes is reassuring, but we cannot afford to be complacent at this stage. While the effective temperature scale of low mass stars is now reasonably well determined for $T_{eff} \geq 3500$ K, it is still poorly defined for M dwarfs and young brown dwarfs with spectral types later than M6, and it will remain so until more complete opacities become available. These opacities must include better treatments of TiO, $H_2O$, and $CH_4$ molecules (including hot bands), grain size distributions, and grain growth time scales appropriate to the high pressures and oxygen-rich conditions found in cool dwarf atmospheres.

While we labor to understand the outer layers of M dwarfs, we must also remember that the interior models that define such sequences suffer their own uncertainties, and that interior and atmospheric properties are intimately coupled in these objects. With fully convective interiors, their radiation fields play the role of an energy valve that regulates both the internal structure and the hydrodynamical (either magnetic or acoustic) heating of their chromospheres. Systematic development of classical atmosphere models (in which molecular opacity calculations and laboratory molecular data are tested) must be tied to (magneto) hydrodynamical studies of chromospheric activity and applied to interior models to truly understand cool dwarfs. The fact that researchers are taking these first steps is an encouraging sign that our field is finally "coming of age" after a long but fruitful adolescence.

ACKNOWLEDGMENTS

We thank Drs. Hugh RA Jones and Tom R Geballe for providing UKIRT near-infrared spectra of M dwarfs and Gl 299B and Sandy K Leggett, Mike S Bessell, John M Brett, T Tsuji, MS Marley, Didier Saumon, DC Monet, Conard C Dahn, and Adrienne M Cool for providing data in electronic form. We would also like to express our gratitude to Suzanne Hawley and Peter Ulmschneider for instructive discussions on activity in M dwarf stars and to Francesca D'Antona, Isabelle Baraffe, and Gilles Chabrier for discussions about VLM star evolution and equations of state, as well as for providing their theoretical isochrones in a numerical form. We are also particularly indebted to Jaymie Matthews for generously proofreading the manuscript and to the Cornell Theory Center (CTC) and the San Diego Supercomputer Center (SDSC) for their allocation of computer time, which made possible some of the calculations and conclusions presented in this review.

This work is funded by grants from the National Science Foundation (NSF) (AST-9217946) to Indiana University, NASA LTSA (NAG5-3435) to Wichita State University, NASA LTSA and ATP to the University of Georgia in Athens, and NASA LTSA (NAGW2628), ATP (NAG53068), and NSF (AST94-17057) to the Arizona State University.



*Literature Cited*

Alcock C, Allsman RA, Axelrod TS, Bennett DP, Cook KH, et al. 1996. *Ap. J.* 461:84–103
Alexander DR. 1975. *Ap. J. Suppl.* 29:363
Alexander DR, Augason GC, Johnson HR. 1989. *Ap. J.* 345:1014–21
Alexander DR, Brocato E, Cassisi S, Castellani V, Ciacio F, DeglÍInnocenti S. 1997. *Astron. Astrophys.* 317:90
Alexander DR, Ferguson JW. 1994a. In *Molecules in the Stellar Environment*, ed. UG Jørgensen, *Lect. Notes Phys.*, pp. 149–62. Berlin/Heidelberg: Springer-Verlag

Alexander DR, Ferguson JW. 1994b. *Ap. J.* 437:879–91
Alexander DR, Rypma RL, Johnson HR. 1983. *Ap. J.* 272:773–80
Allard F. 1990. *Model atmospheres for M dwarfs.* PhD thesis. Ruprecht–Karls University, Heidelberg, Germany
Allard F. 1994. In *Molecular Opacities in the Stellar Environment,* ed. P Thejll, U Jørgensen, *IAU Colloq. 146,* pp. 1–5. Copenhagen: Niels Bohr Inst.; Nordita
Allard F, Hauschildt PH. 1995a. See Tinney 1995, pp. 32–44
Allard F, Hauschildt PH. 1995b. *Ap. J.* 445:433–50
Allard F, Hauschildt PH, Baraffe I, Chabrier G. 1996. *Ap. J. Lett.* 465:L123–27
Allard F, Hauschildt PH, Miller S, Tennyson J. 1994. *Ap. J. Lett.* 426:L39–L41
Aubourg E. 1995. *Astron. Astrophys.* 301:1–5
Auman JJ. 1967. *Ap. J. Suppl.* 14:171
Bahcall JN, Flynn C, Gould A, Kirkhakos S. 1994. *Ap. J.* 435:L51–L54
Baines KH, West RA, Giver LP, Moreno F. 1993. *J. Geophys. Res.* 98:5517–29
Balfour WJ, Klynning L. 1994. *Ap. J.* 424:1049–53
Baraffe I, Chabrier G. 1996. *Ap. J.* 461:L51–L54
Baraffe I, Chabrier G, Allard F, Hauschildt PH. 1995. *Ap. J. Lett.* 446:L35–L38
Baraffe I, Chabrier G, Allard F, Hauschildt PH. 1997. *Ap. J.* In press
Basri G, Marcy GW. 1994. *Ap. J.* 431:844–49
Basri G, Marcy GW. 1995. *Astron. J.* 109:762–73
Basri G, Marcy GW, Graham JR. 1996. *Ap. J.* 458:600–9
Bauschlicher CWJ, Langhoff SR. 1986. *J. Chem. Phys.* 85:5936
Beck HKB, Gail H-P, Henkel R, Sedlmayr E. 1992. *Astron. Astrophys.* 265:626–42
Beck HKB, Hauschildt PH, Gail H-P, Sedlmayr E. 1995. *Astron. Astrophys.* 294:195–205
Becklin EE, Zuckerman B. 1988. *Nature* 336:656–58
Bergeron P, Fontaine G, Brassard P, Lamontagne R, Wesemael F, et al. 1993. *Ap. J.* 106:1987–99
Bergeron P, McGraw JT. 1990. *Ap. J. Lett.* 352:L45–L48
Bernath PF, Black JH, Brault JW. 1995. Preprint.
Bernath PF, Brazier CR. 1985. *Ap. J.* 288:373–76
Berriman G, Reid IN, Leggett SK. 1992. *Ap. J. Lett.* 392:L31–L33
Bessell MS. 1982. *Publ. Astron. Soc. Austr.* 4:417–19
Bessell MS. 1991. *Astron. J.* 101:662
Bessell MS. 1995. See Tinney 1995, pp. 123–31
Bessell MS, Stringfellow GS. 1993. *Astron. Astrophys. Rev.* 31:433–71
Boeshaar PC. 1976. *The spectral classification of M dwarf stars.* PhD thesis. Ohio State Univ., Columbus, OH.
Bøhm-Vitense E. 1958. *Zh. Astrophys.* 46:108
Bohren CF, Huffman DR. 1983. *Absorption and Scattering of Light by Small Particles.* New York: Wiley & Sons. 541 pp.
Borysow A, Frommhold L. 1989. *Ap. J.* 341:549–55
Borysow A, Frommhold L. 1990. *Ap. J. Lett.* 348:L41–L43
Borysow A, Frommhold L, Moraldi M. 1989. *Ap. J.* 336:495–503
Brett JM. 1989. *MNRAS* 241:247–57
Brett JM. 1990. *Astron. Astrophys.* 231:440–52
Brett JM. 1995a. *Astron. Astrophys.* 295:736–54
Brett JM. 1995b. *Astron. Astrophys. Suppl.* 109:263–64
Brett JM, Plez B. 1993. *Publ. Astron. Soc. Austr.* 10:250–53
Bryja C, Humphreys RM, Jones TJ. 1994. *Astron. J.* 107:246–53
Buchholz B. 1995. *Acoustically heated chromospheres of late type stars.* PhD thesis. Univ. Heidelberg. 135 pp.
Burrows A, Hubbard WB, Lunine JI. 1989. *Ap. J.* 345:939–58
Burrows A, Hubbard WB, Saumon W, Lunine JI. 1993. *Ap. J.* 406:158–71
Burrows A, Liebert J. 1993. *Mod. Phys. Rev.* 65:301
Butler RP, Marcy GW. 1996. *Ap. J. Lett.* 464:L153–56
Butler RP, Marcy GW, Williams E, Hauser H, Shirts P. 1997. *Ap. J. Lett.* 474:115–18
Cameron AGW, Pine MR. 1973. *Icarus* 18:377
Canuto VM, Mazzitelli I. 1991. *Ap. J.* 370:295–311
Canuto VM, Mazzitelli I. 1992. *Ap. J.* 389:724–30
Chabrier G, Baraffe I. 1995. *Ap. J. Lett.* 451:L29–L32
Chabrier G, Baraffe I. 1997. *Astron. Astrophys.* In press
Chabrier G, Baraffe I, Plez B. 1996a. *Ap. J. Lett.* 459:L91–L94
Chabrier G, Segretain L, Méra D. 1996b. *Ap. J.* 468:L21–L24
Chan KL, Nordlun A, Steffen M, Stein RF. 1991. *Solar Interior and Atmosphere,* pp. 223–74. Tucson, AZ: Univ. Ariz. Press
Cherchneff I, Barker JR. 1992. *Ap. J.* 394:703–16
Collins JG, Faÿ TDJ. 1974. *J. Quant. Spectrosc. Radiat. Transfer* 14:1259
Comeron F, Rieke GH, Burrows A, Reike MJ. 1993. *Ap. J.* 416:185–203

Cool AM, Piotto G, King IR. 1996. *Ap. J.* 468:655–62
Copenland H, Jensen JO, Jørgensen HE. 1970. *Astron. Astrophys.* 5:12
Cram LE, Mullan DJ. 1985. *Ap. J.* 294:626–33
Dahn CC, Liebert J, Harris HC, Guetter HH. 1995. See Tinney 1995, pp. 239–48
D'Antona F. 1987. *Ap. J.* 320:653–62
D'Antona F. 1990. In *Physical Processes in Fragmentation of Star Formation,* ed. RC Dolcetta, C Choisi, A Di Fazio, pp. 367–79. Dordrecht: Kluwer
D'Antona F. 1995. See Tinney 1995, pp. 13–23
D'Antona F, Mazzitelli I. 1985. *Ap. J.* 296:502–13
D'Antona F, Mazzitelli I. 1994. *Ap. J. Suppl.* 90:467–500
D'Antona F, Mazzitelli I. 1996. *Ap. J.* 456:329–36
Davidge TJ, Boeshaar PC. 1993. *Ap. J. Lett.* 403:L47–L50
Davis SP. 1994. In *Molecules in the Stellar Environment,* ed. UG Jørgensen, *Lect. Notes Phys.,* pp. 397–411. Berlin/Heidelberg: Springer-Verlag
Davis SP, Littleton JE, Phillips JG. 1986. *Ap. J.* 309:449–54
De Jager C, Neven L. 1957. *Liège R. Soc. Mem.* 18:257
Deutsch RW. 1994. PASP 106:1134–37
Dobrijevic M, Goutoulli L, Toublanc D, Parisot JP, Brillet J. 1992. In *Titan, ESA Symp. Ser.,* pp. 329–32, II Univ. Bordeaux, France
Dominik C, Sedlmayr E, Gail H-P. 1989. *Astron. Astrophys.* 223:227–36
Dorman B, Nelson LA, Chau WJ. 1989. *Ap. J.* 342:1003–818
Doverstal M, Weijnitz P. 1992. *Mol. Phys.* 75:1357
Elson RAW, Santiago BX, Gilmore GF. 1996. *New Astron.* 1:1
Farmer CB, Norton RH. 1989. *A High-Resolution Atlas of the Infrared Spectrum of the Sun and the Earth Atmosphere from Space,* Vol. 1: *The Sun,* NASA Ref. Publ. 1224. Washington, DC: NASA
Fegley BJ, Lodders K. 1996. *Ap. J. Lett.* 472:L37–L39
Fleming TA. 1988. *Ap. J.* 331:958–73
Fleming TA, Schmitt JHMM, Giampapa MS. 1995. *Ap. J.* 450:401–10
Flynn C, Gould A, Bahcall J. 1996. *Ap. J. Lett.* 466:L55–L58
Gail H-P, Sedlmayr E. 1988. *Astron. Astrophys.* 206:153–68
Geballe TR, Kulkarni SR, Woodward CE, Sloan GC. 1996. *Ap. J. Lett.* 467:L101–4
Giampapa MS, Liebert J. 1986. *Ap. J.* 305:784–94
Giampapa MS, Worden SP, Linsky JL. 1982. *Ap. J.* 258:740–60
Gliese W, Jahreiss H. 1991. *Astron. Rechen-Inst. Heidelberg Mitt. Ser. A* 224:161–64
Goody RM, Yung YL. 1989. *Atmospheric Radiation.* London/New York: Oxford Univ. Press. 2nd ed.
Gould A, Bahcall JN, Flynn C. 1996. *Ap. J.* 465:759–68
Graff DS, Freese K. 1996. *Ap. J. Lett.* 467:L65–L68
Graham JR, Matthews K, Greenstein JL, Neugebauer G, Tinney CG, Persson SE. 1992. *Astron. J.* 104:2016–21
Gray DF. 1984. *Ap. J.* 277:640–47
Green PJ, Margon B. 1994. *Ap. J.* 423:723–32
Green PJ, Margon B, MacConnell DJ. 1991. *Ap. J. Lett.* 380:L31–L34
Greenstein JL. 1989. *Pac. Astron. Soc. Publ.* 101:787–810
Grossman SA. 1996. *MNRAS* 279:305–36
Guillot T, Gautier D, Chabrier G, Mosser B. 1994. *Icarus* 112:337–53
Gurvich LV. 1978–1981. *Thermodinamicheskie Svoistva Individuak'nikh Vershev,* Vol. 1–3. Moscow: Sovit Acad. Sci.
Gustafsson B, Jørgensen UG. 1994. *Astron. Astrophys. Rev.* 6:19–65
Habets GMHJ, Heintze JRW. 1981. *Astron. Astrophys. Suppl.* 46:193–237
Hambly NC, Hawkins MRS, Jameson RF. 1993. *Astron. Astrophys. Suppl.* 100:607–40
Han C, Gould A. 1996. *Ap. J.* 467:540–45
Hauschildt PH, Wehrse R, Starrfield S, Shaviv G. 1992. *Ap. J.* 393:307–28
Hawley SL, Fisher GH. 1994. *Ap. J.* 426:387–403
Hawley SL, Reid IN, Gizis JE. 1997. *Astron. J.* In press
Heber U, Bade N, Jordan S, Voges W. 1993. *Astron. Astrophys.* 267:L31–L34
Hedgecock LM, Naulin C, Costes M. 1995. *Astron. Astrophys.* 304:667–77
Henry TJ. 1996. *Sky Telescope* 91:24
Henry TJ, Kirkpatrick JD, Ianna PA, Jahreiss H. 1995. *Bull. Am. Astron. Soc.* 187:7022
Henry TJ, Kirkpatrick JD, Simons DA. 1994. *Astron. J.* 108:1437–44
Henry TJ, McCarthy DW. 1993. *Astron. J.* 106:773–89
Houdebine ER, Panagi PM. 1990. *Astron. Astrophys.* 231:459–65
Husson N, Bonnet B, Scott NA, Chedin A. 1992. *J. Quant. Spectrosc. Radiat. Transfer* 48:509
Irwin AW. 1987. *Astron. Astrophys.* 182:348–58
Irwin AW. 1988. *Astron. Astrophys. Suppl.* 74:145–60
Isbell D, Morse D. 1996. NASA press release 96-10
Jensen P. 1989. *J. Mol. Spectrosc.* 133:438
Johns-Krull C, Valenti J. 1996. *Ap. J. Lett.* 459:L95–L98

Jones HRA, Longmore AJ, Allard F, Hauschildt PH. 1996. *MNRAS* 280:77–94
Jones HRA, Longmore AJ, Allard F, Hauschildt PH, Miller S, Tennyson J. 1995. *MNRAS* 277:767–76
Jones HRA, Longmore AJ, Jameson RF, Mountain CM. 1994. *MNRAS* 267:413–23
Jørgensen UG. 1992. *Rev. Mex. Astron. Astrofis.* 23:49–62
Jørgensen UG. 1994. *Astron. Astrophys.* 284:179–86
Jørgensen UG, Jensen P, Sorensen GO. 1994. In *Molecular Opacities in the Stellar Environment,* ed. P Thejll, U Jørgensen, *IAU Colloq. 146,* pp. 51–54. Copenhagen: Bohr Inst., Nordita
Joy AH, Abt HA. 1974. *Ap. J. Suppl.* 28:1–18
Keane TC, Yuan F, Ferris JP. 1996. *Icarus* 122:205
Keenan PC, Schroeder LW. 1952. *Ap. J.* 115:82
Kim YC, Fox PA, Demarque P, Sofia S. 1996. *Ap. J.* 461:499–506
Kippenhan R. 1962. *Proc. XXVIII Course,* Tech. Rep., Int. School Enrico Fermi
Kipper T, Jørgensen UG, Klochkova VG, Panchuk VE. 1996. *Astron. Astrophys.* 306:489
Kirkpatrick JD, Beichman CA. 1995. *Bull. Am. Astron. Soc.* 187:7512
Kirkpatrick JD, Henry TJ, Liebert J. 1993a. *Ap. J.* 406:701–7
Kirkpatrick JD, Henry TJ, Simons DA. 1995. *Astron. J.* 109:797–807
Kirkpatrick JD, Kelly DM, Rieke GH, Liebert J, Allard F, Wehrse R. 1993b. *Ap. J.* 402:643–54
Kirkpatrick JD, McCarthy DWJ. 1994. *Astron. J.* 107:333–49
Kirkpatrick JD, McGraw JT, Hess TR, Liebert J, McCarthy DW J. 1994. *Ap. J. Suppl.* 94:749–88
Kui R. 1991. *Model atmospheres for M-dwarfs.* PhD. thesis. Natl. Univ. Aust.
Kunde V, Hanel R, Maguire W, Gautier D, Baluteau JP, et al. 1982. *Ap. J.* 263:443–67
Kurucz RL. 1970. Smithsonian Astrophys. Obs. Spec. Rep. 309. Washington, DC: Smithsonian Astrophys. Obs.
Kurucz RL. 1992a. *Rev. Mex. Astron. Astrofis.* 23:45
Kurucz RL. 1992b. *Rev. Mex. Astron. Astrofis.* 23:187
Kurucz RL. 1993. *Molecular data for opacity calculations,* Kurucz CD-ROM No. 15
Kurucz RL. 1994. *Atomic data for opacity calculations,* Kurucz CD-ROM No. 1
Kusch HJ. 1958. *Z. Astrophys.* 45:1
Kuznetsova LA, Pazyk EA, Stolyarov AV. 1993. *Russ. J. Phys. Chem.* 67:2046
Lacy CH. 1977. *Ap. J.* 218:444
Laird JB, Carney BW, Latham DW. 1988. *Astron. J.* 95:1843
Langhoff SR, Bauschlicher CW. 1990. *J. Mol. Spectrosc.* 141:243
Langhoff SR, Bauschlicher CW Jr. 1991. *Ap. J.* 375:843
Langhoff SR, Bauschlicher CW Jr. 1994. In *Molecules in the Stellar Environment,* ed. UG Jørgensen, *Lect. Notes Phys.,* pp. 310–25. Berlin/Heidelberg: Springer-Verlag
Larson LE, Mickelson ME. 1997. Preprint
Latham DW, Mazeh T, Stefanik RP, Davis RJ, Carney BW, et al. 1992. *Astron. J.* 104:774–95
Lazarev VV, Pnomarev YN. 1992. *Optics Lett.* 17:1283
Leggett SK. 1992. *Ap. J. Suppl.* 82:351–94
Leggett SK, Allard F, Berriman G, Dahn CC, Hauschildt PH. 1996. *Ap. J. Suppl.* 104: 117
Leggett SK, Harris HC, Dahn CC. 1994. *Astron. J.* 108:944
Leggett SK, Hawkins MRS. 1989. *MNRAS* 238:145
Leinert C, Weitzel N, Richichi A, Eckart A, Taccono-Garman LE. 1994. *Astron. Astrophys.* 291:L47
Liebert J, Kirkpatrick JD, Beichman C, Reid IN, Monet DC, Dahn CC. 1995. *Bull. Am. Astron. Soc.* 187:7502
Liebert J, Probst RG. 1987. *Astron. Astrophys. Rev.* 25:473
Liebert J, Schmidt G, Lesser M, Stephanian JA, Lipovetsky VA, Chaffe FH, Foltz CB, Bergeron P. 1994. *Ap. J.* 421:733
Linsky JL, Wood BE, Brown A, Giampapa MS, Ambruster C. 1995. *Ap. J.* 455:670
Ludwig CB. 1971. *Appl. Opt.* 10(5):1057
Maggazzú A, Martín EL, Rebolo R. 1993. *Ap. J.* 404:L17
Magni G, Mazzitelli I. 1979. *Astron. Astrophys.* 72:134
Marcy G, Basri G, Graham JR. 1994. *Ap. J. Lett.* 428:L57
Marcy GW, Butler RP. 1996a. *Ap. J. Lett.* 464:L147
Marcy GW, Butler RP. 1996b. *Ap. J. Lett.* 464:L151
Marcy GW, Butler RP, Williams E, Bildsten L, Graham J. 1997. *Ap. J.* In press
Marcy GW, Chen GH. 1992. *Ap. J.* 390:550
Marilli E, Catalano S, Trigilio C. 1986. *Astron. Astrophys.* 167:297
Marley MS, Saumon D, Guillot T, Freedman R, Hubbard WB, et al. 1996. *Science* 272:1919
Martín EL, Rebolo R, Magazzú A. 1994. *Ap. J.* 436:262
Martín EL, Rebolo R, Zapatero Osorio MR. 1995. See Tinney 1995, pp. 253–56
Martín EL, Rebolo R, Zapatero Osorio MR. 1996. *Ap. J.* 469:706
Mathioudakis M, Doyle JG. 1992. *Astron. Astrophys.* 262:523

Mathys G, Solanki SK. 1989. *Astron. Astrophys.* 208:189
Matthews K, Nakajima SR, Oppenheiner BR. 1995. *IAU Circ.* 6280:1
Mauas PJD, Falchi A. 1996. *Astron. Astrophys.* 310:245
Mayor M, Queloz D. 1995. *Nature* 378:355
Mazeh T, Latham DW, Stefanik RP. 1996. *Ap. J.* 466:415
Méra D, Chabrier G, Baraffe I. 1996a. *Ap. J.* 459:L87
Méra D, Chabrier G, Schaeffer R. 1996b. *Europhys. Lett.* 33:327
Méra D, Chabrier G, Schaeffer R. 1996c. *Astron. Astrophys.* In press
Merer AJ, Huang G, Cheung AS-C, Taylor AW. 1987. *J. Mol. Spectrosc.* 125:465
Metcalfe TS, Mathieu RD, Latham DW, Torres G. 1996. *Ap. J.* 456:356
Michelis CHP. 1976. *A study of van der Waals broadening for lines of astrophysical interest.* PhD thesis. Texas Univ., Austin
Mickelson ME, Larson LE. 1992. *Bull. Am. Astron. Soc.* 24:990
Mihalas D. 1978. *Stellar Atmospheres.* San Francisco: Freeman. 2nd ed.
Miller S, Tennyson J, Jones HRA, Longmore AJ. 1994. In *Molecules in the Stellar Environment,* ed. UG Jørgensen, *Lect. Notes Phys.*, pp. 296–309. Berlin/Heidelberg: Springer-Verlag
Monet DG, Dahn CC, Vrba FJ, Harris HC, Pier JR, et al. 1992. *Astron. J.* 103:638
Morris S, Wyller A. 1967. *Ap. J.* 150:877
Mould JR. 1975. *Astron. Astrophys.* 38:283
Mould JR. 1976. *Ap. J.* 210:402
Mould JR, Wyckoff S. 1978. *MNRAS* 182:63
Muchmore D. 1987. *Ap. J.* 278:769
Mullan DJ, Cheng QQ. 1993. *Ap. J.* 412:312
Mullan DJ, Cheng QQ. 1994. *Ap. J.* 420:392
Mullan DJ, Dermot J. 1987. In *NASA Wash. M-Type Stars,* pp. 455–79
Mullan DJ, Fleming TA. 1996. *Ap. J.* 464:890
Nakajima T, Oppenheimer BR, Kulkarni SR, Golimowski DA, Matthews K, Durrance ST. 1995. *Nature* 378:463
Neale L, Tennyson J. 1995. *Ap. J. Lett.* 454:L169
Nelson LA, Rappaport S, Chiang E. 1993. *Ap. J.* 413:364
Noyes RW, Hartmann LW, Balinunas SL, Duncan DK, Vaughan AH. 1984. *Ap. J.* 279:763
Oppenheimer BR, Kulkarni SR, Nakajima T, Matthews K. 1995. *Science* 270:1478
Paresce F, De Marchi G, Romaniello M. 1995. *Ap. J.* 468:655
Partridge H, Schwenke DW. 1997. *J. Chem. Phys.* In press
Pavlenko YV, Rebolo R, Martín EL, Garcia-Lopez RJ. 1995. *Astron. Astrophys.* 303:807
Peterson BR. 1980. *Astron. Astrophys.* 82:53
Peterson BR. 1989. *Sol. Phys.* 121:299
Peytremann E. 1974. *Astron. Astrophys.* 33:203
Phillips JG, Davis SP. 1993. *Ap. J.* 409:860
Phillips JG, Davis SP, Lindgren B, Balfour WJ. 1987. *Ap. J. Suppl.* 65:721
Piskunov N, Valenti J, Johns-Krull C. 1996. *Bull. Am. Astron. Soc.* 188:4407
Plez B, Brett JM, Nordlund A. 1992. *Astron. Astrophys.* 256:551
Pollack JB, Hollenbach D, Beckwith S, Simonelli DP, Roush T, Fong W. 1994. *Ap. J.* 421:615
Pollack JB, McKay CP, Christofferson BM. 1985. *Icarus* 64:471
Rebolo R, Martín EL, Basri G, Marcy GW, Zapatero Osorio MR. 1996. *Ap. J.* 469:L53
Rebolo R, Martín EL, Maggazzú A. 1992. *Ap. J.* 389:L83
Rebolo R, Zapatero Osorio MR, Martín EL. 1995. *Nature* 377:129
Reid IN. 1994. *Astrophys. Space Sci.* 217:57
Reid IN, Gilmore RF. 1984. *MNRAS* 206:19
Reid IN, Hawley SL, Gizis JE. 1995a. *Astron. J.* 110:1838
Reid IN, Hawley SL, Mateo M. 1995b. *MNRAS* 272:828
Reid N. 1993. *MNRAS* 265:785
Renzini A, Bragaglia A, Ferraro FR, Gilmozzi R, Ortolani S, et al. 1996. *Ap. J.* 465:23
Richer HB, Fahlman GG, Ibata RA, Stetson PB, Bell RA, et al. 1995. *Ap. J.* 451:17
Robinson RD. 1980. *Ap. J.* 239:961
Rosner R, Golub L, Vaiana G. 1985. *Astron. Astrophys. Rev.* 23:413
Rossi SCF, Marciel WJ, Benevides-Soares P. 1985. *Astron. Astrophys.* 148:93
Rossow WB. 1978. *Icarus* 36:1
Rothman LS, Gamache RR, Tipping RH, Rinsland CP, Smith MAH, et al. 1992. *J. Quant. Spectrosc. Radiat. Transfer* 48:469
Rucinski SM. 1978. *Acta Astron.* 28:167
Russell HM. 1934. *Ap. J.* 79:317
Rutten RGM. 1986. *Astron. Astrophys.* 159:291
Rutten RGM, Zwaan C, Schrijver CJ, Duncan DK, Mewe R. 1989. *Astron. Astrophys.* 219:239
Ryan SG, Norris JE, Bessell MS. 1991. *Astron. J.* 102:303
Saar SH. 1988. *Ap. J.* 324:441
Saar SH. 1994. In *Infrared Solar Physics,* ed. DM Rabin et al, *IAU Colloq. 154,* p. 493. Dordrecht: Kluwer
Saumon D, Bergeron P, Lunine JI, Hubbard WB, Burrows A. 1994. *Ap. J.* 424:333
Saumon D, Burrows A, Hubbard WB. 1995a. See Tinney 1995, pp. 3–12
Saumon D, Chabrier G, Van Horn HM. 1995b. *Ap. J. Suppl.* 99:713
Saumon D, Hubbard WB, Burrows A, Guillot T, Lunine JI, Chabrier G. 1996. *Ap. J.* 460:993
Sauval AJ, Tatum JB. 1984. *Ap. J. Suppl.* 56:193

Schryber HS, Miller S, Tennyson J. 1995. *J. Quant. Spectrosc. Radiat. Transfer* 53:373

Schweitzer A, Hauschildt PH, Allard F, Basri G. 1996. *MNRAS* 283:821–29

Seab GC, Snow TP. 1989. *Ap. J.* 347:479

Seaton JM. 1992. *Rev. Mex. Astron. Astrofis.* 23:180

Seaton JM, Zeippen CJ, Tully JA, Pradhan AK, Mendoza C, et al. 1992. *Rev. Mex. Astron. Astrofis.* 23:19

Sharp CM, Huebner WF. 1990. *Ap. J. Suppl.* 72:417

Simons DA, Becklin EE. 1992. *Ap. J.* 390: 431

Sneden L, Johnson HR, Krupp BM. 1976. *Ap. J.* 204:281

Stauffer J, Hamilton D, Probst R. 1994. *Astron. J.* 108:155

Stauffer JR, Giampapa MS, Herbst W, Vincent JM, Hartmann LW, Stern RA. 1991. *Ap. J.* 374:142

Stauffer JR, Hartmann LW. 1986. *Ap. J. Suppl.* 61:531

Stevenson DJ. 1991. *Astron. Astrophys. Rev.* 29:163

Stone RC, Monet DG, Monet AKB, Walker RL, Ables HD. 1996. *Astron. J.* 111:1721

Strong K, Taylor FW, Calcutt SB, Remedios JJ, Ballard J. 1993. *J. Quant. Spectrosc. Radiat. Transfer* 50:363

Takeda Y, Kato K-I, Watanabe Y, Sadakane K. 1996. *Publ. Astron. Soc. Jpn.* 48:511

Thackrah A, Jones HRA, Hawkins M. 1997. *MNRAS.* In press

Tinney CG. 1993. *The faintest stars.* PhD thesis. Calif. Inst. Technol., Pasadena

Tinney C, ed. 1995. *The Bottom of the Main Sequence—and Beyond. ESO Astrophys. Symp.* Berlin/Heidelberg: Springer-Verlag

Tinney CG. 1996. *MNRAS* 281:644

Tinney CG, Mould JR, Reid IN. 1993. *Astron. J.* 105:1045

Tinney CG, Reid IN, Gizis J, Mould JR. 1995. *Astron. J.* 110:3014

Tsuji T. 1966. *Publ. Astron. Soc. Jpn.* 18:127

Tsuji T. 1973. *Astron. Astrophys.* 23:411

Tsuji T. 1994. In *Molecules in the Stellar Environment,* ed. UG Jørgensen, *Lect. Notes Phys.,* pp. 79–97. Berlin/Heidelberg: Springer-Verlag

Tsuji T, Ohnaka K, Aoki W. 1995. See Tinney 1995, pp. 45–52

Tsuji T, Ohnaka K, Aoki W. 1996a. *Astron. Astrophys.* 305:L1

Tsuji T, Ohnaka K, Aoki W, Nakajima T. 1996b. *Astron. Astrophys.* 308:L29

Tyuterev VG, Babikov YL, Tashkun SA, Perevalov VI, Nikitin A, et al. 1994. *J. Quant. Spectrosc. Radiat. Transfer* 52:459

Unsöld A. 1955. *Physik der Sternatmosphären.* Berlin/Göttingen/Heidelberg: Springer-Verlag. 2nd ed.

Vardya MS. 1966. *MNRAS* 134:347

Veeder GJ. 1974. *Astron. J.* 79:1056

Verner DA, Verner EM, Ferland GJ. 1996. *Bull. Am. Astron. Soc.* 188:5418

Viti S, Jones HRA, Tennyson J, Polyansky OL, Miller S, et al. 1995. *Bull. Am. Astron. Soc.* 187:10314

von Hippel T, Gilmore G, Tanvir N, Robinson D, Jones D. 1996. *Astron. J.* 112:192

Warren SJ, Irwin MJ, Evans DW, Liebert J, Osmer PS, Hewett PC. 1993. *MNRAS* 261:185

Weast RC, ed. 1988. *Handbook of Chemistry and Physics.* Boca Raton, FL: CRC Press. 2nd ed.

Wehrse R. 1972. *Astron. Astrophys.* 19:453

Wehrse R, Liebert J. 1980. *Astron. Astrophys.* 86:139

Weidemann V. 1955. *Z. Astrophys.* 36:101

Williams DM, Boyle RP, Morgan WT, Rieke GH, Stauffer JR, Rieke MJ. 1996. *Ap. J.* 464:238

Williams E, Butler R, Marcy G. 1997. *Ap. J.* In press

Wilson OC, Skumanich A. 1964. *Ap. J.* 140:1401

Zapatero Osorio MR, Rebolo R, Martín EL. 1997. *Astron. Astrophys.* 317:164

Zapatero Osorio MR, Rebolo R, Martín EL, Garcia Lopez RJ. 1996. *Astron. Astrophys.* 305:519

Zheng C, Borysow A. 1994. *Ap. J.* 441:960

Ziurys LM, Barclay WLJ, Anderson MA. 1992. *Ap. J.* 384:63

Ziurys LM, Fletcher DA, Anderson MA, Barclay WLJ. 1996. *Ap. J. Suppl.* 102:425

Zubko VG. 1987. *On the Microstructure Theory of a Condensate Layer in a White Dwarf Atmosphere.* Kiev: Acad. Sci. Ukrainian SSR

Zuckerman B, Becklin EE. 1992. *Ap. J.* 386:260

*Annu. Rev. Astron. Astrophys. 1997. 35:179–215*

# DENSE PHOTODISSOCIATION REGIONS (PDRs)[1]

*D. J. Hollenbach and A. G. G. M. Tielens*
MS 245-3, NASA Ames Research Center, Moffett Field, California 94035;
e-mail: hollenbach@warped.arc.nasa.gov, tielens@dusty.arc.nasa.gov

KEY WORDS: interstellar medium, astrochemistry, infrared astronomy, nebulae, atomic processes

ABSTRACT

All neutral atomic hydrogen gas and a large fraction of the molecular gas in the Milky Way Galaxy and external galaxies lie in PDRs, and PDRs are the origin of most of the nonstellar infrared (IR) and the millimeter CO emission from a galaxy. On the surfaces ($A_V < 1-3$) of interstellar clouds, the absorption of far ultraviolet (FUV) photons ($h\nu < 13.6$ eV) by gas and dust grains leads to intense emission of [C II] 158 $\mu$m, [O I] 63, 146 $\mu$m, and $H_2$ rovibrational transitions, as well as IR dust continuum and polycyclic aromatic hydrocarbon (PAH) emission features. Deeper in PDRs, CO rotational and [C I] 370, 609 $\mu$m lines originate. The transition of H to $H_2$ and $C^+$ to CO occurs within PDRs. Theoretical models compared with observations diagnose such physical parameters as the density and temperature structure, the elemental abundances, and the FUV radiation field in PDRs. Applications include clouds next to H II regions, reflection nebulae, planetary nebulae, red giant outflows, circumstellar gas around young stars, diffuse clouds, the warm neutral medium (WNM), and molecular clouds in the interstellar radiation field: in summary, much of the interstellar medium in galaxies. This review focuses on dense PDRs in the Milky Way Galaxy. Theoretical PDR models help explain the observed correlation of the CO J = 1–0 luminosity with the molecular mass and also suggest FUV-induced feedback mechanisms that may regulate star formation rates and the column density through giant molecular clouds.

# 1. INTRODUCTION

The study of photodissociation regions (PDRs, also called photon-dominated regions) is the study of the importance of FUV (6 eV $< h\nu <$ 13.6 eV) photons on the structure, chemistry, thermal balance, and evolution of the neutral interstellar medium of galaxies. One important aspect of this study is understanding the process of star formation. FUV photons not only illuminate star-forming regions, causing them to glow in infrared (IR) emission diagnostic of the physical conditions, but they also may play an important role regulating the star formation process itself in a galaxy.

Historically, the study of PDRs was stimulated by the early observations of the massive star-forming regions Orion A and M17 in the fine structure lines [C II] 158 $\mu$m and [O I] 63 $\mu$m by Melnick et al (1979), Storey et al (1979), and Russell et al (1980, 1981). These observations pointed to predominantly neutral IR-luminous regions lying outside H II regions. These PDRs are photodissociated and, for elements like carbon with ionization potentials below 13.6 eV, photoionized by the FUV fluxes generated by nearby O stars. The luminosity in the [C II] and [O I] lines, which dominate the cooling of the atomic gas, is on the order of $10^{-3}$–$10^{-2}$ of the IR luminosity from the dust that absorbed the starlight. Figure 1 shows a more recent map of the [C II] 158 $\mu$m and [O I] 63 $\mu$m emission from Orion A (Herrmann et al 1997), showing the large extent ($\gtrsim 5'$ by $5'$ or 0.75 pc by 0.75 pc) and luminosity ($L_{CII} \sim 80\ L_{\odot}$ and $L_{OI} \sim 600\ L_{\odot}$) of the IR-glowing neutral gas associated with the Trapezium stars. Figure 2 (Tielens et al 1993) shows a smaller scale map of the Orion Bar, where the neutral gas outside of the H II region is viewed edge-on. A layered appearance is evident; moving away from the excitation source, the ionization front, H I layer, $H_2$ emission, and CO emission appear in succession.

Despite its historical roots, the study of PDRs is not simply the study of photodissociated gas that lies just outside of dense, luminous H II regions in the Galaxy; it includes as well the pervasive warm neutral medium (WNM), diffuse and translucent clouds, reflection nebulae, the neutral gas around planetary nebulae, photodissociated winds from red giant and asymptotic giant branch (AGB) stars, and the interstellar medium (ISM) in the nuclei of starburst galaxies and galaxies with active galactic nuclei (AGNs). PDRs include all interstellar regions where the gas is predominantly neutral but where FUV photons play a significant role in the chemistry and/or the heating. Figure 3 schematically illustrates the structure of a PDR. The ultraviolet flux from, for example, the interstellar radiation field (ISRF) or from nearby hot stars is incident on a neutral cloud of hydrogen nucleus density *n*. The incident FUV flux $G_0$ (in units of an average interstellar flux of $1.6 \times 10^{-3}$ erg cm$^{-2}$ s$^{-1}$; Habing 1968) can range from the local average ISRF ($G_0 \sim 1.7$, Draine 1978) to $G_0 \gtrsim 10^6$, appropriate,

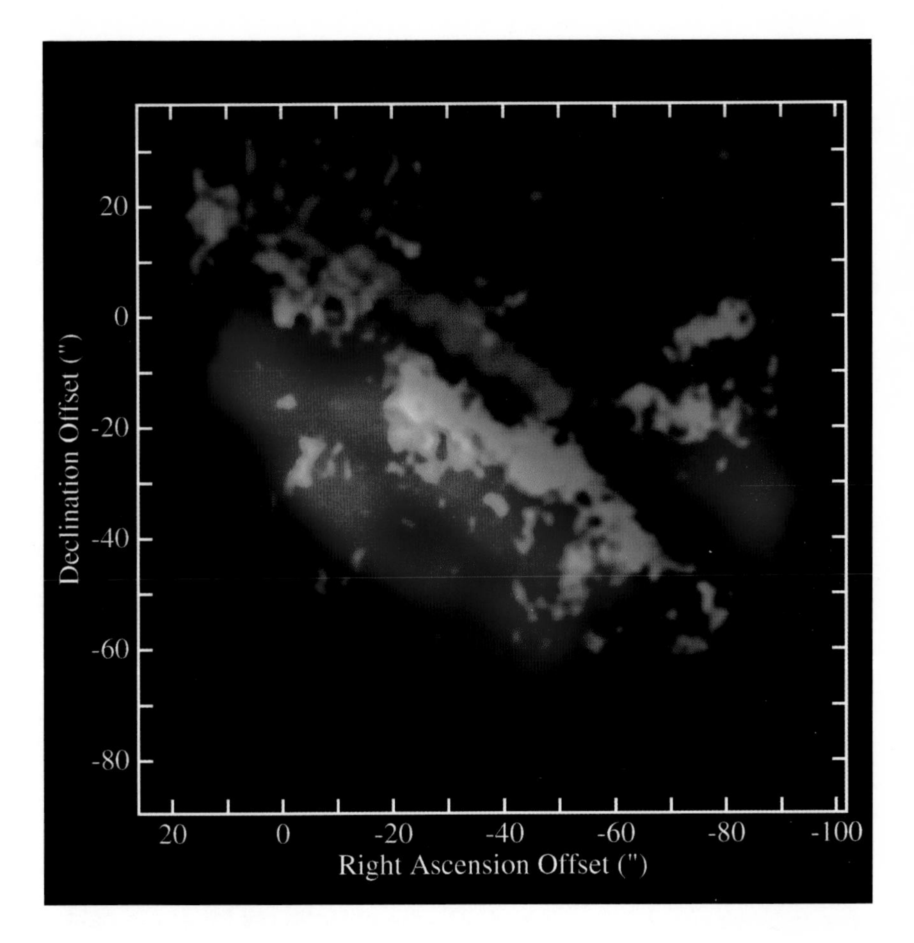


*Figure 2* The Orion Bar region mapped in the 3.3-μm PAH feature *(blue),* $H_2$ 1-0 S(1) emission *(yellow),* and CO J=1-0 emission *(red;* Tielens et al 1993). The (0,0) position corresponds to the (unrelated) star $\theta^2$ A Ori. The illuminating source, $\theta^1$ C Ori and the ionized gas are located to the northwest. For all three tracers, the emission is concentrated in a bar parallel to but displaced to the southeast from the ionization front. The PDR is seen edge on; a separation of $\simeq$ 10" is seen between the PAH emission and the $H_2$ emission, as well as between the $H_2$ emission and the CO emission, as predicted by PDR models (see text).

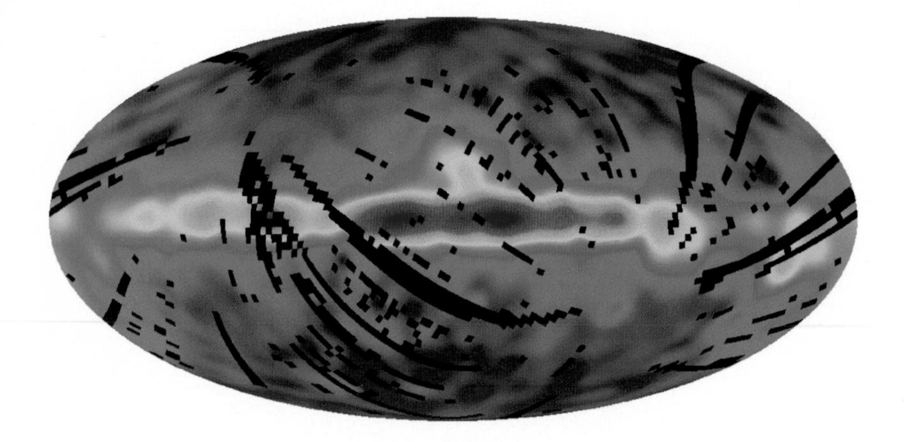

*Figure 4* The [CII] emission of the galaxy observed by the FIRAS instrument on COBE (Bennett et al 1994) (158-μm $C^+$ line intensity). The galactic plane is horizontal with the galactic center at the center.

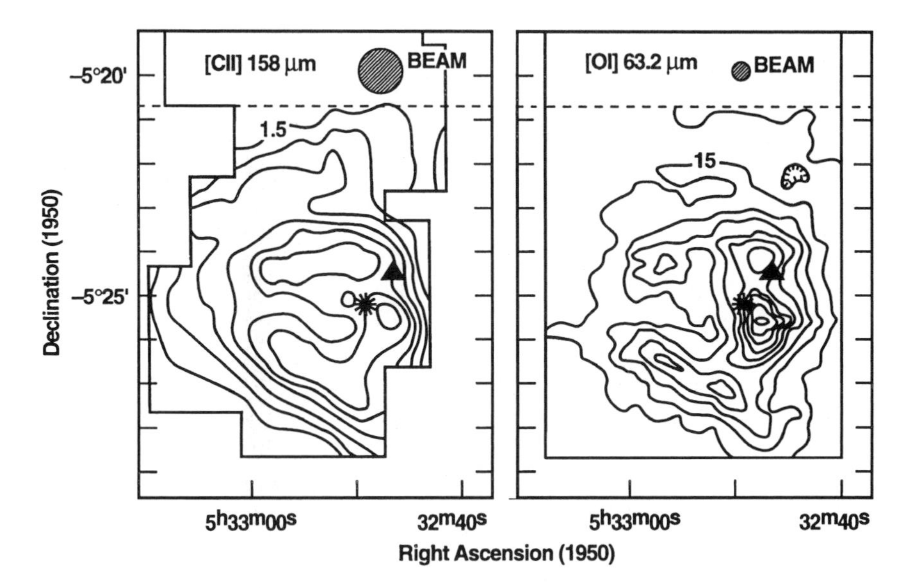


*Figure 1* [C II] 158 $\mu$m and [O I] 63 $\mu$m maps of the Orion A molecular cloud behind the Trapezium (adapted from Herrmann et al 1997). The [C II] contours are in steps of $5 \times 10^{-4}$ erg cm$^{-2}$ s$^{-1}$ sr$^{-1}$ beginning with $1.5 \times 10^{-3}$ erg cm$^{-2}$ s$^{-1}$ sr$^{-1}$. The [O I] are in steps of $10 \times 10^{-3}$ erg cm$^{-2}$ s$^{-1}$ sr$^{-1}$ beginning with $15 \times 10^{-3}$ erg cm$^{-2}$ s$^{-1}$ sr$^{-1}$. A small area of $\simeq 1'$ ($\simeq 0.15$ pc) around IRc 2 has significant [O I] shock emission. The asterisk and triangle indicate the location of $\theta^1$ C Ori and IRc 2, respectively.

for example, to gas closer than 0.1 pc from an O star. Typically, densities $n$ range from $\sim 0.25$ cm$^{-3}$ in the WNM, to $\sim 10$–$100$ cm$^{-3}$ in diffuse clouds, to $\sim 10^3$–$10^7$ cm$^{-3}$ in the PDRs associated with molecular gas. As illustrated in Figure 3, PDRs are often overlaid with H II gas and a thin H II/H I interface that absorb the Lyman continuum photons. Although dependent on the ratio $G_0/n$, the PDR itself is often characterized by a layer of atomic hydrogen that extends to a depth $A_v \sim 1$–$2$ (or a hydrogen nucleus column of $N = 2$–$4 \times 10^{21}$ cm$^{-2}$) from the ionization front, a layer of $C^+$ that extends to a depth $A_v \sim 2$–$4$, and a layer of atomic oxygen that extends to a depth $A_v \sim 10$. The H, $C^+$, and O layers are maintained by the FUV photodissociation of molecules and the FUV photoionization of carbon.

Traditionally, PDRs have been associated with atomic gas. However, with the above definition, PDRs include material in which the hydrogen is molecular and the carbon is mostly in CO, but where FUV flux still strongly affects the chemistry of oxygen and carbon not locked in CO (photodissociating OH, $O_2$, and $H_2O$, for example) and the ionization fraction. With the exception of the

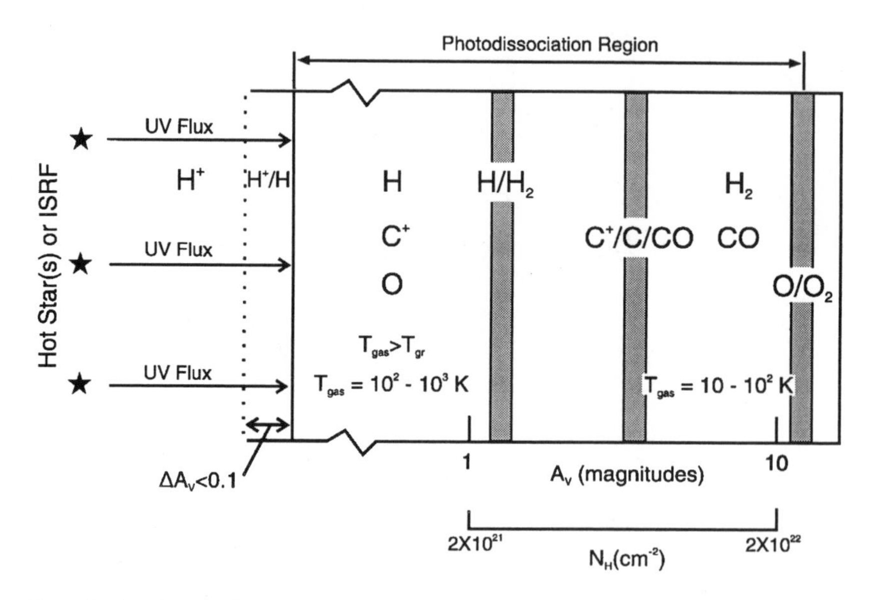


*Figure 3* A schematic diagram of a photodissociation region. The PDR is illuminated from the left and extends from the predominantly atomic surface region to the point where $O_2$ is not appreciably photodissociated ($\simeq$10 visual magnitude). Hence, the PDR includes gas whose hydrogen is mainly $H_2$ and whose carbon is mostly CO. Large columns of warm O, C, $C^+$, and CO and vibrationally excited $H_2$ are produced in the PDR. The gas temperature $T_{gas}$ generally exceeds the dust temperature $T_{gr}$ in the surface layer.

molecular gas in dense star-forming cores, most molecular gas in the Galaxy is found at $A_v \lesssim 10$ in Giant Molecular Clouds (GMCs). Therefore, *all of the atomic and most of the molecular gas in the Galaxy is in PDRs.*

Not only do PDRs include most of the mass of the ISM, but PDRs are the origin of much of the IR radiation from the ISM (the other significant sources are H II regions and dust heated by late-type stars). The incident starlight is absorbed primarily by large carbon molecules (polycyclic aromatic hydrocarbons or PAHs) and grains inside a depth $A_v \sim 1$. Most of the absorbed energy is used to excite the PAHs and heat the grains and is converted to PAH IR features and far-infrared (FIR) continuum radiation of the cooling grains. However, typically 0.1–1% of the absorbed FUV energy is converted to energetic ($\sim$1 eV) photoelectrons that are ejected from PAHs and grains and heat the gas ("photoelectric heating"). Although the gas receives $10^2$–$10^3$ times less heating energy per unit volume than the dust, the gas attains higher equilibrium temperatures because of the much less efficient cooling of the gas (via [C II] 158 $\mu$m and [O I] 63 $\mu$m) relative to the radiative dust cooling.

Much of the [C II], [O I], and [Si II] fine-structure, carbon recombination, the $H_2$ rotational and vibrational, and C I(9850 Å) emission in galaxies originates from depths $A_v \lesssim 4$ in PDRs. Most of the [C I] fine structure and the CO rotational emission in galaxies comes from regions somewhat deeper in PDRs. For example, the COBE 4-$\mu$m to 1000-$\mu$m spectrum of our Milky Way Galaxy (Wright et al 1991) is dominated by PDR emission (see Figure 4 for the [C II] map), with the exception of the [N II] and a fraction of the [C II] fine structure emission, which originates in diffuse H II gas. Evident in Figure 4 are the PDRs associated with the the molecular ring at 3 kpc and the GMCs in Ophiuchus and Orion.

We discuss the physical and chemical processes in PDRs in Section 2, followed by a summary of the theoretical models in Section 3. The observations of dense PDRs in the Milky Way Galaxy (Section 4) span a wide variety of phenomena and diagnose the physical conditions in neutral gas around H II regions, reflection nebulae, planetary nebulae, and the Galactic Center. PDR models explain the $H_2$ spectra, the origin of [C I], the correlation of [C II] to CO J = 1–0, and the correlation of ([C II] + [O I]) to the IR continuum intensity (Section 4). We conclude in Section 5 with brief descriptions of PDR models applied to some interesting problems, including the scaling of CO luminosity with mass, the effect of X rays on molecular gas (X-ray dissociation regions or XDRs), the interstellar conditions of the starburst nucleus of M82, and the regulation of star formation.

Because of space limitations in this review, we have focused on dense PDRs in our Galaxy. Hollenbach & Tielens (1996) present a more comprehensive PDR review that discusses global models of the ISM of the Milky Way and other galaxies, the distribution of H I and $H_2$ in galaxies, H I halos around molecular clouds, the formation of molecular clouds, diffuse and translucent clouds, the neutral phases of the ISM, and more detailed discussion of the problems summarized in Section 5. Glassgold (1996) reviews PDRs associated with red giant outflows. Other PDR reviews include Genzel et al (1989), Jaffe & Howe (1989), Hollenbach (1990), Genzel (1991, 1992), Burton (1992), Sternberg (1992), van Dishoeck (1992), Hollenbach & Tielens (1996), and Sternberg et al (1997).

## 2. PHYSICAL AND CHEMICAL PROCESSES

### 2.1 *The Penetration of FUV Radiation*

One of the keys to understanding PDR structure lies in understanding the attenuation of FUV continuum flux through the PDR. The penetration of FUV radiation is determined by dust absorption and scattering and the geometry and global structure of interstellar clouds. Various authors have studied the penetration of FUV radiation inside homogeneous clouds. In particular, Roberge et al (1981, 1991) solved the radiative transfer equation for plane parallel slabs of various thicknesses and calculated photodissociation and photoionization rates for a variety of astrophysically relevant molecules. Deep inside semi-infinite slabs, the mean intensity is proportional to $\exp[-kA_v]$, where $k$ is a scale factor and $A_v$ the visual extinction measured from the surface. For optically thin clouds, a correction has to be included for photons penetrating from the other side of the slab. Generally, a biexponential fit to the radiative transfer solution suffices (van Dishoeck 1988, Roberge et al 1991).

The intensity of the radiation field inside interstellar clouds depends critically on the adopted absorption and scattering properties of the dust. Generally, theoretical studies rely on either directly measured "average" properties of interstellar dust (cf Savage & Mathis 1979, Mathis 1990) or on models that fit these "average" properties (e.g. Draine & Lee 1984). These "average" dust properties refer exclusively to a (very biased) sample of lines of sight through diffuse interstellar clouds. Moreover, the dust properties are known to vary from one diffuse cloud to another (cf Cardelli et al 1989). Finally, dust in molecular clouds is generally characterized by a high value for R, the ratio of total to selective extinction. If diffuse clouds are a guide, this implies much lower FUV extinction per hydrogen atom than commonly adopted. The uncertainty can easily amount to 50% in $k$.

In recent years, it has become increasingly clear that interstellar clouds are inhomogeneous on all scales (e.g. Falgarone & Phillips 1996 and references therein). This clumpy nature of interstellar clouds can have a profound influence on the penetration of FUV radiation (Stutzki et al 1988, Boissé 1990, Spaans 1996, Hegmann & Kegel 1996). The study by Boissé is particularly instructive and establishes simple scaling laws that can be easily adopted for PDR modeling (cf Tauber & Goldsmith 1990, Howe et al 1991, Hobson & Scheuer 1994, Meixner & Tielens 1993). Depending on the clump filling factor and the density contrast between clump and interclump gas, the scale size for the penetration of FUV radiation can vary by orders of magnitude.

One important characteristic of FUV penetration into clumpy clouds is the existence of large fluctuations in the mean intensity at a given depth (Boissé 1990, Spaans 1996). These fluctuations are particularly important when individual clumps are optically thick (optical depth $\tau_0$), scattering in the clump or interclump gas is unimportant, and the cloud is illuminated by a unidirectional field (i.e. a nearby star). In this limit, the FUV field can fluctuate by $\sim\exp(\tau_0)$ (Monteiro 1991, Störzer et al 1997). When any of these restrictions are relaxed, fluctuations become of lesser importance and may be no more than a factor of a few in realistic situations (Boissé 1990).

## 2.2 *Chemistry in PDRs*

PDR chemistry has been discussed in detail by Tielens & Hollenbach (1985a,c), LeBourlot et al (1993), Hollenbach et al (1991), Fuente et al (1993, 1995), Jansen et al (1995a,b), and Sternberg & Dalgarno (1995). It derives rather directly from the chemistry of those more transparent PDRs, diffuse and translucent clouds (Glassgold & Langer 1974, 1976, Black & Dalgarno 1976, Federman et al 1980, 1984, 1994, Danks et al 1984, van Dishoeck & Black 1986, 1988, 1989, Viala 1986, Viala et al 1988, Federman & Huntress 1989, van Dishoeck 1991, Heck et al 1992, Turner 1996, and references therein). PDR chemistry differs from normal ion-molecule chemistry in a number of ways. Obviously, because of the high FUV flux, photoreactions are very important, as are reactions with atomic hydrogen. Likewise, vibrationally excited $H_2$ is abundant and can play a decisive role in PDR chemistry. If the gas gets very warm ($\gtrsim$500 K), the activation barrier of reactions of atoms and radicals with $H_2$ can be easily overcome and these types of reactions can dominate. Electron recombination and charge exchange reactions are important for the ionization balance. Finally, the FUV flux keeps atomic O very abundant throughout the PDR, and hence burning reactions are effective. Here, we first discuss $H_2$ and CO self-shielding. We follow this with a discussion of the chemistry that can occur when the reactants have non-Maxwellian excitation of the vibrational or translational degrees of freedom. Finally, we briefly summarize chemical networks for PDRs.

PHOTODISSOCIATION AND SELF-SHIELDING OF $H_2$ AND CO Photodissociation and self-shielding of $H_2$ has been discussed by Field et al (1966), Stecher & Williams (1967), Hollenbach et al (1971), Jura (1974), Black & Dalgarno (1977), Shull (1978), Federman et al (1979), deJong et al (1980), van Dishoeck & Black (1986), Abgrall et al (1992), Heck et al (1992), LeBourlot et al (1993), and Lee et al (1996); reviewed by van Dishoeck (1987); and recently studied in detail by Draine & Bertoldi (1996), so a short summary will suffice here. Photodissociation of $H_2$ proceeds through FUV absorption in the Lyman and Werner transitions in the 912- to 1100-Å range, followed by fluorescence to the vibrational continuum of the ground electronic state about 10–15% of the time (see Figure 6 in Section 2.4). The $H_2$ photodissociation rate follows then from a summation over all lines. When the $H_2$ column density exceeds $10^{14}$ cm$^{-2}$, the FUV absorption lines become optically thick and self-shielding becomes important. The photodissociation rate then depends on the $H_2$ abundance and level population distribution as a function of depth in the cloud. Various approximations, appropriate for chemical modeling, have been described by Jura (1974), Federman et al (1979), van Dishoeck & Black (1986), and Draine & Bertoldi (1996). Self-shielding alone dominates $H_2$ dissociation and hence the location of the $H/H_2$ transition when $G_0/n \lesssim 0.01$–$0.1 \text{ cm}^3$. This includes diffuse clouds exposed to the interstellar radiation field and dense clumps in PDRs with higher FUV fluxes. Because of the self-shielding, the $H_2$ column increases rapidly and the $H/H_2$ transition zone is quite sharp. PDRs associated with bright FUV sources typically have $G_0/n \sim 1 \text{ cm}^3$. The location of the $H/H_2$ transition is then dominated by dust absorption and typically occurs at $A_v \simeq 2$. At that point, dust has reduced the $H_2$ photodissociation rate sufficiently that an appreciable column of $H_2$ can build up, $H_2$ self-shielding takes over, and the $H/H_2$ transition will be very rapid again.

Photodissociation of CO has been studied in detail by Bally & Langer (1982), Glassgold et al (1985), van Dishoeck & Black (1988), Viala et al (1988), and Lee et al (1996). High-resolution laboratory studies show that CO photodissociation occurs through discrete absorption into predissociating bound states, implying that CO is also affected by self-shielding (Eidelsberg et al 1992). Further complications arise because of line coincidence with H and $H_2$. CO shielding functions have been tabulated by van Dishoeck & Black (1988) and by Lee et al (1996). For low ratios of $G_0/n$, the effects of CO self-shielding can lead to isotopic fractionation effects at the borders of clouds, where the rarer CO isotopes are preferentially photodissociated. The location of the $C^+$/C/CO transition in bright PDRs is largely governed by dust extinction. Because of the much lower abundance of CO relative to $H_2$, the CO rarely builds up sufficient column to self-shield in the $A_v \lesssim 1$ layer and the $C^+$/C/CO transition is much less sharp than that of $H/H_2$. Like $H_2$, CO abundances can be appreciable near the surfaces of dense clumps. However, that is not a result of self-shielding but rather reflects the high $H_2$ abundance near the surfaces of such clumps, which leads to an enhanced CO formation rate.

NON-MAXWELLIAN CHEMISTRY Neutral-neutral reactions are often endothermic and/or possess appreciable activation barriers because bonds have to be broken or rearranged. In PDRs, reactions of molecular hydrogen with $C^+$, O, N, $S^+$, and $Si^+$ are particularly important in initiating the chemical reactions. FUV pumping of $H_2$ can lead to a vibrational excitation temperature that is considerably higher than the gas temperature. Reactions with vibrational hot $H_2$ ($H_2^*$) have been discussed in an astrophysical context by Wagner & Graff (1987). Non–equilibrium excitation conditions can give rise to reaction rates that are considerably enhanced over the thermal ones (Gardiner 1977, Dalgarno 1985). For example, studies of the O + $H_2^*$, OH + $H_2^*$, and $C^+$ + $H_2^*$ reactions show large enhancements with vibrational excitation of $H_2$ (Light 1978, Schinke & Lester 1979, Schatz & Elgersma 1980, Schatz 1981, Schatz et al 1981, Lee et al 1982, Jones et al 1986). However, state-to-state chemistry is very selective and vibrational excitation of OH does not enhance the reaction rate of OH + $H_2$.

For lack of better knowledge, most studies of the chemistry in PDRs generally assume that exothermic reactions with $H_2^*$ occur at the collision rate.

Translational energy can also be effective in promoting chemical reactions. In particular, the presence of turbulence can lead to non-Maxwellian velocity fields and, hence, non-Maxwellian reaction rates. This effect in PDRs has been shown to be important by Spaans et al (1997) for the reaction $C^+ + H_2$, which is endothermic by 0.4 eV. Because of their larger activation barriers, the reactions $O + H_2$ and $N + H_2$ are much less affected. Spaans & Jansen (in preparation; see Spaans 1995) evaluated in a simplified way, using estimated reaction rates, the effects of turbulence on sulfur chemistry. Finally, Falgarone et al (1995) examined the enhancement of OH, $H_2O$, $CH^+$, and $HCO^+$ in turbulent PDRs.

PDR CHEMICAL NETWORKS The most important reactions in the chemistry of carbon and oxygen compounds are schematically shown in Figure 5. Figure 5 is adapted from Sternberg & Dalgarno (1995) who provide a detailed discussion of PDR chemistry. The PDR surface layer consists largely of neutral or cationic atoms created by photodissociation and ionization reactions. Oxygen-bearing radicals (i.e. OH) are built up through reactions of O with $H_2^*$ and $H_2$. Most of the OH produced is photodissociated again, but a small fraction reacts with

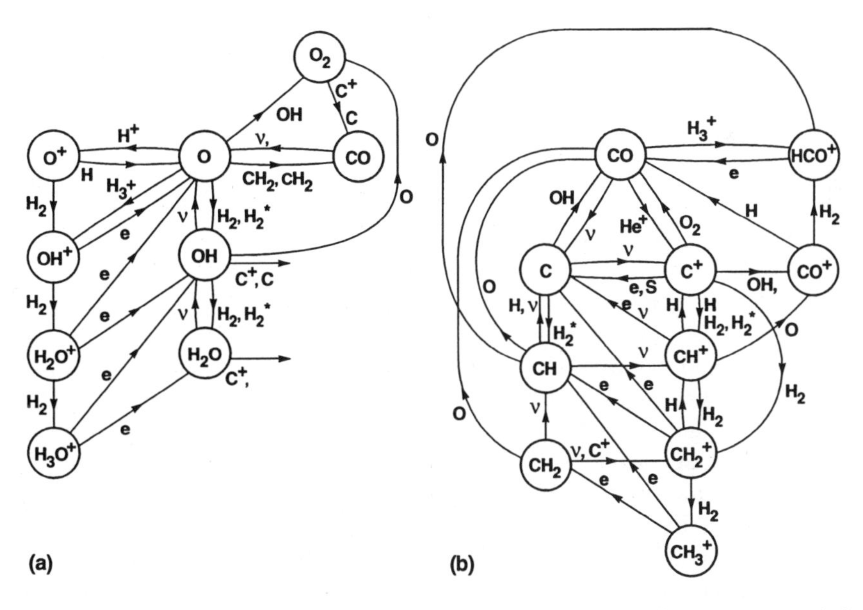


*Figure 5* The most important reactions involved in the PDR chemistry of oxygen-bearing (*left*) and carbon-bearing (*right*) compounds (adapted from Sternberg & Dalgarno 1995).

$C^+$ to form $CO^+$, which charge exchanges with H to form CO. Some $CO^+$ is also formed through the reaction of $CH^+$ with O ("burning"). The CO photodissociates again. The $C^+$/C balance is dominated by photoionization and radiative recombination reactions. A small fraction of the $C^+$ reacts with H and $H_2^*$ to form $CH^+$. Likewise, a small fraction of the neutral C flows to CH through reaction with $H_2^*$. Through reactions with H, $CH^+$ and CH reform $C^+$ and C, respectively. Photoreactions are important for C, OH, and CO but not for the small hydrocarbon radicals and cations. With increasing depth in this so-called radical zone (but recall it is largely atomic), the chemistry involving small radicals, such as the $CH_n^+$-family and OH, becomes more important. In cool low-density PDRs, reactions with FUV-pumped vibrationally excited $H_2^*$ are important, whereas in warm high-density PDRs, reactions with $H_2$ dominate. In this case, reactions of $C^+$ with $H_2$ followed by dissociative electron recombination of $CH^+$ to C can be an important recombination route for $C^+$. As the depth increases, CO formation by burning of small neutral radicals (i.e. CH, $CH_2$) becomes more important than the OH-driven channel. The $CO^+$, produced through the reaction of $C^+$ with OH, reacts with $H_2$ to form $HCO^+$, which dissociatively recombines to CO. This is the start of the $C^+$/C/CO transition zone. Neutralization of $C^+$ through charge transfer with atomic S becomes a dominant source of C. Eventually, PDR chemistry gives way to standard dark cloud, ion-molecule chemistry (Prasad & Huntress 1980, Herbst & Leung 1989). Formation of OH and $H_2O$ is now initiated through the reaction of $H_3^+$ with O. Reactions of atomic O with OH then forms $O_2$.

## 2.3 *Cooling*

The gas in PDRs is cooled by FIR fine structure lines, such as [C II] 158 $\mu$m, [O I] 63, 146 $\mu$m, [Si II] 35 $\mu$m, [C I] 609, 370 $\mu$m, by $H_2$ rovibrational, and by molecular rotational lines, particularly of CO. For high densities and $G_0$, the gas at the surface of the PDR attains temperatures $\gtrsim$5000 K, and significant cooling in [Fe II] (1.26 and 1.64 $\mu$m), [O I] 6300 Å, and [S II] 6730 Å results (Burton et al 1990b). Convenient fitting formulae for the collisional excitation rates have been published by Tielens & Hollenbach (1985a), Hollenbach & McKee (1989), and Spaans et al (1994) for the atomic fine structure and forbidden lines, by Hollenbach & McKee (1979) and McKee et al (1982) for rotational transitions of CO, and by Martin & Mandy (1995) and Martin et al (1996) for rovibrational transitions of $H_2$. At high density, cooling by collisions with the cooler dust grains may be significant (Burke & Hollenbach 1983). The local radiative-cooling rate of a species is also affected by radiative transfer and, hence, depends on the global distribution of the level populations throughout the PDR. Generally, the escape-probability formalism is used to calculate the local cooling rate. As a result, in semi-infinite slabs the PDR temperature structure can be calculated from the outside to the inside without the need for global iterations. However, for clumpy and turbulent clouds, more sophisticated techniques are required when photons escape through $4\pi$ steradians (sr) (Köster et al 1994, Störzer et al 1996) and when line transfer occurs between clumps or between clumps and interclump medium (Hegmann & Kegel 1996, Spaans 1996).

## 2.4 *Heating*

Two main mechanisms couple the gas to the FUV photon energy of stars: the photoelectric effect on PAHs and small dust grains and FUV pumping of $H_2$ molecules. Other heating mechanisms—gas collisions with warm grains, cosmic ray ionization and excitation, ionization of C, pumping of gas particles to excited states by the FIR radiation field of the warm dust followed by collisional deexcitation—play only a limited role in the heating or become important at great depth in the PDR (Tielens & Hollenbach 1985a).

PHOTOELECTRIC HEATING Photoelectric heating is dominated by the smallest grains present in the ISM (Watson 1972, Jura 1976). In recent years, it has become abundantly clear that large PAHs are an important component of the ISM (cf Allamandola et al 1989, Léger & Puget 1989), and these molecular-sized species may play an important role in the heating of interstellar gas (d'Hendecourt & Léger 1987, Lepp & Dalgarno 1988, Verstraete et al 1990, Bakes & Tielens 1994).

Figure 6 schematically shows the physics associated with the photoelectric effect on interstellar grains and PAHs. FUV photons absorbed by a grain create energetic (several electron volts) electrons. These electrons may diffuse in the grain, reach the surface, overcome the work function $W$ of the grain and any Coulomb potential $\phi_c$ if the grain is positively charged, and be injected into the gas phase with excess kinetic energy. The efficiency $\epsilon_{grain}$ of the photoelectric effect on a grain, or the ratio of gas heating to the grain FUV absorption rate, is then given by the yield $Y$, which measures the probability that the electron escapes, times the fraction of the photon energy carried away as kinetic energy by the electron. For large grains and photon energies well above threshold, the photons are absorbed $\sim$100 Å inside the grain and the photoelectrons rarely escape ($Y \sim 0.1$) (Watson 1972, Draine 1978, Bakes & Tielens 1994).

Whereas some of the photon energy may remain behind as electronic excitation energy ($\sim$0.5), the yield is much higher for planar PAH molecules. The limiting factor is now that the ionization potential $IP$ of a charged PAH can be larger than 13.6 eV, and absorbed FUV photons do not lead to the creation of a photoelectron. For example, the second ionization potential of pyrene, $C_{16}H_{10}$, is 16.6 eV, well above the hydrogen ionization limit (Leach 1987). The

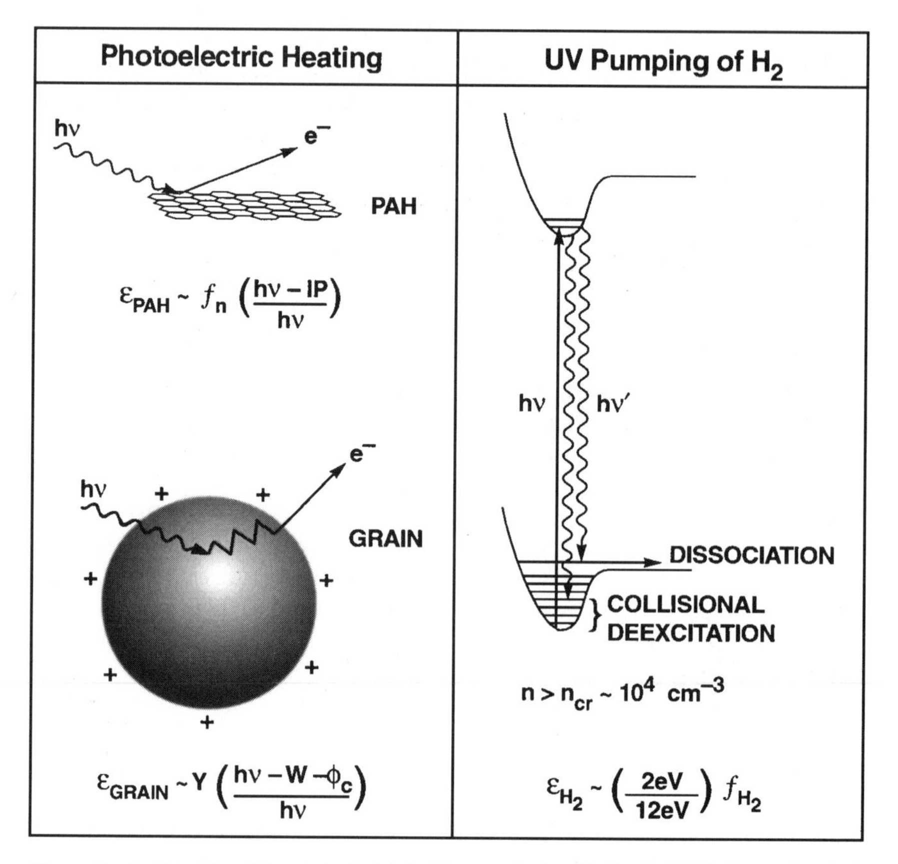


*Figure 6* A schematic of the photoelectric heating mechanism (*left*). An FUV photon absorbed by a dust grain creates a photoelectron that diffuses through the grain until it loses all its excess energy in collisions with the matrix or finds the surface and escapes. For PAHs, the diffusion plays no role. A schematic of the $H_2$ heating mechanism for PDRs (*right*). FUV fluorescence can leave a $H_2$ molecule vibrationally excited in the ground electronic state. Collisional de-excitation of this excited molecule can then heat the gas. In about 10–15% of the pumping cases, the cascade goes to the vibrational continuum of the ground electronic state, which leads to photodissociation of the $H_2$ molecule. Simple expressions for the heating efficiencies, $\epsilon$, of these different processes are indicated. See text for details.

photoelectric heating efficiency, $\epsilon_{PAH}$, for small PAHs is then reduced by the fraction, $f_n$, of PAHs that can still be ionized by FUV photons.

Theoretical calculations show that PAHs and small (<50 Å) grains are more efficient heating agents of the gas than large grains. Interstellar grain size distributions are typically assumed to have number of grains per unit size interval proportional to $a^{-3.5}$, where $a$ is the grain radius (Mathis et al 1977, MRN). For an MRN distribution extending into the molecular PAH domain, about half the gas heating is due to grains with sizes less than 15 Å (Bakes & Tielens 1994). The other half originates in grains with sizes between 15 and 100 Å. Grains larger than 100 Å contribute negligibly to the photoelectric heating of the interstellar gas.

The photoelectric heating efficiency, $\epsilon$, depends on the charge of a grain. A higher charge implies a higher Coulomb barrier (i.e. higher ionization potential) that has to be overcome. Thus, a smaller fraction of the electrons "dislodged" in the grain will escape. Moreover, those that do escape will carry away less kinetic energy (de Jong 1977). For PAHs, the charge determines whether further ionizations can still occur. Hence, the photoelectric heating efficiency will depend on the ratio, $\gamma$, of the photoionization rate over the recombination rate of electrons with grains/PAHs. When $\gamma$ ($\propto G_0/n_e$) is small, grains/PAHs are predominantly neutral and the photoelectric heating has the highest efficiency ($\epsilon \sim 0.05$). When $\gamma$ increases, the grains/PAHs become positively charged and the photoelectric efficiency drops. Based upon extensive theoretical calculations, simple analytical formulae for $\epsilon$ have been derived (Bakes & Tielens 1994).

$H_2$ HEATING As discussed in Section 2.2 the line absorption of an FUV photon will pump $H_2$ molecules to a bound excited electronic state, from which it will fluoresce back to the vibrational continuum of the ground electronic state and dissociate (10–15% of the time) or it will fluoresce back to an excited vibrational state in the electronic ground state (85–90% of the time; Figure 6). At low densities, the excited (bound) vibrational states can cascade down to the ground vibrational state through the emission of IR photons, giving rise to a characteristic far-red and near-IR rovibrational spectrum (see Section 4.2). At high densities, i.e. $n \gtrsim 10^4$ cm$^{-3}$ (Martin & Mandy 1995, Martin et al 1996), collisions with atomic H can also be an important deexcitation mechanism, leading to heating of the gas and thermalization of the rovibrational states. The heating efficiency of this process is then approximately $\epsilon_{H_2} \simeq (E_{vib}/h\nu) f_{H_2} \simeq 0.17 f_{H_2}$. The fraction of the FUV photon flux pumping $H_2$, $f_{H_2}$, depends on the location of the H/$H_2$ transition zone. Thus, when $G_0/n < 10^{-2}$ cm$^3$ (Section 2.2), $H_2$ self-shielding is important, the $H_2$ transition is near the surface, and most of the photons that can pump $H_2$ are absorbed by $H_2$ rather than dust. Under these conditions, $n > 10^4$ cm$^{-3}$, $G_0/n < 10^{-2}$ cm$^3$, $f_{H_2} \simeq 0.25$, and this heating process provides an efficient coupling to the FUV photon flux of the star. Sternberg & Dalgarno (1989) and Burton et al (1990b) describe how $H_2$ heating depends on $G_0$ and $n$.

# 3. PDR MODELS

## 3.1 *Steady-State Stationary PDR Models*

Considerable effort has been expended over the past two decades in constructing PDR models with the assumption of thermal and chemical balance and ignoring any flow through the PDR. Basically, this is equivalent to assuming that the time scale for $H_2$ formation on grains, $\tau_{H_2} \sim (10^9\ \text{cm}^{-3}/n)$ years, which dominates the chemical time scales, is short compared to the dynamical time scales or the time scales for significant change in the FUV flux. Equilibrium PDR models of the transition of H to $H_2$ and $C^+$ to CO, including dust attenuation and self-shielding of the FUV flux, have a long history (Hollenbach et al 1971, Jura 1974, Glassgold & Langer 1975, Langer 1976, Clavel et al 1978, Federman et al 1979, deJong et al 1980, Viala 1986, Abgrall et al 1992, Heck et al 1992, Andersson & Wannier 1993, Draine & Bertoldi 1996). Models focusing on the $H_2$ spectrum are discussed in Section 4.2, and references to diffuse and translucent cloud models can be found in Section 2.2. However, intense FUV fields ($G_0 >> 1$) and high $A_v$ often characterize dense PDRs. Tielens & Hollenbach (1985a), Sternberg & Dalgarno (1989), Hollenbach et al (1991), Abgrall et al (1992), LeBourlot et al (1993), and Diaz et al (1996) model the thermal balance and chemistry in homogeneous PDRs subjected to a range of elevated FUV fluxes.

Generally, these models consider a plane-parallel semi-infinite slab illuminated from one side by an intense FUV field (Figure 7). The penetrating FUV photons create an atomic surface layer. At a depth corresponding to $A_v \simeq 2$, the transition from atomic H to molecular $H_2$ occurs. Because of rapid photodestruction, vibrationally excited molecular hydrogen, $H_2^*$, does not peak until $A_v \sim 2$. Because of dust attenuation, the carbon balance shifts from $C^+$ to C and CO at $A_v \simeq 4$. The second peak in the neutral carbon abundance results from charge exchange between $C^+$ and S. Except for the oxygen locked up in CO, essentially all the oxygen is in atomic form until very deep in the cloud $A_v \sim 8$. Because of their low ionization potential, trace species such as S can remain ionized through a substantial portion of the PDR.

Besides the chemical composition, the FUV photons also control the energy balance of the gas through the photoelectric effect (cf Section 2.4). Typically, about 0.1–1% of the FUV energy is converted into gas heating this way. The rest is emitted as FIR dust continuum radiation. The gas in the surface layer is then much warmer ($\simeq$500 K) than the dust (30–75 K). Somewhat deeper into the PDR ($A_v > 4$), penetrating red and near-IR photons keep the dust warm and gas-grain collisions couple the gas temperature to slightly below the dust temperature. In PDRs, the gas cools through the FIR fine structure lines of mainly [O I] 63 $\mu$m and [C II] 158 $\mu$m at the surface and the rotational lines of CO deeper into the PDR. Figure 7 quantitatively shows this chemical and thermal structure.

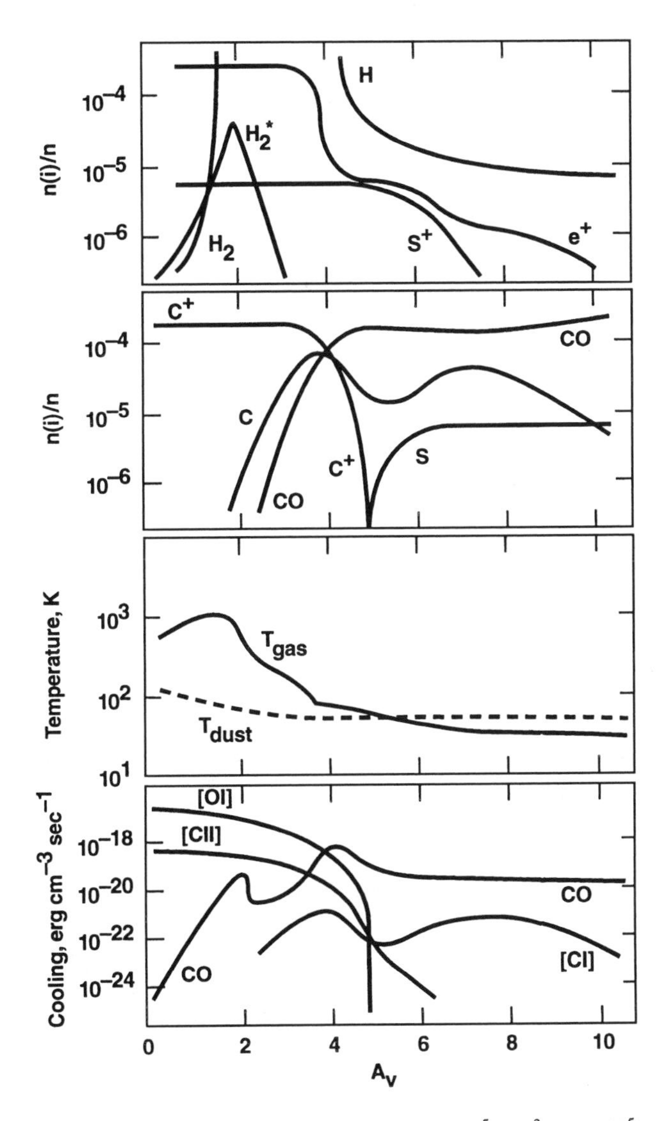


*Figure 7* Calculated structure of the PDR in Orion ($n = 2.3 \times 10^5$ cm$^{-3}$, $G_0 = 10^5$) as a function of visual extinction, $A_V$, into the PDR (Tielens & Hollenbach 1985b). The illuminating source is to the left. *Top two panels*: Abundances relative to total hydrogen. *Third panel*: Gas and dust temperatures. *Bottom panel*: Cooling in the various gas lines.

Specialized aspects of PDR phenomena have been discussed in several papers: the origin of atomic carbon and [C I] 370, 609 $\mu$m emission (Tielens & Hollenbach 1985c, van Dishoeck & Black 1988, Flower et al 1994), PDRs around cooler sources such as F, A, and B stars or the central stars of protoplanetary nebulae (Roger & Dewdney 1992, Spaans et al 1994), the near-IR C I (9850 Å) emission (Escalante et al 1991), carbon radio recombination emission (Natta et al 1994), OH and $CH_3OH$ abundances and masers (Hartquist & Sternberg 1991, Hartquist et al 1995), $CO^+$ abundances (Latter et al 1993, Störzer et al 1995), and molecular chemistry in general (Fuente et al 1993, Sternberg & Dalgarno 1995). Recently, models have been constructed of clumpy PDRs, many of them focused on the $C^+$/C/CO transition and the origin of [C I], $H_2$ 2 $\mu$m, and CO rotational emission (Burton et al 1990b, Tauber & Goldsmith 1990, Gierens et al 1992, Meixner & Tielens 1993, Wolfire et al 1993, Köster et al 1994, Hegmann & Kegel 1996, Spaans 1996, Störzer et al 1996).

## 3.2 *Time-Dependent and Nonstationary PDRs*

There are two basic phenomena that can cause significant time-dependent effects on the chemical abundances and temperatures in a PDR. The first is when $G_0$ or $n$ changes in a smaller time scale than $\tau_{H_2}$. For example, an O or B star suddenly "turns on" in a cloud, clumps moving in PDRs produce rapid shadowing effects, or a PDR shell quickly expands around the central star of a planetary nebula. The second is when, in the frame of the ionization front, the neutral gas flows through the PDR (hence "nonstationary PDR") in a time that is short compared to $\tau_{H_2}$. In this case, the chemical and thermal structure can achieve a steady state, but the resulting structure differs from the structure of a stationary PDR because of the rapid advection of molecular material from the shielded regions into the surface zones.

Hill & Hollenbach (1978), London (1978), and Roger & Dewdney (1992) have considered the time-dependent evolution of the ionization front and $H_2$ dissociation front when O or B stars suddenly turn on in a neutral cloud. Goldshmidt & Sternberg (1995) looked in detail at the enhancement of $H_2$ fluorescent emission when a low ($n \lesssim 10^4$ cm$^{-3}$) constant-density molecular cloud is suddenly illuminated with intense FUV; Hollenbach & Natta (1995) examined the collisionally modified $H_2$ spectra when $n > 10^4$ cm$^{-3}$. Störzer et al (1997) studied the temporary enhancement of $C^0$ abundance and [C I] emission caused by $C^+$ recombination when the FUV flux is suddenly shadowed by moving opaque clumps. Monteiro (1991) found a similar enhancement for rotating clumps in PDRs. Gerola & Glassgold (1978), Tarafdar et al (1985), Prasad et al (1991), Heck et al (1992), Lee et al (1996), and Nelson & Langer (1997) examined the time-dependent evolution of diffuse atomic clouds into molecular clouds. Lee et al (1996) also studied the time-dependent chemistry of an inhomogeneous molecular cloud suddenly exposed to FUV radiation.

The first PDR modeling that incorporated fast flows or advection were chemical models that used rapid turbulent eddies or diffusion to mix molecular gas from opaque molecular cores with surface photodissociated gas (Phillips & Huggins 1981, Boland & deJong 1982, Chieze & Pineau des Forêts 1989,1990, Chieze et al 1991, Williams & Hartquist 1984, 1991, Xie et al 1996). Recently, Bertoldi & Draine (1996) have discussed in detail the time-dependent effect on the PDR structure of the advance of the ionization front (IF) and dissociation front through a cloud. In this case the "flow" velocity is equivalent to the speed of the IF advancing into the PDR. The advected flow is relatively fast when the photoevaporating H II gas is free to flow to lower pressure regions, such as when neutral clumps photoevaporate well inside an H II region. Assuming approximately thermal pressure equilibrium between the photoevaporating H II gas and the neutral PDR gas on the other side of the IF, and assuming a typical PDR temperature of $\sim$1000 K, the PDR density $n_{PDR} \simeq 20 n_{HII}$. Assuming the free-flowing H II gas evaporates from the IF at a speed comparable to its thermal speed, i.e. $c_{HII} \sim 10$ km s$^{-1}$, the flow speed $v_{PDR} \sim (n_{HII}/n_{PDR})c_{HII}$ through the PDR is about 0.5 km s$^{-1}$. The flow time through the PDR is as follows: $\tau_{PDR} \simeq (N_{PDR}/n_{PDR})/v_{PDR}$. Equating $\tau_{PDR}$ to $\tau_{H_2}$, we obtain a critical flow speed of $\sim$0.7 km s$^{-1}$, above which the $H_2$ abundance may be significantly higher than its "stationary" (no-flow) value.

The fact that for free-flowing gas the estimated $v_{PDR}$ is so close to the estimated critical speed indicates that nonstationary PDRs may be important. Bertoldi & Draine (1996) show how $v_{PDR}$ can be related to the EUV flux reaching the IF, because these photons create the new ionizations that push the IF into the PDR. The EUV flux depends on the gas and dust attenuation in the ionized gas and the morphology (e.g. size of clump and distance from star) of the region. This morphology determines the geometric dilution of the flux and the EUV attenuation by recombined atomic hydrogen in the predominantly ionized, evaporative flow. They conclude that for a broad range of conditions, including the famous PDR in Orion (Figures 1 and 2), time-dependent effects may be of some importance.

Much larger effects can be expected when the $H_2$ survives all the way to the IF as it flows from the opaque regions, leading to the merging of the dissociation front (DF) with the IF. Bertoldi & Draine (1996) have quantitatively determined when the IF merges with the DF. Qualitatively, the merger tends to occur for hotter stars (with a high ratio of EUV photons to FUV photons) illuminating clumps or clouds with a geometry such that relatively little attenuation of the EUV flux occurs by recombining hydrogen in the evaporating flow or H II region (e.g. a small, freely evaporating clump near the hot star). Apparently, the condition is not met in the Orion Bar (Figure 2) because the observed layered structure clearly shows the $H_2$ offset from the IF; in fact, Tielens et al (1993) and van der Werf et al (1996) find good agreement with steady-state stationary PDR models.

The consequences of rapid advection through PDRs have not yet been fully determined. Bertoldi & Draine (1996) suggest that $H_3^+$ may be produced in a merged IF/DF and that enhanced $H_2$ IR emission and reduced H I columns will result. Nonstationary PDR models may find interesting application in the photoevaporating protoplanetary disks (proplyds) found in the Orion nebula (Churchwell et al 1987, O'Dell et al 1993).

## 4. OBSERVATIONS OF DENSE PDRs IN THE GALAXY

Dense PDRs are bright in the FIR dust continuum, the PAH emission features, the FIR fine structure lines of [O I] and [C II], and the rotational lines of CO. Besides these dominant cooling processes, PDRs are also the source of (fluorescent or collisionally excited) rovibrational transitions of $H_2$ in the near IR; the atomic fine structure lines of [C I] 609, 370 $\mu$m, and [Si II] 35 $\mu$m; and the recombination lines of C I in the radio (e.g. C91$\alpha$) and the far red (e.g. 9850 and 8727 Å). They are also observed in rotational transitions of trace molecules like $CO^+$, CN, and $C_2H$.

### 4.1 *Physical Conditions*

These observations can be used to determine the physical conditions in the emitting gas. The FIR fine structure lines are particularly useful in that respect because the critical densities of the [C II] and [O I] lines ($3 \times 10^3$–$3 \times 10^5$ cm$^{-3}$) and excitation energies (100–300 K) span the $n, T$ range present in many dense PDRs. The [C II] 158 $\mu$m line, which is (marginally) optically thin, can be used to determine the total mass of emitting gas. These analyses do require assumed elemental abundances (or abundance ratios) for those species that are somewhat uncertain (cf Cardelli et al 1996, Mathis 1997). Often, the physical conditions (e.g. densities) are also constrained by the ratio of the total cooling rate (i.e. the [O I] + [C II] intensity) to the FIR dust continuum. This ratio is approximately equal to the photoelectric heating efficiency, which depends on the local density through the charging of the grains (cf Section 2.4). Of course, this analysis requires a good understanding of the photoelectric heating process itself and hence is also uncertain. Boreiko & Betz (1996a,b) have derived temperatures, column densities, and $^{12}$C/$^{13}$C isotopic ratio ($\simeq$58) in M42 by observing the [O I] 63 $\mu$m and both isotopic [C II] 158 $\mu$m lines with very high velocity resolution. The derived temperatures ($\sim$200 K) are somewhat lower and columns ($\sim 10^{22}$ cm$^{-2}$) somewhat higher than homogeneous PDR models.

The temperature and column density can also be determined directly from the lowest pure rotational lines of $H_2$, which have very low critical densities. To date, that has only been done for the Orion Bar (Parmar et al 1991, Burton et al 1992b) and S140 (Timmermann et al 1996). The intensity distribution of the rotational CO levels also provides density and temperature information, albeit for the molecular component in the PDR ($A_v \simeq 4$) (Harris et al 1985, Jaffe et al 1990, Stutzki et al 1990, Graf et al 1993, Howe et al 1993, Lis et al 1997). Generally, the J = 7–6 transition is quite strong in PDRs, implying densities of $\sim 10^5$ cm$^{-3}$ and temperatures of $\sim$150 K. The presence of CO J = 14–13 emission in some sources (Stacey et al 1993) indicates even higher densities and temperatures ($10^7$ cm$^{-3}$; 500 K).

In Figure 8, observations of the [C II] 158 $\mu$m and [O I] 63 $\mu$m lines are plotted as a function of the intensity of the incident FUV field, $G_0$ ($= I_{IR}/2.6 \times 10^{-4}$ erg cm$^{-2}$ s$^{-1}$ sr$^{-1}$, where $I_{IR}$ is the observed IR intensity). Examining these observations, we conclude that the [C II] line dominates the cooling for $G_0 \lesssim 10^3$. The [O I] line dominates at higher $G_0$. Typically, the total cooling line intensity is in the range $10^{-2}$–$10^{-3}$ of the total FIR continuum intensity, in agreement with predicted grain photoelectric heating efficiencies of $10^{-2}$–$10^{-3}$ (see Section 2.4). This rough correlation extends over a large range in $G_0$ and an even larger range in luminosity ($10^4$–$10^{11}$ $L_\odot$) (Hollenbach 1990).

Also shown in Figure 8 are PDR model calculations for different densities (Hollenbach et al 1991). In the models, the [C II] intensity scales with the incident FUV flux (or FIR intensity) at low $G_0$ where [C II] is the dominant cooling line. The [C II] intensity levels off at high $G_0$ when the temperature exceeds $\sim$100 K. At that point, [O I] 63 $\mu$m becomes a better coolant and its intensity scales with $G_0$. The models are in good agreement with the observations for densities in the range $10^3$–$10^5$ cm$^{-3}$. We note that, as expected, the PDRs in reflection nebulae, such as NGC 7023, and bright rim clouds, such as S140, are characterized by somewhat lower densities ($3 \times 10^3$ cm$^{-3}$) and incident FUV fields than those surrounding H II regions, such as M42 in Orion ($10^5$ cm$^{-3}$).

We summarize the physical conditions in PDRs associated with a number of template objects as derived from observations in Table 1. Typical densities and temperatures are $n \sim 10^3$–$10^5$ cm$^{-3}$ and $T \sim$ 200–1000 K. The atomic mass in the $C^+$ zone, $M_a$, is a significant fraction of the molecular cloud (core) mass, $M_m$ (see also Tielens 1994).

In general, observations reveal pronounced clumping of the emitting gas. First, the [O I], [C II], and [C I] emission is extended on $\gtrsim 5'$ scale, as illustrated in Figure 1 for Orion A. In fact, Stacey et al (1993) trace the [C II] extent even further than Figure 1, and likewise PDR $H_2$ emission has been observed in Orion A on larger scales by Burton & Puxley (1990), Luhman et al (1994), and Usuda et al (1996). All this gas is photoionized, photodissociated, and

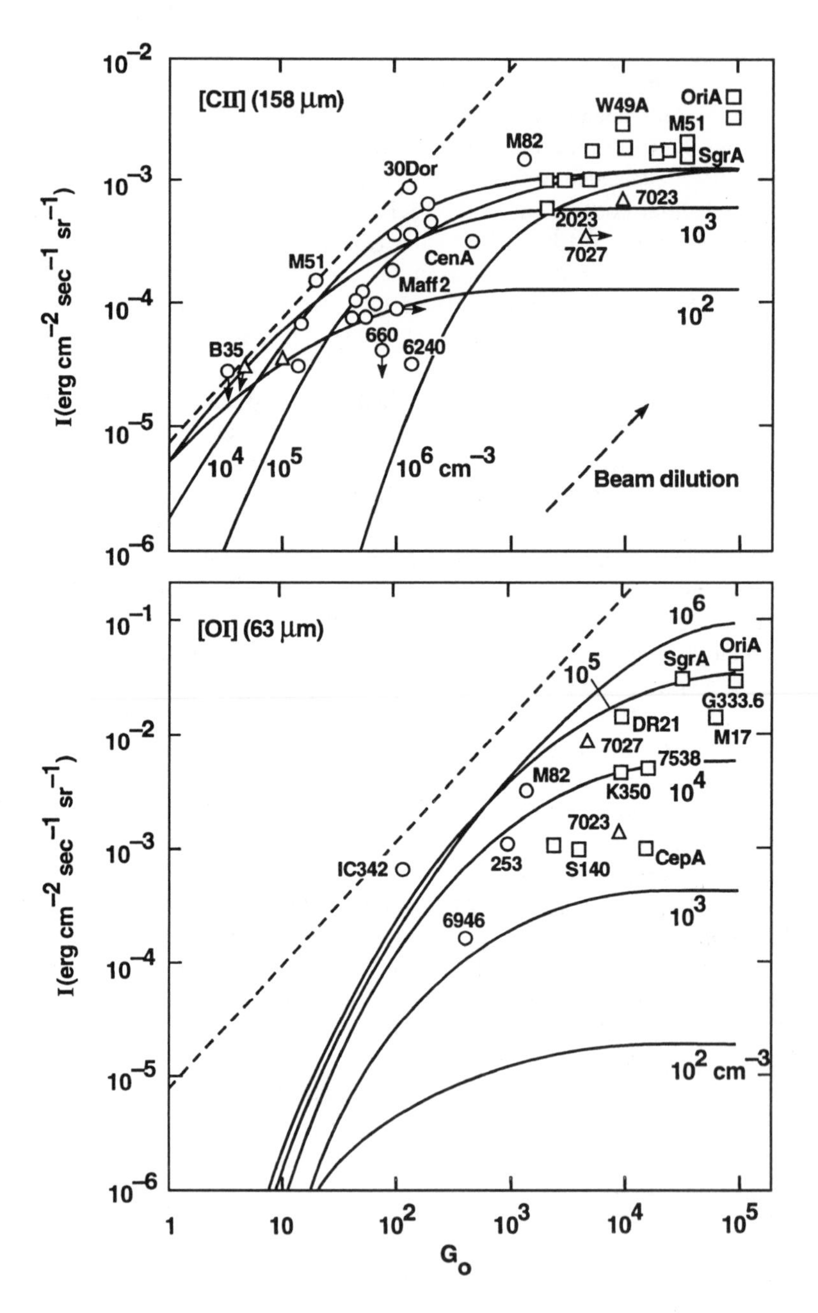

[CII] (158 μm)
W49A
OriA
M51
SgrA
M82
30Dor
7023
2023
7027
CenA
M51
Maff 2
660
6240
B35
10^4
10^5
10^6 cm^-3
10^3
10^2
Beam dilution
I (erg cm^-2 sec^-1 sr^-1)
[OI] (63 μm)
10^6
10^5
SgrA
OriA
G333.6
DR21
M17
7027
7538
M82
K350
10^4
7023
IC342
253
S140
CepA
6946
10^3
10^2 cm^-3
G_o

(photo)heated by $\theta^1$C Ori. Similar results have been obtained for Orion B, M17, S140, NGC 2023, and many other sources (Stutzki et al 1988, Matsuhara et al 1989, Howe et al 1991, White & Padman 1991, Stacey et al 1993, Jaffe et al 1994, Plume et al 1994, Herrmann et al 1997, Luhman & Jaffe 1996, Spaans & van Dishoeck 1997). This emission scale size is much larger than expected for a homogeneous region at the density indicated by the line ratios. Clearly, the gas is organized in filamentary and clumpy structures that allow the penetration of FUV photons to very large distances from the illuminating stars. Second, maps in many molecular species (i.e. $^{13}CO$, $C^{18}O$, CS, $NH_3$, $HCO^+$, HCN, $H_2$, and PAHs) show direct evidence for clumping in PDRs with a surface filling factor of $\simeq$0.1 (Gatley et al 1987, Güsten & Fiebig 1988, Massi et al 1988, Stutzki et al 1988, Stutzki & Güsten 1990, Tauber & Goldsmith 1990, Chrysostomou et al 1992, 1993, Minchin et al 1993, Hobson et al 1994, Tauber et al 1994, Kramer et al 1996, Lemaire et al 1996; R Young Owl, M Meixner, M Wolfire & AGGM Tielens, in preparation). Third, most PDRs show bright molecular lines [e.g. CO J = 14–13, $H_2$ 1–0 and 2–1 S(1)] with high critical densities and high excitation temperatures that attest to the presence of high-density warm molecular gas. The physical conditions of these clumps are not well constrained and likely span a range of values. Thermal $H_2$ 2–1 S(1)/1–0 S(1) ratios require densities $n \gtrsim 10^5$ cm$^{-3}$ and $T \sim 1000$ K. Likewise, C II recombination line studies indicate clumpy high-density ($\sim 10^5$–$10^6$ cm$^{-3}$) regions over a large area of the PDRs in Orion and NGC 2023 (Natta et al 1994, Wyrowski et al 1997). Somewhat higher densities, $n \gtrsim 10^6$–$10^7$ cm$^{-3}$ are required to produce bright CO J = 14–13, 17–16, or 23–22 emission. Table 1 lists estimates for clump physical conditions in several PDRs. We note that homogeneous PDR models fall typically short of observed mid-J (e.g. 6–5 or 7–6) CO intensities, especially from the $^{13}$CO isotope (Graf et al 1993, Lis et al 1997). Clump models may resolve this discrepancy (Köster et al 1994), or it may be that models underestimate the gas temperatures in the $C^+$/C/CO transition regions where these lines originate (Tauber et al 1994).

The inferred thermal pressures for these clumps ($10^9$–$10^{10}$ K cm$^{-3}$) are well in excess of that sustainable by the interclump medium (or the H II region

*Figure 8* Comparison of observations and models of the [C II] 158 $\mu$m and the [O I] 63 $\mu$m line are shown as a function of $G_0$ (Hollenbach et al 1991). Squares are PDRs associated with H II regions; triangles are PDRs associated with dark clouds, reflection nebulae, and planetary nebulae; and circles are PDRs associated with the inner 45–60$''$ of galaxies. Some well-known individual regions have been identified by their catalog numbers. The effects of beam dilution are indicated by the *dashed arrow*. The different models are labeled by their densities. The *dashed line* indicates an efficiency of 5% in the conversion of FUV photons energy into gas cooling—the maximum expected for the photoelectric effect. See text for details.

**Table 1** Characteristics of PDRs

| Object | NGC 2023 | Orion Bar | NGC 7027 | Sgr A | M 82 |
|---|---|---|---|---|---|
| Line intensities[a] | | | | | |
| [OI] 63 $\mu$m | 4. (−3) | 4. (−2) | 1. (−1) | 2. (−2) | 1. (−2) |
| [OI] 145 $\mu$m | 2. (−4) | 2. (−3) | 5. (−3) | 7. (−4) | 1.5 (−4) |
| [SiII] 35 $\mu$m | 2. (−4) | 9. (−3) | — | 2. (−2) | 1. (−2) |
| [CII] 158 $\mu$m | 7. (−4) | 6. (−3) | 1. (−2) | 2. (−3) | 2. (−3) |
| [CI] 609 $\mu$m | — | 5. (−6) | 1.8 (−6) | 4. (−5) | 2. (−5) |
| $H_2$ 1–0 S(1) | 5. (−5) | 2. (−4) | 8. (−4) | 9. (−4) | 5. (−5) |
| CO J = 1–0 | 3. (−8) | 4. (−7) | 1.5 (−6) | 7 (−7) | 2.6 (−7) |
| CO J = 7–6 | 5. (−5) | 2. (−4) | —[b] | 1.5 (−3) | 5. (−5) |
| CO J = 14–13 | — | 3. (−4) | —[b] | 3. (−4) | — |
| FIR[c] | 8. (−1) | 5. (0) | 4. (1) | 5. (1) | 6. (0) |
| PAHs[d] | 9. (−2) | 1.5 (−1) | 1.8 (0) | — | 1.4 (−1) |
| $G_0$ | 1.5 (4) | 4. (4) | 6. (5) | 1. (5) | 1. (3) |
| Physical conditions | | | | | |
| Interclump | | | | | |
| $n$ [cm$^{-3}$] | 7.5 (2) | 5. (4) | 1. (5)[e] | 1. (5) | 1. (4) |
| T [K] | 250 | 500 | 1000[e] | 500 | 250 |
| $M_d/M_m$ | 0.2 | 0.6 | 0.3 | 0.04 | 0.1 |
| Clump | | | | | |
| $n$ [cm$^{-3}$] | 1. (5) | 1. (7) | 1. (7)[e] | 1. (7) | — |
| T [K] | 750[f] | (2000) | (2000)[e] | (2000) | — |
| $f_v$ | 0.1 | 0.005 | 0.05 | 0.06 | 4. (−4) |
| References[g] | 1–5 | 6–8 | 9–13 | 13–16 | 14, 17–19 |

[a]Intensities in units of erg cm$^{-2}$ s$^{-1}$ sr$^{-1}$.
[b] Bright CO J = 17–16 has been observed by Justtanont et al (1997).
[c]Total far IR dust continuum intensity.
[d]Intensity in the PAH emission features.
[e]Interclump and clump refer to halo and torus, respectively (Justtanont et al 1997).
[f]C91$\alpha$ indicates 100–200 K (Wyrowski et al 1997).
[g]References: 1, Steiman-Cameron et al (1997); 2, Wyrowski et al (1997); 3, Sellgren et al (1985); 4, Jaffe et al (1990); 5, Gatley et al (1987); 6, Tielens et al (1993); 7, Tauber et al (1994; 1995); 8, Stutzki et al (1990); 9, Cohen et al (1985); 10, Justtanont et al (1997); 11, Keene (1995); 12, Graham et al (1993); 13, Burton et al (1990a); 14, Wolfire et al (1990); 15, Harris et al (1985); 16, Serabyn et al (1994); 17, Willner et al (1977); 18, Genzel (1992); 19, Schilke et al (1993).

pressure). These clumps may therefore be confined by self-gravity and either are on their way to forming stars or already contain an embedded protostar (Meixner et al 1992). Hence, they may be the PDR counterparts of the small partially ionized (Bok) globules (PIGs) and stellar proplyds observed in H II regions (Churchwell et al 1987, O'Dell et al 1993). These objects seem to be widespread in the molecular clouds.

## 4.2 *The $H_2$ Spectrum*

FUV absorption in the Lyman Werner bands excites molecular hydrogen electronically. Electronic (FUV) fluorescence (Sternberg 1989) leaves the molecule in a vibrationally excited state of the ground electronic state 85–90% of the time (Section 2.4, Figure 6). Molecules in these levels decay through electric quadrupole transitions in the far red and near IR. The intensities of these lines depend then on the detailed distribution of the level populations through the cloud, which themselves depend on the (FUV) line radiative transfer as well as the density and temperature of collision partners. Level populations and IR emission spectra have been calculated for a variety of model clouds by Black & Dalgarno (1976), Black & van Dishoeck (1987), Sternberg [1988, 1990 (for HD, i.e. hydrogen molecule where one of the hydrogens is deuterium)], Sternberg & Dalgarno (1989), Draine & Bertoldi (1996), and Neufeld & Spaans (1996). Pure rotational spectra of $H_2$ have been calculated by Burton et al (1992b) and by Spaans et al (1994).

At low $n$ ($\lesssim 10^4$ cm$^{-3}$), collisional de-excitation is unimportant and the IR spectrum is due to "pure" fluorescence and can be calculated directly from the transition probabilities of the levels involved, taking the radiative transfer into account. For $G_0/n < 10^{-2}$ cm$^3$, $H_2$ self-shielding dominates the opacity in the 912–1100 Å range and the line intensity scales directly with the intensity of the incident FUV radiation field. For $G_0/n \gtrsim 0.1$ cm$^3$, dust opacity is more important and the line intensities become largely independent of $G_0$. At high densities, collisional processes are important, modifying the emitted spectrum to a thermal spectrum in the lowest levels (i.e. v = 1–0 and 2–1) (Sternberg & Dalgarno 1989, Burton et al 1990b, Draine & Bertoldi 1996, Luhman et al 1997). The population of high vibrational (v) levels, though reduced by collisions, is caused by FUV pumping.

Far-red and near-IR $H_2$ fluorescence spectra have been observed in a variety of classical PDRs such as those associated with NGC 2023, S140, NGC 7023, the Orion Bar, Orion A and B, and $\rho$ Oph (Hayashi et al 1985, Gatley et al 1987, Hasegawa et al 1987, Hippelein & Münch 1989, Tanaka et al 1989, Burton et al 1990a, 1992a, Luhman & Jaffe 1996). Figure 9 compares the observed IR spectrum 160″ north of the illuminating star in NGC 2023 (Gatley et al 1987) with model calculations for the pure fluorescence case, illustrating the good fit possible (Black & van Dishoeck 1987). The fit mainly depends on the adopted value for $G_0$ and the detailed geometry of this emission ridge. It is not sensitive to the density for $n \lesssim 10^4$ cm$^{-3}$. The more thermal 1–0 S(1)/2–1 S(1) ratio observed for the bright $H_2$ ridge, 78″ south of the star indicates emission by higher density gas ($10^5$ cm$^{-3}$; Burton et al 1992a, Draine & Bertoldi 1996), in good agreement with studies of the dominant PDR cooling lines and molecular

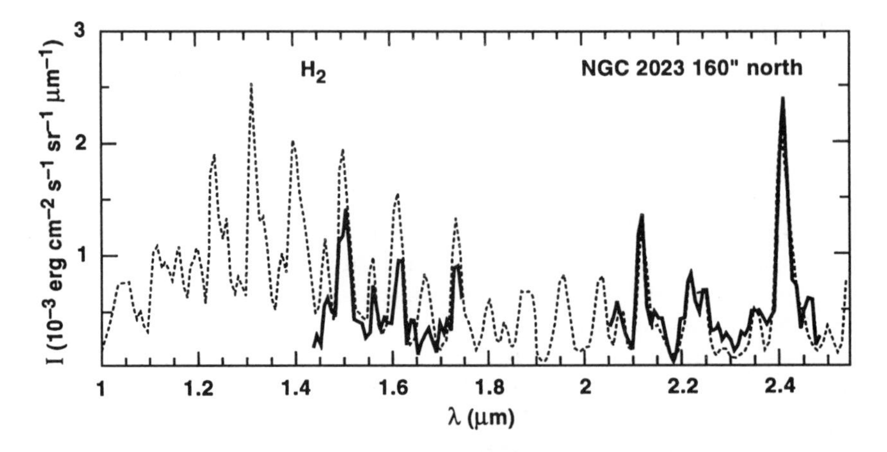


*Figure 9* Comparison of observed and calculated $H_2$ emission spectra, adapted from Black & van Dishoeck (1987). *Solid curve*: spectrum observed 160″ north of the illuminating star in the reflection nebula, NGC 2023 (Gatley et al 1987). *Light curve*: Calculated emission spectrum for $G_0 \simeq 200$ and a density of $10^4$ cm$^{-3}$ convolved to a resolution of $\lambda/\Delta\lambda \simeq 100$ (Black & van Dishoeck 1987). See text for details.

line observations (Steiman-Cameron et al 1997, Jansen et al 1995a, Fuente et al 1995, Wyrowski et al 1997).

## 4.3 *The Molecular Transition Observed in Edge-On PDRs*

The edge-on geometry of the Orion Bar PDR lends itself particularly well to detailed studies of the spatial structure of PDRs (cf Tielens et al 1993, Hogerheijde et al 1995, van der Werf et al 1996; Figure 2). Clearly, the H/$H_2$ transition, as outlined by the emission from vibrationally excited $H_2$, occurs at about 10″ into the cloud from the ionization front (i.e. the location of the Bar in [O I] 6300 and [S II] 6731 Å). The CO 1–0 and [C I] 609 $\mu$m lines peak rather abruptly at a distance of 20″ of the ionization front. These observations agree very well with predictions of the global characteristics of stationary PDR models (Tielens et al 1993, Jansen et al 1995a, van der Werf et al 1996). In such models, the location of the H/$H_2$ transition region and the peak in $H_2^*$ emission is displaced inwards from the ionization front by $A_v \simeq 2$, and the $C^+$/C/CO transition occurs another $\Delta A_v \simeq 2$ deeper into the cloud (Figure 7). Using standard dust parameters, the observed spatial scale translates then into a density of $5 \times 10^4$ cm$^{-3}$ (Tielens et al 1993). Calculated line emission for the fine structure lines of [O I] 63 and 146 $\mu$m, [C II] 158 $\mu$m, [Si II] 35 $\mu$m, [C I] 609 $\mu$m, and CO 1–0 agree well with the observed intensities for this density and $G_0 = 5 \times 10^4$, which is appropriate for $\theta^1$ C Ori at the projected distance of the Orion Bar (Table 1).

The presence of denser clumps ($10^7$ cm$^{-3}$) is indicated by bright emission in CO 7–6 and 14–13 (Stacey et al 1993).

M17-SW is another region where the edge-on geometry allows a detailed study of the structure of PDRs (Stutzki et al 1988, Meixner et al 1992). As in the Orion Bar region, cross scans show a clear separation between the ionized (free-free), atomic ([O I], [C II], [Si II]), and molecular (CO 2–1) gas. M17 is four times farther away than Orion, and the scale size is somewhat larger (40″ rather than 20″); both these factors indicate a much lower average gas density ($3 \times 10^3$ cm$^{-3}$) in the interclump regions of this molecular cloud core. The observed high intensity of the [O I] and [Si II] lines implies then that they mainly originate from the surfaces of the dense ($n \lesssim 10^7$ cm$^{-3}$) clumps evident in the molecular observations (Stutzki et al 1988, Meixner et al 1992). The [C II] and [C I] lines in contrast originate from the lower density interclump gas in the M17-SW core. Besides the M17-SW PDR, the [C II] and [C I] emission extends over a size scale of $\simeq$10 pc (Stutzki et al 1988, Matsuhara et al 1989, Keene et al 1985). This emission may well be associated with clumps illuminated at large distances by FUV penetration through even lower density gas (see Section 4.4).

Edge-on PDRs form an excellent laboratory to study in detail the interaction of FUV photons with gas and dust. One might hope that future studies of, for example, the dominant gas cooling lines and their correlation with the PAH and dust IR emission will allow a semi-empirical evaluation of the cooling and, hence, photoelectric heating process and its dependence on physical conditions. Present and future generations of (sub)millimeter arrays also allow a detailed analysis of the effects of FUV photons on the molecular composition of interstellar gas. The first studies of this kind have already been undertaken for IC 63 (Jansen et al 1995b, 1996), the Orion Bar (Tauber et al 1994, 1995, Hogerheijde et al 1995, van der Werf et al 1996, Fuente et al 1996; R Young Owl, M Meixner, M Wolfire & AGGM Tielens, in preparation), and NGC 2023 (Fuente et al 1995). Finally, limb brightening due to the edge-on geometry also allows the detection of trace species that are otherwise difficult to detect, for example, $CO^+$ (Latter et al 1993, Jansen et al 1995a, Störzer et al 1995).

## 4.4 *The Origin of [C I] Emission*

In retrospect, observations of the [C I] 609 $\mu$m line formed one of the earliest indications that PDRs are a ubiquitous component of molecular clouds (Phillips & Huggins 1981). Atomic carbon has long been recognized to be an abundant species near the surfaces of molecular clouds (Langer 1976, de Jong et al 1980). However, the early observations of the lowest lying transition of [C I] showed more intense emission than expected on the basis of the predicted C I column density. While this discrepancy led to a flurry of papers proposing a variety of schemes to increase the abundance of atomic carbon inside dense clouds, the first detailed PDR models (Tielens & Hollenbach 1985a–c) calculated about a 10-times higher C I column density than the earlier models, partly due to improved chemical schemes (i.e. charge exchange between $C^+$ and S and self-shielding of C I) and partly because of an adopted higher gas-phase carbon abundance. PDR theory (Tielens & Hollenbach 1985a,c, van Dishoeck & Black 1988, Hollenbach et al 1991) is now in reasonable agreement with the observed line intensities (Phillips & Huggins 1981, Keene et al 1985, Genzel et al 1988). Current models explore the effects of the clumpy structure of molecular clouds (Hegmann & Kegel 1996, Spaans & van Dishoeck 1997), the time-dependent effects associated with shadowing by moving clumps (Störzer et al 1997), and the dependence of C I on ionization (Flower et al 1994).

Higher spatial resolution studies with the James Clerke Maxwell Telescope and the Caltech Submillimeter Observatory clearly confirm the association of the [C I] emission with the PDR. In particular, in the Orion Bar, the [C I] 609 and 370 $\mu$m emission forms a bar-like structure at a similar distance from the ionization front as the CO 1–0, in good agreement with theory (White & Padman 1991, White & Sandell 1995, Tauber et al 1995). A similar spatial structure is seen in S140, although an embedded source contributes to the [C I] deeper into the cloud as well (Minchin et al 1993). Similar to the extremely extended [C II] 158 $\mu$m emission, the extended [C I] emission in M17 is thought to originate from the surfaces of clumps illuminated either by deeply penetrating FUV photons or by local B stars (Stutzki et al 1988, Meixner & Tielens 1993, Plume et al 1994). [C I] often correlates with $^{13}$CO 2–1 in spatial extent and line profile (Keene et al 1997), indicating that emission of both species is dominated by clump PDR surfaces (Meixner & Tielens 1993, Spaans & van Dishoeck 1997).

## 4.5 *The [C II]–CO Correlation*

Crawford et al (1985) and Stacey et al (1991) show a linear correlation in the integrated intensities of [C II] 158 $\mu$m and $^{12}$CO J = 1–0 in the observations of bright dense galactic PDRs and PDRs in starburst galactic nuclei. Wolfire et al (1989) explain this correlation with high FUV-field PDR models in which the [C II] 158 $\mu$m emission arises from the warm, $A_v \lesssim 1$–2, outer regions and the $^{12}$CO J = 1–0 originates from the cooler gas somewhat deeper ($A_v \sim 3-4$) in the cloud. However, in low FUV fields, the [C II]/CO ratio is low because the PDR cools and self-shields, which affects [C II] much more than CO. On the other hand, high [C II]/CO ratios are found in CO-deficient (low $A_v$) regions such as diffuse clouds or H I halos of molecular clouds (Jaffe et al 1994). High ratios may also result from geometry effects (Köster et al 1994) or clouds with small molecular cores and large $C^+$ halos such as in the Magellanic Clouds

(Boreiko & Betz 1991, Rubio et al 1993a,b, Mochizuki et al 1994, Poglitsch et al 1995, Israel et al 1996). These trends are noted by Stacey et al (1991).

## 4.6 *Planetary Nebulae (PNe)*

Observations of H I, Na I, $H_2$, CO, and other trace molecules demonstrate that a significant mass of neutral PDR material, on the order of 1 $M_\odot$ compared to $\sim$0.1 $M_\odot$ of ionized gas, is associated with PNe (for references, see recent reviews by Huggins 1992, 1993, Dinerstein 1991, 1995, Tielens 1993). Clearly, studies of these PDRs are crucial for a proper understanding of the dynamic and morphologic evolution of the ejected material. Spherical shells tend to completely ionize in a relatively short time, $t \lesssim 10^3$ year (e.g. Bobrowsky & Zipoy 1989, Tielens 1993, Gussie et al 1995). However, observations of the ionized and neutral gas (e.g. Huggins et al 1992, Graham et al 1993, Latter et al 1995, Kastner et al 1996) clearly show that the ejecta is clumped or nonspherical (e.g. disk-like or torus-like), and PDR models confirm that in evolved ($t \gtrsim 10^3$ year) PNe, atoms and molecules will survive in dense ($n \gtrsim 10^5$ $cm^{-3}$) clumps even though the FUV fluxes are large ($G_0 \sim 10^4$–$10^6$) (Tielens 1993; A Natta & DJ Hollenbach, in preparation). Proper models require a time-dependent calculation of the partial shells or clumps, including the effects of rapid changes in the stellar effective temperature, $G_0$, and $n$, as well as the advance or retreat of the IF with respect to the PDR gas. The $H_2$ emission is often thermal emission of $H_2$ warmed in the PDR to $T \sim 1000$–$2000$ K by grain photoelectric heating, $H_2$ FUV pump heating, and soft X-ray heating. Goldshmidt & Sternberg (1995) suggest that strong $H_2$ emission from young planetary nebulae is due to the $H_2$ enhancement caused by time-dependent PDR $H_2$ chemistry. Shock heating, often invoked in the past, may be less important than previously believed. Considerable work is underway to predict H I, [C II], [O I], $H_2$, and other molecular emission lines and to compare existing observations with the models to determine physical conditions such as the ejected mass (A Natta & DJ Hollenbach, in preparation; WB Latter, AGGM Tielens & DJ Hollenbach, in preparation).

## 4.7 *The Galactic Center*

The center of the Galaxy has been recently reviewed by Morris & Serabyn (1996) and Genzel et al (1994). PDR studies have delineated the structure, dynamics, and excitation mechanisms in the central $\sim$30 pc of the Galaxy. Lugten et al (1986), Genzel et al (1990), and Poglitsch et al (1991) show that a large fraction of the neutral gas, up to 10%, lies in $C^+$ regions, which suggests that most of the neutral gas is in PDRs. The large [C II] intensities from some clouds indicate that they are relatively close ($<$15 pc) to the galactic center, and connecting bridges suggest that the outlying clouds feed material into the center.

A prominent structure in the central 10 pc is the circumnuclear disk (CND) or torus, first observed in FIR continuum by Becklin et al (1982). Subsequent studies in [O I] (Genzel et al 1984, 1985, Jackson et al 1993), [C II] (Genzel et al 1985, Lugten et al 1986, Poglitsch et al 1991), [C I] (Serabyn et al 1994), [Si II] (Herter et al 1986, 1989, Graf et al 1988), and mid–J CO (e.g. Genzel et al 1985, Harris et al 1985) indicate a mass of $10^4$–$10^5$ $M_\odot$, with an inner radius of about 1.5 pc and an outer radius $\gtrsim$8 pc. The atomic carbon is comparable to the CO abundance; this is 10 times its value in local GMCs. The CND seems to consist of several streamers of material and is quite clumpy and turbulent (Jackson et al 1993). Wolfire et al (1990) and Burton et al (1990b) compare PDR models to observations to derive an incident FUV flux $G_0 \sim 10^5$ from the central cavity, which illuminates a clumpy structure with densities ranging from $10^5$–$10^7$ cm$^{-3}$ (Table 1). Overall, there is evidence for a mass infall rate into the central $r < 1.5$-pc cavity of $\sim 10^{-2}$ $M_\odot$ year$^{-1}$, which can feed a central black hole or a future starburst (Jackson et al 1993). The existence of hot massive stars in the central cavity suggests a recent episode of star formation. At least 200 $M_\odot$ of neutral PDR gas has been observed in [O I] in the central cavity (Jackson et al 1993); this gas provides evidence for a possible building reservoir of material available for star formation.

## 5. APPLICATIONS OF PDR MODELS

### 5.1 *CO Line Profiles and the $H_2$ Mass to CO J = 1–0 Luminosity Ratio*

Because the optically thick $^{12}$CO J = 1–0 emission originates largely from FUV heated PDRs on the "surfaces" of opaque molecular clouds, PDR modeling should be able to reproduce the CO line intensities and profiles. Tauber & Goldsmith (1990) and Wolfire et al (1993) construct models that self-consistently calculate the chemistry, thermal balance, and radiative transfer appropriate for a PDR model. They conclude that clumpy models can reproduce the observed centrally peaked profiles; Falgarone et al (1994) point out that turbulent velocity fields with some coherence reproduce the observed smoothness of the profiles better than randomly moving clumps.

There has been considerable discussion over the last decade concerning the correlation of the luminosity of $^{12}$CO J = 1–0 with the molecular mass of a cloud (e.g. Solomon et al 1987). Wolfire et al (1993) have self-consistently calculated the temperature and CO abundance as a function of position in clouds of various mass in order to predict the $^{12}$CO J = 1–0 luminosity. The PDR models match the observed correlation well, largely independent of $G_0$ in the range of $0.1 \lesssim G_0 \lesssim 10^3$. Sodroski et al (1995) present empirical evidence that the correlation factor changes by a factor of 20 when one compares the Galactic center with Galactic molecular clouds at 13 kpc. This change is hard to reconcile with the Wolfire et al model, although the metallicity gradient may enhance the factor gradient in this case.

## 5.2 *X-Ray Dissociation Regions (XDRs)*

Completely analogous to PDRs, XDRs can be defined as predominantly neutral gas in which X rays dominate the chemistry and/or the heating (see Maloney et al 1996, Sternberg et al 1997 for review). X rays can dominate the gas heating by photoionizing atoms and molecules and depositing a significant fraction of the primary and secondary electron energy into heat. X rays can also dominate much of the chemistry through the collisional dissociation and ionization of species by secondary electrons and through the photodissociation and photoionization by FUV photons produced via excitation of H and $H_2$ in collisions with secondary electrons. Molecular gas can be exposed to X rays in a wide range of astrophysical environments: in AGNs, near supernova remnants or fast shocks, around PNe with very hot central stars, and in molecular clouds with embedded X-ray sources such as massive stars, young stellar objects with X-ray active chromospheres, X-ray binaries, or accreting compact objects.

Early work focused on specialized aspects of the chemistry, thermal balance, and $H_2$ excitation in XDRs (Krolik & Kallman 1983, Lepp & McCray 1983, Krolik & Lepp 1989, Draine & Woods 1990, 1991, Voit 1991, Wolfire & Königl 1993, Gredel & Dalgarno 1995, Lepp & Dalgarno 1996, Yan & Dalgarno 1997) Recently, Maloney et al (1996) have done a comprehensive parameter study of equilibrium XDRs, studying the density range $n = 10^3$–$10^5$ cm$^{-3}$ and X-ray ionization rates that range from cosmic ray ionization rates to rates that nearly completely ionize the gas. They derive XDR spectra and show, for example, that [Fe II] 1.26 and 1.64 $\mu$m and $H_2$ 2 $\mu$m emission observed in Seyfert nuclei can originate in XDRs, and not in shocks as had been previously speculated (e.g. Moorwood & Oliva 1990, 1994, Mouri et al 1990a,b). Neufeld et al (1994) and Neufeld & Maloney (1995) applied these XDR models to the higher densities ($n \sim 10^9$ cm$^{-3}$) and X-ray fluxes incident upon the ~0.1–1 pc disks or tori that orbit the central engines of AGN. They showed that luminous water maser emission is characteristically produced under these conditions, explaining the origin of the $H_2O$ megamaser sources. All of the 16 megamasers now known are associated with active galactic nuclei (see Maloney 1997, and references therein).

## 5.3 *The Starburst Nucleus of M82*

Wolfire et al (1990) show how theoretical PDR models can be compared with IR and $^{12}$CO observations to derive numerous interesting average physical parameters that describe the ISM in the central ~1 kpc of relatively nearby IR-bright galaxies. Results have been obtained for M82 by Watson et al (1984), Crawford et al (1985), Lugten et al (1986), Wolfire et al (1990), Harris et al (1991), Schilke et al (1993), White et al (1994), and Lord et al (1996). The derived average gas density in the PDR clouds in the central 500 pc of M82, a relatively typical bright starburst region, is $\sim 10^4$ cm$^{-3}$, and the average incident FUV flux $G_0 \sim 10^3$ (Table 1). The atomic ($C^+$) temperatures are ~250 K, and the mass in the $C^+$ component is very significant, ~10% of the total gas mass. The gas-phase silicon abundances are high, $x_{Si} \sim 1.5 \times 10^{-5}$ (~0.5 solar), which may result from supernova shock destruction of silicate grains. A significant amount of [Si II] emission, and to a lesser extent [C II], may originate from relatively diffuse H II regions (Carral et al 1994, Lord et al 1996). The derived number of clouds and their sizes are surprising; there are numerous ($N \sim 3 \times 10^5$) small ($R \lesssim 1$ pc) clouds present. These "clouds" are individual entities in the sense that they cannot shadow each other from the FUV flux. Nevertheless, they may be clustered together in sheets or filaments. The average conditions in the central 500 pc of M82 are far different from the average ISM conditions in the solar vicinity: The thermal pressures, densities, and FUV fields are higher by factors on the order of $10^3$, $10^2$, and $10^3$ respectively. They resemble the conditions found in PDRs associated with Galactic reflection nebulae such as NGC 2023.

## 5.4 *FUV Regulation of Star Formation in Galaxies*

McKee (1989) and Bertoldi & McKee (1996) explain the observed constancy of $A_v$ (~7.5) in molecular clouds and the regulation of low-mass star formation with a PDR model. They assume that the rate of low-mass star formation is governed by ambipolar diffusion (cf Shu et al 1987) and that newly formed stars inject mechanical energy into the cloud, which supports the cloud against gravitational collapse. The ambipolar diffusion rate is set by the ionization fraction, which depends on the dust shielding of the FUV flux. In the model, the star formation rate increases as the cloud collapses, $A_v$ increases, the FUV-produced ionization level decreases, and the ambipolar diffusion rate increases. However, the cloud collapse is halted as the increased star formation injects turbulent energy. Equilibrium is achieved when $A_v \sim 7.5$. The external FUV flux, in controlling the ionization fraction in most of the cloud, regulates the low-mass star formation rate and the cloud column density.

Parravano (1987, 1988, 1989) proposes that a global feedback mechanism exists in galaxies whereby the galaxy-wide rate of high-mass star formation is also regulated by the interaction of the FUV with the neutral gas. Non–gravitationally bound neutral gas may exist in two phases (cold, ~100 K, and warm, $\sim 10^4$ K) in a galaxy, if the ISM pressure lies in a critical range

$P_{min} < P < P_{max}$ (Field et al 1969). For pressures $P < P_{min}$, only the warm diffuse phase exists. Parravano makes two assumptions in his modeling: (*a*) grain photoelectric heating dominates both phases so that $P_{min}$ monotonically increases with $G_0$, and (*b*) molecular star-forming clouds grow out of the cold phase (e.g. out of the coalescence of cold phase clouds). If the pressure $P$ of the ISM is greater than $P_{min}$, then the cold phase exists, molecular clouds grow, OB stars form, $G_0$ increases, and $P_{min}$ rises. If $P_{min}$ exceeds $P$, however, then the cold phase no longer exists, star formation drops, and $G_0$ and $P_{min}$ both decrease. Thus, the global OB star formation rate is regulated so that $P_{min} \sim P$ in galaxies. Parravano (1988, 1989) offers some observational support of this prediction in external galaxies. Parravano & Mantilla (1991) discuss the radial dependence of the state of the ISM in the Galaxy when both the McKee (1989) and the above self-regulation mechanism are operative.

## 6. CONCLUDING REMARKS

PDRs emit much of the IR radiation (line and continuum) in galaxies. Most of the mass of the gas and dust in the Galaxy resides in PDRs and is significantly affected, either via chemistry or heating, by the FUV flux. Much of the gas is heated by the grain photoelectric heating mechanism. The spectra from PDRs is characterized by luminosity ratios $(L_{CII} + L_{OI})/L_{IR} \sim 10^{-3}$–$10^{-2}$. PDRs are the origin of much of the FIR dust continuum, the near- and mid-IR PAH emission features, and the [C II] 158 $\mu$m, [O I] 63,145 $\mu$m, [Si II] 35 $\mu$m, [C I] 370,609 $\mu$m, low J CO, and C I recombination radiation. PDRs also emit significant $H_2$ rovibrational emission. In regions such as Orion, NGC 2023, NGC 7027, the Galactic center, and the nucleus of M82, the spectra diagnose physical conditions such as the gas density, the temperature, the clumpiness of the clouds, the FUV radiation field, and the elemental abundances. New models of X-ray illuminated molecular clouds may explain near-IR observations of Seyfert nuclei without invoking star formation. PDR models explain the observed correlations of [C II] 158 $\mu$m emission with $^{12}$CO J $=$ 1–0 emission, of $(L_{CII} + L_{OI})$ with $L_{IR}$, and of $^{12}$CO J $=$ 1–0 with $H_2$ mass. The FUV flux in PDRs may regulate low- and high-mass star formation and the column density of gravitationally bound star-forming molecular clouds.

### ACKNOWLEDGMENTS

We would like to acknowledge useful comments on early versions of this paper by D Jaffe, A Sternberg, H Störzer, and J Stutzki. This research was supported in part by the NASA Astrophysical Theory Program, which funds the Center for Star Formation Studies, a consortium of researchers from NASA Ames,

University of California at Berkeley, and University of California at Santa Cruz.



*Literature Cited*

Abgrall H, LeBourlot J, Pineau des Forêts G, Roueff E, Flower D, et al. 1992. *Astron. Astrophys.* 253:525–36
Allamandola LJ, Tielens AGGM, Barker JR. 1989. *Ap. J. Suppl.* 71:733–75
Andersson B-G, Wannier PG. 1993. *Ap. J.* 402: 585–92
Bakes ELO, Tielens AGGM. 1994. *Ap. J.* 427: 822–38
Bally J, Langer WD. 1982. *Ap. J.* 255:143–48
Becklin EE, Gatley I, Werner MW. 1982. *Ap. J.* 258:135–42
Bennett CL, Fixsen DJ, Hinshaw G, Mather JC, Moseley SH, et al. 1994. *Ap. J.* 434:587–98
Bertoldi F, Draine BT. 1996. *Ap. J.* 458:222–32
Bertoldi F, McKee C. 1996. In *Amazing Light*, ed. RY Chiao, pp. 41–53. New York: Springer-Verlag
Black JH, Dalgarno A. 1976. *Ap. J.* 203:132–42
Black JH, Dalgarno A. 1977. *Ap. J. Suppl.* 34:405–23
Black JH, van Dishoeck EF. 1987. *Ap. J.* 322: 412–49
Bobrowsky M, Zipoy DM. 1989. *Ap. J.* 347: 307–24
Boissé P. 1990. *Astron. Astrophys.* 228:483–502
Boland W, de Jong T. 1982. *Ap. J.* 261:110–14
Boreiko RT, Betz AL. 1991. *Ap. J.* 369:382–94
Boreiko RT, Betz AL. 1996a. *Ap. J. Lett.* 464:L83–L86
Boreiko RT, Betz AL. 1996b. *Ap. J. Lett.* 467:L113–16
Burke JR, Hollenbach DJ. 1983. *Ap. J.* 265:223–34
Burton M. 1992. *Aust. J. Phys.* 45:463–85
Burton M, Bulmer M, Moorhouse A, Geballe TR, Brand PWJL. 1992a. *MNRAS* 257:1P–6P
Burton M, Geballe TR, Brand PWJL, Moorhouse A. 1990a. *Ap. J.* 352:625–29
Burton M, Hollenbach D, Tielens AGGM. 1990b. *Ap. J.* 365:620–39
Burton M, Hollenbach D, Tielens AGGM. 1992b. *Ap. J.* 399:563–72
Burton M, Puxley P. 1990. In *The Interstellar Medium in External Galaxies*, ed. DJ Hollenbach, HA Thronson, pp. 238–40. Washington, DC: NASA Conf. Publ. 3084
Cardelli JA, Clayton GC, Mathis JS. 1989. *Ap. J.* 345:245–56
Cardelli JA, Meyer DM, Jura M, Savage BD. 1996. *Ap. J.* 467:334–40
Carral P, Hollenbach DJ, Lord SD, Colgan SWJ, Haas MR, et al. 1994. *Ap. J.* 423:223–36
Chieze JP, Pineau des Forêts G. 1989. *Astron. Astrophys.* 221:89–94
Chieze JP, Pineau des Forêts G. 1990. In *Physical Processes in Fragmentation and Star Formation*, ed. R Capuzzo-Dolcetta, C Chiosi, A DiFazio, pp. 17–25. Dordrecht: Kluwer
Chieze JP, Pineau des Forêts G, Herbst E. 1991. *Ap. J.* 373:110–22
Chrysostomou A, Brand PWJL, Burton MG, Moorhouse A. 1992. *MNRAS* 256:528–34
Chrysostomou A, Brand PWJL, Burton MG, Moorhouse A. 1993. *MNRAS* 265:329–39
Churchwell E, Felli M, Wood DOS, Massi M. 1987. *Ap. J.* 321:516–29
Clavel J, Viala YP, Bel N. 1978. *Astron. Astrophys.* 65:435–48
Cohen M, Allamandola L, Tielens AGGM, Bregman J, Simpson JP, et al. 1985. *Ap. J.* 302:737–49
Crawford MK, Genzel R, Townes CH, Watson DM. 1985. *Ap. J.* 291:755–71
Dalgarno A. 1985. In *Molecular Astrophysics*, ed. GHF Diercksen, WF Huebner, PW Langhof, pp. 281–94. Dordrecht: Reidel
Danks AC, Federman SR, Lambert DL. 1984. *Astron. Astrophys.* 130:62–66
deJong T. 1977. *Astron. Astrophys.* 55:137–45
deJong T, Dalgarno A, Boland W. 1980. *Astron. Astrophys.* 91:68–84
d'Hendecourt L, Léger A. 1987. *Astron. Astrophys.* 180:L9–L12
Diaz RI, Franco J, Shore SN. 1996. *Bull. Am. Astron. Soc.* 188:4010
Dinerstein HL. 1991. *PASP* 103:861–64
Dinerstein HL. 1995. In *Asymmetrical Planetary Nebulae*, ed. N Soker, A Harpaz, pp. 35–54. New York: Am. Inst. Phys.
Draine BT. 1978. *Ap. J. Suppl.* 36:595–619
Draine BT, Bertoldi F. 1996. *Ap. J.* 468:269–89
Draine BT, Lee HM. 1984. *Ap. J.* 285:89–108
Draine BT, Woods DT. 1990. *Ap. J.* 363:464–79
Draine BT, Woods DT. 1991. *Ap. J.* 383: 621–38
Eidelsberg MH, Benayoun JJ, Viala YP, Rostas F, Smith PL, et al. 1992. *Astron. Astrophys.* 265:839–42

Escalante V, Sternberg A, Dalgarno A. 1991. *Ap. J.* 375:630–34
Falgarone E, Lis DC, Phillips TG, Pouquet A, Porter DH, et al. 1994. *Ap. J.* 436:728–40
Falgarone E, Phillips TG. 1996. *Ap. J.* 472:191–204
Falgarone E, Pineau des Forêts G, Roueff E. 1995. *Astron. Astrophys.* 300:870–80
Federman SR, Danks AC, Lambert DL. 1984. *Ap. J.* 287:219–27
Federman SR, Glassgold AE, Jenkins EB, Shaya EJ. 1980. *Ap. J.* 242:545–59
Federman SR, Glassgold AE, Kwan J. 1979. *Ap. J.* 227:466–73
Federman SR, Huntress WT. 1989. *Ap. J.* 338:140–46
Federman SR, Strom CJ, Lambert DL, Cardelli JA, Smith VV, et al. 1994. *Ap. J.* 424:772–92
Field GB, Goldsmith DW, Habing HJ. 1969. *Ap. J. Lett.* 155:L149–52
Field GB, Somerville WB, Dressler K. 1966. *Annu. Rev. Astron. Astrophys.* 4:207–44
Flower DR, LeBourlot J, Pineau des Forêts G, Roueff E. 1994. *Astron. Astrophys.* 282:225–32
Fuente A, Martín-Pintado J, Cernicharo J, Bachiller R. 1993. *Astron. Astrophys.* 276:473–88
Fuente A, Martín-Pintado J, Gaume R. 1995. *Ap. J. Lett.* 442:L33–L36
Fuente A, Rodriguez-Franco A, Martín-Pintado J. 1996. *Astron. Astrophys.* 312:599–609
Gardiner WC. 1977. *Acc. Chem. Res.* 10:326–31
Gatley I, Hasegawa T, Suzuki H, Garden R, Brand P, et al. 1987. *Ap. J. Lett.* 318:L73–L76
Genzel R. 1991. In *The Physics of Star Formation and Early Stellar Evolution*, ed. C Lada, N Kylafis, pp. 155–219. Dordrecht: Kluwer
Genzel R. 1992. In *The Galactic Interstellar Medium: Saas Fe Lectures 1991*, ed. W Burton, R Genzel, BG Elmegreen, pp. 1–85. New York: Springer-Verlag
Genzel R, Harris AI, Jaffe DT, Stutzki J. 1988. *Ap. J.* 332:1049–57
Genzel R, Harris AI, Stutzki J. 1989. In *Infrared Spectroscopy in Astronomy, ESA SP–290*, ed. M Kessler, pp. 115–32. Noordwijk: Eur. Space Agency Publ.
Genzel R, Hollenbach D, Townes CH. 1994. *Rep. Prog. Phys.* 57:417–79
Genzel R, Stacey GJ, Harris AI, Townes CH, Geis N, et al. 1990. *Ap. J.* 356:160–73
Genzel R, Watson DM, Crawford MK, Townes CH. 1985. *Ap. J.* 297:766–86
Genzel R, Watson DM, Townes CH, Dinerstein HL, Hollenbach DJ, et al. 1984. *Ap. J.* 276:551–59
Gerola H, Glassgold AE. 1978. *Ap. J. Suppl.* 37:1–31
Gierens K, Stutzki J, Winnewisser G. 1992. *Astron. Astrophys.* 259:271–82
Glassgold AE. 1996. *Annu. Rev. Astron. Astrophys.* 34:241–78
Glassgold AE, Huggins PJ, Langer WD. 1985. *Ap. J.* 290:615–26
Glassgold AE, Langer WD. 1974. *Ap. J.* 193:73–91
Glassgold AE, Langer WD. 1975. *Ap. J.* 197:347–50
Glassgold AE, Langer WD. 1976. *Ap. J.* 206:85–99
Goldshmidt O, Sternberg A. 1995. *Ap. J.* 439:256–63
Graf P, Herter T, Gull GE, Houck JR. 1988. *Ap. J.* 330:803–8
Graf UU, Eckart A, Genzel R, Harris AI, Poglitsch A, et al. 1993. *Ap. J.* 405:249–67
Graham JR, Serabyn E, Herbst TM, Mathews K, Neugebauer G, et al. 1993. *Astron. J.* 105:250–57
Gredel R, Dalgarno A. 1995. *Ap. J.* 446:852–59
Gussie GT, Taylor AR, Dewdney PE, Roger RS. 1995. *MNRAS* 273:790–800
Güsten R, Fiebig D. 1988. *Astron. Astrophys.* 204:253–62
Habing HJ. 1968. *Bull. Astron. Inst. Netherlands* 19:421–32
Harris AI, Jaffe DT, Silber M, Genzel R. 1985. *Ap. J. Lett.* 294:L93–L96
Harris AI, Hill RE, Stutzki J, Graf UU, Russell APG, et al. 1991. *Ap. J. Lett.* 382:L75–L78
Hartquist TW, Menten KM, Lepp S, Dalgarno A. 1995. *MNRAS* 272:184–88
Hartquist T, Sternberg A. 1991. *MNRAS* 248:48–51
Hasegawa T, Gatley I, Garden RP, Brand P, Ohishi M, et al. 1987. *Ap. J. Lett.* 318:L77–L80
Hayashi M, Hasegawa T, Gatley I, Garden R, Kaifu N. 1985. *MNRAS* 215:31P–36P
Heck EL, Flower DR, LeBourlot J, Pineau des Forêts G, Roueff E. 1992. *MNRAS* 258:377–83
Hegmann M, Kegel WH. 1996. *MNRAS* 283:167–73
Herbst E, Leung CM. 1989. *Ap. J. Suppl.* 69:271–99
Herrmann F, Madden SC, Nikola T, Poglitsch A, Timmermann R, et al. 1997. *Ap. J.* In press
Herter T, Gull GE, Megeath ST, Rowlands N, Houck JR. 1989. *Ap. J.* 342:696–702
Herter T, Houck JR, Graf P, Gull GE. 1986. *Ap. J. Lett.* 309:L13–L16
Hill JK, Hollenbach DJ. 1978. *Ap. J.* 225:390–404
Hippelein HH, Münch G. 1989. *Astron. Astrophys.* 213:323–32
Hobson MP, Jenness T, Padman R, Scott PF. 1994. *MNRAS* 266:972–82

Hobson MP, Scheuer PAG. 1993. *MNRAS* 264:145–60

Hogerheijde MR, Jansen DJ, Van Dishoeck EF. 1995. *Astron. Astrophys.* 294:792–810

Hollenbach DJ. 1988. *Astrophys. Lett. Commun.* 26:191–206

Hollenbach DJ. 1990. In *The Evolution of the Interstellar Medium*, ed. L Blitz, pp. 167–82. San Francisco: Astron. Soc. Pac.

Hollenbach DJ, McKee CH. 1979. *Ap. J. Suppl.* 41:555–92

Hollenbach DJ, McKee CH. 1989. *Ap. J.* 342:306–36

Hollenbach DJ, Natta A. 1995. *Ap. J.* 455:133–44

Hollenbach DJ, Takahashi T, Tielens AGGM. 1991. *Ap. J.* 377:192–209

Hollenbach DJ, Tielens AGGM. 1996. In *The Physics and Chemistry of Interstellar Clouds*, ed. G Winnewisser, GC Pelz, pp. 164–74. Berlin: Springer-Verlag

Hollenbach DJ, Werner M, Salpeter E. 1971. *Ap. J.* 163:165–80

Howe JE, Jaffe DT, Genzel R, Stacey G. 1991. *Ap. J.* 373:158–68

Howe JE, Jaffe DT, Grossman EN, Wall WF, Mangum JG, et al. 1993. *Ap. J.* 410:179–87

Huggins PJ. 1992. In *Mass Loss on the AGB and Beyond*, ed. H Schwarz, pp. 35–54. Noordwijk: Eur. Southern Obs. Publ.

Huggins PJ. 1993. In *Planetary Nebulae, IAU 155*, ed. R Weinberger, A Acker, pp.147–62. Dordrecht: Kluwer

Huggins PJ, Bachiller R, Cox P, Forveille T. 1992. *Ap. J. Lett.* 401:L43–L46

Israel FP, Maloney PR, Geis N, Herrmann F, Madden SC, et al. 1996. *Ap. J.* 465:738–47

Jackson JM, Geis N, Genzel R, Harris AI, Madden S, Poglitsch A, et al. 1993. *Ap. J.* 402:173–84

Jaffe DT, Genzel R, Harris AI, Howe J, Stacey GJ, et al. 1990. *Ap. J.* 353:193–99

Jaffe DT, Howe JE. 1989. *Rev. Mex. Astron. Astrof.* 18:55–63

Jaffe DT, Zhou S, Howe JE, Herrmann F, Madden SC, et al. 1994. *Ap. J.* 436:203–15

Jansen DJ, Spaans M, Hogerheijde MR, van Dishoeck EF. 1995a. *Astron. Astrophys.* 303:541–53

Jansen DJ, van Dishoeck EF, Black JH, Spaans M, Sosin C. 1995b. *Astron. Astrophys.* 302:223–42

Jansen DJ, van Dishoeck EF, Keene J, Boreiko RT, Betz A. 1996. *Astron. Astrophys.* 309:899–906

Jones ME, Barlow SE, Ellison GB, Ferguson EE. 1986. *Chem. Phys. Lett.* 130:218–23

Jura M. 1974. *Ap. J.* 191:375–79

Jura M. 1976. *Ap. J.* 204:12–20

Justtanont K, Tielens AGGM, Skinner CJ, Haas M. 1997. *Ap. J.* In press

Kastner JH, Weintraub DA, Gatley I, Merrill KM, Probst RG. 1996. *Ap. J.* 462:777–85

Keene J. 1995. In *The Physics and Chemistry of Interstellar Molecular Clouds*, ed. G Winnewisser, GC Pelz, pp. 186–94. Berlin: Springer-Verlag

Keene J, Blake GA, Phillips TG, Huggins PJ, Beichman CA. 1985. *Ap. J.* 299:967–80

Keene J, Lis DC, Phillips TG, Schilke P. 1997. In *Molecules in Astrophysics: Probes & Processes*, ed. E van Dishoeck. Dordrecht: Kluwer. In press

Köster B, Störzer H, Stutzki J, Sternberg A. 1994. *Astron. Astrophys.* 284:545–58

Kramer C, Stutzki J, Winnewisser G. 1996. *Astron. Astrophys.* 307:915–35

Krolik JH, Kallman TR. 1983. *Ap. J.* 267:610–24

Krolik JH, Lepp S. 1989. *Ap. J.* 347:179–85

Langer WD. 1976. *Ap. J.* 206:699–712

Latter WB, Kelly DM, Hora JL, Deutsch LK. 1995. *Ap. J. Suppl.* 100:159–67

Latter WB, Walker CK, Maloney PR. 1993. *Ap. J. Lett.* 419:L97–L100

Leach S. 1987. In *Polycyclic Aromatic Hydrocarbons and Astrophysics*, ed. A Léger, L d'Hendecourt, N Boccara, pp. 99–127. Dordrecht: Reidel

LeBourlot J, Pineau des Forêts G, Roueff E, Flower DR. 1993. *Astron. Astrophys.* 267: 233–54

Lee H-H, Herbst E, Pineau des Forêts G, Roueff E, LeBourlot J. 1996. *Astron. Astrophys.* 311:690–707

Lee KT, Bowman JM, Wagner AF, Schatz GC. 1982. *J. Chem. Phys.* 76:3583–96

Léger A, Puget JL. 1989. *Annu. Rev. Astron. Astrophys.* 27:161–98

Lemaire JL, Field D, Gerin M, Leach S, Pineau des Forêts G, et al. 1996. *Astron. Astrophys.* 308:895–907

Lepp S, Dalgarno A. 1988. *Ap. J.* 335:769–73

Lepp S, Dalgarno A. 1996. *Astron. Astrophys.* 306:L21–L24

Lepp S, McCray R. 1983. *Ap. J.* 269:560–67

Light GC. 1978. *J. Chem. Phys.* 68:2831–43

Lis DC, Schilke P, Keene J. 1997. In *CO: 25 Years of Millimeter Wave Spectroscopy*, ed. WB Latter, SJE Radford, PR Jewell, JG Mangum, J Bally. Dordrecht: Kluwer. In press

London R. 1978. *Ap. J.* 225:405–16

Lord SD, Hollenbach DJ, Haas MR, Rubin RH, Colgan SWJ, et al. 1996. *Ap. J.* 465:703–16

Lugten JB, Genzel R, Crawford MK, Townes CH. 1986. *Ap. J.* 306:691–702

Luhman ML, Jaffe DT. 1996. *Ap. J.* 463:191–204

Luhman ML, Jaffe DT, Keller LD, Pak S. 1994. *Ap. J. Lett.* 436:L185–88

Luhman ML, Jaffe DT, Sternberg A, Herrmann F, Poglitsch A. 1997. *Ap. J.* In press

Maloney P. 1997. In *Accretion Phenomena and Related Outflows*, ed. D Wickramasinghe, L Ferrario, G Bicknell. San Francisco: Astron. Soc. Pac. In press

Maloney P, Hollenbach D, Tielens AGGM. 1996. *Ap. J.* 466:561–84

Martin PG, Mandy ME. 1995. *Ap. J.* 455:L89–L92

Martin PG, Schwarz DH, Mandy ME. 1996. *Ap. J.* 461:265–81

Massi M, Churchwell E, Felli M. 1988. *Astron. Astrophys.* 194:116–24

Mathis JS. 1990. *Annu. Rev. Astron. Astrophys.* 28:37–70

Mathis JS. 1997. In *From Stardust to Planetesimals*, ed. Y Pendleton, AGGM Tielens. San Francisco: Astron. Soc. Pac. In press

Mathis JS, Rumpl W, Nordsieck KH. 1977. *Ap. J.* 217:425–33

Matsuhara H, Nakagawa T, Shibai H, Okuda H, Mizutani T, et al. 1989. *Ap. J.* 339:L69–L70

McKee CF. 1989. *Ap. J.* 345:782–801

McKee CF, Storey JWV, Watson DM, Green S. 1982. *Ap. J.* 259:647–56

Meixner M, Haas MR, Tielens AGGM, Erickson EF, Werner M. 1992. *Ap. J.* 390:499–512

Meixner M, Tielens AGGM. 1993. *Ap. J.* 405:216–28

Melnick G, Gull GE, Harwit M. 1979. *Ap. J.* 227:L29–L33

Minchin NR, White GJ, Padman R. 1993. *Astron. Astrophys.* 277:595–608

Mochizuki K, Nakagawa T, Doi Y, Yui YY, Okuda H, et al. 1994. *Ap. J.* 430:L37–L40

Monteiro T. 1991. *Astron. Astrophys.* 241:L5–L8

Moorwood AFM, Oliva E. 1990. *Astron. Astrophys.* 239:78–84

Moorwood AFM, Oliva E. 1994. *Ap. J.* 429:602–11

Morris M, Serabyn E. 1996. *Annu. Rev. Astron. Astrophys.* 34:645–701

Mouri H, Kawara K, Taniguchi Y, Nishida M. 1990a. *Ap. J. Lett.* 356:L39–L42

Mouri H, Nishida M, Taniguchi Y, Kawara K. 1990b. *Ap. J.* 360:55–62

Natta A, Walmsley CM, Tielens AGGM. 1994. *Ap. J.* 428:209–18

Nelson RP, Langer WD. 1997. *Ap. J.* In press

Neufeld DA, Maloney PR. 1995. *Ap. J.* 447:L17–L20

Neufeld DA, Maloney PR, Conger S. 1994. *Ap. J.* 436:L127–30

Neufeld DA, Spaans M. 1996. *Ap. J.* 473:894–902

O'Dell CR, Wen Z, Hu X. 1993. *Ap. J.* 410:696–700

Parmar PS, Lacy JH, Achtermann JM. 1991. *Ap. J.* 372:L25–L28

Parravano A. 1987. In *Supernova Remnants and the ISM, IAU 101*, ed. RS Roger, TL Landecker, pp. 513–18. Cambridge: Cambridge Univ. Press

Parravano A. 1988. *Astron. Astrophys.* 205:71–76

Parravano A. 1989. *Ap. J.* 347:812–16

Parravano A, Mantilla JCh. 1991. *Astron. Astrophys.* 250:70–83

Phillips TG, Huggins PJ. 1981. *Ap. J.* 251:533–40

Plume R, Jaffe DT, Keene J. 1994. *Ap. J.* 425:L49–L52

Poglitsch A, Stacey GJ, Geis N, Haggerty M, Jackson J, et al. 1991. *Ap. J. Lett.* 374:L33–L36

Poglitsch A, Krabbe A, Madden SC, Nikola T, Geis N, et al. 1995. *Ap. J.* 454:293–306

Prasad SS, Heere KR, Tarafdar SP. 1991. *Ap. J.* 373:123–36

Prasad SS, Huntress WT. 1980. *Ap. J. Suppl.* 43:1–35

Roberge WG, Dalgarno A, Flannery BP. 1981. *Ap. J.* 243:817–26

Roberge WG, Jones D, Lepp S, Dalgarno A. 1991. *Ap. J. Suppl.* 77:287–97

Roger RS, Dewdney PE. 1992. *Ap. J.* 385:536–60

Rubio M, Lequeux J, Boulanger F. 1993a. *Astron. Astrophys.* 271:9–17

Rubio M, Lequeux J, Boulanger F, Booth RS, Garay G, et al. 1993b. *Astron. Astrophys.* 271:1–8

Russell RW, Melnick G, Gull GE, Harwit M. 1980. *Ap. J.* 240:L99–L103

Russell RW, Melnick G, Smeyers SD, Kurtz NT, Gosnell TR, et al. 1981. *Ap. J.* 250:L35–L38

Savage BD, Mathis JS. 1979. *Annu. Rev. Astron. Astrophys.* 17:73–111

Schatz GC. 1981. *J. Chem. Phys.* 74:1133–39

Schatz GC, Elgersma H. 1980. *Chem. Phys. Lett.* 73:21–25

Schatz GC, Wagner AF, Walch SP, Bowman JM. 1981. *J. Chem. Phys.* 74:4984–96

Schilke P, Carlstrom JE, Keene J, Phillips TG. 1993. *Ap. J.* 417:L67–L70

Schinke R, Lester WA. 1979. *J. Chem. Phys.* 72:3754–66

Sellgren K, Allamandola LJ, Bregman JD, Werner MW, Wooden DH. 1985. *Ap. J.* 299:416–23

Serabyn E, Keene J, Lis DC, Phillips TG. 1994. *Ap. J.* 424:L95–L98

Shu FH, Adams FC, Lizano S. 1987. *Annu. Rev. Astron. Astrophys.* 25:23–81

Shull JM. 1978. *Ap. J.* 219:877–85

Sodroski TJ, Odegard N, Dwek E, Hauser MG, Franz BA, et al. 1995. *Ap. J.* 452:262–68

Solomon PM, Rivolo AR, Barrett JW, Yahil A. 1987. *Ap. J.* 319:730–41

Spaans M. 1995. *Models of inhomogeneous interstellar clouds*. PhD thesis. University of Leiden, Leiden, The Netherlands
Spaans M. 1996. *Astron. Astrophys.* 307:271–87
Spaans M, Black JH, van Dishoeck EF. 1997. *Ap. J.* In press
Spaans M, Tielens AGGM, van Dishoeck EF, Bakes ELO. 1994. *Ap. J.* 437:270–80
Spaans M, van Dishoeck EF. 1997. *Astron. Astrophys.* In press
Stacey GJ, Geis N, Genzel R, Lugten JB, Poglitsch A, et al. 1991. *Ap. J.* 373:423–44
Stacey GJ, Jaffe DT, Geis N, Genzel R, Harris AI, et al. 1993. *Ap. J.* 404:219–31
Stecher TP, Williams DA. 1967. *Ap. J. Lett.* 149:L29–L32
Steiman-Cameron TY, Haas MR, Tielens AGGM, Burton MG. 1997. *Ap. J.* In press
Sternberg A. 1988. *Ap. J.* 322:400–9
Sternberg A. 1989. *Ap. J.* 347:863–74
Sternberg A. 1990. *Ap. J.* 361:121–31
Sternberg A. 1992. In *Astrochemistry of Cosmic Phenomena*, ed. PD Singh, pp. 329–32. Dordrecht: Kluwer
Sternberg A, Dalgarno A. 1989. *Ap. J.* 338:199–233
Sternberg A, Dalgarno A. 1995. *Ap. J. Suppl* 99:565–607
Sternberg A, Yan M, Dalgarno A. 1997. In *Molecules in Astrophysics: Probes & Processes*, ed. E van Dishoeck. Dordrecht: Kluwer. In press
Storey JWV, Watson DM, Townes CH. 1979. *Ap. J.* 233:109–18
Störzer H, Stutzki J, Sternberg A. 1995. *Astron. Astrophys.* 296:L9–L12
Störzer H, Stutzki J, Sternberg A. 1996. *Astron. Astrophys.* 310:592–98
Störzer H, Stutzki J, Sternberg A. 1997. *Astron. Astrophys.* In press
Stutzki J, Güsten R. 1990. *Ap. J.* 356:513–33
Stutzki J, Stacey GJ, Genzel R, Graf UU, Harris AL, et al. 1990. In *Submillimetre Astronomy*, ed. GD Watt, AS Webster, pp. 269–73. Dordrecht: Kluwer
Stutzki J, Stacey GJ, Genzel R, Harris A, Jaffe D, Lugten J. 1988. *Ap. J.* 332:379–99
Tanaka M, Hasegawa T, Hayashi SS, Brand PWJL, Gatley I. 1989. *Ap. J.* 336:207–11
Tarafdar SP, Prasad SS, Huntress WT, Villere KR, Black DC. 1985. *Ap. J.* 289:220–37
Tauber J, Goldsmith P. 1990. *Ap. J. Lett.* 356:L63–L66
Tauber J, Lis DC, Keene J, Schilke P, Büttgenbach TH. 1995. *Astron. Astrophys.* 297:567–73
Tauber JA, Tielens AGGM, Meixner M, Goldsmith PF. 1994. *Ap. J.* 422:136–52
Tielens AGGM. 1993. In *Planetary Nebulae*, ed. R Weinberger, A Acker, pp.155–62. Dordrecht: Kluwer
Tielens AGGM. 1994. In *Airborne Astronomy Symposium on the Galactic Ecosystem*, ed. MR Haas, JA Davidson, EF Erickson, pp. 3–22. San Francisco: Astron. Soc. Pac.
Tielens AGGM, Hollenbach DJ. 1985a. *Ap. J.* 291:722–46
Tielens AGGM, Hollenbach DJ. 1985b. *Ap. J.* 291:747–54
Tielens AGGM, Hollenbach DJ. 1985c. *ICARUS* 61:40–47
Tielens AGGM, Meixner MM, van der Werf PP, Bregman J, Tauber JA, et al. 1993. *Science* 262:86–89
Timmermann R, Bertoldi F, Wright CM, Drapatz S, Draine BT, et al. 1996. *Astron. Astrophys.* 315:281–85
Turner BE. 1996. *Ap. J.* 461:246–64
Usuda T, Sugai H, Kawabata H, Inoue Y, Kataza H, et al. 1996. *Ap. J.* 464:818–28
van der Werf PP, Stutzki J, Sternberg A, Krabbe A. 1996. *Astron. Astrophys.* 313:633–48
van Dishoeck EF. 1987. In *Astrochemistry*, ed. MS Vardya, SP Tarafdar, pp. 51–65. Dordrecht: Kluwer
van Dishoeck EF. 1988. In *Rate Coefficients in Astrochemistry*, ed. TJ Millar, DA Williams, pp. 49–62. Dordrecht: Kluwer
van Dishoeck EF. 1991. In *Astrochemistry of Cosmic Phenomena*, ed. PD Singh, pp.143–51. Dordrecht: Kluwer
van Dishoeck EF. 1992. In *Infrared Astronomy with ISO*, ed. T Encrenaz, M Kessler, pp. 283–308. New York: Nova
van Dishoeck EF, Black JH. 1986. *Ap. J. Suppl.* 62:109–39
van Dishoeck EF, Black JH. 1988. *Ap. J.* 334: 711–802
van Dishoeck EF, Black JH. 1989. *Ap. J.* 340: 273–97
Verstraete L, Léger A, d'Hendecourt L, Dutuit O, Deforneau D. 1990. *Astron. Astrophys.* 237:436–44
Viala YP. 1986. *Astron. Astrophys. Suppl.* 64:391–437
Viala YP, Letzelter C, Eidelsberg M, Rostas F. 1988. *Astron. Astrophys.* 193:265–72
Voit GM. 1991. *Ap. J.* 377:158–70
Wagner AF, Graff MM. 1987. *Ap. J.* 317:423–31
Watson WD. 1972. *Ap. J.* 176:103–10
Watson D, Genzel R, Townes CH, Werner MW, Storey JWV. 1984. *Ap. J.* 279:L1–L4
White GJ, Ellison B, Claude S, Dent WRF, Matheson PN. 1994. *Astron. Astrophys.* 284:L23–L26
White GJ, Padman R. 1991. *Nature* 354:511–13
White GJ, Sandell G. 1995. *Astron. Astrophys.* 299:179–92

Williams DA, Hartquist TW. 1984. *MNRAS* 210:141–45
Williams DA, Hartquist TW. 1991. *MNRAS* 251:351–55
Willner SP, Soifer BT, Russell RW, Joyce RR, Gillett FC. 1977. *Ap. J.* 217:L121–24
Wolfire M, Hollenbach D, Tielens AGGM. 1993. *Ap. J.* 402:195–215
Wolfire MG, Hollenbach DJ, Tielens AGGM. 1989. *Ap. J.* 344:770–78
Wolfire MG, Königl A. 1993. *Ap. J.* 415:204–17
Wolfire MG, Tielens AGGM, Hollenbach DJ. 1990. *Ap. J.* 358:116–31
Wright EL, Mather JC, Bennett CL, Cheng ES, Shafer RA, et al. 1991. *Ap. J.* 381:200–9
Wyrowski F, Walmsley CM, Natta A, Tielens AGGM. 1997. *Astron. Astrophys.* In press
Xie T, Allen M, Langer WD. 1996. *Ap. J.* 440:674–85
Yan M, Dalgarno A. 1997. *Ap. J.* In press

*Annu. Rev. Astron. Astrophys. 1997. 35:217–66*


# HIGH-VELOCITY CLOUDS

*B. P. Wakker*
Department of Astronomy, University of Wisconsin, 475 N. Charter St., Madison, Wisconsin 53706; e-mail: wakker@astro.wisc.edu

*H. van Woerden*
Kapteyn Instituut, Rijks Universiteit Groningen, Postbus 800, 9700 AV Groningen, The Netherlands; e-mail: hugo@astro.rug.nl

KEY WORDS: interstellar matter, neutral hydrogen, galactic halo, Magellanic Stream, galactic evolution

ABSTRACT

High-velocity clouds (HVCs) consist of neutral hydrogen (HI) at velocities incompatible with a simple model of differential galactic rotation; in practice one uses $|v_{LSR}| \gtrsim 90$ km/s to define HVCs. This review describes the main features of the sky and velocity distributions, as well as the available information on cloud properties, small-scale structure, velocity structure, and observations other than in 21-cm emission. We show that HVCs contain heavy elements and that the more prominent ones are more than 2 kpc from the Galactic plane. We evaluate the hypotheses proposed for their origin and reject those that account for only one or a few HVCs. At least three different hypotheses are needed: one for the Magellanic Stream and possibly related clouds, one for the Outer Arm Extension, and one (or more) for the other HVCs. We discuss the evidence for the accretion and the fountain model but cannot rule out either one.

## 1. INTRODUCTION

When observations by Münch, later published by Münch & Zirin (1961), showed absorption lines of interstellar clouds at distances of 500–1500 pc from the galactic plane, Spitzer (1956) proposed that these clouds might be in equilibrium with an interstellar Galactic Corona having a temperature of about $10^6$ K. In late 1956, Oort suggested to one of us (HvW) that such a corona might contain neutral hydrogen (HI) having high velocities, and that this gas might replenish the gas expanding away from the Galactic Center region



0066-4146/97/0915-0217$08.00

(van Woerden et al 1957). After several years of systematic searches, started by E Raimond at Dwingeloo in 1958, a new receiver brought the first detection of HI at high velocities (Muller et al 1963).

Early on a distinction was made between intermediate-velocity clouds (IVCs) and high-velocity clouds (HVCs), the division being made arbitrarily at a velocity relative to the *local standard of rest* (LSR) of $v_{LSR} = -70$ km/s. Later studies found that some IVCs extend to higher velocities and the division was shifted to $-90$ (or even $-100$) km/s. When HI at high positive velocities was discovered, a similar division was laid at $+90$ km/s. Two problems exist with applying a division in $v_{LSR}$ between IVCs and HVCs. First, it may not have a physical basis. Second, the range of velocities allowed by differential galactic rotation varies strongly with position. In some directions, velocities of 100 km/s are easily understood, while in others velocities of 50 km/s are anomalous. Wakker (1991a) therefore proposed a definition in terms of the *deviation velocity*: HVCs deviate more than 50 km/s from the range allowed by a simple model of differential galactic rotation. At high latitudes this definition preserves the old distinctions. At low latitudes it avoids inclusion of gas with high but regular velocities. A study in terms of the deviation velocity has not been done, but the new Leiden-Dwingeloo survey (Hartmann & Burton 1997) will make this possible. This review uses the old definition.

The negative-velocity HVCs have been studied extensively. In contrast, few major studies exist for the IVCs (but see Wesselius & Fejes 1973, Kuntz & Danly 1996), and studies of positive-velocity HVCs and IVCs are limited to a few clouds. However, all these features are important for a comprehensive understanding of the Galaxy. Although the HVCs represent just a few percent of the observed HI flux, they are relatively distant, and HVCs and related HI clouds may represent up to 10% of the total HI mass in the Galaxy. The precise proportion is not known. The best-studied negative-velocity HVC, complex A, has a mass of $10^{5-6}$ $M_\odot$ (using the distance bracket discussed in Section 4). Assuming a peculiar velocity of 100 km/s, its kinetic energy is $10^{45-46}$ J. For the positive-velocity clouds a mass of $10^4$ $D^2_{kpc}$ $M_\odot$ and kinetic energy $10^{44}$ $D^2_{kpc}$ J are estimated, while the mass and kinetic energy of the largest IVC, the IV Arch, are $5 \times 10^5$ $M_\odot$ and $10^{45}$ J (at its derived distance of about 3 kpc, Kuntz & Danly 1996). Thus, HVCs and IVCs may contain similar amounts of energy and mass, and the bulk motion of the major clouds represents an energy equivalent to that of many supernovae.

A common misconception is that the presence of HVCs with negative velocities must imply an inflow of gas to the Galaxy. However, differential galactic rotation makes an important contribution to the observed velocities. Furthermore, since we do not know the tangential velocity of the HVCs, it is not possible to be certain whether a positive- or negative-velocity cloud moves toward or away from the galactic plane. It is clear, however, that the total picture must be more complicated than a simple inflow.

HVC velocities may be expressed relative to the *galactic standard of rest* ($v_{GSR}$), instead of the LSR, in which case the velocity of the LSR in the direction of the HVC has been taken out: $v_{GSR} = v_{LSR} + 220 \sin l \cos b$. In principle, $v_{GSR}$ is more meaningful than $v_{LSR}$. However, only radial velocities can be observed, and the resulting $v_{GSR}$ will differ from the space velocity of the HVC. Thus, using $v_{GSR}$ instead of $v_{LSR}$ introduces another uncertainty in the analysis, obliterating the gains in physical meaning. Hence, we do not use $v_{GSR}$ in this review.

Few comprehensive reviews of HVCs exist. The first general discussion of their origins by Oort (1966) remains useful. Later reviews updated Oort's discussion (Dieter 1969, Davies 1974, Verschuur 1975, Hulsbosch 1979, de Boer 1983, van Woerden et al 1985, Wakker 1991c). Other overviews of HVCs, also tying them to more general ideas about galaxies, were given at the 1985 Green Bank workshop on gaseous halos of galaxies (Bregman & Lockman 1986) and at IAU Symposium 144 (Bloemen 1991).

We discuss the phenomenology of HVCs in Sections 2–5. In Sections 6 and 7 we evaluate the many hypotheses proposed for their origin, and we give a synthesis in Section 8.

## 2. THE DISTRIBUTION AND STRUCTURE OF HVCs

### 2.1 *Large-Scale Surveys*

For HI at latitudes $b < 10°$ and velocities $v_{LSR} < 90$ km/s, complete coverage was achieved relatively early (grids of 0°.5 with a 0°.5 beam, velocity resolution 1 km/s, and sensitivity $\sim 5 \times 10^{18}$ cm$^{-2}$) (Burton 1985). The general characteristics of HI in the Milky Way have been discussed by Dickey & Lockman (1990). However, most of these surveys miss most HVCs. Separate surveys with less coverage and/or sensitivity were done for high-velocity HI (Table 1).

The combination of the Villa Elisa (Bajaja et al 1985) and Dwingeloo surveys (Hulsbosch & Wakker 1988) covers the whole sky at similar sensitivity and resolution. Together they supersede the older HVC surveys, except for some HVCs that were mapped at higher angular or velocity resolution (e.g. Giovanelli et al 1973, Giovanelli 1980b). A whole-sky map of the HVCs is given in Figure 1. In both surveys the brightness temperature and LSR velocity of the high-velocity profile peaks are given, although near the galactic plane the high-velocity component is often visible only as a wing on a profile whose peak is due to distant disk gas. Since the velocity resolution is 16 km/s, line widths were measured only for very wide profiles (FWHM > 25 km/s). In other cases one must assume a linewidth (e.g. 20 km/s) to derive the column density.

**Table 1** Major surveys for high-velocity clouds[a]

| Reference | Tel[1] | Coverage | Grid | Beam (arcmin) | $v_{LSR}$ range (km/s) | $\Delta v$[2] (km/s) | $\sigma$[3] (K) |
|---|---|---|---|---|---|---|---|
| Muller et al 66 | Dw | $b > 30°$ | $10° \times 10°$, $lb$ | 38.5 | –170:+170 | 6 | 1.2 |
| Hulsbosch & Raimond 66 | Dw | $\delta > -35°$, $\|b\| > 20°$ | $10° \times 10°$, $lb$ | 38.5 | –240: –20 | 16 | 0.4 |
| Hulsbosch 68 | Dw | $\delta > -35°$ | $5° \times 5°$, $lb$[5] | 38.5 | –250: –50 | 16 | 0.4 |
| | | | $10° \times 10°$, $lb$ | | +50:+250 | | |
| Meng & Kraus 70 | OS | $-5° < \delta < +60°$ | $10' \times 1°$, $\alpha\delta$[4] | 10 × 40 | –250:+60 | 20 | 0.5 |
| Dieter 72a, 72b | HC | $\|b\| < 15°$, $l = 10° - 250°$ | $2° \times 2°$, $lb$[6] | 35.2 | –225:+150 | 2.1 | 0.04 |
| | | $\|b\| > 15$, $l = 0°–230°/360°$ | $5° \times 2°$, $lb$[7] | | –225:+150 | 2.1 | |
| van Kuilenburg 72a, 72b | Dw | $\delta > -35°$, $\|l\| > 15°$ | $0.°25/\cos\delta \times 2.°5$ $lb$[4] | 38.5 | –270:+270 | 10.5 | 0.15 |
| Wannier et al 72 | BL | $l = 252°–322°$, $b = 10° - 30°$ | $2° \times 2°$, $lb$ | 120 | +18:+334 | 15.8 | 0.05 |
| Giovanelli et al 73 | GB | A, C, M, 112 + 2–139[8] | $10' \times 10'$, $\alpha\delta$[4] | 9.7 | 130 wide | 1.37 | 0.23[8] |
| Mathewson et al 74 | PK | $\delta < -37.°5$ | $50' \times 2.°5/5$, $\alpha\delta$[4, 9] | 50 | –340:+380 | 10 | 0.3 |
| Giovanelli 80b | GB | $-18° < \delta < 55°$ | $1° \times 2°$, $\alpha\delta$[4] | 9.7 | –900:+900 | 16 | 0.009 |
| Mirabel & Morras 84 | GB | $l = 320° - 50°$, $b > 40°$ | $2° \times 2°$, $\alpha\delta$ | 21 | –950:+950 | 2.1 | 0.03 |
| | | $\delta > -44°$ | | | | | |
| Bajaja et al 85 | VE | $\delta < -10°$ | $2°/\cos\delta \times 2°$, $\alpha\delta$ | 34 | –650:+650 | 16 | 0.025 |
| Hulsbosch & Wakker 88 | Dw | $\delta > -18°$ | $1° \times 1°$, $lb$[10] | 36 | –950:+800 | 16 | 0.01 |
| Stark et al 92 | BL | $\delta > -40°$ | $2° \times 2°$, $\alpha\delta$ | 120 | –325:+325 | 5.3 | 0.017 |
| Hartmann & Burton 97 | Dw | $\delta > -30°$ | $0.°5/\cos b \times 0.°5$, $lb$ | 36 | –450:+400 | 1.5 | 0.07 |

[a] Notes: 1. Telescopes: Dw = Dwingeloo, OS = Ohio State, HC = Hat Creek, BL = Bell Labs, GB = Green Bank, VE = Villa Elisa. 2. $\Delta v$ is the velocity resolution. 3. $\sigma$ is the 1-$\sigma$ rms noise. 4. Drift scans. 5. The discovered HVCs were also observed on a $0.°5$ or $1°$ grid. 6. For $|b| < 15°$, negative velocities are covered on a $2° \times 2°$ grid, but positive velocities on a $10° \times 2°$ grid. 7. For $b > 45°$, the grid is sparser than $5° \times 2°$; the longitude limit is 230° for negative latitudes, 360° for positive latitudes. 8. The sky and velocity coverages consist of irregular regions covering the named HVCs, the sensitivity can also be 0.33 K. 9. Scans spaced $2.°5$ in declination for positive velocities, 5° for negative velocities. 10. Grid in $l$ is 2° for $|b| > 45°$, $2.°5$ for $|b| > 60°$.

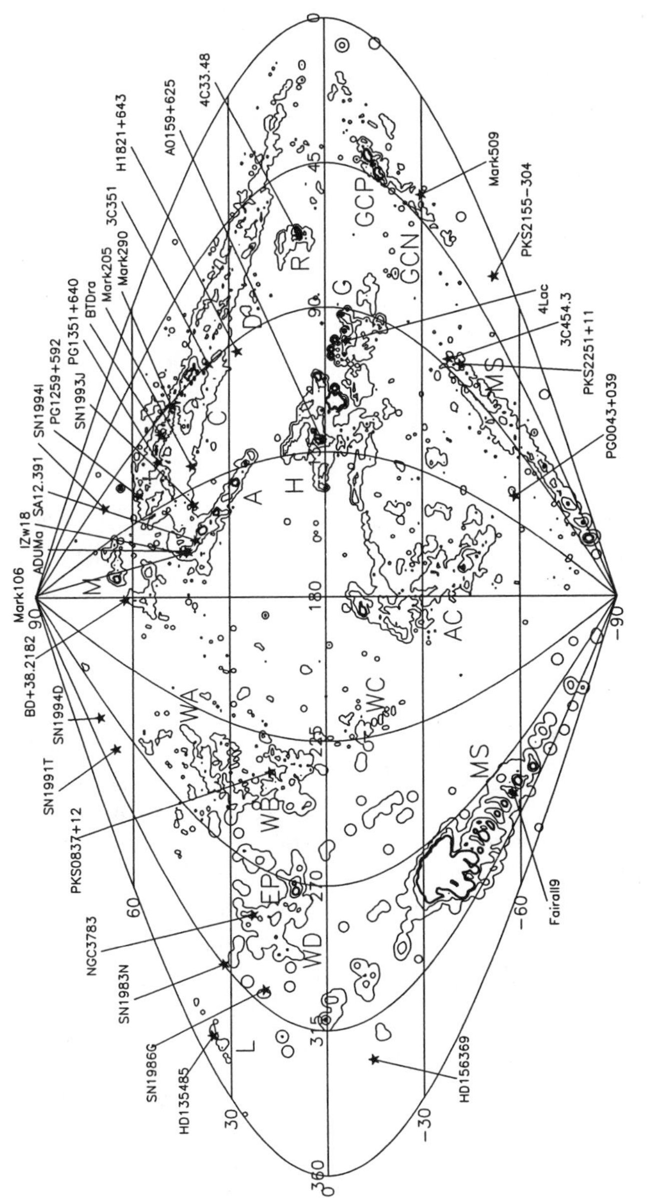


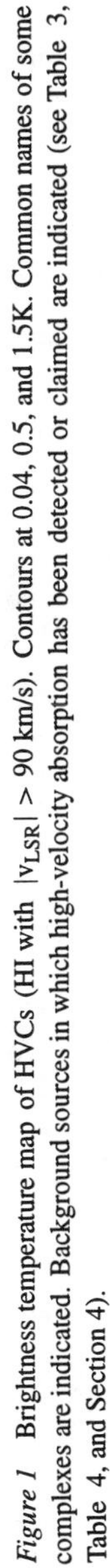
*Figure 1* Brightness temperature map of HVCs (HI with $|v_{LSR}| > 90$ km/s). Contours at 0.04, 0.5, and 1.5K. Common names of some complexes are indicated. Background sources in which high-velocity absorption has been detected or claimed are indicated (see Table 3, Table 4, and Section 4).

The Leiden-Dwingeloo survey (Hartmann & Burton 1997) has a finer grid and higher velocity resolution. At full resolution the 5-$\sigma$ detection limit is 0.35 K. Smoothing to 16 km/s can decrease this limit to 0.09 K, smoothing over 4 grid points to 0.05 K. Thus, for faint HVCs, the sensitivity is comparable to that of the earlier surveys, but more detailed maps of the larger and brighter HVCs can be made, and additional small clouds may be found.

Two HVC catalogues were published recently. That of Wakker & van Woerden (1991) covers the whole sky and was created by manually collecting components at adjacent positions into clouds. Stark et al (1992) used an automated routine to delineate clouds; however, only declinations greater than $-40°$ were covered, and baseline problems may have generated artificial clouds. Since HVCs often show large velocity and brightness gradients, no simple procedure exists to define clouds: an automated procedure may combine or separate clouds in an artificial manner; a manual procedure may be subjective. We use the Wakker & van Woerden (1991) catalogue in this review.

## 2.2 *HVC Complexes*

Before the complete catalogues, a set of more or less well-defined cloud groupings had already been defined, each consisting of a contiguous region on the sky where high-velocity HI at similar velocities is present. The naming of these clouds and cloud complexes was determined historically and does not follow a simple rule. They are identified in Figure 1. HVCs A, B, and C were the three main cores in the first survey (Hulsbosch & Raimond 1966, Hulsbosch 1968). Other complexes were named after their discoverers (M after Mathewson et al 1966 and Meng & Kraus 1970, WA–WD after Wannier et al 1972, the Giovanelli Stream after Giovanelli 1980a, the Cohen Stream after Cohen 1981) or after the person who studied it in detail (H after Hulsbosch 1975). The Magellanic Stream is so named because of its obvious connection to the Magellanic Clouds (Mathewson et al 1974). Other complexes are identified by their position in the sky: ACHV and ACVHV near the Anti-Center, GCN and GCP near the Galactic Center, D and L after the constellation they are in. Finally, cloud R was a feature in the study of Kepner (1970). Wakker & van Woerden (1991) present a table of the properties of these complexes and discuss each in detail.

Figure 2 presents a map of complex A. This map shows that: (*a*) A contains eight cores (A0 through A VI and B), which at medium (10′) resolution have sharp edges (Giovanelli et al 1973, Davies et al 1976, Hulsbosch 1978, Meyerdierks 1991); (*b*) the velocity field shows irregular jumps; and (*c*) A is connected to C by a bridge that was found by Encrenaz et al (1971) but that is seen clearly only in the Dwingeloo survey. The bridge has a complex structure and may be an extension of C, as the velocities are more typical of C than of A.

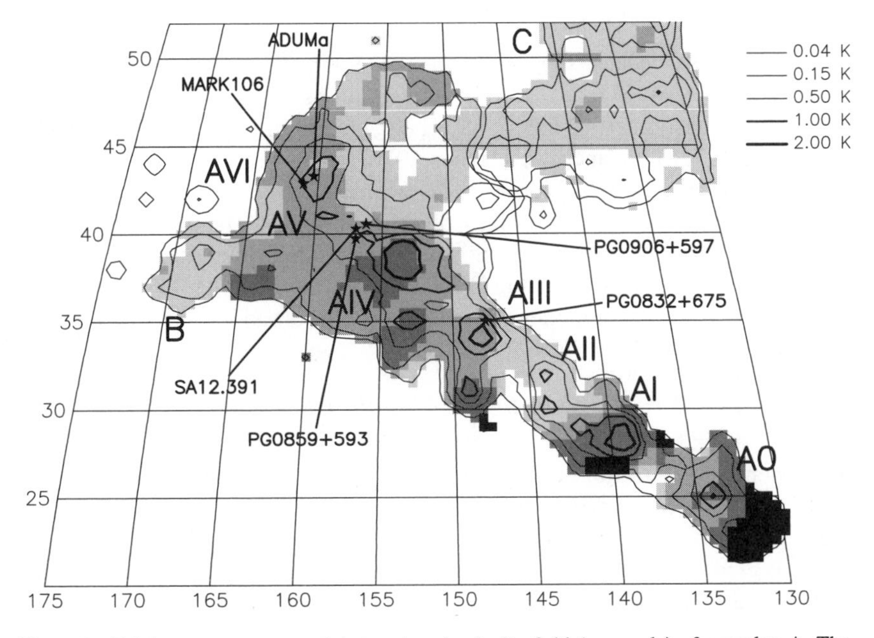


*Figure 2* Brightness temperature (contours) and velocity field (greyscale) of complex A. The darkest greyscale is for velocities less than −190 km/s. It changes at −170, −150, and −130 km/s. Shown are the objects used to derive a distance (Section 4). In the top right corner, part of complex C is included to show the bridge between A and C. Separate contours were drawn for A and C, to show where the artificial boundary was put.

The properties of complex A are often taken to be representative for all HVCs. They include: high galactic latitude, velocities well-separated from low-velocity galactic gas, linear structure, well-defined cores, a two-component structure in velocity, and a velocity field without large-scale structure. However, not all HVCs have the same combinations of these properties, and any interpretation should take that into account.

Apart from the major complexes, many small HVCs have been identified. In the region $l = 0°–90°$, $b < 0°$ there are 34 small HVCs with high negative velocities (Saraber & Shane 1974, Mirabel & Franco 1976, Mirabel 1981a, Mirabel & Morras 1984). Mirabel suggested that these HVCs are a more distant counterpart of the Anti-Center HVCs, that is, they are falling in to the Galactic Center, at distances of about 20 kpc or more. Other HVCs with $v_{LSR} < -250$ km/s concentrate in the southern galactic hemisphere at $l < 180°$ (Wright 1974, Davies 1975, Shostak 1977, Cohen & Mirabel 1978, 1979, Wright 1979a, 1979b, Cohen 1982b, Mirabel & Morras 1984). These clouds are usually called the very-high-velocity clouds or VHVCs.

Many HVCs have positive velocities. They are often ignored in models (cf Section 6), partly because they tend to be fainter than the well-known negative-velocity complexes. Maps at low angular resolutions (35′ or 10′), but sometimes high velocity resolution (2 km/s), have been presented (Smith 1963, Wannier & Wrixon 1972, Giovanelli & Haynes 1976, Mathewson et al 1979, Cohen & Ruelas Mayorga 1980, Morras 1982, Morras & Bajaja 1983, Bajaja et al 1989, Cavarischia & Morras 1989, Colgan et al 1990). The positive-velocity HVCs have dissimilar properties. Some lower-latitude clouds seem connected to galactic spiral structure. Some HVCs close to the Magellanic Clouds have only a broad velocity component, but most show the broad-and-narrow velocity structure seen elsewhere (Section 2.7).

## 2.3 *A HVC Projected onto the Magellanic Clouds?*

IUE and ground-based spectra of stars in the LMC show several absorption components, some from nearby gas around $v_{LSR} = 0$ km/s and some at +200–+300 km/s associated with the LMC. Absorption at velocities between +70 and +170 km/s is also common (Savage & de Boer 1981, Songaila & 1981b, Songaila et al 1986), and it has been interpreted as gas associated with the LMC or as galactic halo gas similar to other HVCs. If the latter were the case, it would give the best opportunity to study abundances in a HVC, as many LMC stars could serve as background probes.

In the 21-cm line these components are often blended with low-velocity or LMC gas and are usually weak. Partial maps have been made by Wayte (1990) and de Boer et al (1990), but the full angular extent remains unknown. Mebold (1991) reviewed these data. The distribution suggests that the +70–+100 km/s gas is not associated with the LMC–SMC system but is a galactic IVC. The gas at +130–+170 km/s is seen only superposed on the HI envelope associated with the LMC ($v_{LSR} \sim 220$–300 km/s). Thus, this gas is most likely related to the Magellanic System, and not to the Galaxy.

## 2.4 *North-South and East-West Asymmetries*

A major finding from the whole-sky surveys is that each of the four quadrants of the sky has different characteristics (Giovanelli 1980b, Wakker & van Woerden 1991). These differences provide major constraints on an acceptable model for the HVCs. Figure 3 shows the distribution of HVCs for eight different latitude ranges. To get a more complete HVC sample, some of the regions between the survey cutoff and the range of allowed velocities should be filled in.

The quadrant $l < 180°$, $b > 0°$ contains four of the large, well-defined and bright complexes (A, C, M, H), all at negative velocities. It also contains most of the bright IVCs, although there does not seem to be a spatial correlation between IVCs and HVCs (Kuntz & Danly 1996).

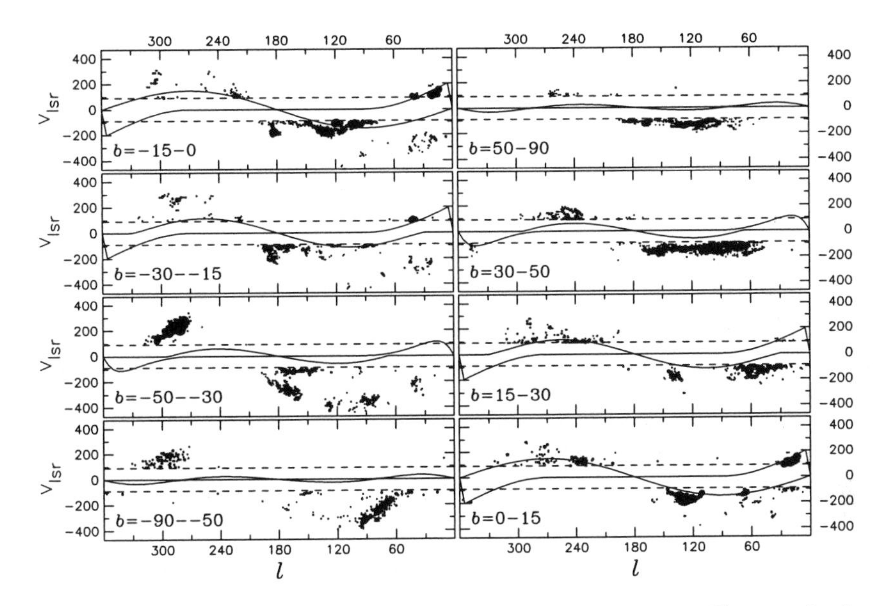


*Figure 3* Longitude-velocity diagrams for eight latitude ranges of equal area. The area of each dot represents the column density in a 1°–by–10 km/s cell. Dashed lines show the survey cutoff of +/–90 km/s. Solid lines show the maximum velocities compatible with differential galactic rotation for the lowest *b* in each latitude range. This velocity range is determined by assuming a simple Galaxy model: solid-body rotation inside 0.5 kpc, a flat rotation curve at 220 km/s elsewhere, and an HI distribution with radius 26 kpc and thickness 4 kpc.

The quadrant $l < 180°$, $b < 0°$ contains the VHVCs, high-negative velocity clouds near the Galactic Center and Anti-Center, and part of the Magellanic Stream. There is no clear counterpart to the large HVCs and IVCs at $b > 0°$, although one might interpret the Anti-Center Clouds as such. IV gas is extensive between $l = 90°$ and $l = 180°$ (Stark et al 1992), but it is not as organized, bright, or well-studied as the northern IVCs.

The quadrant $l > 180°$, $b > 0°$ contains many small positive-velocity clouds and three complexes (WA, WB, and WD) that are not as large, bright or well-connected as A, C, or M. When combined they cover almost 800 square degrees, about a third of the area of complexes A, C, and M. This quadrant also contains clouds at extreme-positive velocities ($v_{LSR} > 200$ km/s), which partly overlap with the other HVCs.

The quadrant $l > 180°$, $b < 0°$ is dominated by the Magellanic Stream, which originates from the Magellanic Clouds. Some faint positive-velocity clouds also occur. However, this part of the sky is comparatively poorly surveyed (only on the 2° × 2° grid of the southern survey), and many undiscovered HVCs may be lurking here.

Giovanelli (1980b) and Wakker & van Woerden (1991) showed that there are at least two classes of HVCs. One class is connected to the Galaxy. The envelope of this class's $l$-$v_{LSR}$ and $b$-$v_{LSR}$ distributions shows the reflection of rotation. Within this envelope the distribution gets denser toward lower velocities, and the clouds in this class are concentrated toward the galactic plane. The second class consists of the VHVCs, which have much higher velocities and which concentrate toward the edge of the $l$-$v_{LSR}$ distribution.

## 2.5 *Sky Coverage and Column Density Spectrum*

The fraction $F$ of sky covered by HVCs depends on the selection criteria for column density and velocity. Applying strict limits of $v_{LSR} > 100$ km/s and $N(HI) > 2 \times 10^{18}\,cm^{-2}$, Wakker (1991a) found $F = 0.18$. Excluding the Outer Arm Extension (Section 6.3) reduces this value to 0.13. Excluding also the Magellanic Stream reduces it further to 0.08. Extrapolating $F$[N(HI)], Wakker estimated $F = 0.2$–$0.4$ for HVCs with $N(HI) > 7 \times 10^{17}\,cm^{-2}$. Murphy et al (1995) confirmed this by searching for HVCs near AGNs, with a 5-$\sigma$ detection limit of $N(HI) = 7 \times 10^{17}\,cm^{-2}$, and found $F = 0.37$. All faint ($N(HI) < 2 \times 10^{18}\,cm^{-2}$) detections are associated with a known complex.

Giovanelli (1980b) found that the distribution of the logarithm of the column density, $F$[log(N(HI))], drops for $N(HI) < 8 \times 10^{18}\,cm^{-2}$, whereas his 5-$\sigma$ detection limit was $3 \times 10^{18}\,cm^{-2}$. However, Wakker (1997, manuscript in preparation) finds that binning N(HI) itself, giving $F$[N(HI)]—instead of binning the logarithm of N(HI), giving $F$[log(N(HI))]—removes the turnover. The cause of this effect is discussed by Heacox (1996). The distribution $F$[N(HI)] based on the Hulsbosch & Wakker (1988) survey, which covers the sky much more completely, continues to rise toward low N(HI). The new data of Murphy et al (1995) connect smoothly with the Dwingeloo survey and seem to indicate that HVCs fainter than $7 \times 10^{17}\,cm^{-2}$ may be widespread. This observation might pose a problem for models predicting that in the Galactic Halo HI with column densities below a few times $10^{18}\,cm^{-2}$ will be ionized by the intergalactic radiation field (e.g. Maloney 1993).

## 2.6 *Small-Scale Structure*

Medium-resolution (10–20′) maps have been made for many HVCs. In complex A these maps show sharp edges and an irregular velocity field, which were attributed to fragmentation after shock compression (Giovanelli & Haynes 1977). In other HVCs sharp edges and velocity jumps occur sometimes. An important result of these medium-resolution observations was the recognition of core-envelope structure, especially pronounced in complex A but seen also in C and positive-velocity clouds.

High-resolution interferometer maps of HVCs have been obtained since the mid-1970s (Greisen & Cram 1976, Schwarz et al 1976, Schwarz & Oort 1981, Wakker 1991b, Wakker & Schwarz 1991, Schwarz et al 1995, Wakker et al 1996a). A resolution of 1′ and 1 km/s has been achieved for the AIV, M I, and H cores, but for other fields lower signal-to-noise ratios limit resolution to 2′ or 3′. Some fields were selected because they are bright at medium resolution, others because they contain a probe star (see Section 4).

A problem with interferometer maps is that large-scale structure is filtered out, so that the final map does not contain the full emission. What is filtered out are features that have smooth angular structure, separately in each velocity channel. The strength of this background can be found by comparing the total flux recovered in the interferometer field to a single-dish observation taken with the same beam size. Schwarz et al (1995) and Wakker et al (1996a) discuss this procedure in detail.

The main results of the high-resolution studies are as follows: (*a*) Structure exists down to the resolution limit, and smaller structure probably exists. (*b*) In cores about 25–50% of the single-dish flux is recovered by the interferometer, while away from cores only about 10% is recovered. (*c*) The steep edges seen at lower resolution break up into a disorganized collection of cloudlets; thus, the suggested evidence for shocks disappears. (*d*) The intensity contrast can be of order 5 to 1 over a few arcmin, taking into account the background. (*e*) The measured line widths decrease with angular size. The median for a single field ranges from 3 to 7 km/s, but there is wide variation, from 2 to 15 km/s. (*f*) The line widths indicate a kinetic temperature of at most a few hundred Kelvin for the fine structure. (*g*) The velocity field is generally irregular, although sometimes there are clear gradients (e.g. Schwarz & Oort 1981). (*h*) Velocity gradients in cores go up to 100 km/s/deg; this contributes 15 km/s to the FWHM when observing the core with a 10′ beam. (*i*) The volume density of the smallest cores is 30–100 $D_{kpc}^{-1}$ cm$^{-3}$.

A word of caution is needed about the volume density estimates. For instance, Verschuur (1993b) studied a set of sizes and densities quoted in the literature and compared these with the beam of each observation. He found that the quoted sizes are usually about three times the beam width and that the quoted volume densities are inversely proportional to the beam size. There are several possible explanations for this effect. If HI clouds are very filamentary and turbulent, then overlapping filaments and velocity crowding will give rise to column density peaks that are misinterpreted as volume density peaks. Alternatively, if the cloud structure can be described by fractals, then the estimated volume density will always depend on the size of the volume over which it is averaged. A fractal dimension of 1.4 is compatible with the structure seen in several HVCs observed at high resolution (Vogelaar & Wakker 1994). Another hint that the cloud structure may be self-similar (fractal) is given by comparing the cumulative distribution of column densities for individual detections (slope −0.5; Wakker 1997, manuscript in preparation) with the cumulative distribution of the fluxes of clouds (slope −0.54; Wakker & van Woerden 1991). In this way, individual detections within a large complex behave as if they were independent clouds. A careful study that accounts for selection effects such as the beam size will be required to interpret the measurements correctly.

## 2.7 *Velocity Structure*

Many HVCs have a two-component velocity profile, composed of a broad (20–25 km/s FWHM) and a narrow (∼8 km/s) component. This property was first emphasized by Cram & Giovanelli (1976), who studied about 100 random directions. The two-component structure seems to occur in HVC cores, but not in the fainter regions. While the data are inhomogeneous and sparse, the following summary can be made. For complexes A, C, and M the average linewidth in the envelopes is about 25 km/s at 36′ resolution and is slightly smaller at 21′. In the cores of A and M the average widths are 5 (at 1′) to 10 km/s (at 10′); the cores of C have not been observed at high resolution. At all resolutions, complex H shows narrower profiles than A, C, and M, down to 3 km/s at 1′. For the Anti-Center Clouds, the line widths are large (>25 km/s), even at 3′ resolution (Mirabel 1982). The Magellanic Stream has only single-component spectra, with typical widths of 30–40 km/s. Positive-velocity HVCs, the GCN clouds, and VHVCs sometimes show two-component (7 + 25 km/s) structure, sometimes just a single component of width 20–40 km/s. This result also depends on the angular resolution of the data. For example, in HVC 267+26+217, Giovanelli & Haynes (1976) (20′ beam) found a 7 + 18 km/s profile, whereas Cavarischia & Morras (1989) (34′ beam) saw only a 24 km/s wide component. In one VHVC (HVC 114−10−440), medium-resolution data (12′) showed two-component structure (Cohen & Mirabel 1979), but high-resolution (1′) data revealed that the broader component is the result of beam smearing of the velocity gradient in the cloud (Wakker & Schwarz 1991). In contrast, another VHVC (HVC 110−7−465) showed only broad profiles at high resolution.

Several problems exist with the study of profile structure. (*a*) The profile sometimes cannot be represented by gaussians. (*b*) Often only an average linewidth is quoted, although a histogram of line widths indicates a wide range of values. (*c*) The quality of fits is variable, and the results depend on whether the authors chose to fit one or two components to their profiles. (*d*) Hardly any information exists on the distribution of line widths across clouds. (*e*) The measured line widths depend on the angular resolution of the data.

The width of a profile arises from a combination of many factors, in unknown proportions. These factors include velocity gradients resulting from projection effects or internal bulk motions, velocity crowding from filaments projected onto the same position, and broadening resulting from internal turbulence or the kinetic temperature of the gas. Verschuur & Magnani (1994) showed that the effect of the beam size on the observed linewidth can be severe. Furthermore, turbulence can produce profiles that are not gaussian but have extended wings (Falgarone et al 1991), simulating a two-component profile.

Nevertheless, the core-envelope structure seen in the angular and velocity distributions is usually interpreted as evidence for a two-phase medium, confined by a hot ($10^6$ K) halo. The cloud would then consist of a warm ($T \sim 10^4$ K) and a cool ($T \sim 10^2$ K) component. A modern version of such a model was worked out by Wolfire et al (1995), who found that the two-phase neutral hydrogen can be in pressure equilibrium with a hot galactic corona for $z > 1$–$5$ kpc, depending on the metallicity and dust content of the cloud.

The envelopes appear not to be in pressure equilibrium with the cores, however. For example, for the envelope of core AIV the radius is 40′, the peak column density $4 \times 10^{19}$ cm$^{-2}$ and the typical linewidth 20 km/s (Davies et al 1976), corresponding to an *average* volume density of $0.6/D_{kpc}$ cm$^{-3}$ and a pressure of $5300/D_{kpc}$ K cm$^{-3}$. For the core, however, the volume density is $40/D_{kpc}$, and the pressure is typically $60000/D_{kpc}$ (Wakker & Schwarz 1991). In fact, the pressures observed in the HVC cores are always much higher than those in the envelopes, suggesting that the cores will expand almost freely and must have been recently formed.

As an alternative to pressure confinement by a hot halo, Dyson & Hartquist (1983) showed that if HVCs are in the outer halo, collisions between randomly moving cloudlets may form high-mass OB stars, whose stellar winds and supernova explosions would maintain the observed velocity dispersion. With 100 HVCs, and 100 stars formed per collision, there would currently be fewer than 1000 such high-z young OB stars, which is compatible with observations. This idea suggests that high-density star-forming cores may exist, in which molecular hydrogen and CO would form. However, no CO emission associated with HVC cores has been detected (Section 3.6). Christodoulou et al (1997) point out that Dyson & Hartquist (1983) used medium-resolution data for their estimate. Using the high-resolution HVC data instead, Christodoulou et al (1997) find that the conditions in the Galactic Halo are not favorable for the formation of high-mass stars.

In summary, not enough is known about the distribution of line widths as a function of position and resolution, or about their origin, to allow a reliable interpretation. A systematic study is sorely needed.

## 2.8 *Time Scales*

Evolution time scales can be calculated from the measured sizes and velocities. Lower limits are given in Table 2, using distance limits discussed in Section 4. The symbols used below are defined in the notes to the table.

**Table 2** Time scales[a]

| HVC | d> (kpc) | $\alpha$ (°) | $\Theta$ (°) | $\theta$ (′) | $v_{dev}$ (km/s) | $\Delta v$ (km/s) | $\Gamma$ (km/s) | $\gamma$ (km/s) | $t_{z=0}$ (Myr) | $t_{sh}$ (Myr) | $t_{cr}$ (Myr) | $t_{e,c}$ (Myr) | $t_{e,s}$ (Myr) |
|---|---|---|---|---|---|---|---|---|---|---|---|---|---|
| A IV | 4.0 | 5 | 1 | 3 | −150 | 20 | 25 | 6.5 | 30 | 20 | 0.5 | 3.0 | 0.5 |
| C IIIB | 2.4 | 15 | 1.5 | — | −80 | 16 | 18 | — | 30 | 40 | 0.8 | 3.5 | — |
| M I | 3.0 | 5 | 0.5 | 3.5 | −110 | 10 | 24 | 4.4 | 30 | 25 | 0.2 | 1.0 | 0.7 |
| H | 5.0 | 20 | 2 | 4 | −90 | — | 22 | 3.2 | — | — | 1.9 | 8.0 | 1.8 |
| AC I | 3* | 10 | 2 | — | −180 | 50 | 25 | — | 15 | 10 | 0.6 | 4.0 | — |
| AC II | 3* | 7 | 1.5 | — | −115 | 20 | 30 | — | 25 | 20 | 0.7 | 2.5 | — |
| HVC 168−46−280 | 3* | 10 | 2.5 | — | −260 | 50 | 30 | — | 10 | 10 | 0.5 | 4.5 | — |
| Cohen Stream @−110 | 3* | 3 | 1 | — | −75 | 20 | 30 | — | 40 | 8 | 0.7 | 1.5 | — |
| 24−20−235 (GCN) | 25* | 20 | 0.5 | — | −180 | 40 | 24 | — | 140 | 220 | 1.2 | 9.0 | — |
| 128−34−392 (#466) | 25* | — | 1.5 | — | −300 | — | 21 | | 80 | — | 2.2 | 31 | — |
| 114−10−440 (#330) | 25* | — | 0.4 | 5 | −310 | — | 23 | 3.9 | 80 | — | 0.6 | 7.5 | 9.3 |
| 244+30+95 (WB) | 5* | 15 | 1.5 | — | +35 | — | 20 | — | 140 | — | 3.7 | 6.5 | — |
| 289+20+250 (EP) | 25* | 4 | 1 | — | +210 | 25 | 23 | — | 120 | 70 | 2.0 | 19 | — |

[a]Distances d are measured or assumed (*) lower limits. The limit for AC I (and by association for the other AC clouds) is based on the possible interaction between AC I and the intermediate-velocity shell at −70 km/s, which has d > 2.5 kpc; the limits for the VHVCs are based on a possible association with the Magellanic Stream. $\alpha$: width of the complex to which the cloud belongs (0.05 K contour); $\Theta$: half-brightness radius of a core; $\theta$: half-brightness radius of a subconcentration; $v_{dev}$: deviation velocity; $\Delta v$: velocity dispersion between different concentrations in a complex; $\Gamma$: FWHM of the broad component at 10′ or 20′ resolution; $\gamma$: FWHM at 1′ resolution. All times are lower limits, scaling linearly with the distance.

The time after which a cloud reaches the galactic plane is found from its z-height and z-velocity. The former is given by $d \sin b$. Although the space velocity is unknown, an estimate of the z-velocity is found by projecting the deviation velocity on the vertical. Thus, this timescale is estimated as $t_{z=0} \sim (d \sin b)/(v_{dev} \sin b)$, and it is several tens of Myr, assuming there is no substantial tangential velocity. Values for positive-velocity clouds represent the time since crossing or leaving the galactic plane.

The time for the cores to shift substantially relative to each other is the ratio of the linear width to the velocity dispersion between cores, as estimated from their relative velocities: $t_{sh} \sim (\alpha d)/\Delta v$. These time scales are at least tens of Myr. Accounting for projection effects would increase the estimates.

The time a core takes to move across its own width is the ratio of the core radius to the deviation velocity: $t_{cr} = (R/v_{dev}) \sim (\Theta d)/v_{dev}$, typically about 1 Myr. Assuming that there are no restraining forces (which is unlikely), a core will double its size in several Myr (the ratio of size and linewidth: $t_{e,c} = (R/\Gamma) \sim (\Theta d)/\Gamma)$). Finally, the time scale for a single subconcentration to expand and dissolve can be found from the sizes and widths measured at high resolution: $t_{e,s} = (r/\gamma) \sim (\theta d)/\gamma$. These time scales are generally shorter than those for the whole core.

Even with the large uncertainties in the distances and the rough estimates of sizes, the time scales for the processes playing a role in determining the fine structure clearly are much shorter than the lifetime of a whole complex. This conclusion was already reached by Oort (1978). Thus, the fine structure will change considerably during the movement of a complex through space, and our present view is only a snapshot of a dynamic process. The medium-scale substructure will change considerably but remain recognizable. The structural integrity of the complexes is mostly preserved while they traverse large distances.

## 2.9 *Interactions Between HVCs and Other Gas*

THEORETICAL CONSIDERATIONS If HVCs are in the lower Galactic Halo, one may suppose that some are at low enough z-heights to show signs of interaction with Disk gas. If drag forces dominate, the velocity becomes proportional to the square root of the column density. Benjamin & Danly (1997) found that low-column density IVCs may reach terminal velocity, but HVCs may be slowed down only slightly. If all clouds would move at their terminal velocity, higher velocities would occur at larger heights, although the distribution also depends on the initial positions and velocities. It may be possible to explain IVC velocities in this way. The data of Hulsbosch & Wakker (1988) have not been analyzed in detail for possible evidence of drag effects, but no obvious cases have been noticed.

Silk & Siluk (1973) claimed $v \propto \mathrm{N(HI)}^n$ with $n = -1$ to $-0.2$ for components with $v_{LSR}<-39$ km/s in the region around $l = 120°$. The exponent is opposite to that expected from drag. However, this correlation, if at all present, seems to exist only for components with $v_{LSR} > -100$ km/s, and the selection mixes complex H with the Disk gas, which is probably unrelated.

The hydrodynamics of collisions of clumps of high-velocity gas with stationary disk gas have been calculated (e.g. Tenorio-Tagle et al 1987, 1988, Comeron & Torra 1992, 1994). Such collisions have been invoked to explain the one-sided and asymmetrical shells seen in low-velocity HI channel maps, as well as star formation in the Orion region (Franco et al 1988, Lepine & Duvert 1994) and Gould's Belt (Comeron & Torra 1994). The calculations show that a small (30–200 pc), dense (0.3–3 $cm^{-3}$), fast (50–300 km/s) cloud will sweep up large amounts of gas and be shocked. The final result of the collision is determined by the position of the cloud at the time that it has been completely shocked. If by that time it has crossed the disk, then it continues on and drills a hole. Otherwise, the shock grows sideways and the remnant grows to a diameter of several hundred parsecs, simulating a superbubble. This result depends non linearly on the initial velocity of the cloud and the density ratio. After several tens of Myr, the disk will restore itself. During the interaction, cold, dense regions may form, in which star formation could be initiated. MacLow et al (1989) point out that such collisions heat the disk gas only to $10^5$ K. Thus, unlike superbubbles caused by (a succession of) supernova explosions, collisions cannot be the source of the X-ray background, which requires gas at $10^6$ K.

OBSERVATIONAL EVIDENCE Signs of interaction between HVCs and IVCs and/or low-velocity gas were reviewed by Mebold et al (1992). The evidence remains inconclusive. Distances to each of the LVCs, IVCs and HVCs in question must be determined to settle the matter.

The most suggestive case is that of HVC ACI, for which Mirabel & Morras (1990) showed that it lies inside an incomplete shell at intermediate velocities (−70 km/s), whose distance is greater than 2.5 kpc (Kulkarni et al 1985, Kulkarni & Mathieu 1986).

In the region around $l = 165°$, $b = -40°$, three HI clouds overlap: a VHVC at $v_{LSR} \sim -280$ km/s, an elongated HVC at $v_{LSR} \sim -110$ km/s (named the Cohen Stream), and a parallel stream at $v_{LSR} \sim -10$ km/s (Cohen & Davies 1975, Cohen 1981). This fact has been used to argue that an infalling VHVC accelerated Disk gas, creating the HVC (Tamanaha 1995). The energetics of such an interaction are consistent with this idea; however, the hypothesis fails to explain why the −110 km/s cloud extends 15° past the supposed interaction point, and why elsewhere there is much more gas with $v_{LSR} \sim -100$ km/s that is not associated with a VHVC.

Meyerdierks (1991, 1992) proposed that cloud A0 is the remainder of a neutral hydrogen bullet that hit the Galactic Disk with a velocity of 325 km/s, creating the IVC in the region. This model was inspired by a possible anticorrelation between the distortions of the IVC and HVC contours. However, because the IVC is at 250 pc distance, the lower limit of 4 kpc for the higher-latitude part of complex A implies that A would point almost straight at the Sun, and be about 16 pc wide at its tip, but >1 kpc wide at the high-latitude end. This geometry is very unlikely, so we doubt that an interaction has taken place. Only distance determinations for different parts of A can resolve the issue.

The most-argued case is that of the Draco Nebula (Goerigk et al 1983, Hirth et al 1985, Rohlfs et al 1989, Herbstmeier et al 1993, 1996), an IVC near $l = 91^\circ, b = 38^\circ$ at $v_{LSR} = -23$ km/s, which overlaps with HVC core CIB. The distance of the Draco Nebula is unknown but could be greater than 0.8 kpc (Goerigk & Mebold 1986). An X-ray shadow has been seen (Burrows & Mendenhall 1991), but no such shadow can be associated with CIB (see Section 3.5). An interaction between the HVC and IVC has been proposed, and although the structural correspondence is suggestive, the evidence remains tentative until the distances of both are measured.

A general claim of interactions between complex C and IVCs has been made by Pietz et al (1996), who found velocity bridges between IVCs and HVCs, including one near the Draco Nebula. They state that these bridges are not related to the IV Arch, although they all occur near its edges, where some faint spurs can be seen in channel maps (Kuntz & Danly 1996). Additional studies are needed to determine the nature of these bridges.

## 3. HVC OBSERVATIONS OTHER THAN IN 21-CM EMISSION

Until recently HVCs had been detected only in 21-cm emission, but much progress has been made in studying the HVCs by other means. Studies of absorption lines resulting from heavy elements merit a separate section (Section 4). The others fall into six categories, discussed here.

### 3.1 *Absorption at 21 cm*

Absorption in the 21-cm line takes place if a sufficiently strong 21-cm continuum source lies projected behind a bright enough area of a HVC. This situation yields a measurement of the hydrogen spin temperature, $T_s$, which is related to the kinetic gas temperature (see e.g. Dickey et al 1979, Liszt 1983). An interferometer is the best instrument for this kind of study, as it minimizes the confusion between absorption and emission in the direction of the source.

Unfortunately, few useful coincidences of background continuum sources and HVCs are known. Real progress will require a systematic search, at reasonably high angular resolution, for bright continuum sources, followed by observing these sources with an interferometer.

Several such studies exist for HVCs. Colgan et al (1990) found $T_s > 20$–70 K for some of the Anti-Center Clouds, $T_s > 20$ K for HVC43−13−309 (cloud 348), and $T_s > 10$ K for the Magellanic Stream. Mebold et al (1991) found $T_s > 50$ K for another direction in the Magellanic Stream. For many other sources the signal-to-noise ratio was insufficient to derive useful limits.

Just two definite results have been reported. Payne et al (1978, 1980) found absorption in 4C 33.48, projected on cloud R, which is centered on $l = 70°$, $b = +12°$ (Kepner 1970) and may be a spiral arm fragment at a galactocentric radius of 25 kpc. Two components occur at −128 km/s, a narrow one with $T_s = 70$ K and a wide one with $T_s = 300$ K. Wakker et al (1991) found absorption in one (possibly two) continuum sources in the Westerbork map of a core of complex H. The derived spin temperature was ~50 K.

In summary, all the derived lower limits are consistent with the two positive results, which yield temperatures on the order of 50–100 K, similar to that of low-velocity cool HI. Since in the core of complex H the line widths of the emission range from about 2.5 to 4 km/s (Wakker & Schwarz 1991), the kinetic temperatures of the gas must be in the range 50–350 K.

## 3.2 *Emission in the Hα line*

Hα emission can be generated by photoionization caused by far-UV radiation or by shock ionization, but one often cannot determine which process dominates (Reynolds 1987, 1992). Hα intensities are usually expressed in Rayleigh (R), corresponding to $10^6/4\pi$ photons $cm^{-2}\,s^{-1}\,sr^{-1}$ or $2.41 \times 10^{-7}$ erg $cm^{-2}\,s^{-1}\,sr^{-1}$.

Only a few attempts have been made to observe Hα emission from HVCs. Several clouds were detected: HVC 168−43−280 (0.08 R; Kutyrev 1986, Kutyrev & Reynolds 1989), cloud M I (0.06 R toward the direction with the brightest HI, 0.26 R just outside the HI contours; Tufte et al 1996), cloud M II (0.1–0.2 R; Münch & Pitz 1990), and the Magellanic Stream (MSII, MSIII, MSIV, and 0.37, 0.21, and 0.20 R, respectively; Weiner & Williams 1996). In the Stream the brightest Hα spot is offset from the brightest HI; between MSII and MSIII the emission is <0.04 R. In complex C two independent measurements for the same position are available. Songaila et al (1989) found 0.03 R; Reynolds et al (1996, manuscript in preparation) instead find 0.09 R.

The detections can be converted to electron density limits if it is assumed that the HI is shock ionized. For HVC 168−43−280, complex C, and the faint directions around M I, I(Hα) ~ 0.08 R, and $n_e \sim 0.04$ $cm^{-3}$. For the bright M I direction and the MS cores, I(H$\alpha$) $\sim$ 0.25 R, or $n_e \sim 0.07$ cm$^{-3}$. In contrast, if the HI were photo-ionized, then $F_{EUV} \sim 1.4 \times 10^5$ photons cm$^{-2}$ s$^{-1}$ sr$^{-1}$ for the directions with faint emission and $F_{EUV} \sim 4.4 \times 10^5$ photons cm$^{-2}$ s$^{-1}$ sr$^{-1}$ for the bright directions. These values are compatible with models of the photo-ionizing radiation in the Galactic Halo (Bregman & Harrington 1986), which predict $F_{EUV} \sim 1 \times 10^5$ photons cm$^{-2}$ s$^{-1}$ sr$^{-1}$. A decision between the ionization mechanisms requires more complete mapping as well as a detailed application of the models.

Mapping the H$\alpha$ emission of HVCs will give a much better grip on their thermal and ionization structure, which in turn will lead to a better understanding of their environment and of their origin. Much progress is expected from the Wisconsin H$\alpha$ mapper (Tufte et al 1996), which is 10 times more sensitive than previous instruments. This work may become one of the more interesting fields of HVC studies in the near future.

## 3.3 *Dust in HVCs*

The IRAS satellite provided the opportunity to search for correlations of HVCs with 100 $\mu$m thermal emission from dust. A major problem is the presence of foreground emission associated with low- (and sometimes intermediate-) velocity HI, which is brighter and nearer. Wakker & Boulanger (1986) searched for 100 $\mu$m emission associated with the brightest HVC cores (M I, AIII, and AIV). The detection limit was 0.1 MJy/sr, but no associated emission was found. A conservative prediction was 0.3 MJy/sr, assuming that I(100 $\mu$m)/N(HI) = 0.6 MJy/sr/10$^{20}$ cm$^{-2}$, as found for low-velocity HI near the HVCs; on average this ratio is higher (Boulanger et al 1996). Since the HVCs were not detected, either there is at least three times less dust than in low-velocity HI, or the dust is too cold. Using a model for the dust extinction and the radiation field, the latter was shown to imply a lower limit of 10 kpc to the distance.

Two other searches also gave a negative result: Fong et al (1987) for MSI and MSII, and Bates et al (1988) for the field around the star 4 Lac. Any future work will remain difficult, as the 100 $\mu$m emission associated with the low- and intermediate-velocity gas must be removed and the ratio I(100 $\mu$m)/N(HI) is not constant.

If the dust in HVCs is very cold, its emission may be relatively enhanced at sub-millimeter wavelengths (Trifalenkov 1993). Further, the abundance ratio of sulphur to iron in HVCs relative to the solar ratio might provide information about the amount and composition of dust.

## 3.4 *Magnetic Fields in HVCs*

Kazès et al (1991) used the Nançay telescope to search for the Zeeman effect in four HVC cores, in order to measure the magnetic field strength. Although they integrated between 36 and 70 hours per field, only one believable detection was found: B $= -11.4 \pm 2.4\ \mu$G for core A 0. Upper limits could be set for three other cores. This field strength is similar to that found for low-velocity galactic HI. Usually, for observations of the Zeeman effect sidelobe effects, beam squint (the possibility that the left and right polarized antennas look in slightly different directions) and velocity gradients are important. However, HVCs are isolated in position and velocity, so these effects can be controlled relatively easily. Small-scale structure is an additional problem.

Despite the expense and difficulty involved, these observations should be pursued further, using more sensitive telescopes. If distances to HVCs can be determined, this may be the only means to directly probe the magnetic field high above the galactic plane.

## 3.5 *X-Ray Shadows and Enhancements*

Observations of X rays can potentially directly trace interactions between HVCs and disk gas (e.g. Bone et al 1983). A diffuse soft (1/4 keV) X-ray emission background is known (McCammon & Sanders 1990). Angular structure in the diffuse X rays is a mixture of (*a*) intrinsic variations in the temperature and density of the continuously distributed emitting medium, (*b*) absorption by foreground HI, and (*c*) possible enhancements resulting from an interaction between HI clouds and their surroundings. In a few cases the emission is clearly shadowed by foreground low- and intermediate-velocity HI (Snowden et al 1991, Burrows & Mendenhall 1991, Lilienthal et al 1992, Benjamin et al 1996). Since HVCs are more distant than low-velocity gas (Section 4), foreground HI will always be present, usually with higher column densities. Thus, the (inhomogeneous) background and any emission generated by interactions will be attenuated (often strongly) by foreground HI, and as a result it will be very hard to show convincingly that HVCs generate X rays or cast shadows. Information on the X-ray spectrum will be necessary to separate out absorption from emission, and definitive proof of an association between X rays and HVCs requires direct determinations of the distance to the various HI structures in the line of sight. Nevertheless, two studies of possible associations have been done, both using ROSAT data.

Herbstmeier et al (1995) analyzed the X-ray distribution near complex M in terms of a simple model of foreground and background X-ray emission by fitting the relation between HI column density and X-ray emission. For fields containing a HVC two fits were made, including as well as excluding the HVC column density. The claim is made that including the HVC reduces the scatter for two of the fields, while at the same time the fitted foreground emission is significantly higher in the HVC fields. However, the improvement of the fits is small, and absent in half the cases. Enhanced X-ray emission near M I and M II was explained as the result of interaction of the HVC with galactic gas, but enhanced foreground emission in fields dominated by low-velocity gas was explained as the result of strong variations in the background.

Kerp et al (1994, 1996) plotted part of complex C on a map of X-ray emission and concluded that a relation exists. An analysis similar to that of Herbstmeier et al (1995) showed that the difference [data minus model] has excess emission where the X rays are bright. These enhancements were interpreted as X rays generated by HVC heating, but deviations from homogeneity of the X-ray–emitting region were not considered as an explanation. Anti-coincidences with the same HVC (i.e. probably at the same distance) are interpreted as shadowing, which seems contradictory. Moreover, the X-ray emission is not compared to the IV Arch, which is present in the same area of sky (Kuntz & Danly 1996).

In our judgment neither study convincingly demonstrates a relation between X rays and HVCs, and a more detailed analysis is necessary, using spectral information for the X rays and still more careful treatment of the low- and intermediate-velocity gas.

### 3.6 *CO Observations of HVCs*

The observed volume density of neutral hydrogen inside HVC cores ranges up to 80 $D_{kpc}^{-1}$ cm$^{-3}$, so molecular hydrogen and CO molecules might form and the $^{12}$CO J = 1–0 115 GHz line might be detectable. Unpublished negative results were mentioned by Hulsbosch (1978) and Giovanelli (1986). Kim et al (1989) and Wakker et al (1997) failed to find CO emission in cores A0, AI, AIV, M I, CIB, H, and HVC 114−10−440, down to a 5-$\sigma$ limit of 0.077 K km/s. Using the ratio between $H_2$ column density and integrated CO intensity for high-latitude translucent clouds, X(CO) $\sim 1.2 \times 10^{20}$ cm$^{-2}$ (K km/s)$^{-1}$ (Magnani & Onello 1995), this corresponds to N($H_2$) $< 9 \times 10^{18}$ cm$^{-2}$, in directions where N(HI) $\sim 3 \times 10^{20}$ cm$^{-2}$. Wakker et al (1997) showed that if the observed HVC cores are nearby ($d < 1$ kpc), one expects that $H_2$ (and CO) would have formed in detectable amounts. If the HVCs are more distant ($d > 3$ kpc), no $H_2$ is expected, consistent with the observed results.

## 4. DISTANCES AND METALLICITIES

The distance is the key unknown for HVCs. Most physical parameters scale with it, for example, size as $d^{+1}$, mass as $d^{+2}$, density and pressure as $d^{-1}$. Only measured temperatures are distance independent. By definition, for HVCs LSR velocities cannot be combined with a model of galactic rotation to derive kinematic distances. The best alternative is the absorption-line method, described in Section 4.1. Nevertheless, several authors have proposed indirect methods to determine distances. These methods are discussed in the next section.

## 4.1 *Indirect Distance Determinations*

All indirect methods depend on one or more key assumptions. Watanabe (1981) assumes that HVC cores are tidally limited in size. Jaaniste (1984) and Espresate et al (1990) assume HVCs are in pressure equilibrium with a hot corona; they calculate the pressure from the observed radius, column density, and temperature, and combine this pressure with a model of the hot halo gas to derive the distance. Haud (1990) assumes that the size-linewidth relation of molecular clouds can be applied to HVCs. Kazès et al (1991) suggest energy equilibrium between motions and magnetic fields. Ferrara & Field (1994) predict a relation between the external halo pressure and amount of ionization, assuming photo-ionization dominates. Benjamin & Danly (1997) derive terminal velocities for drag-dominated vertical motion, finding a relation between z-height and velocity for low-column-density clouds.

These methods give very different answers, ranging from a few hundred pc to a few hundred kpc, and are subject to several major problems: (*a*) the available linewidth data are very sparse and inhomogeneous (Section 2.7); (*b*) quoted line widths, sizes, and column densities depend on the angular resolution (Section 2.6); (*c*) quoted sizes are an undefined average for non circular objects; (*d*) the galactic potential at high z is not well known; (*e*) magnetic fields are difficult to measure, and it is unclear what resolution to use to estimate the internal kinetic energy; and (*f*) if the cloud structure is fractal (self-similar), a method based on line widths or size estimates will always fail.

A statistical approach may also be taken. Kaelble et al (1985) find that the sky and velocity distributions of HVCs are consistent with a population of clouds at $z > 1$ kpc having some radial and vertical infall. This result does not give individual distances, but it sets the scale of the HVC phenomenon. Bajaja et al (1987) reproduce the envelope of the $l$-$v_{LSR}$ distribution by assuming a spherical surface of particles falling in with 75 km/s and moving with 385 km/s toward $l = 110°$, $b = +10°$, similar to the solar motion with respect to the Local Group. They suggest that this result indicates a relation between VHVCs (which are on the envelope) and the Local Group.

## 4.2 *The Absorption-Line Method*

A complete discussion of the absorption-line method and its pitfalls was presented by Schwarz et al (1995) (see Figure 4). In summary, an upper limit to the distance of a HVC can be set by detecting absorption at the velocity of the HVC in the spectrum of a star with known distance. A lower limit is set from stars not showing absorption by the HVC, with a detection limit below the expected line strength, found from an abundance determination based on an extra-galactic object. This method requires high-resolution (better than 15 km/s) spectra of selected ions, as well as high-angular resolution HI data (to

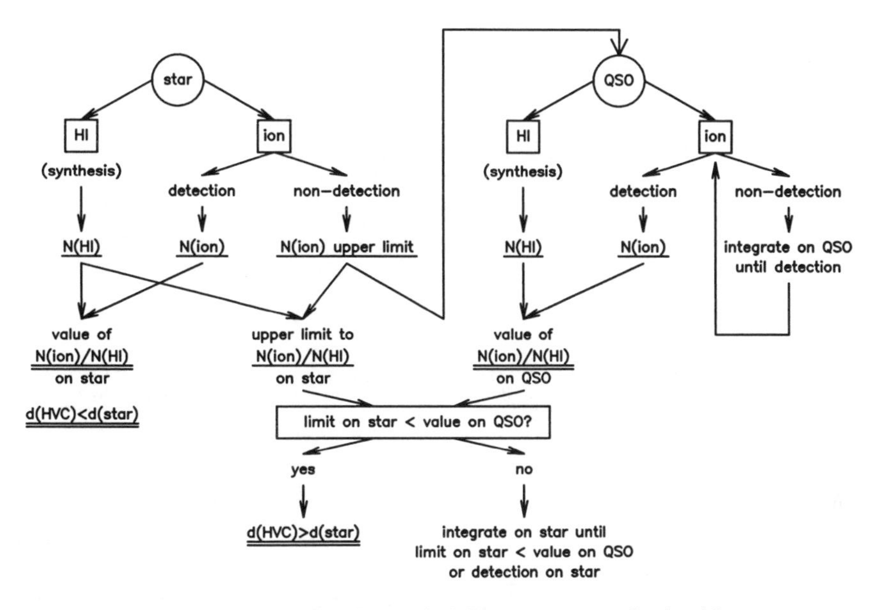


*Figure 4* Procedure of the absorption-line method. For comments see Section 4.2.

account for fine structure in the HI) and also stellar photometry and classification spectra (to determine distances). Further, to derive lower distance limits, one crucial assumption must be made, namely that the abundance does not vary substantially within a cloud or cloud complex. Evidence exists for variations between cloud complexes (Section 4.2), but so far not for variations within one cloud complex.

Until recently, little progress had been made concerning distances and metallicities, but the situation has improved. Table 3 lists all reliable detections of galactic absorption associated with gas at velocities $v_{LSR} > 100$ km/s. Table 4 summarizes the distance limits obtained. It includes only (*a*) stars for which the upper limit on the ion column density is below the abundance derived for the same ion in the same HVC from absorption in an extra-galactic source, and (*b*) stars for which the Si II or Mg II lines give a non detection, as these lines are extremely sensitive. Some less distant stars giving reliable lower limits have been omitted. The columns in Tables 3 and 4 are identical and contain the following data:

1: Probe name.

2,3: Galactic longitude and latitude of probe.

4: Distance in kpc (for stars in Table 4 only).

**Table 3** Absorption lines with $|v_{LSR}| > 100$ km/s

| Probe (1) | l (2) | b (3) | type (5) | HVC (6) | $v_{HI}$ (7) | $N_{HI}$ (8) | tel (9) | ion (10) | $v_{ion}$ (11) | $N_{ion}$ (12) | | $A_{ion}$ (13) | $A/A_{\odot}$ (14) | Ref[a] (15) | Note[b] (16) |
|---|---|---|---|---|---|---|---|---|---|---|---|---|---|---|---|
| **Detections from catalogued HVCs** | | | | | | | | | | | | | | | |
| Mark 106 | 161.14 | 42.88 | Sey | A | –157 | 39±3 | WE | Mg II | –157 | >500 | N | >130 | >0.034 | 24 | |
| | | | | | | | | Ca II | –157 | 6.7±2.3 | N | 1.7±0.6 | 0.0080 | 22 | |
| Mark 290 | 91.49 | 47.95 | Sey | C | –138 | 68±3 | Ef | Ca II | –137 | 20±1 | N | 2.9±0.2 | 0.013 | 23 | 1 |
| PG 1351+640 | 111.89 | 52.02 | QSO | C | –157 | 60±2 | Ef | Ca II | –163 | 9.9±1.0 | N | 1.7±0.2 | 0.0075 | 23 | |
| Mark 205 | 125.45 | 41.67 | Sey | C | –135 | 1.4±0.7 | JB | Mg II | –147 | 55±18 | W | 390±230 | 0.10 | 11 | 2 |
| | | | | 84 | –208 | 3.0±0.7 | | Mg II | –209 | 71±23 | W | 240±100 | 0.062 | 17 | |
| 4C 33.48 | 66.39 | 8.37 | EG? | R | –126 | 140 | Ar | H I | –128 | 0.011 | t | — | | 1 | 3 |
| A 0159+625 | 130.98 | 1.15 | EG? | H | –199 | 110 | Ws | H I | –199 | 0.34 | t | — | | 8 | 4 |
| PKS 0837-12 | 237.17 | 17.43 | QSO | WB | +105 | 14±2 | Pk | Ca II | +106 | 22±4 | N | 16±4 | 0.072 | 7 | 5 |
| NGC 3783 | 287.46 | 22.95 | Sey | EP | +240 | 120±2 | VE | CI | +236 | 240±170 | W | 20±14 | 0.0006 | 14 | |
| | | | | (187) | | | | Si II | +231 | >460 | W | >38 | >0.011 | 14 | |
| | | | | | | | | SII | +236 | 5100±2700 | W | 430±220 | 0.23 | 14 | 6 |
| | | | | | | | | Ca II | +240 | 6.3±2.1 | W | 0.52±0.17 | 0.0024 | 4 | |
| Fairall 9 | 295.07 | –57.83 | QSO | MS | +199 | 69 | Pk | Si II | +195 | >1500 | N | >220 | >0.061 | 15 | 7 |
| | | | | | | | | Ca II | +195 | 33±4 | W | 4.7±0.6 | 0.022 | 2 | |
| **Detections with HVC nearby or uncatalogued** | | | | | | | | | | | | | | | |
| Mark 509 | 35.97 | –29.86 | Sey | (419) | <0.60 | | GB | CIV | –283 | >2100 | W | >34000 | | 20 | 8 |
| | | | | | | | | CIV | –227 | 380±140 | W | >6400 | | 20 | |
| H 1821+643 | 94.01 | 27.41 | QSO | (C?) | <1.8 | | Dw | CIV | –213 | 120±30 | N | >670 | | 18 | 9 |
| HD 212593 | 99.90 | –6.71 | B9Ia | | +110 | 3.0±1 | JB | O I | +110 | >4400 | W | >15000 | >0.20 | 6 | 10 |
| (4Lac) | | | | | | 3.0±1 | | Mg II | +100 | >340 | W | >1100 | >0.30 | 6 | |
| SN 1994 D | 290.07 | 70.13 | SN | | +215 | 2.0 | Ar | Na I | +234 | 1.5±0.2 | L | 7.4±0.9 | 0.036 | 19 | 11 |
| | | | | | | | | Ca II | +232 | 13±1 | L | 63±6 | 0.29 | 21 | |
| SN 1986 G | 309.54 | 19.40 | SN | (219) | | | | Na I | +254 | 2.4±0.05 | W | | | 5 | 12 |
| | | | | | | | | Ca II | +254 | 6.1±0.3 | W | | | 5 | |
| HD 156369 | 328.68 | –14.52 | O9.7 II | (364) | | <2.0 | Pk | C II | +112 | 340±130 | W | >1700 | >0.047 | 9 | 13 |
| | | | | | | | | NV | +128 | 58±22 | N | >290 | | 16 | |

| Detections without known HVC counterpart | | | | | | | | | | | | |
|---|---|---|---|---|---|---|---|---|---|---|---|---|
| PKS 2155-304 | 17.73 | −52.25 | BLLac | | | CIV | −260 | 420 ± 20 L | | | 12 | 14 |
| SN 1994 I | 104.87 | 68.52 | SN | <1.8 | Dw | Na I | +246 | 3.2 ± 1.5 L | >18 | >0.086 | 19 | 15 |
| SN 1993 J | 142.09 | 40.90 | SN | | | Mg II | +231 | 40 ± 12 N | | | 13 | 16 |
| SN 1991 T | 292.59 | 65.18 | SN | <0.30 | HC | Ca II | +215 | 0.74 ± 0.12 L | >25 | >0.11 | 10 | 17 |
| | | | | | | Ca II | +263 | 1.1 ± 0.1 N | >37 | >0.17 | 10 | |
| SN 1983 N | 314.58 | 31.97 | SN | | | Ca II | +248 | 1.6 ± 1.1 W | | | 3 | 18 |

[a]References: (1) Payne et al 1978, 1980; (2) Songaila 1981a; (3) D'Odorico et al 1985; (4) West et al 1985; (5) D'Odorico et al 1989; (6) Bates et al 1990, Gilheany et al 1990; (7) Robertson et al 1991; (8) Wakker et al 1991; (9) Sembach et al 1991; (10) Meyer & Roth 1991; (11) Bowen & Blades 1993; (12) Bruhweiler et al 1993; (13) Bowen et al 1994; (14) Lu et al 1994a; (15) Lu et al 1994b; (16) Sembach et al 1995a; (17) Bowen et al 1995; (18) Savage et al 1995; (19) Ho & Filippenko 1995, 1996; (20) Sembach et al 1995b; (21) King et al 1995; (22) Schwarz et al 1995; (23) Wakker et al 1996b; (24) Wakker et al 1996a.

[b]Notes: 1. Using GHRS on HST, S II associated with complex C has also been detected in Mark 290 but the column density and abundance have not yet been determined. 2. Revised column densities from reference 17; for N(H I) we use an unpublished Jodrell Bank (12′ beam) value. 3. Probably extragalactic; $T_s \sim 70$ K; $T_s\,\tau$ of absorption corresponds to N(H I) $= 6 \times 10^{18}$ cm$^{-2}$. 4. Continuum source in Westerbork map; $T_s = 47 \pm 8$ K. 5. 3C206. 6. Reference 14 uses a linear conversion from equivalent width to find A(S II) = 0.15. 7. H I components separated in Si II, not in Ca II; N(H I) from McGee & Newton (1986). 8. Cloud 419 in pop. GCN ~ 1° away. 10. HVC 100−7+100 is not in Dwingeloo survey; diameter ≪1°; Mg I, Al II, Fe II also detected (N $= 4.3, 97, 300 \times 10^{11}$). 11. In NGC 4526; Faint H I found at Arecibo, possibly a small HVC; Na I also at +214 and +254 ($1.4$ and $0.5 \times 10^{11}$), Ca II also at +204, +216, +254 (N $= 1.7, 3.4$ and $3.5 \times 10^{11}$). 12. In NGC 5128; HVCs 219 and 208 are 1 and 3° away; Na I and Ca II also at +233 (N $= 1.4, 4.3 \times 10^{11}$). 13. Nearest HVC (#364, v = +120) is 6° away, but the region was only mapped on a 2° × 2° grid; Si II, Fe II and Mg II also detected. 14. About 10° away from nearest GCN cloud; CIV 1550 is blended. 15. In NGC 5194; paper assumes an N(Na)/N(H I) ratio, but this implies an N(H I) ≫ limit from the Dwingeloo profile; Na I also at +272 and 301 (N $= 0.8, 1.3 \times 10^{11}$). 16. In M 81. 17. In NGC 4527; N(H I) limit from unpublished Hat Creek spectrum. 18. In M 83.

**Table 4** HVC distance limits (in kiloparsecs)

| Probe (1) | l (2) | b (3) | d (4) | type (6) | $v_{HI}$ (7) | $N_{HI}$ (8) | tel (9) | ion (10) | $N_{ion}$ (12) | | $A_{ion}$ (13) | Ref[a] (15) | Note[b] (16) |
|---|---|---|---|---|---|---|---|---|---|---|---|---|---|
| Complex A [4 < d < 10; 2.5 < z < 6.5] | | | | | | | | Mg II | | | >130 | | 1 |
| | | | | | | | | Ca II | | | 1.7 ± 0.6 | | 1 |
| AD UMa | 160.34 | 43.27 | 9.8 | RRLyr | –158 | 85 ± 2 | Ef | Ca II | See Note 2 | | | | 2 |
| PG 0859+593 | 156.93 | 39.74 | 4.0 | BHB | –166 | 34 ± 1 | WE | Mg II | <23 | N | <6.7 | 8 | |
| PG 0832+675 | 147.75 | 35.01 | 1.6 | sdB | –144 | 30 ± 6 | WE | Ca II | <2.8 | N | <0.93 | 6 | |
| PG 0906+597 | 156.22 | 40.56 | 0.7 | sdO | –153 | 36 ± 1 | WE | Mg II | <12 | N | <3.3 | 8 | |
| Complex C [d > 2.4; z > 1.7] | | | | | | | | Mg II | | | 390 ± 230 | | 1 |
| | | | | | | | | Ca II | | | 2.3 ± 0.6 | | 1 |
| HD 146813 | 85.71 | 43.81 | 2.4 | B1.5 | –109 | 48 ± 6 | GB | Si II | <16 | S | <3.2 | 2 | 3 |
| HD 121800 | 113.01 | 49.75 | 2.2 | B1.5 V | –178 | 2.9 | Dw | Si II | <16 | S | <53 | 2 | 3 |
| | | | | | –127 | 10 ± 6 | GB | Ca II | <0.50 | N | <0.50 | 4 | 4 |
| PG 1705+537 | 81.19 | 36.94 | 2.2 | BHB | –131 | 12 ± 2 | DE | Mg II | <10 | L | <8.3 | 5 | |
| | | | | | | | | FeII | <16 | L | <13 | 5 | |
| Cloud M II/M III [d < 5.0; z < 4.4] | | | | | | | | Mg II | | | 390 ± 230 | | |
| BD+38 2182 | 182.16 | 62.21 | 5.0 | B3 | –93 | 3.5 ± 1.7 | GB | Si II | >200 | N | >570 | 3 | 5 |
| | | | | | | | | Ca II | 1.4 ± 0.3 | N | 4.0 ± 2.1 | 7 | |
| HD 93521 | 183.14 | 62.15 | 1.7 | O9.5 V | –93 | 4.1 ± 1.7 | GB | Si II | <7.3 | S | <18 | 3 | 5, 6 |
| | | | | | | | | Ca II | <0.90 | N | <2.2 | 7 | 7 |

| | | | | | | | | | | | | | |
|---|---|---|---|---|---|---|---|---|---|---|---|---|---|
| Complex H [d > 5.0] | | | | | | | | | | | | | |
| BD+59 367 | 131.10 | −1.28 | 5.0 | O9.5 I | −188 | 17 | Dw | Mg II | <10 | N | <5.9 | 9 | |
| BD+62 338 | 130.91 | 1.07 | 3.6 | B3 II | −199 | 150 | Dw | Mg II | <6.6 | N | <0.44 | 9 | |
| HD 10125 | 128.29 | 1.82 | 3.4 | O9.5 Ib | −203 | 38 ± 2 | Ef | Mg II | <5.2 | N | <1.4 | 9 | |
| | | | | | −147 | 35 ± 2 | Ef | Mg II | <5.2 | N | <1.5 | 9 | |
| HD 12323 | 132.91 | −5.87 | 3.4 | O9 V | −146 | 19 | Dw | Mg II | <8.5 | N | <4.5 | 9 | |
| Cloud 211 (Pop. EP) [d > 6.2; z > 1.9] | | | | | | | | | | | | | |
| HD 86248 | 264.59 | 18.11 | 6.2 | B3 II | +201 | 7.0 ± 2.0 | GB | Si II | <16 | S | <22 | 2 | 3 |

[a]References: (1) Keenan et al 1988; (2) Danly et al 1992; (3) Danly et al 1993; (4) Albert et al 1993; (5) de Boer et al 1994; (6) Schwarz et al 1995 (7) Keenan et al 1995; (8) Wakker et al 1996a; (9) Wakker et al 1997.

[b]Notes: All stellar distances have an uncertainty of typically ±1 kpc. 1. See Table 3 for abundances. 2. Preliminary value for Ca II abundance consistent with that given for the direction of Mark 106 in Table 3 (van Woerden et al 1997, manuscript in preparation). 3. Approximate limit for N(Si II), assuming S/N = 10 in the line. 4. Danly et al (1992) give N(H I) = $13 \times 10^{18}$, while Albert et al (1993) give $6.5 \times 10^{18}$, although both observed the sightline with the Green Bank 140 ft. 5. 12′ Jodrell Bank spectra set 5-$\sigma$ upper limits of $1 \times 10^{18}$ cm$^{-2}$ toward both stars (Sembach 1996, private communication), i.e. no lower distance limit can be derived. 6. Limit to N(Si II) from assuming a S/N = 20 in the continuum. 7. Limit for N($Ca^+$) is implied but not actually given, the errors for BD 31+2182 are used as an estimate.

5: Classification of probe.

6: Common name or number (from Wakker van Woerden 1991) of HVC on which the probe is projected.

7,8: Velocity (relative to the LSR, in km/s) and column density (in units of $10^{18}$ $cm^{-2}$) of the HVC in the direction of the probe. Upper limits are 5-$\sigma$.

9: A code giving the origin of the HI column density: WE, combination of Westerbork and Effelsberg data; Ws, Westerbork (1′); Ef, Effelsberg (9′); GB, Green Bank 140ft (21′); JB, Jodrell Bank MarkII (31′); Pk, Parkes (14′); Ar, Arecibo (3′); VE, Villa Elisa (34′); HC, Hat Creek (36′); Dw, Dwingeloo survey (36′, using a linewidth of 20 km/s to convert $T_B$ to N(HI)); DE, velocity from Dwingeloo, but N(HI) from Effelsberg.

10: Observed ion.

11: The velocity relative to the LSR of the observed ion.

12: Column density (and error) for the observed ion, or (Table 4) its 3-$\sigma$ upper limit, in units of $10^{11}$ $cm^{-2}$. These data are followed by a code showing how the column density was derived: N, column density quoted in reference; L, logarithm of column density given in reference; W, equivalent width in reference, converted to column density, assuming a gaussian profile with width 15 km/s; S, converted from a quoted signal-to-noise level; t, optical depth of 21-cm absorption.

13: Abundance relative to HI, in units of $10^{-8}$.

14: Abundance relative to the solar abundance (Anders & Grevesse 1989).

15: Reference codes, as listed below the table.

16: Notes, listed below the table.

Table 3 contains three subtables. The first shows that absorption associated with known HVCs has been found in the spectra of nine extra-galactic probes, in radio, optical, or UV lines. The derived abundances show that several HVCs are not made of primordial material. The second subtable lists cases where the HVC is absent in the Dwingeloo survey, but a HVC is known a few degrees away or is seen in other H I spectra. The third subtable lists cases with high-velocity absorption in which there is no known associated H I. Most of these cases occur in high S/N spectra of extra-galactic supernovae, so the question remains of whether the absorption is the result of a galactic HVC or is associated with the host galaxy or the supernova. The observed velocities are generally consistent with those of HVCs in the same region of the sky, which argues against association with the external galaxy. For SN 1991 T and SN 1994 D the observed Ca II abundance is greater than 0.15 solar, and this is higher than is usual. This observation implies that either the calcium ionization is unusual, or the calcium abundance is unusually high, or the hydrogen is not neutral. Deeper H I spectra and especially H$\alpha$ observations are needed to settle this question. For objects in the second and third subtables, high-resolution H I maps would allow one to obtain metal ion abundances or better limits.

Not included in the table are the results of Savage et al (1993), who found absorptions with equivalent widths corresponding to 150 km/s or more in low-resolution HST FOS spectra of six sources projected onto the Outer Arm Extension (H 1821+643), complex C (3C 351, PG 1259+593), the Magellanic Stream (PKS 2251+11, 3C 454.3), and cloud #532 (PG 0043+039) (see Figure 1). These observations show that Mg II, C II, and Si II are present, but no column densities or velocities can be derived.

Finally, there are four probes for which high-velocity absorption has been claimed but is dubious. In HD 135485, Ca II absorption at −98 and −127 km/s was reported by Albert et al (1993), and associated with complex L by van Woerden (1993). However, as suspected by Albert et al, these lines are likely to be circumstellar, since the strong UV lines fail to show absorption at the HVC velocity (Danly et al 1995). For IZW18, Kunth et al (1994) reported O I and Si II absorptions in complex A, but the detections are less than 3-$\sigma$. Songaila et al (1985, 1988) claimed a detection of complex C in the spectrum of BT Dra (d = 2 kpc), and (tentatively) of complex A in SA 12.391 (d = 1.6 kpc). However, Lilienthal et al (1990) showed that these absorptions are certainly (for SA 12.391) or probably (for BT Dra) stellar. The implied upper distance limit for complex A would be incompatible with the strong lower limit from HST spectra of more distant stars (Section 4.4). The Ca II abundance for complex C implied by BT Dra would be a factor 5–10 higher than that found from Mark 290 and PG 1351+640 on neighboring sight lines. The absorption at −85 km/s in BT Dra (Songaila et al 1985, 1988) is associated with the IV Arch and is thus compatible with the distance range 0.7–1.7 kpc found for that object (Kuntz & Danly 1996).

## 4.3 *Metallicities*

The intrinsic metallicity is not yet known to better than a factor 3–5 for any HVC. The best ion to observe is $S^+$, because under typical interstellar conditions sulphur is not depleted onto dust and $S^+$ is the dominant ionization stage (Jenkins 1987). SII in HVC 287+23+240 was detected in the spectrum of NGC 3783 (Lu et al 1994a), giving a value of ~0.2 solar (but see Note 5 of Table 3). A limit of S/H $< 0.3$ was set for the Magellanic Stream using Fairall 9 as a probe (Lu et al 1994b). However, in both cases the derived abundance still suffers from the low resolution of the H I data and may be uncertain by a factor of up to 5; a high-resolution H I map is needed. Thus, the conclusions of West et al (1985) and Lu et al (1994a) that HVC 287+23+240 is extra-galactic because of its low abundances may have been premature.

Resolved Mg II lines have been found in three HVCs; Si II lines in two HVCs. In most cases the lines are saturated and abundance determination is not possible. However, the lower limits (>0.01–0.03 solar) are compatible with the depletion seen in low-velocity H I, where the gas-phase abundances of $Mg^+$ and $Si^+$ are $\sim$0.05 solar (Jenkins 1987). The unsaturated detections of HVC#84 and complex C toward Mark 205 give similar abundances [using an improved value for N(H I); see Note 1 in Table 3].

Ca II is the most often detected ion (in six HVCs). Its abundance relative to H I varies between $0.52 \times 10^{-8}$ and $16 \times 10^{-8}$, a factor of 30. For low-velocity gas, Bowen et al (1991) found similar values but with a smaller range ($0.3–1.6 \times 10^{-8}$). Two possible causes of these variations are evident: (*a*) the depletion of calcium onto grains can vary strongly, and (*b*) $Ca^+$ is not the dominant ionization stage under the usual interstellar conditions. In view of these uncertainties, the range of $Ca^+$ abundances found in HVCs is compatible with a solar metallicity.

## 4.4 *Distance Limits*

Two HVCs have been convincingly detected in absorption in a stellar spectrum. The gas between M II and M III causes absorption at $v_{LSR} \sim -95$ km/s, in the spectrum of BD+38 2182 (d $=$ 5 kpc). Less than one degree away is the star HD 93521 (d $=$ 1.7 kpc), the spectrum of which does not show the absorption, providing a distance bracket for this HVC. Two caveats must be mentioned. First, the HVC is detected toward both probes in 20′ resolution H I spectra but not at 12′ resolution, where 5-$\sigma$ upper limits of $1 \times 10^{18}$ cm$^{-2}$ are set (KR Sembach 1996, private communication). Hence, the lower distance limit may be unreliable. Second, clouds M II–M III may not be spatially close to the other clouds in complex M. The complex is discontinuous, the IV Arch also crosses this region, and some M clouds may be part of it. Therefore, the distance bracket might be valid only for M II–M III, not for complex M in general.

In core A VI, Ca II absorption has been found in the spectrum of the Seyfert galaxy Mark106 (Schwarz et al 1995), and the RR Lyrae star AD UMa (distance 10 kpc, van Woerden et al 1997, manuscript in preparation). Mg II absorption is also present in Mark 106 but is absent in the BHB star PG 0859+593 (about 5° from A VI, distance 4 kpc; Wakker et al 1996a). These results give a solid distance bracket for the high-latitude end of complex A, and imply 2.5 kpc $<$ z $<$ 6.5 kpc.

The significant nondetections show that complex C (specifically, concentrations CIA, CIB, and C III B) is beyond 2.4 kpc ($z > 1.7$ kpc). Cloud C II ($l = 110°$, $b = 40°$, $v_{LSR} \sim -80$ km/s) is sometimes considered part of complex C, but may rather be associated with the IV Arch (Kuntz & Danly 1996), and has $z < 1.7$ kpc. Complex H is beyond 3–5 kpc, that is, beyond the Perseus Arm and thus in the outer parts of the Galaxy. The lower limit for population EP (extreme-positive velocity HVCs, as defined by Wakker & van Woerden 1991) is for HVC #211, a different HVC than the one on which NGC 3783 is projected (cloud #187), so that the 6 kpc lower limit cannot be applied to the whole population.

The danger of misinterpreting nondetections as distance limits is illustrated by the results of Welsh et al (1996), who observed Na I and Ca II for stars projected on complex A. The best Ca II abundance limit was less than $1.7 \times 10^{-8}$, which is equal to the value found toward Mark 106 by Schwarz et al (1995). Hence, the absence of absorption in the stellar spectra is not significant and cannot be used to derive a distance limit. Centurión et al (1994) studied the Ca II lines for many stars in the direction of complex H. Although some of the derived abundances are smaller than the values for complex A and C, they cannot yet be used to derive strong distance limits, because the $Ca^+$ abundance of complex H is still unknown. The same problems exist in principle for the strong UV lines (e.g. Mg II, Si II, C II, O I). In practice, however, these lines will be saturated even for low N(H I) in combination with higher-than-usual depletion. Thus, absence of absorption in strong UV lines is more likely to correspond to a lower distance limit, although a check using an extra-galactic source is always required.

Clearly, only limited information exists on HVC distances. Most of the progress has been made since 1992, owing to the recent availability of better instrumentation, including the GHRS on HST and the UES on the Herschel telescope at La Palma, which allow observations at sufficient resolution of blue-horizontal-branch stars and extra-galactic objects to 15th–16th magnitude. A systematic approach to the problem of HVC distances requires directed searches for probe stars in areas with known HVCs. Such searches have not yet been done. Nevertheless, the Palomar-Green catalogue (Green et al 1986) and lists of blue-horizontal-branch candidates (Beers et al 1996) make progress possible.

## 4.5 *Highly Ionized HVCs*

Lines of highly ionized species such as C IV, Si IV, and N V are often seen at low velocities (Sembach & Savage 1992). These show that the Galaxy has a hot gaseous halo. A relation between HVCs and this halo has long been suspected and is required by some models (Section 6.4). This link may be provided by the high-velocity CIV and NV absorption that has been found for four probes (Table 3). In three cases (Mark 509, H 1821+643, and HD

156369), there is a neutral HVC with similar velocities within a few degrees; only PKS 2155−304 is not near any known neutral HVC. Further study of more sightlines (using the new spectrograph, STIS, on HST) may reveal how common such highly ionized clouds are, and whether they are related in any way to the HVCs seen in H I, or whether they represent a completely new component of the Galaxy.

## 5. HVCs IN EXTERNAL GALAXIES

A new area of HVC studies has developed recently, as a result of the increasing sensitivity of both single-dish and synthesis radio telescopes. This area is the detection of gas at anomalous velocities in other galaxies. Wakker et al (1989) showed that if complexes C and A are at distances greater than 20 kpc (implying a mass $\gtrsim 10^7$ $M_\odot$), similar HVCs would be detectable in then-existing interferometer maps of nearby galaxies. Modern maps lower this detection limit to $10^6$ $M_\odot$. Face-on galaxies are the most favorable for such searches, as the HVCs will stand out in velocity. If high-velocity gas occurs at large heights above a galactic plane, it can also be seen in edge-on galaxies. However, the gas may not show unusual velocities, and it can be difficult to separate high-z gas from a warp (see van der Hulst (1996) for a review).

The first extra-galactic anomalous-velocity gas was found in M 101 by van der Hulst & Sancisi (1988), who detected two clouds with masses $10^7$–$10^8$ $M_\odot$ moving at 150 km/s with respect to the disk of M 101. They concluded that the complex is probably the result of a companion cloud now being accreted, rather than the result of the combined effect of many supernovae or spiral dynamics. After more extensive analysis of the same dataset, Kamphuis et al (1991) found more gas with large vertical motions in M 101, including a $3 \times 10^7$ $M_\odot$, 1.5 kpc diameter superbubble expanding at 50 km/s. Many more holes and half-shells with similar velocities were also seen, but only the bubble showed clear expansion. This result is not surprising, since the timescale on which a distinct bubble disappears is comparatively short.

Individual HVCs have been found in NGC 628 (Kamphuis & Briggs 1992), NGC 6946 (Kamphuis & Sancisi 1993), and NGC 5668 (Schulman et al 1996), with masses greater than a few times $10^6$ $M_\odot$. NGC 628 contains two very large HVCs, outside the optical disk and placed symmetrically on opposite sides. NGC 5668 also has two large clouds outside the optical disk. These features may be tidal. However, gas at velocities up to 70 km/s away from galactic rotation is also found in these three galaxies, down to the detection limit of a few times $10^6$ $M_\odot$. It is often seen on top of some deficit in the column density of low-velocity gas. Furthermore, the H I profile is wider inside the optical disk than outside. This observation has been interpreted as the result of superposition of many discrete features below the detection limit. In the edge-on galaxy NGC 891 Swaters et al (1997, manuscript in preparation) have found H I up to z = 6 kpc, but the velocities are not anomalous.

A different approach was taken by Schulman et al (1994, 1996) and Bates & Maddalena (1996), who searched for wings in deep Arecibo or Green Bank 140-ft H I profiles of face-on galaxies, of different Hubble types. Such wings are often found, and the profiles can be fit by a combination of two populations of H I clouds, one with low dispersion and one with relatively high (30–50 km/s) dispersion. An alternative fit of the wings in terms of a warp does not work. More confidence in the interpretation is given because the method also separates out galaxies that are tidally distorted from those with just high-velocity gas. The high-dispersion component can have up to 15% of the total H I mass of a galaxy. Schulman et al (1994) further showed that the far-infrared emission (supposedly a measure of the star formation rate) correlates well with the amount of high-dispersion gas, even after correction for galaxy size. Such a correlation would be expected in a galactic fountain model (see Section 6.4), in which the HVCs are ultimately generated by high-mass stars.

Although this field of study is still in its infancy, it is clear that many galaxies contain gas deviating from differential rotation. The relation to the galactic HVCs remains unclear, but the properties of the individual HVCs found in interferometer maps and the correlation between star formation rate and the presence of high-velocity gas suggest a relation between HVCs and the evolution of massive stars.

## 6. PROPOSED ORIGINS FOR THE HVCs

In this section we discuss the many hypotheses proposed to explain the origins of HVCs. We have ordered them by increasing distance from the Sun. The Magellanic Stream is discussed separately in Section 7.

### 6.1 *Hypotheses That Put HVCs Nearby*

It may seem reasonable to suggest that HVCs are nearby supernova shells. Meaburn (1965), Haslam et al (1971), and Shatsova (1993) proposed a correlation between HVCs and the radio-continuum loops, suggesting that both originated in supernovae. However, no correlation can be seen in the modern data. Heiles (1979) and Weaver (1979) interpreted structures in HVC velocity-channel maps as nearby supernova shells. Oort (1966) already rejected this hypothesis because most supernova shells (*a*) concentrate toward the galactic plane, (*b*) should show both the approaching and the receding half, and (*c*) should show strong H$\alpha$.

The HVCs might also have been generated by many supernovae in a single OB association (Oort 1966), or by a superexplosion in the Perseus Arm (Rickard 1968, Verschuur 1993a). Some of the accelerated gas then would move toward the Sun, and the nearby ($d \sim$ 100–500 pc) bits might become visible as the "infalling" HVCs and IVCs. However, the orbits of such ejecta would result in an $l$-$b$-$v_{LSR}$ distribution that is incompatible with observations (Oort 1966, Wakker & Bregman 1990).

An unusual suggestion was made by Verschuur (1990), who showed that "representative" channel maps of cores in complex A and C have a morphological resemblance to the total CO column-density maps of nearby molecular clouds, if each HVC core is shifted by a different amount. He suggested that this was not a coincidence and implied distances of ~100–800 pc.

These hypotheses are incompatible with the strong lower distance limits for several large HVCs, especially A and C (Section 4). Considering also the indirect arguments, we conclude that HVCs are not nearby supernova or superexplosion debris.

## 6.2 *Hypotheses That Put HVCs at Large Distances*

If HVCs are very distant, one can account for the large line widths without invoking a confining medium and/or short evolution time scales. The virial theorem shows that HVCs will be self-gravitating at distances larger than a few 100 kpc (Verschuur 1969). Alternatively, Eichler (1976) showed that if HVCs were protogalaxies (d > 100 kpc), the orbital energy of a collection of cloudlets would be sufficient to give them internal support, while Doroshkevich & Shandarin (1979) concluded that even after 10 Gyr, thermal instabilities during adiabatic galaxy formation will naturally lead to fragmentation and condensation of cold clouds, yielding HVCs. All these ideas imply extremely large sizes (e.g. >50 kpc for complex A) and unlikely low volume densities ($\ll$0.005 $cm^{-3}$ in the envelopes). Also, abundances are predicted to be much lower than solar, which is not observed.

An interpretation of HVCs as satellites of the Galaxy or as Local Group objects seemed compatible with the data at first (Kerr & Sullivan 1969, Verschuur 1969). However, severe selection effects existed as the distribution of $v_{GSR}$ was plotted, but different $v_{GSR}$ ranges were selected at different sky positions, distorting the histogram.

On a different note, Arp (1985) argued that some small HVCs point toward galaxies near them and hence were shot out of those. This hypothesis ignores the tail on the M 33 cloud (Wright 1979b), as well as the probable relation of the clouds near NGC 300 and NGC 55 to the Magellanic Stream (Haynes & Roberts 1979). The cloud near NGC 520 is HVC 168−46−280, and the map that was used showed only a small part of its core.

An association with Local Group galaxies has been suggested separately for the VHVCs, which concentrate in the same area of the sky. However, Giovanelli (1977, 1978, 1981) showed that this is unlikely because: (*a*) no independent H I clouds have been detected in other galaxy groups; (*b*) the clouds cannot be stable unless the H I represents only a small fraction of the total mass; (*c*) the narrow lines require small velocity dispersion and gradients over tens of kpc; (*d*) the low temperatures (<500 K) implied by the narrow lines require that cooling has taken place, which would be very inefficient at the implied low volume densities; and (*e*) the velocity distribution of the VHVCs is displaced by 200 km/s from that of Local Group galaxies. Recent studies of (*a*) show that $10^6$ $M_\odot$ clouds are detectable out to 1 Mpc but are not common (Zwaan et al 1996). At that distance, the brightest VHVC (HVC 128−34−392) would have a mass of $4 \times 10^8$ $M_\odot$. Wakker & Bregman (1990) show that a population of objects with the same velocity and space distribution as the nearby Local Group galaxies will not reproduce the observed distributions of the VHVCs on the sky and in velocity.

## 6.3 *Hypotheses That Put HVCs in the Outer Galaxy*

HVCs AS OUTER SPIRAL STRUCTURE Although by definition HVCs have velocities incompatible with differential galactic rotation, attempts have been made to understand them as parts of normal spiral structure, invoking systematic peculiar velocities. The "Outer Arm Extension" (not included in Figure 1), which consists of high-velocity gas with $v_{LSR} < -100$ km/s at low latitudes ($0° < b < 15°$) and longitudes 70°–210°, was studied by Habing (1966) and Kepner (1970). This structure has velocities within the range expected from galactic rotation. Hence, the run of velocity with longitude can be understood if the structure is either a high-z extension of the Outer Arm in a warped-up part of the outer Galactic Disk, or a separate high-z spiral arm. Haud (1992) shows that in the former case the velocities can be fitted by assuming a triaxial galactic potential.

Dieter (1971), Davies (1972a, 1972b, 1974), and Verschuur (1973a, 1973b, 1975) proposed a model in which all HVCs are part of a thick ring around the outer Galaxy or an extreme warp. These models do not explain the VHVCs or the positive-velocity HVCs. They might explain complex A and C, but the implied geometry, morphology, and velocities do not fit well (Hulsbosch & Oort 1973). Further, the more extensive data now available show that the structural continuities suggested by Davies (1972a) are not present.

HVCs DRAWN OUT OF THE MAGELLANIC CLOUDS Mathewson et al (1974), Davies & Wright (1977), and Giovanelli (1978) proposed that during a previous passage of the Magellanic Clouds a stream was created, which later pushed a hole in the Galactic Disk and accelerated gas up to 3–5 kpc. Only now is this gas falling back as the northern HVCs. Oort & Hulsbosch (1978) rejected this hypothesis because it implies that the clouds stay coherent for much longer than a Gyr, which is incompatible with the derived time scales (Section 2.8). Also, cloud orbits are expected to yield a different $l$-$v_{LSR}$ distribution (Wakker & Bregman 1990).

HVCs AS A POLAR RING Haud (1988) and Mathewson et al (1987) suggested that the Magellanic Stream, complex C, and some positive-velocity HVCs form part of a polar ring around our Galaxy, at a distance of 90 kpc. Haud (1988) also included some lower-velocity clouds near the North Galactic Pole but did not justify this inclusion. This model has several problems: (*a*) complex C makes a sizable angle with the purported plane; (*b*) complexes A, M, H, and many positive-velocity HVCs are not included; (*c*) the GSR velocity was plotted against longitude along the polar ring, but the criterion to include clouds was their LSR velocity, leading to an artificial sine wave; and (*d*) the IV Arch would also lie approximately in the polar ring plane, but its velocities do not fit at all. This hypothesis assigns major significance to the fact that complex C and the Magellanic Stream are linear structures at opposite sides of the sky, even though they do not mirror each other in any significant detail. It further suppresses several properties of the HVC and IVC distributions. Thus, although the Magellanic Stream may well be in a polar orbit and although there is no strong direct evidence against a similar polar orbit for complex C, this hypothesis cannot provide a general explanation for the HVCs.

A CONNECTION WITH GLOBULAR CLUSTERS Attempts to connect high-velocity gas with globular clusters have been made a few times. At first sight the $l$-$b$-$v_{LSR}$ distributions of Galactic Halo objects and of HVCs are similar (Shatsova 1984, Mathewson et al 1987). However, this is just a reflection of galactic rotation (Wakker & Bregman 1990), and if HVCs were distributed like halo objects there would be more high-latitude HVCs. A physical connection between HVCs and the nearest dwarf galaxies and distant globular clusters (Lynden-Bell 1976) is also inconsistent with the data (see Lin & Lynden-Bell 1977). Neither is a physical connection required to explain the coincidence of the velocities of H I emission and the cluster for M92 (Kerr & Knapp 1972) and M 56 (Birkinshaw et al 1982), as 23 globular clusters are projected onto a HVC, and thus a few velocity coincidences are not unexpected. Lynden-Bell & Lynden-Bell (1995) discuss the possibility of streams caused by tidal interactions between the Galaxy and the Magellanic Clouds, involving distant globular clusters and nearby dwarf spheroidals. Some of these hypothetical streams might be identified with HVCs or VHVCs.

VHVCs AND THE MAGELLANIC STREAM VHVCs lie in the same region of the sky as the tip of the Magellanic Stream, which is fragmentary. Thus, Giovanelli (1981) argued that the VHVCs are loose shreds of the Stream. He thought that this would work only if the Stream is leading the Magellanic Clouds, which is not the case (Section 7). However, Wakker & Bregman (1990) found that a large scatter can also be produced if the Stream is trailing. Detailed models of the Stream (see Section 7) are not yet sophisticated enough to produce the VHVCs, however.

If the VHVCs are related to the tip of the Magellanic Stream, they must be distant ($d \gtrsim 50$ kpc), and their masses ($M = 1 \times 10^6\,M_\odot$ for HVC 128−34−392 at 50 kpc) and evolution time scales become comparable to those for the northern HVCs. Unlike the Magellanic Stream, however, many VHVCs show two-component structure in their profiles, possibly indicating that not all are in the same environment or at the same distance as the Stream.

A relation has also been proposed between VHVCs and the most-negative velocity clouds in the Anti-Center and Galactic Center regions (Mirabel 1981a, 1982, Mirabel & Morras 1984), because these three populations are contiguous in the $l$-$v_{LSR}$ diagram (Figure 3). The main problem with this idea is that models for the Magellanic Stream (Section 7) do not predict gas to show up near these regions.

## 6.4 *Hypotheses That Put HVCs in the Lower Galactic Halo*

A PLASMA CHANNEL CREATED BY THE GALACTIC NUCLEUS If between 80 and 150 Myr ago our Galaxy was a Seyfert, the Galactic Center might have ejected the HVCs (Oort 1966, Saraber & Shane 1974). Oort rejected this hypothesis because very low angular momentum is expected for the gas, contrary to observations. Also, the velocities have to be fine tuned for the current situation to arise. Kundt (1987) suggested instead that the gas in the HVCs was not ejected directly, but that relativistic electrons ejected from the Galactic Center created a plasma channel, whose walls are compressed and cool faster, so that H I clouds condense and rain down ballistically. This process would explain both complexes C and A as well as the Magellanic Stream. Kundt stated that it would also generate a regular velocity field, as seen in the Stream. However, the velocity fields of complexes C and A are irregular, and the Anti-Center Clouds, the positive-velocity clouds, and complex H remain unexplained.

PRIMORDIAL GAS BEING ACCRETED BY THE GALAXY This hypothesis was favored by Oort (1966), Oort (1970), and Oort & Hulsbosch (1978), and could also apply to IVCs. The hypothesis suggests that some gas left over from the formation of the Galaxy may only now be reaching the Disk. During its approach, high-z (1–3 kpc), cold, low-density H I is swept up, shock-heated, accelerated and then recools, forming the HVCs. The resulting mixture would be about 70% swept-up gas. Most of the velocity comes from the overpressure provided by the infalling cloud. This model raises two main questions: Could clouds survive for ~10 Gyr before impacting the Galaxy, and would the gas cool quickly enough after being shocked so that it would be visible at z-distances of order 1 kpc?

The accretion hypothesis has several testable consequences. First, it implies z-heights on the order of a kpc. Second, metal abundances would be (slightly) lower than normal. Third, *all* HVCs should show signs of interaction. Currently, distance limits are not strong enough to exclude that most HVCs are at z ~ 1–2 kpc, except for complex A which has z $\gtrsim$2.5 kpc. An extended vertical component is known to exist for the neutral hydrogen (scale height 400 pc; Dickey & Lockman 1990) as well as the ionized hydrogen (scale height ~1 kpc; Reynolds 1996). It remains to be seen whether the volume density of the extended component is large enough at the z-heights of the HVCs to cause adequate interaction. Metal abundances in HVCs are still mostly unknown (Section 4.2), so they cannot be used to argue for or against this hypothesis.

THE SIGNATURE OF A GALACTIC FOUNTAIN The idea that has gained the most ground as an explanation of HVCs and IVCs has been that of the galactic fountain (Shapiro & Field 1976, Bregman 1980, Kahn 1981, 1990, Houck & Bregman 1990, Li & Ikeuchi 1992). A review of the subject is given by Bregman (1996). In the fountain model, hot gas is produced by the cumulative effect of supernovae in the disk; thermal expansion lifts it to large z-heights, where it becomes unstable; cool clouds condense and fall back ballistically. In effect, this process sets up a galactic-scale convection. The uplifting force may also be provided by spiral shocks (Suchkov et al 1974, Schchekinov 1982) or cosmic rays (Breitschwerdt et al 1993). In contrast, Ferrara & Einaudi (1992) considered the effects of convection and thermal conduction and argued that HVCs cannot form as a result of thermal instability in a hot galactic flow, but that dynamic instability may develop.

How the hot gas behaves depends on the ratio of cooling time to sound crossing time (Houck & Bregman 1990), and its temperature and density at $z = 0$. If $t_{cool} < t_{sound}$ ($T_o > 3 \times 10^5$ K, $n_o > 0.0015$ cm$^{-3}$), a transonic fountain develops in which clouds form in the upward flow, grow and then fall back. For $t_{sound} < t_{cool} < 10t_{sound}$ ($2 \times 10^5 < T_o < 10^6$ K, $0.0005 < n_o < 0.0015$ cm$^{-3}$), a subsonic fountain develops in which all clouds form at similar z-height. Finally, if $t_{cool} > 10t_{sound}$ ($T_o > 10^6$ K, $n_o < 0.0005$ cm$^{-3}$), convection develops and clouds form all over, with both downward and upward velocities. If a transonic fountain developed in the Galaxy, it may provide an explanation for the east-west asymmetry. The positive-velocity clouds would be still-forming clouds in the upward flow; when they slow down and start falling they will become lost in the low-velocity gas because of projection effects. The negative-velocity clouds at $l < 180°$ are fully developed falling sheets. Evidence for a hot galactic corona is provided by X-ray observations (e.g. Snowden et al 1993) and CIV absorption lines (Sembach & Savage 1992). This corona has a scaleheight of several kpc, temperature $5 \times 10^5$–$10^6$ K, and density at the base of about $n_o = 0.006$ cm$^{-3}$, corresponding to a transonic fountain.

The energy input from supernovae turns out to be sufficient to set up a fountain. The requirement that superbubble blowout occurs may be fulfilled in the absence of magnetic fields (MacLow et al 1989). Under some circumstances magnetic fields may prevent break-out (Tomisaka 1991, 1992), but only if the field is mostly aligned with the Galactic Disk. Even then the bubble can still burst, if the supernova rate is above some critical value ($L_{mech} > 3 \times 10^{31}$ W if $B = 5\ \mu$G). This rate corresponds to a time of 1000 yr between supernovae in a single bubble. For lower energy inputs, a field of 5 $\mu$G confines the bubble to $R < 160$pc.

Predictions for the angular and velocity structure are similar to those of the accretion model. However, the fountain model differs in the following aspects: (*a*) the ultimate source of the gas is in the Galactic Disk instead of intergalactic space; (*b*) many HVCs will not be interacting with low-z gas; (*c*) the predicted z-heights are in the range 0–10 kpc (Wakker & Bregman 1990), instead of 1–3 kpc; and (*d*) metallicities will be normal instead of slightly reduced (although the distinction may be difficult to make in practice). The maximum vertical velocity in the fountain model is ~70–100 km/s, far less than the observed velocities of many HVCs. However, Wakker & Bregman showed that projection effects resulting from differential galactic rotation increase the apparent velocities sufficiently. Another difference between the models might lie in the predicted sky distribution of velocities, as a result of the motion of the Galaxy in the Local Group, but no predictions are available.

Wakker & Bregman (1990) derived that in a galactic fountain the HVCs would represent a flow rate on the order of 5 $M_\odot$/year. This rate is comparable to the flow rate favored by Sembach & Savage (1992) to interpret the C IV, Si IV, and N V absorption lines toward halo stars in terms of a fountain. It is also similar to the detailed predictions of fountain models (Edgar & Chevalier 1986, Shapiro & Benjamin 1991), showing that theory and observations of the fountain model are compatible. Since the total H I mass of the Galaxy is estimated to be about $6 \times 10^9\ M_\odot$ (Giovanelli & Haynes 1989), such a flow rate implies that all interstellar matter circulates through the Halo each Gyr.

## 7. THE MAGELLANIC STREAM

### 7.1 *Observations*

One of the best studied HVCs is the Magellanic Stream, reviewed by Mathewson et al (1987), Wayte (1989), and Gardiner & Noguchi (1996). Maps of the whole Stream or parts of it were presented in several papers (Mathewson 1976, Mathewson et al 1979, Mirabel et al 1979, Haynes 1979, Erkes et al 1980, Mirabel 1981b, Cohen 1982a, Morras 1983, 1985, Morras & Bajaja 1986, Wayte 1989). Mathewson & Ford (1984) gave an overview map. The Stream breaks up into six main concentrations, named MS I through MS VI, with fainter extended emission between. Mathewson et al (1974) present a suggestive picture in which the very-high positive-velocity HVCs around $l = 270°$, $b = +20°$ (population EP of Wakker & van Woerden 1991) are part of the Stream, but they give no evidence for this. The line profiles in the Stream are usually wide, while those in population EP show the two-component structure seen elsewhere. Whether this represents a physical difference or is a resolution effect due to different distances, it casts doubt upon the association suggested by Mathewson et al (1974).

Several unsuccessful searches have been made for an excess of stars associated with the Stream. Philip (1976) and Mathewson et al (1979) presented no data, but stated a negative result. Searches for A stars by Recillas-Cruz (1982), Tanaka & Hamajima (1982), and Brück & Hawkins (1983) reached down to magnitude 20.5, corresponding to 90 kpc for an A0V star and 40 kpc for an A9V star. If the Stream has a tidal origin, the stellar density is expected to increase from an average of 3500 to about 5500 per square degree. The negative results can be explained if the Stream is more distant than 50 kpc, or if it is older than 1 Gyr, in which case the A stars have evolved away from the main sequence. Deeper searches may then turn up an excess of F stars. However, in Gardiner et al's (1994) tidal model the Stream originates in the outer parts of the SMC, and a lower star-to-gas ratio is expected than would follow from an origin in the body of the SMC.

## 7.2 *Discarded Models*

Currently, there are two main models for the Stream: the tidal and the ram-pressure model. Some other proposals have fallen by the wayside because of various incompatibilities with the data: (*a*) Primordial gas in orbit around the Magellanic Clouds (Mathewson et al 1974, Mathewson & Schwarz 1976, Einasto et al 1976, Fujimoto 1979) seems unlikely, especially since metals have been detected in the Stream (Songaila 1981a). (*b*) A relation with the dwarf spheroidals had also been proposed (Lynden-Bell 1976, Kunkel 1979). (*c*) The idea that the Stream is a wake (i.e. cooled halo gas), caused by the passage of the Clouds through the hot Galactic Halo (Mathewson et al 1977), was refuted by Bregman (1979).

## 7.3 *Tidal Tail Models*

Originally there was a controversy between models in which the Stream was a trailing tidal tail (Murai & Fujimoto 1980, Lin & Lynden-Bell 1982; predicting a distance (d) of ~50 kpc for the tip) or a leading tidal bridge (Davies & Wright 1977, Lin & Lynden-Bell 1977, Tanaka 1981; predicting d ~ 10 kpc for the tip). Major problems existed with these early tidal models (e.g. Gingold 1984, Mathewson et al 1987). All leading-bridge models can now be discarded, since the tangential velocity of the LMC has been shown to be 290 km/s in the direction away from the Stream (Jones et al 1994).

The most recent good tidal model was presented by Gardiner et al (1994) and updated by Gardiner & Noguchi (1996) and Lin et al (1995). It represents the SMC by a self-gravitating collection of particles, the LMC by a point mass. The results show that LMC and SMC have been a binary galaxy for a Hubble time but had a close (7 kpc) encounter 1.5–2 Gyr ago. Tidal forces on the SMC produced its large depth and some gas was freed to form the Stream. The leading bridge is swallowed by the LMC, except for some shreds that can be identified with low-velocity H I clouds. A second close encounter 0.2 Gyr ago produced what is now the Inter-Cloud region, and its structure is also partly explained.

## 7.4 *Ram-Pressure Models*

Two versions of the ram-pressure model exist: diffuse and discrete. Both try to explain the column density gradient and the presence of concentrations along the Stream. The diffuse ram-pressure model postulates that diffuse hot galactic or intergalactic gas (density $\sim 10^{-4}$ cm$^{-3}$) exerts a drag on the outer parts of the Magellanic Clouds (Mathewson et al 1977, 1987, Sofue 1994, Heller & Rohlfs 1994). With proper adjustments the velocity field of the Stream can be reproduced. However, with a realistic drag coefficient the predicted transverse velocities are too low (Meurer et al 1985, Gardiner et al 1994). In the discrete ram-pressure model, inter-cloud gas is decelerated by random collisions with cool clouds in the Galactic Halo, possibly represented by the HVCs with $v_{LSR} > +200$ km/s (Mathewson et al 1987, Sofue & Wakamatsu 1991). This uses the conclusion that HVC 287+23+240 is intergalactic because of its low abundances; however, this question has not been settled (see Section 4.2).

Moore & Davis (1994) listed the following arguments in favor of the drag model: (*a*) The gas with the lowest column density has lost the most angular momentum and is now at the Stream's tip. (*b*) The drag model does not require the leading bridge that tidal models predict. [However, in the best tidal models the bridge is swallowed by the LMC.] (*c*) Most of the 0.5- to 1-keV X-ray background originates from beyond 60 pc (McCammon & Sanders 1990), which is extrapolated to imply a $2 \times 10^6$ K galactic corona. [However, this conclusion is uncertain.] (*d*) Drag is invoked to explain the absence of gas in nearby dwarf spheroidals and globular clusters. [However, this property might have an evolutionary explanation.]

Against the tidal model Moore & Davis (1994) presented the following arguments: (*a*) Gas tidally stripped from the LMC would not show the observed density contrasts. [However, in the best tidal models the gas originates in the SMC.] (*b*) The observed Stream is trailing, which happens naturally in a drag-dominated model. [However, this is not a problem in tidal models.] (*c*) The observed velocity field is inconsistent in tidal models. [However, Gardiner et al (1994) reproduced it fairly well.] (*d*) No stars are observed in the Stream. [However, the available observations may have been insufficient to settle this question.] Thus, the arguments presented by Moore & Davis (1994) against the tidal model are not very strong, while those in favor of the drag model have their own, bigger, problems.

### 7.5 *A Wake Model and Summary*

Liu (1992) proposed that a wake in the hot gas is filled by cold LMC gas, drawn out by drag forces and accelerated down this tube by the gravity of the Galaxy. The Stream then outlines the Cloud's orbit. This is the only model that can naturally account for both the velocity gradient and the density gradient along the Stream (the latter as result of the changing balance between pressure gradient and gravity).

In general, however, we conclude that the tidal model seems to have the stronger case. But further exploration of the tidal, drag and wake models is needed.

## 8. SUMMARY

### 8.1 *Recent Progress and New Insights*

The past ten years have seen publication of the first complete sky surveys for high-velocity clouds (HVCs). These surveys have yielded the distributions of HVCs on the sky and in velocity, have shown that more than 10% of the sky is covered with HVCs and have clarified their relationships to the Galaxy. They have also provided homogeneous catalogues of the properties of HVCs.

Radio synthesis maps show that many HVCs contain arcmin-size structures, with densities of 10–100 $cm^{-3}$ and temperatures between 50 and 350 K. The lifetimes of these condensations are on the order of 1 Myr, much shorter than those of the larger complexes as a whole. H$\alpha$ emission and X-ray shadows have been reported for a few HVCs, but the evidence for interactions with the Galactic Halo or Disk remains circumstantial. Searches for dust and for CO molecules have set significant upper limits.

Visual band and/or ultraviolet absorption by metal ions has been measured in at least seven HVCs, showing that these are not primordial, and opening the road to distance determinations. Conversion of ion abundances into metallicities still requires understanding of small-scale structure, ionization, and depletion conditions. Good lower distance limits (several kpc) now exist for several major HVC complexes; part of complex M is found to lie below $z = 4.5$ kpc; complex A lies between $z = 2.5$ and 6 kpc.

These recent results on distance and composition, along with dynamic models, now allow a critical evaluation of the various theories for HVCs.

## 8.2 *The Origin of HVCs*

In Section 6 we questioned hypotheses that may give an acceptable explanation for one or a few HVCs, but then do not account for others nor for the large-scale distribution in $l$, $b$ and $v_{LSR}$. In this, we follow the principle of Occam's razor, which says that it is best to minimize the number of hypotheses. Yet, it is obvious that the observational selection criteria catch a range of phenomena and at least three hypotheses are needed: one for the Magellanic Stream and possibly related clouds, one for the Outer Arm Extension, and at least one for the other HVCs. It is unclear whether the Anti-Center and Galactic Center Complexes require a fourth hypothesis.

The Magellanic Stream is obviously connected to the Magellanic Clouds. Whether it was formed tidally or by ram pressure remains unclear. Tidal models seem more natural and give good fits to most properties but still do not account well for the presence of concentrations along the Stream. The distribution of the VHVCs suggests that they are related to the Stream, although neither model predicts them, and alternative explanations are not completely excluded. The characteristics of the Outer Arm Extension are compatible with its being a high-z extension of the Outer Arm (Section 6.3). That other HVCs can be explained in either manner appears unlikely.

The key to an understanding of the other HVCs lies in a direct determination of their distances, but these are still largely lacking. However, the lower limits that have been set for some HVCs allow us to exclude all hypotheses that put these HVCs nearby.

The following evidence indicates that the major HVCs are not distant ($d > 100$ kpc) protogalactic objects: (*a*) large ($10^8\ M_\odot$) independent H I clouds are not seen in other galaxy groups; and (*b*) several HVCs do not consist of primordial gas, as shown by the detection of heavy elements.

There are also several arguments that many HVCs are related to the Galactic Halo: (*a*) The lower distance limits for some clouds put them far above the Disk (Section 4); (*b*) The sky and velocity distributions are consistent with a population of clouds at $z = 2$–$5$ kpc having some radial and vertical infall (Kaelble et al 1985, Wakker & Bregman 1990); and (*c*) The level of H$\alpha$ emission and the absence of dust and CO emission all are compatible with high z but not with low z (Section 3).

Only two of the hypotheses discussed in Section 6 provide a unified explanation for most HVCs with $|v_{LSR}| < 200$ km/s: the accretion model (Section 6.4.2) and the fountain model (Section 6.4.3). Both models can account for the observed fine structure, since the HVCs would consist of cooled hot gas, as suggested by the irregular velocity field, the presence of clumpiness, and the core-envelope structure. The accretion model might account for the Anti-Center Clouds as the result of the Galaxy's motion through the Local Group, but it has difficulties with the positive-velocity clouds. The fountain model provides a natural explanation for the positive-velocity clouds as clouds that formed in the upward flow, but it has difficulties accounting for the north-south asymmetry and possibly the Anti-Center Clouds. Both models require that IVCs are present too, and the properties of the IVCs also fit. A choice between the two models could be made on the following basis: (*a*) $z \sim 1$–3 kpc for all HVCs in the accretion model, $z = 0$–10 kpc in the fountain model; (*b*) all HVCs should show signs of interaction in the accretion model, only some in the fountain model; and (*c*) a detailed prediction for the sky and velocity distributions might show differences. So far, none of these possible differentiations can be applied. Considering phenomena observed in other galaxies, it appears possible that both accretion and a fountain are important. If a fountain occurs, the HVCs represent a flow rate of $\sim$5 $M_{\odot}$/yr, similar to theoretical predictions. Such a flow circulates all interstellar matter through the Halo in about a Gyr. In the accretion model a smaller rate ($\sim$1 $M_{\odot}$/yr) is derived and no circulation takes place.

## 8.3 *Problems, Needs, and Prospects*

The physical conditions in HVCs are still poorly known, and the processes that are going on are little understood. Synthesis maps have shown a wealth of small-scale structure in several HVCs (Section 2.6); however, other major complexes (notably complex C, the Anti-Center Complexes, and the Magellanic Stream) and many other HVCs have not yet been studied at high resolution. Searches for radio sources behind HVCs should make spin-temperature determinations possible (Section 3). Mapping of H$\alpha$ emission, which is now becoming possible (Section 3.1), will reveal the thermal and ionization structure of HVCs and provide information on their interactions with Galactic Halo and Disk. Further studies of distribution and spectrum of soft X rays may also contribute (Section 3.4). Analyses of dynamic processes are sorely needed in order to clarify the orbits of HVCs, their interactions with the environment, and the evolution of small-scale structure. Further exploration of the possibility that HVCs colliding with the Disk can trigger star formation is also needed.

Studies of HVCs in other galaxies may provide crucial information to decide between the accretion and fountain models, or suggest alternatives. A larger sample of face-on and edge-on spirals should be analyzed at high sensitivity, and the statistics of broad line wings (Section 5) should be studied. Also, a thorough analysis of existing material on the nearest spirals, M 31 and M 33, would be worthwhile.

Analysis of the vast material in the Leiden-Dwingeloo survey will bring rich rewards. It will clarify the relations between HVCs and IVCs, and allow redefinition of the HVCs by inclusion of gas with anomalous velocities below the $v_{LSR} < 90$ km/s limit. Also, it will make possible a systematic study of line widths and of line profile shapes. Finally, it will allow searches for small HVCs missed in the previous surveys.

As emphasized earlier, knowledge of distances is crucial to our understanding. Direct determinations are needed separately for each HVC. The metal ions required for absorption-line studies are present, and modern spectrographs on HST and large ground-based telescopes can reach the faint magnitudes ($m > 15$) involved. Searches for blue horizontal-branch stars in fields selected to contain HVCs should provide probes over a suitable range of magnitudes and distances. Although various ion abundances in HVCs have been measured (Section 4.2), determination of intrinsic metallicities will require observation of specific ultraviolet lines (e.g. S II, Zn II) in HST spectra of faint extra-galactic probes, combined with high-resolution H I maps. Combining many ultraviolet lines can provide information on ionization and dust composition.

Acquisition of these various data on temperature, ionization, structure, metallicity, and distance will reveal the origins of HVCs and their role in the evolution of our Galaxy.

We thank UJ Schwarz, BD Savage, and RJ Reynolds for valuable comments and discussions, and JN Bregman, F Briggs, JM van der Hulst, and K Sembach for providing reviews and data before publication.



*Literature Cited*

Albert CE, Blades JC, Morton DC, Lockman FJ, Proulx M, Ferrarese L. 1993. *Ap. J. Suppl.* 88:81–117

Anders N, Grevesse E. 1989. *Geochim. Cosmochim. Acta* 53:197–205

Arp H. 1985. *Astron. J.* 90:1012–18

Bajaja E, Cappa de Nicolau CE, Cersosimo JC, Martin MC, Loiseau N, et al. 1985. *Ap. J. Suppl.* 58:143–65

Bajaja E, Cappa de Nicolau CE, Martin MC, Morras R, Olano CA. 1989. *Astron. Astrophys. Suppl.* 78:345–62

Bajaja E, Morras R, Pöppel WGL. 1987. *Publ. Astron. Inst. Czech. Acad. Sci.* 69:237—39

Bates B, Catney MG, Keenan FP. 1988. *Astrophys. Space Sci.* 146:195–200

Bates B, Catney MG, Keenan FP. 1990. *MNRAS* 242:267–70

Bates N, Maddalena RJ. 1996. *Bull. Am. Astron. Soc.* 28:839 (Abstr.)

Beers TC, Wilhelm R, Doinidis SP, Mattson CJ. 1996. *Ap. J. Suppl.* 103:433–66

Benjamin RA, Danly L. 1997. *Ap. J.* In press

Benjamin RA, Venn KA, Hiltgen DD, Sneden C. 1996. *Ap. J.* 464:836–41

Birkinshaw M, Ho PTP, Baud B. 1982. *Astron.*

*Astrophys.* 125:271–75
Bloemen H, ed. 1991. *The Interstellar Disk-Halo Connection in Galaxies, IAU Symp. No. 144,* pp. 1–448. Dordrecht: Reidel
Bone DA, Hartquist TW, Sanford P. 1983. *Astrophys. Space Sci.* 89:173–76
Boulanger F, Abergel A, Bernard JP, Burton WB, Désert FX, et al. 1996. *Astron. Astrophys.* 312:256–62
Bowen DV, Blades JC. 1993. *Ap. J.* 403:L55–58
Bowen DV, Blades JC, Pettini M. 1995. *Ap. J.* 448:662–67
Bowen DV, Pettini M, Penston MV, Blades JC. 1991. *MNRAS* 249:145–58
Bowen DV, Roth KC, Blades JC, Meyer DM. 1994. *Ap. J.* 420:L71–74
Bregman JN. 1979. *Ap. J.* 229:514–23
Bregman JN. 1980. *Ap. J.* 236:577–91
Bregman JN. 1996. In *The Interplay Between Massive Star Formation, the ISM and Galaxy Evolution, 11th IAP Meeting.* Paris: Inst. d'Astrophys. Paris
Bregman JN, Harrington JP. 1986. *Ap. J.* 309:833–45
Bregman JN, Lockman FJ, eds. 1986. *Proc. NRAO Workshop No. 12, Gaseous Halos of Galaxies,* pp. 1–284. Green Bank: Natl. Radio Astron. Oobs.
Breitschwerdt D, McKenzie JF, Völk HJ. 1993. *Astron. Astrophys.* 269:54–66
Bruhweiler FC, Boggess A, Norman DJ, Grady CA, Urry CM, Kondo Y. 1993. *Ap. J.* 409:199–204
Brück MRS, Hawkins MT. 1983. *Astron. Astrophys.* 124:216–22
Burrows DN, Mendenhall JA. 1991. *Nature* 351:629–31
Burton WB. 1985. *Astron. Astrophys. Suppl.* 62:365–643
Cavarischia GA, Morras R. 1989. *Astron. Astrophys. Suppl.* 78:437–40
Centurión M, Vladilo G, de Boer KS, Herbstmeier U, Schwarz UJ. 1994. *Astron. Astrophys.* 292:261–70
Christodoulou DM, Tohline JE, Keenan FP. 1997. *Ap. J.* In press
Cohen RJ. 1981. *MNRAS* 196:835–44
Cohen RJ. 1982a. *MNRAS* 199:281–93
Cohen RJ. 1982b. *MNRAS* 200:391–405
Cohen RJ, Davies RD. 1975. *MNRAS* 170:23p–27p
Cohen RJ, Mirabel IF. 1978. *MNRAS* 182:395–99
Cohen RJ, Mirabel IF. 1979. *MNRAS* 186:217–30
Cohen RJ, Ruelas Mayorga RA. 1980. *MNRAS* 193:583–91
Colgan SWJ, Salpeter EE, Terzian Y. 1990. *Ap. J.* 351:503–14
Comeron F, Torra J. 1992. *Astron. Astrophys.* 261:94–104
Comeron F, Torra J. 1994. *Astron. Astrophys.* 281:35–45
Cram TR, Giovanelli R. 1976. *Astron. Astrophys.* 48:39–47
D'Odorico S, Pettini M, Ponz D. 1985. *Ap. J.* 299:852–64
D'Odorico S, di Serego Alighieri S, Pettini M, Magain P, Nissen PE, Panagia N. 1989. *Astron. Astrophys.* 215:21–32
Danly L, Albert CE, Kuntz KD. 1993. *Ap. J.* 416:L29–31
Danly L, Lee YP, Albert CE. 1995. *Bull. Am. Astron. Soc.* 27:860 (Abstr.)
Danly L, Lockman FJ, Meade MR, Savage BD. 1992. *Ap. J. Suppl.* 81:125–61
Davies RD. 1972a. *MNRAS* 160:381–406
Davies RD. 1972b. *Nature* 237:88–91
Davies RD. 1974. In *Galactic Radio Astronomy, IAU Symp. No. 60,* ed. FJ Kerr, SC Simonson III, pp. 599–616. Dordrecht: Reidel
Davies RD. 1975. *MNRAS* 170:45p–49p
Davies RD, Buhl D, Jafolla J. 1976. *Astron. Astrophys. Suppl.* 23:181–204
Davies RD, Wright AE. 1977. *MNRAS* 180:71–88
De Boer KS. 1983. *Highlights Astron.* 6:657–63
De Boer KS, Altan AZ, Bomans DJ, Lilienthal D, Moehler S, et al. 1994. *Astron. Astrophys.* 286:925–34
De Boer KS, Morras R, Bajaja E. 1990. *Astron. Astrophys.* 233:523–26
Dickey JM, Lockman FJ. 1990 *Annu. Rev. Astron. Astrophys.* 28:215–61
Dickey JM, Terzian Y, Salpeter EE. 1979. *Ap. J.* 228:465–74
Dieter NH. 1969. *Publ. Astron. Soc. Pacific* 81:186–223
Dieter NH. 1971. *Astron. Astrophys.* 12:59–75
Dieter NH. 1972a. *Astron. Astrophys. Suppl.* 5:21–80
Dieter NH. 1972b. *Astron. Astrophys. Suppl.* 5:313–68
Doroshkevich AG, Shandarin SF. 1979. *Sov. Astron.* 23:265–71
Dyson JE, Hartquist TW. 1983. *MNRAS* 203:1233–38
Edgar RJ, Chevalier RA. 1986. *Ap. J.* 310:L27
Eichler D. 1976. *Ap. J.* 208:694–700
Einasto J, Haud U, Jôeveer M, Kaasik A. 1976. *MNRAS* 177:357–75
Encrenaz PJ, Penzias AA, Wrixon GT, Gott R III, Wilson RW. 1971. *Astron. Astrophys.* 12:16–20
Erkes JW, Philip AGD, Turner KC 1980. *Ap. J.* 238:546–53
Espresate J, Franco J, Cantó J. 1990. *Rev. Mex. Astron. Astrophys.* 18:185 (Abstr.)
Falgarone E, Phillips TG, Walker CK. 1991. *Ap. J.* 378:186–201
Ferrara A, Einaudi G. 1992. *Ap. J.* 395:475–83
Ferrara A, Field GB. 1994. *Ap. J.* 423:665–73

Fong R, Jones LR, Shanks T, Stevenson PRF, Strong AW, Dawe JA, Murray JD. 1987. *MNRAS* 224:1059–72

Franco J, Tenorio-Tagle G, Bodenheimer P, Różyczka, Mirabel IF. 1986. *Ap. J.* 333:826–39

Fujimoto M. 1979. In *The Large-Scale Characteristics of the Galaxy, IAU Symp. No. 84*, ed. WB Burton, pp. 557–66. Dordrecht: Reidel

Gardiner LT, Noguchi M. 1996. *MNRAS* 278:191–208

Gardiner LT, Sawa T, Fujimoto M. 1994. *MNRAS* 266:567–82

Gilheany S, Bates B, Catney MG, Keenan FP, Davies RD. 1990. In *Evolution in Astrophysics: IUE Astronomy in the Era of New Space Missions,* pp. 489–93, ESA

Gingold RA. 1984. *Proc. Astron. Soc. Austr.* 5:469–71

Giovanelli R. 1977. *Astron. Astrophys.* 55:395–400

Giovanelli R. 1978. In *Structure and Properties of Nearby Galaxies, IAU Symp. No. 77*, ed. EM Berkhuijsen, R Wielebinski, pp. 293–98. Dordrecht: Reidel

Giovanelli R. 1980a. *Ap. J.* 238:554–59

Giovanelli R. 1980b. *Astron. J.* 85:1155–81

Giovanelli R. 1981. *Astron. J.* 86:1468–79

Giovanelli R. 1986. *Proc. NRAO Workshop No. 12, Gaseous Halos of Galaxies,* pp. 99–114. Green Bank: Natl. Radio Astron. Obs.

Giovanelli R, Haynes MP. 1976. *MNRAS* 177:525–30

Giovanelli R, Haynes MP. 1977. *Astron. Astrophys.* 54:909–13

Giovanelli R, Haynes MP. 1989. In *Galactic and Extragalactic Radioastronomy*, ed. GL Verschuur, KI Kellerman, pp. 522–62. Berlin: Springer

Giovanelli R, Verschuur GL, Cram TR. 1973. *Astron. Astrophys. Suppl.* 12:209–62

Goerigk W, Mebold U. 1986. *Astron. Astrophys.* 162:279–82

Goerigk W, Mebold U, Reif K, Kalberla PMW, Velden L. 1983. *Astron. Astrophys.* 120:63–73

Green RF, Schmidt M, Liebert J. 1986. *Ap. J. Suppl.* 61:305–52

Greisen EW, Cram TR. 1976. *Ap. J.* 203:L119–21

Habing HJ. 1966. *Bull. Astron. Inst. Neth.* 18: 323–52

Hartmann D, Burton WB. 1997. *Atlas of Galactic Neutral Hydrogen,* pp.1–247. Cambridge: Univ. Press

Haslam CGT, Kahn FD, Meaburn J. 1971. *Astron. Astrophys.* 12:388–97

Haud U. 1988. *Astron. Astrophys.* 198:125–34

Haud U. 1990. *Astron. Astrophys.* 230:145–52

Haud U. 1992. *MNRAS* 257:707–14

Haynes MP. 1979. *Astron. J.* 84:1173–80

Haynes MP, Roberts MS. 1979. *Ap. J.* 227:767–75

Heacox WD. 1996. *Publ. Astron. Soc. Pacific* 108:591–93

Heiles C. 1979. *Ap. J.* 229:533–44

Heller P, Rohlfs K. 1994. *Astron. Astrophys.* 291:743–53

Herbstmeier U, Heithausen A, Mebold U. 1993. *Astron. Astrophys.* 272:514–32

Herbstmeier U, Kalberla PMW, Mebold U, Weiland H, Souvatzis I, et al. 1996. *Astron. Astrophys. Suppl.* 117:497–518

Herbstmeier U, Mebold U, Snowden SL, Hartmann D, Burton WB, et al. 1995. *Astron. Astrophys.* 298:606–23

Hirth W, Mebold U, Müller P. 1985. *Astron. Astrophys.* 153:249–52

Ho LC, Filippenko AV. 1995. *Ap. J.* 444:165–74

Ho LC, Filippenko AV. 1996. *Ap. J.* 463:818 (erratum)

Houck JC, Bregman JN. 1990. *Ap. J.* 352:506–21

Hulsbosch ANM. 1968. *Bull. Astron. Inst. Neth.* 20:33–39

Hulsbosch ANM. 1975. *Astron. Astrophys.* 40:1–25

Hulsbosch ANM. 1978. *Astron. Astrophys. Suppl.* 33:383–406

Hulsbosch ANM. 1979. In *The Large-Scale Characteristics of the Galaxy, IAU Symp. No. 84*, ed. WB Burton, pp. 525–33. Dordrecht: Reidel

Hulsbosch ANM, Oort JH. 1973. *Astron. Astrophys.* 22:153–54

Hulsbosch ANM, Raimond E. 1966. *Bull. Astron. Inst. Neth.* 18:413–20

Hulsbosch ANM, Wakker BP. 1988. *Astron. Astrophys. Suppl.* 75:191–236

Jaaniste J. 1984. In *Structure and Evolution of the Magellanic Clouds, IAU Symp. No. 108,* ed. S van den Bergh, KS de Boer, pp. 141–44. Dordrecht: Reidel

Jenkins EB. 1987, In *Interstellar Processes*, ed. DJ Hollenbach, HA Thronson, pp. 533–39. Dordrecht: Reidel

Jones BF, Klemola AR, Lin DNC. 1994. *Astron. J.* 107:1333–37

Kaelble A, de Boer KS, Grewing M. 1985. *Astron. Astrophys.* 143:408–12

Kahn FD. 1981. In *Investigating the Universe*, ed. FD Kahn, pp. 1–28. Dordrecht: Reidel

Kahn FD. 1990. In *Structure and Dynamics of the Interstellar Medium, IAU Coll. No. 120*, ed. G Tenorio-Tagle, M Moles, J Melnick, pp. 474–83. Berlin: Springer

Kamphuis J, Briggs F. 1992. *Astron. Astrophys.* 253:335–48

Kamphuis J, Sancisi R. 1993. *Astron. Astrophys.* 273:L31–34

Kamphuis J, Sancisi R, van der Hulst T. 1991. *Astron. Astrophys.* 244:L29–32

Kazès I, Troland TH, Crutcher RM. 1991. *Astron. Astrophys.* 245:L17–19
Keenan FP, Conlon ES, Brown PJF, Dufton PL. 1988. *Astron. Astrophys.* 192:295–98
Keenan FP, Shaw CR, Bates B, Dufton PL, Kemp SN. 1995. *MNRAS* 272:599–604
Kepner ME. 1970. *Astron. Astrophys.* 5:444–69
Kerp J, Lesch H, Mack K. 1994. *Astron. Astrophys.* 286:L13–16
Kerp J, Mack KH, Egger R, Pietz J, Zimmer F, et al. 1996. *Astron. Astrophys.* 312:67–73
Kerr FJ, Knapp GR. 1972. *Astron. J.* 77:354–59
Kerr FJ, Sullivan WT III. 1969. *Ap. J.* 158:115–22
Kim K-T, Minh YC, Hasegawa TI. 1989. *J. Korean Astron. Soc.* 22:25–30
King DL, Vladilo G, Lipman K, de Boer KS, Centurión M, et al. 1995. *Astron. Astrophys.* 300:881–89
Kulkarni SR, Dickey JM, Heiles C. 1985. *Ap. J.* 291:716–21
Kulkarni SR, Mathieu R. 1986. *Astrophys. Space Sci.* 118:531–33
Kundt W. 1987. *Astrophys. Space Sci.* 129:195–201
Kunkel WE. 1979. *Ap. J.* 228:718–33
Kunth D, Lequeux J, Sargent WLW, Viallefond F. 1994. *Astron. Astrophys.* 282:709–16
Kuntz KD, Danly L. 1996. *Ap. J.* 457:703–17
Kutyrev AS. 1986. *Astron. Tsirkular* 1396:3 (Abstr.)
Kutyrev AS, Reynolds RJ. 1989. *Ap. J.* 344:L9–11
Lepine JRD, Duvert G. 1994. *Astron. Astrophys.* 286:60–71
Li F, Ikeuchi S. 1992. *Ap. J.* 390:405–22
Lilienthal D, Hirth W, Mebold U, de Boer KS. 1992. *Astron. Astrophys.* 255:323–49
Lilienthal D, Meyerdierks H, de Boer KS. 1990. *Astron. Astrophys.* 240:487–96
Lin DNC, Jones BF, Klemola AR. 1995. *Ap. J.* 439:652–71
Lin DNC, Lynden-Bell D. 1977. *MNRAS* 181:59–81
Lin DNC, Lynden-Bell D. 1982. *MNRAS* 198:707–21
Liszt HS. 1983. *Ap. J.* 275:163–74
Liu Y. 1992. *Astron. Astrophys.* 257:505–10
Lu L, Savage BD, Sembach KR. 1994a. *Ap. J.* 426:563–76
Lu L, Savage BD, Sembach KR. 1994b. *Ap. J.* 437:L119–22
Lynden-Bell D. 1976. *MNRAS* 174:695–710
Lynden-Bell D, Lynden-Bell RM. 1995. *MNRAS* 275:429–42
MacLow M, McCray R, Norman ML. 1989. *Ap. J.* 337:141–54
Magnani L, Onello J. 1995. *Ap. J.* 443:169–80
Maloney P. 1993. *Ap. J.* 414:41–50
Mathewson DS. 1976. *R. Greenwich Obs. Bull.* 182:217–32
Mathewson DS, Cleary MN, Murray JD. 1974. *Ap. J.* 190:291–96
Mathewson DS, Ford VL. 1984. In *Structure and Evolution of the Magellanic Clouds, IAU Symp. No. 108*, ed. S van den Bergh, KS de Boer, pp. 125–36. Dordrecht: Reidel
Mathewson DS, Ford VL, Schwarz MP, Murray JD. 1979. In *The Large-Scale Characteristics of the Galaxy, IAU Symp. No. 84,* ed. WB Burton, pp. 547–56. Dordrecht: Reidel
Mathewson DS, Meng SY, Brundage WD, Kraus JD. 1966. *Astron. J.* 71:863 (Abstr.)
Mathewson DS, Schwarz MP. 1976. *MNRAS* 176:47p–51p
Mathewson DS, Schwarz MP, Murray JD. 1977. *Ap. J.* 217:L5–8
Mathewson DS, Wayte SR, Ford VL, Ruan K. 1987. *Proc. Astron. Soc. Austr.* 7:19–25
McCammon D, Sanders WT. 1990. *Annu. Rev. Astron. Astrophys.* 28:657–85
McGee RX, Newton LM. 1986. *Proc. Astron. Soc. Austr.* 6:358–85
Meaburn J. 1965. *Nature* 207:179–80
Mebold U. 1991. In *The Magellanic Clouds, IAU Symp. No. 148* , ed. R Haynes, D Milne, pp. 463–68. Dordrecht: Reidel
Mebold U, Greisen EW, Wilson W, Haynes RF, Herbstmeier U, Kalberla PMW. 1991. *Astron. Astrophys.* 251:L1–4
Mebold U, de Boer KS, Wennmacher A. 1992. In *New Windows on the Universe II*, 11th Eur. Reg. Meet. IAU. 11:413–31
Meng SY, Kraus JD. 1970. *Astron. J.* 75:535–62
Meurer GR, Bicknell GV, Gingold RA. 1985. *Proc. Astron. Soc. Austr.* 6:195–98
Meyer DM, Roth KC. 1991. *Ap. J.* 383:L41–44
Meyerdierks H. 1991. *Astron. Astrophys.* 251:269–75
Meyerdierks H. 1992. *Astron. Astrophys.* 253:515–20
Mirabel IF. 1981a. *Ap. J.* 247:97–103
Mirabel IF. 1981b. *Ap. J.* 250:528–33
Mirabel IF. 1982. *Ap. J.* 256:112–19
Mirabel IF, Cohen RJ, Davies RD. 1979. *MNRAS* 186:433–51
Mirabel IF, Franco ML. 1976. *Astrophys. Space Sci.* 42:483–94
Mirabel IF, Morras R. 1984. *Ap. J.* 279:86–92
Mirabel IF, Morras R. 1990. *Ap. J.* 356:130–34
Moore B, Davis M. 1994. *MNRAS* 270:209–21
Morras R. 1982. *Astron. Astrophys.* 115:249–52
Morras R. 1983. *Astron. J.* 88:62–66
Morras R. 1985. *Astron. J.* 90:1801–6
Morras R, Bajaja E. 1983. *Astron. Astrophys. Suppl.* 51:131–34
Morras R, Bajaja E. 1986. *Rev. Mex. Astron. Astrophys.* 13:69–71
Muller CA, Oort JH, Raimond E. 1963. *C. R. Acad. Sci. Paris* 257:1661
Muller CA, Raimond E, Schwarz UJ, Tolbert

CR. 1966. *Bull. Astron. Inst. Neth. Suppl.* 1:213–44
Murai T, Fujimoto M. 1980. *Publ. Astron. Soc. Jpn.* 32:581–603
Murphy EM, Lockman FJ, Savage BD. 1995. *Ap. J.* 447:642–45
Münch G, Pitz E. 1990. In *Structure and Dynamics of the Interstellar Medium, IAU Coll. No. 120,* ed. G Tenorio-Tagle, M Moles, J Melnick, pp. 373–82. Berlin: Springer
Münch G, Zirin H. 1961. *Ap. J.* 133:11–28
Oort JH. 1966. *Bull. Astron. Inst. Neth.* 18:421–38
Oort JH. 1970. *Astron. Astrophys.* 7:381–404
Oort JH. 1978. In *Problems of Physics and the Evolution of the Universe,* ed. L Mirzoyan, pp. 259–80, Yerevan: Acad. Sci. Armenian SSR
Oort JH, Hulsbosch ANM. 1978. In *Astronomical Papers Dedicated to B. Strömgren,* ed. A Reiz, T Anderson, pp. 409–28. Copenhagen: Copenhagen Univ. Observatory
Payne HE, Dickey JM, Salpeter EE, Terzian Y. 1978. *Ap. J.* 221:L95–98
Payne HE, Salpeter EE, Terzian Y. 1980. *Ap. J.* 240:499–513
Philip AGD. 1976. *Bull. Am. Astron. Soc.* 8:352 (Abstr.)
Pietz J, Kerp J, Kalberla PMW, Mebold U, Burton WB, Hartmann D. 1996. *Astron. Astrophys.* 308:L37–40
Recillas-Cruz E. 1982. *MNRAS* 201:473–78
Reynolds RJ. 1987. *Ap. J.* 323:553–56
Reynolds RJ. 1992. *Ap. J.* 392:L35–38
Reynolds RJ. 1996. In *The Physics of Galactic Halos, 156th WE-Heraeus-Sem.,* ed. H Lesch, RJ Dettmar, U Mebold, R Schlickeiser. Berlin: Akademie Verlag
Rickard JJ. 1968. *Ap. J.* 152:1019–42
Robertson JG, Schwarz UJ, van Woerden H, Murray JD, Morton DC, Hulsbosch ANM. 1991. *MNRAS* 248:508–14
Rohlfs R, Herbstmeier U, Mebold U, Winnberg A. 1989. *Astron. Astrophys.* 211:402–8
Saraber MJM, Shane WW. 1974. *Astron. Astrophys.* 36:365–68
Savage BD, Lu L, Bahcall JN, Bergeron J, Boksenberg A, et al. 1993. *Ap. J.* 413:116–36
Savage BD, Sembach KR, Lu L. 1995. *Ap. J.* 449:145–55
Savage BD, de Boer KS. 1981. *Ap. J.* 243:460–84
Schchekinov YA. 1982. *Astrophysics* 16:159–63
Schulman E. 1996. *Publ. Astron. Soc. Pacific* 108:460
Schulman E, Bregman JN, Brinks E, Roberts MS. 1996. *Astron. J.* 112:960–71
Schulman E, Bregman JN, Roberts MS. 1994. *Ap. J.* 423:180–89
Schwarz UJ, Oort JH. 1981. *Astron. Astrophys.* 101:305–14
Schwarz UJ, Sullivan WT III, Hulsbosch ANM. 1976. *Astron. Astrophys.* 52:133–38
Schwarz UJ, Wakker BP, van Woerden H. 1995. *Astron. Astrophys.* 302:364–81
Sembach KR, Savage BD. 1992. *Ap. J. Suppl.* 83:147–201
Sembach KR, Savage BD, Lu L. 1995a. *Ap. J.* 439:672–81
Sembach KR, Savage BD, Lu L, Murphy EM. 1995b. *Ap. J.* 451:616–32
Sembach KR, Savage BD, Massa D. 1991. *Ap. J.* 372:81–96
Shapiro PR, Benjamin RA. 1991. *Publ. Astron. Soc. Pacific* 103:923–27
Shapiro PR, Field GB. 1976. *Ap. J.* 205:762–65
Shatsova RB. 1984. *Astrophysics* 19:441–48
Shatsova RB. 1993. *Astrophys. Space Sci.* 201:91–105
Shostak GS. 1977. *Astron. Astrophys.* 54:919–24
Silk J, Siluk RS. 1973. *Astrophys. Letters* 13:143–45
Smith GP. 1963. *Bull. Astron. Inst. Neth.* 17:203–8
Snowden SL, McCammon D, Verter F. 1993. *Ap. J.* 409:L21–24
Snowden SL, Mebold U, Hirth W, Herbstmeier U, Schmitt JHMM. 1991. *Science* 252:1529–32
Sofue Y. 1994. *Publ. Astron. Soc. Jpn.* 46:431–40
Sofue Y, Wakamatsu K. 1991. *Publ. Astron. Soc. Jpn.* 43:L57–64
Songaila A. 1981a. *Ap. J.* 243:L19–22
Songaila A. 1981b. *Ap. J.* 248:945–55
Songaila A, Blades JC, Hu EM, Cowie LL. 1986. *Ap. J.* 303:198–215
Songaila A, Bryant W, Cowie LL. 1989. *Ap. J.* 345:L71–73
Songaila A, Cowie LL, Weaver HF. 1988. *Ap. J.* 329:580–88
Songaila A, York DG, Cowie LL, Blades JC. 1985. *Ap. J.* 293:L15–18
Spitzer L. 1956. *Ap. J.* 124:20–34
Stark AA, Gammse CF, Wilson RW, Bally J, Linke RA, et al. 1992. *Ap. J. Suppl.* 79:77–104
Suchkov AA, Schchekinov YA. 1974. *Astrophysics* 10:159–62
Tamanaha CM. 1995. *Ap. J.* 450:638–62
Tanaka KI. 1981. *Publ. Astron. Soc. Jpn.* 33:247–64
Tanaka KI, Hamajima K 1982. *Publ. Astron. Soc. Jpn.* 34:417–22
Tenorio-Tagle G, Franco J, Bodenheimer P, Różyczka M. 1987. *Astron. Astrophys.* 179:219–30
Tenorio-Tagle G, Franco J, Bodenheimer P,

Różyczka M. 1988. *Astron. Astrophys.* 193:372 (erratum)
Tomisaka K. 1991. In *The Interstellar Disk-Halo Connection in Galaxies, IAU Symp. No. 144*, ed. H Bloemen, pp. 407–16. Dordrecht: Reidel
Tomisaka K. 1992. *Publ. Astron. Soc. Jpn.* 44:177–91
Trifalenkov IA. 1993. *Astron. Rep.* 37:327–28
Tufte SL, Reynolds RJ, Haffner LM, Jaehring K. 1996. *Bull. Am. Astron. Soc.* 128:890 (Abstr.)
van der Hulst JM. 1996. In *The Physics of Galactic Halos, 156th WE-Heraeus-Sem.*, ed. H Lesch, RJ Dettmar, U Mebold, R Schlickeiser. Berlin: Akademie Verlag
van der Hulst T, Sancisi R. 1988. *Astron. J.* 95:1354–59
van Kuilenburg J. 1972a. *Astron. Astrophys.* 16:276–81
van Kuilenburg J. 1972b. *Astron. Astrophys. Suppl.* 5:1–20
van Woerden H. 1993. In *Luminous High-Latitude Stars*, ASP Conf. Ser. 45, ed. DD Sasselov, pp. 11–29, San Francisco:ASP
van Woerden H, Rougoor GW, Oort JH. 1957. *C. R. Acad. Sci. Paris* 244:1691
van Woerden H, Schwarz UJ, Hulsbosch ANM. 1985. In *The Milky Way Galaxy, IAU Symp. No. 106*, ed. H van Woerden, RJ Allen, WB Burton, pp. 387–408. Dordrecht: Reidel
Verschuur GL. 1969. *Ap. J.* 156:771–77
Verschuur GL. 1973a. *Astron. Astrophys.* 22:139–51
Verschuur GL. 1973b. *Astron. Astrophys.* 27:407–11
Verschuur GL. 1975. *Annu. Rev. Astron. Astrophys.* 13:257–93
Verschuur GL. 1990. *Ap. J.* 361:497–510
Verschuur GL. 1993a. *Ap. J.* 409:205–33
Verschuur GL. 1993b. *Astron. J.* 106:2580–86
Verschuur GL, Magnani L. 1994. *Astron. J.* 107:287–97
Vogelaar M, Wakker B. 1994. *Astron. Astrophys.* 291:557–68
Wakker BP. 1991a. *Astron. Astrophys.* 250:499–508
Wakker BP. 1991b. *Astron. Astrophys. Suppl.* 90:495–511
Wakker BP. 1991c. In *The Interstellar Disk-Halo Connection in Galaxies, IAU Symp. No. 144*, ed. H Bloemen, pp. 27–40. Dordrecht: Reidel
Wakker BP, Boulanger F. 1986. *Astron. Astrophys.* 170:84–90
Wakker BP, Bregman JN. 1990. In *Interstellar Neutral Hydrogen at High Velocities*, PhD thesis BP Wakker, Rijks Univ. Groningen, ch. 5
Wakker BP, Broeils AH, Tilanus RPJ, Sancisi R. 1989. *Astron. Astrophys.* 226:57–58
Wakker BP, Howk C, van Woerden H, Schwarz UJ, Beers TC, et al. 1996a. *Ap. J.* 473:834–48
Wakker BP, Murphy EM, van Woerden H, Dame T. 1997. *Astron. Astrophys.* In press
Wakker BP, Schwarz UJ. 1991. *Astron. Astrophys.* 250:484–98
Wakker BP, Vijfschaft B, Schwarz UJ. 1991. *Astron. Astrophys.* 249:233–38
Wakker BP, van Woerden H. 1991. *Astron. Astrophys.* 250:509–32
Wakker BP, van Woerden H, de Boer KS, Kalberla PMW. 1997. *Astron. Astrophys.* Submitted
Wakker BP, van Woerden H, Schwarz UJ, Peletier RF, Douglas NG. 1996b. *Astron. Astrophys.* 306:L25–28
Wannier P, Wrixon GT. 1972. *Ap. J.* 173:L119–23
Wannier P, Wrixon GT, Wilson RW. 1972. *Astron. Astrophys.* 18:224–31
Watanabe T. 1981. *Astron. J.* 86:30–35
Wayte SR. 1989. *Proc. Astron. Soc. Austr.* 8:195–203
Wayte SR. 1990. *Ap. J.* 355:473–95
Weaver H. 1979. In *The Large-Scale Characteristics of the Galaxy, IAU Symp. No. 84*, ed. WB Burton, pp. 295–300. Dordrecht: Reidel
Weiner BJ, Williams TB. 1996. *Astron. J.* 111:1156–63
Welsh BY, Craig N, Roberts B. 1996. *Astron. Astrophys.* 308:428–32
Wesselius PR, Fejes I. 1973. *Astron. Astrophys.* 24:15–34
West KA, Pettini M, Penston MV, Blades JC, Morton DC. 1985. *MNRAS* 215:481–97
Wolfire MG, McKee CF, Hollenbach D, Tielens AGGM. 1995. *Ap. J.* 453:673–84
Wright MCH. 1974. *Astron. Astrophys.* 31:317–22
Wright MCH. 1979a. *Ap. J.* 233:35–38
Wright MCH. 1979b. *Ap. J.* 234:27–32
Zwaan M, Briggs F, Sprayberry D. 1996. *Proc. Astron. Soc. Austr.* In press

*Annu. Rev. Astron. Astrophys. 1997. 35:267–307*


# LOW SURFACE BRIGHTNESS GALAXIES

*Chris Impey*
Steward Observatory, University of Arizona, Tucson, Arizona 85721;
e-mail: cimpey@as.arizona.edu

*Greg Bothun*
Physics Department, University of Oregon, Eugene, Oregon 97403;
e-mail: nuts@moo2.uoregon.edu

KEY WORDS: physical properties, galaxy morphology, cosmology, luminosity function, quasar absorption

ABSTRACT

The properties of galaxies that are lower in surface brightness than the dark night sky are reviewed. There are substantial selection effects against the discovery of galaxies that are unevolved or diffuse; these systems are missing from most wide field catalogs. Low surface brightness galaxies make up a significant amount of the luminosity density of the local universe. They contribute substantial but poorly determined amounts to the census of baryons and dark matter. Low surface brightness galaxies are also relevant to the interpretation of quasar absorption lines and to the understanding of rapidly evolving galaxy populations in the more distant universe. Theories of galaxy formation and evolution must accomodate the properties of these diffuse stellar systems.

## INTRODUCTION

The Cosmological Principle states that the universe will appear homogeneous and isotropic to a typical observer, but one of the deeper implications of this principle is rarely examined. Observational cosmology is usually probed via catalogs of galaxies. Although much of the universe is dark, galaxies are the prime repositories of shining baryonic matter. Galaxy properties are used to measure the size and shape of the universe; deviations from Hubble flow and image distortions through lensing are used to map out the dark matter distribution. If galaxies are to be used as effective cosmological probes, then our



0066-4146/97/0915-0267$08.00

catalogs must be complete and homogeneous, in accord with the Cosmological Principle. Yet the detectability of galaxies depends very much on the cosmic environment. An observer whose star was in a giant molecular cloud or near the center of an elliptical galaxy would have difficulty discovering external galaxies and so would perceive the universe quite differently from us. Put simply, we only catalog the galaxies we can see.

Observational bias in the selection of galaxies dates back to Messier and Herschel. Galaxies are diffuse objects selected in the presence of a contaminating signal: the brightness of the night sky. Below a certain percentage of the night sky brightness, no galaxy can be detected. Above this limiting isophote, a galaxy must present a large enough angular size to be distinguished from a star. For a given luminosity and radial profile, a galaxy will be visible to the maximum distance at a surface brightness level substantially higher than the limiting isophote. High surface brightness (HSB) galaxies are small because they are intrinsically compact, and low surface brightness (LSB) galaxies are small because they mostly fall below the limiting isophote. The night sky essentially acts as a filter, which, when convolved with the true population of galaxies, gives the population of galaxies we observe. This censorship due to surface brightness was first commented on by Zwicky (1957) and was further investigated by Arp (1965) and Disney (1976).

This review deals primarily with LSB galaxies in the local universe ($z \lesssim 0.1$). There is no convention for defining LSB; discussion is mostly restricted to galaxies with central surface brightness fainter than 23 $B$ mag arcsec$^{-2}$. Stellar systems more luminous than $M_B = -14$ are considered; for reviews of LSB dwarf galaxies in the Local Group and beyond, see Ferguson & Binggeli (1994), Irwin & Hatzidimitriou (1995), and Mateo (1996). First, the structural properties of galaxies are summarized, and the influence of surface brightness selection is described. Next, the potential incompleteness of galaxy luminosity functions is considered. Surveys for LSB galaxies are reviewed. We then establish the significance of LSB galaxies for the census of light and matter in the universe, for the formation and evolution of galaxies, and for the statistics of quasar absorption. This historically neglected population has important implications for virtually every aspect of observational cosmology. Unless otherwise noted, $H_0 = 100$ km s$^{-1}$ Mpc$^{-1}$ and $q_0 = 0.5$ is assumed, or results are expressed in terms of $h_{100} = H_0/100$.

# THE TRUE POPULATION OF GALAXIES

## *Surface Brightness Selection and Galaxy Visibility*

For selecting and cataloging galaxies, the simplest assumption is that galaxies form a univariate distribution in apparent brightness. Figure 1*a* shows the

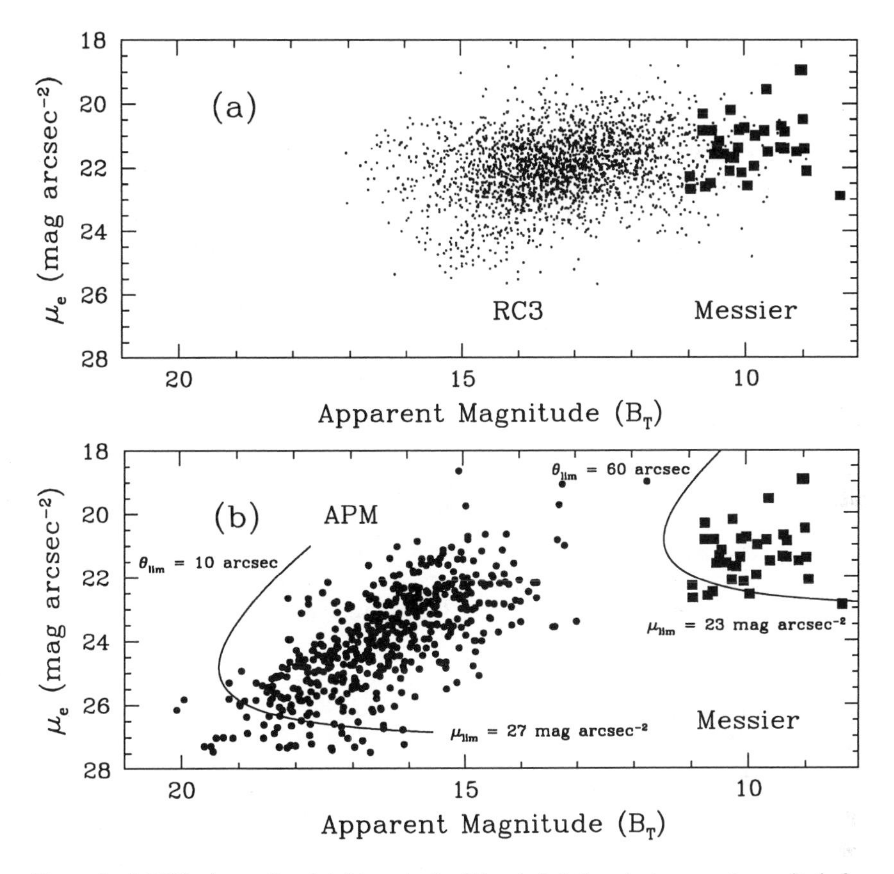


*Figure 1* (*a*) Effective surface brightness in the B band plotted against apparent magnitude for Messier galaxies (*filled squares*) and RC3 galaxies (*small dots*). (*b*) Effective surface brightness plotted against apparent magnitude for Messier galaxies (*filled squares*) and LSB galaxies with redshifts from the APM survey (*filled circles*). The curves show the approximate selection functions for each sample. (From de Vaucouleurs et al 1991, Impey et al 1996.)

distribution in apparent brightness and effective surface brightness in *B* of the galaxies from the Messier catalog and a much larger number of galaxies from the *Third Reference Catalog of Bright Galaxies* (hereafter RC3) (de Vaucouleurs et al 1991).

Messier produced a catalog of the most prominent nebulae in the sky as a guide to stationary objects that comet hunters should avoid. The intervening two centuries have yielded photographic all-sky galaxy samples that reach apparent brightness that is hundreds of times fainter but not substantially fainter in average surface brightness. Despite the inhomogeneous nature of these samples, these data support the idea of a characteristic surface brightness for galaxies.

SPECIFYING A GALAXY CATALOG Any galaxy catalog is specified by three parameters: the surface brightness of the limiting isophote ($\mu_{\rm lim}$), the isophotal magnitude limit ($m_{\rm lim}$), and the isophotal size limit in arcseconds at the limiting isophote ($\theta_{\rm lim}$). Assuming symmetric galaxies with no internal absorption, the fraction of the galaxy flux detected above the limiting isophote is

$$S_{\rm lim}/S_{\rm tot} = 1 - [1 + 0.4 \ln 10(\mu_{\rm lim} - \mu_0)]\, {\rm dex}[-0.4(\mu_{\rm lim} - \mu_0)] \quad (1)$$

for spiral disks and

$$S_{\rm lim}/S_{\rm tot} = 1 - \left\{1 + \sum_{n=1}^{7} [0.4 \ln 10(\mu_{\rm lim} - \mu_0)]^n/n!\right\} + {\rm dex}[-0.4(\mu_{\rm lim} - \mu_0)] \quad (2)$$

for ellipticals and bulges. This formalism was first presented by Disney & Phillipps (1983). These relations have maxima that occur at $\mu_{\rm lim} - \mu_0 = 5.37$ for spiral disks and $\mu_{\rm lim} - \mu_0 = 10.93$ for ellipticals, where $\mu_0$ is the central surface brightness.

The observed distribution of galaxy central surface brightness has long been known to be peaked. In the earliest work on the subject, the maximum value for spirals was at $\mu_0 = 21.6 \pm 0.4$ mag arcsec$^{-2}$ in $B$ (Freeman 1970), and the maximum value for ellipticals was at $\mu_0 = 14.8 \pm 0.9$ mag arcsec$^{-2}$ in $B$ (Fish 1964). Subsequently, the distributions have been found to be somewhat broader, with a dispersion of $\sigma(\mu_0) = 2$ mag arcsec$^{-2}$ for spirals (Boroson 1981, van der Kruit 1987, Bosma & Freeman 1993, Courteau 1996) and a dispersion of $\sigma(\mu_0) = 1.5$ mag arcsec$^{-2}$ for ellipticals (Kormendy 1977). The distributions may reflect an intrinsic property of the galaxy population, a selection effect, or both.

At this point, we make the simplifying assumption that LSB galaxies are suitably described by exponential radial profiles. Late-type galaxies are composite stellar systems, where the ratio of disk to bulge flux is $D/B = 0.28\,(s/r_e)^2\,[(I_0)_{\rm disk}/(I_e)_{\rm bulge}]$, where the disk exponential scale length $s$ and the bulge effective radius $r_e$ are measured in arcseconds. The assumption of exponential profiles is only valid for disk-dominated systems, which is appropriate for most deep galaxy surveys. The Hubble Space Telescope (HST) Medium Deep Survey finds ~80% of the $I$ band selected sample to be well fit by exponentials (Im et al 1995b). Among low luminosity galaxies, dwarf ellipticals and dwarf irregulars have long been known to have exponential profiles (Binggeli et al 1984, Gallagher & Hunter 1984). More recently, the low luminosity LSB galaxies in clusters have been shown to be well fit by exponential profiles (e.g. Caldwell

& Bothun 1987, Impey et al 1988, Davies et al 1990, Bothun et al 1991). All of the LSB galaxies in Figure 1*b* have redshifts, and their effect on the luminosity function is considered in the next section.

In Figure 1*b*, the Messier galaxies are compared with the LSB galaxies from a recent survey carried out using Automated Plate Measuring (APM) machine scans of UK Schmidt sky survey plates. This survey is used here to illustrate the issues of surface brightness selection effects because it is the largest set of field LSB galaxies for which redshifts have been published (Impey et al 1996). The superimposed curves represent the spiral selection function described by Equation 1. As surveys probe to smaller fractions of the night sky brightness, the accessible parameter space for detecting galaxies expands. Objects are missed at faint $\mu_0$ owing to the poor contrast with the sky brightness. Objects are missed at faint $m_{\rm tot}$ because they become smaller than the limiting angular size at the limiting isophote. Objects are missed (in principle) at bright $\mu_0$ and faint $m_{\rm tot}$ because they become difficult to resolve from stars.

Practical limitations also exist in the detection of galaxies. For the APM survey, Sprayberry et al (1996) found 50% completeness for $\mu_0 = 24$ $B$ mag arcsec$^{-2}$ and $s = 10$ arcsec, or $\mu_0 = 23$ $B$ mag arcsec$^{-2}$ and $s = 3$ arcsec. Photographic amplification of UK Schmidt plates reaches a limiting isophote that is a factor of two deeper (Malin 1978, Impey et al 1988). With the new Kodak Technical Pan 4415 emulsion, these limits will become $\mu_0 = 25.5$ mag arcsec$^{-2}$ for $s = 10$ arcsec in the red, or $\mu_0 = 25$ $R$ mag arcsec$^{-2}$ for $s = 3$ arcsec, again in the $R$ band (Schwartzenberg et al 1995a). Digital coaddition of scanned films should yield an additional improvement factor of 2–3. The detection of faint, extended objects with CCDs is limited by the photon flux and the quality of the flat field. Small-scale variations in background can be caused by the detector (i.e. fringing at long wavelengths) or the sky (i.e. faint sources in the sky flat). In practice, it is difficult to achieve flat fields better than 0.1–0.2% of the sky brightness (Tyson 1988a). In addition, Capaccioli & de Vaucouleurs (1983) have argued that systematic errors preclude reliable surface photometry much fainter than 28 mag arcsec$^{-2}$. A typical limit for CCD surveys is $\mu_0 = 27$ mag arcsec$^{-2}$ and $s = 10$ arcsec, or $\mu_0 = 26.5$ mag arcsec$^{-2}$ and $s = 3$ arcsec (Schwartzenberg et al 1995b), although galaxy classification is extremely difficult at that level. Phillipps & Disney (1985) have argued that the limit for detecting LSB objects by indirect methods, such as star and galaxy counts, can be as low as 29–30 mag arcsec$^{-2}$. The lowest backgrounds for the detection of LSB objects are in the vacuum ultraviolet and the near infrared, as seen from space (Wright 1985, O'Connell 1987, O'Neil et al 1996).

GALAXY VISIBILITY The visibility of galaxies is governed by the point at which they are lost from the sample by either falling below the angular size or the flux limit of the survey. Using the expression for the radial intensity profile, the limiting angular diameter at the isophotal limit can be calculated (see Allen & Shu 1979):

$$\theta_{\rm lim} = 1.84s(\mu_{\rm lim} - \mu_0), \tag{3}$$

and the case of a flux limit is

$$S_{\rm lim} = 2\pi s^2 \mu_0 \{1 - [1 + 0.92(\mu_{\rm lim} - \mu_0)]\, e^{-0.92(\mu_{\rm lim} - \mu_0)}\}. \tag{4}$$

We now assume that the number of galaxies observed is proportional to the volume sampled. This implies a survey large enough to include a fair sample of galaxies in the local universe. It also implies that galaxies that cover the full range of structural parameters can be observed at some point in the volume (i.e. no correction factor can be applied for galaxy types that are not detected). In this case $n(\mu_0) \propto \theta_{\rm lim}^{-3}$ for the angular size limit, and $n(\mu_0) \propto S_{\rm lim}^{-3/2}$ for the flux limit. The inevitable consequence of this visibility argument is that the detectability of galaxies is a strong function of surface brightness. The corollary is that the limiting distance to which a galaxy can be observed is a strong function of surface brightness, since $d_{\rm lim} \propto \theta_{\rm lim}^{-1}$ for an angular size limit and $d_{\rm lim} \propto S_{\rm lim}^{-1/2}$ for a flux limit. Davies (1990) has modeled the visibility of inclined two-component bulge/disk galaxies. Although the selection function no longer has a simple analytical form, the basic features of the more simple treatment are confirmed.

For a catalog with an angular size cutoff, the selection criteria are the isophotal size, $\theta_{\rm lim}$, and the limiting isophote at which that angular size is measured, $\mu_{\rm lim}$. The volume sampled is

$$V(h, \mu_0) \propto s^3(\mu_{\rm lim} - \mu_0), \tag{5}$$

where $s$ and $\mu_0$ are the structural parameters that completely define an exponential profile. For a catalog with a flux limit, the selection criteria are the magnitude limit, $m_{\rm lim}$, and the limiting isophote within which the flux is measured. Because catalogs limited by total flux do not exist, we must also specify $S_{\rm lim}/S_{\rm tot}$, the fraction of the total flux that is measured down to the limiting isophote. Then the visibility volume is

$$V(h, \mu_0) \propto s^3 10^{-0.6(\mu_{\rm lim} - \mu_0^*)} \times \{1 + [1 + 0.92(\mu_{\rm lim} - \mu_0)] \exp[-0.92(\mu_{\rm lim} - \mu_0)]^{3/2}\}, \tag{6}$$

where $\mu_0^*$ is a fiducial luminosity analogous to $L^*$, conveniently taken to be the Freeman value. The last two equations reflect the formalism of McGaugh et al (1995); Disney & Phillipps (1983) had previously defined visibility functions in terms of luminosity and surface brightness. Because galaxies are not selected by total magnitude, and the proportion of the galaxy flux detected depends on the surface brightness, it is simpler to deal with the visibility functions defined in terms of structural parameters. As expected, the range of surface brightness for nearby galaxies is substantially larger than for more distant galaxies (Impey et al 1996). This is exactly as expected under the influence of surface brightness selection, where galaxies of intermediate surface brightness can be seen to the largest distances (see also Phillipps et al 1990). Regardless of whether the catalog is limited by angular size or isophotal flux, the visibility increases monotonically with larger scale length and brighter surface brightness. The volume sampling functions are very steep; the volume over which low surface brightness galaxies can be observed is small, so the correction to the space density is correspondingly large.

THE INTRINSIC SURFACE BRIGHTNESS DISTRIBUTION The intrinsic surface brightness distribution can be recovered only if a survey has selection parameters $\mu_{\rm lim}$, $m_{\rm lim}$, and $\theta_{\rm lim}$, which have been rigorously applied to the catalog. McGaugh (1996) has applied these methods to several samples, notably the diameter-limited Fornax catalog of Davies (1990), and has shown that the presumption of a preferred surface brightness at the Freeman value is clearly invalid. The distribution function cuts off sharply at high surface brightness (HSB), but it has a long tail that is populated down the faintest detectable galaxies. The lack of HSB disk galaxies is real and not a selection effect. Below $\mu_0 \approx 21.5$ mag arcsec$^{-2}$, the numbers fall to about one fourth to one fifth of the peak at $\mu_0 \approx 23.5$ mag arcsec$^{-2}$, but there is no sign of a sharp turndown. No cutoff in the surface brightness distribution has been found down to the limit of deep CCD surveys, $\mu_0 \sim 26$ mag arcsec$^{-2}$ (Schwartzenberg et al 1995b, Dalcanton 1995).

The studies just referred to do not have redshifts, so they must make the implicit assumption of a separable bivariate luminosity function, $\Phi(s, \mu)$, i.e. that scale length and central surface brightness are not correlated. The surveys of de Jong (1995) and Impey et al (1996) do not have to make this assumption because redshifts allow absolute scale lengths to be calculated. For the APM sample, the detection probability as a function of $\mu_0$ and $s$ was estimated, and the observed distribution $n(\mu_0)$ was corrected for this incompleteness. For the raw data, $\langle V/V_{\rm max}\rangle = 0.18 \pm 0.06$. After weighting by the inverse probability of detection, $\langle V/V_{\rm max}\rangle = 0.44 \pm 0.06$, which is consistent with completeness brighter than $\mu_0 = 25$ $B$ mag arcsec$^{-2}$, above which the corrections become very large and uncertain. The result is a distribution with a broad peak at $\approx$21.5 mag arcsec$^{-2}$, which falls by a factor of 4–5 by $\mu_0 \approx 23.5$ mag arcsec$^{-2}$, but then continues with no sign of a cutoff (Sprayberry et al 1996). The result is inconsistent with the traditional description of a Gaussian distribution of surface brightness. The number of disk galaxies falls slowly as a function of $\mu_0$ but with no limit apart from that imposed by observational selection.

### *The Local Galaxy Luminosity Function*

The luminosity function of galaxies is fundamental to observational cosmology. As emphasized by Binggeli et al (1988), there is no universal luminosity function; the space density of galaxies is a function of Hubble type and the density of the environment. Accurate knowledge of the luminosity function is required to test cosmological world models and to understand galaxy evolution. Presumably, the shape of the luminosity function also contains "frozen in" clues to the process of galaxy formation. For example, if Freeman's law for spiral disks is correct, there is only one parameter relevant to galaxy selection, as only variations in size act to modulate variations in luminosity. This would require all the physical processes of galaxy formation and evolution to conspire to result in one specific value of central surface brightness for all galaxies.

THE GALAXY LUMINOSITY FUNCTION The parameterization of the number of galaxies per unit volume according to Schechter (1976) is

$$\phi(L)dL = \phi^*(L/L^*)^\alpha e^{-L/L^*} d(L/L^*), \tag{7}$$

where $\phi^*$ characterizes the space density of galaxies, $L^*$ is the luminosity above which galaxies are rare, and $\alpha$ is the asymptotic slope of the faint end of the luminosity function.

Field luminosity functions in the local universe were first reviewed by Felten (1977) and Binggeli et al (1988). More recently, large photographic surveys have been used to define the local luminosity function with better statistics. Efstathiou et al (1988) found $\phi^* = 0.016$, $M^* = -19.7$, and $\alpha = -1.07$. Loveday et al (1992) found $\phi^* = 0.014$, $M^* = -19.5$, and $\alpha = -0.97$ for the Stromlo-APM survey. Marzke et al (1994b) found $\phi^* = 0.010$, $M^* = -18.8$, and $\alpha = -1.07$ for the Center for Astrophysics survey, and da Costa et al (1994) found $\phi^* = 0.015$, $M^* = -19.5$, and $\alpha = -1.20$ for its southern equivalent, the Southern Sky Redshift Survey. The first extensive field galaxy survey to be selected from CCD images is the Las Campanas Redshift Survey; Lin et al (1996) present the parameters $\phi^* = 0.019$, $M^* = -20.3$, and $\alpha = -0.70$. For the photographic surveys, $M^*$ is measured in absolute $B$ magnitudes; the Las Campanas survey used Gunn-$r$ band selection. All surveys quote $L^*$ in the equivalent in solar units, and the normalization $\phi^*$ in units of $h_{100}^3$ Mpc$^{-3}$ mag$^{-1}$. Note that these recent determinations differ by more than a factor of two in both normalization, $\phi^*$, and characteristic luminosity, $M^*$. See also the discussion by Ellis (1997) in this volume.

THE FAINT END SLOPE IN CLUSTERS AND THE FIELD The faint end slope of the galaxy luminosity function is most easily studied in nearby clusters, where cluster membership can be decided with reasonable reliability in the absence of redshifts, using morphology and two-dimensional spatial distribution as a guide.

Sandage et al (1985) revealed the existence of large numbers of faint galaxies in the Virgo cluster. Ferguson & Sandage (1988) followed that work with a similar survey of the Fornax cluster, deriving a faint end slope of $\alpha \approx -1.3$ for both clusters. Correcting for surface brightness selection effects, Impey et al (1988) deduced $\alpha \approx -1.6$ for the Virgo cluster, and Bothun et al (1991) found $\alpha \approx -1.5$ for the Fornax cluster. Tyson & Scalo (1988) postulated a large population of gas-rich dwarfs that could substantially steepen the luminosity function. In both clusters, the population that steepens the luminosity function has low luminosity ($M_B > -16$), moderate scale length (3–10 kpc), and low surface brightness (LSB) ($\mu_e > 25$ $B$ mag arcsec$^{-2}$). These studies were all based on photographic material. CCD surveys have advanced our census of intrinsically faint galaxy populations. The dwarf luminosity function in Abell 963 has an overall slope of $\alpha \approx -1.8$ (Driver et al 1994a), Bernstein et al (1995) observed a faint end slope of $\alpha \approx -1.4$ in the core of the Coma cluster, and Trentham (1997) found slopes of $-1.8 < \alpha < -1.6$ for three spiral-rich, poor clusters. The latter two results reach down to the luminosity of the Local Group dwarf spheroidals. With smaller samples and a less secure background correction, de Propris et al (1995) found even steeper slopes for four rich Abell clusters, $\alpha \approx -2.2$.

A magnitude-limited redshift survey must sample several thousand galaxies to include a few dozen fainter than $M^* + 5$. Our knowledge of the field galaxy luminosity function fainter than $M_B = -16$ is poor for two reasons. Shallow, wide field surveys have small effective volumes for the detection of low luminosity systems. Also, regardless of the exact relationship between luminosity and surface brightness, magnitude-limited redshift surveys must be increasingly censored by surface brightness selection effects at $M_B > -16$.

Driver & Phillipps (1996) find the Stromlo-APM survey to be entirely consistent with a faint end slope of $\alpha = -1.5$. A reanalysis of the CfA redshift survey by Marzke et al (1994a) suggests that the faint end slope is as high as $-1.85$ for low luminosity spirals and irregulars. Marzke & da Costa (1997) deduce a Schechter function with $\alpha = -1.5$ for the blue galaxies ($B$–$R < 1.3$) in the Southern Sky Redshift Survey, and the deepest part of the sample yields a slope as steep as $\alpha = -1.7$. The ESO Slice Survey has also been compared directly to the Stromlo-APM survey by Zucca (1997), who finds a higher normalization than Loveday et al (1992), and a faint end with $\alpha = -1.6$ due entirely to LSB and compact star-forming galaxies. An extension of the Las

Campanas Redshift Survey to correct for excluded LSB galaxies also results in a steeper faint end slope (JJ Dalcanton & SA Schectman, in preparation). Bershady et al (1997) have carried out a local ($z < 0.1$) *UBVRI* survey down to $\mu_B = 24.5$ *B* mag arcsec$^{-2}$ over 1 deg$^2$ and find blue field galaxies to have a steep slope of $\alpha = -1.6$. The Texas Deep Sky Survey is a *UBVRI* survey covering 50 deg$^2$ that will reach $\mu_B = 23.5$ *B* mag arcsec$^{-2}$. These multicolor CCD surveys are deep enough to detect LSB galaxies that are missing from most photographic catalogs.

The data in both clusters and the field are best described by a composite luminosity function. Giant galaxies, especially those of early type with HSB, have either a conventional Schechter function with $\alpha = -1$ to $-1.1$ or a Gaussian luminosity function. Dwarf galaxies, especially those of late type with LSB, have steep Schechter slopes of $\alpha = -1.5$ to $-2$, which begin to dominate at $-15 < M_B < -17$. A single Schecter function is a poor description of almost any survey that spans more than 6 magnitudes (mag) in luminosity. The steep faint end slope has so far been most clearly seen in poor, spiral-rich clusters or in samples of blue field galaxies. We note that in the case where $\alpha = -1.5$, $L_{tot} \approx 1.8\phi^* L^*$ and the number of galaxies per luminosity interval is $dN = \phi^* \exp(-L/L^*) dL/L^*$. The steep luminosity function at the faint end for gas-rich dwarfs was first proposed by Tyson & Scalo (1988), and it has been argued for on different grounds by Schade & Ferguson (1994).

RECOVERING THE TRUE GALAXY POPULATION Correct calculation of a luminosity function requires a measurement of the bivariate distribution of luminosity and surface brightness. The luminosity should be calculated from a total rather than an isophotal magnitude. Most constructions of the luminosity function implicitly assume a delta function surface brightness distribution, and no published luminosity function has made corrections for surface brightness selection effects. Figure 2 shows how significant the addition of diffuse galaxies can be. The RC3 galaxies form a broad distribution around a characteristic value of $\mu_e = 22$ mag arcsec$^{-2}$ in *B*; optical and/or 21-cm redshifts have been measured for all LSB galaxies plotted in Figure 2. With a fainter limiting isophote, the APM survey recovered previously uncataloged galaxies down to $\mu_e = 26$ mag arcsec$^{-2}$. The correlation between total luminosity and surface brightness has a very large scatter. Over the range $23 < \mu_e < 25$, galaxies are being discovered over the entire range $-14 > M_B > -21$. The far left solid curve of Figure 2 represents the selection function at 50% completeness.

The derivation of a luminosity function is complicated by the correlation between surface brightness and luminosity. Rather than calculate $n(L, \mu)$, it is possible to calculate $n(I_0, s)$ because $I_0$ and $s$ uniquely define luminosity and

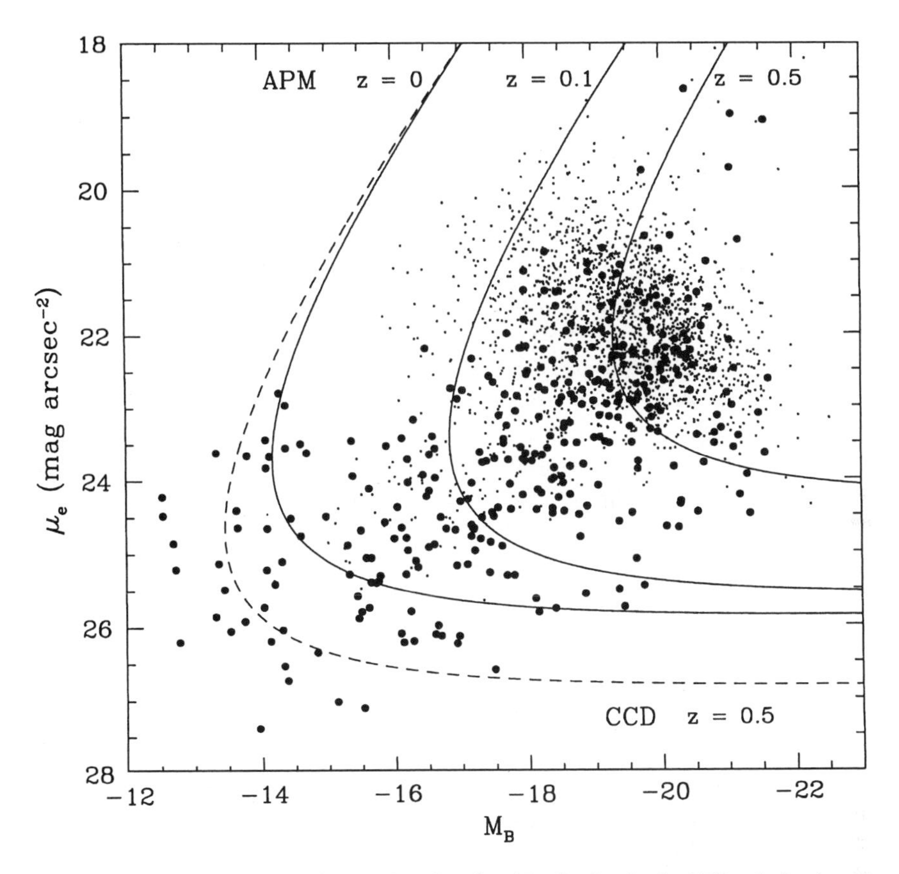


*Figure 2* Effective surface brightness plotted against blue luminosity for RC3 galaxies (*small dots*) and LSB galaxies with redshifts from the APM survey (*large dots*). The correlation between luminosity and surface brightness for the LSB galaxies is weak. (From de Vaucouleurs et al 1991, Impey et al 1996.) The three solid curves show the APM selection function, assuming $\mu_{\mathrm{lim}} = 26$ $B$ mag arcsec$^{-2}$ and $\theta_{\mathrm{lim}} = 10$ arcsec. The dashed curve is the selection function from the pencil-beam CCD survey of Lilly (1993), assuming limits of $\mu_{\mathrm{lim}} = 28.6$ $B$ mag arcsec$^{-2}$ and $\theta_{\mathrm{lim}} = 2$ arcsec.

they are observed to be uncorrelated for LSB galaxies fit by exponentials. In this case,

$$n(L)dL = 2\pi \int_{s_{\mathrm{min}}}^{s_{\mathrm{max}}} \int_{I_{\mathrm{min}}}^{I_{\mathrm{max}}} I_0^{-p} s^{-2q} dI_0\, ds, \tag{8}$$

where the distributions of structural parameters, after correction for selection effects, are $n(I_0) \propto I_0^{-p}$ and $n(s) \propto s^{-q}$ (Phillipps & Driver 1995).

Assuming that this double integral is bounded by the observed upper limits on central surface brightness and scale length, $n(L) \propto [1 - (L/L_{\max})^{1-p}][1 - (L/L_{\max})^{(1-2q)/2}]$, which reduces to $n(L) \propto L^{-1.5}$, if $p \approx 1$ and $q \approx 2$, as observed. As noted by Phillipps & Driver (1995), $n(L)dL \propto \int I_0^{-1}(I_0/L)(I_0L)^{-1/2}dI_0dL$, which reduces to $n(L) \propto L^{-3/2}(I_{\max}^{1/2} - I_{\min}^{1/2})$, and so the fraction of galaxies above $I_{\min}$ is $1 - (I_{\min}/I_{\max})^{1/2}$, independent of luminosity.

Sprayberry et al (1997) have compared the APM survey of LSB galaxies directly to the CfA redshift survey. By measuring the surface brightness properties of the CfA galaxies, a volume-limited sample of LSB galaxies with redshifts can be defined that has no overlap with the CfA survey (i.e. $\mu_0 > 22$ mag arcsec$^{-2}$). The predominantly late-type LSB galaxies in the range $22 < \mu_0 < 25$ $B$ mag arcsec$^{-2}$ are at least as numerous as the late-type CfA galaxies presented by Marzke et al (1994a). The amount of extra luminosity density is 25–30%. Because the number distribution is nearly flat down to the lowest surface brightness reached by the APM survey, these are lower bounds to the missing LSB population. Using the maximum likelihood estimator of Sandage et al (1979), a Schechter function for the giants, and a power-law function for the dwarfs, the luminous galaxies are fit by a standard slope of $\alpha = -0.9$, and the dwarfs have a steeper (but more uncertain) slope of $\alpha = -2.2$. It is not known how this analysis would be affected by the addition of galaxies at even fainter surface brightness levels of $\mu \sim 27$ mag arcsec$^{-2}$, such as have been discovered in CCD surveys (Schwartzenberg et al 1995a, Dalcanton 1995).

To summarize, the local galaxy luminosity function cannot be derived without correcting for surface brightness selection effects. These effects are more severe for late-type and gas-rich galaxies than for early-type and gas-poor galaxies because the latter have generally higher surface brightness and a tighter correlation between luminosity and surface brightness (Binggeli et al 1984). Isophotal magnitudes generally measure a smaller fraction of the total galaxy flux as the surface brightness decreases. Procedures that assume that each luminosity interval has a similar distribution of surface brightness, or that $S_{\text{lim}}/S_{\text{tot}}$ is not a function of surface brightness, may be in error. For galaxy types where the correlation between luminosity and surface brightness has large scatter, the corrections are substantial. The volume corrections are largest for the faintest galaxies, where the numbers in local surveys are fewest and where the incompleteness corrections are largest. When these corrections are made to shallow wide-angle surveys, or when considering deep pencil-beam surveys that tend to have fainter isophotal limits, the slope of the tail of the luminosity function is at least as steep as $\alpha = -1.4$. By far the most ubiquitous type of galaxy in the universe is the LSB dwarf. These galaxies are almost completely missing from published local luminosity functions.

## SURVEYS OF LSB GALAXIES

Over the past 15 years, a number of surveys have succeeded in locating LSB galaxies. The distribution in surface brightness is continuous, but operationally we choose to define galaxies with $\mu_0 \geq 23$ mag arcsec$^{-2}$ as low surface brightness. In terms of the narrow surface brightness distribution of Freeman (1970), a disk galaxy this diffuse should be extremely rare. In practice, LSB galaxies are a mixed bag, including objects as diverse as giant gas-rich disks and dwarf spheroidals. This of course is the reason for their importance: Such galaxies offer a new window onto the diversity of galaxy morphology and evolution.

### *Photographic Surveys*

Unsurprisingly for someone who left his idiosyncratic mark on much of extragalactic astronomy, Zwicky (1957) was one of the first to speculate on the existence of LSB galaxies. His claim of a steeply increasing tail of faint galaxies was at odds with Hubble's (1936) earlier Gaussian form for the luminosity function. In retrospect, both were correct; Hubble had identified mostly galaxies of high surface brightness, Zwicky had discovered an exponential tail of mostly LSB dwarfs. The David Dominion Observatory (DDO) catalog (van den Bergh 1959) was the first catalog to contain significant numbers of diffuse galaxies, although most of them had low mass and so were not representative of the full range of LSB types. Meanwhile, Reaves (1956) and Arp (1965) had identified the selection effect that might lead LSB galaxies to be missed. This selection bias was first clearly formulated by Disney (1976).

The discovery of LSB galaxies advanced considerably in the 1980s. In a prescient piece of work, Longmore et al (1982) obtained optical and 21-cm data on a sample of 151 LSB galaxies selected by visual inspection of UK Schmidt plates. Many early studies were based on the diameter-limited Uppsala General Catalog of Galaxies (UGC) (Nilson 1973). With no explicit surface brightness selection, the UGC contains, for example, an order of magnitude more LSB galaxies than the catalog of Fisher & Tully (1981). The LSB galaxies from the UGC catalog were subsequently studied by Romanishin et al (1982), who noted that they had relatively large amounts of gas for their luminosity.

Large numbers of LSB dwarfs were detected in the monumental photographic survey of the Virgo cluster by Binggeli et al (1985) using plates taken with the Du Pont 100-inch telescope. Another large survey was carried out in the nearby Fornax cluster using both Du Pont plates (Ferguson & Sandage 1988, Ferguson 1989) and sky survey plates from the UK Schmidt Telescope (Phillipps et al 1987). The surface brightness limits of both these surveys were $\mu_{\rm lim} \approx 25$ $B$ mag arcsec$^{-2}$, however the spatial resolution and morphological classification is superior on the Du Pont plates, which gives an advantage in defining cluster membership in the absence of redshifts.

The next improvement was offered by visual searches of the POSS-II plates, which reach to a deeper limiting isophote, $\mu_{\rm lim} \approx 26$ $B$ mag arcsec$^{-2}$ (Schombert & Bothun 1988, Schombert et al 1992). Binggeli et al (1990) used deep Palomar plates to identify several hundred predominantly LSB dIm and dE galaxies. Impey et al (1988) used photographically amplified images of Virgo to push the limiting isophote down to $\mu_{\rm lim} \approx 27.5$ $B$ mag arcsec$^{-2}$, and a similar surface brightness can be reached by automated scans of UK Schmidt plates (Irwin et al 1990b). These studies yielded new samples of extremely LSB galaxies in Virgo and Fornax (Davies et al 1988, Bothun et al 1991). A further gain in sensitivity can be achieved by digitally stacking scans of existing sky survey plates or by using Tech Pan emulsions; large-scale surveys are currently under way with limits of $\mu_{\rm lim} \approx 27$ $R$ mag arcsec$^{-2}$ (Schwartzenberg et al 1995a).

### *CCD Surveys*

Digital detectors can survey for LSB galaxies in the field down to a much lower limiting isophote but over much smaller areas (Schwartzenberg et al 1995b). CCD surveys of nearby (Turner et al 1993, Bernstein et al 1995) and more distant clusters have been undertaken (Driver et al 1994a). The Texas Survey for field LSB galaxies adds a new dimension with red selection down to $\mu_{\rm lim} \approx 27.5$ $R$ mag arcsec$^{-2}$ (O'Neil et al 1997). Dalcanton et al (1997b) have used strip scans made with the Palomar 200-inch telescope operating in transit mode to find galaxies with $23 < \mu_0 < 25$ $V$ mag arcsec$^{-2}$. This approaches the limit below which individual galaxies cannot be distinguished from distant clusters of galaxies (Schectman 1973). At the limit of deep surveys, LSB galaxies are being mined from the WFPC2 images of the HST Deep Field Study by several groups. The next major step forward in large area surveys will come with the Sloan Digital Sky Survey (Gunn & Knapp 1993).

The diversity of galaxies uncovered by these surveys is striking. An early surprise was the accidental discovery of the giant LSB disk galaxy Malin 1 in a survey of the Virgo cluster (Bothun et al 1987). This remarkable galaxy is the prototype of systems that have low surface-mass density stellar disks, large physical sizes, and enormous amounts of neutral hydrogen (Impey & Bothun 1989, Knezek 1993). These galaxies are extreme cousins of the gas-rich LSB galaxies discussed by Longmore et al (1982), characterized by large exponential scale lengths ($s > 10$ kpc) and low central surface brightnesses ($\mu_0 > 25$ $B$ mag arcsec$^{-2}$). Further examples have recently been found (Bothun et al 1990, Sprayberry et al 1993, 1995b). Dwarf spirals have also been detected (Schombert et al 1995); both the smallest and the largest spiral galaxies known have low surface brightness (LSB). At the other extreme are LSB dwarfs, with similar surface brightness but much smaller scale lengths ($s \sim 1$–2 kpc). Surface brightness selection accounts for the fact that dwarf members of the Local Group continue to be discovered (Irwin et al 1990a). Even at the modest distance of the Coma cluster, many Local Group dwarf spheroidals would be too low in surface brightness to be detected in a shallow survey and too compact to be distinguished from stars in a deep survey.

## THE SIGNIFICANCE OF LSB GALAXIES

Our ignorance of the true galaxy population affects virtually every aspect of observational cosmology. Study of the detailed properties of galaxies is confined to nearby prominent examples. Beyond the Milky Way, most of our information on the stellar populations, kinematics, dark matter content, star formation history, and large-scale clustering of galaxies is based on studies of high surface brightness objects. Without accounting for LSB galaxies, we cannot complete a census of the luminous density and matter content of the universe.

### *Light and Matter in the Universe*

There are reasons to believe that the existing census of galaxies is incomplete. Figure 3 shows current constraints on luminous and nonluminous material in the universe. The left-hand panel shows the integrated surface intensity of the galaxy counts down to $V = 27$ (Tyson 1995). The sum is dominated by counts well above the magnitude limit of the survey, and this is nearly an order of magnitude lower than the upper bound on the extragalactic part of the diffuse sky brightness (Mattila 1990). The center panel of Figure 3 shows HI column density, where bright spirals have a peak HI column density in the range $5 \times 10^{20} < N_{\rm HI} < 2 \times 10^{21}$ atoms cm$^{-2}$ (Warmels 1986, Cayette et al 1993). At lower HI column densities, $3 \times 10^{20} < N_{\rm HI} < 8 \times 10^{20}$ atoms cm$^{-2}$, the efficiency of star formation is greatly reduced (Kennicutt 1989, van der Hulst et al 1993). At still lower column densities, disks may be ionized, and such diffuse gas clouds may not be detectable with HI surveys (Maloney 1993).

The right-hand panel of Figure 3 shows various measures of the mean mass density, $\Omega_0$. The agreement between the observed abundances of light elements and the predictions of primordial nucleosynthesis is one of the great successes of standard cosmology (Walker et al 1991). Yet, a careful accounting of the visible material in and between galaxies shows that it falls short of matching the amount of baryonic material predicted by the Big Bang model (Persic & Salucci 1992). For an allowed range of nucleon densities, $2.8 \times 10^{10} < n_{\rm baryon}/n_\gamma < 4.0 \times 10^{10}$, Walker et al (1991) find the constraint on the baryon density parameter to be $0.010 < \Omega_{\rm baryon} h_{100}^2 < 0.015$.

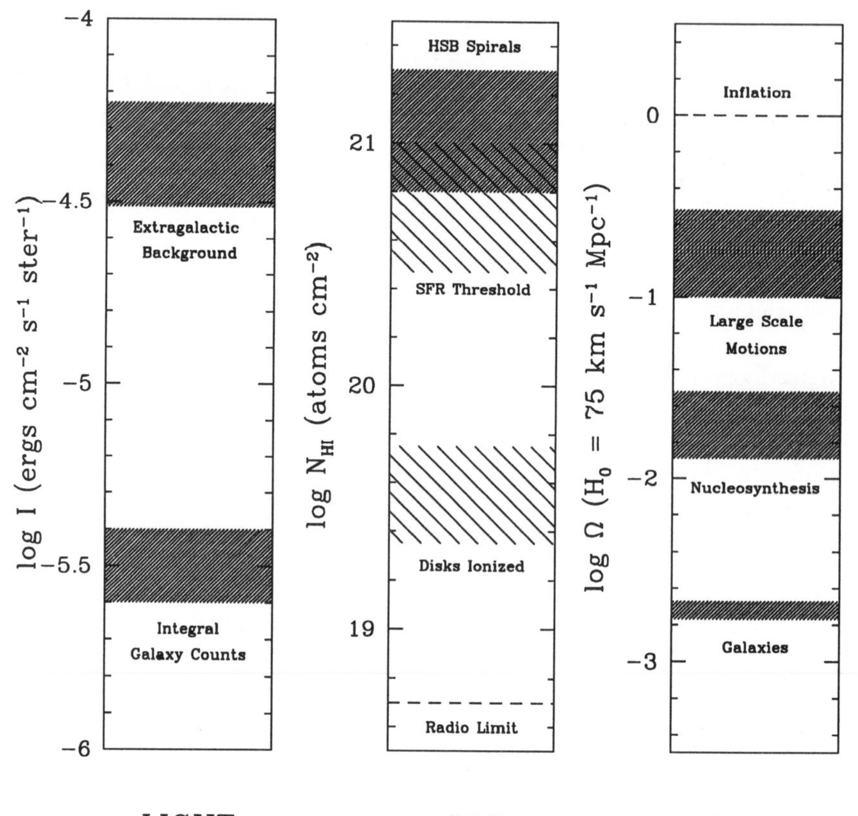


*Figure 3* The left panel shows the intensity of diffuse extragalactic light. The range of integral of the faint galaxy counts (Cowie et al 1995b, Tyson 1995) is nearly an order of magnitude below the observations' upper bound (Mattila et al 1991). The center panel shows the HI column density of gas in galaxies. Superimposed are the range of column densities for normal luminous spirals (Cayette et al 1993), the range of the threshold column density below which star formation is inefficient (Kennicutt 1989), and the probable column density at which disks in the local universe are ionized (Maloney 1993). The right panel shows mass components in the universe as a fraction of the critical density. The upper band is the range from large-scale motions (e.g. Peebles 1993), the middle band is the range from nucleosynthesis arguments (Walker et al 1991), and the lower band is the observed contribution in luminous galaxies (Persic & Salucci 1992).

The visible contribution from baryons is given by the sum $\Omega_{\text{baryon}} = \Omega_{\text{E/S0}} + \Omega_{\text{Sp}} + \Omega_{\text{clusters}} + \Omega_{\text{groups}}$, where the results of the Gunn-Peterson test put a very low limit on any diffuse component of cold intergalactic hydrogen. The inventory by Persic & Salucci (1992) yields a total of $\Omega_{\text{baryon}} = (2.2 + 0.6h_{100}^{-3/2}) \times 10^{-3}$, so that most baryons must be dark for $0.5 < h_{100} < 1$. X-ray observations of rich clusters imply $\Omega_{\rm baryon}/\Omega_{\rm tot} \sim 0.1$–0.3 (e.g. White et al 1993). Because inhomogeneous nucleosynthesis cannot be used to raise the baryon density above $\Omega_{\rm baryon} \sim 0.1h_{100}^2$ (Malaney & Mathews 1993), this observation is difficult to understand unless the standard scenario of a flat universe is incorrect; see the extensive discussion of baryonic dark matter by Carr (1994).

LUMINOSITY DENSITY For cosmological purposes, the integrated luminosity density is more important than the integrated number density. It is also a more robust way of comparing galaxy samples than by Schechter function parameters (assuming that the integral has converged 3–4 mag below $L^*$, when the incompleteness of most samples becomes large). For the CfA survey (Marzke et al 1994b),

$$L_{\rm tot} = \int\int \phi(L, \mu_0) L \, dL d\mu_0 = 11 \pm 4 \times 10^7 h_{100} \, L_\odot \, {\rm Mpc}^{-3}. \tag{9}$$

A variation of nearly a factor of 2 is seen between this and other published values for the luminosity density: Efstathiou et al (1988) found $19 \pm 7 \times 10^7 h_{100} \, L_\odot \, {\rm Mpc}^{-3}$, Loveday et al (1992) found $15 \pm 3 \times 10^7 h_{100} \, L_\odot \, {\rm Mpc}^{-3}$, and Lin et al (1996) found $19 \pm 1 \times 10^7 h_{100} \, L_\odot \, {\rm Mpc}^{-3}$. From the APM survey of late-type LSB galaxies, Sprayberry et al (1997) derived $3 \pm 0.5 \times 10^7 h_{100} \, L_\odot \, {\rm Mpc}^{-3}$. This exceeds the luminous density of late-type irregulars found by Marzke et al (1994b) and is 15–30% of the luminous density for all morphological types found by all investigators. Dalcanton et al (1997a) reached a similar conclusion based on a smaller number of redshifts from a CCD survey; LSB galaxies contribute 10–100% of the luminous density seen in HSB galaxies. The LSB correction to the luminosity density is only valid over the range $22 < \mu_0 < 25$ mag arcsec$^{-2}$, and this correction must be a lower limit because the surface brightness distribution is nearly flat down to the limits of current observations.

EXTRAGALACTIC BACKGROUND LIGHT The extragalactic background light (EBL) is an integral sum of the star formation history of the universe, more sensitive in practice to galaxy evolution than to the parameters of the cosmological model (Harrison 1964). It also provides a fundamental limit to the potential profusion of LSB galaxies. The observed upper limit on the diffuse EBL is subject to the accurate elimination of foreground components that are several hundred times larger; Mattila et al (1991) summarize a number of measurements that yield upper limits in the range 4.5–10 $\times$ $10^{-9}$ erg cm$^{-2}$ s$^{-1}$ sr$^{-1}$ Å$^{-1}$. The integral of the number counts of galaxies down to a $B$ band isophote of 30 mag arcsec$^{-2}$ (Tyson 1995) is

$$I_{\text{EBL}} = \int_m^\infty N(m) 10^{-0.4(m+20.45)}\, dm$$
$$= 6.4^{+0.2}_{-0.1} \times 10^{-10} \text{ ergs cm}^{-2}\text{ s}^{-1}\text{ sr}^{-1}\text{ Å}^{-1}. \quad (10)$$

The corresponding number from the survey of Cowie et al (1995b) is $4 \times 10^{-10}$ erg cm$^{-2}$ s$^{-1}$ sr$^{-1}$ Å$^{-1}$.

The existing limits on the EBL allow for large populations of LSB galaxies; note, however, that they are not present in large numbers in the deep CCD data (Tyson 1995). Väisänen (1996) concluded that populations of LSB galaxies permitted by the number counts can raise the EBL by a factor of 2–3. Models that include large numbers of LSB dwarfs lead to a predicted EBL within a factor of 2 of the current limit (Ferguson & McGaugh 1995, Morgan & Driver 1995). Depending on the evolution model assumed, the steep faint end tail of the local luminosity function discussed previously would not contribute to the pencil-beam counts until $B \sim 26$, which is substantially fainter than the level of the peak contribution of the counts to the EBL, $B \sim 24$. Because LSB galaxies are largely quiescent, their existence in large numbers would not violate the constraint that we have already identified the sources of most of the metal production in the universe (Cowie 1988). The directly measured contribution of LSB galaxies from the APM survey is $2 \times 10^{-10}$ erg cm$^{-2}$ s$^{-1}$ sr$^{-1}$ Å$^{-1}$, or 30–50% of the amount from the integral number counts.

A separate constraint comes from the fluctuations in the EBL (Schectman 1974). Note that LSB galaxies can add to the level of the EBL without increasing the amplitude of the fluctuations because they are observed to be weakly clustered (Mo et al 1994). It is possible to use the correlation properties of the extragalactic background to constrain galaxy evolution models (Cole et al 1992). However, this type of analysis must take into account that a significant fraction of the background from discrete sources is at much lower redshifts than the galaxies seen in deep pencil-beam surveys, and that LSB galaxies are more readily detected in pencil-beam surveys than in wide field surveys.

GAS MASS DENSITY Galaxy evolution proceeds by the conversion of gas into stars. There is ample evidence that the census of diffuse galaxies as measured by their light is incomplete. However, the search for gas through the 21-cm emission line offers a complementary approach. Radio telescopes have been used not only to measure the gas contents of cataloged galaxies, but to place limits on the space density of intergalactic clouds of neutral hydrogen (Fisher

& Tully 1977). For a source larger than the beam,

$$(N_{\rm HI})_{\rm lim} > 2 \times 10^{16}\ (S/N)\ T_{\rm sys} \left(\frac{\Delta V}{t}\right)^{1/2} \ \text{atoms cm}^{-2}, \tag{11}$$

where $t$ is the integration time, $\Delta V$ in kilometers per second is the bandwidth in terms of the antenna equation or the velocity dispersion of the gas for a galaxy, and $(N_{\rm HI})_{\rm lim}$ is independent of telescope size because telescope area and beam size cancel (Disney & Banks 1996). For sources smaller than the beam, the visibility volume is given by

$$V(M_{\rm HI}, \Delta V) \propto n_{\rm beam} D t^{-1/4} \left(\frac{M_{\rm HI}}{\sqrt{\Delta V}}\right), \tag{12}$$

where $t$ is the integration time per pointing or beam, which means that the typical observing strategy is to maximize the observable volume with short integrations and a shallow survey. As a consequence, very little is known about the HI content below a column density of $N_{\rm HI} \sim 10^{19}$ atoms cm$^{-2}$. There is a rough scaling $N_{\rm HI} \sim 10^{20}(M_{\rm HI}/L_{\rm B})10^{0.4(27-\mu_{\rm B})}$ atoms cm$^{-2}$ (Disney & Banks 1996), such that galaxies from most HI surveys should be readily detectable on POSS II or UKST survey plates, and galaxies with either $N_{\rm HI} < 10^{19}$ atoms cm$^{-2}$ or $\mu_e > 27$ mag arcsec$^{-2}$ are extremely difficult to detect by any technique.

Published radio surveys have a number of limitations. The sensitivity to HI masses below $10^8\ M_\odot$ is poor, and the small beam sizes and limited bandwidths allow relatively small volumes to be probed, with correspondingly weak limits on rare objects. Interferometers gain over single dishes by the number of beams per field of view, but they lose bandwidth due to the need for $n(n-1)$ correlators, where $n$ is the number of telescopes in the array. Equally important is the region of space targeted. Most surveys have targeted optically selected galaxies, which involves a bias towards those galaxies that have been most efficient in forming stars. Other surveys target optically bright galaxies but are sensitive to HI signals elsewhere in the velocity bandwidth or in the "off" beams. This prejudices the sample to the immediate environments of bright galaxies, which may be atypical. Many HI surveys are confined to the Local Supercluster, a volume that is over-dense by a factor of ~2.3 compared to the cosmic average (Felten 1977). There have been relatively few unbiased HI surveys.

Schneider et al (1990, 1992) have published extensive HI surveys, including observations of many dwarf and LSB galaxies (mostly from the UGC catalog). Rao & Briggs (1993) have calculated the HI mass function of late-type galaxies over the range $10^7 < M_{\rm HI} < 10^{10}\ M_\odot$ using the surveys of Fisher & Tully (1981) and Hoffman et al (1989). Other "blind" surveys have been used to quantify the space density of gas-rich dwarfs (Weinberg et al 1991, Szomoru et al 1994) and giants (Briggs 1990). The HI mass function has an analogous form to the Schechter luminosity function

$$\Theta(M_{\rm HI})d(M_{\rm HI}) = \left(\frac{\phi^*}{\beta+1}\right)\left(\frac{M_{\rm HI}}{M^*_{\rm HI}}\right)^{\frac{-(\alpha+\beta)}{(1+\beta)}} \exp\left(\frac{-M_{\rm HI}}{M^*_{\rm HI}}\right)^{\frac{1}{(1+\beta)}} d\left(\frac{M_{\rm HI}}{M^*_{\rm HI}}\right), \quad (13)$$

where the extra component is the trend of HI richness with luminosity $M_{\rm HI}/L \propto L^{\beta}$. Studies of spiral and irregular galaxies indicate $\beta \approx -0.1$ (Fisher & Tully 1975), and when $\beta$ is small the shape of the HI mass function and the optical luminosity function are very similar (Rao & Briggs 1993). Figure 4*a* shows a direct comparison between the Rao & Briggs mass function and the count of LSB galaxies from the APM survey, normalized at $M_{\rm B} = -21$ or $M_{\rm HI} = 3 \times 10^9\ M_\odot$. Both functions are consistent with a faint end slope corresponding to $\alpha = -1.4$ in a Schechter parameterization.

Figure 4*b* shows the mass function from a sensitive Arecibo HI survey (Schneider 1996), with the luminosity function of the LSB galaxies from the APM survey superimposed. Optical luminosity is mapped onto HI mass for the APM sample by the relation $M_{\rm HI} = 10^{10-(M_{\rm B}+a)/b}\ M_\odot$, with $a = 21.7$ and $b = 3.16$, which accounts for the fact that the galaxies are optically selected, and so the distribution is censored at low values of $M_{\rm HI}$ for a given luminosity. This relationship agrees well with the slope (to 10%) and the normalization at $10^8\ M_\odot$ (to 50%) of the relationships adopted by Briggs (1990), $a = 20.3$ and $b = 2.78$, and Tyson & Scalo (1988), $a = 20.9$ and $b = 3.12$. LSB galaxies are a factor of $\sim$2 more gas-rich at a given luminosity than HSB galaxies. The two functions are normalized at $M_{\rm HI} = 3 \times 10^9\ M_\odot$ where the HI survey is reasonably complete.

The HI mass function shows a clear upturn at $M_{\rm HI} \sim 10^8\ M_\odot$, corresponding to $M_{\rm B} \sim -16$, where the APM luminosity function turns up with a faint end slope of $\alpha = -1.4$. At the high end of the HI mass function, Briggs (1990) has put a limit on the space density of giant gas disks like Malin 1 (also see Bothun 1985). With 95% confidence assuming Poisson statistics, $\rho < 4.1 \times 10^{-4}$ Mpc$^{-3}$ [Weinberg et al (1991) have a weaker limit of $\rho < 0.029$ Mpc$^{-3}$]. This upper bound is weak enough that objects like Malin 1 could contribute as much integrated gas density as HI-rich dwarfs with $M_{\rm HI} = 10^8\ M_\odot$. Several thousand such galaxies could lie closer than the prototype. Nevertheless, the cosmological conclusion from Figure 4 is that gas-rich LSB galaxies do not dominate the local gas mass density of the universe.

MASS DENSITY IN BARYONS Even after correcting for surface brightness selection, the luminosity density is one step removed from the function of

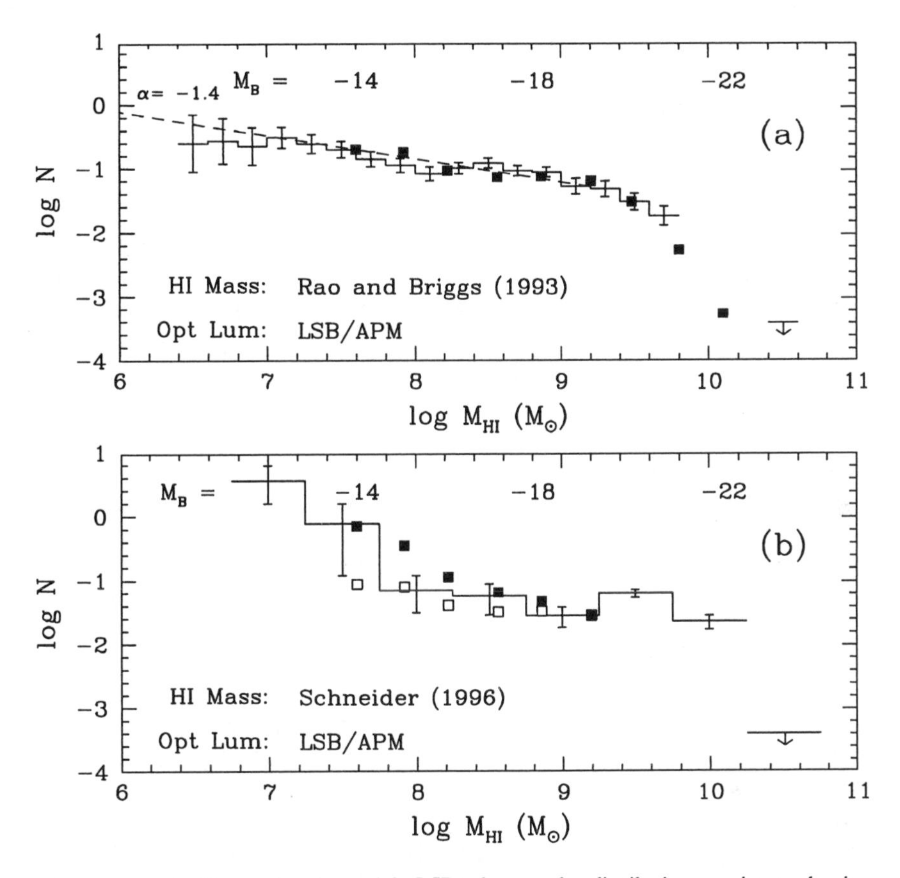


*Figure 4* (*a*) The HI mass function and the LSB galaxy number distribution superimposed, using the mean relationship between light and gas mass from the APM LSB survey, normalized at $M_B = -21$. (From Rao & Briggs 1993, Impey et al 1996, Sprayberry et al 1997.) (*b*) As above, with the HI mass function taken from a deep Arecibo survey. The open squares are the APM counts of LSB galaxies as in (*a*); the filled squares show the counts using a correction for censored regions in the surface brightness–luminosity plane. (Sources as above; in both cases the limit on high HI mass Malin 1-type galaxies is from Briggs 1990.)

cosmological interest. The baryonic density is

$$\rho_{\text{baryon}} = \sum \int \phi(L, \mu)\Gamma(L, \mu)Ld\, Ld\mu, \tag{14}$$

where both the luminosity function and the baryonic mass-to-light ratio $\Gamma(L, \mu) = (M/L)_{\text{baryon}} = A(\mu)(L/L_*)^\eta$ are functions of surface brightness. A large amount of uncertainty is hidden in the normalization factor $A(\mu)$ because there is very little data on the dependence of baryonic $M/L$ as a function of surface brightness. To simplify, we can factor $A$ out of the integral

$$\rho_{\text{baryon}} = \phi^* L^* A \int_{L_{\text{min}}}^{L_{\text{max}}} (L/L^*)^{1+\alpha+\eta} \exp(-L/L^*) d(L/L^*), \qquad (15)$$

where the surface brightness corrections to the luminosity function are accounted for in terms of an increased normalization, $\phi^*$, and a steeper faint end slope, $\alpha$. Following Persic & Salucci (1992), we take the bounds of the integral as $0.01 < (L/L^*) < 8$ . Bristow & Phillipps (1994) plot the integrated mass function for different values of $\alpha$, $\eta$, and $L_{\text{min}}$. To sharpen up the cosmological comparison, we use recent nucleosynthesis constraints (Copi et al 1995) and a Hubble constant bound of $0.5 < h_{100} < 0.8$, which encompasses 95% of the published values since 1995 (RC Kennicutt, private communication), including the recent HST Key Project result of $H_0 = 73 \pm 10$ km s$^{-1}$ Mpc$^{-1}$ (Freedman et al 1996). Although we use a scaling to $H_0 = 100$ km s$^{-1}$ Mpc$^{-1}$ in this review, it is almost certain that $h_{100} < 1$, and a number of direct distance scale measurements indicate $h_{100} < 0.7$ (e.g. Saha et al 1996, Grogin & Narayan 1996). The result is $0.014 < \Omega_{\text{baryon}} < 0.080$, which may increase by a factor of 2–3 if the $D/H$ ratio proves to be as low as $1\text{–}2 \times 10^{-5}$. Most matter is nonbaryonic, and most baryons have not yet been detected.

Given the uncertainties in this calculation, we quote only illustrative results. For spirals, Persic & Salucci (1990) derive $A = 2.4h_{100}(M_\odot/L_\odot)$, whereas McGaugh (1992) derives $A \approx 5h_{100}(M_\odot/L_\odot)$ for LSB spirals, a ratio of a factor of 2.1. Based on the APM survey (Impey et al 1996), we increase the normalization by a factor of 1.3. The product of these two factors raises the spiral contribution by a factor of 2.7, and the overall Persic & Salucci census by 40% to $\Omega_{\text{baryon}} = 0.0042$. The biggest leverage in the mass density integral comes from faint and low mass galaxies. Persic & Salucci (1990) assume values that make the exponent $1+\alpha+\eta > 0$, but if $1+\alpha+\eta < -1$, the mass integral rises rapidly. Adopting a faint end slope of $\alpha = -1.4$, and assuming $\eta \approx 0$ (because LSB galaxies appear to have higher $M/L$ and luminosity and surface brightness are correlated), $\Omega_{\text{baryon}} = 0.0062$. If either $\alpha = -1.80$ or $\eta = -0.5$ (baryons scale with dark matter), the correction factor becomes 3.6, with a total $\Omega_{\text{baryon}} = 0.013$. Bristow & Phillipps (1994) use more extreme, but not implausible, values to deduce $\Omega_{\text{baryon}} = 0.025$. We conclude that LSB galaxies can easily be the sites of much of the missing baryonic matter in the universe.

TOTAL MASS-TO-LIGHT RATIOS Dynamical measures of the total mass-to-light ratios of LSB galaxies are difficult to obtain. A small number of HI rotation curves of LSB giants have been published (de Blok et al 1996), but velocity dispersions for the LSB dwarfs are beyond the capabilities of existing telescopes. There is nonetheless indirect evidence that LSB galaxies have higher

$M/L$ than HSB galaxies of the same size. LSB disks have been found to follow the same Fisher-Tully relation as normal bright spirals but with increased scatter (Zwaan et al 1995, Sprayberry et al 1995a). If the mass $M$ is proportional to $v_{\max}^2 h$, then $v_{\max}^4 \propto M^2/h^2 \propto M^2 I_0/L$, since $L \propto I_0 h^2$. Zwaan et al (1995) found that LSB galaxies from the surveys of McGaugh & Bothun (1994), Knezek (1993), and de Blok et al (1995) have the same luminosities at a fixed line width as the HSB galaxies observed by Broeils (1992). This implies $M/L \propto I_0^{-1/2}$. Although the scatter is large, luminosity and central surface brightness are correlated for the APM sample, with a dependence that scales on $L \propto I_0^{1/2}$. Using a projection of the luminosity–surface brightness distribution that accounts for censored galaxies, this flattens to $L \propto I_0$. With this latter dependence, $M/L \propto L^{-1/2}$, which gives LSB galaxies considerable leverage in the mass census of disk systems.

Less can be said about the $M/L$ of dwarf LSB galaxies in the field because the galaxies are small and the HI line widths are narrow (10–30 km s$^{-1}$), so the rotation curves are usually poorly sampled. The low surface mass density of the (mostly gas-poor) LSB dwarfs in clusters like Virgo and Fornax suggests a stability constraint. Assuming isotropic velocities, the mass density in the core of a rich cluster is

$$\rho_c = \frac{9\sigma^2}{4\pi G r_c^2} = 3 \times 10^{-3} h_{100}^2 \, M_\odot \, \mathrm{pc}^3, \tag{16}$$

where $\sigma = 870$ km s$^{-1}$ and $r_c = 0.2 h_{100}^{-1}$ Mpc (Peebles 1993). This number can be compared with the stellar mass density of a typical LSB galaxy:

$$\rho_{\mathrm{LSB}} = \frac{L(r)}{V(r)} (M/L)_{\mathrm{LSB}} = 3 \left\{ \frac{L_{\mathrm{tot}}[1 - (1+n)e^{-n}]}{4\epsilon\pi (nh)^3} \right\} (M/L)_{\mathrm{LSB}}, \tag{17}$$

where $n$ is the number of scale lengths considered and $\epsilon$ is a shape parameter that describes the departure from spherical symmetry. Adopting $n = 3$ (which includes 80% of the total light) and $\epsilon = 0.3$ leads to a condition for the stability of the diffuse galaxy in the tidal field of a cluster core $(M/L)_{\mathrm{LSB}} > 0.13(h^3/L_{\mathrm{tot}})$, where $h$ is in parsecs and $L_{\mathrm{tot}}$ is in solar units. The envelope of this distribution is of course defined by selection effects, which mitigate against the combination of large scale length and low luminosity. However, cluster LSB samples (Impey et al 1988, Bothun et al 1991, Turner et al 1993) include galaxies whose long-term existence implies larger than expected mass-to-light ratios. The condition $M/L > 3$ corresponds to $h > 3$ kpc at $M_B = -17$ or $h > 10$ kpc at $M_B = -21$. Finally, we note that the innocuous Local Group dwarf spheroidals have some of the highest $M/L$ ratios of any galaxies (Mateo 1996). The galaxies that have been undercounted in all existing surveys are those with the highest mass fractions of gas and the highest $M/L$ ratios.

### *Galaxy Evolution*

Our understanding of how galaxies form and evolve is highly incomplete. Ironically, we know as much about the linear and high-temperature physics during the first few minutes of the universe, through the successes of the hot Big Bang model (e.g. Peebles et al 1991), as we do about the billion years after density perturbations became nonlinear. Our ignorance of galaxy evolution has hampered tests of the deceleration parameter and the curvature of the universe (Sandage 1988). Reliable measures of ages and stellar populations can only be obtained for nearby prominent galaxies. These same HSB galaxies are used as probes of large-scale structure in redshift surveys. The study of LSB galaxies opens a new window onto galaxy formation and evolution.

THE COLORS OF LSB GALAXIES The LSB dwarf galaxies in clusters are distinguished by their blue colors and low gas contents. Imaging of sufficient resolution reveals mostly dIm and dE morphologies. In Virgo and Fornax combined, the mean colors are $B - V = 0.58 \pm 0.02$ for 31 objects, and $V - I = 1.00 \pm 0.03$ for 23 objects (Impey et al 1988, Bothun et al 1991). Davies et al (1990) found $B - R \approx 1.4$ in Fornax, corresponding to a slightly redder $B - V \approx 0.7$. There is no simple explanation for these colors, which are similar to the colors of the most metal-poor galactic globular clusters. The lack of a correlation between color and surface brightness argues against LSB dwarfs being faded remnants of more gas-rich dwarfs. One scenario identifies the diffuse stellar component as a fossilized metal-poor remnant of a galaxy with a steadily softening potential, where most of the gas was driven away long ago (Dekel & Silk 1986). Although the blue colors are consistent with a young median age, spectroscopy is required to disentangle effects of age and metallicity.

Although fading is not indicated as a cause for the low surface brightness, a red population of LSB galaxies cannot be ruled out. All of the large area surveys have been carried out in the blue, and a starbursting dwarf could rapidly redden and fade below threshold imposed by the night sky. A $10^7$ year starburst with a conventional IMF would redden from $B - V = -0.30$ to $B - V = 0.50$ after $10^9$ years, whereas the total light fades by $\sim$5 mag (Bruzual & Charlot 1993). After $10^{10}$ years, the color reddens further to $B - V = 0.85$ and the total light fades another $\sim$2 mag. Assuming the galaxy fades with constant scale length, a galaxy fading by $\Delta\mu$ magnitudes is lost more rapidly owing to falling below the angular size limit ($n \propto \Delta\mu^3$) than to falling below the flux limit ($n \propto 10^{\Delta\mu/2.5}$). Each factor of 2 of fading reduces the visibility volume by a factor of 8. In addition, the sky contrast improves towards the blue for all objects with $B - I < 1.3$. The combined result is a formidable selection effect against post-starburst LSB dwarfs.

LSB disk galaxies also have blue colors. After excluding galaxies with significant bulges, McGaugh & Bothun (1994) found $U - B = -0.17$, $B - V =$

$0.49 \pm 0.04$, and $V - I = 0.89$, similar to the colors of an actively star-forming Sc galaxy. Romanishin et al (1983) found averages of $B - V = 0.43 \pm 0.04$ and $V - R = 0.60 \pm 0.02$. This compares with $B - V = 0.75 \pm 0.03$ and $V - R = 0.53$ for normal spirals that are on average 2 mag brighter in $\mu_B$ (de Jong & van der Kruit 1994). For seven giant LSB disks, Sprayberry et al (1995b) found $B - V = 0.73 \pm 0.05$ and $V - R = 0.50 \pm 0.04$. LSB galaxies in general are about 0.25 mag bluer in $B - V$ than HSB galaxies, and both populations show a correlation between increasing redness and larger scale length. This effect, plus the fact that there is no correlation between color and surface brightness, can be used to rule out fading as a cause of the low surface brightness for both LSB giants and dwarfs. The HII region oxygen abundances are also uncorrelated with color, which means that the blue colors are not caused by low metallicity. Metallicities are typically one third solar (McGaugh 1994a).

Sprayberry et al (1995b) found four Seyfert 1 nuclei and one Seyfert 2 nuclei in a study of 10 LSB giants, as indicated by broad H$\alpha$ emission and sometimes by [NII]/H$\alpha$ ratios indicative of excitation by a power law (see also Knezek 1993). Although the sample is small, the Poisson probability of drawing so many Seyferts from a population of normal galaxies with the same luminosity (Meurs & Wilson 1984) is small, $\sim 2 \times 10^{-6}$. It is possible that the kinematics of a low surface mass density disk can facilitate mass transfer into the central parsec, where it can fuel nuclear activity.

GAS CONTENT AND EVOLUTION Combining radio and optical data leads to a better understanding of the evolution of LSB galaxies. The HI properties of LSB galaxies as a class were first studied by Hawarden et al (1981). More recently, radio synthesis telescopes have been used to derive rotation curves and to map out the gas in the disk. LSB disks have low star formation rates, despite their normal HI contents and luminosities. Kennicutt (1989) has shown that the star formation rate in HSB disks has a threshold, below which widespread star formation does not occur. The critical HI surface mass density is

$$\Sigma_{\rm crit} = \frac{\eta \sigma_{\rm v} V}{2.38 G R} \left(1 + \frac{R}{V}\frac{dV}{dR}\right)^{1/2}, \tag{18}$$

where $\eta$ is a dimensionless constant around unity, $\sigma_{\rm v}$ is the velocity dispersion in the gas disk, and the circular velocity of the gas is $V$ at radius $R$. Using HI synthesis data from the Westerbork Telescope, van der Hulst et al (1993) have found that LSB disks have HI surface mass densities of 3–6 $M_\odot$ pc$^{-2}$, a factor of 2 lower than HSB disks (Cayette et al 1993). This puts most of the disk below the threshold for star formation, given by $\Sigma_{\rm crit} = 0.059\sigma_{\rm v}(V/R)$ $M_\odot$ pc$^{-2}$, where $\sigma_{\rm v}$ is in kilometers per second, and where $V$ is the rotation velocity in kilometers per second at a distance $R$ in kiloparsecs.

The physical basis for this threshold is the onset of gravitational instability in a thin rotating disk (Toomre 1964, Quirk 1972). Star formation is therefore governed by local physics; despite the low overall rate, the star formation in isolated regions can be prodigious (Impey & Bothun 1989). A plausible hypothesis is that the evolution of all galaxies is driven by surface mass density. Because mass is roughly given by $M \propto v_{\rm max}^2 h$, and because both HSB and LSB galaxies follow a Fisher-Tully relation ($L \propto v_{\rm max}^4$), it follows that $(M/L) \propto v_{\rm max}^2 h/(Mh)^2 \propto \bar{\Sigma}$, where $\bar{\Sigma}$ is the mean surface mass density (de Blok et al 1995). Low surface mass density galaxies evolve slowly, forming few stars, which results in low surface brightness (LSB) and high values of $M/L$.

It is important to reiterate the facility with which quiescent and low mass galaxies can escape detection by either optical or radio surveys. The neutral disks of spiral galaxies are observed to truncate sharply below a column density of 2–3 $\times$ $10^{19}$ atoms cm$^{-3}$ (e.g. van Gorkom et al 1996). As the column density of HI falls off, galaxy disks can be ionized, which rapidly reduces the detectable HI mass (Maloney 1993, Corbelli & Salpeter 1993, Dove & Shull 1994). As parameterized by Dalcanton et al (1997), the observable HI mass is $M_{\rm HI} = 50\pi\,\Sigma_{\rm HI}h^2[1 - 0.01\Sigma_{\rm HI}^{-1.43}(5.61 + 1.43\ln\Sigma)]\ M_{\odot}$, where $h$ is in kiloparsecs and $\Sigma_{\rm HI}$ is in units of $10^{21}$ atoms cm$^{-2}$. This trend is illustrated in the left panel of Figure 5. Galaxies with scale lengths under 5 kpc or central surface HI densities under $10^{20}$ atoms cm$^{-2}$ drop below the sensitivity limit of a typical Arecibo survey. Galaxies with scale lengths under 2 kpc or central surface HI densities under $3 \times 10^{19}$ atoms cm$^{-2}$ are difficult to detect in any HI survey.

There is a similarly rapid drop in the detectability of newly formed stars as a function of the total (HI plus $H_2$) hydrogen surface density. Using a conversion from H$\alpha$ flux to star formation rate from Kennicutt (1983), the dependence of H$\alpha$ surface brightness on total hydrogen surface density for Sc galaxies (Kennicutt 1989) can be parameterized in a quasi-Schechter form as $\Sigma_{\rm SFR} = 2.72\ \exp(-5/\Sigma_{\rm H})(5/\Sigma_{\rm H}^{-0.78})\ M_{\odot}\ {\rm pc}^{-2}\ {\rm Gyr}^{-1}$, where $\Sigma_{\rm H}$ is in units of $M_{\odot}$ pc$^{-2}$. Note that this is only a rough scaling, as the threshold surface densities range from 2–10 $M_{\odot}$ pc$^{-2}$, and the H$\alpha$ surface brightness at a given hydrogen surface density ranges over more than a factor of 10. The right panel of Figure 5 shows this dependence, where the dashed line represents a steady star formation rate over a Hubble time. Below a few $M_{\odot}$ pc$^{-2}$, the star formation rate drops so low that H$\alpha$ emission is not detectable, and the rate of gas consumption is so low that the galaxy is essentially quiescent.

### *Implications for Quasar Absorption*

The narrow absorption lines seen in the spectra of quasars are powerful probes of dim and low column density material along the line of sight. The neutral

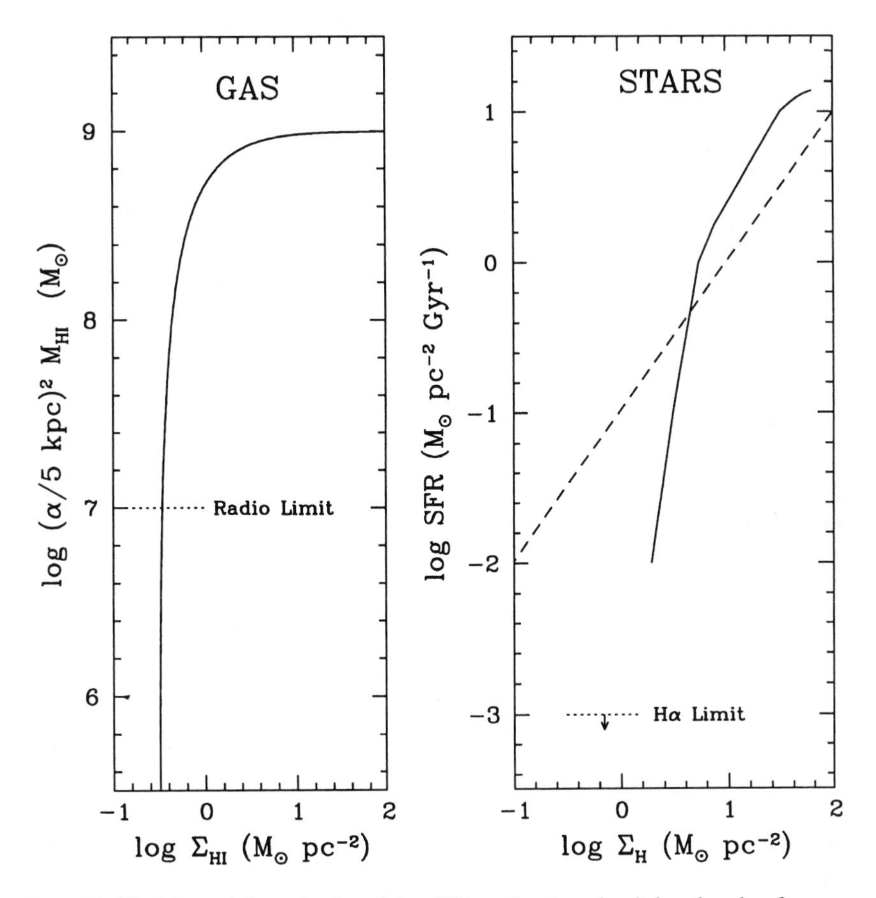


*Figure 5* The left panel shows the detectivity of HI as a function of scale length and surface mass density of gas. (From Dalcanton et al 1997a.) The right panel shows the dependence of disk star formation rate on the surface mass density of gas, HI plus $H_2$. The dashed line corresponds to a constant star formation rate over a Hubble time. (From Kennicutt 1989.)

hydrogen column densities of the absorbers range over 10 orders of magnitude, $10^{12} < n_{HI} < 10^{22}$ cm$^{-2}$. The high column density or "damped" Lyman-$\alpha$ systems have associated metals, and they may be the progenitors of normal spiral galaxies. The low column density systems of the Lyman-$\alpha$ "forest" have much lower metallicity and are only weakly associated with galaxies. Intermediate HI column densities are probed by the CIV $\lambda\lambda$1548,1550 and MgII $\lambda\lambda$2797, 2803 absorption doublets, which are strongly clustered in velocity space and are presumed to originate in the halos of normal galaxies, which are sometimes located in clusters. For a recent review, see the proceedings edited by Meylan (1995).

The number of absorbers per unit redshift down to a particular rest equivalent width line is given by

$$\frac{dN}{dz} = \frac{c\sigma n}{H_0}(1+z)^{\gamma}(1-q_0 z)^{-1/2}\int_0^{\infty}\phi_g(L)A(L)dl, \tag{19}$$

where $\phi_g(L)$ is the conventional Schechter function multiplied by $f_g$, the fraction of gas-rich galaxies, basically spirals and irregulars. We assume no evolution in the redshift path density, i.e. $\gamma = 1$, which is appropriate for both the MgII absorbers (Steidel & Sargent 1992) and the low column density Lyman-$\alpha$ absorbers at low redshift (Bahcall et al 1993). It is also likely to be appropriate for the high column density, damped Lyman-$\alpha$ absorbers (Lanzetta et al 1991, but see Rao et al 1995). The cross-sectional area of the absorbers is $A(L/L^*) = (\pi R_*^2/2)(L/L^*)^{2\beta}\eta^2$, where $R_*$ is a fiducial radius, typically the Holmberg radius for an $L^*$ galaxy, and $\eta$ is a factor that relates the optical size of a galaxy to the HI size at the column density of interest. This yields

$$\frac{dN}{dz} = \left(\frac{\pi c}{2H_0}\right)(\eta R_*)^2(f_g\phi^*)(1+z)(1-q_0 z)^{-1/2}\Gamma(1+2\beta+\alpha). \tag{20}$$

The absorption path length depends on the galaxy luminosity, the faint end slope of the luminosity function, and the relationship between absorption cross section and luminosity. Assuming no evolution of the galaxy luminosity function (either $\phi^*$ or $\alpha$) over the path length, the integral over redshift gives the fraction of the absorption path length that is caused by galaxies brighter than a certain luminosity. The conventional Holmberg (1975) relation is $(R/R_*) \propto (L/L^*)^{0.4}$. Using this scaling and the conventional $\alpha = -1$, dwarfs with $M_B > -16$ should contribute $\sim$10% of the absorption cross section.

METAL LINE ABSORBERS MgII absorbers are equivalent to absorbers selected to be optically thick in the Lyman continuum, $\tau(912\ \text{Å}) > 1$ or $N_{HI} > 3 \times 10^{17}$ atoms cm$^{-2}$ (Sargent et al 1988). There is strong statistical evidence linking MgII absorption to the presence of a bright galaxy near the line of sight (Bergeron & Boisse 1991, Steidel & Dickinson 1992). The traditional Holmberg scaling is almost certainly not appropriate. Impey et al (1988) showed that for a Virgo sample of galaxies all at roughly the same distance, the relation $R \propto L^{0.4}$ is partly a consequence of selection effects caused by sky brightness. Steidel (1993) studied the impact parameters of galaxies causing 56 MgII absorptions over the range $0.2 < z < 2.2$. The results were inconsistent with $\beta = 0.4$ and had a maximum likelihood fit of $\beta = 0.2$ and $R_* = 35h_{100}^{-1}$ kpc. The scaling $R \propto L^{0.2}$ with $\alpha = -1$ implies that dwarfs contribute $\sim$75% of the absorption cross section. Alternatively, we can adopt $\beta = 0.4$ and consider the effect of a steep faint end slope to the luminosity function. Dwarfs contribute 35% of the cross section if $\beta = 0.4$ and $\alpha = -1.4$ and 85% of the cross section if $\beta = 0.4$ and $\alpha = -1.7$. The integral $\int \phi(L)A(L)dL$ diverges and dwarfs dominate the cross section for any combination $\alpha + 2\beta \leq -1$.

Equation 20 uses local galaxy properties to predict the demographics of the quasar absorbers. This can be compared with the observed properties of individual absorbers, which are detected as galaxies with the appropriate redshift lying at small impact parameters from the quasar sightline. Steidel (1993) found no absorbers with $M_B > -19.4 + 5\ \log h_{100}$, as opposed to 80% predicted if $\beta = 0.2$ and 50% predicted if $\alpha = -1.4$. This sharp difference implies that copious gas-rich dwarfs (York et al 1986, Tyson & Scalo 1988) or unseen LSB galaxies (Phillipps et al 1993) cannot be a major contributor to MgII absorption. We also consider it unlikely that $f_g$ declines with decreasing luminosity; a number of field galaxy luminosity functions have a rising tail of primarily gas-rich dwarfs. However, the gas in low mass galaxies is probably ionized, so that dwarfs are not responsible for HI absorption in the range $3 \times 10^{17} < N_{HI} < 3 \times 10^{19}$ atoms cm$^{-2}$. Maloney (1993) has shown that the column density $N_{crit}$ below which gas is ionized, and galaxies therefore are mostly invisible in the 21-cm line, is a function of halo surface mass density, $N_{crit} \propto \Sigma_{halo}^{0.6}$. Low mass or low surface mass density galaxies will therefore be ionized at larger total gas column densities.

HIGH COLUMN DENSITY HYDROGEN ABSORBERS The most massive gas disks produce damped Lyman-$\alpha$ absorption with HI columns in the range $2 \times 10^{20} < N_{HI} < 6 \times 10^{22}$ atoms cm$^{-2}$. At a mean redshift of $z = 2.5$, the absorbers have $dN/dz \approx 0.2$ and contribute $\Omega = 1.45 \times 10^{-3} h_{100}^{-1}$ ($q_0 = 0.5$) to the mass density of the universe (Lanzetta et al 1991). Impey & Bothun (1989) made the point that gas-rich LSB disks have many similar properties to the progenitors of spiral galaxies seen at high redshift, including size, HI column density, and a low mean star formation rate. Bergeron (1995) has directly observed LSB galaxy counterparts to damped Lyman-$\alpha$ absorbers at intermediate redshift.

For the large LSB disks, the observational bounds are $2.8 \times 10^{-5} < dN/dz < 1.9 \times 10^{-2}$. The upper bound comes from the diameter, $140 h_{100}^{-1}$ kpc at the $3 \times 10^{19}$ atoms cm$^{-2}$ contour, and the limit on the space density of less than $4.1 \times 10^{-4}$ Mpc$^{-3}$ for objects like Malin 1. The much less restrictive lower bound comes from the minimum space density of galaxies in the tail of the size distribution of UGC galaxies, those with diameter greater than 50 kpc. Rao & Briggs (1993) derived $dN/dz = 0.015 \pm 0.004$ for the gas cross section of normal galaxies at the present epoch, so it is possible that giant disks like Malin 1 are the largest contributor to high column density HI absorption. The upper bound on the incidence of damped Lyman-$\alpha$ systems at low redshift is

$dN/dz < 0.05$ (Storrie-Lombardi et al 1994). The contribution of damped Lyman-$\alpha$ systems to the density parameter at $z \sim 0$ is found to be $\Omega = 1.9 \times 10^{-4}$ (Lanzetta et al 1995). This can be compared to the bounds from the gas and stars in giant LSB galaxies: $\Omega_{\rm gas} = \rho_{\rm gas}/\rho_{\rm crit} < 1.5 \times 10^{-4}$, where $\rho_{\rm crit} = 2.78 \times 10^{11}\ h_{100}^2\ M_\odot\ {\rm Mpc}^{-3}$, and an HI mass of $10^{11}\ M_\odot$ is adopted for Malin 1, and $\Omega_{\rm stars} = \rho_{\rm stars}/\rho_{\rm crit} > 1.1 \times 10^{-4}(M/L)$, where we use the luminosity density of LSB galaxies, $3 \pm 0.5 \times 10^7 h_{100}\ L_\odot\ {\rm Mpc}^{-3}$. Therefore, large gas disks can be a substantial contributor to the local population of damped Lyman-$\alpha$ absorbers.

LOW COLUMN DENSITY HYDROGEN ABSORBERS The last comparison concerns the low HI column density absorbers of the Lyman-$\alpha$ forest, which have $dN/dz \approx 100$ for $N_{\rm HI} > 10^{13}$ atoms cm$^{-2}$ (Morris et al 1991), and $dN/dz \approx 15$ for $N_{\rm HI} > 10^{14}$ atoms cm$^{-2}$ (Bahcall et al 1993). Tyson (1988b) speculated that the copious low column density absorbers were associated with gas-rich dwarf galaxies. The space density of the absorbers is $\rho = 2.9 \times 10^{-2}\ {\rm Mpc}^{-3}$, where we adopt the large characteristic absorber size of $500h_{100}^{-1}$ kpc at $z \sim 0.7$, measured by Dinshaw et al (1995) using common absorption in a quasar pair to define a size. The maximum likelihood technique yields a 95% confidence interval, which translates to $\rho_{\rm Ly\alpha} = 0.3\text{–}3.7 \times 10^{-2}\ {\rm Mpc}^{-3}$. This overlaps with the observational range of the number density of low luminosity galaxies $N_{\rm gal} = 1.1\text{–}6.3 \times 10^{-2}\ {\rm Mpc}^{-3}$, which makes use of the density of $M_{\rm B} = -14$ galaxies from Loveday et al (1992) as a lower bound and the surface brightness corrected density of $M_{\rm B} < -14$ galaxies from the APM LSB survey as an upper bound.

LSB galaxies share the space density and clustering properties of Lyman-$\alpha$ absorbers seen at somewhat higher redshifts. Rauch & Haehnelt (1995) showed that for uniform clouds of thickness $D$ and temperature $T$ in a UV ionizing background $I$ (in units of $10^{-21}$ erg Hz$^{-1}$ sr$^{-1}$ s$^{-1}$ cm$^{-2}$), $\Omega_{\rm baryon} \approx 0.027h_{100}^{-1}$ $(T/3 \times 10^4\ {\rm K})^{1/3}\ I^{1/2}\ (D/100\ {\rm kpc})^{1/2}$. If the clouds are large and elongated, most baryons may be in this form. Direct association of low redshift absorbers with LSB galaxies is unlikely; it has been ruled out in some cases (Rauch et al 1996). On the other hand, the association of the absorbers with bright galaxies is difficult to prove; all that the observations provide is an impact parameter, and there may be undetected galaxies closer to the line of sight. Large samples are required before the kinematics and peculiar velocities of the absorbers can be compared with similar information obtained for galaxies along the line of sight. Simulations that incorporate gas dynamics have shed some light on the situation. The low column density HI absorbers form a filamentary network that traces out the dark matter potential of large-scale structure (Cen et al 1994, Petitjean et al 1995, Hernquist et al 1996). The copious number of observed

LSB dwarfs might be associated with collapsed regions within a more extensive network of diffuse hydrogen.

## *Galaxy Formation*

LSB GALAXIES AT COSMOLOGICAL DISTANCES The discussion so far has concentrated on the properties of LSB galaxies in the local universe. However, these unassuming stellar systems have great relevance for surveys of galaxies at cosmological distances and significant look-back times.

The visibility of galaxies is defined by the isophotal limits of the survey, but it is also a strong function of galaxy redshift and type. As first discussed by Phillipps et al (1990), galaxies in deep surveys are more likely to be blue and of high surface brightness (HSB), since those types have a large accessible volume. The cosmological corrections give

$$(\mu_{\rm lim} - \mu_0)_{\rm obs} = (\mu_{\rm lim} - \mu_0)_{z=0} - 10\log(1+z) - k(z) \tag{21}$$

and

$$M_{\rm B} = m_{\rm B} - 5\log\{(2c/H_0)(1+z)[1-(1+z)^{-1/2}]\} - 25 - k(z), \tag{22}$$

where $k(z)$ are the $k$ corrections. As we consider galaxies at higher and higher redshifts, the visibility is reduced by a combination of $k$ corrections and the Tolman $(1+z)^4$ cosmological dimming of the surface brightness (Phillipps et al 1990). Figure 6 shows the visibility distance for spirals (panel *a*) and ellipticals (panel *b*), assuming an angular diameter–limited survey with three different values of $\theta_{\rm lim}$ plotted. The dashed curves show the selection functions with cosmological effects ($k$ corrections, Tolman surface brightness dimming) included. The peak visibility of ellipticals at zero redshift is 30% smaller than that of spirals, and ellipticals have larger $k$ corrections. The result is that ellipticals are much less visible than spirals in deep optical surveys.

The isophotal limits of the pencil-beam CCD surveys are so deep that those surveys are sensitive to galaxies that are absent from local wide field surveys. The curves superimposed on Figure 2 illustrate this important point (see also McGaugh 1994b). The far left solid curve shows the selection function of the APM LSB survey (Sprayberry et al 1996), assuming $\mu_{\rm lim} = 26$ $B$ mag arcsec$^{-2}$ and $\theta_{\rm lim} = 10$ arcsec. The two curves to the right show the equivalent selection function at $z = 0.1$ and $z = 0.5$. Only the most luminous and HSB galaxies at $z = 0.5$ are detected on Schmidt photographic plates. Whereas about 30% of RC3 galaxies fit this criterion, only about 10% of the APM sample do. The dashed curve shows the selection function at $z = 0.5$ for the deep CCD survey of Lilly (1993), with limits of $\mu_{\rm lim} = 28.6$ $B$ mag arcsec$^{-2}$ and $\theta_{\rm lim} = 2$ arcsec. The CCD survey encompasses a larger region of ($\mu$, $L$) space at $z = 0.5$ than the APM survey does in the local universe. In other words, deep CCD surveys

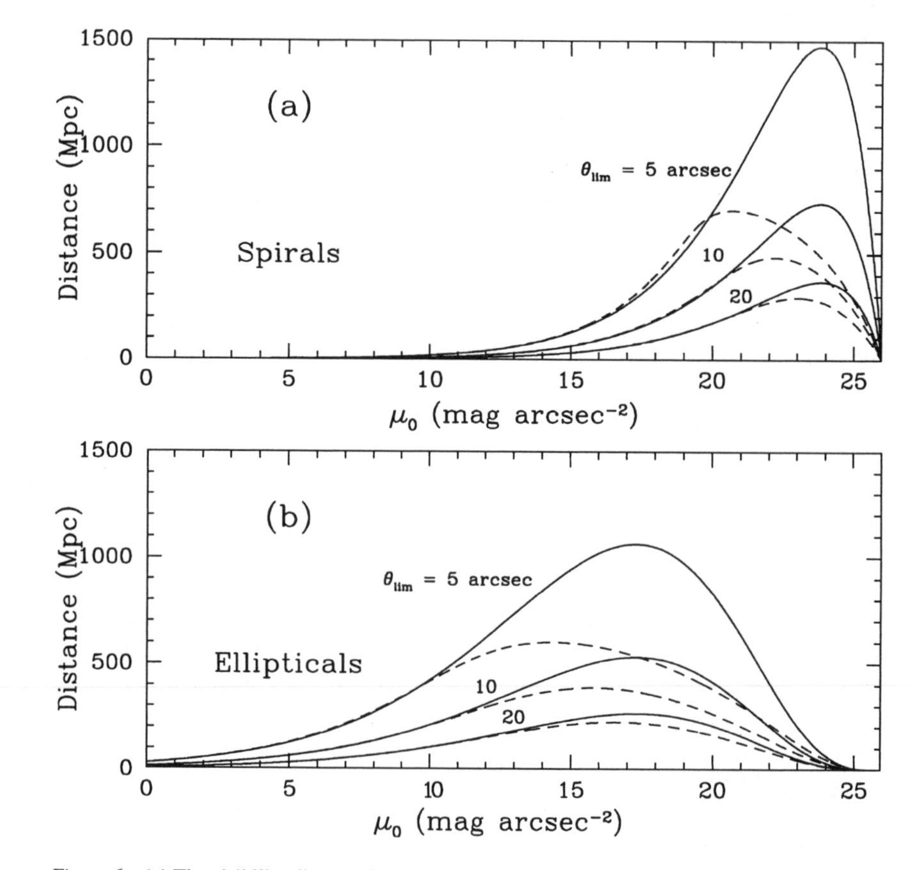


*Figure 6* (*a*) The visibility distance for spirals assuming three different values of the angular size limit in arcseconds. The dashed curves show the effect of cosmological corrections to the visibility distance. (*b*) As above, for elliptical galaxies. In both cases, $M_B = -21$ has been assumed; the curves for other cases scale with the luminosity distance. (From Phillipps et al 1990.)

are sensitive to a potentially large number of galaxies (not necessarily dwarfs) that are essentially absent from published wide field surveys.

RELATION TO DEEP GALAXY SURVEYS The number of faint galaxies in the $B$ and $I$ bands exceeds the expectations of all cosmological models that do not include evolution (Tyson & Jarvis 1979, Kron 1982, Tyson 1988a, Lilly et al 1991). The excess of blue galaxies must however be reconciled with the observed redshift distribution at $B = 24$, which is consistent with a no-evolution prediction. Many of the solutions that have been proposed so far are problematic. The excess population at $z = 0.3$–$0.5$ cannot be removed by merging because the faint blue galaxies are not strongly enough clustered (Efstathiou et al 1991), and the requisite merger remnants cannot be found (Dalcanton 1993). Selective luminosity evolution has also been proposed, where a high space density of bright dwarfs evolves strongly to become a common but faint population locally (Broadhurst et al 1988, Lilly 1993). This rapid and strong evolution has not been observed to occur. The distasteful option of a cosmological constant has probably been ruled out by the statistics of gravitational lensing (Fukagita & Turner 1991). More conservative solutions involve adjustments of the local luminosity functions of the different galaxy types (Koo et al 1993). For an extensive discussion, see the article by Ellis (1997) in this volume.

McGaugh (1994b) has shown that LSB disks with $\mu_0 \sim 23.5$ mag arcsec$^{-2}$ and $s \sim 3$ kpc can be readily detected by ultra-deep surveys out to $z \sim 0.8$, because the high sensitivity of CCD surveys more than compensates for cosmological dimming and $k$ corrections. However, a disk galaxy with the properties just described will only be detectable in the wide area survey of Loveday et al (1992) out to $z = 0.02$, with only one third of the flux detected above the limiting isophote. Both the numbers and fluxes of LSB galaxies are underestimated by large area photographic surveys. To summarize, the optimum surface brightness for galaxy detection moves to higher *intrinsic* surface brightness with redshift, to counteract the effects of $k$ corrections and cosmological dimming. These distance-dependent selection effects can mimic evolutionary changes in the galaxy population and distort the observed luminosity function at high redshift.

Sprayberry et al (1997) carried out a morphological separation of the LSB galaxies with redshifts from the APM survey, of which ~50% are spirals and ~40% are irregulars (the remainder have peculiar morphologies). Most of the rise in the numbers fainter than $M_B = -16$ is accounted for by irregulars. This joins a growing list of studies that find gas-rich irregulars to be a steeply rising component of the field galaxy luminosity function. Although the statistics are poor, Marzke et al (1994a) found a faint end slope of $\alpha = -1.87$ for the irregular Sm-Im types. Schade & Ferguson (1994) deduced $\alpha = -1.63$ for low luminosity star-forming galaxies, based on a reinterpretation of the survey of Salzer (1989, see also Boroson et al 1993). Driver et al (1994b) have suggested a dwarf-rich model to explain deep *BVRI* galaxy counts. The model with $\alpha = -1.8$ for the dwarfs can be used to fit the $B$ band counts down to $B = 27.5$, but such a model predicts lower redshifts than are observed by spectroscopy down to $B = 24$ (Cowie et al 1991). More recently, Driver et al (1995a) have measured the morphologies of galaxies from a single ultra-deep WFPC2 image. At a depth corresponding to $B \approx 26$, half the galaxies are late-type spirals or irregulars, and some combination of rapid evolution and a steep faint end slope for the local dwarf luminosity function is required to account for the data.

The local low luminosity LSB population can be compared with the faint blue galaxies in the Medium Deep Survey (MDS) of Driver et al (1995b), which are taken to be at a typical redshift of $z \sim 0.3$. LSB field galaxies with $M_B > -16$ are at typical distances of $10 < d < 40h_{100}^{-1}$ Mpc. They have central surface brightnesses in the range $22 < \mu < 25$ $B$ mag arcsec$^{-2}$ and effective angular radii of $6 < r_{\rm eff} < 20$ arcsec (Impey et al 1996). If they are related to the LSB dwarfs in clusters, they will have $B-V \sim 0.5$ (Impey et al 1988). The late-type and irregular (Sdm/Irr) MDS galaxies have median effective radii of 0.4 arcsec (Im et al 1995b), which would scale to 20 arcsec for a local population. The central surface brightnesses of the MDS sample convert into a range $23 < \mu < 24$ $B$ mag arcsec$^{-2}$ locally, assuming no evolution. The Sdm/Irr MDS galaxies have colors $V - I \sim 1$, also consistent with a local color of $B - V \sim 0.5$, again assuming no evolution. In addition, Dalcanton & Schectman (1996) showed that the faint blue "chain" galaxies found by Cowie et al (1995b) can be plausibly related to edge-on LSB disk galaxies. Both the faint blue galaxies (Efstathiou et al 1991) and the local LSB galaxies (Mo et al 1994) are weakly clustered.

Although local LSB galaxies can be shown to have similar properties to many of the faint blue galaxies, it is not clear that the high and low redshift populations can be related with a consistent evolutionary scheme. All such comparisons must take account of the change in morphology with luminosity at any redshift, from luminous E/S0/Sp types down to dwarf E/S0/Im types. Detailed modeling is hampered by the factor of 2 disagreement between the normalizations of the deep survey luminosity functions (Ellis et al 1996, Cowie et al 1996), which are probably caused by large-scale structure along the pencil-beam surveys. Generally speaking, deep redshift surveys point to two populations at $z \sim 0.5$: red galaxies that evolve very little and a rapidly evolving (and presumably fading) blue population (Lilly et al 1996, Ellis et al 1996). Ferguson & McGaugh (1995) have demonstrated that LSB galaxies could be a substantial contributor to the faint galaxy counts. They considered two rather extreme populations in the $(\mu, L)$ plane: galaxies with $L$ and $\mu$ uncorrelated and galaxies with a tight correlation between $L$ and $\mu$. Redshifts for the APM sample (Impey et al 1996) indicate that the truth lies between these two situations (see Figure 2). Babul & Ferguson (1996) have used plausible evolutionary models to make detailed predictions of faint blue galaxies that fade and redden, based on the earlier model of Babul & Rees (1992). There are a number of difficulties in identifying local LSB galaxies as the faded remnants of the faint blue galaxy population. If faded red galaxies are present in sufficient profusion, they should appear in the low redshift bins of the deep redshift surveys (Lilly et al 1995, Bouwens & Silk 1996), and they should contribute to the deep counts around $B = 26$ (Driver & Phillipps 1996). In addition, the nearby LSB population is too blue to fit a fading scenario, and the disks appear to be evolving slowly not rapidly (McGaugh 1994a, Ferguson & McGaugh 1995). Fading scenarios usually invoke supernova-driven gas ejection during a burst of star formation, which conflicts with the large gas content of the local sample. The existing uncertainty strongly motivates new searches for dim and diffuse galaxies.

FORMATION OF LSB GALAXIES Finally, LSB galaxies can be placed in the context of galaxy formation and evolution in the broadest sense. Figure 7*a* shows a schematic version of the bivariate luminosity function of galaxies in stars

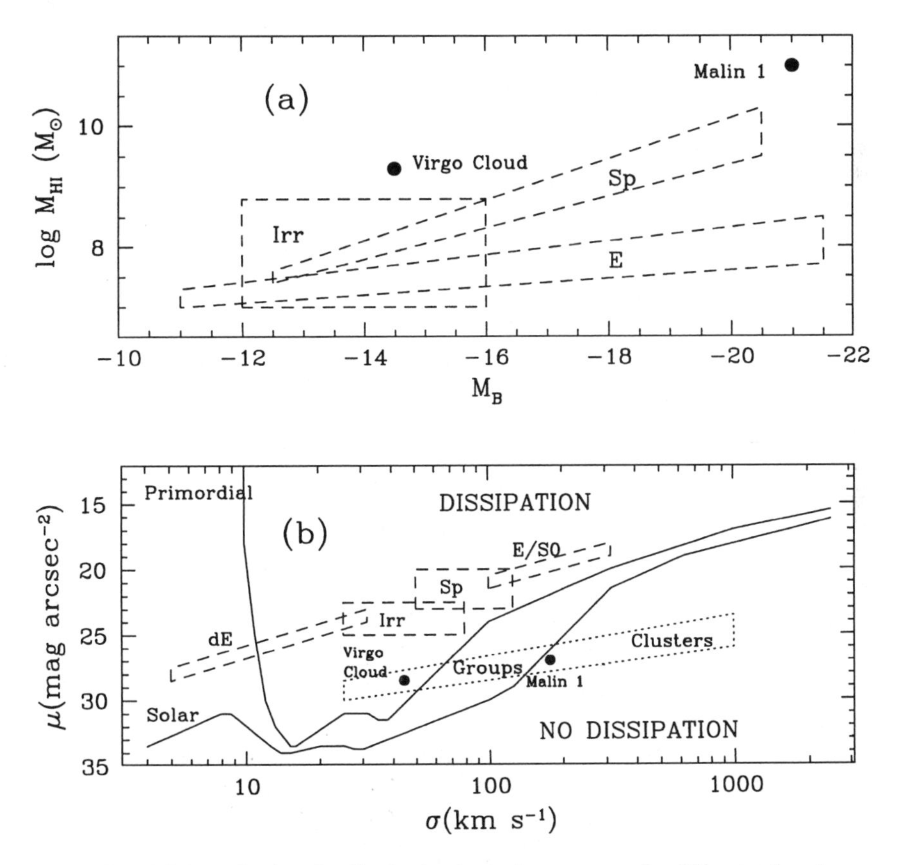


*Figure 7* (*a*) Schematic plot of stellar luminosity against gas mass for different galaxy types. LSB galaxies of various types lie not far off the relationship defined by most galaxies. (From Giovanelli & Haynes 1989, Bothun et al 1987.) (*b*) Plot of surface brightness in the B band or surface mass density against velocity dispersion for various cosmic structures, assuming dissipation from intercloud collisions. The giant LSB disk galaxies have the dynamic properties of groups of galaxies. (From Efstathiou & Silk 1983; see also Burstein et al 1995.)

measured by luminosity and surface brightness and gas measured by the 21-cm line. Extreme LSB galaxies like Malin 1 and the Virgo cloud (Giovanelli & Haynes 1989) are formidable. Because starbursting dwarfs do not have the gas supply to sustain their star formation for long, we might wonder how to detect the larger number of quiescent dwarfs. Our understanding of what drives the star formation history and visibility of galaxies, through the conversion of $M_{\rm HI}$ into $M_{\rm B}$, is still very incomplete.

It appears that the formation of large numbers of LSB galaxies is an unavoidable consequence of hierarchical schemes for the formation of large-scale structure. In the standard picture of the spherical collapse of a top-hat density perturbation (Gunn & Gott 1972), smaller perturbations reach their maximum extents more slowly and collapse at later times. Assuming Gaussian fluctuations, small amplitude peaks will be found preferentially in under-dense regions, such that galaxies that result from small amplitude peaks are less correlated than galaxies that collapse earlier from large amplitude peaks (Kaiser 1984). It is natural to associate dwarf galaxies with the 1–2$\sigma$ peaks (Dekel & Silk 1986). Mo et al (1994) showed that the clustering properties of LSB galaxies were consistent with late formation. The giant gas-rich galaxies like Malin 1 are more likely to originate from rare 3$\sigma$ peaks in under-dense regions because the gas disks are vulnerable to interactions in over-dense regions (Hoffman et al 1992).

In their application of the formalism of Press & Schechter (1974), Dalcanton et al (1997) find that the mean baryonic surface mass density $\bar{\Sigma} \propto F r_* \Lambda^3(\lambda)$ $(\delta\rho/\rho)^3$ in the case $\Omega = 1$. The mean surface density is defined with a region of size $r_*$, which has a collapse factor $\Lambda(\lambda)$, where $\lambda$ is the dimensionless spin parameter. The time scale of violent relaxation and virialization is $\tau_{\rm vir} \propto [(\delta\rho/\rho) + 1]^{-1/2}$, and the virialized radius is $R_{\rm vir} = R_{\rm init}(1+z)^{-1}(1 + 178\ \Omega^{-0.6})^{-1/3}$. Furthermore, the amplitude of the correlation function traced by galaxies is $\bar{\Sigma} \propto \Lambda^3(\lambda)\ \xi(r)^{3/2}$, so that the correlation amplitude of galaxies only falls as a weak function of surface mass density. This is consistent with the clustering amplitude of LSB galaxies measured by Mo et al (1994). The total mass density of LSB galaxies in hierarchical clustering schemes can exceed the mass density of normal, high surface brightness galaxies.

Figure 7*b* shows the role that dissipation plays in structure formation in the universe (e.g. Efstathiou & Silk 1983). In this plot of surface brightness (or equivalently, surface mass density) versus velocity dispersion, the solid lines show the demarcation between structures that have undergone dissipation and those that have not. This simple model assumes that star formation occurs by radiative shocks in intercloud collisions. If either the surface mass density is too low or the velocity dispersion is too high, the cooling is inefficient and so there is no compression and no subsequent star formation. The stellar velocity dispersion is therefore a fossilized measure of the cloud velocity dispersion at the epoch when most of the dissipation (and star formation) occurred.

In hierarchical clustering, the relationship between surface mass density and velocity dispersion for nondissipative structures is $\bar{\Sigma} \propto \sigma^{-4(n+2)/(1-n)}$, where $n$ is the spectral index of the initial fluctuations (Efstathiou & Silk 1983). Clustering goes towards increasing $\sigma$ if $n < 1$ and towards increasing $\bar{\Sigma}$ if $n > -2$. The eventual goal is to understand LSB systems well enough to interpret them in terms of an "HR" diagram for galaxies, a goal that is being approached for dynamically hot systems (Bender et al 1993). This goal is plausible as long as the structure of the dissipative baryonic matter is directly related to the structure of the dark halo (Faber 1982). Figure 7*b* illustrates that extreme LSB galaxies like Malin 1 and the Virgo cloud have similar properties to small groups or clusters of galaxies (Burstein et al 1995). They are displaced by a factor of 100 in surface mass density from the conventional galaxies of the Hubble sequence. This suggests an alternate evolutionary history that involves late collapse and dissipation, slow evolution, and a low density cosmic environment.

LSB galaxies play a key role in this cosmogony. Navarro et al (1996) argue that dark matter halos that formed through dissipationless hierarchical clustering have a universal mass profile. Each halo mass is specified solely by its characteristic overdensity, which depends on the mean density at the epoch of collapse. HSB galaxies have fast-rising rotation curves that flatten off to outer regions of dark matter domination. LSB disks have slowly rising rotation curves that are dark-matter dominated at almost all radii. Pickering et al (1997) show examples that can be traced out to $\sim 100 h_{100}^{-1}$ kpc. In general, LSB galaxies have high angular momentum disks, with larger scale lengths at the same circular velocity (Zwaan et al 1995). Dalcanton et al (1997) describe the evolution of high angular momentum and low surface mass density disks, and they provide a context for understanding the broad relationship between luminosity and surface brightness and for understanding the properties of the local luminosity function. Moreover, the Cold Dark Matter model predicts far too many dark halos to match the cataloged galaxy population (White & Frenk 1991). Many dark halos may have failed to form galaxies or may have formed galaxies that have so far escaped detection. Observational limitations and theoretical expectations align to point to the importance of LSB galaxies.

## ACKNOWLEDGMENTS

We acknowledge inspiration and provocation from our former students, Stacy McGaugh and David Sprayberry; our current students, Tim Pickering and Karen O'Neil; and our longtime collaborators, Jim Schombert and David Malin. A number of colleagues contributed greatly to the results and ideas described in this review—Bruno Binggeli, Frank Briggs, Julianne Dalcanton, Simon Driver, Harry Ferguson, Mike Irwin, Pat Knezek, Ron Marzke, Allan Sandage, and Steve Schneider. We are grateful to the Welsh "dragons," Mike Disney, Steve Phillipps, and Jon Davies, for breathing life and fire into this subject over a number of years. We acknowledge support from the NSF under grant AST-9003158.



*Literature Cited*

Allen R, Shu F. 1979. *Ap. J.* 227:67
Arp HC. 1965. *Ap. J.* 145:402
Babul A, Ferguson HC. 1996. *Ap. J.* 458:100
Babul A, Rees MJ. 1992. *MNRAS* 255:346
Bahcall JN, Bergeron J, Boksenberg A, Hartig GF, Jarmuzi BT, et al. 1993. *Ap. J. Suppl.* 87:1
Bender R, Burstein D, Faber SM. 1993. *Ap. J.* 411:153
Bergeron J. 1995. In *ESO Workshop on Quasar Absorption Lines*, ed. G Meylan, pp. 127–36. Dordrecht: Springer-Verlag
Bergeron J, Boisse P. 1991. *Astron. Astrophys.* 243:344
Bernstein GM, Nichol RG, Tyson AJ, Ulmer MP, Wittman D. 1995. *Astron. J.* 110:1507
Bershady MA, SubbaRao MU, Szalay A, Koo DC, Kron RG, Hereld M. 1997. *Ap. J.* Submitted
Binggeli B, Sandage A, Tammann GA. 1985. *Astron. J.* 90:1681
Binggeli B, Sandage A, Tammann GA. 1988. *Annu. Rev. Astron. Astrophys.* 26:509
Binggeli B, Sandage A, Tarenghi M. 1984. *Astron. J.* 89:64
Binggeli B, Tarenghi M, Sandage A. 1990. *Astron. Astrophys.* 228:42
Boroson TA. 1981. *Ap. J. Suppl.* 46:177
Boroson TA, Salzer JJ, Trotter A. 1993. *Ap. J.* 412:524
Bosma A, Freeman KC. 1993. *Astron. J.* 106:1394
Bothun GD. 1985. *Astron. J.* 90:1982
Bothun GD, Impey CD, Malin DF. 1991. *Ap. J.* 376:404
Bothun GD, Impey CD, Malin DF, Mould JR. 1987. *Astron. J.* 94:23
Bothun GD, Schombert JM, Impey CD, Schneider SE. 1990. *Ap. J.* 360:427
Bouwens RJ, Silk J. 1996. *Ap. J. Lett.* 471:L19
Briggs FH. 1990. *Astron. J.* 100:999
Bristow PD, Phillips S. 1994. *MNRAS* 267:13
Broadhurst TJ, Ellis RS, Shanks TS. 1988. *MNRAS* 235:827
Broeils AH. 1992. *Dark and visible mass in spiral galaxies.* PhD thesis. Univ. Groningen
Bruzual GA, Charlot S. 1993. *Ap. J.* 405:538
Burstein D, Bender R, Faber SM, Nolthenius R. 1995. *Astron. Lett. Comm.* 31:95
Caldwell N, Bothun GD. 1987. *Astron. J.* 94:1126
Capaccioli A, de Vaucouleurs G. 1983. *Astron. Astrophys. Suppl.* 52:465
Carr BJ. 1994. *Annu. Rev. Astron. Astrophys.* 32:531
Cayette V, Kotanyi C, Balkowski C. 1993. *Astron. J.* 107:1003
Cen R, Miralda-Escude J, Ostriker JP, Rauch M. 1994. *Ap. J. Lett.* 437:L9
Cole S, Treyer M, Silk J. 1992. *Ap. J.* 385:9
Copi CJ, Schramm DN, Turner MS. 1995. *Nature* 267:192
Corbelli E, Salpeter EE. 1993. *Ap. J.* 419:104
Courteau S. 1996. *Ap. J. Suppl.* 103:363
Cowie LL. 1988. In *The Post-Recombination Universe*, ed. N Kaiser, A Lazenby, pp. 1–18. Dordrecht: Kluwer
Cowie LL, Hu EM, Songaila A. 1995a. *Nature* 377:603
Cowie LL, Hu EM, Songaila A. 1995b. *Astron. J.* 109:1522
Cowie LL, Songaila A, Hu EM. 1991. *Nature* 354:460
Cowie LL, Songaila A, Hu EM, Cohen JG. 1996. *Astron. J.* 112:839
da Costa G, Nicolaci L, Geller MJ, Pellegrini PS, Latham DW, et al. 1994. *Ap. J. Lett.* 424:1
Dalcanton JJ. 1993. *Ap. J. Lett.* 415:L87
Dalcanton JJ. 1995. *The nature of ultra-low surface brightness objects.* PhD thesis. Univ. Princeton
Dalcanton JJ, Schectman SA. 1996. *Ap. J. Lett.* 465:L9
Dalcanton JJ, Spergel DN, Summers FJ. 1997a. *Ap. J.* Submitted
Dalcanton JJ, Spergel DN, Gunn JE, Schmidt M, Schneider DP. 1997b. *Ap. J.* Submitted

Davies JI. 1990. *MNRAS* 244:8
Davies JI, Phillips S, Cawson MGM, Disney MJ, Kibblewhite EJ. 1988. *MNRAS* 232: 239
Davies JI, Phillips S, Disney MJ. 1990. *MNRAS* 244:385
Dekel A, Silk J. 1986. *Ap. J.* 303:39
de Blok WJG, McGaugh SS, van der Hulst JM. 1996. *MNRAS* 283:18
de Blok WJG, van der Hulst JM, Bothun GD. 1995. *MNRAS* 274:235
de Jong RS. 1995. *Spiral galaxies: the light and color distributions in the optical and near-infrared.* PhD thesis. Univ. Groningen
de Jong RS, van der Kruit PC. 1994. *Astron. Astrophys. Suppl.* 106:451
de Propris R, Pritchet CJ, Harris WE, McClure RD. 1995. *Ap. J.* 450:534
de Vaucouleurs G, de Vaucouleurs A, Corwin HG, Buta R, Paturel G, Fouque P. 1991. *Third Reference Catalog of Bright Galaxies.* New York: Springer-Verlag
Dinshaw N, Foltz CB, Impey CD, Weymann RJ, Morris SL. 1995. *Nature* 373:223
Disney MJ. 1976. *Nature* 263:573
Disney MJ, Banks G. 1997. *MNRAS.* Submitted
Disney MJ, Phillips S. 1983. *MNRAS* 205: 1253
Dove JB, Shull JM. 1994. *Ap. J.* 423:196
Driver SP, Phillips S. 1996. *Ap. J.* 469:529
Driver SP, Phillips S, Davies JI, Morgan I, Disney MJ. 1994a. *MNRAS* 268:393
Driver SP, Phillips S, Davies JI, Morgan I, Disney MJ. 1994b. *MNRAS* 266:155
Driver SP, Windhorst RA, Ostrander EJ, Keel WC, Griffiths RE, Ratnatunga KU. 1995a. *Ap. J. Lett.* 449:L23
Driver SP, Windhorst RAW, Griffiths RE. 1995b. *Ap. J.* 453:48
Efstathiou G, Bernstein G, Tyson JA, Katz N, Guhathakurta P. 1991. *Ap. J. Lett.* 380:L47
Efstathiou G, Ellis RS, Peterson BA. 1988. *MNRAS* 232:431
Efstathiou G, Silk J. 1983. *Fundam. Cosm. Phys.* 9:1
Ellis RS. 1997. *Annu. Rev. Astron. Astrophys.* 35:389
Ellis RS, Colless M, Broadhurst T, Heyl J, Glazebrook K. 1996. *MNRAS* 280:235
Faber SM. 1982. In *Astrophysical Cosmology,* ed. HA Bruck, GV Coyne, MS Longair, pp. 191–215. Vatican: Pontif. Acad. Sci.
Felten JE. 1977. *Astron. J.* 82:861
Ferguson HC. 1989. *Astron. J.* 98:367
Ferguson HC, Binggeli B. 1994. *Astron. Astrophys. Rev.* 6:67
Ferguson HC, McGaugh SS. 1995. *Ap. J.* 440:470
Ferguson HC, Sandage A. 1988. *Astron. J.* 96:1520
Fish RA. 1964. *Ap. J.* 139:284
Fisher JR, Tully RB. 1975. *Astron. Astrophys.* 44:151
Fisher JR, Tully RB. 1977. *Ap. J. Lett.* 243:23
Fisher JR, Tully RB. 1981. *Ap. J. Suppl.* 47:139
Freedman WL, Madore BF, Kennicutt RC. 1997. In *The Extragalactic Distance Scale,* ed. M Livio, M Donohue, N Panagia. Cambridge: Cambridge Univ. Press. In press
Freeman KC. 1970. *Ap. J.* 160:811
Fukagita M, Turner EL. 1991. *MNRAS* 253: 99
Gallagher JS, Hunter DA. 1984. *Annu. Rev. Astron. Astrophys.* 22:37
Giovanelli R, Haynes MP. 1988. *Astron. J.* 97:633
Giovanelli R, Haynes MP. 1989. *Ap. J. Lett.* 396:L5
Grogin NA, Narayan R. 1996. *Ap. J.* 464:92
Gunn JE, Gott JR. 1972. *Ap. J.* 176:1
Gunn JE, Knapp GR. 1993. In *Sky Surveys: Protostars to Protogalaxies,* ed. BT Soifer, pp. 267–80. San Francisco: Astron. Soc. Pac.
Harrison ER. 1964. *Nature* 204:271
Hawarden TG, Longmore AJ, Goss WM, Mebold U, Tritton SB. 1981. *MNRAS* 196:175
Hernquist L, Katz N, Weinberg DH, Miralda-Escude J. 1996. *Ap. J. Lett.* 457:L51
Hoffman GL, Lewis BM, Helou G, Salpeter EE, Williams HL. 1989. *Ap. J. Suppl.* 69:65
Hoffman Y, Silk J, Wyse RFG. 1992. *Ap. J. Lett.* 388:13
Holmberg E. 1975. In *Stars and Stellar Systems,* Vol. 9, ed. A Sandage, M Sandage, J Kristian, p. 123. Chicago: Univ. Chicago Press
Hubble E. 1936. *The Realm of the Nebulae.* New Haven, CT: Yale Univ. Press
Im M, Casretano S, Griffiths RE, Ratnatunga JU, Tyson JA. 1995a. *Ap. J.* 441:494
Im M, Ratnatunga KU, Griffiths RE, Casertano S. 1995b. *Ap. J. Lett.* 445:15
Impey CD, Bothun GD. 1989. *Ap. J.* 341:89
Impey CD, Bothun GD, Malin DF. 1988. *Ap. J.* 330:634
Impey CD, Sprayberry D, Irwin MJ, Bothun GD. 1996. *Ap. J. Suppl.* 105:209
Irwin MJ, Bunclark PS, Bridgeland MT, McMahon RG. 1990a. *MNRAS* 244:16P
Irwin MJ, Davies JI, Disney MJ, Phillips S. 1990b. *MNRAS* 245:289
Irwin MJ, Hatzidimitriou D. 1995. *MNRAS* 277:1354
Kaiser N. 1984. *Ap. J. Lett.* 284:L9
Kennicutt RC. 1983. *Ap. J.* 272:54
Kennicutt RC. 1989. *Ap. J.* 344:685
Knezek PM. 1993. *The stellar and gaseous content of giant low surface brightness galaxies.* PhD thesis. Univ. Massachusetts
Koo DC, Gronwall C, Bruzual G. 1993. *Ap. J. Lett.* 415:L21
Kormendy J. 1977. *Ap. J.* 217:406

Kormendy J. 1985. *Ap. J.* 295:73
Kron RG. 1982. *Vistas in Astronomy* 26:37
Lanzetta KM, Wolfe AM, Turnshek DA. 1995. *Ap. J.* 440:435
Lanzetta KM, Wolfe AM, Turnshek DA, Lu L, McMahon RG, Hazard C. 1991. *Ap. J. Suppl.* 77:1
Lilly SJ. 1993. *Ap. J.* 411:502
Lilly SJ, Cowie LL, Gardner JP. 1991. *Ap. J.* 369:79
Lilly SJ, LeFevre O, Hammer F, Crampton D. 1996. *Ap. J. Lett.* 460:L1
Lin H, Kirshner RP, Schectman SA, Landy SD, Oemler A. 1996. *Ap. J.* 464:60
Longmore AJ, Hawarden TG, Goss WM, Mebold U, Webster BL. 1982. *MNRAS* 200:325
Loveday J, Peterson BA, Efstathiou G, Maddox SJ. 1992. *Ap. J.* 390:338
Malaney P, Mathews G. 1993. *Phys. Rep.* 229:147
Malin DF. 1978. *Nature* 276:591
Maloney P. 1993. *Ap. J.* 414:41
Marzke RO, da Costa LN. 1997. *Astron. J.* 113:185
Marzke RO, Geller MJ, Huchra JP, Corwin HG Jr. 1994a. *Astron. J.* 108:437
Marzke RO, Huchra JP, Geller MJ. 1994b. *Ap. J.* 428:43
Mateo M. 1996. In *Formation of the Galactic Halo, Inside and Out*, ed. H Morrison, A Sarajedini, p. 434. San Francisco: Astron. Soc. Pac.
Mattila K. 1990. In *The Galactic and Extragalactic Background Radiation*, ed. S Bowyer, C Leinert, p. 257. Dordrecht: Kluwer
Mattila K, Leinart Ch, Schnur G. 1991. In *The Early Observable Universe from Diffuse Backgrounds*, ed. B Rocca-Volmerange, JM Deharvang, J Tran Thanh Van, p. 133. France: Ed. Front.
McGaugh SS. 1992. *The physical properties of low surface brightness galaxies.* PhD thesis. Univ. Michigan, Ann Arbor
McGaugh SS. 1994a. *Ap. J.* 426:135
McGaugh SS. 1994b. *Nature* 367:538
McGaugh SS. 1996. *MNRAS* 280:337
McGaugh SS, Bothun GD. 1994. *Astron. J.* 107:530
McGaugh SS, Bothun GD, Schombert JM. 1995. *Astron. J.* 110:573
Meurs EJA, Wilson AS. 1984. *Astron. Astrophys.* 136:206
Meylan G, ed. 1995. *ESO Workshop on Quasar Absorption Lines.* Dordrecht: Springer-Verlag
Mo HJ, McGaugh SS, Bothun GD. 1994. *MNRAS* 267:129
Morgan I, Driver SP. 1995. In *Extragalactic Background Radiation*, ed. D Calzetti, M Livio, P Madau, pp. 285–88. Cambridge: Cambridge Univ. Press
Morris SL, Weymann RJ, Savage BD, Gilliland RL. 1991. *Ap. J. Lett.* 377:L21
Nilson PN. 1973. *Uppsala General Catalog of Galaxies.* Uppsala, Finland: Uppsala Astron. Obs.
O'Connell RW. 1987. *Astron. J.* 94:876
O'Neil K, Bothun GD, Cornell M, Impey CD. 1997. *Astron. J.* Submitted
O'Neil K, Bothun GD, Smith EP, Stecher TP. 1996. *Astron. J.* 112:431
Peebles PJE. 1993. *Principles of Physical Cosmology*. Princeton: Princeton Univ. Press. 391 pp.
Peebles PJE, Schramm D, Turner E, Kron R. 1991. *Nature* 352:769
Persic M, Salucci P. 1990. *MNRAS* 245:57
Persic M, Salucci P. 1992. *MNRAS* 258:14P
Petitjean P, Mucket JP, Kates RE. 1995. *Astron. Astrophys.* 295:L9
Phillipps S, Davies JI, Disney MJ. 1990. *MNRAS* 242:235
Phillipps S, Disney MJ. 1985. *Astron. Astrophys.* 148:234
Phillipps S, Disney MJ, Davies JI. 1993. *MNRAS* 260:453
Phillipps S, Disney MJ, Kibblewhite EJ, Cawson MGM. 1987. *MNRAS* 229:505
Phillipps S, Driver SP. 1995. *MNRAS* 274:832
Phillipps S, Fong R, Shanks T. 1981. *MNRAS* 194:49
Pickering TE, Impey CD, van Gorkom J, Bothun GD. 1997. *Ap. J.* Submitted
Press WH, Schechter PL. 1974. *Ap. J.* 330:579
Quirk WJ. 1972. *Ap. J. Lett.* 176:9
Rao SM, Briggs FH. 1993. *Ap. J.* 419:515
Rao SM, Turnshek DA, Briggs FH. 1995. *Ap. J.* 449:488
Rauch M, Haehnelt MG. 1995. *MNRAS* 275:76
Rauch M, Weymann RJ, Morris SL. 1996. *Ap. J.* 458:518
Reaves G. 1956. *Astron. J.* 61:69
Romanishin W, Krumm N, Salpeter E, Knapp G, Strom KM, Strom SE. 1982. *Ap. J.* 263:94
Romanishin W, Strom KM, Strom SE. 1983. *Ap. J.* 263:94
Saha A, Sandage A, Labhardt L, Tammann GA, Macchetto FD, Panagia N. 1996. *Ap. J. Suppl.* 107:693
Salzer JJ. 1989. *Ap. J.* 347:152
Sandage A. 1988. *Annu. Rev. Astron. Astrophys.* 26:561
Sandage A, Binggeli B, Tammann GA. 1985. *Astron. J.* 90:1759
Sandage A, Tammann G, Yahil A. 1979. *Ap. J.* 232:352
Sargent WLW, Steidel CC, Boksenberg A. 1988. *Ap. J.* 334:22
Schade DJ, Ferguson HC. 1994. *MNRAS* 267:889

Schechter P. 1976. *Ap. J.* 203:297
Schectman SA. 1973. *Ap. J.* 188:233
Schectman SA. 1974. *Ap. J.* 179:681
Schombert JM, Bothun GD. 1988. *Astron. J.* 95:1389
Schombert JM, Bothun GD, Schneider SE, McGaugh SS. 1992. *Astron. J.* 103:1107
Schombert JM, Pildis RA, Eder JA, Oemler A. 1995. *Astron. J.* 110:2067
Schneider SE. 1996. In *Minnesota Lectures on Extragalactic HI*, ed. E Skillman, pp. 323–48. San Francisco: Astron. Soc. Pac.
Schneider SE, Thuan TX, Magnum JG, Miller J. 1992. *Ap. J. Suppl.* 81:5
Schneider SE, Thuan TX, Magri C, Wadiak JE. 1990. *Ap. J. Suppl.* 72:245
Schwartzenberg JM, Phillips S, Parker QA. 1995a. *Astron. Astrophys.* 293:332
Schwartzenberg JM, Phillips S, Smith RM, Couch WJ, Boyle BJ. 1995b. *MNRAS* 275:171
Sprayberry D, Bernstein GM, Impey CD, Bothun GD. 1995a. *Ap. J.* 438:72
Sprayberry D, Impey CD, Bothun GD, Irwin MJ. 1995b. *Astron. J.* 109:558
Sprayberry D, Impey CD, Irwin MJ. 1996. *Ap. J.* 463:535
Sprayberry D, Impey CD, Irwin MJ, Bothun GD. 1997. *Ap. J.* In press
Sprayberry D, Impey CD, Irwin MJ, McMahon RG, Bothun GD. 1993. *Ap. J.* 417:114
Steidel CC. 1993. In *The Evolution of Galaxies and Their Environment, 3rd Teton Summer School*, ed. JM Shull, H Thronson, pp. 263–94. Dordtrecht: Kluwer
Steidel CC, Dickinson M. 1992. *Ap. J* 394:81
Steidel CC, Sargent WLW. 1992. *Ap. J. Suppl.* 80:1
Storrie-Lombardi LJ, McMahon RG, Irwin MJ, Hazard C. 1994. *Ap. J. Lett.* 427:L13
Szomoru A, Guhathakurta P, van Gorkom JH, Knapen JH, Weinberg DH, Fruchter AS. 1994. *Astron. J.* 108:491
Toomre A. 1964. *Ap. J* 139:1217
Trentham N. 1997. *MNRAS*. In press
Turner JA, Phillips S, Davies JI, Disney MJ. 1993. *MNRAS* 261:39
Tyson AJ. 1988a. *Astron. J.* 96:1
Tyson AJ. 1995. In *Extragalactic Background Radiation*, ed. D Calzetti, M Livio, P Madau, pp. 103–28. Cambridge: Cambridge Univ. Press
Tyson AJ, Jarvis JF. 1979. *Ap. J. Lett.* 230: L153
Tyson ND. 1988b. *Ap. J. Lett.* 329:L57
Tyson ND, Scalo JM. 1988. *Ap. J.* 329:618
Väisänen P. 1996. *Astron. Astrophys.* 315:21
van den Bergh S. 1959. *Publ. David Dunlap Obs.* 2:147
van der Hulst JM, Skillman, ED, Smith TR, Bothun GD, McGaugh SS, de Blok WJG. 1993. *Astron. J.* 106:548
van der Kruit PC. 1987. *Astron. Astrophys.* 173:59
van Gorkom J, Cornwell T, van Albada TS, Sancisi R. 1997. *Ap. J.* In press
Walker TP, Steigman G, Schramm DN, Olive KA, Kang H-S. 1991. *Ap. J.* 376:51
Warmels RH. 1986. *HI properties of spiral galaxies in the Virgo Cluster.* PhD thesis. Univ. Groningen
Weinberg DH, Szomoru A, Guhathakurta P, van Gorkom JH. 1991. *Ap. J. Lett.* 372:L13
White SDM, Frenk CS. 1991. *Ap. J.* 379:52
White SDM, Navarro JF, Evrard AE, Frenk CS. 1993. *Nature* 366:429
Wright EL. 1985. *Publ. Astron. Soc. Pac.* 97: 451
York DG, Dopita M, Green RF, Bechtold JB. 1986. *Ap. J.* 311:610
Zucca E. 1997. In *37th Herstmonceux Conference on HST and the High Redshift Universe*, ed. M Pettini. Cambridge: Cambridge Univ. Press. In press
Zwaan MA, van der Hulst JM, de Blok WJG, McGaugh SS. 1995. *MNRAS* 273:35
Zwicky F. 1957. *Morphological Astronomy.* New York: Springer-Verlag

*Annu. Rev. Astron. Astrophys. 1997. 35:309–55*


# OPTICAL SPECTRA OF SUPERNOVAE

*Alexei V. Filippenko*
Department of Astronomy, University of California, Berkeley, California 94720-3411;
e-mail: alex@astro.berkeley.edu

KEY WORDS: spectroscopy, stellar evolution, supernovae

ABSTRACT

The temporal evolution of the optical spectra of various types of supernovae (SNe) is illustrated, in part to aid observers classifying supernova candidates. Type II SNe are defined by the presence of hydrogen, and they exhibit a very wide variety of photometric and spectroscopic properties. Among hydrogen-deficient SNe (Type I), three subclasses are now known: those whose early-time spectra show strong Si II (Ia), prominent He I (Ib), or neither Si II nor He I (Ic). The late-time spectra of SNe Ia consist of a multitude of blended emission lines of iron-group elements; in sharp contrast, those of SNe Ib and SNe Ic (which are similar to each other) are dominated by several relatively unblended lines of intermediate-mass elements. Although SNe Ia, which result from the thermonuclear runaway of white dwarfs, constitute a rather homogeneous subclass, important variations in their photometric and spectroscopic properties are undeniably present. SNe Ib/Ic probably result from core collapse in massive stars largely stripped of their hydrogen (Ib) and helium (Ic) envelopes, and hence they are physically related to SNe II. Indeed, the progenitors of some SNe II seem to have only a low-mass skin of hydrogen; their spectra gradually evolve to resemble those of SNe Ib. In addition to the two well-known photometric subclasses (linear and plateau) of SNe II, which may exhibit minor spectroscopic differences, there is a new subclass (SNe IIn) distinguished by relatively narrow emission lines with little or no P Cygni absorption component and slowly declining light curves. These objects probably have unusually dense circumstellar gas with which the ejecta interact.

## 1. INTRODUCTION

The study of supernovae (SNe) has expanded tremendously during the past decade. A major motivation, of course, was provided by SN 1987A, by far



0066-4146/97/0915-0309$08.00

the most thoroughly observed supernova (SN) in history. Advances in the field have also been driven by technology: The advent of sensitive detectors, especially charge-coupled devices (CCDs), and the proliferation of moderately large telescopes made it possible to obtain excellent photometry and spectroscopy of large numbers of SNe. In addition to their intrinsic interest, SNe are relevant to nucleosynthesis and galactic chemical evolution, the production of neutron stars and black holes, the origin of cosmic rays, the physical state of the interstellar medium, and induced star formation; thus, they have been investigated from a wide range of perspectives. Finally, the enormous potential of SNe as cosmological distance indicators is inspiring many new studies.

This review concentrates primarily on the observed optical spectra of SNe, illustrating the temporal evolution of the major classes and subclasses. When combined with other observations and properly interpreted, such data can reveal the chemical composition of the ejecta, the nature of the progenitors, the explosion mechanisms, and even the distances of SNe. Optical light curves are briefly summarized; it is difficult to entirely decouple discussions of the photometric and spectral evolution of SNe. Details concerning light and color curves can be found in Kirshner (1990), Ford et al (1993), Leibundgut (1994, 1996), Patat et al (1994), Suntzeff (1996), Richmond et al (1996b), and other articles.

Conference proceedings or collections of reviews devoted to many different aspects of SNe (in some cases largely to SN 1987A) include those edited by Wheeler (1980), Rees & Stoneham (1982), Bartel (1985), Danziger (1987), Kafatos & Michalitsianos (1988), Brown (1988), Proust & Couch (1988), Wheeler et al (1990), Petschek (1990), Woosley (1991), Danziger & Kjär (1991), Ray & Velusamy (1991), Audouze et al (1993), Clegg et al (1994), Bludman et al (1995), McCray & Wang (1996), and Ruiz-Lapuente et al (1997). Additional reviews include those of Oke & Searle (1974), Trimble (1982, 1983), Woosley & Weaver (1986), Dopita (1988), Weiler & Sramek (1988), Arnett et al (1989), Imshennik & Nadëzhin (1989), Hillebrandt & Höflich (1989), Wheeler & Harkness (1990), Branch et al (1991), Branch & Tammann (1992), McCray (1993), Chevalier (1981, 1995), and Arnett (1996, especially Chapter 13). Infrared (IR) spectra of SNe are discussed by Meikle et al (1993, 1997) and others. A thorough atlas of International Ultraviolet Explorer (IUE) spectra of SNe, together with some optical spectra, light curves, and many useful references, has been published by Cappellaro et al (1995c). Recently, Wheeler & Benetti (1997) have concisely summarized many of the basic observed properties of SNe.

The most extensive catalogs of SNe are those of the Asiago Observatory (Barbon et al 1989, with an update by van den Bergh 1994) and the Sternberg Astronomical Institute. These are now regularly maintained and available electronically on the World Wide Web (http://athena.pd.astro.it/~supern/snean.txt and http://www.sai.msu.su/groups/sn/sncat/sn.cat, respectively). The Palomar catalog of SNe (Kowal & Sargent 1971) is no longer updated.

Unless otherwise noted, the optical spectra illustrated here were obtained by the author or his collaborators, primarily with the 3-m Shane reflector at Lick Observatory. They have been shifted to their rest frame; in each case the adopted redshift is listed in the caption. Telluric lines were generally removed, but the spectra were not dereddened. Although the relative spectrophotometry (i.e. the shape of each spectrum) is accurate, the absolute scale is arbitrary. Universal time (UT) dates are used throughout this review. When referring to phase of evolution, the variables $t$ and $\tau$ denote time since maximum brightness (usually in the B passband) and time since explosion, respectively.

## 2. GENERAL OVERVIEW OF SUPERNOVAE

### 2.1 *Spectra*

Supernovae occur in at least three, and possibly four or more, spectroscopically distinct varieties. The two main classes, Types I and II, were firmly established by Minkowski (1941, but see Popper 1937). Type I SNe are defined by the absence of obvious hydrogen in their optical spectra, except for possible contamination from superposed H II regions. SNe II all prominently exhibit hydrogen in their spectra, yet the strength and profile of the H$\alpha$ line vary widely among these objects. Until recently, most spectra of SNe have been obtained near the epoch of maximum brightness, but in principle the classification can be made at any time, as long as the spectrum is of sufficiently high quality. Only occasionally (Section 5.5) do SNe metamorphose from one type to another, suggesting the use of hybrid designations.

The early-time ($t \approx 1$ week) spectra of SNe are illustrated in Figure 1. The lines are broad owing to the high velocities of the ejecta, and most of them have P Cygni profiles formed by resonant scattering above the photosphere. SNe Ia are characterized by a deep absorption trough around 6150 Å produced by blueshifted Si II $\lambda\lambda$6347, 6371 (collectively called $\lambda$6355). Members of the Ib and Ic subclasses do not show this line. The presence of moderately strong optical He I lines, especially He I $\lambda$5876, distinguishes SNe Ib from SNe Ic (Wheeler & Harkness 1986, Harkness & Wheeler 1990).

The late-time ($t \gtrsim 4$ months) optical spectra of SNe provide additional constraints on the classification scheme (Figure 2). SNe Ia show blends of dozens of Fe emission lines, mixed with some Co lines. SNe Ib and Ic, on the other hand, have relatively unblended emission lines of intermediate-mass elements such as O and Ca. Emission lines in SNe Ib are narrower (Filippenko et al 1995b) and perhaps stronger (Wheeler 1990) than those in SNe Ic, but these

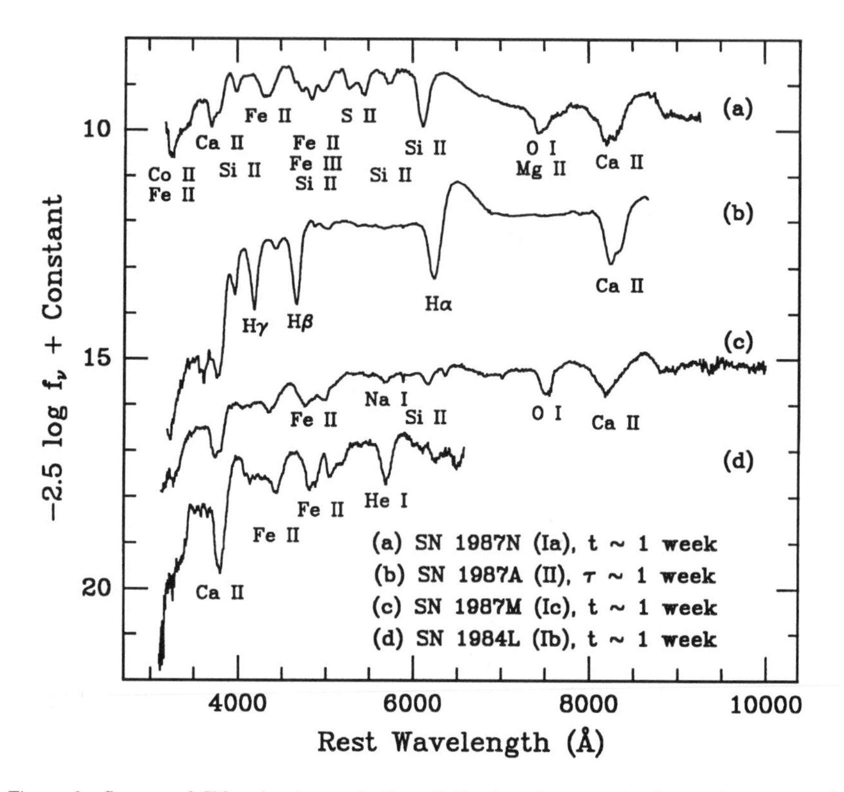


*Figure 1* Spectra of SNe, showing early-time distinctions between the four major types and subtypes. The parent galaxies and their redshifts (kilometers per second) are as follows: SN 1987N (NGC 7606; 2171), SN 1987A (LMC; 291), SN 1987M (NGC 2715; 1339), and SN 1984L (NGC 991; 1532). In this review, the variables $t$ and $\tau$ represent time after observed B-band maximum and time after core collapse, respectively. The ordinate units are essentially "AB magnitudes" as defined by Oke & Gunn (1983).

conclusions are based on the few existing late-time spectra of SNe Ib, and no other possibly significant differences have yet been found. At this phase, SNe II are dominated by the strong H$\alpha$ emission line; in other respects, most of them spectroscopically resemble SNe Ib and Ic, but the emission lines are even narrower and weaker (Filippenko 1988). The late-time spectra of SNe II show substantial heterogeneity, as do the early-time spectra.

At ultraviolet (UV) wavelengths, all SNe I exhibit a very prominent early-time deficit relative to the blackbody fit at optical wavelengths (e.g. Panagia 1987). This is due to line blanketing by multitudes of transitions, primarily those of Fe II and Co II (Branch & Venkatakrishna 1986). The spectra of SNe Ia (but not of SNe Ib/Ic) also appear depressed at IR wavelengths (Meikle

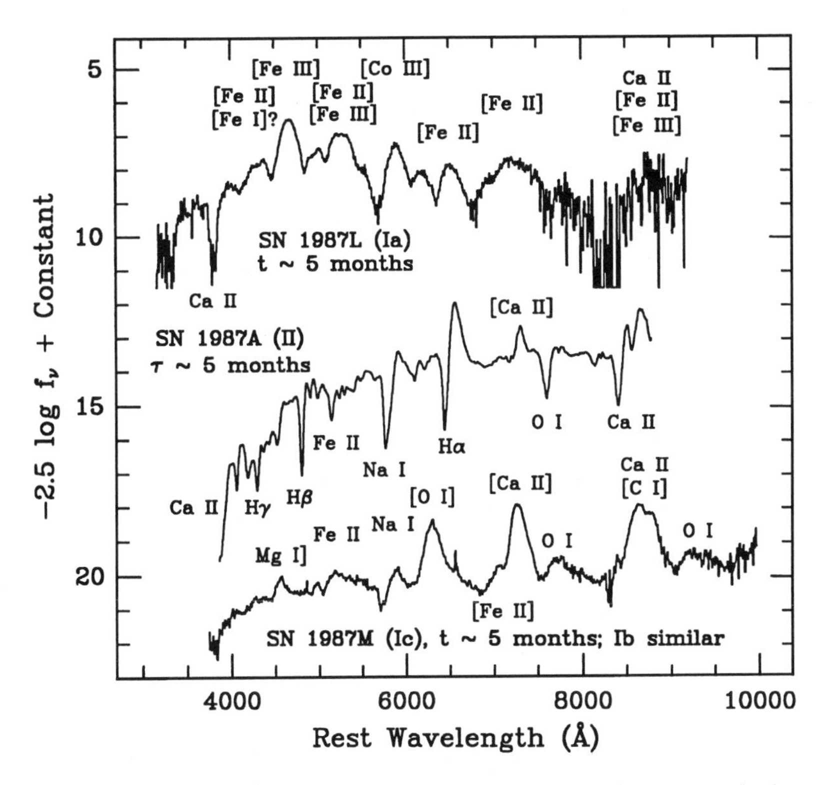


*Figure 2* Spectra of SNe, showing late-time distinctions between various types and subtypes. Notation is the same as in Figure 1. The parent galaxy of SN 1987L is NGC 2336 ($cz$ = 2206 km s$^{-1}$); others are listed in the caption of Figure 1. At even later phases, SN 1987A was dominated by strong emission lines of H$\alpha$, [O I], [Ca II], and the Ca II near-IR triplet, with only a weak continuum.

et al 1997). The early-time spectra of most SNe II, in contrast, approximate the single-temperature Planck function from UV through IR wavelengths, with occasionally even a slight UV excess. SN 1987A was an exception: The earliest IUE spectra showed a strong UV deficit relative to the blackbody curve defined at optical wavelengths (Danziger et al 1987), as in SNe I.

## 2.2 *Light Curves*

Some representative optical light curves of SNe are shown by Minkowski (1964); much more complete atlases of SNe I and SNe II are given by Leibundgut et al (1991c) and Patat et al (1993), respectively. The most obvious conclusion is that to first order, the light curves of SNe I are all broadly similar, whereas those of SNe II exhibit much dispersion.

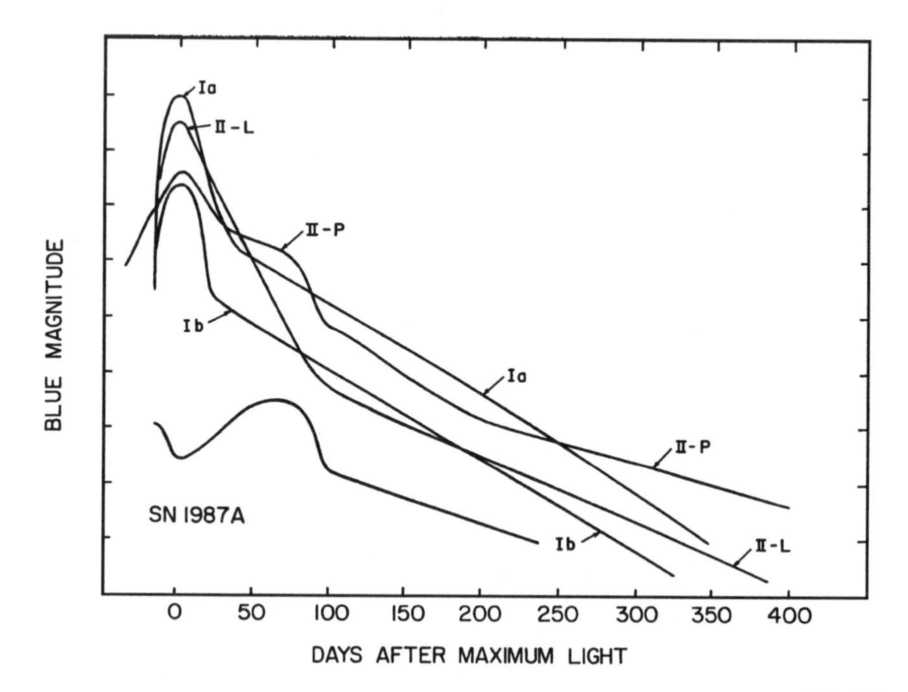


*Figure 3* Schematic light curves for SNe of Types Ia, Ib, II-L, II-P, and SN 1987A. The curve for SNe Ib includes SNe Ic as well, and represents an average. For SNe II-L, SNe 1979C and 1980K are used, but these might be unusually luminous. From Wheeler 1990; reproduced with permission.

Despite their bewildering variety, the majority of early-time light curves of SNe II ($t \lesssim 100$ days) can be usefully subdivided into two relatively distinct subclasses (Barbon et al 1979, Doggett & Branch 1985). The light curves of SNe II-L ("linear") generally resemble those of SNe I, whereas SNe II-P ("plateau") remain within ~1 mag of maximum brightness for an extended period (Figure 3). The degree to which there is a continuity between SNe II-L and SNe II-P is still debated, and other possible classification schemes are being considered (Patat et al 1994). The plateau of SN 1992H, for example, was somewhat shorter than usual and declined with time in the VRI passbands, and it was barely visible or nonexistent in B and U, leading Clocchiatti et al (1996a) to call this a hybrid object.

The peak absolute magnitudes of SNe II-P show a very wide dispersion (Schmitz & Gaskell 1988, Young & Branch 1989), almost certainly due to differences in the radii of the progenitor stars. Most SNe II-L, on the other hand, have a nearly uniform peak absolute magnitude (Young & Branch 1989,

Gaskell 1992), ~2.5 mag fainter than SNe Ia, although a few exceptionally luminous SNe II-L (SN 1979C, and to a lesser extent SN 1980K) are known. At late times ($t \gtrsim 150$ days), the light curves of most SNe II resemble each other, both in shape and absolute flux (Turatto et al 1990, Patat et al 1994). The decline rate is close to that expected from the decay of $Co^{56}$ to $Fe^{56}$ (0.98 mag/100 days), especially in V.

The light curve of SN 1987A (Figure 3), although unusual, was generically related to those of SNe II-P; the initial peak was low because the progenitor was a blue supergiant, much smaller than a red supergiant (Arnett et al 1989 and references therein). Additional objects of a possibly similar nature are SN 1909A (Young & Branch 1988) and SNe 1923A, 1948B, and 1965L (Schmitz & Gaskell 1988). Some SNe II decline very slowly at early times (Section 5.4), probably because of energy radiated during the interaction of the ejecta with circumstellar gas (e.g. SN 1987F, Chugai 1991; SN 1988Z, Turatto et al 1993b, Chugai & Danziger 1994); they do not fit into the two main photometric subclasses. Indeed, they appear to constitute a subclass that is also spectroscopically distinct (SN IIn; Section 5.4).

Barbon et al (1973, their Figure 1) illustrate the B-band light curves of 38 SNe I on a single plot, after having adjusted them in both coordinates to minimize the dispersion. There is considerable scatter, some of which is due to intrinsic differences among SNe Ia (Section 3.3). Moreover, three SNe now known to have been SNe Ib (1962L, 1964L, 1966J), as well as some number of unrecognized SNe Ib/Ic, were included in this compilation, yet their light curves are not identical to those of SNe Ia. (Note that the archaic "SNe Ia" and "SNe Ib" of Barbon et al, defined according to photometric properties, should not be confused with the modern, spectroscopic designations.) SNe Ic 1987M and 1994I, for example, declined markedly faster than SNe Ia (Filippenko et al 1990, Richmond et al 1996b), whereas the decline rate of the few SNe Ib that have been studied appears to be slower than that of SNe Ia (Schlegel & Kirshner 1989, Kirshner 1990). Clocchiatti & Wheeler (1997) recently found that the light curves of SNe Ic cluster into two different categories: Some fall more slowly than SNe Ia (like SNe Ib), and others more rapidly, even though they have nearly identical spectra. This serves as an important reminder that light curves can provide physical diagnostics not available from spectra alone.

It is appropriate at this point to mention Zwicky's (1965) SN Types III, IV, and V, all of which had peculiar light curves. Each of these classes has only a few known examples, and in all cases their spectra showed hydrogen. As stated by Oke & Searle (1974) and Doggett & Branch (1985), they should therefore be considered as SNe II (or peculiar SNe II), at least until additional objects that have similar characteristics are found and examined in detail. SN 1961V in NGC 1058 (Type V) had the most bizarre light curve ever recorded. The less luminous SN 1954J, also known as Variable 12 in NGC 2403, had a light curve shape similar to that of SN 1961V; see Humphreys & Davidson (1994). SNe V may not even be genuine SNe, a conclusion reached by Goodrich et al (1989) and Filippenko et al (1995a) for the specific case of SN 1961V; rather, they may be super-outbursts of luminous blue variables such as $\eta$ Carinae.

### 2.3 *Environments and Progenitors*

The locations at which SNe occur provide important clues to their nature and to the mass of their progenitor stars. SNe II, Ib, and Ic have never been seen in elliptical galaxies and rarely if ever in S0 galaxies. They are generally in or near spiral arms and H II regions (Huang 1987, Porter & Filippenko 1987, Van Dyk 1992, Van Dyk et al 1996a), implying that their progenitors must have started their lives as massive stars ($\gtrsim$8–10 $M_{\odot}$). SNe Ia, on the other hand, occur in all types of galaxies, including ellipticals, and in spirals there is no strong preference for spiral arms (Maza & van den Bergh 1976, Van Dyk 1992, McMillan & Ciardullo 1996; but see Bartunov et al 1994b). Because SNe Ia occur most frequently in spiral galaxies, with the rate per unit K-band (2.2 $\mu$m) luminosity increasing from early to late Hubble types (Della Valle & Livio 1994), the majority of SNe Ia probably come from intermediate-age ($\sim$0.1–0.5 billion years), moderately massive stars (4–7 $M_{\odot}$); see Oemler & Tinsley (1979) for an interesting early discussion.

The progenitors of SNe Ia are carbon-oxygen white dwarfs that accrete matter from a companion star and undergo thermonuclear runaway (Nomoto et al 1984, Woosley & Weaver 1986, and references therein). Although the white dwarfs probably reach the Chandrasekhar limit prior to exploding, this is not yet certain (Woosley et al 1994, Livne & Arnett 1995, but see Höflich et al 1996a). SNe II are thought to arise from evolved, massive progenitors (initial mass $\gtrsim$ 8–10 $M_{\odot}$) that suffer core collapse (generally iron) and subsequently rebound (e.g. Arnett et al 1989), leaving a neutron star or perhaps in some cases a black hole (Brown & Bethe 1994). Most workers now believe that SNe Ib/Ic are produced by the same mechanism as SNe II, except that the progenitors were stripped of their hydrogen (SN Ib) and possibly helium (SN Ic) envelopes prior to exploding, either via mass transfer to companion stars (Nomoto et al 1994, Woosley et al 1995) or through winds (Woosley et al 1993, Swartz et al 1993b, and references therein). White dwarf models have been discussed (e.g. Branch & Nomoto 1986) but are implausible.

## 3. TYPE Ia SUPERNOVAE

### 3.1 *Spectral Evolution*

The first thorough long-term set of spectra of a SN Ia was that of SN 1937C, obtained photographically by Minkowski (1939), with an intensity calibration provided by Greenstein & Minkowski (1973). The use of linear detectors having high quantum efficiency (especially CCDs) led to the publication of better sequences of spectra of "normal SNe Ia" (Section 3.3), specifically those of SN 1972E (Kirshner et al 1973, Kirshner & Oke 1975), SN 1981B (Branch et al 1983), SN 1989B (Barbon et al 1990, Wells et al 1994), and SN 1994D (Meikle et al 1996, Patat et al 1996, Filippenko 1997a). Wells et al (1994) discuss in some detail the temporal changes in their densely sampled spectra of SN 1989B. A representative set of spectra of SN 1994D is illustrated in Figure 4.

The early-time spectra of SNe Ia exhibit prominent broad peaks and valleys. Extensive computer modeling of the expanding ejecta has resulted in reliable identifications for most features, after decades of uncertainty. In general, at very early times they are attributed to lines of neutral and singly ionized intermediate-mass elements (O, Mg, Si, S, Ca), with some contribution from iron-peak elements (Fe, Co) especially at near-UV wavelengths (Branch et al 1983, 1985, Harkness 1986, 1991, Kirshner et al 1993, Mazzali et al 1993). The strongest features are Si II $\lambda$6355 and Ca II H&K $\lambda\lambda$3934, 3968. Convincing evidence for the presence of helium has never been shown, although the results of Meikle et al (1996) are intriguing and warrant further scrutiny.

The relative contribution of iron-group elements quickly increases as the photosphere recedes into the ejecta. By $t \approx 2$ weeks the spectrum is dominated by lines of Fe II, which is consistent with an iron-rich core (Harkness 1991), but some lines of intermediate-mass elements are still present (e.g. Si II, Ca II). Thereafter the spectral changes are more gradual, although forbidden emission lines of Fe (and some Co, most prominent at $\sim$5900 Å) eventually dominate in the nebular phase (Axelrod 1980), which begins roughly one month past maximum brightness. Nevertheless, Ca II remains visible, primarily in absorption (Ca II H&K and the near-IR triplet of $\lambda\lambda$8498, 8542, 8662). Qualitatively, Axelrod (1980) was able to show that the cobalt lines decrease with time in a manner suggestive of radioactive decay. More recently, compelling evidence that the late-time tail of SNe Ia is powered by the decay of radioactive $Co^{56}$ (initially from radioactive $Ni^{56}$) was found from the temporal changes in the intensity ratio of two relatively unblended [Co III] and [Fe III] emission lines (Kuchner et al 1994; see also Varani et al 1990 for the case of SN 1987A). Also, Spyromilio et al (1992) demonstrated the presence of a large amount (0.4–0.7 $M_{\odot}$) of iron in the ejecta of SN 1991T. Incidentally, SN 1991T is the only SN Ia to have been observed spectroscopically more than two years past maximum, but the data were dominated by an echo produced by foreground dust reflecting the near-maximum spectrum (Schmidt et al 1994c).

Early-time photospheric expansion velocities, determined by measuring the positions of Doppler-shifted absorption minima in the P Cygni profiles of strong, fairly unblended lines such as Si II $\lambda$6355, are typically $\gtrsim$10,000 km s$^{-1}$

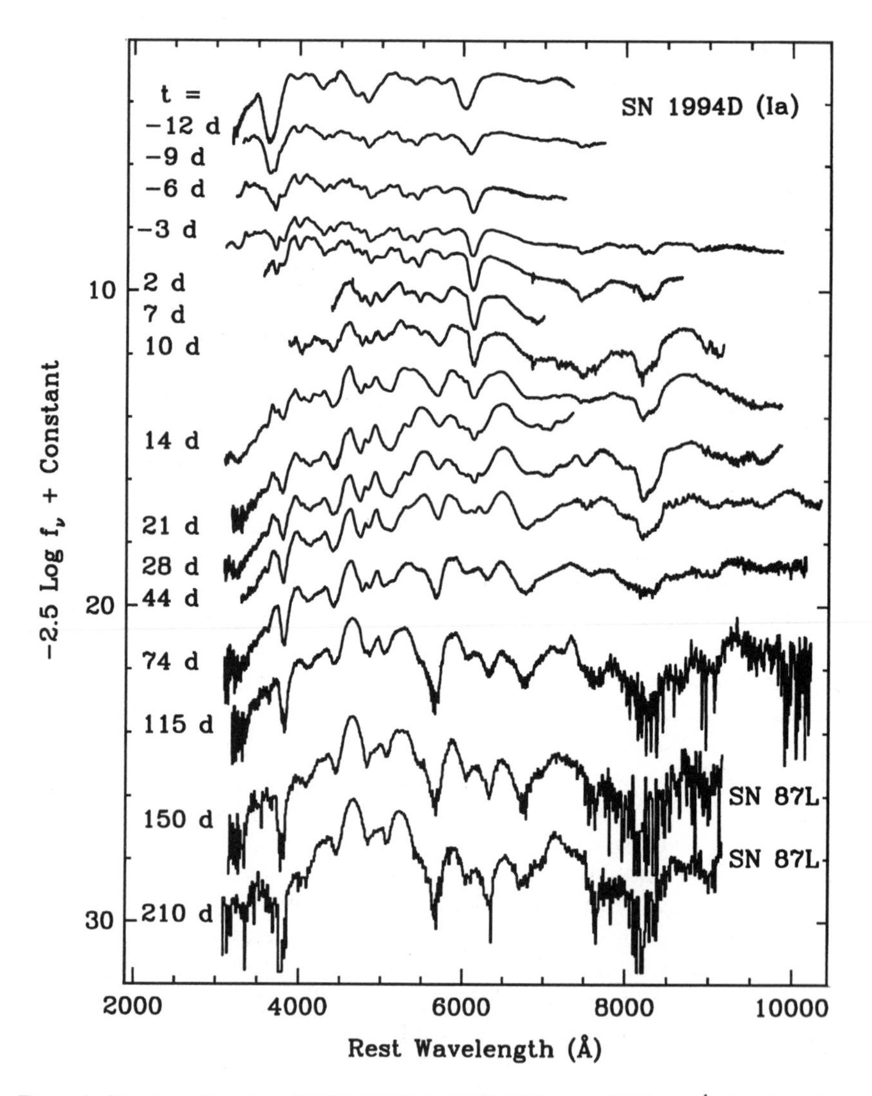


*Figure 4* Montage of spectra of SN Ia 1994D in NGC 4526 ($cz = 850$ km s$^{-1}$), based on data from Patat et al (1996; reproduced with permission) and Filippenko (1997a). Epochs (days) are given relative to maximum B brightness (March 20.5, 1994). The last two spectra are of the similar SN Ia 1987L in NGC 2336.

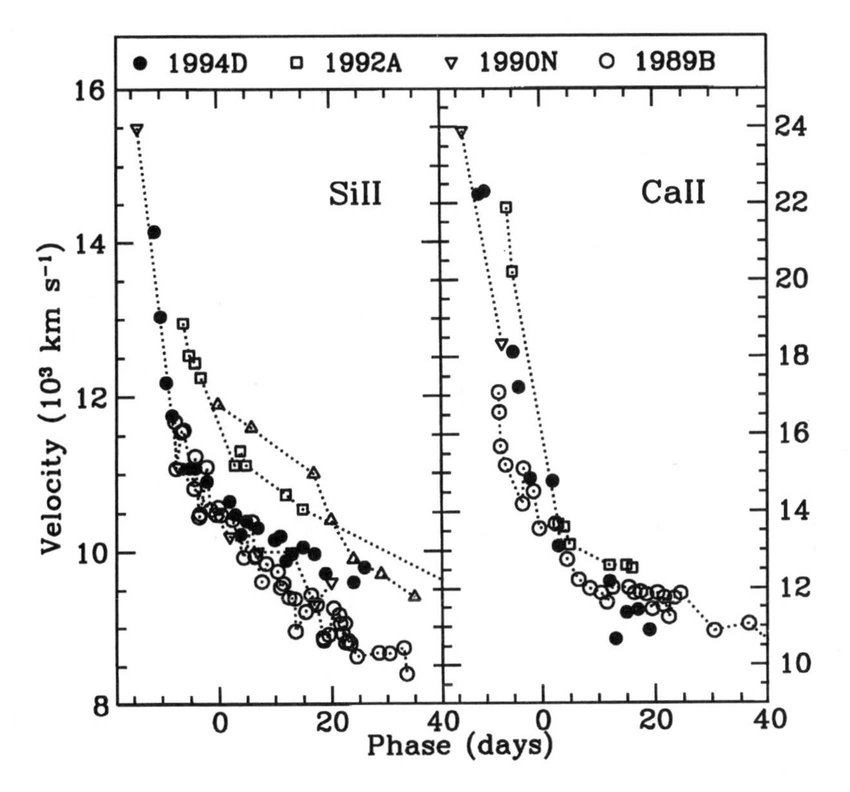


*Figure 5* Evolution of the expansion velocity as deduced from the minima of the Si II λ6355 (*left panel*) and Ca II H&K (*right panel*) absorption troughs for SNe Ia 1994D, 1992A, 1990N, 1989B, and 1981B. From Patat et al (1996); reproduced with permission.

(Pskovskii 1977, Branch 1981). Different lines do not have identical velocities: for example, Si II (10,000–12,000 km s$^{-1}$) and Ca II H&K (13,000–15,000 km s$^{-1}$) at maximum brightness. Initially the velocity obtained from the Si II and Ca II H&K lines decreases rapidly with time, as shown most recently by Wells et al (1994) and Patat et al (1996); see Figure 5. Patat et al (1996) note that there is a sudden break in the decline rate around $t = -6$ days, most easily visible in the Si II line (Figure 5).

Note that the broad emission near 6500 Å, which begins to develop around $t = 2$ weeks and remains thereafter with some changes in shape (Figure 4), is not H$\alpha$ but rather Fe II and later [Fe II]. It is occasionally incorrectly identified as H$\alpha$ by observers attempting to classify spectra of SN candidates. Similarly, the strong Si II λ6355 line at early times is sometimes attributed to H$\alpha$. In both cases, erroneous classifications of Type II are made, especially when uncalibrated spectra are quickly examined at the time of observation.

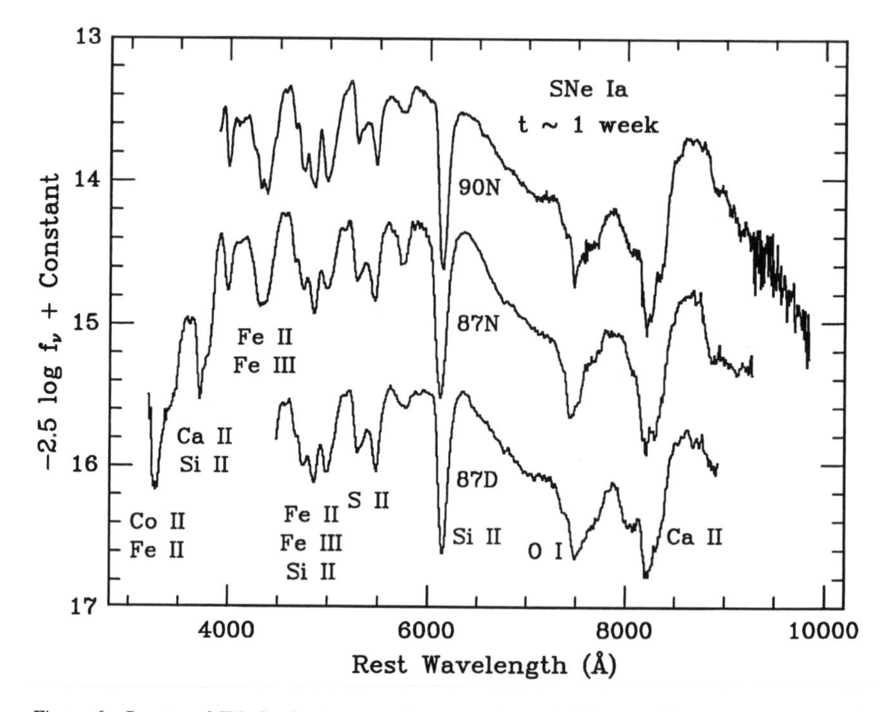


*Figure 6* Spectra of SNe Ia about one week past maximum brightness. The parent galaxies and their redshifts (kilometers per second) are as follows: SN 1990N (NGC 4639; 970), SN 1987N (NGC 7606; 2171), and SN 1987D (MCG+00-32-01; 2227).

## 3.2 *Homogeneity*

It was noticed long ago that the optical spectra of SNe Ia are usually quite homogeneous, if care is taken to compare objects at similar times relative to maximum brightness (Oke & Searle 1974, and references therein). One can even deduce a fairly accurate age of a "normal" SN Ia at the time of observation by comparison of its spectrum with a series of template spectra, such as those of SNe 1937C, 1972E, 1981B, 1989B, and 1994D mentioned above. A good example of this homogeneity is provided by spectra (Figure 6) of SN 1987D, SN 1987N, and SN 1990N, each about one week after maximum brightness. Especially impressive are the small "notches" visible in all three spectra near 4550 Å, 4650 Å, and 5150 Å.

The optical light curve shapes of many SNe Ia also closely resemble each other (e.g. Hamuy et al 1991). By examining a large quantity of data, Leibundgut (1988) constructed "template" light curves in several bandpasses. It is striking that the B and V light curves of SN 1990N (Leibundgut et al 1991a), discovered two weeks before maximum brightness, very closely match the previously determined templates.

Branch & Miller (1993), Vaughan et al (1995), and others have quantified the dispersion in the peak absolute magnitude of unreddened SNe Ia. By including only those SNe Ia whose B–V color at maximum brightness is within the range $-0.25$ to 0.25 mag, Vaughan et al found that $\langle M_B \rangle = -18.54 \pm 0.06 + 5 \log (H_0/85)$ mag and $\langle M_V \rangle = -18.59 \pm 0.06 + 5 \log (H_0/85)$ mag, where the units of $H_0$ are kilometers per second per megaparsec. Much of the measured dispersion (only 0.30 mag in both cases) may be due to observational errors, incorrect relative distances, and residual reddening; the intrinsic dispersion is certainly smaller.

## 3.3 *Heterogeneity*

Nevertheless, careful inspection of high-quality data demonstrates that differences among SNe Ia do indeed exist. In Figure 6, for example, the depths of the features near 5750 Å and 8050 Å differ. Moreover, the ejecta of SNe Ia do not always have the same velocity at a given phase (Branch 1987, Branch et al 1988; see also Figures 5 and 6). Remarkably, the smallest ejection velocities are generally found among SNe Ia in elliptical galaxies (Filippenko 1989b, Branch & van den Bergh 1993). A spectroscopic distinction between SNe Ia in spiral and elliptical galaxies clearly indicates that there are real physical differences among SNe Ia; the dissimilar ejection velocities cannot be a consequence of viewing an asymmetric explosion from different angles. Also, after many tenuous suggestions (e.g. Barbon et al 1973, Rust 1974, Pskovskii 1977, 1984, Branch 1981), variations in the light curve shapes among SNe Ia have finally been confirmed beyond doubt and shown to be correlated with luminosity: Intrinsically bright SNe Ia rise and decline more slowly than dim ones (Phillips 1993, Hamuy et al 1995, 1996a, Riess et al 1995a, 1996). The most luminous SNe Ia seem to occur in young stellar populations (Hamuy et al 1995, Branch et al 1996), which is an important result.

Spectroscopic and photometric peculiarities have been noted with increasing frequency in well-observed SNe Ia during the past decade. One of the best examples is SN 1986G: Phillips et al (1987) and Cristiani et al (1992) observed anomalies in the optical spectra as well as a rapid postmaximum decline in the UBV bands, Frogel et al (1987) reported clearly discrepant IR (JHK) light curves, and the early-time UV spectrum was unusual (Panagia & Gilmozzi 1991). Another interesting object, SN 1990N (Leibundgut et al 1991a, Phillips et al 1992, Mazzali et al 1993), had significantly weaker Si II absorption (at 6150 Å) at $t = -1$ week than did the normal SNe Ia 1989B and 1994D (Figure 7), but its postmaximum evolution was typical. A striking case is SN 1991T; its premaximum spectrum did not exhibit Si II or Ca II

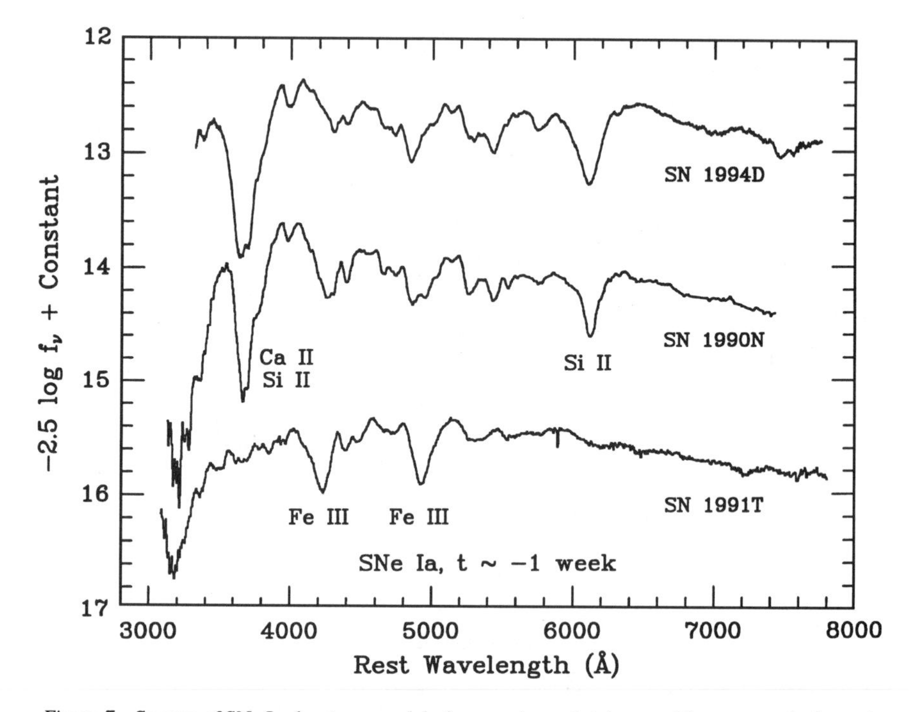


*Figure 7* Spectra of SNe Ia about one week before maximum brightness. The parent galaxies and their redshifts (kilometers per second) are as follows: SN 1994D (NGC 4526; 850), SN 1990N (NGC 4639; 970), and SN 1991T (NGC 4527; 1740).

absorption lines at all (Figure 7), yet two months past maximum the spectrum was nearly indistinguishable from that of a classical SN Ia (Filippenko et al 1992b, Ruiz-Lapuente et al 1992, Phillips et al 1992, Jeffery et al 1992, Mazzali et al 1995). Superior photometry reported by Phillips et al (1992) shows that the light curves of SN 1991T were slightly broader than the SN Ia template curves, and the object was probably somewhat more luminous than average at maximum (Filippenko et al 1992b, Phillips 1993, Riess et al 1996). Spectroscopically similar objects were subsequently identified by Filippenko & Leonard (1995; SN 1995ac) and by Garnavich et al (1995b; SN 1995bd).

SN 1991bg, in the E1 galaxy NGC 4374, is the reigning champion of well-observed peculiar SNe Ia (Filippenko et al 1992a, Leibundgut et al 1993, Turatto et al 1996, Mazzali et al 1997). At maximum brightness, SN 1991bg was subluminous by 1.6 mag in V and 2.5 mag in B, compared with normal SNe Ia. The colors were unusually red at maximum, but the object was not significantly reddened by dust; indeed, at late times the object may have been slightly bluer than normal SNe Ia (Leibundgut et al 1993). The decline from maximum was certainly quite steep; Filippenko et al (1992a) measured an initial linear V fading of 0.10 mag day$^{-1}$, rather than the typical value of 0.06 mag day$^{-1}$ for SNe Ia, and a late-time decline of 0.034 mag day$^{-1}$, rather than 0.026 mag day$^{-1}$. Also, the "knee" in the V light curve of SN 1991bg occurred only ~20 days past maximum, in contrast to the usual value of ~35 days. Unlike the case in normal SNe Ia, the I-band light curve did not exhibit a secondary maximum, and the R-band light curve showed no sign of a plateau. Furthermore, the spectrum of SN 1991bg at maximum had a deep trough around 4200 Å produced by Ti II (Filippenko et al 1992a); lines of Ti II and other intermediate-mass elements were present elsewhere in the spectrum as well, whereas Fe II was weak or absent. The expansion velocity (~10,000 km s$^{-1}$) was slightly lower than average (11,000–13,000 km s$^{-1}$) for luminous SNe Ia. A spectrum obtained three weeks past maximum showed a relatively narrow absorption line attributed to Na I D, as well as the emergence of forbidden emission lines; apparently the nebular phase began very early in SN 1991bg. Three months past maximum the [Ca II] $\lambda\lambda$7291, 7324 blend began dominating the spectrum.

Although "normal" SNe Ia (defined by objects such as SNe 1937C, 1972E, 1981B, 1989B, and 1994D) constitute a majority ($\gtrsim$80%) of SNe Ia observed thus far (Branch et al 1993), very subluminous and peculiar SNe Ia like SN 1991bg might be intrinsically as common per unit volume but are more difficult to find (e.g. Schaefer 1996). The second known example is SN 1992K (Hamuy et al 1994), whose luminosity, spectral characteristics, light curves, and color curves were nearly identical to those of SN 1991bg. Note that the host galaxy of SN 1992K is a spiral, whereas that of SN 1991bg is an elliptical, though SN 1992K may have been associated with the older bulge population. Yet another probable member of this subclass is SN 1991F in the lenticular galaxy NGC 3458 (Gómez & López 1995): Its late-time spectrum closely resembled that of SN 1991bg, but its photometric properties are unknown. In all cases where the requisite data have been obtained, the subluminous objects are intrinsically red compared with most normal SNe Ia (Hamuy et al 1995, Vaughan et al 1995).

Maza et al (1994) present a striking example of the photometric heterogeneity of SNe Ia. As part of the Calán/Tololo search for SNe, they discovered two SNe Ia at redshift $z = 0.02$, presumably at nearly the same distance because the peculiar velocities of their host galaxies are likely to be small relative to the Hubble flow. SN 1992bc was 0.69-mag brighter than SN 1992bo in B; in addition, its light curves declined more slowly than those of SN 1992bo, and in fact closely resembled those of the overluminous SN 1991T (Phillips et al 1992, Phillips 1993). Differences in the shapes of high-quality CCD light curves of numerous SNe Ia are illustrated by Riess (1996) and Hamuy et al (1996c).

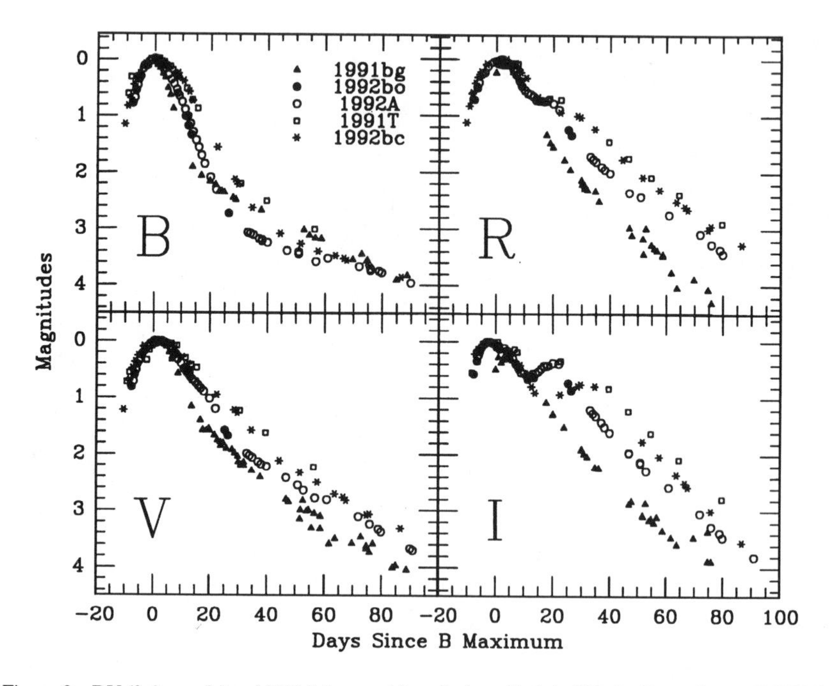


*Figure 8* BV (*left panels*) and RI (*right panels*) evolution of bright SNe Ia. From Suntzeff (1996); reproduced with permission.

The photometric heterogeneity among extreme examples of SNe Ia is perhaps best demonstrated by Suntzeff (1996) with five objects that have excellent BVRI light curves, all scaled to the same maximum brightness (Figure 8). The differences in decline rates at early times are largest in B, followed by V; they are relatively minor at R and I, except for those of SN 1991bg. At late times, however, the R and I light curves exhibit more scatter than the B and V curves. Two months past maximum, for example, SN 1991T and SN 1991bg differ by 1.3 mag in R (but very little in B) after normalizing to the same peak. Suntzeff (1996) goes on to show the difference in bolometric (near-UV through near-IR) luminosity of four SNe Ia as a function of time. At maximum brightness, SN 1991T was a factor of 5 more luminous than SN 1991bg, and this grew to a factor of 9 by one month past maximum.

## 3.4 *Cosmological Uses of SNe Ia*

Despite not being truly "standard" candles (cf Branch & Tammann 1992, and references therein), SNe Ia are still exceptionally useful for cosmological studies (e.g. Riess et al 1995a,b, 1996, Hamuy et al 1995, 1996a,b, Perlmutter et al

1995, 1997, Goobar & Perlmutter 1995, Leibundgut et al 1996, Goldhaber et al 1997, Filippenko 1997b, Kim et al 1997, Schmidt 1997). The key point is that luminosity correlates strongly with light curve shape; given enough high-quality observations, the luminosity of each object can therefore be calibrated. Even the extinction can be determined reliably from observations through several filters (Riess et al 1996), obviating the need to use other, much more uncertain methods such as those based on the strength of the Na I D interstellar absorption lines (e.g. Richmond et al 1994).

There are additional, potentially tight correlations that could enable the peak luminosity of individual SNe Ia to be calibrated even more accurately. Fisher et al (1995) found that the velocity of the red edge of the Ca II H&K absorption line at $t \gtrsim 60$ days in spectra of SNe Ia correlates reasonably well with absolute visual magnitude. Nugent et al (1995) discuss other spectral trends and begin to explore their physical basis. For example, the ratio of the Si II absorption line at 5750 Å to that at 6150 Å increases with decreasing luminosity, as does the ratio of the two peaks on either side of the Ca II H&K absorption trough. The latter could be especially useful when classifying high-redshift SNe Ia. Similarly, Branch et al (1996) found that the U–B color at maximum brightness correlates with absolute magnitude; intrinsically luminous SNe Ia generally exhibit the largest UV excess. Further work is required to see whether at least two independent methods yield the same corrections (within the uncertainties) to the derived peak luminosities of SNe Ia.

## 3.5 *Hydrogen in Spectra of SNe Ia*

The presence or absence of circumstellar material can shed light on the nature and evolution of SN Ia progenitors, as discussed by Branch et al (1995, and references therein). One way in which this gas can reveal itself is through transient, narrow H$\alpha$ emission or absorption lines in early-time spectra of SNe Ia. Branch et al (1983), for instance, gave tentative evidence for a weak, narrow H$\alpha$ emission line in a spectrum of SN 1981B obtained six days after maximum brightness, but Cumming et al (1996) showed that this interpretation is unlikely to be correct. Similarly, Polcaro & Viotti (1991) claimed to have detected H$\alpha$ absorption in a spectrum of SN 1990M obtained four days after maximum brightness, but Della Valle et al (1996) argued that this was probably an artifact of the reduction procedure.

Calculations by Cumming et al (1996) indicate that circumstellar emission in H$\alpha$ will drop rapidly after explosion; detection is not possible unless very early observations are made. Sensitive high-resolution spectroscopy is starting to set useful limits on H$\alpha$ absorption or emission and in turn on the amount of circumstellar hydrogen around SN Ia progenitors. Cumming et al (1996) did not detect H$\alpha$ in a spectrum of SN 1994D obtained at $t = -10$ days; under the assumption of spherical symmetry for the progenitor's wind, they find an upper limit of $\dot{M}_\odot \approx 2.5 \times 10^{-5}$ M year$^{-1}$ (Lundqvist & Cumming 1997) if the wind speed is 10 km s$^{-1}$. Unfortunately, this limit can exclude only the most extreme symbiotic systems as progenitors of SNe Ia. Later (at t $\approx$ 23 days), Ho & Filippenko (1995; see also Filippenko 1997a) used the Keck telescope to carry out a more sensitive search for H$\alpha$ in SN 1994D, though they did not detect any absorption or emission features (equivalent width 2$\sigma$ upper limits of $\sim$3 mÅ) within $\pm$100 km s$^{-1}$ of the SN's systemic velocity. Early-time observations at this sensitivity should be able to reveal narrow H$\alpha$ from nearby SNe Ia, if it is present.

Thus, to date there have been no convincing detections of narrow, transient H$\alpha$ in early-time spectra of any SNe Ia, though the sample is still very small. (Also, no such helium lines have been reported, but few if any careful searches have been attempted.) Note, however, that weak hydrogen in spectra of SNe Ia, if ever detected, will not necessarily be of circumstellar origin. For example, at the time of explosion the surface of the white dwarf may contain some hydrogen, presumably donated by the secondary star. If so, it should be a broad feature, as it is in the spectra of classical novae (e.g. Williams et al 1994) but much more subtle. In progenitors consisting of main-sequence or subgiant donors (e.g. cataclysmic variables), the ejecta can strip and ablate gas from the secondary star, thereby contaminating the early-time spectrum with hydrogen (Applegate & Terman 1989, Wheeler 1992), but this has never actually been seen.

For certain progenitor models, H$\alpha$ emission might be expected in the late-time spectra of SNe Ia. Chugai (1986b) predicted that most of the hydrogen-rich material stripped from a red-giant secondary during the explosion is trapped within the ejecta, subsequently expanding at relatively low speeds. Two-dimensional hydrodynamic calculations supported this hypothesis (Livne et al 1992). The hydrogen becomes visible only after the photosphere recedes substantially, and the expected line width is small: Full width at half maximum (FWHM) $\approx$ 2000 km s$^{-1}$. Such a feature may have been detected in SN 1991bg by Ruiz-Lapuente et al (1993; see also Turatto et al 1996 and Garnavich & Challis 1997), but there are other possible interpretations if it is real (e.g. [Fe II]; Turatto et al 1996).

# 4. TYPE Ib AND Ic SUPERNOVAE

## 4.1 *Historical Development*

Bertola (1964) and Bertola et al (1965) noticed that the early-time spectra of some SNe I (specifically SNe 1962L and 1964L) lack the deep 6150-Å absorption trough. For two decades few, if any, new examples of such objects existed, and they were simply labeled as "peculiar SNe I" (SNe Ip). Interest in them was revitalized in the mid-1980s by the studies of several newly discovered SNe Ip by Uomoto & Kirshner (1985), Wheeler & Levreault (1985), Elias et al (1985), and Panagia et al (1986b). Particularly influential (but unpublished) was an optical, UV, and IR investigation of SN 1983N done by Panagia et al (1986a); subsets of these data have been discussed by Panagia (1985) and Gaskell et al (1986), and radio observations were presented by Sramek et al (1984).

As summarized by Porter & Filippenko (1987), SNe Ip seemed to constitute a distinct subclass, characterized by their (*a*) lack of the 6150-Å Si II absorption trough, (*b*) preference for galaxies having Hubble types Sbc or later, (*c*) proximity to H II regions, (*d*) rather low luminosity, typically 1.5-mag fainter than classical SNe I, (*e*) distinct IR light curves having no secondary maximum around one month past primary maximum, (*f*) reddish colors, and (*g*) emission of radio waves within a year past maximum. The subclass was coined "Type Ib" (Elias et al 1985) to distinguish it from normal SNe Ia. At least one of the earliest studies (Wheeler & Levreault 1985) concluded that the explosion mechanism might be more closely related to that of SNe II than to SNe Ia, but this was not yet certain because the spectroscopic appearance of SNe Ib near maximum seemed to resemble that of somewhat older SNe Ia ($t \approx 1$ month).

The serendipitous discovery of SN 1985F (Filippenko & Sargent 1985, 1986) initially compounded the confusion. Its spectrum was dominated by very strong, broad emission lines of neutral and singly ionized species such as [O I] $\lambda\lambda$6300, 6364, [Ca II] $\lambda\lambda$7291, 7324, the Ca II near-IR triplet, Mg I] $\lambda$4571, and Na I D. The strength of the forbidden lines suggested that SN 1985F was an old SN, as did the exponential decline of the derived light curve (Filippenko et al 1986). This was later confirmed by Tsvetkov (1986), whose inspection of prediscovery plates showed that SN 1985F had reached maximum brightness at $B = 12.1$ mag (one of the brightest SNe in many years!) about 260 days prior to discovery. The complete absence of hydrogen led to a formal classification of SN I, although no known spectra of SNe I at any stage of development resembled that of SN 1985F. The dominance of intermediate-mass elements and other factors suggested that SN 1985F was the explosion of a massive star that had rid itself of hydrogen prior to exploding (Filippenko & Sargent 1986, Begelman & Sarazin 1986, Schaeffer et al 1987), somewhat like the progenitor long ago proposed for Cas A by Chevalier (1976). Was this yet another type of SN I, distinct from SNe Ia and SNe Ib?

An important "unification" occurred when Gaskell et al (1986) showed that a spectrum of the Type Ib SN 1983N, obtained eight months past maximum, was very similar to that of SN 1985F at the time of its discovery. Moreover, Kirshner (quoted in Chevalier 1986; see also Schlegel & Kirshner 1989) found that a late-time spectrum of the Type Ib SN 1984L also resembled that of SN 1985F. Thus, SN 1985F was probably a SN Ib discovered long after maximum, and, conversely, SNe Ib eventually turn into objects whose spectra really are vastly different from those of SNe Ia. Interestingly, Chugai (1986a) had, in fact, already suggested that SN 1985F might be a SN Ib discovered long after maximum. Note, however, that there are no early-time spectra of SN 1985F; thus, it may have been a SN IIb (Section 5.5) or perhaps even a SN Ic, although the latter is unlikely given the relatively slow decline of its light curve (Wheeler & Harkness 1990).

The link between SN 1985F and SNe Ib 1983N/1984L, as well as the convincing discovery of He I lines in early-time spectra of the latter (Harkness et al 1987), provided substantial evidence that SNe Ib are a physically separate subclass of SNe I, probably driven by core collapse of an initially massive star (Wheeler & Levreault 1985). Detailed analysis of the nebular spectrum of SN 1985F (Fransson & Chevalier 1989) strongly supported this hypothesis; in their model, the progenitor had initial and final masses of 25 $M_{\odot}$ and 8 $M_{\odot}$, respectively, and $\sim$2 $M_{\odot}$ of oxygen was ejected. Large departures from local thermodynamic equilibrium (LTE) were invoked by Harkness et al (1987) to produce the observed He I lines.

Gradually it became clear that SNe Ib constitute a heterogeneous subclass, with substantial variations in the observed He I strengths in spectra obtained around maximum brightness. Wheeler & Harkness (1986; see also Harkness et al 1987) suggested that SNe Ib should actually be divided into two separate categories: SNe Ib are those showing strong He I absorption lines (especially He I $\lambda$5876) in their early-time photospheric spectra, whereas SNe Ic are those in which He I is not easily discernible. However, they modeled SNe Ic in the same physical way as SNe Ib (Wheeler et al 1987) but with different relative concentrations of He and O in the envelope. Although this nomenclature has been adopted by most authors, use of two subtypes (Ib and Ic) might not be observationally warranted. Few objects have been studied in detail; it is possible that a continuum of helium strengths exists among SNe Ib and that He-rich objects are not fundamentally different from He-poor objects in terms of physical origin (Wheeler et al 1987). Clocchiatti & Wheeler (1997), on the other hand, have recently argued that SNe Ib and Ic show a roughly bimodal distribution of He I strengths and that their progenitors may have significantly different evolutionary phases.

The amount of ejected iron in SNe Ib/Ic is not yet clear. If radioactive $Ni^{56}$ powers the optical display, and if SNe Ia produce 0.6 $M_{\odot}$ of this isotope (Woosley & Weaver 1986), then the corresponding mass for SNe Ib might be only $\sim$0.15 $M_{\odot}$ because they are roughly four times fainter than SNe Ia (Wheeler & Levreault 1985). Graham et al (1986) proposed the presence of [Fe II] $\lambda$1.644 $\mu$m emission in a late-time spectrum of SN Ib 1983N, and they calculated an ejected iron mass of 0.3 $M_{\odot}$, but Oliva (1987) suggested that the proper identification is Si I.

## 4.2 *SNe Ib: Spectral Evolution*

Genuine helium-rich SNe Ib appear to be rather rare objects. At present, the most complete series of published spectra of a SN Ib is that of Harkness & Wheeler (1990) for SN 1984L, and even this case includes only the first two months past maximum brightness. As shown in Figure 9, there is no strong evidence of hydrogen, and the 6150-Å trough of SNe Ia is weak or absent at all times. Although some of the absorption lines can be attributed to He I in the earliest spectra, other alternatives exist (e.g. Na I D for the feature near 5800 Å). However, the gradual strengthening of blueshifted ($\sim$7500 km s$^{-1}$) lines corresponding to He I $\lambda\lambda$4471, 5876, 6678, and 7065 makes the helium identification unambiguous a few weeks past maximum brightness (e.g. the spectrum at $t$ = 20 days). This differs markedly from the spectral development of SNe Ia (Figure 4).

The seventh spectrum in Figure 9, obtained about two months past maximum brightness, appears to exhibit weak [O I] $\lambda\lambda$6300, 6364 and especially [O I] $\lambda$5577, indicating an early onset of the nebular phase. Strong, broad (FWHM $\approx$ 4500 km s$^{-1}$) lines of [O I] and [Ca II] $\lambda\lambda$7291, 7324 are visible in the spectra of SN 1984L obtained by Schlegel & Kirshner (1989) 13–14 months past maximum, as is somewhat narrower Mg I] $\lambda$4571, but the data are noisy. The last plot in Figure 9 instead shows the late-time spectrum ($t \approx$ 8 months) of a similar SN Ib, SN 1983N (from Gaskell et al 1986).

Early-time spectra of SN 1983N were presented by Richtler & Sadler (1983), before the SN Ib subclass had been recognized. The blueshifted minimum of the He I $\lambda$5876 line corresponds to a photospheric velocity of 18,200 km s$^{-1}$ about 15 days prior to maximum brightness, and 10 days later this decreased to 13,100 km s$^{-1}$ (Wheeler & Harkness 1990). The velocity given by this line was only $\sim$10,000 km s$^{-1}$ at maximum brightness (Harkness et al 1987), further evidence of the retreat of the photosphere into deeper, more slowly moving layers.

## 4.3 *Type Ic Spectral Evolution*

The first relatively complete set of spectra illustrating the evolution of a SN Ic (Figure 10) was obtained by Filippenko et al (1990). Shortly after it was discovered, SN 1987M showed spectral characteristics typical of SNe Ic. No lines of hydrogen were visible, and the 6150-Å absorption trough was much weaker than in normal SNe Ia. The strongest features were the P Cygni profile of the Ca II near-IR triplet, O I $\lambda$7774 absorption, and Ca II H&K absorption. As the object aged, Fe II lines became prominent (e.g. near 4900 Å and 5500 Å). Strong He I lines did not appear, unlike the case in SN 1984L (Figure 9), and this is the basis for identifying SN 1987M as a SN Ic rather than a SN Ib.

Nebular [O I] $\lambda\lambda$6300, 6364 emission first emerged at $t$ = 1–2 months, considerably earlier than had been expected. The two lines initially had roughly comparable strength, rather than the usual intensity ratio of three to one, because

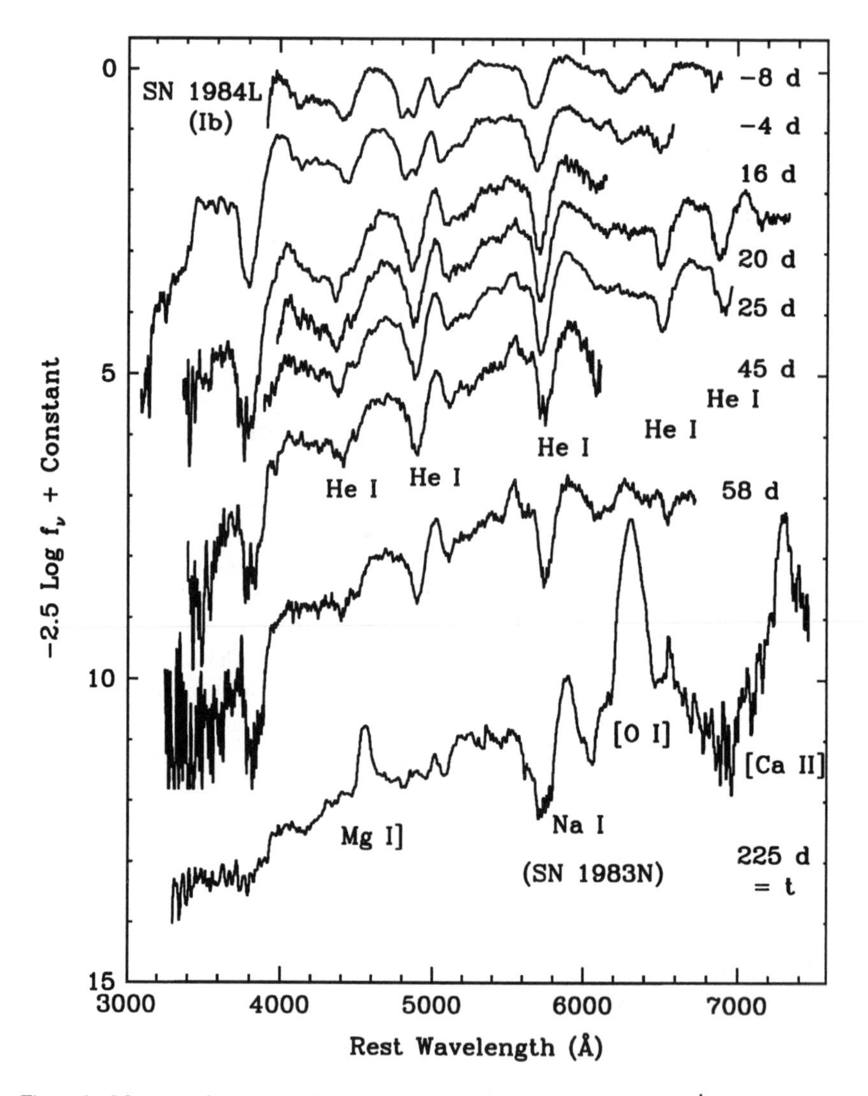


*Figure 9* Montage of spectra of SN Ib 1984L in NGC 991 ($cz = 1532$ km s$^{-1}$), from Harkness et al (1987). The last spectrum is of SN Ib 1983N in NGC 5236 ($cz = 516$ km s$^{-1}$), from Gaskell et al (1986). Epochs (days) are given relative to maximum B brightness (September 7, 1984, for SN 1984L; July 17, 1983, for SN 1983N). Reproduced with permission.

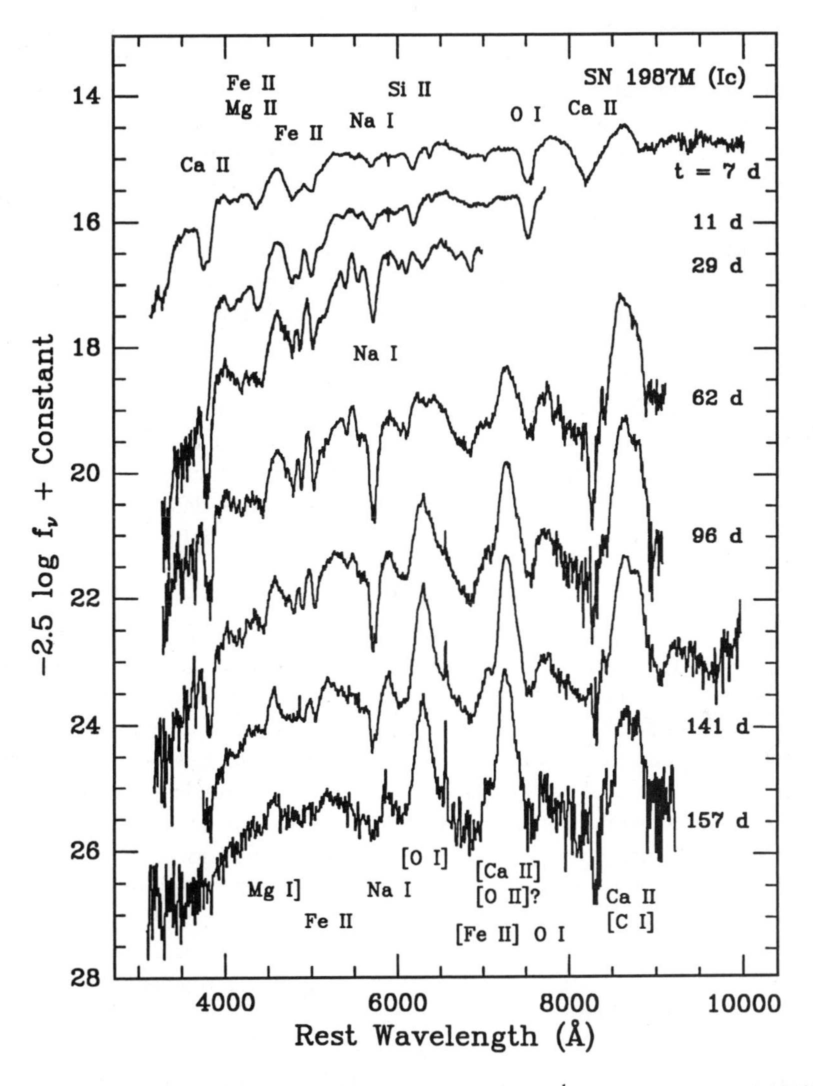


*Figure 10* Spectra of SN Ic 1987M in NGC 2715 ($cz = 1339 \, \mathrm{km \, s^{-1}}$), from Filippenko et al (1990), showing the development of the nebular phase. Epochs (days) are given relative to maximum B brightness (estimated to be September 21, 1987).

of self-absorption. The intensity ratio of [O I] and [Ca II] emission to the Ca II near-IR triplet increased with time as a consequence of the steadily decreasing electron density. For a while Na I D absorption grew deeper, but after 2–3 months it began to fade. By 5 months past maximum the nebular emission completely dominated the spectrum.

Despite their superficial similarities at early times, the spectra of SNe Ia and SNe Ic evolve in very different manners. The nebular spectra of SNe Ia consist of broad emission-line blends of many forbidden transitions of singly and doubly ionized Fe and Co (Figure 2). SNe Ic (and SNe Ib), on the other hand, are dominated by a few strong, broad, relatively unblended emission lines of neutral oxygen and singly ionized calcium, together with weaker lines of C I, Mg I, Na I, and other intermediate-mass elements (Figure 10).

## 4.4 *Is There Helium in SNe Ic?*

What evidence do we have that helium is truly absent from the spectra of genuine SNe Ic? Do some (or even most) SNe Ic actually have weak He I lines? [Filippenko (1991b) demonstrated how difficult it can be to distinguish between SNe Ib and SNe Ic, especially if the explosion date is unknown.] If so, what does the helium tell us about the progenitors and their evolutionary histories?

The bright, nearby SN Ic 1994I in NGC 5194 (M51) has begun to shed light on these issues. As shown by Filippenko et al (1995b), strong He I $\lambda$10,830 absorption was visible during the first month past maximum brightness (Figure 11). Moreover, the Na I D $\lambda$5892 absorption line may have been somewhat contaminated by He I $\lambda$5876 (see also Clocchiatti et al 1996c), as evidenced by the weak notch in the blue wing of the Na I D absorption on April 18, 1994. Based on the optical region alone, SN 1994I is clearly a SN Ic; nevertheless, its atmosphere cannot be completely devoid of helium, as is most convincingly demonstrated by the He I $\lambda$10,830 line. Clocchiatti et al (1996c) and Clocchiatti & Wheeler (1997) showed that weak optical He I lines appear to be present in several other classical SNe Ic, including SN 1987M (see also Jeffery et al 1991). Indeed, the line at ~5520 Å in the first four spectra of Figure 10 (most easily visible in the third spectrum) is identified as He I $\lambda$5876. Thus, the presence of at least some helium is a common property of the progenitors of SNe Ic.

Nomoto et al (1994; see also Iwamoto et al 1994) suggested that the progenitor of SN 1994I was a 2.2-$M_\odot$ C-O core formed as a consequence of two stages of mass transfer in a binary system; during the second stage, helium was lost to the close companion (most likely an O-Ne-Mg white dwarf). Woosley et al (1995), on the other hand, invoked only the first stage of mass transfer, during which the hydrogen envelope is lost to the companion. Depending on the initial mass of the star, the resulting helium star has a mass in the range

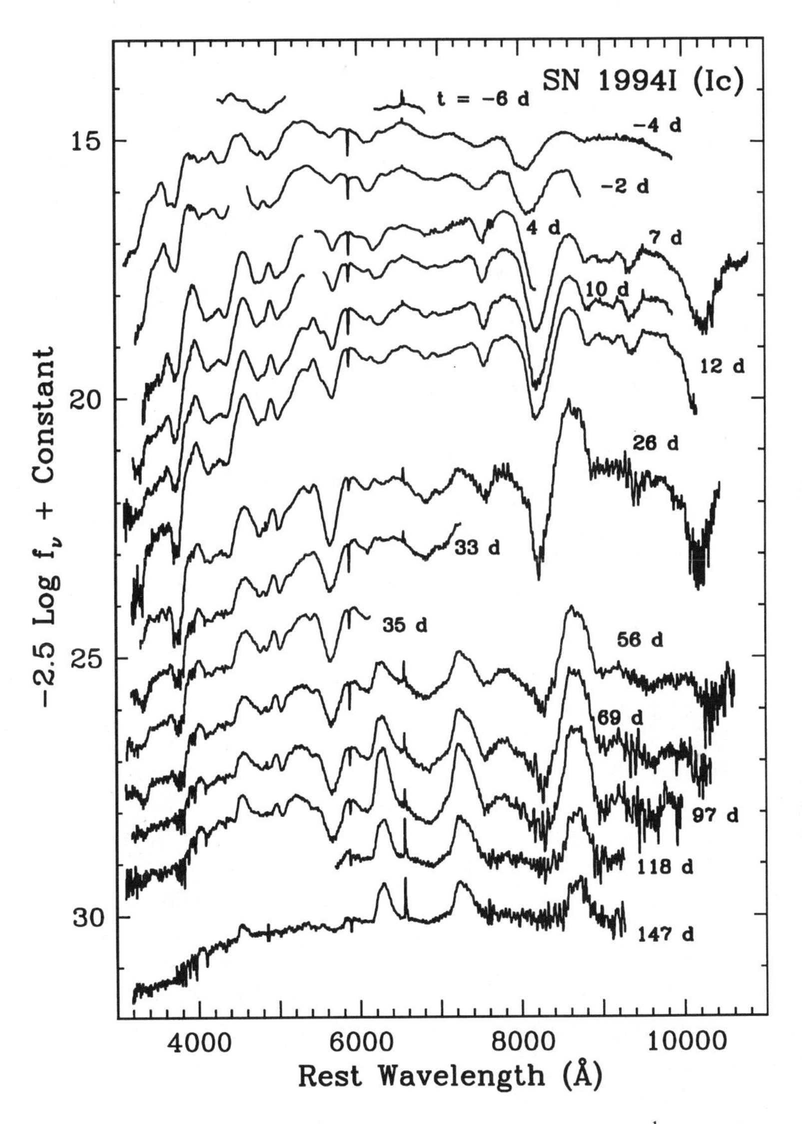


*Figure 11* Montage of spectra of SN 1994I in NGC 5194 ($cz = 500$ km s$^{-1}$), from Filippenko et al (1995b). Epochs (days) are given relative to maximum B brightness (April 8, 1994). The late-time spectra are significantly contaminated by gas and early-type stars in the host galaxy; note the blue continuum, as well as the Balmer absorption and emission lines. Blueshifted He I λ10,830 is prominent at early times, and the transition to the nebular phase is rapid.

4–20 $M_{\odot}$, but subsequent mass loss through winds is very efficient and makes the final mass of the C-O star always converge to the narrow range 2.26–3.55 $M_{\odot}$. In both cases, the explosion mechanism was iron core collapse, as in SNe II, but the mass of ejected helium is rather different: $\sim$0.01 $M_{\odot}$ (Nomoto et al 1994) or 0.1–0.3 $M_{\odot}$ (Woosley et al 1995).

Baron et al (1996) found no direct evidence for helium (upper limit of $\sim$0.1 $M_{\odot}$) in a preliminary analysis of the optical spectra of SN 1994I, but they did not include the He I $\lambda$10,830 line and non-LTE effects. Naively, if 0.1 $M_{\odot}$ of helium is moving at 16,500 km s$^{-1}$ (as observed for the He I $\lambda$10,830 absorption minimum), the corresponding kinetic energy ($\sim 0.3 \times 10^{51}$ erg) already seems excessive; this may favor the Nomoto et al (1994) model, which requires a factor of 10 less helium. On the other hand, Nomoto et al made a specific prediction about the late-time spectrum of SN 1994I: Emission lines of calcium should exhibit velocities of up to $\sim$10,000 km s$^{-1}$, and the oxygen lines would be even broader because oxygen is concentrated in the outermost gas layers (Iwamoto et al 1994). This is not the case (Filippenko et al 1995b); despite consisting of the more closely spaced doublet, the [Ca II] line is broader than [O I], and the most rapidly moving oxygen has $v \lesssim 7000$ km s$^{-1}$.

# 5. TYPE II SUPERNOVAE

## 5.1 *Type II-P Supernovae*

Excellent examples of SNe II-P are SN 1969L (Ciatti et al 1971), SN 1986I (Pennypacker et al 1989), SN 1988A (Turatto et al 1993a), SN 1990E (Schmidt et al 1993, Benetti et al 1994), and SN 1991G (Blanton et al 1995). At very early times, the spectrum is nearly featureless and quite blue, indicating a high color temperature ($\gtrsim$10,000 K). Very weak hydrogen Balmer lines and He I $\lambda$5876 are often visible. Initially, the widths of the Balmer lines and the blueshifts of their P Cygni absorption minima decrease noticeably in some objects (e.g. SN 1987A; Menzies 1991), as the photosphere quickly recedes to the inner, more slowly moving layers of the homologously expanding ejecta. The temperature rapidly decreases with time, reaching $\sim$5000 K within a few weeks, as expected from the adiabatic expansion and associated cooling of the ejecta. It remains roughly constant at this value during the plateau, while the hydrogen recombination wave moves through the massive ($\sim$10 $M_{\odot}$) hydrogen ejecta and releases the energy deposited by the shock. At this stage, strong Balmer lines and Ca II H&K with well-developed P Cygni profiles appear, as do weaker lines of Fe II, Sc II, and other iron-group elements. Subsequently, as the light curve drops to the late-time tail, the spectrum gradually takes on a nebular appearance; the continuum fades, but H$\alpha$ becomes very strong, and prominent emission lines of [O I], [Ca II], and Ca II also appear.

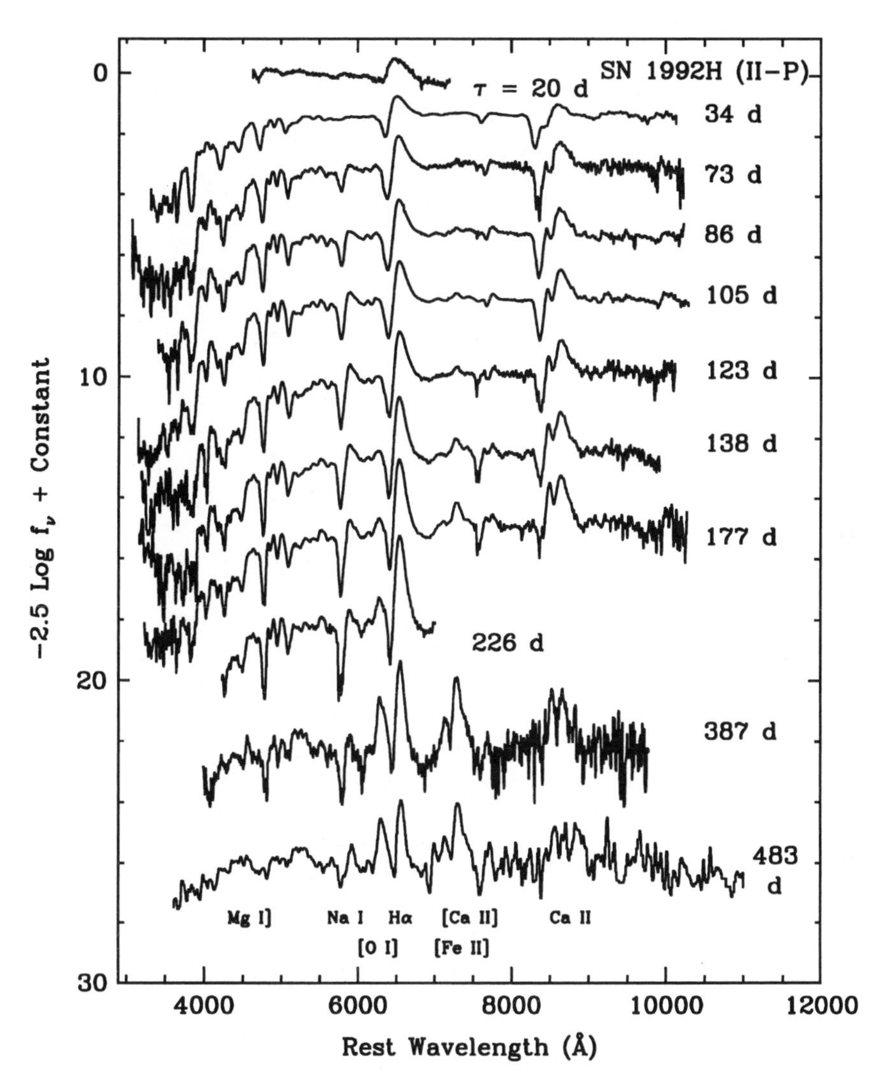


*Figure 12* Montage of spectra of SN 1992H in NGC 5377 ($cz = 1793$ km s$^{-1}$). Epochs (days) are given relative to the estimated date of explosion, February 8, 1992.

This behavior is well illustrated in Figure 12 with SN 1992H (see also Clocchiatti et al 1996a, who estimated the explosion date to be February 8, 1992). Although its V-band plateau ($\tau = 40$–100 days, or perhaps 50–90 days, depending on one's definition) was somewhat shorter than that of the most famous SNe II-P mentioned above and declined slowly with time, SN

1992H can still be considered a SN II-P, and its spectral development was quite typical. Weak He I $\lambda$5876 was superposed on a blue continuum on day 20. H$\beta$ absorption was present, but the corresponding component of H$\alpha$ was weak or absent; H$\alpha$ emission, on the other hand, was obvious. The H$\alpha$ absorption line must have developed very rapidly, as it was strong by day 34, and the continuum was redder. The spectrum changed little between days 34 and 123, with the absorption lines gradually growing stronger; Na I D became very prominent, and many lines of singly ionized metals were present. The emergence of weak forbidden emission lines (day 105), most notably [Ca II] $\lambda\lambda$7291, 7324, roughly coincided with the end of the plateau phase; the Ca II near-IR triplet and Na I D emission also became more prominent. By day 138, and certainly by day 177, [O I] $\lambda$5577 and [O I] $\lambda\lambda$6300, 6364 were unmistakable. At $\tau \approx 1$ year, when the continuum was faint, the spectrum was dominated by H$\alpha$, [Ca II] $\lambda\lambda$7291, 7324, and [O I] $\lambda\lambda$6300, 6364; weaker [Fe II] $\lambda$7155, Na D, Mg I] $\lambda$4571, the Ca II near-IR triplet, and blends of Fe II lines (especially near 5300 Å) were also present.

SNe II-P are excellent distance indicators, using the "Expanding Photosphere Method" (a variant of the Baade-Wesselink method) described by Kirshner & Kwan (1974); see Schmidt et al (1994a,b), Eastman et al (1996), Filippenko (1997b), and references therein. This technique is independent of the various uncertain rungs in the cosmological distance ladder: It relies only on an accurate measurement of the effective temperature (from the measured colors, with appropriate modeling of deviations from a blackbody spectrum) and the velocity of the photosphere (from the wavelengths of weak absorption lines such as those of Sc II) during the plateau phase. An important check is that the object's derived distance should be independent of time.

## 5.2 *Supernova 1987A*

SN 1987A, a peculiar variant of SNe II-P, is described extensively in many other reviews (see Section 1); it is mentioned here only briefly. In general, its spectral evolution resembled that of SN 1992H (Figure 12), as can be seen, for example, in Menzies (1991). Jeffery & Branch (1990) presented an analysis that showed that to a considerable extent the evolution of the line spectrum during the first 100 days could be understood on the basis of simplifying assumptions such as resonant-scattering line source functions and LTE line optical depths. One important aspect of the SN 1987A spectrum is that narrow emission lines from the circumstellar ring were present (Wampler & Richichi 1989), and these became dominant at $\tau \gtrsim 3$ years (Wang et al 1996).

Danziger et al (1988) found that around day 530, the peaks of several emission lines (most notably [O I] $\lambda\lambda$6300, 6364) shifted rapidly to bluer wavelengths. This was probably due to the formation of dust in the ejecta (Lucy et al 1991), which is consistent with the nearly simultaneous increase in the decline rate of the optical light curves and the rapid growth of an IR excess. The lines remained blueshifted even in the very late-time spectra obtained with the Hubble Space Telescope ($\tau \approx 2000$ days; Wang et al 1996), which shows that the dust was still present.

The optical spectra of SN 1987A provide considerable evidence for the formation of clumps and mixing of different layers in the ejecta, as had already been deduced from other studies (e.g. X-ray emission; Arnett et al 1989, and references therein). 1. Very early, Hanuschik & Dachs (1987; see also Phillips & Heathcote 1989) drew attention to the "Bochum event," an asymmetry in the H$\alpha$ and other hydrogen-line profiles. One possibility is that a blob of $Ni^{56}$ was ejected asymmetrically (Chugai 1992, Utrobin et al 1995). 2. Stathakis et al (1991) showed that the [O I] $\lambda\lambda$6300, 6364 profile was serrated in a manner similar to that found for SN 1985F by Filippenko & Sargent (1989), with FWHM typically 80 km $s^{-1}$ for the emission-line peaks. The interpretation is that the [O I]-emitting material is clumpy, probably owing to the formation of Rayleigh-Taylor instabilities at the boundary of the oxygen-rich and helium-rich layers. Chugai (1994a) estimated the mass of clumpy oxygen to be 1.2–1.5 $M_{\odot}$. 3. Hanuschik et al (1993) showed that the H$\alpha$ profile exhibited peaks with a somewhat larger velocity scale: FWHM = 160–400 km $s^{-1}$. Subsequently, Spyromilio et al (1993) demonstrated that at least one of the H$\alpha$ clumps was also visible in the [Ca II] $\lambda\lambda$7291, 7324 and [Fe II] $\lambda$7155 emission lines, directly demonstrating that small-scale mixing of hydrogen and radioactive products occurred in the ejecta.

## 5.3 *Type II-L Supernovae*

Few SNe II-L have been observed in as much detail as SNe II-P. Figure 13 shows the spectral development of SN 1979C (Branch et al 1981), an unusually luminous member of this subclass (Gaskell 1992). Near maximum brightness the spectrum was very blue and almost featureless, with a slight hint of H$\alpha$ emission. A week later, H$\alpha$ emission was more easily discernible, and low-contrast P Cygni profiles of Na I, H$\beta$, and Fe II appeared. By $t \approx 1$ month, the H$\alpha$ emission line was very strong but still devoid of an absorption component, while the other features clearly had P Cygni profiles. Strong, broad H$\alpha$ emission dominated the spectrum at $t \approx 7$ months, and [O I] $\lambda\lambda$6300, 6364 emission was also present.

SN 1980K (Uomoto & Kirshner 1986), another extensively studied (and somewhat overluminous) SN II-L, also did not exhibit an absorption component of H$\alpha$ at any phase of its development. Several authors (e.g. Wheeler & Harkness 1990, Filippenko 1991a) have speculated that the absence of H$\alpha$ absorption spectroscopically differentiates SNe II-L from SNe II-P, but the small

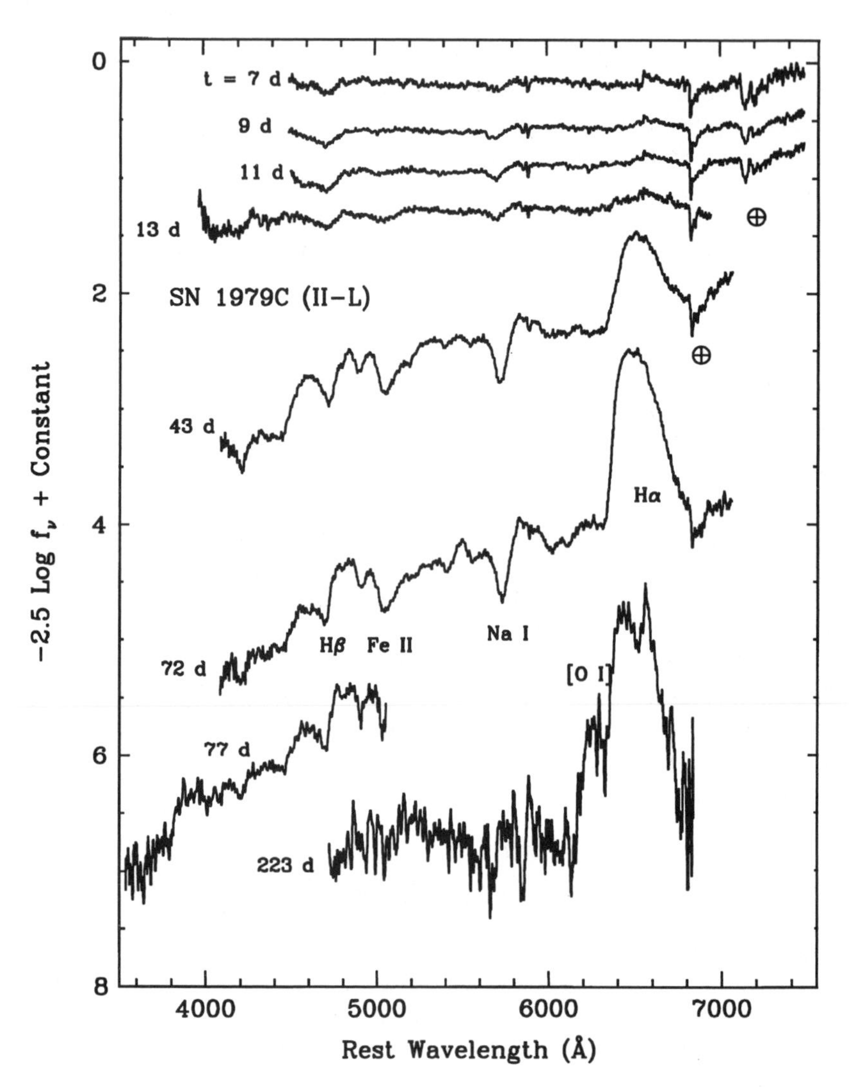


*Figure 13* Montage of spectra of SN 1979C in NGC 4321 ($cz = 1571$ km s$^{-1}$). From Branch et al (1991); reproduced with permission. Epochs (days) are given relative to the date of maximum brightness, April 15, 1979.

size of the sample of well-observed objects precluded definitive conclusions. Recently, Schlegel (1996) collected a somewhat larger set of data and formally proposed that SNe II-L and SNe II-P can be spectroscopically separated in this manner, but it is not yet clear that this is justified. As he himself admitted, the conclusion is based heavily on spectra of SNe 1979C and 1980K, yet these are not necessarily typical SNe II-L. More data are needed to convincingly demonstrate a spectroscopic difference between SNe II-L and SNe II-P.

The progenitors of SNe II-L are generally believed to have relatively low-mass hydrogen envelopes (a few solar masses); otherwise, they would exhibit distinct plateaus, as do SNe II-P. On the other hand, perhaps their envelopes are very extended, or the progenitors have more circumstellar gas than do SNe II-P, and this could give rise to the emission-line dominated spectra (see Section 5.4). They are often radio sources (Sramek & Weiler 1990); moreover, Fransson (1982, 1984) suggested that the UV excess (at $\lambda \lesssim 1600$ Å) seen in SNe 1979C and 1980K is produced by inverse Compton scattering of photospheric radiation by high-speed electrons in shock-heated ($T \approx 10^9$ K) circumstellar material. Finally, the light curves of some SNe II-L reveal an extra source of energy: After declining exponentially for several years after outburst, the H$\alpha$ flux of SN 1980K reached a steady level, showing little if any decline thereafter (Uomoto & Kirshner 1986, Leibundgut et al 1991b). The excess almost certainly came from the kinetic energy of the ejecta that was thermalized and radiated owing to an interaction with circumstellar matter (Chevalier 1990, Leibundgut 1994, and references therein). Note that Swartz et al (1991) explored the possibility that SNe II-L may result from electron-capture–induced collapse of an O-Ne-Mg core, rather than by collapse of an Fe core due to photodissociation.

Fesen & Becker (1990), Leibundgut et al (1991b), and Fesen & Matonick (1993, 1994; see also Fesen et al 1995) illustrated very late-time spectra of SNe II-L 1980K and 1979C. Additional objects having similar characteristics include SN 1970G (Fesen 1993) and SN 1986E (Cappellaro et al 1995b). The spectra consist of a few strong, broad emission lines such as H$\alpha$, [O I] $\lambda\lambda$6300, 6364, and [O III] $\lambda\lambda$4959, 5007. Their relative intensities and temporal changes are generally consistent with the circumstellar interaction models of Chevalier & Fransson (1994), and they can be used to further constrain the nature of the progenitor star.

## 5.4 *Type IIn Supernovae*

During the past decade, there has been the gradual emergence of a new, distinct subclass of SNe II (Filippenko 1991a,b, Schlegel 1990, Leibundgut 1994) whose ejecta are believed to be strongly interacting with dense circumstellar gas (see Chevalier 1990 for an overview of this process). The derived mass-loss rates for the progenitors can exceed $10^{-4} M_{\odot}$ year$^{-1}$ (Chugai 1994b). In these objects, the broad absorption components of all lines are weak or absent throughout their evolution. Instead, their spectra are dominated by strong emission lines, most notably H$\alpha$, that have a complex but relatively narrow profile. Although the details differ among objects, H$\alpha$ typically exhibits a very narrow component (FWHM $\lesssim 200$ km s$^{-1}$) superposed on a base of intermediate width

(FWHM $\approx$ 1000–2000 km s$^{-1}$; sometimes a very broad component (FWHM $\approx$ 5000–10,000 km s$^{-1}$) is also present. Schlegel (1990) christened this subclass "Type IIn," the "n" denoting "narrow" to emphasize the presence of the intermediate-width or very narrow emission components. Representative spectra of five SNe IIn are shown in Figure 14, with two epochs for SN 1994Y.

The early-time continua of SNe IIn tend to be bluer than normal. Occasionally He I emission lines are present in the first few spectra [e.g. SN 1994Y in Figure 14 and SN 1987B (see Figure 1.22 of Harkness & Wheeler 1990)]. Very narrow Balmer absorption lines are visible in the early-time spectra of some of these objects, often with corresponding Fe II, Ca II, O I, or Na I absorption as well (e.g. SNe 1994W and 1994ak in Figure 14). Some of them are unusually luminous at maximum brightness, and they generally fade quite slowly, at least at early times. The equivalent width of the intermediate H$\alpha$ component can grow to astoundingly high values at late times.

One of the first extensively observed SNe IIn was SN 1987F (Filippenko 1989a, Wegner & Swanson 1996). Initially, broad H$\alpha$ emission was superposed on a luminous ($M_V \approx -19.3$ mag), nearly featureless continuum, but its profile did not have the characteristic P Cygni shape, and its centroid was blueshifted by $\gtrsim$1500 km s$^{-1}$ with respect to the systemic velocity of the parent galaxy. Many months later, the broad H$\alpha$ in SN 1987F was more luminous and had much larger equivalent width; Fe II, Ca II, and O I emission were detected as well (see also SN 1994Y in Figure 14). Forbidden lines, normally prominent at this phase, were very weak; Filippenko (1989a, 1991a) concluded that the ejecta had high electron density ($n_e \gtrsim 10^9$ cm$^{-3}$). The narrow component of H$\alpha$, initially quite luminous, was now much weaker. At early times it may have been produced by material previously lost from the progenitor, but this gas was eventually engulfed by the expanding SN ejecta. At 10 months after maximum, SN 1987F was $\sim$2-mag more luminous than typical SNe II-P (Cappellaro et al 1990). Chugai (1991) modeled the data according to an interaction of the SN ejecta with dense circumstellar matter.

Another example is SN 1988Z (Filippenko 1991a,b, Stathakis & Sadler 1991, Turatto et al 1993b, Chugai & Danziger 1994). At early times, SN 1988Z showed very narrow (FWHM $\lesssim$ 100 km s$^{-1}$) [O III] $\lambda$4363 and [O III] $\lambda\lambda$4959, 5007 emission lines whose relative intensities indicated $n_e \gtrsim 10^7$ cm$^{-3}$. They were almost certainly produced by circumstellar gas released by the progenitor prior to exploding and then photoionized by the intense flash of UV radiation emitted at the time of shock breakout. A resolved, intermediate-width (FWHM $\approx$2000 km s$^{-1}$) component of H$\beta$ appeared less than two months after discovery and steadily grew stronger. At H$\alpha$, this component was superposed on a much broader emission line (FWHM $\approx$ 15,000 km s$^{-1}$; see Figure 14). Nearly a year later, the intermediate-width component completely dominated the optical

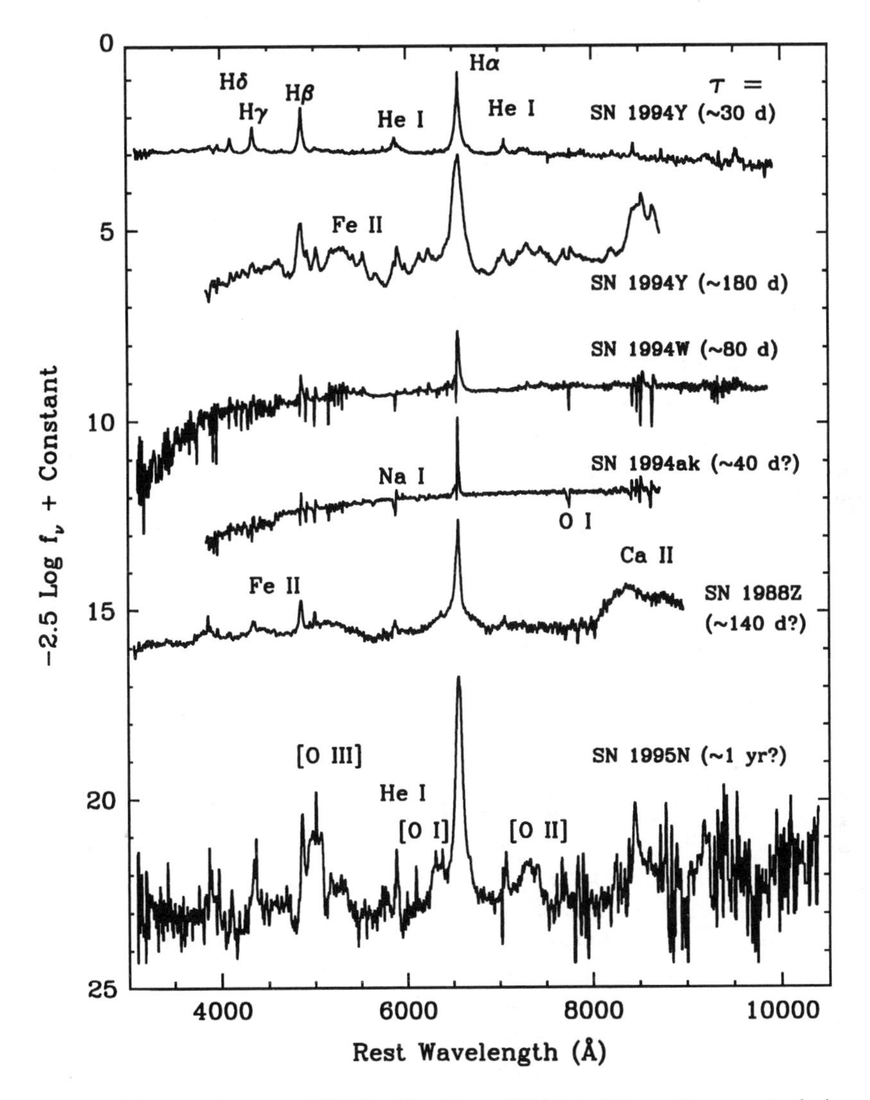


*Figure 14* Montage of spectra of SNe IIn. The objects, UT dates of observation, parent galaxies, and adopted redshifts (kilometers per second) are as follows: SN 1994Y (September 1, 1994 and January 26, 1995; NGC 5371; 2553), SN 1994W (October 1, 1994; NGC 4041; 1234), SN 1994ak (January 26, 1995; NGC 2782; 2562), SN 1988Z (April 27, 1989; MCG+03-28-022; 6595), and SN 1995N (May 24, 1995; MCG +02-38-017; 1534). Epochs are given relative to the estimated dates of explosion rather than maximum brightness; the rise times to maximum can differ substantially among SNe IIn.

spectrum (Filippenko 1991a,b, Turatto et al 1993b). Its Balmer decrement was very steep, possibly indicating "Case C" recombination conditions (e.g. Xu et al 1992) in which the gas is optically thick to the Lyman and Balmer series, although collisional excitation may have also contributed to the peculiar line intensity ratios. We were probably seeing shock-induced emission from dense clumps in a wind emitted by the progenitor star (Chugai & Danziger 1994). As in SN 1987F, forbidden lines were weak or absent, and very strong lines of Fe II, Ca II, and O I emerged (Figure 14). The blend of O I $\lambda$8446 and the Ca II near-IR triplet, in particular, became stronger than the very broad component of H$\alpha$, yet little or no [Ca II] $\lambda\lambda$7291, 7324 was present, indicating high density. However, Chugai & Danziger (1994) argued that the envelope was not massive, and hence the progenitor itself may have had a relatively low mass, in contrast with the conclusion of Stathakis & Sadler (1991).

The late-time optical spectra of SN 1988Z closely resembled those of SN 1986J, an object that was discovered at radio wavelengths long after its optical outburst (Rupen et al 1987, Leibundgut et al 1991b). Accordingly, Filippenko (1991a) predicted that SN 1988Z should eventually become very luminous at radio wavelengths, as did SN 1986J. SN 1988Z was indeed subsequently detected at radio wavelengths with a luminosity comparable to that of SN 1986J, and analysis of the radio light curves suggested a high mass-loss rate (Van Dyk et al 1993). SN 1988Z was also detected as an X-ray source (Fabian & Terlevich 1996). Another similar object is SN 1978K (Ryder et al 1993, Chugai et al 1995), which was luminous at radio and X-ray energies, although its Balmer decrement was not unusually steep and suggests Case B recombination.

Type IIn supernovae exhibit considerable heterogeneity. For example, objects like SNe 1986J, 1988Z, 1993N (Filippenko & Matheson 1993, 1994), and 1995N (Pollas et al 1995, Garnavich et al 1995a, Van Dyk et al 1996b), whose spectra were for many years completely dominated by H$\alpha$ emission of FWHM $\approx$ 1000 km s$^{-1}$, became strong radio and X-ray sources. They seem to have the densest circumstellar material. Of these objects, the ones observed at early times (SNe 1988Z and 1993N) had relatively featureless blue continua with almost no H$\alpha$ emission. Other SNe IIn, however, are distinct from the SN 1988Z flavor; they exhibit strong H$\alpha$ emission right from the start (e.g. SN 1994Y in Figure 14), and they don't become luminous radio sources (Van Dyk et al 1996c). Even among these latter objects there is considerable heterogeneity: Witness the presence of narrow absorption lines in SN 1994W (and also SN 1994ak) but not in SN 1994Y (Figure 14). Moreover, as illustrated by Cumming & Lundqvist (1997), the brightness of SN 1994W dropped precipitously after an age of four months, while SN 1994Y remained quite bright for several years after outburst (AV Filippenko, unpublished data). As another example, the early-time spectrum of SN 1987B (Harkness & Wheeler 1990, Filippenko 1991b, Schlegel et al 1996) closely resembled that of SN 1994Y

(Figure 14), with Balmer emission lines and He I that had broad bases; on the other hand, a few months later, the spectrum of SN 1987B exhibited only a hint of broad emission and was instead dominated by relatively narrow absorption lines (Filippenko 1991b), while that of SN 1994Y showed broad Balmer and Fe II emission lines (Figure 14).

SN 1983K (Niemela et al 1985), despite its classification as a SN II-P by Phillips et al (1990), might also be a variant of SNe IIn and further illustrates the diversity of this subclass: Emission lines of N III λ4651 and He II λ4686, as well as hydrogen Balmer emission lines, were superposed on a very blue continuum in spectra obtained about 10 days prior to maximum brightness, but by maximum brightness the spectrum showed only a few weak and narrow absorption lines. In the case of SN 1984E (Henry & Branch 1987), there was evidence that circumstellar material had been ejected from the progenitor in a relatively discrete event less than 30 years before the explosion (Gaskell & Keel 1988, but see Dopita et al 1984). Recently there have been a substantial number of other SNe IIn discovered (as documented in *IAU Circulars*), and they seem to exhibit a great variety of properties that should provide clues to the nature of mass loss in evolved massive stars.

## 5.5 *Links Between SNe II and SNe Ib/Ic*

5.5.1 SN 1987K SN 1987K (Figure 15) appears to be a link between SNe II and SNe Ib (Filippenko 1988). Near maximum brightness, it was undoubtedly a SN II but with rather weak photospheric Balmer and Ca II lines. Many months after maximum, the broad Hα emission that dominates the late-time spectra of other SNe II was weak or absent. Instead, broad emission lines of [O I] λλ6300, 6364, [Ca II] λλ7291, 7324, and the Ca II near-IR triplet were the most prominent features, just as in SNe Ib. Such a metamorphosis was, at that time, unprecedented in the study of SNe.

SN 1987K provides very strong evidence for a physical continuity between the progenitors and explosion mechanisms of SNe II and at least some SNe Ib/Ic. The simplest interpretation is that SN 1987K had a meager hydrogen atmosphere at the time it exploded (but see Harkness & Wheeler 1990); it would naturally masquerade as a SN II for a while, and as the expanding ejecta thinned out the spectrum would become dominated by emission from deeper and denser layers. It is noteworthy that, aside from the Hα line, the early-time spectrum of SN 1987K was actually quite similar to those of SNe Ic 1983I and 1983V (Wheeler et al 1987). The progenitor was probably a star that, prior to exploding via iron core collapse, lost almost all of its hydrogen envelope either through mass transfer onto a companion or as a result of stellar winds.

5.5.2 SN 1993J The data for SN 1987K (especially its light curve) were rather sparse, making it difficult to model in detail; moreover, it was an apparently

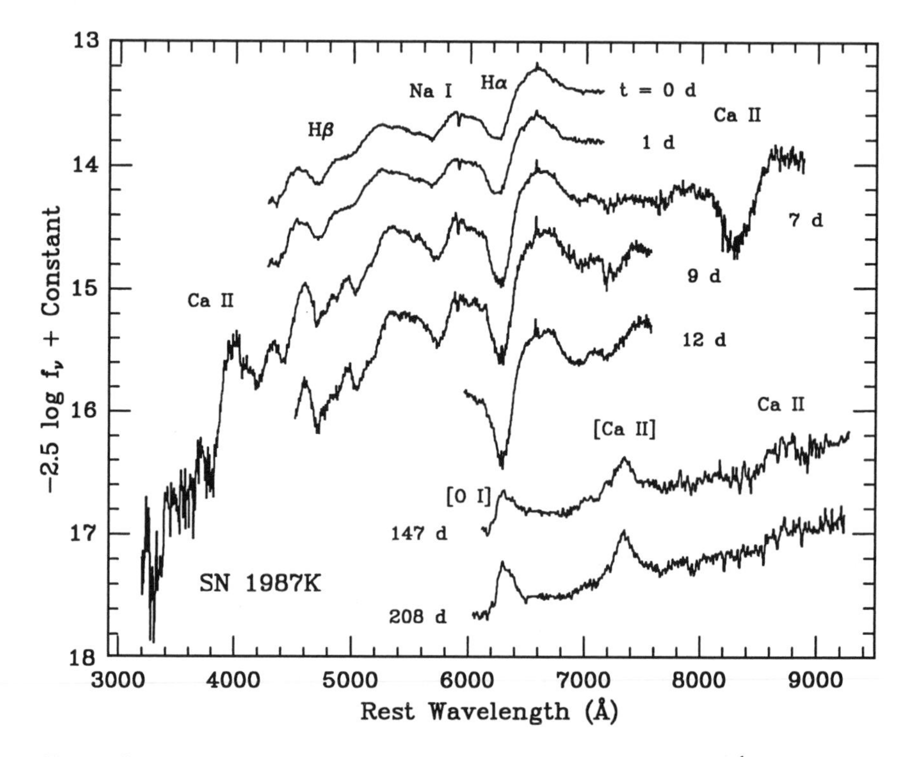


*Figure 15* Montage of spectra of SN 1987K in NGC 4651 ($cz = 817$ km s$^{-1}$), showing the dramatic transformation from a SN II to a SN Ib. Narrow emission lines, produced by superposed H II regions, have been excised in the late-time spectra. Adapted from Filippenko (1988). Epochs are given relative to the date of maximum brightness, July 31, 1987.

unique object and hence perhaps somewhat of a fluke. Consequently, there may have been some room for skepticism regarding the connection between SNe II and SNe Ib (dubbed "SNe IIb" by Woosley et al 1987, who had proposed a similar preliminary model for SN 1987A before it was known to have a massive hydrogen envelope; see Arnett et al 1989). Fortunately, the Type II SN 1993J in NGC 3031 (M81) came to the rescue. Bright ($V \gtrsim 10.7$ mag), nearby ($d =$ 3.6 Mpc; Freedman et al 1994), and well placed in the night sky (circumpolar for many northern observatories), it was studied in greater detail than any SN since SN 1987A. Its early history of observations and modeling is documented by Wheeler & Filippenko (1996).

The unusual nature of SN 1993J first became apparent through its light curves (Benson et al 1994, Lewis et al 1994, Richmond et al 1994, 1996a, Barbon et al 1995, Prabhu et al 1995). At visual wavelengths, it rose to maximum brightness in just a few days, then plummeted by $\sim$1.3 mag over the next week, and subsequently climbed for two weeks to a second peak of brightness comparable to the first. Thereafter it declined, initially about as rapidly as the second rise. By ~40 days after the explosion, it had settled onto an exponential tail of ~0.02 mag day$^{-1}$. This behavior differed substantially from that of either the "plateau" or "linear" SNe II (Barbon et al 1979, Doggett & Branch 1985). The rapid initial rise and the twin peaks were reminiscent of SN 1987A, but the time scale of SN 1993J's second peak was much shorter.

A number of independent groups quickly concluded that the progenitor of SN 1993J probably had a relatively low-mass hydrogen envelope (Nomoto et al 1993, Podsiadlowski et al 1993, Ray et al 1993, Bartunov et al 1994a, Utrobin 1994, Woosley et al 1994); otherwise, the second peak would have resembled the more typical plateau of SNe II, since stored energy slowly diffuses out of a massive envelope. Indeed, Nomoto et al (1993) pointed out that the second rise and decline closely resembled the light curve of SN 1983N (Clocchiatti et al 1996b), a prototypical SN Ib. Most groups found that the light curve could be modeled well by assuming that the progenitor was a ~4 $M_{\odot}$ He core having a low-mass (0.1–0.6 $M_{\odot}$) "skin" of hydrogen; the explosion mechanism was the standard iron core collapse of SNe II.

The likely progenitor of SN 1993J was identified as a G8I-K5I star with a bolometric magnitude of about −7.8 (Aldering et al 1994, Garnavich et al 1997). The general consensus (for a dissenting view, see Höflich et al 1993) is that its initial mass was ~15 $M_{\odot}$. A star of such low mass cannot shed nearly its entire hydrogen envelope without the assistance of a companion star. Thus, the progenitor of SN 1993J probably lost most of its hydrogen through "Case C" mass transfer (in the asymptotic giant phase, after core He burning) to a bound companion 3- to 20-AU away. In addition, part of the gas may have been lost from the system.

A specific prediction made by Nomoto et al (1993) and Podsiadlowski et al (1993) was that the spectrum of SN 1993J should evolve to resemble those of SNe Ib/Ic, as in the case of SN 1987K described in Section 5.5.1. Nearly simultaneously with the submission of these papers, Filippenko et al (1993) identified prominent absorption lines of He I in SN 1993J, confirming the prediction. Instead of growing progressively more prominent with time (relative to other features), the emission component of H$\alpha$ developed a distinct notch identified as blueshifted He I $\lambda$6678 (see Figure 16). Swartz et al (1993a) came to essentially the same conclusion: The spectrum of SN 1993J was transforming itself into that of a SN Ib together with a bit of hydrogen.

Filippenko et al (1993) suggested that many months after the explosion, the spectrum of SN 1993J would closely resemble the late-time spectra of SNe Ib—dominated by strong emission lines of [O I], [Ca II], and Ca II, with H$\alpha$ weak or absent (cf SN 1987K). This was confirmed by Filippenko et al (1994; see

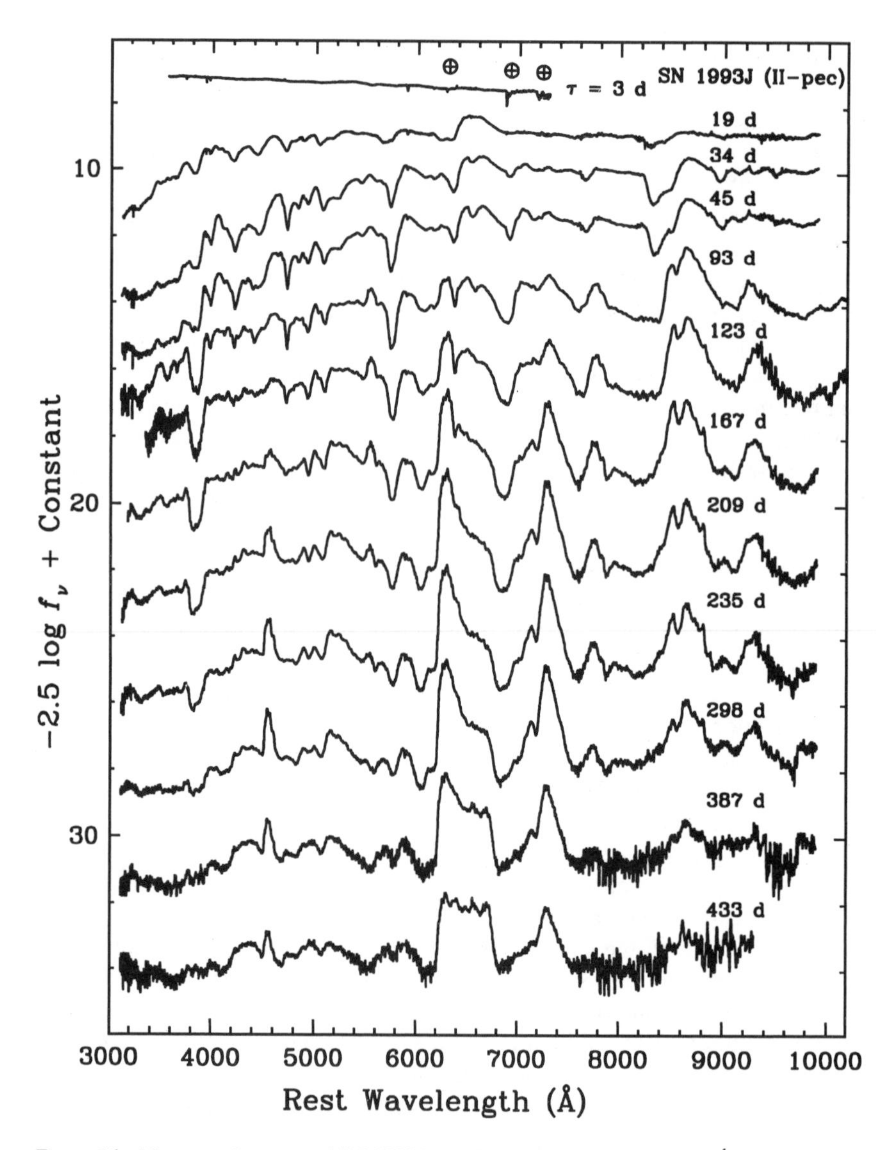


*Figure 16* Montage of spectra of SN 1993J in NGC 3031 ($cz = -34$ km s$^{-1}$), adapted from Filippenko et al (1994). Epochs (days) are given relative to the estimated date of explosion, March 27.5, 1993. There are a few telluric features in the first spectrum.

also Finn et al 1995), as illustrated in Figure 16. Although H$\alpha$ remained visible throughout the evolution of SN 1993J, it was weak relative to the neighboring [O I] line at $\tau$ = 6–10 months, whereas in normal SNe II it dominates the optical spectrum at these epochs (e.g. Figure 12). The emission lines were considerably broader than those of normal SNe II at comparable phases, which is consistent with the progenitor having lost a majority of its hydrogen envelope prior to exploding.

The photometric behavior during the first four months (twin peaks; 0.02-mag day$^{-1}$ exponential decline), together with the spectral evolution, strongly suggests that the progenitor of SN 1993J had only a low-mass skin of hydrogen. Moreover, the V-band decline rate at very late times ($\sim$0.014 mag day$^{-1}$ at $\tau \approx$ 300 days; Richmond et al 1996a) was substantially steeper than that of normal SNe II (0.009 mag day$^{-1}$) and comparable to that of SNe I (Turatto et al 1990), again indicating a low mass for the ejecta. Had the progenitor lost essentially all of its hydrogen prior to exploding, it would have had the optical characteristics of SNe Ib. Consequently, there is now little doubt that most SNe Ib, and probably SNe Ic as well, result from core collapse in stripped massive stars rather than from the thermonuclear runaway of white dwarfs. In addition, it seems likely that a good fraction of the progenitors lost their mass largely via transfer to a bound companion rather than with winds. These conclusions have broad implications for the chemical evolution of galaxies, the expected number density of compact remnants, the potential detectability of neutrino bursts, and the origin of X-ray binary stars.

SN 1993J held several more surprises, however. Observations at radio (Van Dyk et al 1994) and X-ray (Suzuki & Nomoto 1995) wavelengths revealed that the ejecta were interacting with relatively dense circumstellar material (Fransson et al 1996), probably ejected from the system during the course of its pre-SN evolution. Optical evidence for this interaction also began emerging at $\tau \gtrsim$ 10 months: The H$\alpha$ emission line grew in relative prominence, and by $\tau \approx$ 14 months it had become the dominant line in the spectrum (Filippenko et al 1994, Finn et al 1995, Patat et al 1995), consistent with the model of Chevalier & Fransson (1994). Its profile was very broad (FWHM $\approx$ 17,000 km s$^{-1}$; Figure 16) and had a flat top, but with prominent peaks and valleys whose likely origin is Rayleigh-Taylor instabilities in the cool, dense shell of gas behind the reverse shock (Chevalier et al 1992). (The late-time emergence of H$\alpha$ is the main reason SN 1993J is listed as a "SN II-pec" in Figure 16; otherwise, it is a prime example of SNe IIb.) Radio VLBI measurements showed that the ejecta are circularly symmetric but with significant emission asymmetries (Marcaide et al 1995), which are possibly consistent with the asymmetric H$\alpha$ profile.

5.5.3 IS THERE HYDROGEN IN OTHER SNe Ib/Ic? In view of the prominence of H$\alpha$ in the first few spectra of SNe 1987K and 1993J, and the resemblance between the late-time spectra of these two SNe II and SNe Ib/Ic, the search for hydrogen in the spectra of SNe Ib/Ic is of interest. Perhaps it is present in some cases but at a lower level than in the "SN IIb" prototypes; this would set more constraints on the nature of the progenitors.

Wheeler et al (1994) pointed out a possible H$\alpha$ absorption line in spectra of the prototypical SNe Ib 1983N and 1984L. Similarly, Branch (1972) noted that SN Ib 1954A may have exhibited H$\alpha$. The presence of weak H$\alpha$ in SNe Ib is not unexpected, given the examples of SNe 1987K and 1993J; the entire hydrogen envelope need not be expelled prior to core collapse, regardless of whether the progenitor loses gas primarily through winds (very massive star, either isolated or in a wide binary) or via transfer to a companion.

Filippenko (1988, 1992) suggested that SNe Ic 1987M, 1988L, and 1991A had weak H$\alpha$ emission; SN 1987M may also have exhibited H$\alpha$ absorption (Filippenko et al 1990, Jeffery et al 1991). If hydrogen is indeed present in some SNe Ic, then SNe Ic cannot be the explosions of isolated type-WC Wolf-Rayet stars—i.e. those that have lost their entire H envelope and much of their He layer as well. [Note that Van Dyk et al (1996a) also argued against the WR progenitor model, based on a study of the association of SNe Ib/Ic with H II regions.] The binary scenario of Shigeyama et al (1990), on the other hand, may be consistent with a weak H$\alpha$ line, since a helium layer (perhaps with some remaining hydrogen on it, or mixed with it) is present. However, there are other problems with this model—especially the absence of prominent He I lines (Baron 1992). Indeed, the spectral synthesis of Swartz et al (1993b) demonstrated that the progenitors of SNe Ic cannot even have much helium, let alone hydrogen, in their outer layers—if they did, then the signature of these elements would easily be visible at early times.

## 6. FINAL REMARKS

This review has emphasized the optical spectral classification of SNe. The approach is largely taxonomical—or, in the style of Zwicky (1965), "morphological." Like botanists and zoologists, we wish to find observable characteristics that eventually provide a deeper physical understanding of the objects under consideration. Naturally, certain properties may turn out to be more useful than others, and it is our goal to identify these.

For example, a clue to whether SNe Ib and SNe Ic are physically different might be provided by the degree to which the He I line strengths form a continuous sequence among the two subclasses; a roughly bimodal distribution could indicate distinct progenitors or evolutionary paths. Similarly, if the observed properties of SNe IIn really do result from unusually dense circumstellar media, they will provide information on stellar mass loss under different circumstances.

The apparent similarity of the early-time spectra of SNe Ic and subluminous SNe Ia (Filippenko et al 1992a), on the other hand, may end up being nothing more than an indication of comparable primordial iron abundances in the atmospheres of the progenitors, as suggested by Clocchiatti & Wheeler (1997).

There is a tendency to assign each object to a new pigeonhole, based on small variations in the spectra or light curves. This should generally be resisted, unless there are clear physical grounds for doing so (as in the case of SNe IIb), because the proliferation of subtypes having few known members generally does not enhance our understanding of the nature of the phenomenon. For instance, the unusual strength of Ba II lines in early-time spectra of SN 1987A (Williams 1987), though interesting and important to understand, does not justify the creation of a new subclass. Of course, deviants often give valuable clues that are otherwise difficult to notice or entirely unavailable, and considerable attention should be paid to them. Good examples are provided by the peculiar SNe Ia 1991T and 1991bg, which dramatically illustrate the heterogeneity of SNe Ia and considerably strengthen the apparent correlation between luminosity and decline rate.

The greatest benefits, of course, are often achieved when many different types of observations are combined. For example, optical polarimetry and spectropolarimetry, though not discussed here owing to space limitations, can give information on asymmetries in the ejecta of SNe (e.g. Jeffery 1991, Höflich et al 1996b). Similarly, the recent work of Clocchiatti & Wheeler (1997) showed that SNe Ic with essentially identical spectra can be distinguished by their optical light curves, perhaps indicating rather important differences in the internal structure of the ejecta. IR emission is prominent throughout the evolution of SNe and thoroughly dominates at late times. Conditions at the time of shock breakout are best studied at UV wavelengths, X-ray and radio observations provide clues to the circumstellar environment of SNe, gamma-ray data are used to study the products of explosive nucleosynthesis, and neutrinos can indicate whether a neutron star was formed. The observational study of SNe is a rich and active field; this review was necessarily restricted to only a small subset of the relevant data.

ACKNOWLEDGMENTS

Many colleagues have contributed to my knowledge of the spectra of SNe, sharing their insights in spirited conversations and stimulating lectures, as well as in comments on a draft of this review. Some of my students and former students (AJ Barth, LC Ho, DC Leonard, T Matheson, and JC Shields) helped obtain and calibrate the Lick spectra shown here; special thanks go to DC Leonard for his assistance with many of the figures. S Benetti and A Clocchiatti kindly sent their data in digital format. My research on SNe has been supported by the National Science Foundation, most recently through grant AST–9417213.



*Literature Cited*


Aldering G, Humphreys RM, Richmond M. 1994. *Astron. J.* 107:662–72

Applegate JH, Terman JL. 1989. *Ap. J.* 340:380–83

Arnett D. 1996. *Supernovae and Nucleosynthesis.* Princeton: Princeton Univ. Press

Arnett WD, Bahcall JN, Kirshner RP, Woosley SE. 1989. *Annu. Rev. Astron. Astrophys.* 27:629–700

Audouze J, Bludman S, Mochkovitch R, Zinn-Justin J, eds. 1993. *Supernovae.* Les Houches Sess. LIV. Amsterdam: Elsevier

Axelrod TA. 1980. *Late time optical spectra from the $Ni^{56}$ model for type I supernovae.* PhD thesis. Univ. Calif., Santa Cruz. 250 pp.

Barbon R, Benetti S, Cappellaro E, Patat F, Turatto M, Iijima T. 1995. *Astron. Astrophys. Suppl.* 110:513–19

Barbon R, Benetti S, Cappellaro E, Rosino L, Turatto M. 1990. *Astron. Astrophys.* 237:79–90

Barbon R, Cappellaro E, Turatto M. 1989. *Astron. Astrophys.* 81:421–43

Barbon R, Ciatti F, Rosino L. 1973. *Astron. Astrophys.* 25:241–48

Barbon R, Ciatti F, Rosino L. 1979. *Astron. Astrophys.* 72:287–92

Baron E. 1992. *MNRAS* 255:267–68

Baron E, Hauschildt PH, Branch D, Kirshner RP, Filippenko AV. 1996. *MNRAS* 279:799–803

Bartel N, ed. 1985. *Supernovae as Distance Indicators. Lect. Notes Phys. Vol. 224.* Berlin: Springer-Verlag. 226 pp.

Bartunov OS, Blinnikov SI, Pavlyuk NN, Tsvetkov DY. 1994a. *Astron. Astrophys.* 281:L53–L55

Bartunov OS, Tsvetkov DY, Filimonova IV. 1994b. *Publ. Astron. Soc. Pac.* 106:1276–84

Begelman MC, Sarazin CL. 1986. *Ap. J. Lett.* 302:L59–L62

Benetti S, Cappellaro E, Turatto M, Della Valle M, Mazzali P, Gouiffes C. 1994. *Astron. Astrophys.* 285:147–56

Benson P, Herbst W, Salzer JJ, Vinton G, Hanson, GJ, et al. 1994. *Astron. J.* 107:1453–60

Bertola F. 1964. *Ann. Ap.* 27:319–26

Bertola F, Mammano A, Perinotto M. 1965. *Asiago Contr.* 174:51–61

Blanton EL, Schmidt BP, Kirshner RP, Ford CH, Chromey FR, Herbst W. 1995. *Astron. J.* 110:2868–75

Bludman S, Feng DH, Gaisser T, Pittel S, eds. 1995. *Phys. Rep.* 256:1–235

Branch D. 1972. *Astron. Astrophys.* 16:247–51

Branch D. 1981. *Ap. J.* 248:1076–80

Branch D. 1987. *Ap. J. Lett.* 316:L81–L83

Branch D, Doggett JB, Nomoto K, Thielemann F-K. 1985. *Ap. J.* 294:619–25

Branch D, Drucker W, Jeffery DJ. 1988. *Ap. J. Lett.* 330:L117–18

Branch D, Falk SW, McCall ML, Rybski P, Uomoto AK, Wills BJ. 1981. *Ap. J.* 224:780–804

Branch D, Fisher A, Nugent P. 1993. *Astron. J.* 106:2383–91

Branch D, Lacy CH, McCall ML, Sutherland PG, Uomoto A, et al. 1983. *Ap. J.* 270:123–29

Branch D, Livio M, Yungelson LR, Boffi FR, Baron E. 1995. *Publ. Astron. Soc. Pac.* 107:1019–29

Branch D, Miller DL. 1993. *Ap. J. Lett.* 405:L5–L8

Branch D, Nomoto K. 1986. *Astron. Astrophys.* 164:L13–L15

Branch D, Nomoto K, Filippenko AV. 1991. *Comm. Astrophys.* 15:221–37

Branch D, Romanishin W, Baron E. 1996. *Ap. J.* 465:73–76. Erratum 467:473

Branch D, Tammann GA. 1992. *Annu. Rev. Astron. Astrophys.* 30:359–89

Branch D, van den Bergh S. 1993. *Astron. J.* 105:2231–35

Branch D, Venkatakrishna KL. 1986. *Ap. J. Lett.* 306:L21–L23

Brown GE, ed. 1988. *Phys. Rep.* 163:1–204

Brown GE, Bethe HA. 1994. *Ap. J.* 423:659–64

Cappellaro E, Danziger IJ, Turatto M. 1995b. *MNRAS* 277:106–12

Cappellaro E, Della Valle M, Iijima T, Turatto M. 1990. *Astron. Astrophys.* 228:61–68

Cappellaro E, Turatto M, Fernley J. 1995c. *IUE–ULDA Access Guide No. 6. Supernovae.* Noordwijk: ESA

Chevalier RA. 1976. *Ap. J.* 208:826–28

Chevalier RA, 1981. *Fundam. Cosmic Phys.* 7:1–58

Chevalier RA. 1986. *Highlights Astron.* 7:599–609
Chevalier RA. 1990. See Petschek 1990, pp. 91–110
Chevalier RA. 1995. *Space Sci. Rev.* 74:289–98
Chevalier RA, Blondin JM, Emmering RT. 1992. *Ap. J.* 392:118–30
Chevalier RA, Fransson C. 1994. *Ap. J.* 420:268–85
Chugai NN. 1986a. *Sov. Astron. Lett.* 12:192–95
Chugai NN. 1986b. *Sov. Astron.* 30:563–67
Chugai NN. 1991. *MNRAS* 250:513–18
Chugai NN. 1992. *Sov. Astron. Lett.* 18:50–53
Chugai NN. 1994a. *Ap. J. Lett.* 428:L17–L19
Chugai NN. 1994b. See Clegg et al 1994, pp. 148–52
Chugai NN, Danziger IJ. 1994. *MNRAS* 268:173–80
Chugai NN, Danziger IJ, Della Valle M. 1995. *MNRAS* 276:530–36
Ciatti F, Rosino L, Bertola F. 1971. *Mem. Soc. Astron. Italia* 42:163–84
Clegg RES, Stevens IR, Meikle WPS, eds. 1994. *Circumstellar Media in the Late Stages of Stellar Evolution.* Cambridge: Cambridge Univ. Press
Clocchiatti A, Benetti S, Wheeler JC, Wren W, Boisseau J, et al. 1996a. *Astron. J.* 111:1286–1303
Clocchiatti A, Wheeler JC. 1997. See Ruiz-Lapuente et al 1997, pp. 863–83
Clocchiatti A, Wheeler JC, Benetti S, Frueh M. 1996b. *Ap. J.* 459:547–54
Clocchiatti A, Wheeler JC, Brotherton MS, Cochran AL, Wills D, et al. 1996c. *Ap. J.* 462:462–68
Cristiani S, Cappellaro E, Turatto M, Bergeron J, Bues I, et al. 1992. *Astron. Astrophys.* 259:63–70
Cumming RJ, Lundqvist P. 1997. In *Advances in Stellar Evolution,* ed. RT Rood, A Renzini. Cambridge: Cambridge Univ. Press. In press
Cumming RJ, Lundqvist P, Smith LJ, Pettini M, King DL. 1996. *MNRAS.* 283:1355–61
Danziger IJ, ed. 1987. *ESO Workshop on the SN 1987A,* Conf. Proc. No. 26. Garching: Eur. Southern Obs.
Danziger IJ, Bouchet P, Fosbury RAE, Gouiffes C, Lucy LB, et al. 1988. See Kafatos & Michalitsianos 1988, pp. 37–50
Danziger IJ, Fosbury RAE, Alloin D, Cristiani S, Dachs J, et al. 1987. *Astron. Astrophys.* 177:L13–L16
Danziger IJ, Kjär K, eds. 1991. *Supernova 1987A and Other Supernovae, Conf. Proc. No. 37.* Garching: Eur. Southern Obs.
Della Valle M, Benetti S, Panagia N. 1996. *Ap. J. Lett.* 459:L23–L25
Della Valle M, Livio M. 1994. *Ap. J. Lett.* 423:L31–L33
Doggett JB, Branch D. 1985. *Astron. J.* 90:2303–11
Dopita MA. 1988. *Space Sci. Rev.* 46:225–71
Dopita MA, Evans R, Cohen M, Schwartz RD. 1984. *Ap. J. Lett.* 287:L69–L71
Eastman RG, Schmidt BP, Kirshner RP. 1996. *Ap. J.* 466:911–37
Elias JH, Matthews K, Neugebauer G, Persson SE. 1985. *Ap. J.* 296:379–89
Fabian AC, Terlevich R. 1996. *MNRAS* 280:L5–L8
Fesen RA. 1993. *Ap. J. Lett.* 413:L109–12
Fesen RA, Becker RH. 1990. *Ap. J.* 351:437–42
Fesen RA, Hurford AP, Matonick DM. 1995. *Astron. J.* 109:2608–10
Fesen RA, Matonick DM. 1993. *Ap. J.* 407:110–14
Fesen RA, Matonick DM. 1994. *Ap. J.* 428:157–65
Filippenko AV. 1988. *Astron. J.* 96:1941–48
Filippenko AV. 1989a. *Astron. J.* 97:726–34
Filippenko AV. 1989b. *Publ. Astron. Soc. Pac.* 101:588–93
Filippenko AV. 1991a. See Woosley 1991, pp. 467–79
Filippenko AV. 1991b. See Danziger & Kjär 1991, pp. 343–62
Filippenko AV. 1992. *Ap. J. Lett.* 384:L37–L40
Filippenko AV. 1997a. See Ruiz-Lapuente et al 1997, pp. 1–32
Filippenko AV. 1997b. In *18th Texas Symp. Relativistic Astrophys.,* ed. A Olinto, J Frieman, D Schramm. Singapore: World Sci. In press
Filippenko AV, Barth AJ, Bower GC, Ho LC, Stringfellow GS, et al. 1995a. *Astron. J.* 110:2261–73. Erratum 112:806
Filippenko AV, Barth AJ, Matheson T, Armus L, Brown M, et al. 1995b. *Ap. J. Lett.* 450:L11–L15
Filippenko AV, Leonard DC. 1995. *IAU Circ.* No. 6237
Filippenko AV, Matheson T. 1993. *IAU Circ.* No. 5788
Filippenko AV, Matheson T. 1994. *IAU Circ.* No. 5924
Filippenko AV, Matheson T, Barth AJ. 1994. *Astron. J.* 108:2220–25
Filippenko AV, Matheson T, Ho LC. 1993. *Ap. J. Lett.* 415:L103–6
Filippenko AV, Porter AC, Sargent WLW. 1990. *Astron. J.* 100:1575–87
Filippenko AV, Porter AC, Sargent WLW, Schneider DP. 1986. *Astron. J.* 92:1341–48
Filippenko AV, Richmond MW, Branch D, Gaskell CM, Herbst W, et al. 1992a. *Astron. J.* 104:1543–56
Filippenko AV, Richmond MW, Matheson T, Shields JC, Burbidge EM, et al. 1992b. *Ap. J. Lett.* 384:L15–L18
Filippenko AV, Sargent WLW. 1985. *Nature* 316:407–12

Filippenko AV, Sargent WLW. 1986. *Astron. J.* 91:691–96
Filippenko AV, Sargent WLW. 1989. *Ap. J. Lett.* 345:L43–L46
Finn RA, Fesen RA, Darling GW, Thorstensen JR. 1995. *Astron. J.* 110:300–7
Fisher A, Branch D, Höflich P, Khokhlov A. 1995. *Ap. J. Lett.* 447:L73–L76
Ford CH, Herbst W, Richmond MW, Baker ML, Filippenko AV, et al. 1993. *Astron. J.* 106:1101–12
Fransson C. 1982. *Astron. Astrophys.* 111:140–50
Fransson C. 1984. *Astron. Astrophys.* 133:264–84
Fransson C, Chevalier RA. 1989. *Ap. J.* 343:323–42
Fransson C, Lundqvist P, Chevalier RA. 1996. *Ap. J.* 461:993–1008
Freedman WL, Hughes SM, Madore BF, Mould JR, Lee GL, et al. 1994. *Ap. J.* 427:628–55
Frogel JA, Gregory B, Kawara K, Laney D, Phillips MM, et al. 1987. *Ap. J. Lett.* 315:L129–34
Garnavich P, Challis P, Berlind P. 1995a. *IAU Circ.* No. 6174
Garnavich P, Höflich P, Kirshner RP, Kurucz RL, Challis P, Filippenko AV. 1997. *Ap. J.* Submitted
Garnavich P, Riess A, Kirshner R. 1995b. *IAU Circ.* No. 6278
Garnavich PM, Challis PM. 1997. See Ruiz-Lapuente et al 1997, pp. 711–13
Gaskell CM. 1992. *Ap. J. Lett.* 389:L17–L20
Gaskell CM, Cappellaro E, Dinerstein HL, Garnett D, Harkness RP, Wheeler JC. 1986. *Ap. J. Lett.* 306:L77–L80
Gaskell CM, Keel WC. 1988. See Kafatos & Michalitsianos 1988, pp. 13–15
Goldhaber G, Deustua S, Gabi S, Groom D, Hook I, et al. 1997. See Ruiz-Lapuente et al 1997, pp. 777–84
Gómez G, López R. 1995. *Astron. J.* 109:737–41
Goobar A, Perlmutter S. 1995. *Ap. J.* 450:14–18
Goodrich RW, Stringfellow GS, Penrod GD, Filippenko AV. 1989. *Ap. J.* 342:908–16
Graham JR, Meikle WPS, Allen DA, Longmore AJ, Williams PM. 1986. *MNRAS* 218:93–102
Greenstein JL, Minkowski R. 1973. *Ap. J.* 182:225–43
Hamuy M, Phillips MM, Maza J, Suntzeff NB, Della Valle M, et al. 1994. *Astron. J.* 108:2226–32
Hamuy M, Phillips MM, Maza J, Suntzeff NB, Schommer RA, Avilés R. 1995. *Astron. J.* 109:1–13
Hamuy M, Phillips MM, Maza J, Wischnjewsky M, Uomoto A, et al. 1991. *Astron. J.* 102:208–17
Hamuy M, Phillips MM, Schommer RA, Suntzeff NB, Maza J, Avilés R. 1996a. *Astron. J.* 112:2391–97
Hamuy M, Phillips MM, Suntzeff NB, Schommer RA, Maza J, et al. 1996b. *Astron. J.* 112:2398–407
Hamuy M, Phillips MM, Suntzeff NB, Schommer RA, Maza J, et al. 1996c. *Astron. J.* 112:2438–47
Hanuschik RW, Dachs J. 1987. *Astron. Astrophys.* 182:L29–L30
Hanuschik RW, Spyromilio J, Stathakis R, Kimeswenger S, Gochermann J, et al. 1993. *MNRAS* 261:909–20
Harkness RP. 1986. In *Radiation Hydrodynamics in Stars and Compact Objects,* ed. D Mihalas, K-HA Winkler, pp. 166–81. Berlin: Springer-Verlag
Harkness RP. 1991. See Danziger & Kjär 1991, pp. 447–56
Harkness RP, Wheeler JC. 1990. See Petschek 1990, pp. 1–29
Harkness RP, Wheeler JC, Margon B, Downes RA, Kirshner RP, et al. 1987. *Ap. J.* 317:355–67
Henry RBC, Branch D. 1987. *Publ. Astron. Soc. Pac.* 99:112–15
Hillebrandt W, Höflich P. 1989. *Rep. Prog. Phys.* 52:1421–73
Ho LC, Filippenko AV. 1995. *Ap. J.* 444:165–74. Erratum 1996. 463:818
Höflich P, Khokhlov A, Wheeler JC, Phillips MM, Suntzeff NB, Hamuy M. 1996a. *Ap. J. Lett.* 472:L81–L84
Höflich P, Langer N, Duschinger M. 1993. *Astron. Astrophys.* 275:L29–L32
Höflich P, Wheeler JC, Hines DC, Trammell SR. 1996b. *Ap. J.* 459:307–21
Huang Y-L. 1987. *Publ. Astron. Soc. Pac.* 99:461–66
Humphreys RM, Davidson K. 1994. *Publ. Astron. Soc. Pac.* 106:1025–51
Imshennik VS, Nadëzhin DK. 1989. *Sov. Astrophys. Space Phys. Rev.* 8:1–147
Iwamoto K, Nomoto K, Höflich P, Yamaoka H, Kumagai S, Shigeyama T. 1994. *Ap. J. Lett.* 437:L115–18
Jeffery DJ. 1991. *Ap. J.* 375:264–87
Jeffery D, Branch D. 1990. See Wheeler et al, pp. 149–247
Jeffery DJ, Branch D, Filippenko AV, Nomoto K. 1991. *Ap. J. Lett.* 377:L89–L92
Jeffery DJ, Leibundgut B, Kirshner RP, Benetti S, Branch D, Sonneborn G. 1992. *Ap. J.* 397:304–28
Kafatos M, Michalitsianos A, eds. 1988. *Supernova 1987A in the Large Magellanic Cloud.* Cambridge: Cambridge Univ. Press
Kim A, Gabi S, Goldhaber G, Groom DE, Hook IM, et al. 1997. *Ap. J. Lett.* 476:L63–L66
Kirshner RP. 1990. See Petschek 1990, pp. 59–75

Kirshner RP, Jeffery DJ, Leibundgut B, Challis PM, Sonneborn G, et al. 1993. *Ap. J.* 415:589- 615

Kirshner RP, Kwan J. 1974. *Ap. J.* 193:27–36

Kirshner RP, Oke JB. 1975. *Ap. J.* 200:574–81

Kirshner RP, Oke JB, Penston MV, Searle L. 1973. *Ap. J.* 185:303–22

Kowal CT, Sargent WLW. 1971. *Astron. J.* 76:756–64

Kuchner MJ, Kirshner RP, Pinto PA, Leibundgut B. 1994. *Ap. J. Lett.* 426:L89–L92

Leibundgut B. 1988. *Light curves of supernovae type I.* PhD thesis. Univ. Basel. 137 pp.

Leibundgut B. 1994. See Clegg et al 1994, pp. 100–11

Leibundgut B. 1996. See McCray & Wang 1996, pp. 11–18

Leibundgut B, Kirshner RP, Filippenko AV, Shields JS, Foltz CB, et al. 1991a. *Ap. J.* 371:L23–L26

Leibundgut B, Kirshner RP, Phillips MM, Wells LA, Suntzeff NB, et al. 1993. *Astron. J.* 105:301–13

Leibundgut B, Kirshner RP, Pinto PA, Rupen MP, Smith RC, et al. 1991b. *Ap. J.* 372:531–44

Leibundgut B, Schommer R, Phillips M, Riess A, Schmidt B, et al. 1996. *Ap. J. Lett.* 466:L21–L24

Leibundgut B, Tammann GA, Cadonau R, Cerrito D. 1991c. *Astron. Astrophys. Suppl.* 89:537–79

Lewis JR, Walton NA, Meikle WPS, Martin R, Cumming RJ, et al. 1994. *MNRAS* 266:L27–L39

Livne E, Arnett D. 1995. *Ap. J.* 452:62–74

Livne E, Tuchman Y, Wheeler JC. 1992. *Ap. J.* 399:665–71

Lucy LB, Danziger IJ, Gouiffes C, Bouchet P. 1991. See Woosley 1991, pp. 82–94

Lundqvist P, Cumming RJ. 1997. In *Advances in Stellar Evolution,* ed. RT Rood, A Renzini. Cambridge: Cambridge Univ. Press. In press

Marcaide JM, Alberdi A, Ros E, Diamond P, Shapiro II, et al. 1995. *Science* 270:1475–78

Maza J, Hamuy M, Phillips MM, Suntzeff NB, Avilés R. 1994. *Ap. J. Lett.* 424:L107–10

Maza J, van den Bergh S. 1976. *Ap. J.* 204:519–29

Mazzali PA, Chugai NN, Turatto M, Lucy LB, Danziger IJ, et al. 1997. *MNRAS* 284:151–71

Mazzali PA, Danziger IJ, Turatto M. 1995. *Astron. Astrophys.* 297:509–34

Mazzali PA, Lucy LB, Danziger IJ, Gouiffes C, Cappellaro E, Turatto M. 1993. *Astron. Astrophys.* 269:423–45

McCray R. 1993. *Annu. Rev. Astron. Astrophys.* 31:175–216

McCray R, Wang Z, eds. 1996. *Supernovae and Supernova Remnants.* Cambridge: Cambridge Univ. Press

McMillan R, Ciardullo R. 1996. *Ap. J.* 473:707–12

Meikle WPS, Bowers EJC, Geballe TR, Walton NA, Lewis JR, Cumming RJ. 1997. See Ruiz-Lapuente et al 1997, pp. 53–64

Meikle WPS, Cumming RJ, Geballe TR, Lewis JR, Walton NA, et al. 1996. *MNRAS* 281:263–80

Meikle WPS, Spyromilio J, Allen DA, Varani G-F, Cumming RJ. 1993. *MNRAS* 261:535–72

Menzies JW. 1991. See Danziger & Kjär 1991, pp. 209–15

Minkowski R. 1939. *Ap. J.* 89:156–217

Minkowski R. 1941. *Publ. Astron. Soc. Pac.* 53:224–25

Minkowski R. 1964. *Annu. Rev. Astron. Astrophys.* 2:247–66

Niemela VS, Ruíz MT, Phillips MM. 1985. *Ap. J.* 289:52–57

Nomoto K, Suzuki T, Shigeyama T, Kumagai S, Yamaoka H, Saio H. 1993. *Nature* 364:507–9

Nomoto K, Thielemann F-K, Yokoi K. 1984. *Ap. J.* 286:644–58

Nomoto K, Yamaoka H, Pols OR, van den Heuvel EPJ, Iwamoto K, et al. 1994. *Nature* 371:227–29

Nugent P, Phillips MM, Baron E, Branch D, Hauschildt P. 1995. *Ap. J. Lett.* 455:L147–50

Oemler A, Tinsley BM. 1979. *Astron. J.* 84:985–92

Oke JB, Gunn JE. 1983. *Ap. J.* 266:713–17

Oke JB, Searle L. 1974. *Annu. Rev. Astron. Astrophys.* 12:315–29

Oliva E. 1987. *Ap. J. Lett.* 321:L45–L49

Panagia N. 1985. See Bartel 1985, pp. 14–33

Panagia N. 1987. In *High Energy Phenomena Around Collapsed Stars,* ed. F Pacini, pp. 33–49. Dordrecht: Reidel

Panagia N, Gilmozzi R. 1991. See Danziger & Kjär 1991, pp. 575–94

Panagia N, Meikle WPS, Allen DA, Andrews PL, Barr P, et al. 1986a. Preprint

Panagia N, Sramek RA, Weiler KW. 1986b. *Ap. J. Lett.* 300:L55–L58

Patat F, Barbon R, Cappellaro E, Turatto M. 1993. *Astron. Astrophys. Suppl.* 98:443–76

Patat F, Barbon R, Cappellaro E, Turatto M. 1994. *Astron. Astrophys.* 282:731–41

Patat F, Benetti S, Cappellaro E, Danziger IJ, Della Valle M, et al. 1996. *MNRAS* 278:111–24

Patat F, Chugai N, Mazzali PA. 1995. *Astron. Astrophys.* 299:715–23

Pennypacker CR, Burns MS, Crawford FS, Friedman PG, Graham JR, et al. 1989. *Astron. J.* 97:186–93

Perlmutter S, Gabi S, Goldhaber G, Groom DE, Hook IM, et al. 1997. *Ap. J.* In press

Perlmutter S, Pennypacker CR, Goldhaber G,

Goobar A, Muller RA, et al. 1995. *Ap. J. Lett.* 440:L41–L44
Petschek AG, ed. 1990. *Supernovae.* New York: Springer-Verlag
Phillips MM. 1993. *Ap. J. Lett.* 413:L105–8
Phillips MM, Hamuy M, Maza J, Ruiz MT, Carney BW, Graham JR. 1990. *Publ. Astron. Soc. Pac.* 102:299–305
Phillips MM, Heathcote SR. 1989. *Publ. Astron. Soc. Pac.* 101:137–46
Phillips MM, Phillips AC, Heathcote SR, Blanco VM, Geisler D, et al. 1987. *Publ. Astron. Soc. Pac.* 99:592–605
Phillips MM, Wells LA, Suntzeff NB, Hamuy M, Leibundgut B, et al. 1992. *Astron. J.* 103:1632–37
Podsiadlowski P, Hsu JJL, Joss PC, Ross RR. 1993. *Nature* 364:509–11
Polcaro VF, Viotti R. 1991. *Astron. Astrophys.* 242:L9–L11
Pollas C, Albanese D, Benetti S, Bouchet P, Schwarz H. 1995. *IAU Circ.* No. 6170
Popper DM. 1937. *Publ. Astron. Soc. Pac.* 49:283
Porter AC, Filippenko AV. 1987. *Astron. J.* 93:1372–80
Prabhu TP, Mayya YD, Singh KP, Rao NK, Ghosh KK, et al. 1995. *Astron. Astrophys.* 295:403–12
Proust KM, Couch WJ, eds. 1988. *Astron. Soc. Aust.* 7:343–562
Pskovskii YuP. 1977. *Sov. Astron.* 21:675–82
Pskovskii YuP. 1984. *Sov. Astron.* 28:658–64
Ray A, Singh KP, Sutaria FK. 1993. *J. Astrophys. Astron.* 14:53–63
Ray A, Velusamy T, eds. 1991. *Supernovae and Stellar Evolution.* Singapore: World Sci.
Rees MJ, Stoneham RJ, eds. 1982. *Supernovae: A Survey of Current Research,* NATO Adv. Study Inst. Dordrecht: Reidel
Richmond MW, Treffers RR, Filippenko AV, Paik Y. 1996a. *Astron. J.* 112:732–41
Richmond MW, Treffers RR, Filippenko AV, Paik Y, Leibundgut B, Schulman E, et al. 1994. *Astron. J.* 107:1022–40
Richmond MW, Van Dyk SD, Ho W, Peng C, Paik Y, et al. 1996b. *Astron. J.* 111:327–39
Richtler T, Sadler EM. 1983. *Astron. Astrophys.* 128:L3–5
Riess AG. 1996. *Type Ia multicolor light curve shapes.* PhD thesis. Harvard Univ. 150 pp.
Riess AG, Press WH, Kirshner RP. 1995a. *Ap. J. Lett.* 438:L17–L20
Riess AG, Press WH, Kirshner RP. 1995b. *Ap. J. Lett.* 445:L91–L94
Riess AG, Press WH, Kirshner RP. 1996. *Ap. J.* 473:88–109
Ruiz-Lapuente P, Canal R, Isern J, eds. 1997. *Thermonuclear Supernovae.* Dordrecht: Kluwer
Ruiz-Lapuente P, Cappellaro E, Turatto M, Gouiffes C, Danziger IJ, et al. 1992. *Ap. J. Lett.* 387:L33–L36
Ruiz-Lapuente P, Jeffery DJ, Challis PM, Filippenko AV, Kirshner RP, et al. 1993. *Nature* 365:728–30
Rupen MP, van Gorkom JH, Knapp GR, Gunn JE, Schneider DP. 1987. *Astron. J.* 94:61–70
Rust BW. 1974. *The use of supernovae light curves for testing the expansion hypothesis and other cosmological relations.* PhD thesis. Univ. Ill. 388 pp.
Ryder S, Staveley-Smith L, Dopita M, Petre R, Colbert E, et al. 1993. *Ap. J.* 416:167–81
Schaefer BE. 1996. *Ap. J.* 464:404–11
Schaeffer R, Cassé M, Cahen S. 1987. *Ap. J. Lett.* 316:L31–L35
Schlegel EM. 1990. *MNRAS* 244:269–71
Schlegel EM. 1996. *Astron. J.* 111:1660–67
Schlegel EM, Kirshner RP. 1989. *Astron. J.* 98:577–89
Schlegel EM, Kirshner RP, Huchra JP, Schild RE. 1996. *Astron. J.* 111:2038–46
Schmidt BP. 1997. See Ruiz-Lapuente et al 1997, pp. 765–75
Schmidt BP, Kirshner RP, Eastman RG, Hamuy M, Phillips MM, et al. 1994a. *Astron. J.* 107:1444–52
Schmidt BP, Kirshner RP, Eastman RG, Phillips MM, Suntzeff NB, et al. 1994b. *Ap. J.* 432:42–48
Schmidt BP, Kirshner RP, Leibundgut B, Wells LA, Porter AC, et al. 1994c. *Ap. J. Lett.* 434:L19–L23
Schmidt BP, Kirshner RP, Schild R, Leibundgut B, Jeffery D, et al. 1993. *Astron. J.* 105:2236–50
Schmitz MF, Gaskell CM. 1988. See Kafatos & Michalitsianos 1988, pp. 112–15
Shigeyama T, Nomoto K, Tsujimoto T, Hashimoto M. 1990. *Ap. J. Lett.* 361:L23–L27
Spyromilio J, Meikle WPS, Allen DA, Graham JR. 1992. *MNRAS* 258:53p–56p
Spyromilio J, Stathakis RA, Meurer GR. 1993. *MNRAS* 263:530–34
Sramek RA, Panagia N, Weiler KW. 1984. *Ap. J. Lett.* 285:L59–L62
Sramek RA, Weiler KW. 1990. See Petschek 1990, pp. 76–90
Stathakis RA, Dopita MA, Cannon RD, Sadler EM. 1991. See Woosley 1991, pp. 95–101
Stathakis RA, Sadler EM. 1991. *MNRAS* 250:786–95
Suntzeff NB. 1996. See McCray & Wang 1996, pp. 41–48
Suzuki T, Nomoto K. 1995. *Ap. J.* 455:658–69
Swartz DA, Clocchiatti A, Benjamin R, Lester DF, Wheeler JC. 1993a. *Nature* 365:232–34
Swartz DA, Filippenko AV, Nomoto K, Wheeler JC. 1993b. *Ap. J.* 411:313–22

Swartz DA, Wheeler JC, Harkness RP. 1991. *Ap. J.* 374:266–80
Trimble V. 1982. *Rev. Mod. Phys.* 54:1183–224
Trimble V. 1983. *Rev. Mod. Phys.* 55:511–63
Tsvetkov DYu. 1986. *Sov. Astron. Lett.* 12:328–29
Turatto M, Benetti S, Cappellaro E, Danziger IJ, Della Valle M, et al. 1996. *MNRAS* 283:1–17
Turatto M, Cappellaro E, Barbon R, Della Valle M, Ortolani S, Rosino L. 1990. *Astron. J.* 100:771–81
Turatto M, Cappellaro E, Benetti S, Danziger IJ. 1993a. *MNRAS* 265:471–85
Turatto M, Cappellaro E, Danziger IJ, Benetti S, Gouiffes C, Della Valle M. 1993b. *MNRAS* 262:128–40
Uomoto A, Kirshner RP. 1985. *Astron. Astrophys.* 149:L7–L9
Uomoto A, Kirshner RP. 1986. *Ap. J.* 308:685–90
Utrobin V. 1994. *Astron. Astrophys.* 281:L89–L92
Utrobin VP, Chugai NN, Andronova AA. 1995. *Astron. Astrophys.* 295:129–35
van den Bergh S. 1994. *Ap. J. Suppl.* 92:219–27
Van Dyk SD. 1992. *Astron. J.* 103:1788–803
Van Dyk SD, Hamuy M, Filippenko AV. 1996a. *Astron. J.* 111:2017–27
Van Dyk SD, Sramek RA, Weiler KW, Montes MJ, Panagia N. 1996b. *IAU Circ.* No. 6386
Van Dyk SD, Weiler KW, Sramek RA, Panagia N. 1993. *Ap. J. Lett.* 419:L69–L72
Van Dyk SD, Weiler KW, Sramek RA, Rupen MP, Panagia N. 1994. *Ap. J. Lett.* 432:L115–18
Van Dyk SD, Weiler KW, Sramek RA, Schlegel EM, Filippenko AV, et al. 1996c. *Astron. J.* 111:1271–77
Varani G-F, Meikle WPS, Spyromilio J, Allen DA. 1990. *MNRAS* 245:570–76
Vaughan TE, Branch D, Miller DL, Perlmutter S. 1995. *Ap. J.* 439:558–64
Wampler EJ, Richichi A. 1989. *Astron. Astrophys.* 217:31–34
Wang L, Wheeler JC, Kirshner RP, Challis PM, Filippenko AV, et al. 1996. *Ap. J.* 466:998–1010
Wegner G, Swanson SR. 1996. *MNRAS* 278:22–38
Weiler KW, Sramek RS. 1988. *Annu. Rev. Astron. Astrophys.* 26:295–341
Wells LA, Phillips MM, Suntzeff NB, Heathcote SR, Hamuy M, et al. 1994. *Astron. J.* 108:2233–50
Wheeler JC, ed. 1980. *Proc. Texas Workshop on Type I Supernovae.* Austin: Univ. Texas
Wheeler JC. 1990. See Wheeler et al 1990, pp. 1–93
Wheeler JC. 1992. In *Evolutionary Processes in Interacting Binary Stars,* ed. Y Kondo, RF Sisteró, RS Polidan, pp. 225–34. Dordrecht: Kluwer
Wheeler JC, Benetti S. 1997. In *Astrophysical Quantities.* In press. 4th ed.
Wheeler JC, Filippenko AV. 1996. See McCray & Wang 1996, pp. 241–76
Wheeler JC, Harkness RP. 1986. In *Galaxy Distances and Deviations from Universal Expansion,* ed. BF Madore, RB Tully, pp. 45–54. Dordrecht: Reidel
Wheeler JC, Harkness RP. 1990. *Rep. Prog. Phys.* 53:1467–557
Wheeler JC, Harkness RP, Barker ES, Cochran AL, Wills D. 1987. *Ap. J. Lett.* 313:L69–L73
Wheeler JC, Harkness RP, Clocchiatti A, Benetti S, Brotherton MS, et al. 1994. *Ap. J. Lett.* 436:L135–38
Wheeler JC, Levreault R. 1985. *Ap. J. Lett.* 294:L17–L20
Wheeler JC, Piran T, Weinberg S, eds. 1990. *Supernovae.* Singapore: World Sci.
Williams RE. 1987. *Ap. J. Lett.* 320:L117–20
Williams RE, Phillips MM, Hamuy M. 1994. *Ap. J. Suppl.* 90:297–316
Woosley SE, ed. 1991. *Supernovae.* New York: Springer-Verlag
Woosley SE, Eastman RG, Weaver TA, Pinto PA. 1994. *Ap. J.* 429:300–18
Woosley SE, Langer N, Weaver TA. 1993. *Ap. J.* 411:823–39
Woosley SE, Langer N, Weaver TA. 1995. *Ap. J.* 448:315–38
Woosley SE, Pinto PA, Martin PG, Weaver TA. 1987. *Ap. J.* 318:664–73
Woosley SE, Weaver TA. 1986. *Annu. Rev. Astron. Astrophys.* 24:205–53
Xu Y, McCray R, Oliva E, Randich S. 1992. *Ap. J.* 386:181–89
Young TR, Branch D. 1988. *Nature* 333:305–6
Young TR, Branch D. 1989. *Ap. J. Lett.* 342:L79–L82
Zwicky F. 1965. In *Stars and Stellar Systems,* ed. LH Aller, DB McLaughlin, 8:367–423. Chicago: Univ. Chicago Press

*Annu. Rev. Astron. Astrophys. 1997. 35:357–88*


# COMPACT GROUPS OF GALAXIES

*Paul Hickson*
Department of Physics and Astronomy, University of British Columbia, 2219 Main Mall, Vancouver, British Columbia V6T1Z4, Canada; email: paul@astro.ubc.ca

KEY WORDS: clusters, evolution

ABSTRACT

Compact groups of galaxies have posed a number of challenging questions. Intensive observational and theoretical studies are now providing answers to many of these and, at the same time, are revealing unexpected new clues about the nature and role of these systems. Most compact groups contain a high fraction of galaxies having morphological or kinematical peculiarities, nuclear radio and infrared emission, and starburst or active galactic nuclei (AGN) activity. They contain large quantities of diffuse gas and are dynamically dominated by dark matter. They most likely form as subsystems within looser associations and evolve by gravitational processes. Strong galaxy interactions result and merging is expected to lead to the ultimate demise of the group. Compact groups are surprisingly numerous and may play a significant role in galaxy evolution.

## 1. INTRODUCTION

As their name suggests, compact groups are small systems of several galaxies in a compact configuration on the sky. The first example was found over one hundred years ago by Stephan (1877) who observed it visually using the 40-cm refractor of the Observatoire de Marseille. Stephan's Quintet is a small group of five galaxies, three of which show strong tidal distortions due to gravitational interaction. A second example was found 71 years later by Seyfert (1948) from a study of Harvard Schmidt plates. Seyfert's Sextet (Figure 1) is one of the densest groups known, having a median projected galaxy separation of only $6.8h^{-1}$ kpc (the Hubble Constant $H_0 = 100h$ km s$^{-1}$ Mpc$^{-1}$).

The Palomar Observatory Sky Survey (POSS) provided a new and extensive resource for the systematic investigation of small groups of galaxies. Two catalogs, the *Atlas of Interacting Galaxies* (Vorontsov-Velyaminov 1959, 1975) and



0066-4146/97/0915-0357$08.00

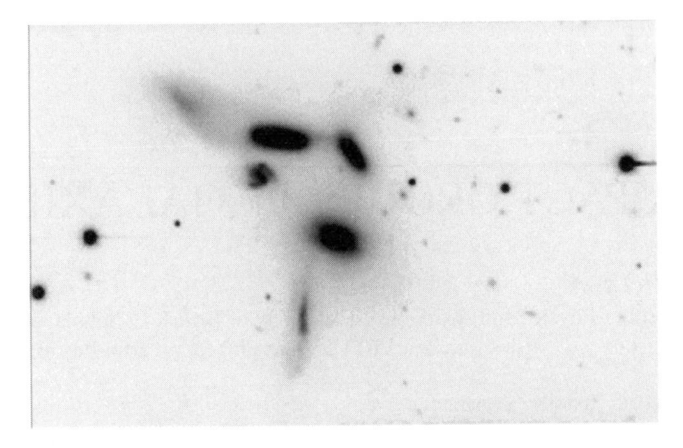

*Figure 1* Seyfert's Sextet. Discovered in 1948, this group of five galaxies is one of the densest known. The sixth object appears to be a tidal plume. The small face-on spiral galaxy has a redshift that is more than four times larger than those of the other galaxies.

the *Atlas of Peculiar Galaxies* (Arp 1966), contain galaxies or galaxy groups selected on the basis of visible signs of interaction or peculiar appearance. In addition to Stephan's Quintet and Seyfert's Sextet, these include many new compact groups, including a striking chain of five galaxies, VV 172. Prior to these, Shakhbazian (1957) had discovered a small dense cluster of 12 faint red galaxies that appeared so compact that they were initially mistaken for stars. Over the next two decades, Shakhbazian and collaborators examined over 200 POSS prints covering 18% of the sky and cataloged 376 additional "compact groups of compact galaxies" (Shakhbazian 1973, Shakhbazian & Petrosian 1974, Baier et al 1974, Petrosian 1974, 1978, Baier & Tiersch 1975–1979). Apart from occasional photographic or spectroscopic observations (e.g. Mirzoian et al 1975, Tiersch 1976, Massey 1977, Shakhbazian & Amirkhanian 1979, Vorontsov-Velyaminov et al 1980, Vorontsov-Velyaminov & Metlov 1980), these systems initially received little attention. However, interest in them is growing. Although the majority seem to be small clusters, they share some of the same properties and pose some of the same questions as do compact groups.

When redshifts were measured for galaxies in the first compact groups (Burbidge & Burbidge 1959, 1961a), surprises were found. Both Stephan's Quintet and Seyfert's Sextet contain a galaxy with a discordant redshift. It seemed unlikely that a foreground or background galaxy would appear so often projected within such compact systems (Burbidge & Burbidge 1961a). This impression was further reinforced with the discovery of yet another discordant redshift in VV 172 (Sargent 1968). Are these examples of physical association between objects of widely different redshifts, as has been advocated for many years by Arp (1987)?

Even if the discordant galaxies are ignored, the velocity dispersions of these systems are generally higher than would be expected given their visible mass (Burbidge & Burbidge 1959, 1961b, 1961c, Burbidge & Sargent 1971). It was argued that such groups must be unbound and disrupting (e.g. Ambartsumian 1961), although Limber & Mathews (1960) showed that the virial theorem could be satisfied for Stephan's Quintet, given the uncertainties in the projection factors, if the individual galaxy masses were considerably larger than those of isolated galaxies. The observations can of course also be explained if the bulk of the mass is in a nonvisible form. In hindsight, this was one of the earliest indications of the possible existence of dark matter in galactic systems.

A new problem emerged with the realization that bound groups would be unstable to orbital decay resulting from gravitational relaxation processes (Peebles 1971). A simple calculation indicated that the dynamical-friction time scale was much shorter than the Hubble time (Hickson et al 1977), as was soon confirmed by numerical simulations (Carnevali et al 1981). At the same time, it became increasingly clear that mergers played an important role in the evolution of many, if not all, galaxies (Press & Schechter 1974, Ostriker & Tremaine 1975). Compact groups emerged as prime locations for investigations of the dynamical evolution of galaxies.

Motivated by the desire for a homogeneous sample that could be subject to statistical analysis, Rose (1977) and, later, Hickson (1982) produced the first catalogs of compact groups having specific, quantitative, selection criteria. Subsequent detailed investigation, at many wavelengths, has produced a large body of observational data for the Hickson catalog. As a result, it has now become possible to address some of the outstanding questions concerning the nature of compact groups and their role in galaxy evolution. Not surprisingly, new questions have been raised and new controversies have appeared. However, much progress has been made in resolving both old and new issues.

This review is organized as follows: In Section 2, the definition of a compact group is discussed, along with methods of identification and surveys that have been made. In Sections 3–5, observed properties of these systems are summarized and discussed. Sections 6 and 7 focus mainly on interpretation of the observations and on implications of these results. All work on compact groups of galaxies cannot possibly be discussed in this short paper, although an attempt is made to touch upon most current topics. Other recent reviews of compact groups and closely related subjects include those by White (1990), Hickson (1990, 1997), Whitmore (1992), Kiseleva & Orlov (1993), Sulentic (1993), and Mamon (1995).

## 2. IDENTIFICATION AND SURVEYS

By "compact group," we mean a small, relatively isolated system of typically four or five galaxies in close proximity to one another. Such groups do not necessarily form a distinct class but may instead be extreme examples of systems having a range of galaxy density and population. Because of this, the properties of the groups in any particular sample may be strongly influenced by the criteria used to define the sample. The early surveys used qualitative criteria that, while successful in finding many interesting individual objects, do not easily allow one to draw broad conclusions about the groups as a whole. Thus, the focus in recent years has been on samples selected using specific, quantitative, criteria. These criteria define the minimum number and magnitude range of the galaxies and also consider the galaxy spatial distribution.

The use of quantitative selection criteria was pioneered by Rose (1977), who searched for groups that have three or more galaxies that are brighter than a limiting magnitude of 17.5 and that have a projected surface density enhancement of a factor of 1000 compared to the surrounding background galaxy density. Searching an area of 7.5% of the sky, he found 170 triplets, 33 quartets, and 2 quintets. Unfortunately, the sample received little follow-up study. Sulentic (1983) reexamined the 35 Rose groups that contain four or more galaxies and found that only a third actually satisfied the selection criteria. This is testimony to the difficulty of visual searches. A more fundamental problem is that the fixed magnitude limit in the selection criteria makes the sample susceptible to strong distance-dependent biases.

In an attempt to reduce such effects, Hickson (1982, 1993, 1994) adopted a relative magnitude criterion, selecting systems of four or more galaxies whose magnitudes differ by less than 3.0. A distance independent (to first order) compactness criterion was employed: $\bar{\mu}_G < 26$, where $\bar{\mu}_G$ is the mean surface brightness of the group calculated by distributing the flux of the member galaxies over the smallest circular area containing their geometric centers. To avoid including the cores of rich clusters, an isolation criterion was necessary so as to reject the group if a nonmember galaxy, not more than 3 mag fainter than the brightest member, occurred within three radii of the center of the circle. (A nonmember galaxy is a galaxy that, if included in the group, would cause the group to fail one or more of the selection criteria.) From a search of 67% of the sky (all the POSS prints), and using magnitudes estimated from the POSS red prints, exactly 100 groups were found satisfying these criteria (hereafter HCG's). As the HCG sample is now the most widely studied, it is important to examine the biases introduced by the criteria and by the visual search procedure.

Any sample selected on the basis of surface density will suffer from geometric and kinematic biases. The former occurs because nonspherical systems will be preferentially selected if they are oriented to present a smaller cross-sectional area (e.g. prolate systems pointed towards us); the latter because we will preferentially select systems that, owing to galaxy orbital motion, are momentarily in a more compact state (transient compact configurations). Thus, a compact group might result from a chance alignment or transient configuration within a loose group (Mamon 1986). This question will be considered in more detail in Section 6.2.

Other biases arise from the subjective nature of the search procedure. The original catalog contains a few misidentifications, such as compact galaxies mistaken for stars and marginal violations of the isolation criteria. In addition, when photometry was obtained for the galaxies in the catalog (Hickson et al 1989), it was found that some groups would not satisfy the selection criteria if photometric magnitudes are used. Attempts to rectify these problems have been made by Hickson et al (1989) and Sulentic (1997). However, it should be emphasized that changes, such as using photometric magnitudes in the selection criteria, are not corrections to the catalog, but are actually the imposition of *additional* a postiori selection criteria. The resulting subsample is by no means complete because the new criteria are applied only to the visually selected HCG catalog and not to the entire sky.

Because of the difficulty of identifying faint groups, the HCG catalog starts to become significantly incomplete at an integrated magnitude of about 13 (Hickson et al 1989, Sulentic & Rabaça 1994). A more subtle effect results from the difficulty of recognizing low-surface–brightness groups. Because of this, the catalog also becomes incomplete at surface brightnesses fainter than 24 (Hickson 1982). Yet another effect is that groups may be more noticeable if the magnitude spread of their members is small. Thus, the catalog may also be incomplete for magnitude intervals greater than about 1.5 (Prandoni et al 1994). These effects are of critical importance in statistical analyses of the sample. One immediate conclusion is that the actual number of groups that satisfy the selection criteria may be considerably larger than the number found by a subjective search.

It has recently become feasible to find compact groups by automated techniques. Mamon (1989) used a computer to search Tully's (1987) catalog of nearby galaxies and identified one new compact group, satisfying Hickson's criteria, in the Virgo Cluster. Prandoni et al (1994) applied similar criteria to digital scans of $\sim$1300 deg$^2$ around the southern galactic pole and detected 59 new southern compact groups (SCGs). Observations are presently underway to obtain accurate photometry and redshifts for this sample (A Iovino, private communication). The digitized Palomar Sky Survey II also offers new opportunities for the identification of compact groups (De Carvalho & Djorgovski 1995).

An alternative approach is to identify groups of galaxies from redshift information, as was first done by Humason et al (1956). With the advent of large-scale redshift surveys, it has become possible to identify a reasonably large sample of compact groups in this way. Barton et al (1996) have compiled a catalog of 89 redshift-selected compact groups (RSCGs) found by means of a friends-of-friends algorithm applied to a complete magnitude-limited redshift survey. Galaxies having projected separations of $50h^{-1}$ kpc or less and line-of-sight velocity differences of 1000 km $s^{-1}$ or less are connected, and the sets of connected galaxies constitute the groups. The numerical values were chosen to best match the characteristics of the HCG sample, and indeed, many of those RSCGs that have at least four members are also HCGs. There are some significant differences, however: Because foreground and background galaxies are automatically eliminated by the velocity selection criteria, this technique is more effective at finding groups in regions of higher galaxy density, which would fail the HCG isolation criterion. This criterion requires that the distance to the nearest neighbor be at least as large as the diameter of the group. The RSCG criteria, on the other hand, require only that the nearest-neighbor distance be larger than the threshold distance ($50h^{-1}$ kpc for the RSCGs), which may be considerably smaller. It will therefore allow the inclusion of groups that are physically less isolated (and therefore less physically distinct) than would the HCG criterion. One would also expect that the numbers of groups found in a given volume by the less restrictive RSCG criteria to be larger than by the HCG criteria, as seems to be the case. While the redshift-selection method compliments the HCG angular-selection technique, one should keep in mind that it also selects groups according to apparent (projected) density—the velocity information serves only to reject interlopers. Thus it will be subject to some of the biases discussed above. Also, because the galaxy sample used is magnitude limited, rather than volume limited, there will be redshift-dependent biases in the RSCGs. However, the well-defined selection criteria and the completeness of the sample should allow a quantitative determination of the effects of this bias.

## 3. SPACE DISTRIBUTION AND ENVIRONMENT

The space distribution and environment of compact groups provide important clues to their nature. The median redshift of the HCGs is $z = 0.030$, placing most of them at distances well beyond the Virgo Cluster (Hickson et al 1992). A cursory inspection reveals that they are fairly uniformly distributed and show no preference for rich clusters. This is at least partly due to the isolation criterion. However, galaxies in rich clusters have rather different kinematical and morphological properties than do those in compact groups, so one might justifiably argue that small clumps of galaxies within clusters are not compact groups.

A natural question is whether or not compact groups are associated with loose groups. Rood & Struble (1994) observed that 70% of the HCGs are located within the bounds of cataloged loose groups and clusters. Studies of the distribution of galaxies in redshift space (Vennik et al 1993, Ramella et al 1994, Sakai et al 1994, Garcia 1995, Barton et al 1996) indicate that compact subcondensations do occur within loose groups and filaments. Vennik et al (1993) and Ramella et al (1994) find that most HCGs are indeed associated with loose groups.

While the above studies show that compact groups trace large-scale structure, it is also clear that they prefer low-density environments. Sulentic (1987), Rood & Williams (1989), Kindl (1990), and Palumbo et al (1995) have examined the surface density of galaxies surrounding the groups. They generally agree that about two thirds of the groups show no statistically significant excess of nearby neighbors. This is not inconsistent with the redshift-space results because most of the HCG associations identified by Ramella et al (1996) contain fewer than five excess galaxies. Thus while compact groups are associated with loose groups and filaments, these tend to be low-density and sparsely populated systems.

Are the galaxies in compact groups in any way distinct from those in their immediate environments? Rood & Williams (1989) and Kindl (1990) both found that compact groups, including those in rich environments, contain a significantly smaller fraction of late-type (spiral and irregular) galaxies than do their neighborhoods. This result is of particular importance to the question of the physical nature of compact groups, discussed below in Section 6. In addition, many independent studies have found significant differences between galaxies in compact groups and those in other environments. These are examined in Section 5.

## 4. DYNAMICAL PROPERTIES

The first studies of individual compact groups (Burbidge & Burbidge 1959, 1961b, 1961c, Burbidge & Sargent 1971, Rose & Graham 1979, Kirshner & Malumuth 1980) indicated short dynamical times and mass-to-light ratios intermediate between those of galaxies and rich clusters. However, because of the small number of galaxies, estimates of the space velocities and physical separations of galaxies in individual groups are highly uncertain. Meaningful dynamical conclusions about systems containing only four or five galaxies requires statistical analysis of large homogeneous samples.

By 1992, velocities had been measured for almost all 462 galaxies in the HCG catalog (Hickson et al 1992). The distribution of galaxy velocities, relative to the median of each group, is shown in Figure 2. It can be seen that the majority of velocities fall within a roughly Gaussian distribution of standard deviation $\sim$250 km $s^{-1}$ (the sharp peak at zero velocity results from the use of

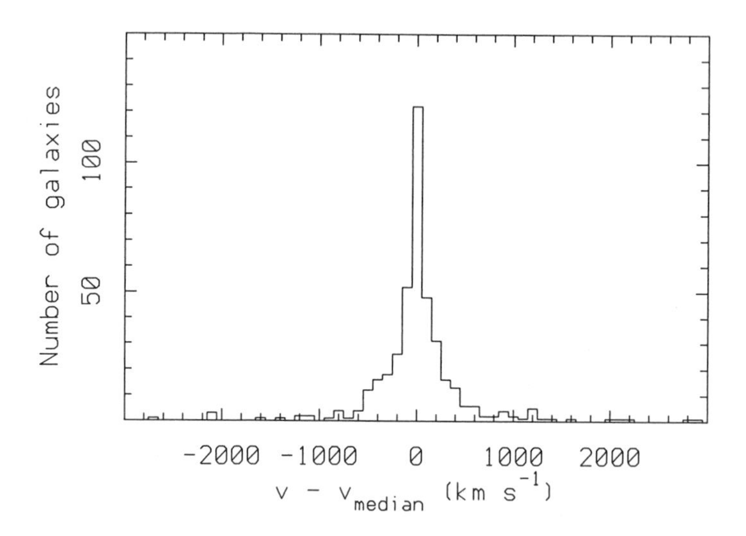


*Figure 2* Velocity distribution of galaxies in compact groups. The figure shows the distribution of the difference between the observed galaxy radial velocity and the median velocity of galaxies in the group to which it belongs, for 410 galaxies in the HCG catalog. Most galaxies (77%) have velocity differences less than 500 km $s^{-1}$ from the median.

the median). This characteristic velocity is quite similar to velocity dispersions found in loose groups and much smaller than typical velocity dispersions in rich clusters. In addition to the Gaussian core, a flat component is seen in the velocity distribution of Figure 2. This is expected, as some galaxies that are not physically related to the group will appear projected on the group by chance. This component contains about 25% of the total number of galaxies. Whether chance projection can account for such a large number of "discordant" galaxies is still a matter of some debate and is discussed further in Section 6.1.

For a system of characteristic linear size $R$ and internal velocity $V$, a characteristic dynamical time is $t_d = R/V$. A characteristic mass density is $\rho = 1/Gt_d^2$, from which one can estimate the total mass within the region occupied by the galaxies. After removing galaxies whose velocities differ from the group median by more that 1000 km $s^{-1}$, Hickson et al (1992) found that $t_d \sim 0.02\ H_0^{-1}$ and obtained a mean mass-to-light ratio of $50h$ (solar units) for the HCGs. Similar values were found for several Shakhbazian groups (Tikhonov 1986, Amirkhanian & Egikian 1987, Amirkhanian 1989, Lynds et al 1990, Amirkhanian et al 1991). Since the mass-to-light ratios of individual HCG galaxies are on the order of $7h$ (Rubin et al 1991), the galaxies appear to contain only $\sim$15% of the total mass.

# 5. STRUCTURE AND MORPHOLOGY

The spatial distribution and luminosities of the member galaxies provide further clues to the nature of compact groups. If they are primarily projections or transient configurations, the luminosity function should be the same as that of the parent systems, and the spatial distribution of the galaxies should be consistent with a random distribution. If they are bound physical systems, the luminosity function and spatial distribution might show features that reflect the origin or subsequent evolution of compact groups.

## 5.1 *Shapes and Orientations*

The shapes of compact groups were first investigated by Arp (1973), who concluded that galaxy "chains" were unusually predominant. However, Rose (1977) determined that the ellipticities of his groups were consistent with a random distribution of galaxies. Using the larger HCG sample, Hickson et al (1984) and Malykh & Orlov (1986) reached the same conclusion as Arp—the groups are typically more elongated than would be a random distribution of galaxies. An immediate consequence of this result is that compact groups cannot easily be explained as random projections or chance crossings, as this would largely erase any inherent ellipticity of a parent loose group. From static simulations, Hickson et al (1984) concluded that the observed ellipticities are best matched by three-dimensional shapes that are intrinsically prolate. The same result was found by Oleak et al (1995) in a recent study of the shapes of 95 Shakhbazian compact groups. These conclusions, however, are not unique. Hickson et al (1984) also found the shapes to be consistent with those seen in dynamical simulations of compact groups seen in projection as subgroups within loose groups. In addition, one must always be concerned about possible selection biases. It may be that highly elongated groups (such as VV 172) are more easily noticed in visual searches. It will be interesting to see if these results are confirmed by studies of groups found by automated searches.

If the intrinsic shapes of compact groups are related to their formation process, one might expect to see a relationship between the orientation angle of a group and the environment. Palumbo et al (1993) examined the environments of the HCGs and found that the orientations of the major axes of the groups were consistent with an isotropic distribution.

If compact groups are not simply projection effects, they might be expected to show a centrally concentrated surface density profile, as is seen in clusters of galaxies. Although the number of galaxies in individual compact groups is small, with a large sample it is possible to estimate a mean profile. By scaling and superimposing the HCGs, Hickson et al (1984) found evidence for central concentration. Mendes de Oliveira & Girard (1994), using a similar analysis, concluded that the mean surface density profile is consistent with a King (1962) model with typical core radius of $15h^{-1}$ kpc. Most recently, Montoya et al (1996) have analyzed the profiles of the 42 HCG quartets that have accordant redshifts. Their technique uses the distribution of projected pair separations and thus avoids assumptions about the location of the group center. They find a smaller core radius ($6h^{-1}$ kpc for a King model). The fact that Montoya et al (1996) find a consistent density profile for all groups, without any scaling, is particularly interesting. This would not be expected if most groups are chance alignments within loose groups. It also implies that compact groups have a unique scale, which seems counter to the concept of hierarchical clustering. Montoya et al (1996) suggest that this arises as a result of a minimum mass density and velocity dispersion required for the groups to be virialized (Mamon 1994).

## 5.2 *Compact Group Galaxies*

There have been several studies of the morphological types of galaxies in compact groups (Hickson 1982, Williams & Rood 1987, Sulentic 1987, Hickson et al 1988b). Most studies agree that the fraction $f_s$ of late type galaxies is significantly less in compact groups than in the field. Hickson et al (1988b) find $f_s = 0.49$ for the HCGs; Prandoni et al (1994) obtain $f_s = 0.59$ for the SCGs. Both these values are substantially lower than those found for field galaxy samples ($f_s \simeq 0.82$, Gisler 1980, Nilson 1973).

Also well established is morphological type concordance, observed in both the HCGs and SCGs (Sulentic 1987, Hickson et al 1988b, Prandoni et al 1994). A given compact group is more likely to contain galaxies of a similar type (early or late) than would be expected for a random distribution. White (1990) has pointed out that such concordance could result from a correlation of morphological type with some other property of the group. The strongest such correlation found to date is between morphological type and velocity dispersion (Hickson et al 1988b). As Figure 3 shows, groups with higher velocity dispersions contain fewer late-type (gas-rich) galaxies. They also tend to be more luminous. The importance of velocity dispersion, in addition to local density, on the galaxy morphology had previously been emphasized by De Souza et al (1982). A crucial clue is that the morphology-density relation seen in clusters and loose groups (Dressler 1980, Postman & Geller 1984, Whitmore & Gilmore 1992) is not the dominant correlation in compact groups (Hickson et al 1988b), although some effect is present (Mamon 1986). This suggests that the velocity dispersion is more fundamental, i.e. of greater physical relevance to the formation and evolution of galaxies in compact groups, than is apparent physical density.

There is much evidence that interaction is occurring in a large fraction of galaxies in compact groups. The strongest direct support comes from kinematical

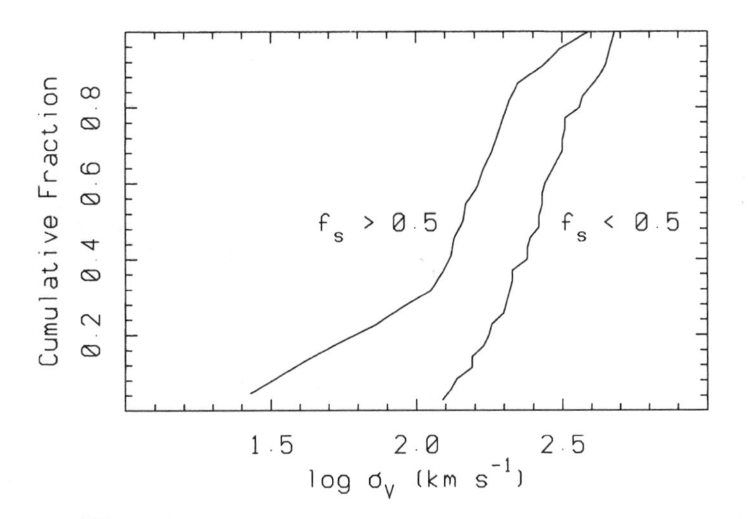


*Figure 3* Morphology-velocity correlation for compact groups. The figure shows the cumulative distributions of velocity dispersion for spiral-rich ($f_s > 0.5$) and spiral-poor ($f_s < 0.5$) groups. The former have typically half the velocity dispersion than the latter and a broader velocity range.

studies. Rubin et al (1991) found that two thirds of the 32 HCG spiral galaxies that they observed have peculiar rotation curves. These show asymmetry, irregularity, and in some cases extreme distortion, characteristic of strong gravitational interaction. This result has recently been challenged by Mendes de Oliveira et al (preprint), who obtained $H_\alpha$ velocity maps for 26 HCG spiral galaxies and found that only one third showed abnormal rotation curves. They suggest that the difference is due to the more complete spatial sampling of their data.

In their study, Rubin et al (1991) observed 12 HCG elliptical galaxies and detected nuclear emission in 11 of them. This high fraction suggests that interactions and mergers may be supplying gas to these galaxies. This idea received independent support from radio observations in which neutral hydrogen emission was detected in three compact groups which contain only elliptical galaxies (Huchtmeier 1994).

Zepf & Whitmore (1993) found that elliptical galaxies in compact groups tend to have lower internal velocity dispersions than do ellipticals in other environments having the same effective radii, absolute magnitudes, and colors. They therefore do not lie on the fundamental plane defined by other elliptical galaxies. This discrepancy correlates with isophote shape in that those galaxies that have "disky" or irregular isophotes tend to have lower velocity dispersion. Both Zepf & Whitmore (1993) and Bettoni & Fasano (1993, 1995, 1996, Fasano

& Bettoni 1994) report that HCG elliptical galaxies are less likely to have "boxy" isophotes and more likely to have irregular isophotes. Such effects are consistent with results of simulations of tidal encounters (Balcells & Quinn 1990).

## 5.3 *Optical Luminosity Function*

The luminosity function of compact groups was first estimated by Heiligman & Turner (1980). They examined a sample consisting of Stephan's Quintet, Seyfert's Sextet, and eight more compact groups from the Arp and Vorontsov-Velyaminov catalogs, and they concluded that compact groups contain relatively fewer faint galaxies than does a comparable field galaxy sample. Analysis of the relative luminosities within individual HCGs (Hickson 1982), and studies of several Shakhbazian groups (Kodaira et al 1991), showed a similar effect, although Tikhonov (1987) found a luminosity function similar to that of field and cluster galaxies.

The larger HCG sample allows the question of the galaxy content of compact groups to be addressed with greater certainty. The standard technique for determination of the luminosity function weights each galaxy by $V_m / V$, where $V$ is the volume of the smallest geocentric sphere containing the group, and $V_m$ is the volume of the largest such sphere within which the group could have been detected. Using this approach Sulentic & Rabaça (1994) obtained a luminosity function for HCG galaxies similar to that of field galaxies. However, Mendes de Oliveira & Hickson (1991) argued that the standard calculation does not address the selection effects of the HCG sample. For example, the luminosity range within an individual group is limited by the 3-mag range of the selection criteria. Because of this, fainter galaxies within compact groups are not included in the catalog. In order to account for such biases, they used a modeling technique in which galaxies were drawn from a trial luminosity function and assigned to groups. Redshifts were given to each group according to the observed distribution, and groups that failed to meet the HCG selection criteria were rejected. The luminosity distribution of the resulting galaxy sample was then compared to the observations and the process repeated with different trial luminosity functions. Their best-fit luminosity function is deficient in faint galaxies, although a normal field-galaxy luminosity function is not excluded.

To avoid the selection problem, Ribeiro et al (1994) obtained deeper photometry for a subsample of the HCGs in order to include the fainter galaxies explicitly. Since redshifts are not known for these galaxies, a correction for background contamination was made statistically. The luminosity function that they obtained is similar to that of field galaxies. Most recently, the luminosity function for the RSCGs has been computed by Barton et al (1996). They find it to be mildly inconsistent with that of field galaxies, in the same

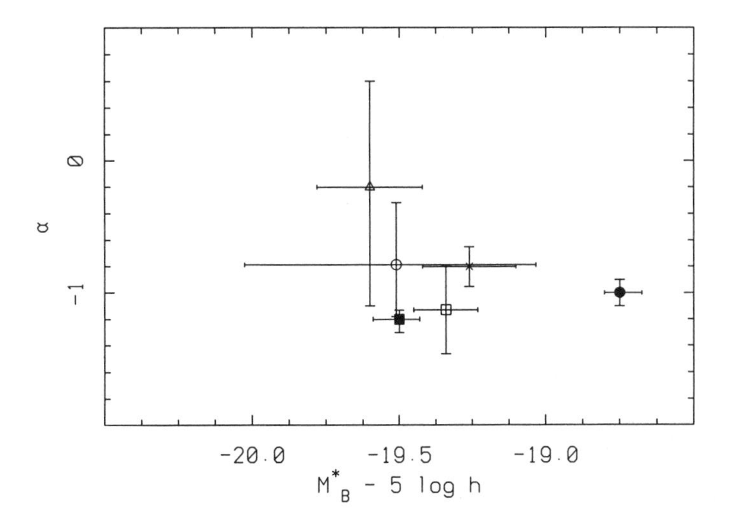


*Figure 4* Optical luminosity function parameters of compact groups. *Triangle*: Mendes de Oliveira & Hickson (1991), *open square*: Sulentic & Rabaça (1994), *cross*: Ribeirao et al (1994), *open circle*: mean of the three RSCG samples (Barton et al 1996). For comparison, the *filled circle and square* indicate luminosity function parameters for galaxies in the CFA-Combined (Marzke et al 1994) and SSRS2 (da Costa et al 1994) surveys respectively, representing galaxies in lower density environments.

sense as that of Mendes de Oliveira & Hickson (1991) for the HCGs. Figure 4 summarizes these estimates of the luminosity function of compact galaxies in terms of the Schechter (1976) parameters $M^*$ and $\alpha$.

How do we interpret these apparently conflicting results? Prandoni et al (1994) have argued that the HCG catalog is biased toward groups with a small magnitude range $\Delta m$, because the SCGs have a larger fraction of high $\Delta m$ groups. However, it is not known what fraction of such groups are physically real, as few redshifts have yet been obtained. Such a bias could affect the luminosity function of Mendes de Oliveira & Hickson (1991), particularly at the faint end, but it is not evident that the bias is sufficient to account for the apparent faint-galaxy deficit. On the other hand, the small sample used by Ribeiro et al (1994) may not be representative of compact groups in general. Hickson (1997) points out that the Ribeiro et al (1994) sample has a spiral fraction of 0.60, substantially higher than that of the whole HCG catalogue, and contains 7 of the 16 HCGs found to be in a high density environment by Palumbo et al (1995). This suggests that the their sample has more than the usual amount of field galaxy contamination. De Carvalho et al (1994), note that the faint galaxies form a more extended distribution than do the brighter galaxies.

Thus they may be a dynamically distinct component, or simply unrelated field galaxies. Finally, the redshift-selected RSCG sample also shows mild evidence for a faint galaxy deficiency.

While the faint end of the LF in compact groups appears to be depleted, there is evidence that the bright end may be enhanced. Limber & Matthews (1960) were the first to remark that "the members of Stephan's Quintet are to be classed among the brightest of galaxies." It is possible that this may be in part due to interaction-induced star formation, at least for the spiral galaxies. On the other hand, Mendes de Oliveira & Hickson (1991) compared their luminosity function of elliptical galaxies in compact groups with those in the Virgo and Coma Clusters (as reported by Sandage et al 1985 and Thompson & Gregory 1980) and found that the compact group ellipticals have a luminosity enhancement of more than 1 mag compared to cluster ellipticals. Sulentic & Rabaça (1994) find a similar enhancement in their morphological-type–specific luminosity function. This suggests that compact group elliptical galaxies may have a unique formation mechanism.

From the luminosity function, one can estimate the contribution of compact groups of galaxies $\mathcal{L}_{CG}$ to the total galaxian luminosity density $\mathcal{L}$. Mendes de Oliveira & Hickson (1991) obtained a ratio of $\mathcal{L}_{CG}/\mathcal{L} \simeq 0.8\%$. Applying the same analysis to the luminosity function of Ribeiro et al (1994) gives a ratio of 3.3%. For the RSCGs, the figure is comparable: for groups of four or more galaxies, Barton et al (1996) obtain a compact group abundance of $1.4\times10^{-4}h^{-3}$ $\mathrm{Mpc}^{-1}$, which leads to a luminosity density ratio of approximately 3%. These are surprisingly high figures considering the short dynamical times of most compact groups.

## 5.4 *Star Formation and Nuclear Activity*

Evidence is accumulating that tidal interactions play an important role in triggering starburst activity in galaxies (e.g. Maccagni et al 1990, Campos-Aguilar & Moles 1991, Kormendy & Sanders 1992, Sanders & Mirabel 1996). Compact groups, with their high galaxy density and evident signs of galaxy interaction should be ideal systems in which to study such effects. Many HCGs do in fact contain galaxies showing starbursts or harboring active galactic nuclei (AGN). For example, HCG 16 is found to contain a Seyfert 2 galaxy, two LINERs, and three starburst galaxies (Ribeiro et al 1996). HCG 31 contains five galaxies showing signs of recent starburst activity (Rubin et al 1990, Iglesias-Páramo & Vílchez preprint). Seyfert galaxies are also found in HCG 77, 92, 93, and 96.

The general degree of star formation activity in compact group galaxies can be determined from infrared observations. To date, studies have been based primarily on data from the IRAS satellite. Hickson et al (1989) found sources in 40 HCG from a search of the Point Source Catalog. They concluded that the ratio of far infrared–to-optical luminosity is greater by about a factor of two in compact group galaxies, compared to that of isolated galaxies. This result was disputed by Sulentic & De Mello Rabaça (1993), who argued that the low spatial resolution of the data made the assignment of infrared flux to individual galaxies ambiguous. They concluded that redistribution of the flux could result in little or no infrared enhancement, a conclusion echoed by Venugopal (1995). However, in cases of doubt, Hickson et al (1989) identified the infrared galaxy on the basis of radio emission. The well-known correlation between infrared and radio continuum emission makes it unlikely that the results are much in error. Analysis of improved data (e.g. Allam et al 1996) should soon resolve questions about the identifications and infrared fluxes.

The resolution problem can be avoided by considering the infrared colors of the sources instead of the infrared/optical ratio. Zepf (1993) compared the ratio of 60 $\mu$m to 100 $\mu$m fluxes of compact group galaxies with those of isolated galaxies and also with those of galaxies believed to be currently merging. He found that the compact group sample was significantly different from both other samples, and estimated that about one third of the compact group galaxies had warm colors (larger 60/100 $\mu$m ratios) similar to those of merging galaxies.

Another approach to interpreting the infrared results was taken by Menon (1991) who emphasized that the strong correlation between radio and infrared radiation indicates that these likely originate from a common region. In compact group spirals, the radio emission is primarily nuclear whereas in isolated spirals it originates in the disk. If this is also true for the infrared flux, there must be an enhancement of the infrared/optical ratio, in the nuclear region, of more than an order of magnitude. This idea is supported by recent millimeter-wavelength observations (Menon et al 1996) in which CO emission is detected in 55 of 70 IRAS-selected HCG galaxies. The inferred ratio of infrared luminosity-to-$H_2$ mass showed an enhancement which correlates with the projected nearest-neighbor distance.

Further clues are provided by radio continuum studies. Nonthermal emission from spiral galaxies can arise from both disk and nuclear sources. Disk emission is predominantly due to supernova remnants and is thus related to the star formation rate. Nuclear emission can arise both from star formation and from an active nucleus. Menon (1995) observed 133 spiral galaxies in 68 HCG and found that overall they typically show less continuum emission than those in isolated environments, which is consistent with the neutral hydrogen observations. However, when considering the nuclear regions alone, the radio emission is found to be an order of magnitude higher compared to isolated spirals. The implication is that star formation and/or AGN activity is substantially enhanced in the nuclear regions of many compact group spiral galaxies. This is generally consistent with a picture in which galaxy interactions remove gas from the outer regions of galaxies, while simultaneously allowing gas to flow inwards toward the nucleus, resulting in enhanced star formation in the nuclear region and possibly fueling an active nucleus.

Although there is a clear example of tidal interaction stimulating disk radio emission in at least one compact group (Menon 1995a), statistical evidence for a link between interactions and radio emission in compact groups is only now accumulating. If interactions are stimulating nuclear radio emission, one would expect the radio luminosity to be correlated with some index describing the degree of interaction such as the projected distance to the nearest neighbor. Evidence in support of this was found by Vettolani & Gregorini (1988) who observed that early-type galaxies that have a high ratio of radio-to-optical emission show an excess of nearby neighbors. A similar effect was observed by Malumian (1996) for spiral galaxies in groups. Examining compact group galaxies, Menon (1992) found that elliptical and S0 galaxies detected at a wavelength of 20 cm had closer neighbors than the undetected galaxies. The effect was not found for spiral galaxies, but if one considers only the detected galaxies, there is a significant correlation between radio-to-optical luminosity and nearest neighbor distance for both early- and late-type galaxies (TK Menon, private communication).

Continuum radio emission has also been detected in a number of HCG elliptical galaxies. Unlike those found in cluster ellipticals, the radio sources are low luminosity and compact. Where spectral indices are available, they indicate that the radio emission arises from an AGN rather than from starburst activity (TK Menon, private communication). In the HCG sample, there is a significant preference for radio-loud elliptical galaxies to be first-ranked optically (Menon & Hickson 1985, Menon 1992). The probability of radio emission does not correlate with absolute luminosity, but instead correlates with *relative* luminosity within a group. Spiral HCG galaxies do not show this effect. Although the tendency of radio galaxies in rich clusters to be first-ranked has been known for many years, it is surprising to find a similar effect in small groups, where the number of galaxies and luminosity range is small, the gravitational potential well is much less clearly defined, and it is unlikely that any individual galaxy holds a central location. The effect of optical rank on radio emission had been previously noted in other small groups by Tovmasian et al (1980), although these authors made no distinction between elliptical and spiral galaxies. It is difficult to imagine any explanation for this result in which the compact group is not a true physical system. It would appear that, regardless of absolute luminosity, only the first-ranked (presumably the most massive in the group) elliptical galaxy can develop a radio source.

## 5.5 *Diffuse Light*

Stars stripped from galaxies by tidal forces should accumulate in the potential well of the group and may be detectable as diffuse light. In an early photographic study, Rose (1979) found no evidence for diffuse light in his groups and was lead to the conclusion that most of his groups must be transient configurations. However, Bergvall et al (1981) were successful in detecting ionized gas and a common halo around a compact quartet of interacting early-type galaxies, and evidence for a common halo in VV 172 was reported by Sulentic & Lorre (1983). Diffuse light can clearly be seen in HCG 94 and has been found in HCG 55 (Sulentic 1987), but Pildis et al (1995b) did not detect any in seven other compact groups. Analysis by Mamon (1986) indicated that although the expected diffuse light should be detectable with modern techniques, it would generally be very faint. Estimates of the total amount of diffuse light in the detected groups are rather uncertain as they depend sensitively on subtraction of the galactic light and the sky background. Deeper photometry and improved image processing techniques may yet reveal diffuse light in other compact groups (Sulentic 1997).

## 5.6 *Cool Gas*

The mass and distribution of cool galactic and intergalactic gas can be obtained from observations of the 21-cm line of neutral hydrogen. The first such study of a large sample of compact groups is that of Williams & Rood (1987), who found a median HI mass of $2.2 \times 10^{10}\ M_{\odot}$. They concluded that compact groups are typically deficient in neutral hydrogen by about a factor of two compared to loose groups. This effect is consistent with similar deficit in continuum radio emission seen in the disks of compact group spiral galaxies (Menon 1995) and suggests that interactions in compact groups have removed much of the gas from the galaxies. Simulations suggest that in addition to an outflow of gas, inflow also occurs, which may fuel nuclear star formation, as suggested by the strongly enhanced radio emission seen in the nuclear regions of compact group spiral galaxies (Menon 1995).

High-resolution studies of individual groups (Williams & van Gorkom 1988, Williams et al 1991) showed clearly that the gas is not confined to the galaxies. In two of three groups studied, the radio emission originates from a common envelope surrounding the group, and in the third group there are signs of tidal distortion. These results strongly indicate that at least these compact groups are physically dense systems and not chance alignments or transient configurations in loose groups. They also show that many groups have evolved to the point that gas contained within individual galaxies has been distributed throughout the group.

In contrast to the HI results, initial CO-line observations of 15 compact group galaxies (Boselli et al 1996) indicated a normalized molecular gas content similar to that of isolated spiral galaxies. However, this result is based on normalizing the flux by the optical area of the galaxy, rather than by the infrared luminosity, and may be biased by the relatively small sizes of compact group galaxies. Further CO studies, currently in progress, should soon settle this question.

## 5.7 *Hot Gas*

X-ray observations of hot gas in clusters of galaxies can reveal the amount, distribution, temperature, and metallicity of the gas, as well as the relative amount and distribution of the total gravitating mass. Temperature, metallicity (fraction of solar abundance), and bolometric luminosities are estimated by fitting a spectral model, such as that of Raymond & Smith (1977) to the data. X-ray emission from compact "poor clusters" was first reported by Schwartz et al (1980), who concluded that their X-ray properties were similar to those of rich clusters. Using the Einstein observatory, Bahcall et al (1984) first detected X-ray emission from Stephan's Quintet. The X-ray map revealed that the emission is diffuse and not centered on individual galaxies. However, we now know that most of this emission is associated with a shock front rather than gas trapped in the group potential well (Sulentic et al 1995). Although several other groups (Bahcall et al 1984, Biermann & Kronberg 1984) were detected by the Einstein observatory, further progress required the improved sensitivity of the ROSAT X-ray observatory.

Pointed ROSAT observations revealed massive hydrogen envelopes surrounding the NGC 2300 group, a bright elliptical-spiral pair with two fainter members (Mulchaey et al 1993), and HCG 62, a compact quartet of early-type galaxies (Ponman & Bertram 1993), and showed that these systems are dominated by dark matter. Subsequent investigations detected X rays from 18 additional compact groups, either from individual galaxies or from diffuse gas (Ebeling et al 1994, Pildis et al 1995a, Sarraco & Ciliegi 1995, Sulentic et al 1995). These studies showed that the physical properties of individual systems span a wide range, but that the ratio of gas-to-stellar mass is significantly lower than in rich clusters. Moreover, the detected compact groups all contained a majority of early-type galaxies. No spiral-rich groups were detected (although Mulchaey et al 1996b pointed out that they might be found from QSO absorption spectra). This result is consistent with the fact that X-ray–selected groups (Henry et al 1995) and loose groups (Mulchaey et al 1996a) tend to be spiral poor, and it led to the suggestion that spiral-rich compact groups might not be physically dense systems at all.

The most extensive X-ray study of compact groups to date is that of Ponman et al (1996). These authors combined pointed and survey-mode observations of a complete sample of 85 HCGs and detected diffuse emission in 22 groups. They conclude that, when the detection limits are considered, diffuse emission is present in at least 75% of the systems. Significantly, they detected diffuse

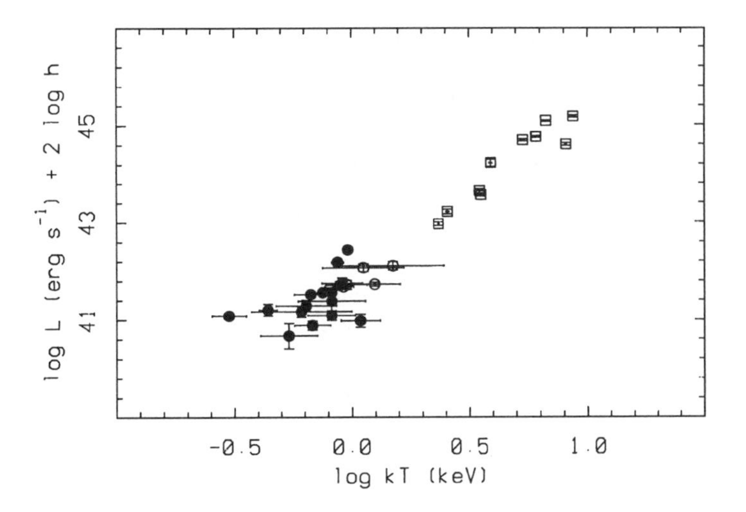


*Figure 5* Bolometric X-ray luminosity vs temperature. *Filled circles* indicate compact groups, *open circles* indicate X-ray selected groups (Henry et al 1995), and *squares* indicate clusters. X-ray data for the compact groups and clusters are taken from Ponman et al (1996). A single relation is consistent with clusters, groups, and compact groups.

emission in several spiral-rich groups. In these, the surface brightness is lower, and the X-ray emission has a lower characteristic temperature, as would be expected given the lower velocity dispersions of spiral-rich compact groups. The diffuse X-ray luminosity was found to correlate with temperature, velocity dispersion, and spiral fraction, but not with optical luminosity. The last result suggests that the gas is mostly primordial and not derived from the galaxies. The correlations with temperature and velocity dispersion appear to be consistent with a single relation for clusters and groups (Figure 5).

The total mass in compact groups typically exceeds the stellar and gas mass by an order of magnitude. Pildis et al (1995) derived baryon fractions of 12–19%. Davis et al (1996) obtains 10–16% for the NGC 2300 group. These are comparable to the fractions found for poor clusters (Dell'Antonio et al 1995) and are about half the typical values found for rich clusters. However, the derived baryon fraction depends sensitively on the radius within which it is measured and on the assumed background level (Henriksen & Mamon 1994). Both the total mass, which is dominated by dark matter, and the gas mass continue to increase with radius. Consequently, both the baryon fraction and the gas fraction are poorly determined.

The contribution of compact groups to the X-ray luminosity function has been estimated by PBEB, who find that on the order of 4% of the total luminosity in the range $10^{41}$–$10^{43}$ erg s$^{-1}$ comes from HCGs. This is higher than the contribution of HCG galaxies to the local optical luminosity density estimated at 0.8% by Mendes de Oliveira & Hickson (1991), but it is comparable to the value found by Ribeiro et al (1994).

The metallicity inferred for the X-ray–emitting gas in compact groups is relatively low. PBEB obtain a mean metallicity of 0.18 solar, compared to the value, 0.3–0.4 solar, found in rich clusters. This is comparable to the low value ($<0.11$ solar) found for the NGC 2300 group (Davis et al 1996). These figures suggest that the gas is largely primordial, a result supported by the absence of a correlation between X-ray and optical luminosity. However, given the limited spectral resolution of ROSAT, these low metalicities cannot yet be considered secure. The higher spectral resolution and sensitivity of the ASCA satellite should provide more definitive results. Recent observations of HCG 51 and the NGC 5044 group found metal abundances comparable to those of clusters (Fukazawa et al 1996).

# 6. PHYSICAL NATURE

## 6.1 *Discordant Redshifts*

The nature of the discordant redshift members of compact groups has been a subject of debate for many years (e.g. Burbidge & Burbidge 1961a, Burbidge & Sargent 1971, Nottale & Moles 1978, Sulentic 1983). If the frequency of discordant galaxies is inconsistent with the statistics of chance projection, it might signify the need for new physical theories (Arp 1987) or for gravitational amplification of background galaxies (Hammer & Nottale 1986). Initial estimates of the chance probability of finding discordant galaxies in groups like Stephan's Quintet, Seyfert's Sextet, and VV 172 were very small (Burbidge & Sargent 1971). However, such probabilities were recognized to be difficult to determine reliably because the a priori probability of *any* particular configuration of galaxies is also very small (Burbidge & Sargent 1971). Only with a well-defined sample of groups and a complete characterization of selection effects, can meaningful estimates of the probabilities be made. The explicit selection criteria of the HCG catalog in principal make this sample suitable for a quantitative statistical investigation of the discordant-redshift question. Sulentic (1987) first concluded that the number of discordant redshifts in the catalog is too large to explain by chance. On the other hand, Hickson et al (1988a) and Mendes de Oliveira (1995), applying the selection criteria more rigorously, found no strong statistical evidence for this. Their result, however, may be biased by incompleteness in the HCG catalog: There seem to be too few low surface brightness groups in the catalog, and the "missing" groups may have a higher fraction of discordant redshifts (Sulentic 1997).

In order to address the incompleteness issue, Iovino & Hickson (1996) combined observational results from both the HCG and SCG catalogs with Monte-Carlo simulations. Their technique exploits the unbiased nature of the SCG catalog and the complete redshift coverage of the HCG sample. They conclude that for all except the two highest surface-brightness quintets (Stephan's Quintet and Seyfert's Sextet), the number of discordant redshifts is consistent with chance projections. For these two, the chance probabilities are low. However, for both of these systems, there is independent physical evidence that the discordant galaxies are at the cosmological distances that correspond to their redshifts and are therefore not group members (Kent 1981, Wu et al 1994).

One should not assume that the situation is now completely settled. Further studies will be possible when redshifts have been obtained for the SCG galaxies. There are still other questions that have not been adequately addressed, such as reported redshift quantization (Cocke & Tifft 1983). However, at this point it appears that the frequency of discordant galaxies does not require a new interpretation of galaxy redshifts. In fact, physical evidence suggests the opposite. The discordant galaxies all have physical properties consistent with a cosmological distance. For example those with higher redshift tend to be smaller and fainter than other members of the group, and vice versa (Mendes de Oliveira 1995).

## 6.2 *Physical Association and Density*

Because we can measure only three phase-space dimensions for galaxies in compact groups (two components of position and one of velocity), the groups are subject to projection effects. Because of this, they may not be physically dense or even physically related systems.

The following interpretations have so far been suggested for compact groups:

1. transient dense configurations (Rose 1977)
2. isolated bound dense configurations (Sulentic 1987, Hickson & Rood 1988)
3. chance alignments in loose groups (Mamon 1986, Walke & Mamon 1989, Mamon 1995)
4. filaments seen end-on (Hernquist et al 1995)
5. bound dense configurations within loose groups (Diaferio et al 1994, Governato et al 1996)

Evidence for and against physical association and high density in the HCG sample, to 1988, was summarized by Hickson & Rood (1988) and by Walke & Mamon (1989), respectively. Since that time, several new results have emerged.

From an analysis of optical images, Mendes de Oliveira & Hickson (1994) concluded that 43% of all HCG galaxies show morphological features indicative of interaction and/or merging and that 32% of all HCGs contain three or more interacting galaxies. These percentages are likely to rise with more detailed studies and sophisticated image analysis (Longo et al 1994). This high frequency of interactions observed in compact groups is difficult to reconcile with the chance alignment and filament hypotheses, even if the alignments contain physical binaries (Mamon 1995).

The high fraction of HCGs showing diffuse X-ray emission is very strong evidence that a large fraction of these systems are physically dense and are not transient configurations or projection effects. Although the exact numbers are not final, owing to the faintness of the sources and the problems of contamination by sources associated with the individual galaxies, it seems evident that many groups are dense bound systems. The correlations seen between X-ray and optical properties, and the fact that the X-ray properties of compact groups are not inconsistent with those of clusters reinforces this conclusion.

Ostriker et al (1995) have argued that the relatively low X-ray luminosities of compact groups might not be due to a low gas fraction but instead could be understood if the groups are filaments seen in projection (Hernquist et al 1995). However, Ponman et al (1996) point out that in order to explain even the fainter compact groups, gas temperatures $T \sim 1$ keV and densities $n \sim 10^{-4}$ cm$^{-3}$ would be required. These appear to be ruled out by both observations (Briel & Henry 1995) and simulations (Diaferio et al 1995, Pildis et al 1996).

Even if compact groups are physically dense, they may not be as dense as they appear. As mentioned in Section 2, a sample of groups selected on the basis of high apparent density will be biased by the inclusion of looser systems that appear more compact owing to geometrical or kinematic effects. Is this bias large? Its magnitude can be estimated as follows: Consider $n$ galaxies randomly located within a circle of radius $R$ on the sky. What is the probability $f(x, n)$ that they will fall within some circular subarea of radius $xR$? The answer can be obtained using analytic expressions derived by Walke & Mamon (1989). From their equations 1 and 6 (setting $N = n$ and $\mathcal{N}_{ext} = 1$), we obtain

$$f(x, n) = \frac{n!}{\pi^n} \int_0^x dr \int_0^{1-r} d\rho \frac{d\mathcal{N}}{dr\, d\rho}, \tag{1}$$

where

$$d\mathcal{N} = \frac{2\pi^{n-1} n r^{2n-3}}{(n-2)!} 2\pi\rho\, d\rho\, dr \tag{2}$$

is the number of possible configurations with radius between $r$ and $r + dr$ and distance from the center between $\rho$ and $\rho + d\rho$. Here we have neglected a small edge contribution that is unimportant for small values of $x$ [Walke & Mamon's (1989) case 3]. This gives

$$f(x,n) = n^2 x^{2n-2}\left[1 - 4\frac{n-1}{2n-1}x + \frac{n-1}{n}x^2\right] \tag{3}$$

Now, an observer would infer a galaxy space density that is higher by a factor $\beta = x^{-3}$, so the average apparent space density enhancement is

$$\begin{aligned}\langle\beta\rangle &= \int_0^1 \beta\frac{df(x,n)}{dx}dx \\ &= \int_0^1 n^2 x^{2n-5}\left[1 - 4\frac{n-1}{2n-1}x + \frac{n-1}{n}x^2\right]dx \\ &= \frac{2n^2(n-1)}{(2n-5)(2n-3)(n-2)}.\end{aligned} \tag{4}$$

Thus we expect to typically overestimate the space density by about a factor of 12.0 for triplets (or quartets containing a physical binary), 3.2 for true quartets, and 2.0 for quintets.

# 7. COSMOLOGICAL IMPLICATIONS

## 7.1 *Clustering and Large-Scale Structure*

A key question that remains is the position of compact groups in the clustering hierarchy. Are compact groups distinct entities (Sulentic 1987) or an intermediate stage between loose groups and triplets, pairs, and individual galaxies (Barnes 1989, White 1990, Cavaliere et al 1991, Rampazzo & Sulentic 1992, Diaferio et al 1994)? Some compact groups are purely projection effects, others may be small clusters (Ebeling et al 1995), but most appear to be real. It seems that they can arise naturally from subcondensations in looser groups, but further studies are needed to better determine both the observed space density of groups as a function of population and the time scales involved in the evolutionary process.

This question is related to that of the formation mechanism of compact groups. Two mechanisms have been discussed in the literature. Diaferio et al (1994) conclude that compact groups form continually from bound subsystems within loose groups. This gains some support from the observation (see Section 3) that most HCGs are embedded in loose groups, although it is not obvious that these loose groups are sufficiently rich (Sulentic 1997). Governato et al (1996) proposed a model in which merging activity in compact groups is accompanied by infall of galaxies from the environment. This naturally explains the observed mix of morphological types, and it allows compact groups to persist for longer times.

Where do the Shakhbazian groups fit in this picture? Recent studies (Tikhonov 1986, Amirkhanian & Egikian 1987, Amirkhanian et al 1988, 1991, Amirkhanian 1989, Kodaira et al 1988–1991, Stoll et al 1993–1996) show that these objects are typically compact clusters or groups of early-type galaxies. Although the systems were selected on the basis of red colors and compact appearance of their galaxies, both of these factors result from their large distances because K-corrections and contrast effects become significant. The galaxies are in fact relatively normal, although luminous (Del Olmo et al 1995). However, the number of blue (gas rich) galaxies in these systems does seem to be very small. Thus it appears that the Shakhbazian groups are mostly small clusters, possibly intermediate in physical properties between classical compact groups and clusters.

## 7.2 *Galaxy Evolution and Merging*

If the groups are dynamically bound, galaxy mergers should commence within a few dynamical times (Carnevali et al 1981, Ishizawa et al 1983, Barnes 1985, Ishizawa 1986, Mamon 1987, 1990, Zheng et al 1993). Both N-body and hydrodynamic simulations indicate that the dark matter halos of individual galaxies merge first, creating a massive envelope within which the visible galaxies move (Barnes 1984, Bode et al 1993). Kinematic studies of loose groups (e.g. Puche & Carignan 1991) indicate that the dark matter is concentrated around the individual optical galaxies. In contrast, the X-ray observations indicate that in most compact groups, the gas and dark matter is more extended and is decoupled from the galaxies. This may explain the observation that galaxies in compact groups typically have mass-to-light ratios 30% to 50% lower than more isolated galaxies (Rubin et al 1991).

Is there any observational evidence that galaxies in compact groups are merging? By 1982 it was evident that first-ranked galaxies in compact groups did not appear to be merger products, because the fraction of first-ranked galaxies that are type E or S0 is the same as for the general population of HCG galaxies (Hickson 1982). If mergers were a dominant effect, the first-ranked galaxies would be expected to be more often elliptical. The same conclusion was reached by Geller & Postman (1983) who found that the luminosities of first-ranked galaxies were consistent with a single luminosity distribution for all group galaxies. Of course this may just mean that in small groups, the first-ranked galaxy is not necessarily the most evolved. Rather, one should ask if *any* galaxies in compact groups show indications of merging. The relative paucity of merging galaxies in compact groups was first noted by Tikhonov (1987), from a visual inspection of optical images. Zepf & Whitmore (1991) realized that elliptical galaxies formed by recent mergers of gas-rich systems should have bluer colors than normal. Examining the HCGs, they found only a small enhancement in the fraction of early-type galaxies having blue colors, a conclusion reinforced by an independent study by Moles et al (1994). On the other hand, Caon et al (1994) argued that the large effective radii of compact group elliptical galaxies is indicative of an origin by merging or accretion of companions.

Zepf (1993) estimated that roughly 7% of the galaxies in compact groups are in the process of merging. This conclusion was based on roughly consistent frequencies of (*a*) optical signatures of merging, (*b*) warm far-infrared colors, and (*c*) sinusoidal rotation curves. However, few galaxies show all of these effects simultaneously. The merging fraction may thus be as high as 25% if one allows that any one of these criteria would be considered to be sufficient to indicate a merger (Hickson 1997). Given the small numbers of objects in these studies, it is fair to say that the fraction of merging galaxies is highly uncertain at present. It seems safe to conclude that current observations do not rule out a significant amount of merging in compact groups.

Detailed studies of individual compact groups can be quite revealing. Many galaxies that at first appear normal are revealed to have peculiar morphology or spectra when examined more closely. Many, perhaps most, compact groups clearly contain galaxies that are dynamically interacting. However, the groups likely span a range of evolutionary states. At the extreme end are high-density groups like Seyfert's Sextet, HCG 31, HCG 62, HCG 94 (Pildis 1995), and HCG 95 (Rodrigue et al 1995), in which we find strong gravitational interactions. At the other end are lower density compact groups, such as HCG 44, which most likely are in a less advanced stage of evolution. This picture is supported by radio observations: Seyfert's Sextet and HCG 31 are both embedded in extended HI clouds, whereas in HCG 44 the HI is associated with individual galaxies (Williams et al 1991).

It seems clear that the groups as we now see them can persist for only a fraction of a Hubble time. Simulations indicate that merging should destroy the group on a time scale $t_m$ that is typically an order of magnitude larger than $t_d$, depending on the distribution of dark matter (Cavaliere et al 1983, Barnes 1984, Navarro et al 1987, Kodaira et al 1990) and initial conditions (Governato et al 1991). Assuming that the groups are in fact bound dynamical systems, we can draw two conclusions: (*a*) There must be an ongoing mechanism for forming or replacing compact groups, and (*b*) there must be a significant population of relics of merged groups.

What are the end-products of compact groups? It is tempting to identify them with field elliptical galaxies, following a suggestion first made by Toomre (1977). Simulations (Weil & Hernquist 1994) indicate that multiple mergers in small groups of galaxies best reproduce the observed kinematical properties of elliptical galaxies. The resulting galaxies are predicted to possess small kinematic misalignments, which can be detected by detailed spectroscopic and photometric studies. Neverthess, it remains to be demonstrated that these merger remnants can reproduce the tight correlation between size, luminosity, and velocity dispersion found in present-day elliptical galaxies.

If compact groups have lifetimes on the order of $t_m$, and form continuously, then the number of relics, per observed group, is expected to be on the order of $(H_0 t_m)^{-1}$. Thus, the number of relics could exceed that of present day groups by as much as an order of magnitude. Mamon (1986) estimated that, if all HCGs are real, then the relics would account for about 25% of luminous field elliptical galaxies. As we have seen, the true space density of compact groups is uncertain by at least a factor of two and may be underestimated because of selection biases. There is then the potential problem of producing too many relics.

A second problem is the fact that the integrated luminosities of compact groups are typically a factor of three to four times greater than luminosities of isolated elliptical galaxies (Sulentic & Rabaça 1994). It is possible that interaction-induced star formation has boosted the luminosities of some compact group galaxies, and that some degree of fading of the merger product is expected. However, at this point it is not clear whether or not the relics can be identified with isolated elliptical galaxies.

Despite these problems, a fossil compact group may have actually been found. Ponman et al (1995) have detected a luminous isolated elliptical galaxy surrounded by diffuse X-ray emission which is consistent with the expected end-product of a compact group. If more objects like this are found, it may be possible to compare their space density with that expected for compact group relics.

## 7.3 *Role in Galaxy Formation and Evolution*

Interactions are often implicated in the development of active nuclei in galaxies (e.g. Freudling & Almudena Prieto 1996). The HCG catalog includes several examples of compact groups containing both starburst galaxies and AGN. Several recent examples of associations between starburst galaxies or AGN and what appear to be compact groups have been reported: Del Olmo & Moles (1991) have found a broad-line AGN in Shakhbazian 278; Zou et al (1995) found that the luminous infrared source IRAS 23532 coincides with a compact group that includes a Seyfert 1 as well as a starburst galaxy. If this association extends to QSOs, one would expect to find numerous compact groups at redshifts $z \sim 2$, where the comoving number density of QSOs peaks (e.g. Hartwick & Schade 1984). The tendency for QSOs to have close companions has been known for some time (e.g. Stockton 1982, Bahcall et al 1997). Recently, several examples of compact groups associated with luminous infrared galaxies, AGN, and QSOs at $z \simeq 2$ have been found using HST (Pascarelle et al 1996, Francis et al 1996, Matthews et al 1994, Tsuboi & Nakai 1994, Hutchings 1995, Hutchings et al 1995).

These observations provide support to the idea that tidally triggered star formation is a predominant factor in the galaxy formation process (Lacey & Silk 1991, Lacey et al 1993). In this model, disk star formation occurs relatively late, after the compact group has formed and tidal interactions are strong. This seems at least qualitatively consistent with the fragmentary nature of high-redshift galaxies observed with the Hubble Space Telescope (Schade et al 1995), although these fragments appear to be much less luminous and more irregular than most present-day compact group galaxies. The model also offers a possible explanation for the excess numbers of faint blue galaxies found in field galaxy counts as dwarf galaxies undergoing star formation at a redshift of $z \simeq 1$.

Compact groups may possibly play a role in the formation of other systems. We have seen that giant galaxies may be formed as the end product of compact group evolution. At the other end of the scale, dwarf galaxies have physical properties distinct from normal galaxies, which suggests a unique formation mechanism. One possibility is that they form during gravitational interactions from tidal debris (Duc & Mirabel 1994). If this is the case, one would expect to find evidence for this in compact groups of galaxies. From an examination of condensations in tidal tails, Hunsberger et al (1996) concluded that the fraction of dwarf galaxies produced within tidal debris in compact groups is not negligible. There is also evidence that star clusters form from tidal debris. Longo et al (1995) have found an excess population of unresolved blue objects around HCG 90 that appear to be recently formed star clusters. These may be similar to the population of new star clusters recently reported in the merger remnant NGC 7252 (Whitmore et al 1993).

## 7.4 *Gravitational Lensing*

Because compact groups have a high galaxy surface density, they may form effective gravitational lenses. Gravitational amplification of background field galaxies was proposed by Hammer & Nottale (1986) as a possible explanation for the presence of the high-redshift discordant member of this group. Mendes de Oliveira & Giraud (1994) and Montoya et al (1996) find that most HCGs are too nearby to produce strong lensing effects. However, because the critical mass density required for strong lensing depends reciprocally on distance, analogous systems 5–10 times more distant should produce a nonnegligible fraction of giant arcs.

### 7.5 *Cosmology*

Studies of small groups may provide clues to the overall structure of the universe. The baryon fractions found in clusters of galaxies appear to be inconsistent with a density parameter $\Omega = 1$, unless the dark matter is more prevalent outside clusters (White 1992, Babul & Katz 1993). Compact groups provide a means to study dark matter in such regions. The baryon fractions found for compact groups do appear to be lower than those for clusters. David et al (1995) argued that the gas is the most extended component, with galaxies the most compact and the dark matter intermediate. They concluded that the baryon fraction approaches 30% on large enough scales, which is comparable to the values found for clusters. Given the constraints of standard Big Bang nucleosynthesis, this would imply that the density parameter $\Omega$ is at most 0.2. On the other hand, the infall picture of compact group evolution (Governato et al 1996) requires a high-density $\Omega \sim 1$ universe. In a low-density universe the infall rate is insufficient. As there is at present no other clear mechanism for avoiding the overproduction of relics by merging compact groups, this may be a strong argument for a high-density universe.

During the last two decades we have seen a resurgence of interest in compact groups. Although initially little more than a curiosity, these systems are now viewed as potentially important sites of dynamical evolution, shaping the structure of many galaxies. It now seems clear that while many compact groups are contaminated by projections, a large fraction of at least the high surface-brightness HCGs are physically dense. They form by gravitational relaxation processes within looser associations of galaxies. The densest are generally in an advanced stage of evolution characterized by strong interactions, starburst and AGN activity, stripping of stellar and dark matter halos, and merging. They contain large amounts of dark matter and primordial X-ray–emitting gas trapped within the gravitational potential well.

Despite this progress, many questions remain unanswered. What are the end products of compact group evolution, and do they have properties consistent with any known population of objects? What is the space density of such relics? Where do compact groups fit in the overall clustering hierarchy? What is their role in the evolution of galaxies both past and present? Given the current interest and research activity in this area, it is likely that many of these questions may soon be addressed.

## ACKNOWLEDGMENTS

It is a pleasure to thank the Observatories of Brera and Capodimonte for hospitality during the initial work on this review. I have benefitted from discussions with many individuals, but I would like to acknowledge particularly the contributions of A Iovino, E Kindl, G Longo, G Mamon, C Mendes de Oliveira, TK Menon, G Palumbo, H Rood, and J Sulentic. I thank G Mamon, A Sandage, and J Sulentic for providing helpful comments on an earlier version of the manuscript. Financial support was provided by the Natural Sciences and Engineering Research Council of Canada and NATO.



*Literature Cited*


Allam S, Assendorp R, Longo G, Braun M, Richter G. 1996. *Astron. Astrophys. Suppl.* 117:39–82

Ambartsumian BA. 1961. *Astron. J.* 66:536–40

Amirkhanian AS. 1989. *Akad. Nauk Arm. SSR Bjurak. Obs. Soobšč.* 61:25–28

Amirkhanian AS, Egikian AG. 1987. *Astrofizika* 27:395–97

Amirkhanian AS, Egikian AG, Sil'chenko OK. 1988. *Pis'ma Astron. Zh.* 14:404–8

Amirkhanian AS, Eghikian AG, Tikhonov NA, Shakhbazian RK. 1991. *Astrofizika* 35:67–75

Arp H. 1966. *Ap. J. Suppl.* 14:1–20

Arp H. 1973. *Ap. J.* 185:797–808

Arp H. 1987. *Quasars, Redshifts and Controversies.* Berkeley: Interstellar Media

Babul A, Katz N. 1993. *Ap. J. Lett.* 406:L51–L54

Bahcall JN, Kirharos S, Saxe DH, Schneider DP. 1997. *Ap. J.* 479: In press

Bahcall NA, Harris DE, Rood HJ. 1984. *Ap. J. Lett.* 284:L29–L33

Baier FW, Petrosian MB, Tiersch H, Shakhbazian RK. 1974. *Astrofizika* 10:327–35

Baier FW, Tiersch H. 1975. *Astrofizika* 11:221–28

Baier FW, Tiersch H. 1976a. *Astrofizika* 12:7–12

Baier FW, Tiersch H. 1976b. *Astrofizika* 12:409–15

Baier FW, Tiersch H. 1978. *Astrofizika* 14:279–82

Baier FW, Tiersch H. 1979. *Astrofizika* 15:33–35

Balcells N, Quinn PJ. 1990. *Ap. J.* 361:381–93

Barnes J. 1984. *MNRAS* 208:873–85

Barnes JE. 1985. *MNRAS* 215:517–36

Barnes JE. 1989. *Nature* 338:123–26

Barton E, Geller MJ, Ramella M, Marzke RO, da Costa LN. 1996. *Astron. J.* 112:871–86

Bergvall N, Ekman A, Lauberts A. 1981. *Astron. Astrophys.* 95:266–77

Bettoni D, Fasano G. 1993. *Astron. J.* 105:1291–307

Bettoni D, Fasano G. 1995. *Astron. J.* 109:32–55

Bettoni D, Fasano G. 1996. *Astron. Astrophys. Suppl.* 118:429–39

Biermann P, Kronberg PP. 1984. *Phys. Scr.* T7:169

Bode PW, Cohn HN, Lugger PM. 1993. *Ap. J.* 416:17–25

Boselli A, Mendes de Oliveira C, Balkowski C, Cayatte V, Casoli F. 1996. *Astron. Astrophys.* 314:738–44

Briel UG, Henry JP. 1995. *Astron. Astrophys.* 302:L9–L12

Burbidge EM, Burbidge GR. 1959. *Ap. J.* 130:23–25

Burbidge EM, Burbidge GR. 1961a. *Astron. J.* 66:541–50

Burbidge EM, Burbidge GR. 1961b. *Ap. J.* 134:244–47

Burbidge EM, Burbidge GR. 1961c. *Ap. J.* 134:248–56

Burbidge EM, Sargent WLW. 1971. *Pontif. Acad. Sci. Scr. Varia* 35:351–78

Campos-Aguilar A, Moles M. 1991. *Astron. Astrophys.* 241:358–64

Caon N, Capaccioli M, D'Onofrio M, Longo G. 1994. *Astron. Astrophys.* 286:L39–L42

Carnevali P, Cavaliere A, Santangelo P. 1981. *Ap. J.* 249:449–61

Cavaliere A, Colafrancesco S, Scaramella R. 1991. *Ap. J.* 380:15–23

Cavaliere A, Santangelo P, Tarquini G, Vittorio N. 1983. In *Clustering in the Universe*, ed D Gerbal, A Mazure, pp. 25–33. Gif-sur-Yvette: Ed. Front.

Cocke WJ, Tifft WG. 1983. *Ap. J.* 268:56–59

da Costa LN, Geller MJ, Pellegrini PS, Latham DW, Fairall AP, et al. *Ap. J. Lett.* 424:L1–L4

David LP, Jones C, Forman W. 1995. *Ap. J.* 445:578–90

Davis DS, Mulchaey JS, Mushotzky RF, Burstein D. 1996. *Ap. J.* 460:601–11
De Carvalho RR, Djorgovski SG. 1995. *Bull. Am. Astron. Soc.* 187:53.02
De Carvalho RR, Ribeiro ALB, Zepf SE. 1994. *Ap. J. Suppl.* 93:47–63
Dell'Antonio IP, Geller MJ, Fabricant DG. 1995. *Astron. J.* 110:502–12
Del Olmo A, Moles M. 1991. *Astron. Astrophys.* 245:27–30
Del Olmo A, Moles M, Perera J. 1995. In *Groups of Galaxies, ASP Conf. Ser.* 70:117–26
De Souza RE, Capelato HV, Arakaki L, Logullo C. 1982. *Ap. J.* 263:557–63
Diaferio A, Geller MJ, Ramella M. 1994. *Astron. J.* 107:868–79
Diaferio A, Geller MJ, Ramella M. 1995. *Astron. J.* 109:2293–304
Dressler A. 1980. *Ap. J.* 236:351–65
Duc P-A, Mirabel IF. 1994. *Astron. Astrophys.* 289:83–93
Ebeling H, Mendes de Oliveira C, White DA. 1995. *MNRAS* 277:1006–32
Ebeling H, Voges W, Boehringer H. 1994. *Ap. J.* 436:44–55
Fasano G, Bettoni D. 1994. *Astron. J.* 107:1649–67
Francis PJ, Woodgate BE, Warren SJ, Moller P, Mazzolini M, et al. 1996. *Ap. J.* 457:490–99
Freudling W, Almudena Prieto M. 1996. *Astron. Astrophys.* 306:39–48
Fukazawa Y, Makishima K, Matsushita K, Yamasaki N, Ohashi T, et al. 1996. *Publ. Astron. Soc. Jpn.* 48:395–407
Garcia A. 1995. *Astron. Astrophys.* 297:56–60
Geller MJ, Postman M. 1983. *Ap. J.* 274:31–38
Gisler G. 1980. *Astron. J.* 85:623–25
Governato F, Chincarini G, Bhatia R. 1991. *Ap. J.* 371:L15–L18
Governato F, Tozzi P, Cavaliere A. 1996. *Ap. J.* 458:18–26
Hammer F, Nottale L. 1986. *Astron. Astrophys.* 155:420–22
Heiligman GM, Turner EL. 1980. *Ap. J.* 236:745–49
Henriksen MJ, Mamon GA. 1994. *Ap. J. Lett.* 421:L63–L66
Henry JP, Gioia IM, Huchra JP, Burg R, McLean B, et al. 1995. *Ap. J.* 449:422–30
Hernquist L, Katz N, Weinberg DH. 1995. *Ap. J.* 442:57–60
Hickson P. 1982. *Ap. J.* 255:382–91
Hickson P. 1990. In *Paired and Interacting Galaxies*, Int. Astron. Union Colloq. No. 124, pp. 77–91. Huntsville, AL: NASA Marshall Space Flight Center
Hickson P. 1993. *Ap. Lett. Commun.* 29:1–207
Hickson P. 1994. *Atlas of Compact Groups of Galaxies.* Basel: Gordon & Breach
Hickson P. 1997. *Ap. Lett. Commun.* In press
Hickson P, Kindl E, Huchra JP. 1988a. *Ap. J.* 329:L65–L67
Hickson P, Kindl E, Huchra JP. 1988b. *Ap. J.* 331:64–70
Hickson P, Kindl E, Auman JR. 1989. *Ap. J. Suppl.* 70:687–98
Hickson P, Mendes de Oliveira C, Huchra JP, Palumbo GG. 1992. *Ap. J.* 399:353–67
Hickson P, Menon TK, Palumbo GGC, Persic M. 1989. *Ap. J.* 341:679–84
Hickson P, Ninkov Z, Huchra JP, Mamon GA. 1984. In *Clusters and Groups of Galaxies*, ed. F Mardirossian, G Giuricin, M Mezzetti, pp. 367–73. Dordrecht: Reidel
Hickson P, Richstone DO, Turner EL. 1977. *Ap. J.* 213:323–26
Hickson P, Rood HJ. 1988. *Ap. J.* 331:L69–L72
Huchtmeier W. 1994. *Astron. Astrophys.* 286: 389–94
Hunsberger SD, Charlton JC, Zaritsky D. 1996. *Ap. J.* 462:50–56
Humason ML, Mayall NU, Sandage AR. 1956. *Astron. J.* 61:97
Hutchings JB. 1995. *Astron. J.* 109:928–34
Hutchings JB, Crampton D, Johnson A. 1995. *Astron. J.* 109:73–80
Iovino A, Hickson P. 1996. *MNRAS.* In press
Ishizawa T. 1986. *Astrophys. Space Sci.* 119: 221–25
Ishizawa T, Matsumoto R, Tajima T, Kageyama H, Sakai H. 1983. *Publ. Astron. Soc. Jpn.* 35:61–76
Kent SM. 1981. *Publ. Astron. Soc. Pac.* 93:554–57
Kindl E. 1990. *A photometric and morphological study of compact groups of galaxies and their environments.* PhD thesis. Univ. British Columbia, Vancouver, Canada. 168 pp.
Kirshner RP, Malumuth EM. 1980. *Ap. J.* 236:366–72
Kiseleva LG, Orlov VV. 1993. *Vistas Astron.* 36:1–30
Kodaira K, Doi M, Ichikawa S-I, Okamura S. 1990. *Jpn. Natl. Astron. Obs. Publ.* 1:283–95
Kodaira K, Iye M, Okamura S, Stockton A. 1988. *Publ. Astron. Soc. Jpn.* 40:533–45
Kodaira K, Sekiguchi M, Sugai H, Doi M. 1991. *Publ. Astron. Soc. Jpn.* 43:169–76
Kormendy J, Sanders DB. 1992. *Ap. J.* 390:L53–56
Lacey C, Guiderdoni B, Rocca-Volmerange B, Silk J. 1993. *Ap. J.* 402:15–41
Lacey C, Silk J. 1991. *Ap. J.* 381:14–32
Limber DN, Mathews WG. 1960. *Ap. J.* 132:286–305
Longo G, Busarello G, Lorenz H, Richter G, Zaggia S. 1994. *Astron. Astrophys.* 282:418–24
Longo G, Grimaldi A, Richter G. 1995. *Astron. Astrophys.* 299:L45–L48
Lynds CR, Khachikian EE, Amirkhanian AS.

1990. *Pis'ma Astron. Zh.* 16:195–98
Maccagni D, Gioia IM, Henry JP, Maccacaro T, Vettolani GP. 1990. *Astron. J.* 100:1461–67
Malumian VH. 1996. *Astron. Nachr.* 317:101
Malykh SA, Orlov VV. 1986. *Astrofizika* 24:445–51
Mamon GA. 1986. *Ap. J.* 307:426–30
Mamon GA. 1987. *Ap. J.* 321:622–44
Mamon GA. 1989. *Astron. Astrophys.* 219:98–100
Mamon GA. 1990. In *Paired and Interacting Galaxies*, Int. Astron. Union Colloq. No. 124, pp. 77–91. Huntsville, AL: NASA Marshall Space Flight Center
Mamon GA. 1995. In *Groups of Galaxies*, ASP Conf. Ser. 70:83–94
Marzke RO, Huchra JP, Geller MJ. 1944. *Ap. J.* 428:43–50
Massey P. 1977. *Publ. Astron. Soc. Pac.* 89:13–18
Matthews K, Soifer BT, Nelson J, Boesgaard H, Graham JR, et al. 1994. *Ap. J. Lett.* 420:L13–L16
Mendes de Oliveira C. 1995. *MNRAS* 273:139–45
Mendes De Oliveira C, Giraud E. 1994. *Ap. J.* 437:L103–6
Mendes De Oliveira C, Hickson P. 1991. *Ap. J.* 380:30–38
Mendes De Oliveira C, Hickson P. 1994. *Ap. J.* 427:684–95
Menon TK. 1991. *Ap. J.* 372:419–23
Menon TK. 1992. *MNRAS* 255:41–47
Menon TK. 1995a. *Astron. J.* 110:2605–9
Menon TK. 1995b. *MNRAS* 274:845–52
Menon TK, Hickson P. 1985. *Ap. J.* 296:60–64
Menon TK, Leon S, Combes F. 1996. *Bull. Am. Astron. Soc.* 188:4502
Mirzoian LV, Miller JS, Osterbrock DE. 1975. *Ap. J.* 196:687–88
Moles M, Del Olmo A, Perea J, Masegosa J, Marquez I, Costa V. 1994. *Astron. Astrophys.* 285:404–14
Mulchaey JS, Davis DS, Mushotzky RF, Burstein D. 1993. *Ap. J. Lett.* 404:L9–L12
Mulchaey JS, Davis DS, Mushotzky RF, Burstein D. 1996a. *Ap. J.* 456:80–97
Mulchaey JS, Mushotzky RF, Burstein D, Davis DS. 1996b. *Ap. J. Lett.* 456:L5–L8
Navarro JF, Mosconi M, Garcia Lambas D. 1987. *MNRAS* 228:501–11
Nilson PN. 1973. *Uppsala General Catalogue of Galaxies,* Uppsala Obs. Ann. 6
Nottale L, Moles M. 1978. *Astron. Astrophys.* 66:355–58
Oleak H, Stoll D, Tiersch H, Macgillivray H. 1995. *Astron. J.* 109:1485–89
Ostriker JP, Lubin LM, Hernquist L. 1995. *Ap. J.* 444:L61–L64
Ostriker JP, Tremaine SD. 1975. *Ap. J. Lett.* 202:L113–17
Palumbo GGC, Saracco P, Hickson P, Mendes de Oliveira C. 1995. *Astron. J.* 109:1476–84
Palumbo GGC, Saracco P, Mendes de Oliveira C, Hickson P, Tornatore V, Baiesi-Pillastrini GC. 1993. *Ap. J.* 405:413–18
Pascarelle SM, Windhorst RA, Driver SP, Ostrander EJ, Keel WC. 1996. *Ap. J. Lett.* 456:L21–L24
Peebles PJE. 1971. *Physical Cosmology.* Princeton: Princeton Univ. Press
Petrosian MB. 1974. *Astrofizika* 10:471–76
Petrosian MB. 1978. *Astrofizika* 14:631–35
Pildis RA. 1995. *Ap. J.* 455:492–96
Pildis RA, Bregman JN, Evrard AE. 1995a. *Ap. J.* 443:514–26
Pildis RA, Bregman JN, Schombert JM. 1995b. *Astron. J.* 110:1498–506
Pildis RA, Evrard AE, Bregman JN. 1996. *Astron. J.* 112:378–87
Ponman TJ, Allan DJ, Jones LR, Merrifield M, McHardy IM, et al. 1995. *Nature* 369:462–64
Ponman TJ, Bertram D. 1993. *Nature* 363:51–54
Ponman TJ, Bourner PDJ, Ebeling H, Bohringer H. 1996. *MNRAS* 283:690–708
Postman M, Geller MJ. 1984. *Ap. J.* 281:95–99
Prandoni I, Iovino A, Macgillivray HT. 1994. *Astron. J.* 107:1235–44
Press WH, Schechter P. 1974. *Ap. J.* 187:425–38
Puche D, Carignan C. 1991. *Ap. J.* 378:487–95
Ramella M, Diaferio A, Geller MJ, Huchra JP. 1994. *Astron. J.* 107:1623–28
Rampazzo R, Sulentic JW. 1992. *Astron. Astrophys.* 259:43–60
Raymond JC, Smith BW. 1977. *Ap. J. Suppl.* 35:419–39
Ribeiro ALB, De Carvalho RR, Zepf SE. 1994. *MNRAS* 267:L13–L16
Ribeiro ALB, De Carvalho RR, Coziol R, Capelato HV, Zepf SE. 1996. *Ap. J. Lett.* 463:L5–L8
Rodrigue M, Schultz A, Thompson J, Colegrove T, Spight LD, et al. 1995. *Astron. J.* 109:2362–67
Rood HJ, Struble MF. 1994. *Publ. Astron. Soc. Pac.* 106:413–16
Rood HJ, Williams BA. 1989. *Ap. J.* 339:772–82
Rose JA. 1977. *Ap. J.* 211:311–18
Rose JA. 1979. *Ap. J.* 231:10–22
Rose JA, Graham JA. 1979. *Ap. J.* 231:320–26
Rubin VC, Ford WKJ, Hunter DA. 1990. *Ap. J.* 365:86–92
Rubin VC, Hunter DA, Ford WKJ. 1991. *Ap. J. Suppl.* 76:153–83
Sakai S, Giovanelli R, Wegner G. 1994. *Astron. J.* 108:33–43
Sandage A, Binggeli B & Tammann GA. 1985. *Astron. J.* 90:1759–71
Sanders DB, Mirabel IF. 1996. *Annu. Rev. As-*

*tron. Astrophys.* 34:749–92
Saracco P, Ciliegi P. 1995. *Astron. Astrophys.* 301:348–58
Sargent WLW. 1968. *Ap. J. Lett.* 153:L135–37
Schade D, Lilly SJ, Crampton D, Hammer F, Le Fevre O, Tresse L. 1995. *Ap. J.* 451:L1–L4
Schechter P. 1976. *Ap. J.* 203:297–306
Schwartz DA, Schwarz J, Tucker W. 1980. *Ap. J.* 238:L59–L62
Seyfert CK. 1948. *Astron. J.* 53:203–4
Shakhbazian RK. 1957. *Astron. Tsirk.* 177: 11
Shakhbazian RK. 1973. *Astrofizika* 9:495–501
Shakhbazian RK, Amirkhanian AS. 1979. *Astrofizika* 14:455–66
Shakhbazian RK, Petrosian MB. 1974. *Astrofizika* 10:13–20
Stephan ME. 1877. *MNRAS* 37:334–39
Stockton A. 1982. *Ap. J.* 257:33–39
Stoll D, Tiersch H, Oleak H, Baier F, Macgillivray HT. 1993a. *Astron. Nachr.* 314:225–67
Stoll D, Tiersch H, Oleak H, Baier F, Macgillivray HT. 1993b. *Astron. Nachr.* 314:317–60
Stoll D, Tiersch H, Oleak H, Macgillivray HT. 1994a. *Astron. Nachr.* 315:11–61
Stoll D, Tiersch H, Oleak H, Macgillivray HT. 1994b. *Astron. Nachr.* 315:97–150
Stoll D, Tiersch H, Braun M. 1996a. *Astron. Nachr.* 317:239–57
Stoll D, Tiersch H, Braun M. 1996b. *Astron. Nachr.* 317:315–32
Sulentic JW. 1983. *Ap. J.* 270:417–21
Sulentic JW. 1987. *Ap. J.* 322:605–17
Sulentic JW. 1993. In *Progress in New Cosmologies: Beyond the Big Bang*, ed. HC Arp et al, pp. 49–64. New York: Plenum
Sulentic JW. 1997. *Ap. J.* In press
Sulentic JW, De Mello Rabaça DF. 1993. *Ap. J.* 410:520–25
Sulentic JW, Lorre JJ. 1983. *Astron. Astrophys.* 120:36–52
Sulentic JW, Pietsch W, Arp H. 1995. *Astron. Astrophys.* 298:420–26
Sulentic JW, Rabaça CR. 1994. *Ap. J.* 429:531–39
Thompson LA, Gregory SA. 1980. *Ap. J.* 242:1–7
Tiersch H. 1976. *Astron. Nachr.* 297:301–3
Tikhonov NA. 1986. *Soobšč. Spets. Astrofiz. Obs.* 49:69–84
Tikhonov NA. 1987. *Astrofizika* 27:253–64
Tovmasian GM, Shakhbazian ETs, Shirbakian MS. 1980. *Astrofizika* 15:513–14
Toomre A. 1977. In *The Evolution of Galaxies and Stellar Populations*, ed. BM Tinsley, RB Larson, pp. 401–16. New Haven: Yale Univ. Obs.
Tsuboi M, Nakai N. 1994. *Publ. Astron. Soc. Jpn.* 46:L179–82
Tully RB. 1987. *Nearby Galaxies Catalog.* Cambridge: Cambridge Univ. Press
Vennik J, Richter GM, Longo G. 1993. *Astron. Nachr.* 314:393
Venugopal VR. 1995. *MNRAS* 277:455–57
Vettolani G, Gregorini L. 1988. *Astron. Astrophys.* 189:39–41
Vorontsov-Velyaminov BA. 1959. *Atlas and Catalog of Interacting Galaxies,* Vol 1. Moscow: Sternberg Inst.
Vorontsov-Velyaminov BA. 1975. *Atlas of Interacting Galaxies*, Part II, *Astron. Astrophys. Suppl.* 28:1
Vorontsov-Velyaminov BA, Dostal VA, Metlov VG. 1980. *Soviet Astron. Lett.* 6:394–97
Vorontsov-Velyaminov BA, Metlov VG. 1980. *Soviet Astron. Lett.* 6:199–201
Walke DG, Mamon GA. 1989. *Astron. Astrophys.* 225:291–302
Weil ML, Hernquist L. 1994. *Ap. J. Lett.* 431:L79–L82
White SDM. 1990. In *Dynamics and Interactions of Galaxies*, ed. R Weilen, pp. 380–88. Heidelberg: Springer-Verlag
White SDM. 1992. In *Clusters and Superclusters of Galaxies*, pp. 1–15. NATO Adv. Study Inst.
Whitmore BC. 1992. In *Physics of Nearby Galaxies: Nature or Nurture?*, ed. TX Thuan, C Balkowski, J Tran Thanh Van, *12th Moriond Astrophys. Meet.*, pp. 351–66. Gif-sur-Yvette: Ed. Front.
Whitmore BC, Gilmore DM. 1992. *Ap. J.* 367:64–68
Whitmore BC, Schweizer F, Leitherer C, Borne K, Robert C. 1993. *Astron. J.* 106:1354–70
Williams BA, Mcmahon PM, van Gorkom JH. 1991. *Astron. J.* 101:1957–68
Williams BA, Rood HJ. 1987. *Ap. J. Suppl.* 63:265–94
Williams BA, Van Gorkom JH. 1988. *Astron. J.* 95:352–55
Wu W, Rabaça CR, Sulentic JW. 1994. *Bull. Am. Astron. Soc.* 26:1494
Zepf SE. 1993. *Ap. J.* 407:448–55
Zepf SE, Whitmore BC. 1991. *Ap. J.* 383:542–49
Zepf SE, Whitmore BC. 1993. *Ap. J.* 418:72–81
Zepf SE, Whitmore BC, Levison HF. 1991. *Ap. J.* 383:524–41
Zheng J, Valtonen MJ, Chernin AD. 1993. *Astron. J.* 105:2047–53
Zou Z-L, Xia X-Y, Deng Z-G, Wu H. 1995. *Astron. Astrophys.* 304:369–73

*Annu. Rev. Astron. Astrophys. 1997. 35:389–443*


# FAINT BLUE GALAXIES

*Richard S. Ellis*
Institute of Astronomy, University of Cambridge, Madingley Road,
Cambridge CB3 0HA, England; e-mail: rse@ast.cam.ac.uk

KEY WORDS: distant field galaxies, evolution, galaxy formation, cosmology

ABSTRACT

The physical properties of the faint blue galaxy population are reviewed in the context of observational progress made via deep spectroscopic surveys and Hubble Space Telescope imaging of field galaxies at various limits and theoretical models for the integrated star formation history of the universe. Notwithstanding uncertainties in the properties of the local population of galaxies, convincing evidence has emerged from several independent studies for a rapid decline in the volume-averaged star-formation rate of field galaxies since a redshift $z \approx 1$. Together with the small angular sizes and modest mean redshift of the faintest detectable sources, these results can be understood in hierarchical models where the bulk of the star formation occurred at redshifts between $z \approx 1-2$. The physical processes responsible for the subsequent demise of the faint blue galaxy population remain unclear. Considerable progress will be possible when the evolutionary trends can be monitored in the context of independent physical parameters such as the underlying galactic mass.

## 1. INTRODUCTION

The nature of the faintest galaxies detectable with our large telescopes has a long and intriguing history and remains one of astronomy's grand questions. The quest to measure galaxy counts faint enough to verify the cosmological principle and to constrain world models motivated the construction of a series of ever larger telescopes, and this, in turn, led to a renaissance of observational cosmology that has progressed apace during the twentieth century (Hubble 1926, 1936). The interest in the deep universe as probed by faint galaxy statistics and the puzzling results obtained with modern instrumentation have been major driving forces in extragalactic astronomy through the 1980s and 1990s.



0066-4146/97/0915-0389$08.00

The most recent chapter in this observational story is the Hubble Deep Field (HDF) project (Williams et al 1996)—an unprecedented long exposure of an area of undistinguished sky observed with the Hubble Space Telescope (HST) undertaken with the purpose of extending the known limits of the faint universe as delineated by field galaxies.

Although the original motivation for what has always been difficult observational work was an attempt to quantify the cosmological world model, the bulk of this review is concerned with the study of faint galaxies as a way of probing their evolutionary history. The evolutionary possibilities were, of course, recognized by the early pioneers (Hubble 1936, Humason et al 1956). However, galaxy evolution was surprisingly slow to emerge as the dominant observational motivation (see Sandage 1995 for a historical account). Even in the 1970s much of the relevant time on the Palomar 200-inch telescope was aimed at estimating cosmological parameters by using galaxies as tracers of the cosmic deceleration (Tammann 1984). Evolution was considered primarily as a necessary "correction" to apply in the grander quest for the nature of the world model; the term evolutionary correction remains (cf Poggianti 1997). Galaxy evolution only rose to prominence when quantitative predictions for the star-formation histories of normal galaxies became available (Sandage 1961, Tinsley 1972). An equally important theoretical motivation for studying evolution was the development of the statistical machinery to understand how physical structures form in the expanding universe (see e.g. Peebles 1971). Both propelled observers to construct catalogs of faint field galaxies and to consider comparing their statistical properties with theoretical predictions.

An appropriate birthplace for the modern era of faint field galaxy studies was the symposium Evolution of Stellar Populations, held at Yale University (Larson & Tinsley 1977). The theoretical ingredients were largely in place in time for the first deep optical surveys made possible by a new generation of wide-field 4-m telescope prime focus cameras. In an era in which the photographic plate is so often disregarded, it is salutary to realize that many of the basic observational results that generated the discussion that follows were first determined from photographic surveys exploited by the new generation of computer-controlled measuring machines (Kron 1978, Peterson et al 1979, Tyson & Jarvis 1979, Koo 1981).

The question of interpreting faint galaxy data in the context of the evolutionary history of field galaxies was comprehensively reviewed by Koo & Kron (1992). Their article summarized the developments from the early photographic results through to deeper charge-coupled device (CCD) galaxy counts and, most importantly, provided a critical account of the interpretation of the first deep redshift surveys. As they stressed, the addition of spectroscopic data provided a valuable first glimpse at the distributions in redshift, luminosities, and star-formation characteristics of representative populations of field galaxies a few billion years ago. Several puzzling features emerged that have collectively been referred to as the faint blue galaxy problem (Kron 1978). In its simplest manifestation, an apparent excess of faint blue galaxies is seen in the source counts over the number expected on the basis of local galaxy properties. A more specific version of the problem that attracted much attention followed the results of the first faint redshift surveys (Broadhurst et al 1988, Colless et al 1990). The count-redshift data from these surveys did not solve the number problem by revealing a redshift range (at either low or high redshift) where this additional population could be logically placed. Relatively complex evolutionary hypotheses were then proposed to reconcile these results, including luminosity-dependent evolution, galaxy merging, and the existence of a new population of source present at modest redshift but, mysteriously, absent locally. Koo & Kron (1992) reviewed the material at hand but considered many of these hypotheses to be premature. They stressed the relatively poor knowledge of the local galaxy population and urged a more cautious approach. Meanwhile, however, the fundamental question remains as to how to account for the high surface density of faint blue sources seen to limits well beyond those of the spectroscopic surveys.

Although it is only five years since Koo & Kron's review, it is timely to address the question anew for several reasons. First, there has been an explosion of interest in the subject as evidenced by a number of diagnostics including large numbers of articles and the statistics of telescope time applications. Clearly, in many minds, the controversy remains. Second, considerable observational progress has been made through larger, more comprehensive, ground-based redshift surveys including those from the first generation of 8- to 10-m telescopes. The redshift boundary for statistically complete field surveys has receded from $z = 0.7$ (Colless et al 1990, 1993) to 1.6 (Cowie et al 1996). Third, significant new data bearing on the question has arrived from the refurbished HST including the Medium Deep Survey (MDS) (Griffiths et al 1994, Windhorst et al 1996) and the HDF (Williams et al 1996). Reliable morphologies and sizes of faint field galaxies have become available for the first time, providing a surge of new data similar to that provided by the first redshift surveys discussed by Koo & Kron.

The major questions addressed in this review are as follows: What are the faint blue galaxies seen in the deep optical galaxy counts and what role do they play in the evolution and formation of normal field galaxies; is there convincing evidence for recent evolution of the forms proposed; and, if so, what are the physical processes involved? "Faint" in this context is defined, for convenience, to be at or beyond the limiting magnitude of the Schmidt sky surveys (i.e. $V > 21$). "Faint" need not necessarily imply "distant," although a major motivation is the hope of learning something fundamental about distant young galaxies. The adjective "blue" is not applied rigorously in most of the articles in this area, and the term primarily reflects the significance of the excess population when observed at optical wavebands sensitive to changes in the short-term star-formation rate. "Field" implies selection without regard to the local environment (Koo & Kron 1992). Normally this is in the context of systematic surveys of randomly chosen areas that span large cosmic volumes. The distinction between field and cluster observations is straightforward to apply, but the physical significance of the different results remains unclear. The 1977 Yale symposium also saw the first quantitative evidence for an increase with redshift in the blue galaxy population in rich clusters (Butcher & Oemler 1978). Recent work (Couch et al 1994, Dressler et al 1994) interprets this evolution as produced via a changing star-formation rate in certain types of cluster galaxies, presumably as a result of environmental processes. However, until the physical cause of these activities is better understood, a connection between field and cluster evolutionary processes should not be ruled out.

Two points should be made about the scope of this review and its intended audience. First, a quantitative review of the evolution of galaxies would require a critical discussion of many related issues such as the stellar and dynamical history of nearby galaxies, indirect probes of the high redshift universe such as QSO absorption line statistics, primeval galaxy searches, the growth of large-scale structure, uncertainties in stellar evolution, and the reliability of evolutionary modeling used to interpret the wealth of data now available. All of these are active areas that impact heavily on galaxy evolution and merit reviews of their own. The strategy here is to focus solely on the faint blue galaxy question, drawing on additional evidence from these fields where appropriate. Second, the subject is developing rapidly with many new observational claims, some of which appear to contradict one another. A detailed resolution of these issues is usually highly technical and in many cases not yet possible. The review is therefore primarily aimed at a fresh graduate student entering the field rather than at the expert who is active in the subject. The hope is to bring out the key issues without getting overly bogged down in the numerical detail.

## 2. BASIC METHODOLOGIES

The primary observational material consists of a number of measurements for every galaxy in a survey that can be regarded as statistically complete. Completeness means that the results of one survey can be compared with those of other observers and with the predictions of contemporary models that take account of the various selection criteria. Raw measurements for each galaxy in a given survey typically include isophotal or pseudototal magnitudes, aperture colors, spectroscopic redshifts, and line strengths (such as [O II] 3727-Å emission line fluxes or equivalent widths); dynamical line widths; and, from HST, images, sizes, and shapes. Derived pseudophysical parameters for each galaxy include its luminosity in some rest-frame bandpass, the star-formation rate as inferred from emission line characteristics or ultraviolet (UV) continuum flux, and some form of classification. Simple considerations (Struck-Marcell & Tinsley 1978) suggest that galaxies of similar Hubble types have had similar star-formation histories, providing the strong incentive to classify faint samples. This has been done variously via a morphological type from the HST image (Glazebrook et al 1995b, Driver et al 1995a,b), by examination of the color or spectral energy distribution (SED) (Lilly et al 1995), or via cross-correlation of detailed spectral features against local spectroscopic data (Heyl et al 1997, Kennicutt 1992). Assuming completeness, population statistics can then be derived, such as the galaxy luminosity function and the volume-averaged star-formation rate, both as a function of redshift and galaxy type. These data can be used with other indicators, such as probes of the gas content, to infer the global history of star formation and metal production (Songaila et al 1990, Fall et al 1996, Madau et al 1996).

Before discussing how observers transform their raw measurements into population statistics, an important point of principle needs to be discussed. In the above methodology, the population statistics are used only to present an empirical description of galaxy evolution; no extensive modeling is usually involved, although occasionally extrapolations are made into territories where no data exist. Recent examples of the empirical approach include redshift survey articles by Lilly et al (1996), Ellis et al (1996a), and Cowie et al (1996). An alternative approach, which we call the ab initio approach, starts from a cosmogonic theory. A particular initial power spectrum of density fluctuations is adopted and gas-consumption time scales and morphological types are assigned to assemblies of dark and baryonic matter, which grow hierarchically. Star-formation histories for model galaxies are then used to predict observables directly in the context of particular surveys. The ab initio approach has its origins in the evolutionary predictions made by Tinsley (1980), Bruzual (1983), Guiderdoni & Rocca-Volmerange (1987), and Yoshii & Takahara (1988) on the basis of simple assumptions that did not rely on particular cosmogonic models. However, more elaborate calculations are now possible in the physical context of hierarchical dark matter cosmologies, such as in recent articles by Kauffmann et al (1994), Cole et al (1994a), and Baugh et al (1996).

Both the empirical and ab initio approaches have their place in observational cosmology. The empirical approach attempts to encapsulate the data in the simplest way and has the advantage of presenting results in a way that is not tied to any particular cosmogonic theory. However, it is not always clear how uncertainties in the raw data plane transform to those in the derived physical plane. The complexity of various selection effects is increasingly used in many areas of astronomy as an argument for the ab initio approach (cf simulations of the Lyman alpha forest discussed by Miralda-Escude et al 1996). Numerical simulations are now sufficiently sophisticated that remarkably realistic predictions can be made that take into account complexities such as merging, starbursts, feedback, and ionization effects that affect model galaxies according to their physical situation rather than their observable quantities. In such cases, intuitive approaches are not possible. Ultimately the ab initio approach may become the preferred way to interpret data, but this will only occur when such models have strong predictive capabilities. The difficulty of relying entirely on the ab initio approach is that, at most, the current methodology reveals models that are only consistent with the observational datasets; one might argue this is the minimum required of any model! In no sense does agreement of a complex model with data imply that that particular model is correct.

## 2.1 *Predictions in the Absence of Galaxy Evolution*

The "no-evolution prediction" originated as a null hypothesis used to address the most basic question: Is evolution detected? Despite its frequent use, it clearly describes an unphysical situation. The passive evolution of stellar populations (defined as that arising from stellar population changes occurring in the absence of continued star formation) alone can lead to detectable luminosity and color changes over the past few billion years (Tinsley 1972). However, the look-back time to a distant galaxy (which makes the entire subject of galaxy evolution observationally possible) is accompanied by redshift and related observational selection biases that seriously affect the empirical approach of monitoring evolution as a function of redshift in flux-limited catalogs of galaxies. For redshifts of $z < 1$–$2$, the no-evolution prediction has acted as a valuable standard baseline that incorporates these biases from which the various evolutionary differences can be compared.

To predict the observables in the empirical approach, consider a local galaxy population with a luminosity function (LF) whose form is defined by the Schechter function (1976):

$$\phi(\mathrm{L})\mathrm{dL} = \phi^*(\mathrm{L}/\mathrm{L}^*)^{-\alpha} \exp(-\mathrm{L}/\mathrm{L}^*)\, \mathrm{dL}/\mathrm{L}^*. \tag{1}$$

Here, $\phi^*$ is a normalization related to the number of luminous galaxies per unit volume; $\mathrm{L}^*$ is a characteristic luminosity; and $\alpha$ represents the shape of the function, which determines the ratio of low luminosity to giant galaxies. The mean luminosity density for the Schechter function is

$$\rho_\mathrm{L} = \int \mathrm{L}\phi(\mathrm{L})\, \mathrm{dL} = \phi^*\mathrm{L}^*\Gamma(\alpha + 2). \tag{2}$$

Assuming that the LF is independent of location and galaxies are distributed homogeneously, the integrated source count to apparent magnitude m in the nonrelativistic case is given by

$$N(<m) \propto d^{*3}(m) \int dL\, \phi(L)(L/L^*)^{3/2} \propto \phi^* L^{*3/2} \Gamma\left(\alpha + \frac{5}{2}\right), \tag{3}$$

where $d^*(m)$ is the limiting depth of the survey for a $L^*$ galaxy.

Equation 3 shows that the count normalization depends not only on $\phi^*$ but also on a good understanding of $L^*$ and $\alpha$. Recent determinations of the local field galaxy LF (Loveday et al 1992, Marzke et al 1994, Lin et al 1996, Zucca et al 1997) are somewhat discrepant in all three parameters. It is also important to determine the LF as a function of morphology or color because the visibility of the different galaxy types is affected by redshift bandpass effects. Color-based local LFs have been discussed by Metcalfe et al (1991) and morphological-based ones by King & Ellis (1985), Loveday et al (1992), and Marzke et al (1994).

The Schechter formula need not be adopted to make predictions. One could simply use the raw luminosity-redshift data for local surveys to predict the appearance of the faint universe. Surprisingly, this has not been done. Indeed growing evidence shows that the simple Schechter expression fails to describe the full observational extent of the local field galaxy LF (Ferguson & McGaugh 1995, Zucca et al 1997). Studies of nearby clusters (Binggeli et al 1988, Bernstein et al 1995), where volume-limited samples can be constructed, also indicate significant departures from the Schechter form at luminosities $M_B > -16 + 5 \log h$ (h denotes Hubble's constant in units of 100 km $s^{-1}$ $Mpc^{-1}$). An upturn at the faint end of the field galaxy LF could make a significant Euclidean contribution (i.e. $N \propto 10^{0.6m}$) to the faint number counts. Marzke et al's (1994) type-dependent LF suggests that such a contribution would also be quite blue.

Given a LF for type j, $\phi(M, j)$, the type-dependent source counts N(m, j) and redshift distributions N(z, j) in flux-limited samples are calculable once the redshift visibility functions are known. In addition to the cosmological distance modulus, the type-dependent k-correction must be determined using the appropriate SED, $f(\nu, j)$. Following Humason et al (1956),

$$k(z, j) = -2.5 \log(1+z) \int f(\nu, j) S(\nu)\, d\nu \Big/ \int f\left(\frac{\nu}{1+z}, j\right) S(\nu)\, d\nu, \tag{4}$$

where $S(\nu)$ is the detector response function.

Progress has been limited in measuring the integrated SEDs of galaxies of different types over a wide wavelength range since the early studies of Pence (1976), Wells (1978), and Coleman et al (1980) used by King & Ellis (1985), because there has been no appropriate UV satellite with which integrated large aperture SEDs can be determined. The lack of reliable k-corrections particularly affects the blue counts because, at high redshift, UV luminosities enter the visible spectral region. Kinney et al (1996) have obtained IUE-aperture spectrophotometry of 15 galaxies from $\lambda\lambda 1200$ Å to 1 $\mu$m (a technique attempted also by Ellis et al 1982) and claim that the small physical size of the areas sampled may nonetheless be useful if the SEDs span the range representative of the integrated galaxy light.

Ideally, the k-correction should be averaged for galaxies of a given rest-frame color rather than binned by morphology, because the correlation between morphology and color is actually quite poor (Huchra 1977) and there are added complications for spheroidal galaxies arising from luminosity dependencies (Sandage & Visvanathan 1978). Recognizing these difficulties, Bruzual & Charlot (1993) and Poggianti (1997) advocated determining the k-correction from model SEDs known to reproduce the integrated broad-band colors of real galaxies. A comprehensive analysis by Bershady (1995) indicates such a technique can match five-color data of a large sample of galaxies with $z < 0.3$, with typical errors of only 0.04 mags. At higher redshifts, alternative routes to the k-correction have included interpolating SEDs from rest-frame colors (Lilly et al 1995) and matching spectroscopic data against local samples (Heyl et al 1997).

To what extent could k-correction uncertainties seriously affect our perceived view of the distant universe? The uncertainty in the optical k-correction for the bulk of the Hubble sequence is probably fairly small to $z = 1$–$1.5$ (Figure 1*a*). However, this conclusion is based on optically selected samples. Although Bershady (1995) found a few galaxies with colors outside the range expected on the basis of his modeling, for high redshift work the test should be extended into the UV. Serious errors of interpretation at fairly modest redshifts could occur if there existed a population of galaxies whose UV-optical colors did not match the model SEDs. Using balloon-borne instrumentation, Donas et al (1995) presented a large aperture UV (2000-Å) optical color distribution of a sample of galaxies selected in the UV; a significant proportion of this sample at all optical colors shows UV excesses compared to the conventional range of SEDs (Figure 2). Redshifts are needed to derive luminosity densities for

→

*Figure 1* (*a*) Type-dependent k-corrections for the B band using two different approaches. *Bold lines* indicate corrections based on model SEDs that reproduce integrated broadband colors (Poggianti 1997); *nonbold lines* indicate corrections derived from aperture spectrophotometry analyzed in conjunction with color data (Kinney et al 1996). (*b*) The k-correction for ellipticals (*solid lines*), Sa (*short-dash*), and Sc (*long-dash*) galaxies for the UBIK photometric bands from the models of Poggianti (1997).

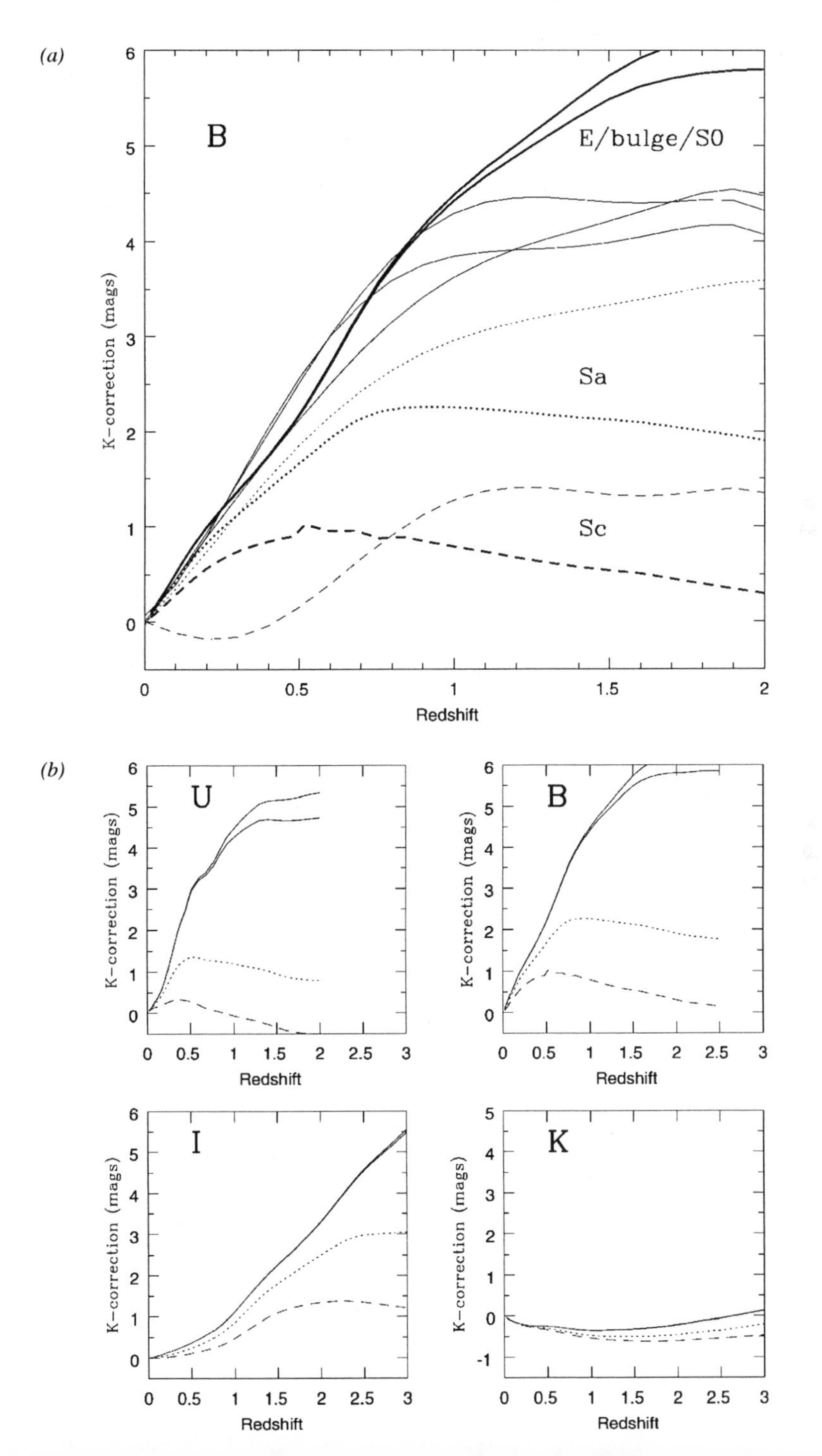
(a)
B
E/bulge/S0
Sa
Sc
K-correction (mags)
Redshift
(b)
U
B
I
K
K-correction (mags)
Redshift

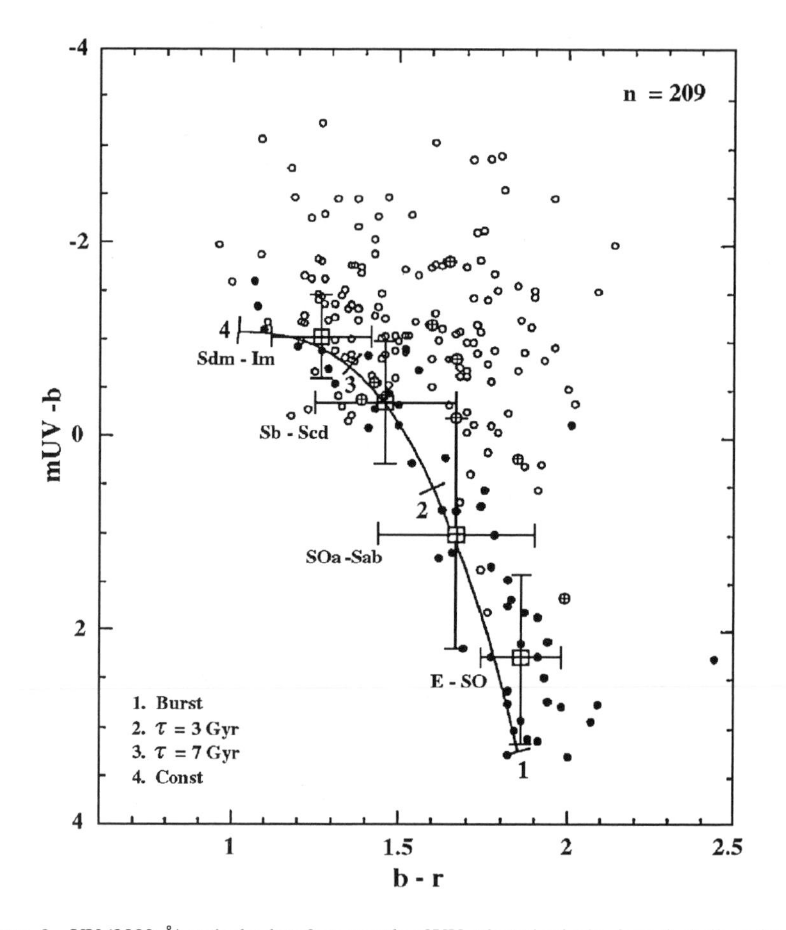


*Figure 2* UV (2000-Å) optical colors for a sample of UV-selected galaxies from the balloon-borne imaging study of Donas et al (1995). A significant proportion of the sample appears to have normal **b-r** colors but UV excesses of 1–2 mag compared to standard models (*solid curve*; Bruzual & Charlot 1993) used to infer k-corrections for the field galaxy population.

these galaxies in order to quantify their possible effect on the k-corrections in use. Such inconsistencies in our understanding of the UV continua of galaxies indicate much work remains to be done in this area and also highlights the continued need for survey facilities at these wavelengths.

Inspection of the k-correction as a function of observed passband (Figure 1*b*) reveals the important point that the variation with type or optical color is considerably reduced in the near-infrared (Bershady 1995). There is also the benefit that, to high redshifts, the required SEDs are based on large aperture optical data. The smaller k-corrections reflect the dominant contribution of red giants at longer wavelengths in old stellar populations, and this has been used to justify a number of faint K-band surveys for evolution and cosmology (Gardner et al 1993, Songaila et al 1994, Glazebrook et al 1995c, Djorgovski et al 1995, Cowie et al 1996), on the grounds that their interpretation is less hampered by k-correction uncertainties. By implication, the observed mixture of types at faint limits is less distorted to star-forming types than is the case in optical surveys. However, if the primary motivation of a faint survey is to learn about the star-formation history of field galaxies, the gains of an infrared-selected survey are somewhat illusory. The insensitivity of the K-band light to young stars means that a deep infrared survey is primarily tracing the mass distribution of distant galaxies. When optical-infrared colors are introduced to aid the interpretation (Cowie et al 1996), the poorly understood UV continua return to plague the analysis.

The cosmological sensitivity that originally motivated the topic of galaxy counts (Hubble & Tolman 1935, Hubble 1936, Sandage 1961) enters through the luminosity distance, $d_L(z)$ (Weinberg 1974, Peebles 1994), and the volume element, $dV(z)$. The apparent magnitude of a faint galaxy of absolute magnitude M is determined by both the luminosity distance, $d_L(z, \Omega_M, \Omega_\Lambda = \Lambda/3H_0^2)$ in megaparsecs (Carroll et al 1992, see equation 25), and the type-dependent k-correction. Because the flux received from an apparent magnitude m scales as $10^{-0.4m}$, the contribution that the differential counts N(m) make to the extragalactic background light (EBL) is obtained via the following (Mattila et al 1991, Vaisanen 1996):

$$I_{EBL} \propto \int N(m) 10^{-0.4m} \, dm. \tag{5}$$

Given the local properties of galaxies, the joint distribution N(m, z, j) can be predicted and compared with faint datasets. As Koo & Kron (1992) discussed, the availability of redshift data enables far more sensitive tests than for photometric data alone. Some information on the redshift distribution of faint sources can be derived from multiband colors. However, redshifts based on colors suffer from similar uncertainties to those discussed above for the k-correction. If local SEDs do not fully span those sampled in faint data, systematic errors may occur. The technique has a long history beginning with Baum (1962), Koo (1985), and Loh & Spillar (1986; who sought to constrain cosmological models via volume tests). In recent years, more sophisticated techniques have been introduced that rely on spectroscopic confirmation for some subset of the sample, on the use of magnitude information to break various degeneracies (Connolly et al 1995), or on strong discontinuities in the spectral energy distributions (Guhathakurta et al 1990, Steidel & Hamilton 1992, Madau et al 1996).

For extremely faint galaxies, color-based redshifts are increasingly used as a way of locating information for sources beyond the spectroscopic limits (see Section 5).

## 2.2 *Evolutionary Predictions*

The no-evolution models defined above have been very useful as a baseline for comparing the predictions and observations made by different workers, but, as the datasets available probe to higher redshift, evolutionary modeling has become increasingly important. These models take as ingredients the stellar evolutionary tracks (normally for a restricted metallicity range), the initial mass function, and a gas-consumption time scale adjusted to give the present range of colors across the Hubble sequence. Assuming each galaxy evolves as an isolated system, the rest-frame SED can be predicted at a given time, and thus an evolutionary correction can be determined with respect to the no-evolution equations above, or the predictions can be incorporated in ab initio models to generate simulated data sets. The cosmological model is a crucial, but often overlooked, variable in linking time and redshift. For $H_0 = 70$ and $\Lambda = 0$, the redshift corresponding to a look-back time of, say, 7 Gyr, varies from $z = 1$–$3$ depending on $\Omega$.

A discussion of the reliability of these evolutionary predictions and the differences between the various approaches is beyond the scope of this review. Physical difficulties arise from the degenerate effects of age and metallicity and the uncertainties of post–main sequence stellar evolution. Results for various models have been intercompared by Mazzei et al (1992), Bruzual & Charlot (1993), and, most recently for populations of the same input age and metallicity, by Charlot et al (1996). Recent efforts have concentrated on improving the stellar tracks and spectral libraries (Bruzual & Charlot 1993) and including the effects of chemical evolution (Arimoto et al 1992). An all-inclusive database of progress in this area is presented by Leitherer et al (1996).

For single burst populations presumed appropriate for early-type galaxies, Charlot et al (1996) discuss surprisingly large discrepancies in the predicted behavior of such populations at times after 1 Gyr. The differences amount to at least 0.03 mag in rest-frame B-V, 0.13 mag in rest-frame V-K, and a 25% dispersion in the V-band mass/light ratio. The large uncertainties in the predicted optical-infrared colors and visual luminosity evolution imply a significant age range (4–13 Gyr) that is permissible even for the simplest case of a passively evolving red galaxy, emphasizing the continuing need to compare these models with representative high redshift data, as well as the need to improve our knowledge of post–main sequence stellar evolution.

For populations with constant star formation, both the evolutionary corrections and the discrepancies between the available models are less. This is because the same main sequence stellar types dominate the spectra at most times and their theoretical behavior is considerably better understood. Unfortunately, precise predictions are required for both types of model galaxy (as well as the large range in between) because faint blue galaxies could be either passively evolving systems seen at an early stage, systems of constant star formation, or bursts of star formation imposed on a quiescent system. A comparison of the predicted evolutionary behavior for two of these cases is shown in Figure 3. Such uncertainties represent a formidable obstacle to detailed modeling in the ab initio approach.

## 2.3 *Images from Hubble Space Telescope*

Since the review by Koo & Kron (1992), multicolor HST images have become available for galaxies to the limits achievable with ground-based spectroscopy and beyond. The images are spectacular and provide morphological data for large numbers of distant galaxies, a growing number of which have redshifts. The potential for directly tracking the evolution of the Hubble sequence of types is clearly an exciting opportunity. However, interpretation of the HST data in this way has not been straightforward (cf discussion by Abraham et al 1996a,b). Although many workers have concluded that the faint field population is dominated by "irregular" and "merging" galaxies (Glazebrook et al 1995b, Driver et al 1995b), both the reliability of this result and a physical understanding of what it signifies require careful study. When comparing distant images with possible local counterparts, the reduced signal-to-noise ratio, relative increase in background, and cosmological surface brightness dimming tend to accentuate the visibility of high contrast features. Furthermore, bandpass shifting effects arising from the k-correction mean that low redshift optical images are being compared with high redshift UV images (O'Connell & Marcum 1996). An illustration of these effects is given in Figure 4, following the work described by Abraham et al (1996b). Such studies indicate that, although the broad classification system (E/Spiral/Irr) is reliable to $z \approx 1$ for most large regular galaxies, any unfamiliarity in the images of higher z systems may be spurious.

The concept of comparing HST images of galaxies grouped by class at different redshifts raises a very important problem. Most of the above discussion has centered on statistical results from surveys. By examining the entire field population accessible at each redshift, robust statements are made about the global changes in, for example, the galaxy LF and volume-averaged star-formation rates (SFRs). Yet, to physically understand these changes in terms of the elaborate models now available, we are encouraged to divide the samples into "subclasses" in the hope of determining the evolutionary behavior as a function of type. This can only be done if there is some fundamental slowly changing property of a galaxy that can act as a label. Clearly, both color and spectral

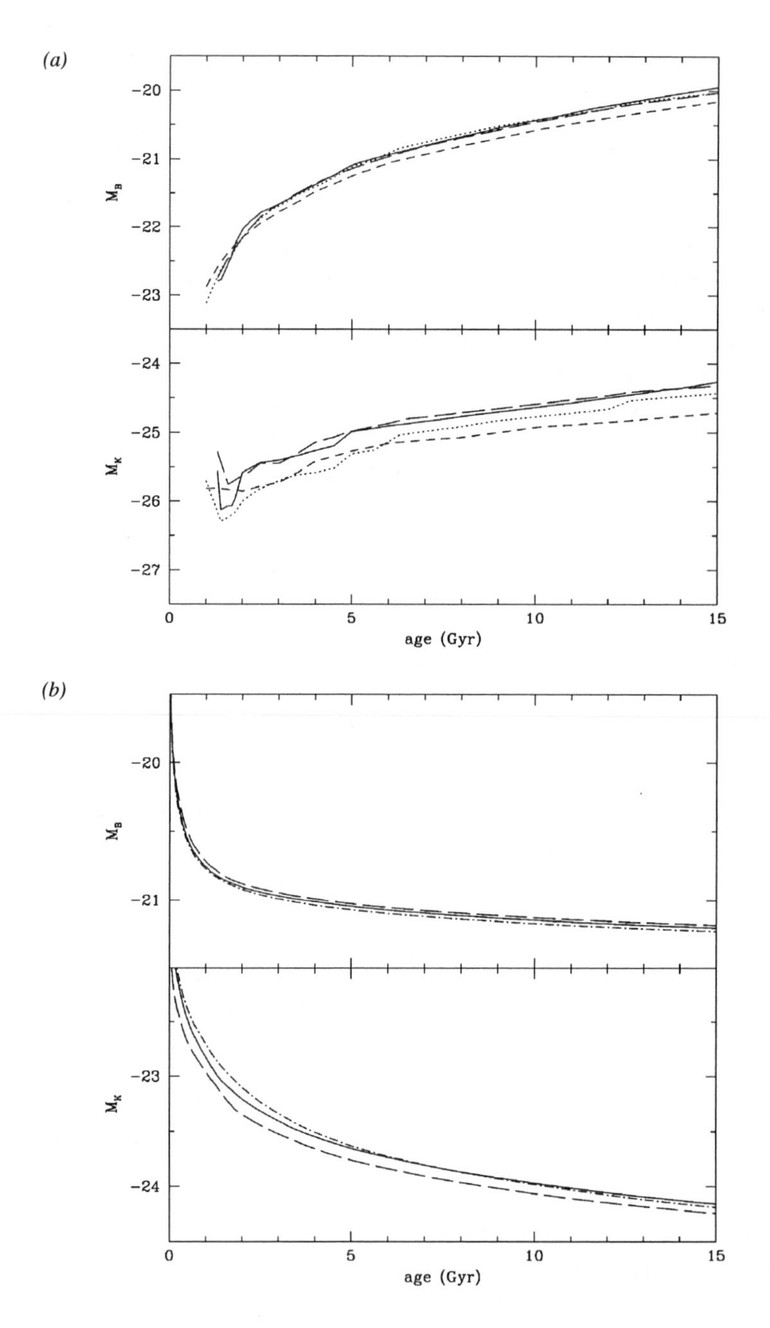

(a)
-20
-21
-22
-23
$M_B$
-24
-25
-26
-27
$M_K$
0
5
10
15
age (Gyr)
(b)
-20
-21
$M_B$
-23
-24
$M_K$
0
5
10
15
age (Gyr)

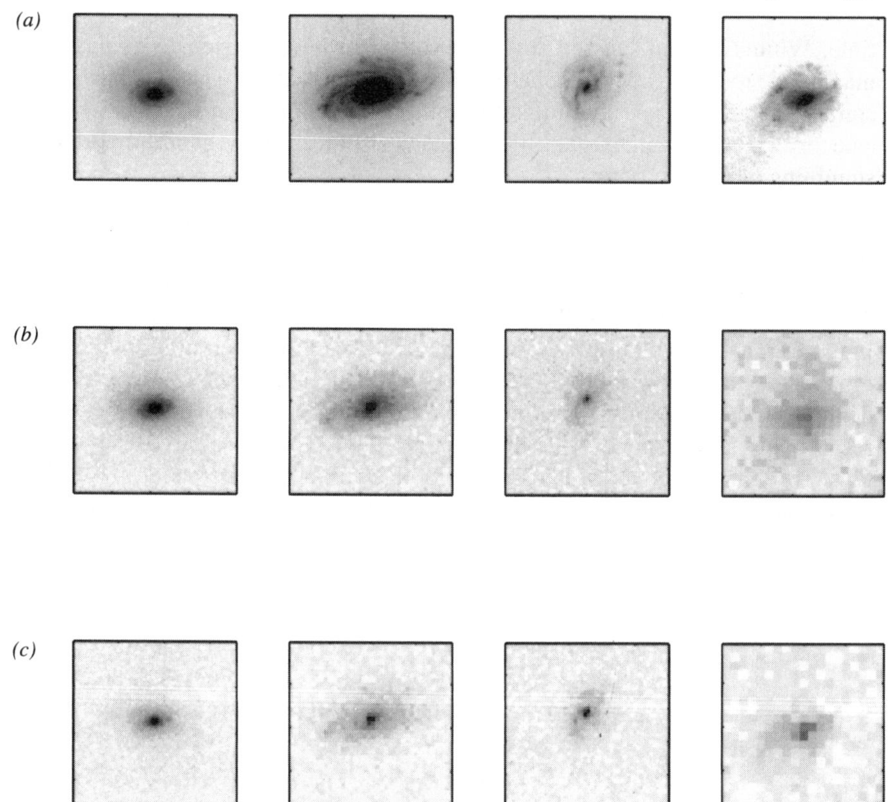


*Figure 4* Simulated appearance of NGC 4450 (Sab), 3953 (Sbc), 5669 (Scd), and 4242 (Sdm) at various redshifts. With the exception of NGC 4242, which is subluminous, all galaxies have L $\approx$ $L^*$. From top to bottom the panels show (*a*) the rest-frame B band images, (*b*) the appearance as viewed at z = 0.75 in a three-orbit F814W WFPC-2 image, and (*c*) the appearance at z = 1.5 in an exposure equivalent to the HDF. (Courtesy of R Abraham.)

class (cf Lilly et al 1995, Heyl et al 1997) could be transient properties affected by short-term changes in star formation induced, for example, by merging or other processes. Likewise galaxy morphology, even if correctly assigned in HST images that sample a range of rest-frame wavelengths, may change with

⟵

*Figure 3* Evolution of the rest-frame B and K band luminosity with time, according to recent models by Bruzual & Charlot (1993; *solid line*), Bertelli et al (1994; *dotted*), and Worthey (1994; *short-dash*); long dash represents a Bruzual & Charlot model based on alternative evolutionary stellar tracks (see Charlot et al 1996 for details). (*a*) Comparison for a passively evolving galaxy following an instantaneous burst with solar metallicity (see Charlot et al 1996 for details); (*b*) comparison for a constant star formation model (see Charlot 1996 for details).

time. White (1996) discussed physical situations where gas-rich disk galaxies may merge to form gas-poor spheroidals, only to accrete more gas and become spirals again. The possible migration of galaxies in and out of a faint blue category by some unidentified time-dependent process may be an important stumbling block to progress.

# 3. RECENT OBSERVATIONAL RESULTS

## 3.1 *Number Magnitude Counts*

Large format CCD and infrared-sensitive detectors have been used to extend ground-based number counts since Koo & Kron's (1992) review. The greater depth now available has confirmed a break in the count slope that occurs, respectively, at around B $\approx$ 25 (Lilly et al 1991, Metcalfe et al 1995b) and K $\approx$ 18 (Gardner et al 1993, Djorgovski et al 1995, Moustakas et al 1997) shown in Figure 5*a*. The slope, $\gamma$ = dlogN/dm, ranges from $\gamma_B$ = 0.47 to

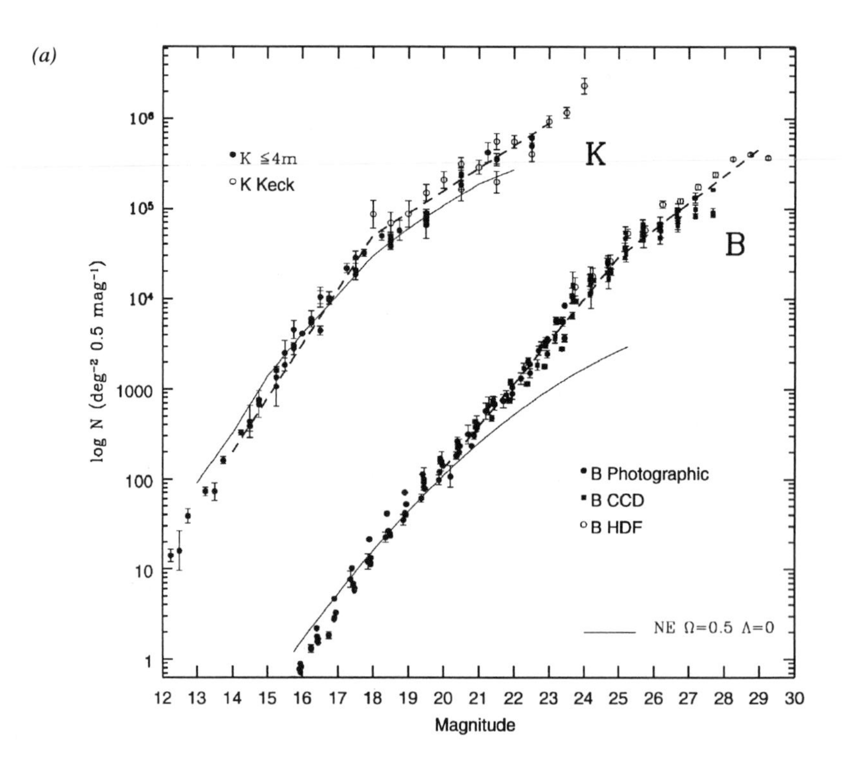


*Figure 5* (*continued*)

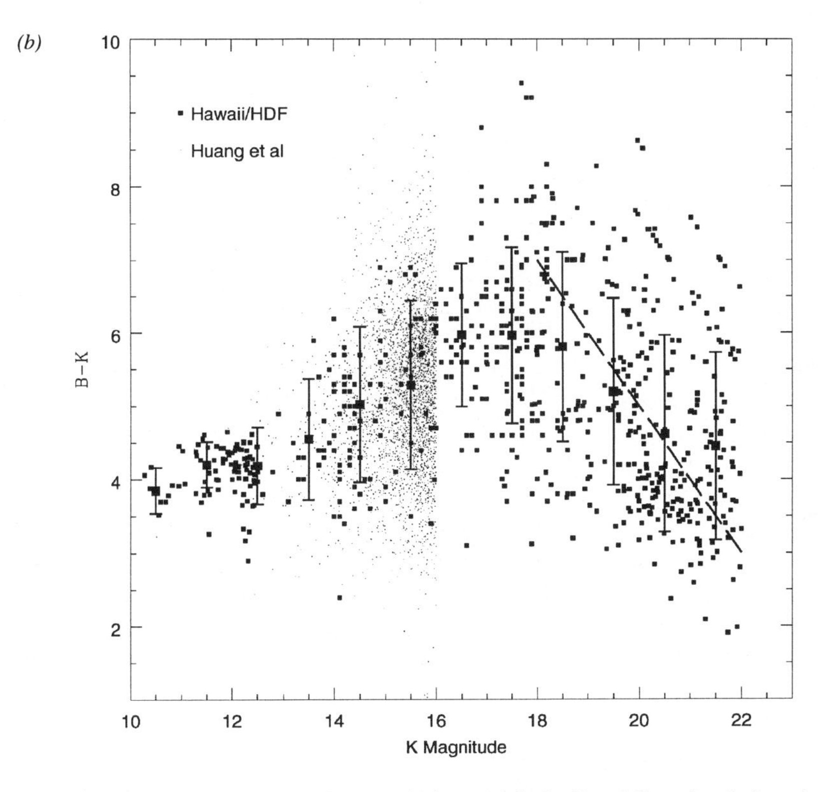


*Figure 5* (*a*) Differential galaxy number magnitude counts in the B and K passbands from the compilation of Metcalfe et al (1996) augmented with the Keck K counts of Moustakas et al (1997). The K counts have been offset by +1 dex for clarity. The two power law slopes (*dashed lines*) drawn have $\gamma$ (= dlog N/dm) = 0.47,0.30 around B = 25 (Metcalfe et al 1995b) and 0.60,0.25 around K = 18 (Gardner et al 1993). The *solid curve* indicates a no-evolution prediction for an Einstein–de Sitter universe based on King & Ellis' (1985) k-corrections and luminosity functions from local surveys with a normalization raised by 50% (see Section 4 for discussion). Beyond the limits drawn, the no-evolution prediction ceases to be reliable. (*b*) B-K versus K for galaxies from the Hawaii Bright Survey (*fine dots*, Huang et al 1996), the Hawaii Deep Surveys (Cowie et al 1996 and references therein), and HDF (Cowie 1996). *Points with error bars* refer to mean colors for various magnitude slices. The *dashed line* indicates the current limit for optical spectroscopy.

0.30 and from $\gamma_K = 0.60$ to 0.25, but the surface density at the B break is ≈30 times higher than that at the K break. If the two effects were manifestations of the same phenomenon, e.g. a decline in volume density beyond some redshift limit, then the mean B-K color should not change significantly across the break points. In contrast, the change in slope is accompanied by a marked increase in the number of galaxies with colors B-K < 5, which explains the steeper $\gamma_B$ at fainter limits. Such blue galaxies were originally referred to as flat spectrum galaxies by Cowie et al (1989) because their SEDs approximate ones with constant f($\nu$).

The break in slope and, more importantly, the disparate behavior between the B and K counts with respect to the no-evolution predictions makes it unlikely that a major portion of the excess counts arises via a nonzero cosmological constant, $\Lambda$, as postulated by Fukugita et al (1990) and Yoshii & Peterson (1991) (see Carroll et al 1992 for a full discussion). That such a dramatic excess should be seen in B but not K could only be consistent with the hypothesis of a nonzero cosmological constant if the B $>$ 25 sources were significantly more distant than the K $>$ 19 ones, which seems unlikely (Section 5). Note again that the K counts, by virtue of their insensitivity to the k-correction (Figure 1*b*), remain an interesting cosmological probe, although a satisfactory conclusion concerning $\Omega$ and $\Lambda$ will remain elusive so long as the evolutionary behavior is poorly determined (Djorgovski et al 1995).

The break in slope more probably signals a transition in galaxy properties at some redshift, with the bulk of the fainter sources being drawn from intrinsically less luminous sources at similar redshifts, a conclusion favored by Lilly et al (1991), Gardner et al (1993), and Metcalfe et al (1995b). In this case, the abundance of fainter blue sources would suggest a high volume density of low mass galaxies and the slope of the B and K counts would directly reflect their relative contribution to the luminosity function at that time. Indeed, from Equation 5, clearly the break indicates the apparent magnitude beyond which the contribution of galaxies to the extragalactic background begins to converge. Specifically, in the B band that samples the rest-frame UV at the appropriate redshifts, the location of the break defines, albeit qualitatively, that era in which galaxies contribute most in terms of short-term star formation and associated metal production (Songaila et al 1990, Lilly et al 1996; Section 6).

### 3.2 *Faint Redshift Surveys*

Ground-based redshift surveys have not yet convincingly penetrated past the break point in the optical counts to demonstrate the suggestions above. The bulk of the published work has been concerned with tracking the evolution of the galaxy LF up to B $=$ 24, I $=$ 22, and K $=$ 18; these are the effective limits in reasonable integration times on 4-m class telescopes. Although redshifts are available beyond these limits from the first Keck exposures (Koo et al 1996, Illingworth et al 1996), with the exception of the Cowie et al (1996, 1997) surveys, these do not yet constitute a controlled magnitude-limited sample. A summary of the published (and where known, unpublished) faint field redshift data is given in Table 1. The Kitt Peak survey discussed provisionally by Koo & Kron in their 1992 review is to be published in a series of forthcoming papers commencing with Munn et al (1997).

**Table 1** Recent faint galaxy redshift surveys

| Reference | Survey[1] | Magnitude range | Redshifts |
|---|---|---|---|
| Lilly et al (1995) | CFRS | $17.5 < I < 22$ | 591 |
| Ellis et al (1996a) | Autofib/LDSS | $11.5 < b_J < 24$ | 1726 |
| Cowie et al (1996) | Keck LRIS | $14 < K < 20$ | 346 |
| | | $I < 23$ | 287 |
| Cowie et al (1997) | Keck LRIS | $B < 24.5$ | 166 |
| Lin et al (1997) | CNOC1 | $r < 22$ | 389 |
| Koo et al (1996), Illingworth et al (1996) | Keck DEEP | $I < 25$ | ≈115 |
| Munn et al (1997) | KPGRS | $R < 20+$ | 739 |

[1]CFRS, Canada France Redshift Survey (CFHT); Autofib, multi-fiber spectrograph (AAT); LDSS, Low Dispersion Survey Spectrograph (AAT/WHT); CNOC1, Canadian Network for Observational Cosmology (field component of cluster survey, CFHT); KPGRS, Kitt Peak Galaxy Redshift Survey; DEEP, Deep Extragalactic Evolutionary Probe (Keck).

The survey data suggest a mean redshift $z \approx 0.8$ at the break point $B = 25$, although no complete survey this faint yet exists. The surveys are still incomplete at some level beyond $B \approx 23$ and $I \approx 21$, and this can seriously affect the inferred redshift distribution. There is, for example, a significant difference between the faint $B < 24$ LDSS-2 redshift distribution (Glazebrook et al 1995a) used by Ellis et al (1996a) and that presented by Cowie et al (1996), arising presumably from a loss of higher redshift galaxies whose [O II] emission was redshifted out of the LDSS-2 spectroscopic window. Such incompleteness will remain a concern until infrared spectrographs are available that can systematically track [O II] and H$\alpha$ emission to $z \approx 2$ (Ellis 1996b).

The most significant results from the redshift surveys are the redshift and type-dependent LFs. Furthermore, provided the spectroscopic surveys sample representative regions that are sufficiently faint to probe the apparent excess population, physical models that explain the counts to fainter apparent magnitude limits can also be tested. By combining data that sample a wide range of apparent magnitude, a range in luminosity is available for $z < 0.5$ [for the B-selected survey of Ellis et al (1996a)] and $z < 1$ [for I- and K-selected surveys of Lilly et al (1995) and Cowie et al (1996)]. The LFs so derived indicate little change in the volume density of luminous galaxies to $-z = 1$, whereas less luminous galaxies appear to evolve more rapidly. Ellis et al (1996a) presented evidence for a steepening with redshift of the faint end slope of the overall LF consistent with a rising contribution of less luminous galaxies. The effect is particularly strong when parameterized according to [O II] emission line strength (Ellis et al 1996a) and UV-optical color (Cowie et al 1996). The bulk of the excess population to B $=$ 24 compared to the no-evolution predictions (Figure 5*a*) may be largely due to this phenomenon (Glazebrook et al 1995a).

Such luminosity-dependent evolution was originally proposed by Broadhurst et al (1988) to reconcile their redshift survey with photometric data at brighter limits (B $<$ 21.5). However, strong evolution at these limits, corresponding to $z \approx 0.25$, now seems less likely because of growing evidence for a higher normalization of the local counts. The early B-selected surveys (Broadhurst et al 1988, Colless et al 1990, 1993) used the Durham-AAT Redshift Survey (DARS; Peterson et al 1986, Efstathiou et al 1988) and APM-Stromlo (Loveday et al 1992) data sets as local benchmarks. As Ellis et al (1996a) discuss, the low-z LF derived from the fainter surveys indicates a higher normalization. An upward revision, consistent with the galaxy counts at B $\approx$ 19, would reduce the original dilemma presented by Broadhurst et al and Colless et al, whereby excess galaxies are seen in the counts but within the same redshift range as expected in the no-evolution case. The implication of this change may be that some local data is unrepresentative, possibly because of photometric difficulties at bright apparent magnitudes or because the survey regions used are deficient in galaxies for some reason. The effect is apparent also in the source counts (Maddox et al 1990).

Given the uncertainties in the normalization of the local benchmark samples, Lilly et al (1995) argue that greater reliance should be placed on evolutionary trends determined internally from self-consistent data sets. The CFRS survey has the considerable advantage of probing $0.2 < z < 1$ on the basis of a single well-defined photometric scale. The limited apparent magnitude range sampled offers less conclusive results on possible evolution in the shape of the LF (Figure 6*a*), but the luminosity-dependent trends are similar to those in the other surveys. Whether the shape is changing or the overall LF is brightening can be judged to a limited extent by examining the contrasting behavior of the CFRS and Autofib/LDSS LFs in the redshift ranges where both have reasonable sample sizes (Figure 6*b*). The CFRS results do admit some luminosity evolution at the bright end, whereas the changes in the Autofib/LDSS LF occur solely for the less luminous objects.

The underlying trend supported by all the major redshift surveys is a marked increase in the volume density of star-forming galaxies with redshift. That this evolution occurs primarily in subluminous galaxies is less clear. Lilly et al (1995) support the distinction in the evolutionary behavior of galaxies redder and bluer than the rest-frame color of a typical Sbc spiral. Cowie et al (1996) conclude the evolution proceeds according to a "down-slicing" trend in progressively less massive systems. Ellis et al (1996a) propose a two-component "passively evolving giant plus rapidly evolving dwarf model" as the simplest empirical description of their data.

## 3.3 *Other Probes of Luminosity Evolution*

One of the most satisfactory developments in galaxy evolution has been the verification of the redshift survey trends from completely independent approaches and, in particular, from the study of QSO absorbers. A redshift survey of 55 galaxies with $0.3 < z < 1$ producing Mg II absorption in the spectra of background QSOs indicates little change in the volume density, characteristic luminosity, or rest-frame colors of typical $L^*$ galaxies with established metallic halos (Steidel et al 1994). Of importance, there appears a marked distinction between these systems and the rapidly evolving later-type galaxies that do not

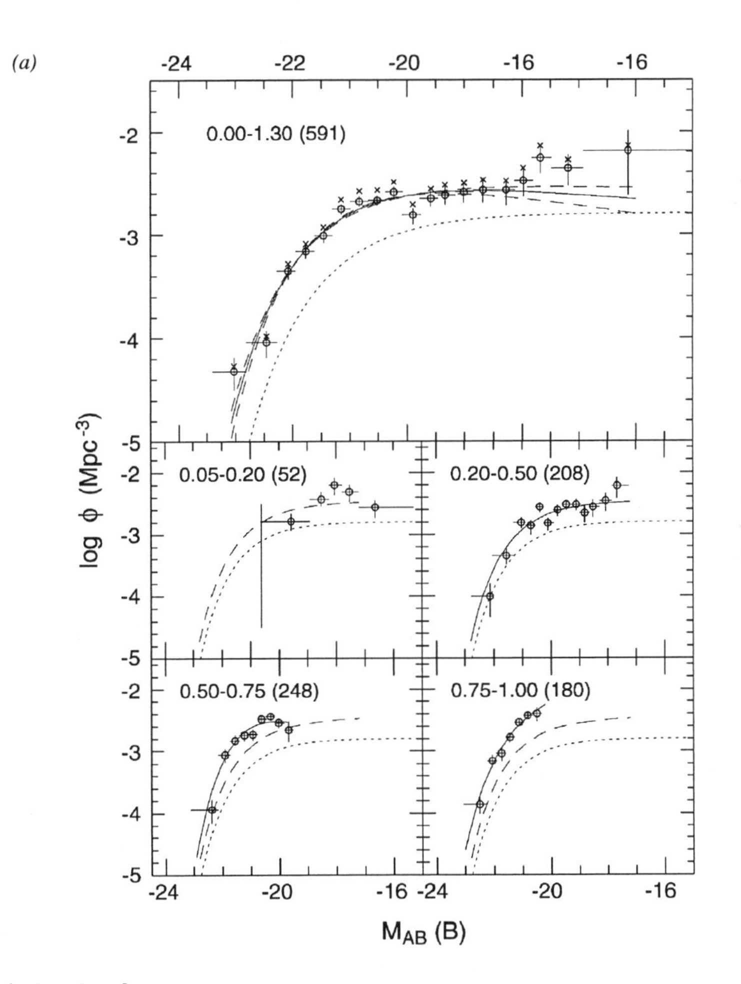


*Figure 6 (continued)*

*(b)*

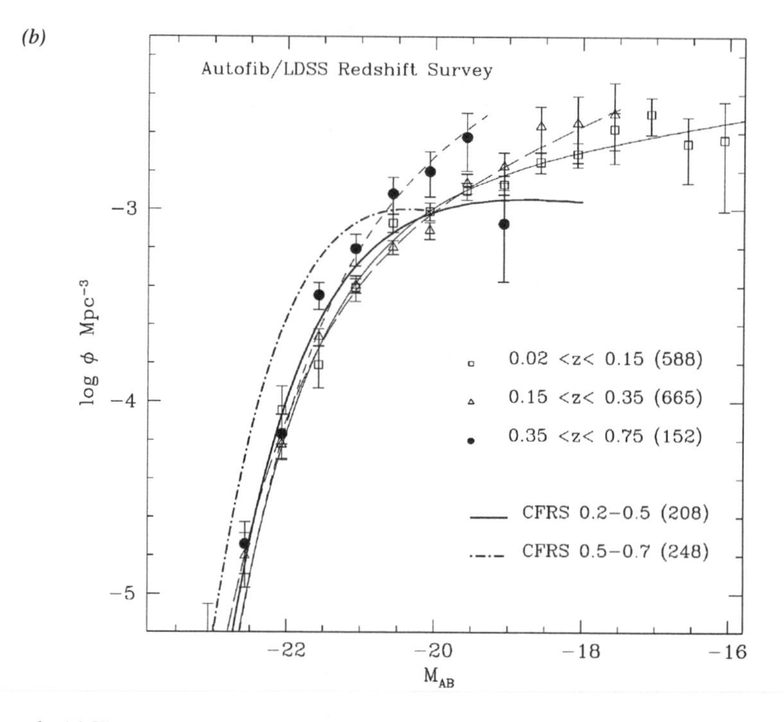


*Figure 6* (*a*) The rest-frame B(AB) luminosity function for various redshift ranges derived from the CFRS (Lilly et al 1995); numbers in parentheses denote sample size. *Solid* and *dashed curves* represent fitted Schechter functions (best fit and 1-$\sigma$ fits); the *dotted curve* represents Loveday et al's (1992) local LF. In each panel where $z > 0.2$, the *dashed curve* represents the fit for $0.2 < z < 0.5$. With a single photometric scale, clear evidence of luminosity evolution is apparent. (*b*) Comparison of the Autofib/LDSS (Ellis et al 1996a) and CFRS LFs in the range $0.2<z<0.7$ in the CFRS B(AB) system. For convenience, both panels adopt $H_0 = 50$ km s$^{-1}$ Mpc$^{-1}$ and $\Omega = 1$, as in Lilly et al (1995). Numbers denote sample sizes in the appropriate redshift ranges for the two surveys. The CFRS results admit modest evolution for luminous galaxies, whereas the Autofib/LDSS results suggest significant changes occur only for $L < L^*$ galaxies.

seem capable of producing significant Mg II absorption (Steidel et al 1993, 1996a). Although the straightforward physical interpretation has been questioned by Charlton & Churchill (1996) because of a poor correlation between the galaxy-QSO impact parameter and the Mg II absorption structure (Churchill et al 1996), and the paucity of nonabsorbing galaxies found at small impact parameters, these problems may relate to the difficulty of extrapolating from kinematics on small scales to the large scale properties of the absorbing galaxies.

Progress has also been made in studying possible evolution in the surface photometry and dynamics of various classes of distant galaxies. The crucial point in these studies is that evolution is normally inferred via a comparison with local data. Consequently it is important to understand exactly how the distant samples are selected. On the basis of luminosities and scale lengths for a magnitude-limited sample of galaxies selected to have faint bulge components drawn largely from the CFRS, Schade et al (1996) claimed a mean rest-frame B-disk brightening of $\approx$1 mag, compared to Freeman's (1970) law by $z = 0.55$. Although this seems difficult to reconcile with LF studies from the same survey (Figure 6*a,b*), it should be noted the LF trends are $\Omega$-dependent, whereas the surface brightness test is not. Moreover, Schade et al applied their test globally to all systems with weak bulges, and presumably this incorporated the rapidly evolving lower luminosity systems discussed in Section 3.2.

Vogt et al (1996) obtained spatially resolved rotation curves for nine galaxies imaged with HST and, using the Tully-Fisher relation, obtained more modest changes of $\approx$0.6 mag to $z \approx 1$ but warn of possible biases toward more luminous, larger star-forming galaxies at high z. Rix et al (1997) and Simard & Pritchet (1997) sought to address the question of whether a faint blue galaxy is rotationally supported and thus selected the bluest or strongest emission line sources at $z = 0.25$–$0.45$, considered to be typical of the excess population. Rix et al measured line widths indicative of rotational support and, as with Simard & Pritchet, claimed to demonstrate that their galaxies are at least 1.5-mag brighter than expected from the local Tully-Fisher relation. The difficulty is finding an appropriate local calibration for these galaxies. Rix et al presented various alternatives, but more work is needed to define a self-consistent dynamic data set over a range in redshift in conjunction with HST images.

## 3.4 *Imaging Surveys with Hubble Space Telescope*

Given the heroic efforts to resolve the faint blue population from ground-based telescopes (Giraud 1992, Colless et al 1994), one might imagine rapid progress in understanding their nature would be possible from the first refurbished HST images that became available in 1994. In practice, interpretation of the HST data has been hindered by a number of factors introduced in Section 2. The following discussion concentrates on post-refurbishment data. Discussion of the HDF is below in Section 5.

Two survey techniques have been used to image the faint population. The MDS (Griffiths et al 1994) has utilized parallel WFPC-2 data that sample random high-latitude fields in F785W or F814W and, where possible, F555W. The typical primary exposure time has seriously limited long parallel exposures and thus the bulk of the useful data comes from exposures of $\approx$1 h in F814W. The primary advantage of the MDS is the total survey field area. Morphological data and image parameters have been presented to $I_{814} = 22$–$23$ (Glazebrook et al 1995b, Driver et al 1995a) for over 300 sources to $I_{814} = 22$ within 13

WFPC-2 fields (0.02 $deg^2$). Driver et al (1995b) also analyze 227 galaxies from a single deeper pointed exposure of 5.7 h to $I_{814} = 24.5$. The main disadvantage of the MDS is that the small field of view of each WFPC-2 image is poorly matched to ground-based multiobject spectrographs, and thus follow-up redshift work is rather inefficient. The MDS survey is thus best viewed as a deep 2-D survey of the faint sky. Windhorst et al (1996) provides a good summary of the overall results. The alternative technique discussed by Ellis (1995), Schade et al (1995), Cowie et al (1995a,b), Koo et al (1996), and LeFevre et al (1996a) involves taking HST images in primary mode of ground-based redshift survey fields. By arranging WFPC-2 exposures in a contiguous strip, an effective match is obtained with existing redshift data. A variant here is the exploitation of the "Groth strip," a GTO exposure in F606W and F814W that consists of 28 overlapping WFPC-2 fields (Koo et al 1996).

Glazebrook et al (1995b) and Driver et al (1995a,b) have classified the MDS and related samples visually into spheroidal/compact, spiral, and "irregular/peculiar/merger" categories. They claimed the number of regular galaxies (spheroidals and spirals) to I = 22–24.5 is approximately as expected on the basis of the local passively evolving populations, whereas the irregular/peculiar/merger population is considerably in excess of expectations with a much steeper count slope. Abraham et al (1996b) derived the same conclusion from an automated treatment of image morphology based on the concentration of the galaxy light (which correlates closely with the bulge/disk ratio) and the asymmetry (used to locate irregulars). Odewahn et al (1996) investigate the use of artifical neural networks to classify the data and reach similar conclusions. Note that the latter techniques do not yet, as presented, include corrections for bandpass shifting biases. In each of these studies, galaxies were located after smoothing the MDS images to ground-based resolution to avoid double-counting close pairs of galaxies, and the conclusions were based on the increased normalization of the LF discussed earlier.

It is tempting to connect the rapidly evolving blue galaxies in the redshift surveys with the irregular/peculiar/merger systems seen in the MDS data (Ellis 1995), but central to the MDS results is the physical interpretation of the irregular/peculiar/merger class. Could this category of objects not simply be an increasing proportion of sources rendered unfamiliar by redshift or other effects? A convincing demonstration that this is not the case requires a detailed analysis of the inferred morphology, when viewed at likely redshifts, of a representative multicolor CCD sample of nearby galaxies of known type. Although the machinery is available to conduct such simulations (Abraham et al 1996b) (Figure 4), it is not yet clear whether the available local samples are properly represented in all classes. However, thus far it seems unlikely such a large bias could occur within the redshift range appropriate for $I < 22$ samples as defined from the CFRS, because the F814W images correspond typically to rest-frame B. However, the precise distinction between late-type spiral and irregular/peculiar/merger may remain uncertain (Ellis 1996a).

Within the limited sample sizes available through HST images of the redshift survey galaxies, the identification of such a high fraction of irregular galaxies is less clear. Schade et al (1995) studied 32 CFRS galaxies with $z > 0.5$ using WFPC-2 images in F450W and F814W and commented on the high proportion (30%) of blue nucleated galaxies. Many appear asymmetric and some shows signs of interaction. Cowie et al (1995a,b) discussed the unfamiliar nature of their distant sample and, for the faintest bluest sources with $z > 1$, introduced the terminology of chain galaxies—multiple systems apparently merging along one dimension (but see Dalcanton & Schectman 1996). The varied interpretation of these HST images again underlines the difficulty of connecting high redshift morphology with local ground-based data. Possibly the most robust conclusion so far is that the bulk of the faint blue population are galaxies comparable in size to the remainder of the population, a result that is apparent in earlier ground-based efforts (Colless et al 1994). Few are truly compact systems of the kind selected for detailed study by Koo et al (1995) and Guzman et al (1996).

### 3.5 *Evolutionary Constraints from Galaxy Correlation Functions*

The analysis of positional data in deep galaxy catalogs in terms of 2-D angular correlations or 3-D redshift space correlations is primarily useful as a probe of the evolution of clustering with look-back time (Phillipps et al 1978). The role of correlation functions in constraining galaxy evolution, as stressed by Koo & Szalay (1984), has, to a large extent, been overtaken by the redshift surveys, which give a much clearer and less ambiguous indicator of luminosity changes. Nonetheless, a bewildering amount of 2-D data has been analyzed in recent years, mostly from panoramic photographic and CCD-based surveys in various bands, and the results arising from this work have featured prominently in the debate on faint blue galaxies.

The angular correlation function, $w(\theta)$, of faint galaxies is linked to its spatial equivalent, $\xi(r)$, by terms that depend on the redshift distribution (see Peebles 1994 for definitions). The connecting relationship (Limber 1953, Phillipps et al 1978) takes into account both the angular diameter distance, $d_A = d_L/(1+z)^2$, and dilution from uncorrelated pairs distributed along the line of sight. In practice, as $\xi(r)$ is locally type-dependent (Davis & Geller 1976), $w(\theta)$ is weighted by the type-dependent redshift distribution N(z, j) determined by the k-correction and evolutionary behavior of each type. A further factor is the likely evolution in spatial clustering conventionally parameterized in proper space as

$$\xi(r, z) = (r/r_o)^{-\gamma}(1+z)^{-(3+\varepsilon)}, \tag{6}$$

where $r_o$ (= 5 $h^{-1}$ Mpc; Peebles 1980) is the current scale length of galaxy clustering and $\varepsilon = 0$ corresponds to clustering fixed in proper coordinates. For a correlation function $\xi(r) = (r/r_o)^{-1.8}$, clustering fixed in comoving coordinates yields $\varepsilon = -1.2$, whereas linear theory with $\Omega = 1$ indicates growth equivalent to $\varepsilon = +0.8$ (Peacock 1997).

Broadly speaking, the angular correlation function results for the faint surveys are consistent with a decline in the spatial correlation function with increasing redshift (Roche et al 1993, 1996b, Infante & Pritchet 1995, Brainerd et al 1995, Hudon & Lilly 1996, Villumsen et al 1996), but any quantitative interpretation that would allow statements to be made on luminosity evolution would have to make careful allowance for the change in apparent mix of types with redshift. Because of the morphology-density relation, the k-correction works in the direction of suppressing the visibility of correlated spheroidal galaxies, and therefore an apparent decrease in the amplitude of clustering is inevitable. Using color-selected subsamples, some workers (Efstathiou et al 1991, Neuschaefer et al 1991, Neuschaefer & Windhorst 1995) have addressed this issue and claimed that a large fraction of the faintest blue samples must be weakly correlated unless they are at very high redshift ($z > 3$). The surprisingly large decrease in angular clustering with apparent magnitude has even led to the suggestion that the bulk of the faintest sources represent a population unrelated to normal galaxies (Efstathiou 1995).

The dramatic evolution claimed from the 2-D data sets contrasts somewhat with those studies based on 3-D data from the redshift surveys. Cole et al (1994b) derived spatial correlation functions from the Autofib redshift survey, and Bernstein et al (1994) analyzed panoramic 2-D data within magnitude limits where the redshift range is likewise known. Neither found significant changes in the clustering scale to $z = 0.3$ other than can be accounted for if the bluer star-forming galaxies were somewhat less clustered, as seen locally in IRAS-selected samples. Estimates of spatial clustering at higher redshift in the CFRS redshift surveys give rather uncertain, though modest, growth estimates ($0 < \varepsilon < 2$) (LeFevre et al 1996b). The results from the deeper Keck LRIS survey (Carlberg et al 1997) also indicate modest evolution (a 60% decrease in $r_o$ from $z = 0.6$ to 1.1) with a possible strong cross-correlation between high and low luminosity galaxies on 100 $h^{-1}$ kpc scales.

The apparent conflict between the weak angular clustering of faint sources and the modest decline in the spatial clustering to $z = 1$ might be understood if an increasing proportion of the detected population is (*a*) less clustered and (*b*) itself evolving at a physically reasonable rate. Villumsen et al (1997) argue that over $20 < R < 29$, $w(\theta)$ can be explained via the linear growth of a component whose present correlation length is comparable to that of local IRAS galaxies ($r_o \approx 4\ h^{-1}$ Mpc). Smaller correlation lengths ($\approx 2\ h^{-1}$ Mpc) are claimed by Brainerd et al (1995), but the differences may relate to the adopted redshift distributions for the faint populations as well as possible sampling variations in the small fields available to date. Peacock (1997) concludes the evolution in $\xi(r)$ seen by LeFevre et al can be simply interpreted via a single population of sources evolving with $\varepsilon \approx 1$—corresponding to $\Omega \approx 0.3$. That such diverse conclusions can emerge from these basic observational trends gives some indication of the degeneracies involved in analyzing the data.

## 4. ADDRESSING THE MAIN ISSUES

The aim in this section is to distill the observational phenomena discussed in Section 3 into a few key physical issues that will assist in understanding the evolutionary role of faint blue galaxies. As the greatest progress has been made in the redshift interval to $z = 1$, the discussion of fainter samples and the HDF is deferred until Section 5.

### 4.1 *Uncertainties in the Local Field Galaxy Luminosity Function*

The nature of the local LF was comprehensively reviewed by Bingelli et al (1988). Since then, considerable progress has been achieved through deeper, more representative local optical and infrared surveys, such as the the Stromlo-APM (Loveday et al 1992), Las Campanas (Lin et al 1996), CfA (Marzke et al 1994), SSRS2 (da Costa et al 1994), and DARS (Peterson et al 1986, Efstathiou et al 1988, Mobasher et al 1993) redshift surveys. These are utilized both as photometric databases for normalizing the galaxy counts at the bright end (Maddox et al 1990) and in deriving LFs as a function of type. Colless (1997) provides a recent summary of local LF determinations (Figure 7) to which is added the recent LF analysis of the ESO Slice Project (ESP) (Zucca et al 1997), a survey of 3342 galaxies to $b_J = 19.4$.

The local normalization depends on a combination of the LF parameters (not just $\phi^*$, see Equation 3), and its value is central to the question of whether there is significant evolution in galaxy properties to $z = 0.3$–$0.5$. Early faint galaxy workers (Broadhurst et al 1988, Colless et al 1990) assumed either the Efstathiou et al (1988) or Loveday et al (1992) normalization, $\phi^* = (1.40 \pm 0.17) \times 10^{-2}\ h^3\ \mathrm{Mpc}^{-3}$, which, if too low, would increase the evolution inferred, assuming $L^*$ and $\alpha$ were correctly determined. Although this might be explained

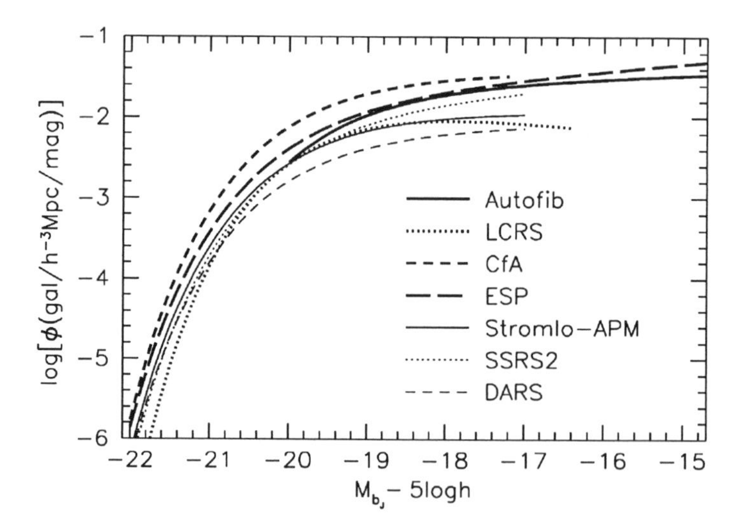


*Figure 7* A comparison of the Schechter function fits to various determinations of the local galaxy LF from the compilation of Colless (1997), updated to include the results of Zucca et al (1997). All LFs are placed on the $b_J$ system using transformations given by Colless (1997).

via a local minimum in the southern galaxy distribution, there is no convincing indication of this in the redshift distribution, and so the anomalously steep count slope found by Maddox et al (1990) brighter than $B = 18$ has been taken to indicate that a problem may exist elsewhere.

The Stromlo-APM survey limited at $b_J = 17.2$ is the largest and most well-documented local galaxy survey, but its photometry has been questioned in several recent papers. Metcalfe et al (1995a) and Bertin & Dennefeld (1997) compared APM magnitudes with their own CCD-calibrated measures and detected a scale error and considerable scatter in the region $17 < b_J < 19$. Bertin & Dennefeld explained this, in part, to unmeasured light fainter than the relatively bright APM isophote. However, Metcalfe et al found that high surface brightness galaxies lose more light than those of low surface brightness because of the limited dynamic range of the plate-measuring machines involved. Bertin & Dennefeld presented B-band counts whose dlogN/dm slope, $\gamma_{16:19}$ in the $16 < b_J < 19$ range, is 0.53 (cf 0.59 for Maddox et al 1990), a value more consistent with no-evolution expectations. It should be noted that any photometric limitations of the APM data may possibly apply also to the COSMOS-based photometry used by Zucca et al (1997) because Metcalfe et al (1995a) and Rousseau et al (1996) found similar behavior in their COSMOS data. However, not all photographic photometry may be affected in this way. Weir et al (1995) present photographic counts whose slope is flatter that of Maddox et al.

The above calibrations of nonlinear photographic survey data using linear CCD detectors are the forerunners of a new generation of photometric surveys based entirely on CCDs or infrared arrays. Gardner et al (1996) presented counts in BVIK for two large northern fields covering 10 deg$^2$ and found a B-band slope $\gamma_{16:19} = 0.50 \pm 0.03$ in reasonable agreement with Bertin & Dennefeld (1997). Huang et al (1997) presented counts in BIK for a similarly sized equatorial area. Although both Gardner et al and Huang et al claim to disagree, the difference lies mostly in the bright K data. Huang et al's Table 3 gives $\gamma_{16:19} = 0.53$. Although these survey areas are very small in comparison with the 4300 deg$^2$ of the APM survey, they nonetheless support the above claims for a photometric revision. For the $\alpha = -1$ Schechter function adopted by Loveday et al (1992), Gardner et al's counts indicate $\phi_B^* = 2.02 \times 10^{-2}$ h$^3$ Mpc$^{-3}$, (i.e. a normalization about 50% higher than the APM result). A similarly high normalization emerges from the ESP results limited at $b_J = 19.4$ (Zucca et al 1997).

The LF and counts are also available from the extensive Las Campanas Redshift Survey (LCRS) (Lin et al 1996). Galaxies were selected in r according to a relatively bright isophote over an area of 720 deg$^2$. When allowance is made for the different photometric scale, Lin et al found a low normalization (as did Loveday et al 1992) and somewhat discrepant $\alpha$ and $L^*$ in comparison to those of other surveys. The latter point is significant given that the count normalization depends on all three parameters. The differences are not understood but may arise from the poor match between their actual LF and the Schechter function, possible inaccuracies in the single type-invariant k-correction used (a criticism also applicable to the ESP survey), and the outer redshift limit at which the final normalization is made.

In summary, these issues, although not entirely resolved, support a normalization increase of at least 50% over that adopted by Loveday et al (1992) and indicate that at least some photographic data brighter than $B = 19$ may be inaccurate in both photometric scale and possibly even in source detection (see Section 4.2 below). The normalization derived for galaxies with $z < 0.1$ observed to be fainter than $B = 19$ (Ellis et al 1996a, Cowie et al 1996) supports this upward revision, which has been incorporated in Figure 5*a*.

Turning to the shape of the LF and its dependence on morphology or spectral class, Driver & Phillipps (1996) have demonstrated the difficulty in estimating the faint end slope reliably from magnitude-limited redshift surveys. Regardless of size, such surveys are optimized to define the LF in the region around $M^*$. An additional difficulty with the deeper LCRS and ESP surveys (Lin et al 1996, Zucca et al 1997) is their reliance on a single k-correction formula for all galaxies. If, as suggested by Marzke et al (1994), the mean spectral type is a function of luminosity, systematic errors may be made in the volume correction, leading to a distortion in the derived shape.

Recognizing these local uncertainties, Koo et al (1993) and Gronwall & Koo (1995) postulated the existence of a population of intrinsically faint blue star-forming galaxies whose LF has a much steeper faint end slope; this serves to minimize the evolution necessary to explain the faint galaxy counts. Limited evidence in support of an underestimated component of blue dwarf galaxies has been presented by Metcalfe et al (1991), Marzke et al (1994), and Zucca et al (1997). The hypothesis was central to the "conservative" model proposed in Koo & Kron's (1992) review.

Although the faint end of the local field LF will inevitably remain uncertain to some unsatisfactory degree, to make a fundamental difference in the interpretation of the faint counts without affecting the redshift distributions at the limits now reached is fairly difficult. At the time of Koo & Kron's review, the faint redshift data only extended to $B = 22.5$. However, by extending the redshift surveys to $B = 24$, Glazebrook et al (1995a), Ellis et al (1996a), and Cowie et al (1996) found only a minimal contribution of local dwarfs to the faint counts. Few $22 < B < 24$ galaxies are observed with $z < 0.1$, in marked contrast to Gronwall & Koo's (1995) predictions (Glazebrook et al 1995a, Cowie et al 1996). Although some incompleteness remains at these limits, it is difficult to understand how this could apply to blue systems at such low redshift where the normal spectroscopic diagnostics are readily visible. Driver & Phillipps (1996) correctly pointed out that such inferences should be drawn from a comparison of an absolute number of sources seen rather than by matching N(z) distributions normalized to evolving numbers. Even so, taking the $z < 0.1$ redshift survey data fainter than $B = 17$, no convincing evidence exists to support a faint end slope steeper than $\alpha = -1.2$ down to at least $M_B = -15 + 5 \log h$ (Ellis et al 1996a, Cowie et al 1996).

Zucca et al (1997) presented evidence for an upturn in their local LF fainter than $M_B = -17 + 5 \log h$, similar to trends observed for cluster LFs (Bernstein et al 1995). Although there is no reason to suppose the field and cluster LFs should be the same, as the faint end of the cluster galaxy LF is more reliably determined (given the volume-limited nature of that data), the suggestion is an important one. However, the upturn identified by Zucca et al is based on 37 galaxies fainter than $-16$ and only 14 fainter than $-15$. The implications of such a LF upturn in the apparent magnitude range probed by the redshift surveys would be very small. Thus there is no conflict between this result and the local LF limits provided by the faint redshift surveys. However, a significant contribution to the counts from this upturn would not commence until $B = 26$ [cf model (c) of Driver & Phillipps 1996]. The sub-Euclidean number count slope beyond $B = 25$ (Figure 5*a*) provides a further constraint on such a dwarf population.

In conclusion, much work is still needed to verify the detailed form of the local LF. The large-scale redshift surveys underway at the AAT and shortly with the SDSS to B = 19–20 will make a big improvement, provided that their photometric scales are robust. However, despite these uncertainties, the deeper redshift surveys now available indicate that a poorly understood faint end slope fainter than $M_B = -15 + 5 \log h$ is unlikely to seriously distort our understanding of the faint counts to $B \approx 26$.

## 4.2 *The Role of Low Surface Brightness Galaxies*

Although intimately connected with uncertainties in the local LF, the presence (or otherwise) of an abundant population of low surface brightness galaxies (LSBGs) is best considered as a separate issue affecting the interpretation of faint data. McGaugh (1994, 1996) has shown how the presence of systems with central surface brightnesses fainter than $\mu_B = 23$ arcsec$^{-2}$ covering a wide range of luminosities would seriously affect determinations of the LF particularly if the local data were plagued by isophotal losses, as discussed in Section 4.1. McGaugh (1994), Phillipps & Driver (1995), Ferguson & McGaugh (1995), and Babul & Ferguson (1996) have explored this uncertainty and proposed the existence of an abundant population of local LSBGs that could be faded remants of blue star-forming systems identified fainter than $B > 22$. The idea stems from Babul & Rees' (1992) suggestion that the excess seen in the counts might arise from a separate population of dwarfs whose initial star-formation era is delayed until the UV ionizing background drops below a critical value. Rapid fading would produce a large present-day abundance of red LSBGs. The role of LSBGs in these suggestions is thus twofold; first, as an additional uncertainty in the local LF and second, as remnants of the faint blue galaxies.

Quantifying the contribution of LSBGs to the local LF will remain controversial until a suitable catalog exists for which rigorous selection criteria have been applied and redshifts determined. Most of the available field data is angular-diameter limited from photographic plates (Impey et al 1988, 1996, Schombert et al 1992, Sprayberry et al 1996) and, although illustrating the range of scale lengths and surface brightnesses possible, cannot easily be converted into volume-limited data. McGaugh (1996) convincingly argues that the presence of only a small number of LSBG examples implies a significant correction must be made to the faint end slope, although several assumptions are made in calculating the survey volume as a function of limiting surface brightness.

Although CCDs have been used for LSBG searches in clusters, Dalcanton et al (1996) recently analyzed 17.4 deg$^2$ of deep transit scan CCD data and identified seven LSBGs with $\mu_v > 23$ arcsec$^{-2}$. Of importance, spectroscopic data for this sample have provided an estimate of the volume density of LSBGs of known physical size. The large mean distance for these LSBGs indicates a LSBG volume density, though still uncertain, that is comparable to that of normal galaxies. As expected, the contribution to the overall luminosity density is very small. Dalcanton et al (1997) argued, as did McGaugh (1996), that part of the normalization change in the local LF (discussed in Section 4.1) may arise from the selective loss of these systems in the bright photographic data.

Although LSBGs undoubtedly exist, perhaps quite abundantly and with a range of properties, it seems unlikely that they dominate the luminosity density or that the bulk of them represent the faded remants of a faint blue population. The fading required to push a typical faint blue galaxy below typical local detection thresholds would have to be 2.5–4 mag, depending on the true faint end slope of the local LF (Phillipps & Driver 1995). To be effective, given the narrow time interval involved, the end of the star-forming era would also have to be very abrupt. Moreover, most well-studied LSBGs are gas-rich and blue (de Blok et al 1996) and quite unlike the postulated faded remnants. Neither is there an obvious correlation between LSBG central surface brightness and color (McGaugh & Bothun 1994). An abundant population of faded remants would be detectable as a significant Euclidean upturn in the number of red and infrared sources at faint limits, which has not yet been seen (Babul & Ferguson 1996).

## 4.3 *Is Number Evolution Required?*

Are simple number-conserving models adequate? Or are additional populations of star-forming sources required with no detectable local counterpart, as has been suggested in hypotheses based on a postulated population of fading dwarfs (Babul & Rees 1992, Cole et al 1992) or recent galaxy-galaxy mergers (Rocca-Volmerange & Guiderdoni 1990)? Although the upward revision in the normalization of the local LF and the possibility of a steeper faint end slope of blue sources (discussed in Section 4.1) reduces the no-evolution N(m, z) paradox originally introduced by Broadhurst et al (1988), even internally within the CFRS data (Lilly et al 1995) there are signs of strong evolution from $z = 0.3$ to 1, i.e. neglecting any reference to local data. Similar conclusions can be drawn from the internally presented LFs in the Ellis et al (1996a) and Cowie et al (1996) samples, although, as these data comprise a number of individual surveys with slightly different photometric selection criteria, the conclusions are perhaps less compelling.

To understand the extent to which this evolution is number-conserving, it is important to revisit the nature of the LF changes presented in Figure 6 in the context of simple, number-conserving, luminosity evolution. Whereas Lilly et al (1995) claim at most a 0.5-mag brightening in the luminosity scale of their redder population over $0.3 < z < 0.6$, they detect brightening of more than 1 mag in the luminosity scale for the bluer population in the same interval. Further brightening is seen at high redshifts, particularly for less luminous sources. Because of the limited magnitude range sampled in the CFRS, the luminosity overlap at various redshifts is relatively small, but, given the homogeneity of the survey, the data by Lilly et al provide incontrovertible evidence for evolution in the blue sources corresponding to a factor of 2–2.5 increase in the comoving rest-frame 4400-Å luminosity density over $0.3 < z < 0.8$, depending on $\Omega$ (Lilly et al 1996).

Although the Autofib/LDSS redshift survey does not penetrate as deep as the CFRS, the large number (548) of fiber-based redshifts in the intermediate interval $19.7 < b_J < 22.0$ (selected primarily from 4-m prime focus plates that are presumably unaffected by the photometric difficulties discussed in Section 4.1) is a valuable component of this survey. This enables the LF to be probed at all redshifts in a consistent manner to $M_{bJ} = -18 + 5 \log h$. By virtue of the blue selection, the overall LF is more affected by the evolutionary changes than in those selected in I and K. Ellis et al (1996a) reject an unchanging local LF as a fit to their data to $z = 0.75$ with very high significance. Furthermore, statistical tests reveal that the evolution begins strongly beyond $z = 0.3$ in the sense that there is a steeper faint end slope; no convincing shift is seen in $L^*$ for the overall population. As with the CFRS survey, the bulk of these changes occur in the star-forming component selected with strong [O II] emission. The luminosity decline or fading of the [O II]-strong population is roughly comparable to that seen for the blue galaxies in CFRS.

The question of whether the observed changes in the LF can be understood principally as a luminosity shift for some subset of the population [pure luminosity evolution (PLE) models] or whether more complex models are required (e.g. ones that violate number conservation) is difficult to determine from LF data alone because of the statistical nature of the observations. A galaxy may enter the [O II]-strong or blue class only temporarily and thus a variety of physical scenarios might be compatible with the trends observed. However, although there seems a natural reluctance among some workers to contemplate "exotic" interpretations of the data (Shanks 1990, Koo et al 1993, Metcalfe et al 1996), luminosity-independent evolution could be even harder to understand physically, particularly when extending such models to very high redshift.

The traditional PLE models date from Baade (1957) and were explored in detail by Tinsley (1972, 1980). They assume, for simplicity, that all galaxies of a certain class change their luminosities with redshift by the same amount in magnitudes irrespective of their luminosity. Although the LF would maintain the same shape at all redshifts, because different subclasses are allowed to evolve at different rates with different LF shapes, the integrated form may show evolution that is luminosity-dependent (as observed). In such models (e.g. Pozzetti et al 1996), evolutionary corrections are taken from synthesis codes that aim to reproduce the present-day SEDs via star-formation histories of a simple form (Bruzual 1983, Bruzual & Charlot 1993). An unavoidable corollary of those models that can successfully account for the present range of Hubble sequence colors is much stronger evolution for early-type galaxies than for the later types. According to Pozzetti et al, a significant proportion of the excess blue $B = 25$ population are distant young ellipticals with redshifts generally higher than the surveys indicate.

The PLE models represent a subset of an entire class of models whose evolutionary codes aim to satisfy the joint number-magnitude-color-redshift data sets. They can be considered as ab initio models (Section 2). Many fit the data to some degree but do not define a unique interpretation. An important conclusion from the LFs discussed in Section 3 is the qualitatively different pattern of evolution that is seen compared with that expected in the standard PLE models. Luminous early-type galaxies appear to evolve very little to $z = 1$, and the bulk of the bluing observed appears to occur by virtue of an increased proportion of star-forming $L^*$ galaxies, few of which have spheroidal forms.

Although number evolution cannot be verified to the limits of the redshift surveys, the situation at fainter limits is much clearer. Figure 8 updates a useful diagram originally published by Lilly et al (1991). Adopting a local LF upwardly normalized with a steeper faint end slope ($\alpha = -1.3$), to take into account possible uncertainties discussed in Section 4.1, the calculations in the Figure show the maximum redshift to which a source that is brighter than a given luminosity would have to be visible so as to account for the integrated number of sources to $B = 28$, as defined using the ground-based data of Metcalfe et al (1996). Because the majority of the faintest sources cannot lie at very high redshifts by virtue of the absence of a Lyman limit (Section 5), and it is unreasonable to suppose low luminosity galaxies could shine continuously for a Hubble time, strong number evolution seems unavoidable in order to account for the integrated population beyond the break at $B \approx 25$.

## 4.4 *A Distinct Population of Rapidly Fading Objects?*

The difficulties encountered with the traditional PLE models (Section 4.3) can be overcome by invoking a separate population of blue objects that undergo strong luminosity evolution (Phillipps & Driver 1995). Unlike the traditional PLE models, this hypothesis invokes a delayed star formation in a subset of the galaxy population followed by a remarkably rapid decline in activity thereafter. The present-day remnant of this population would presumably occupy the faint end of the local LF. Although physical models have been proposed to support this picture (Cowie et al 1991, Babul & Rees 1992, Babul & Ferguson 1996), the original hypothesis was proposed only to fit the data. Nevertheless, there is a strong theoretical motivation because ab initio models tend to overpopulate

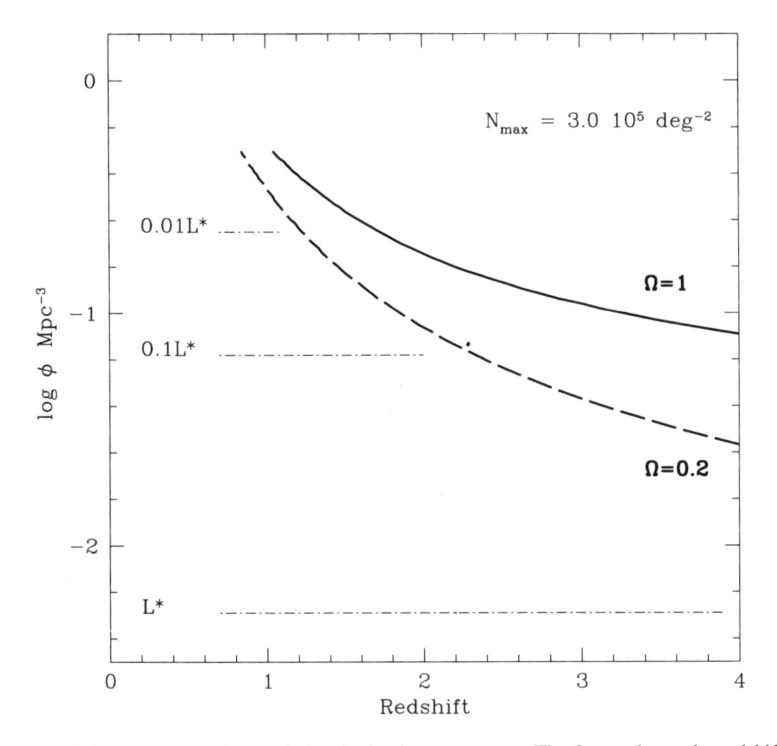


*Figure 8* Evidence for number evolution in the deepest counts. The figure shows the redshift to which a population of galaxies brighter than a given luminosity would have to be seen to account for the integrated number of sources seen in ground-based images to B = 28 (Figure 5*a*). The calculation adopts a high normalization $\alpha = -1.3$ local LF similar to that observed at z = 0.3 in the Autofib/LDSS survey (Ellis et al 1996a).

the faint end of the local LF (Cole et al 1994a). Evidence in support of this empirical picture includes the very different evolutionary behavior of field LFs characterized by color and spectral features (Lilly et al 1995, Heyl et al 1997) and the rapid increase in the number of irregulars in the MDS data (Glazebrook et al 1995b, Driver et al 1995b). A high fraction of the HST irregulars are also blue [O II]-strong sources (LeFevre et al 1996a). However, the key to testing this model further would be to verify that a distinct population exists and to demonstrate that it fades according to the rate determined from the LF studies.

Some progress might be made by studying the supposedly rapidly evolving component in more detail, as well as by finding local examples. A typical, rapidly evolving blue galaxy at $z > 0.3$ has $M_B \approx -19.5 + 5\log h$, B-I $\approx$ 1–1.5, a rest-frame equivalent width of [O II] of 20–40 Å, and an irregular appearance with knots and occasionally a compact core. Although a few giant irregulars are found locally (Gallagher et al 1989), the bulk of the blue sources with comparable [O II] strengths at low redshift are gas-rich metal-poor dwarfs believed to have suffered sporadic bursts of star formation (Telles et al 1997, Gallego et al 1995, Ellis et al 1996a). Leitherer (1996) provides a comprehensive review of the physical processes responsible for sudden star formation in galaxies. There are two broad categories: (*a*) H II region–like spectra superimposed on that of an older population (Telles et al 1997) and (*b*) nuclear starbursts. Both may be induced by interactions.

Surprisingly little is known about the astrophysical properties of the faint blue galaxies, which at first sight seems remarkable given spectra are available for several thousand such galaxies! With the exception of Broadhurst et al's (1988) original data taken at a spectral resolution of 4 Å (which they interpreted as supporting signs of short-term star formation in the distant population), both the signal/noise and resolution of more recent, fainter data are optimized for little more than measuring redshifts; a resolution of 20–40 Å is more typical. Heyl et al (1997) and Hammer et al (1997) discuss the spectral properties of co-added spectral datasets in the Autofib/LDSS and CFRS surveys, respectively, but disagree on the degree to which short-term star formation may be occurring in their samples. Higher quality spectra are required to make progress.

A tentative connection between the faint blue population and active galaxies is reported by many groups. On the basis of emission line ratios, Tresse et al (1996) claimed between 8 and 17% of the $z < 0.3$ CFRS sample could be Seyfert 2 or LINERs, depending on uncertain corrections for underlying stellar absorption. This is about four times higher than the local rate and would suggest a dramatic evolution in the proportion of active sources over quite recent epochs. Likewise, Treyer & Lahav (1996) claimed $B < 23$ galaxies with $z < 0.3$ could contribute 22% of the 0.5- to 2-keV X-ray background. At fainter limits, Windhorst et al (1995) found that the micro-Jy radio sources overlap significantly with the faint blue population. Windhorst et al argue that the nonthermal activity is more likely a by-product of interactions and mergers rather than the direct output of classical active galactic nuclei.

How might fading in a given population be demonstrated convincingly? We return to the question of selecting some physical parameter that does not change in the process. Koo et al (1995) and Guzman et al (1996) have studied a particular subclass of compact narrow emission line galaxies (CNELG) selected on the basis of very small effective radii (1 kpc) and have suggested that these may be the ancestors of local dwarf spheroidals. Although only a minority of the faint blue population lies in this category, this suggestion is important because, if correct, it would demonstrate for a subset of the faint population that significant fading has occurred. Spectroscopic velocity dispersions from the Keck HIRES are typically 30–50 km $s^{-1}$, which indicates low mass/light ratios and bursting star–formation histories. Substantial fading (4–7 mag in $M_B$ in 3 Gyr) would be needed to reduce such systems to the present luminosities of dwarf spheroidals.

The CNELG study illustrates an important way forward because, if local dwarf galaxies are faded versions of their high z cousins, the surface brightnesses within some fixed metric scale must also have declined by the same amount as indicated in the LF studies. The crucial test for the fading hypothesis is thus to examine the surface brightness distributions of various subsets of the field population, taking due care to allow for possible selection biases. The availability of HST images for subsets of the large redshift surveys (Ellis 1995, Schade et al 1995, LeFevre et al 1996a) makes this an important direction for future study.

## 4.5 *Evidence for Mergers?*

An increase with redshift in the rate of galaxy merging is an attractive way to satisfy the changing shape of the LF. Such an explanation is a natural consequence of hierarchical pictures (Carlberg 1992) and features prominently in many of the ab initio models discussed earlier (Baugh et al 1996). The importance of merging in the faint counts has a long but somewhat inconclusive history (see Carlberg 1996 for a recent summary). The optical LF data has been particularly difficult to interpret in this picture, primarily because the rest-frame light indicates the history of star formation, whereas the merger predictions are based on the evolution of the mass. Rocca-Volmerange & Guiderdoni (1990) introduced a self-similar, mass-conserving evolutionary model in which the comoving number density is required to increase as $(1 + z)^{1.5}$ in a $\Omega = 1$ universe. Broadhurst et al (1990), analyzing redshift and number count data in the context of a low local normalization, required a mass growth rate such that a typical galaxy became 4–6 fragments by $z = 1$. Eales (1993) incorporated a more physically based model and predicted changes in LF shape not dissimilar to those seen in the recent redshift surveys. To overcome uncertainties in relating mass and light, Broadhurst et al (1990) advocated conducting deep K-limited surveys to test the merger hypothesis and predicted a turnover in the mean redshift at faint limits consistent with the absence of large mass objects at high z. Detailed predictions in this context have been made by Carlberg & Charlot (1992) and the latest available K-selected data provide some support for such a picture (Cowie et al 1996); over $K = 18–20$, the mean population redshift hardly increases with apparent magnitude. However, incompleteness in the $z > 1.5$ range remains a concern.

A more direct approach might be to estimate the interaction rate by searching for close pairs. Barnes & Hernquist (1995) discussed the importance of "major" and "minor" mergers with respect to the morphology of the host galaxy, and energy arguments suggest a strong redshift dependence in the interaction rate (Carlberg 1996). However, even the local interaction rate remains uncertain because not all of the diagnostic features of a merger are expected to be easily visible (Mihos 1995). One of the earliest observational studies attempting to define the merger rate at large look-back times was that of Zepf & Koo (1989), who found 20 close pairs in a faint photographic sample limited at $B = 22$ and concluded that the pair merger rate increases with redshift as $(1 + z)^m$ where $m = 2$–$4$. The difficulty here lies in correcting for a large number of biases that may artificially raise the apparent interaction rate. Such biases include the dissimilar tactics in analyzing low and high redshift data, a possible boosting in luminosity of satellite galaxies whose star formation is triggered by a merger, and the k-correction, which leads to an increase in the number of late-type spirals at high redshift that often have peripheral H II regions; the latter are rendered more visible in the rest-frame UV. A recent analysis using ground-based data is discussed by Woods et al (1995).

The arrival of HST images has led to renewed interest in this area. Neuschaefer et al (1997) presented a comprehensive analysis of the number of close pairs (<3 arcsec) to $I = 23.5$ in 56 MDS fields and analyzed their results in the context of HST and ground-based angular correlation functions. In hierarchical merging, one need not expect to find a significant excess in the angular correlation function on small scales. Neuschaefer et al's data supercedes the earlier MDS study of Burkey et al (1994) [which claimed an increase in the merger rate comparable to Zepf & Koo (1989)] and gave $m = 1.2 \pm 0.4$, as did Woods et al (1995). No convincing excess of pairs with separations less than 3 arcsec was found in comparison with an extrapolation of the angular correlation function.

Greater progress will be possible when redshifts are available. The physical scale around each host galaxy can then be defined and, more importantly, satellites can be constrained to include only those above a fixed luminosity limit (providing some estimate of the k-correction is made). Patton et al (1997) use the field galaxies located via the CNOC1 cluster galaxy survey (Carlberg et al 1996) to define a sample of close pairs within 20 $h^{-1}$ kpc. Redshifts are available for half of the secondary images and demonstrate that a sizable fraction are true physical associations. By $z = 0.33$, $4.7 \pm 0.9\%$ of the faint population are claimed to be merging, and comparison with local data suggests $m = 2.8 \pm 0.9$. A number of corrections are required in this analysis to allow for the idiosyncrasies of the observing strategy that produced the primary sample and contamination from cluster galaxies in the secondary sample. LeFevre et al (1996a) had both the advantage of HST images and a very wide redshift range, which means that the evolutionary trend can be examined without reliance to any local data. Moreover, LeFevre et al sampled to much fainter limits around each primary galaxy. More than half of the major mergers in the LeFevre et al sample have $z > 0.8$; indeed no strong trend is seen until the redshift is quite large, in qualitative agreement with the predictions of hierarchical clustering (Baugh et al 1996). However, the absolute rate remains uncertain when determined by pair counts, and so whether merging is the dominant process driving the evolution of the LF remains unclear.

The above studies demonstrate the difficulties of verifying the merger hypothesis quantitatively. However, some important points can be made. First, following the upward revision in the local LF (Section 4.1), there is less need for rapid number evolution at low z, and this considerably reduces the difficulties concerning the abundance of recent merger products (Dalcanton 1993). Second, notwithstanding the uncertainties, there is growing observational evidence from HST images of galaxies of known redshift and the modest depth of the faintest K-limited surveys that merging is of increasing importance at high redshift.

# 5. SURVEYS BEYOND THE SPECTROSCOPIC LIMIT

Various workers have considered ways in which the mean redshift of the very faint population (say $B \approx 27$, $R \approx 26$, $I \approx 26$, i.e. beyond normal spectroscopic limits) could be estimated statistically. One expectation might be that the bulk of such a faint population lies beyond $z = 1$. A redshift of unity has represented a significant barrier to the systematic study of normal galaxy populations for many years. It corresponds to the 4-m telescope limit for which spectroscopic redshifts are possible for normal $L^*$ field galaxies, and it also marks an important transition from analyses that are, broadly speaking, independent of the cosmological framework to ones where the volume element and time-redshift relation depend critically on $\Omega$ and $\Lambda$. A lesson that emerges from Section 4 is that a fundamental difficulty in making progress is the need to ensure that local and high z data sets are treated similarly. As we turn to the next logical step in the observational challenge, this becomes even more the case.

## 5.1 *Constraints from Gravitational Lensing*

The suggestion that the mean redshift of the $B = 27$ and $I = 26$ population might be rather low, consistent with discussion of the break in the count slope in Section 3, first arose from the pioneering CCD exposures conducted by Tyson and collaborators (Tyson 1988, Guthakurta et al 1990). Tyson also developed the first practical applications of gravitational lensing by rich clusters as a tool for estimating the mean statistical distance to the background population (Tyson et al 1990).

At faint magnitudes, lensing by foreground masses affects source properties by an amount that depends on the nature of the intervening lens, the relative distances to the lens and source, and the cosmological model (Blandford & Narayan 1992, Fort & Mellier 1994). The phenomenon manifests itself in several ways, depending on the geometrical configuration and lens scale. In the case of giant arcs, dense concentrated clusters of galaxies beyond $z = 0.1$ magnify faint sources considerably, extending spectroscopic and photometric detections to fainter limits. Certain clusters have well-constrained mass distributions, either from giant arcs and multiple images of known redshift or from indirect probes such as X-ray luminosities and velocity dispersions. The lensing shear field viewed through these clusters can provide a statistical estimate of the mean distance to sources that are too faint for conventional spectroscopy. As the technique is purely geometric in nature, it provides an independent probe of the distances to faint galaxies.

Recognizing the need to separate the dependence of the weak lensing signal on both the relative distances of source and lens and the nature of the lensing cluster, Smail et al (1994, 1995) compared the shear seen in a background population to $I = 25$ as measured through three clusters at different redshifts. The relatively weak shear found by Smail et al (1994) through an X-ray luminous cluster at $z = 0.54$ suggested a relatively low mean redshift for the population, but this is in marked contrast to the conclusions of Luppino & Kaiser (1997), who detected shear through a similar cluster at $z = 0.83$. Difficulties arise because this technique relies critically on measuring the absolute shear, as well as understanding the properties of the lensing cluster. A comparison of the various techniques used to estimate the shear by different workers on the same clusters is badly needed, and deep HST data of more distant clusters is also required to correct for seeing and other effects that may affect ground-based images.

Considerable progress is possible if redshifts are available for some of the lensed features, because this reduces the dependence on absolute measurements of the lensing signal (Fort & Mellier 1994). Giant arcs are strong lensing events of high magnification but offer a rather unreliable glimpse of the high redshift population. Spectroscopy is now available for about 20 cases; a few have redshifts beyond 1. Although these are galaxies found serendipitously by virtue of their location behind unrelated foreground clusters and their unlensed magnitudes are quite faint, important selection effects operate in their recognition and thus they are an unreliable statistical probe of the background redshift distribution. Strong lensing is optimal when the background source is around two to three times the angular diameter distance of the lens, and so a low frequency of high redshift arcs may simply reflect the paucity of concentrated high z clusters. At the moment, few convincing arcs have been seen in clusters beyond $z = 0.5$. Second, the arcs are, without exception, found as high surface brightness features in optical CCD images often by virtue of their contrasting blue color as compared with the red cluster population. Smail et al (1993, 1996) have examined the optical-infrared colors and HST angular sizes of giant arc samples and, not surprisingly, deduced that many are representative of late-type galaxies undergoing vigorous but extended star formation.

The more important role of arc spectroscopy is to further constrain the properties of the lensing cluster. The lensing geometry can then be used together with the image shapes and orientations for the fainter population of less distorted "arclets" to yield a statistical redshift for each one, assuming, on average, that they are intrinsically round sources. This "lensing inversion" technique was first developed by the Toulouse group (Kneib et al 1994) using ground-based data for the well-studied cluster Abell 370, and it has now been extended using HST images and more comprehensive ground-based spectroscopy for Abell 2218 and Abell 2390 (Kneib et al 1996). These studies suggest that the mean redshift of $I < 25.5$ populations cannot significantly exceed 1. In several cases, the constraints on the arclet redshifts are sufficiently tight that the predicted redshifts are worth verifying spectroscopically. In a number of such cases, Ebbels (1996), Ebbels et al (1996), and Bezecourt & Soucail (1997) presented convincing evidence that the inversion technique works well. However, the construction of genuine magnitude-limited samples appears to be difficult using this method. Not only are the lensed images magnified by different amounts, but inversion is unreliable for the smallest sources that increasingly dominate the faint counts (Roche et al 1996).

To overcome difficulties inherent in analyses of distorted images, Broadhurst (1997) proposed the use of "magnification bias" or gravitational convergence. A lensing cluster enlarges the background sky, and this produces a diminution in the surface density of sources depending on the relative distances involved. For galaxy counts, the effect is in the opposite direction to the magnification described above (Tyson et al 1984, Broadhurst et al 1995). The background source counts viewed at radius r from the center of a foreground lens become

$$N(<m, r) = N_0(<m)\mu(r)^{2.5\gamma-1}, \tag{7}$$

where $N_0(<m)$ represents the true counts, $\mu(r)$ is the magnification at angular radius r from the center of the lens, and $\gamma = \mathrm{dlogN/dm}$ is the slope of the number-magnitude counts. For $\gamma = 0.4$, the magnification and dilution effects cancel out and no effect is seen. However, when $\gamma < 0.4$, the counts decrease, particularly near the critical angular radius. The location of this point depends on the relative angular diameter distances of the sources and the lens, as well as on the cosmological model. Fort et al (1997) and Mellier (1997) described promising applications of this technique. As the method relies only on source counting rather than a reliable measurement of image shapes, the technique can probe to very faint limits in a controlled manner.

## 5.2 *The Hubble Deep Field*

The Hubble Deep Field (HDF) project (Williams et al 1996) and related Keck spectroscopic programs (Steidel et al 1996c, Illingworth et al 1996, Cohen et al 1997, Lowenthal et al 1997) have already given a tremendous boost to studies of the universe beyond the limits of the 4-m telescope redshift surveys. The investment of a large amount of HST Director's Discretionary time enabled the first comprehensive multicolor study of a deep field with WFPC-2. The relatively poor efficiency of HST blueward of 500 nm had previously prevented even the most adventurous workers from attempting deep UV and blue imaging in normal guest observer allocations. Furthermore, the public availability of the HDF image has led to a rapid delivery of scientific results concerning the nature and redshift distribution of galaxies well beyond spectroscopic limits.

The most striking result, from a consideration of the high surface density of $B = 28$ galaxies in ground-based data (cf Metcalfe et al 1995b, 1996) (Figure 5*a*), is the large fraction of blank sky in the HDF image. This simple observation reflects the fact that the bulk of the $I > 25$ population has very small angular sizes, continuing trends identified first in the MDS (Mutz et al 1994, Roche et al 1996a). Another result that represents a continuation of MDS work is the increasing proportion of faint irregular structures from $I = 22$ to $I = 25$ (Abraham et al 1996a). Although the widely distributed HDF color images emphasize irregularity via blue features that sample the rest-frame UV, the morphological analysis of Abraham et al is based on the F814W image, which, at $z \approx 1{-}1.5$, is equivalent to rest-frame U or B. Beyond $I = 25$, the resolution is insufficient for detailed study; indeed, Colley et al (1996) raise the important question of the definition of a galaxy in this regime. The small angular sizes and high abundance compared to $z < 1$ LF estimators, together with the declining slope of the counts (Figure 5*a*), suggest that many of these sources may be subgalactic components at an early stage of formation.

The availability of images in four passbands, supplemented by deep ground-based data at infrared wavelengths (Cowie 1996), has led to a surge of interest in estimating redshifts from colors. In its most elementary form, a set of discontinuities (Lyman or Ca II 4000-Å break) is located in individual galaxies via imaging through a set of filters. The HDF observing strategy was chosen to extend earlier work by Guhathakurta et al (1990) and Steidel & Hamilton (1992) and isolate those sources whose Lyman limit discontinuities are redshifted into the optical. Similar techniques have also been used via ground-based images to detect higher redshift sources around QSOs (Giallongo et al 1996). The validity of the technique is reviewed comprehensively by Madau et al (1996) in the context of what is known about the UV SEDs of galaxies and the attenuation of UV light by intervening HI clouds. They claim the strategy is robust to quite considerable uncertainties in the precise shape of the SEDs near the Lyman limit and possible effects of dust.

Lists of candidate high z galaxies in the HDF selected on the basis of the Lyman discontinuity were first published by Abraham et al (1996a) and Clements & Couch (1996), but most of the progress has been achieved through spectroscopic exploitation of the technique in the HDF and other fields using the Keck telescope (Steidel et al 1996a,b, 1997, Giavilisco et al 1996, Lowenthal et al 1997). The spectra demonstrate a success rate of 100% for the selection of high z sources using the Lyman limit; not a single redshift has been confirmed outside the expected range. The Keck surveys are still in progress but already provide a significant constraint on the proportion of $R < 25$ star-forming galaxies beyond $z = 2.3$ (the redshift corresponding to the limit entering the F300W filter). These results have offered a valuable glimpse at the nature of a population of galaxies with $2.3 < z < 3.5$, and perhaps the most significant results are their relatively modest star-formation rates (SFR, 1–6 $M_o$ $year^{-1}$) and volume densities comparable to those of local $L^*$ galaxies ($\approx 8 \times 10^{-4} h^3 Mpc^{-3}$) (Steidel et al 1996a,b, Madau et al 1996). (However, note that these figures refer to a $\Omega = 1, \Lambda = 0$ world model and would be much larger if $\Omega \ll 1$.) Comparable SFR for high z galaxies have been estimated from independent studies by Ebbels et al (1996), Djorgovski et al (1996), and Hu & McMahon (1996). The weak emission lines in these high z star-forming galaxies may explain the null results of many years of primeval galaxy searching based on the assumption of intense photoionized Lyman $\alpha$ emission (Djorgovski & Thompson 1992).

A minimum signal-to-noise ratio in the UV-optical SED is required to convincingly detect the presence or otherwise of the Lyman limit, and thus the question remains as to whether a larger fraction of the sources fainter than $R = 25$ have higher redshift. As the longest wavelength band (F814W) still samples the rest-frame UV at $z > 4$, only star-forming objects above some threshold can be visible at high z in the HDF. To eliminate a large population of sources with low SFR would require K-band imaging to much deeper limits than are currently possible (Cowie 1996, Moustakas et al 1997).

The multicolor HDF data has also been analyzed by numerous workers with respect to various template SEDs in an attempt to secure statistical redshift distributions (Table 2). The critical uncertainty in these studies is the form of the UV SED sampled by the optical HDF data for $z > 2$. Lanzetta et al (1996) grafted large aperture optical SEDs from Coleman et al (1980) with much smaller aperture UV data from Kinney et al (1996), whereas the other workers generally adopted model SEDs in the UV. Gwyn & Hartwick (1996) used model SEDs throughout. With the exception of Gwyn & Hartwick, all fitting thus far was done on the basis of present-day SEDs in the (unjustified) hope that an evolving SED must somehow move along the locus of those observed today.

**Table 2** Color-based redshift surveys in the HDF

| Reference | Magnitude limit | Template SED |
| --- | --- | --- |
| Gwyn & Hartwick (1996) | $I_{814} < 28$ | Bruzual-Charlot models |
| Lanzetta et al (1996) | $I_{814} < 28$ | z = 0 observed + Ly∞[1] |
| Mobasher et al (1996) | $I_{814} < 28$ | z = 0 observed |
| Sawicki et al (1997) | $I_{814} < 27$ | z = 0 observed + UV models + Ly∞ |
| Cowie (1996) | $H + K < 22.5$ | z = 0 observed |

[1]Lyman limit.

Although Bershady (1995) discussed this assumption in the context of six-color low z data, it remains unclear what systematic effects this will have at high z with four-color data sampling the UV.

A further difference among the observers is the aperture they have used to measure the colors. Sawicki et al (1997) determined an "optimum aperture" that varies for each source; Gwyn & Hartwick (1996) used a fixed aperture of 0.2 arcsec, whereas Mobasher et al (1996) experimented with apertures of 0.5–3 arcsec depending on magnitude. Given the signal/noise and irregular structure of the images, it seems reasonable to expect that the results obtained may depend on the chosen aperture (FDA Hartwick, private communication).

As an illustration of the uncertainties arising from analyses that differ only in the aperture, fitting algorithm, and template SEDs used, the redshift estimates of Mobasher et al (1996), Lanzetta et al (1996), and Cowie (1996) are compared in Figure 9. Cowie (1996) tabulates redshifts for a significant sample of $K < 20$ galaxies from the compilations of groups at the California Technical Institute (Caltech) and the University of Hawaii. When compared with the photometric redshifts, the agreement is only satisfactory for the optical + infrared data analyzed by Cowie; a surprising fraction of spectroscopically confirmed low z galaxies are considered to be at high z by Lanzetta et al (Figure 9). Nonetheless, for the $H + K < 22$ sample discussed by Cowie (1996), the distribution of photometric redshifts obtained by Mobasher et al and Lanzetta et al are quite similar (Figure 9). Cowie's distribution reveals a somewhat unphysical gap for

*Figure 9* A study of various photometric redshift catalogs available for the HDF (see Table 2). (*a*) Photometric redshifts for three groups compared with currently available Keck spectroscopic values. Note that Cowie's (1996) estimates take advantage of deep infrared photometry. (*b*) Distribution of photometric redshifts to a limit of $H + K < 22$ [Cowie: *filled*, Lanzetta et al (1996): *solid line*, Mobasher et al (1996): *dashed line*]. (*c*) A comparison of photometric redshifts to $I < 26$. (*d*) Distribution of photometric redshifts to $I < 26$ (Mobasher et al: *shaded*, Lanzetta et al: *solid line*).

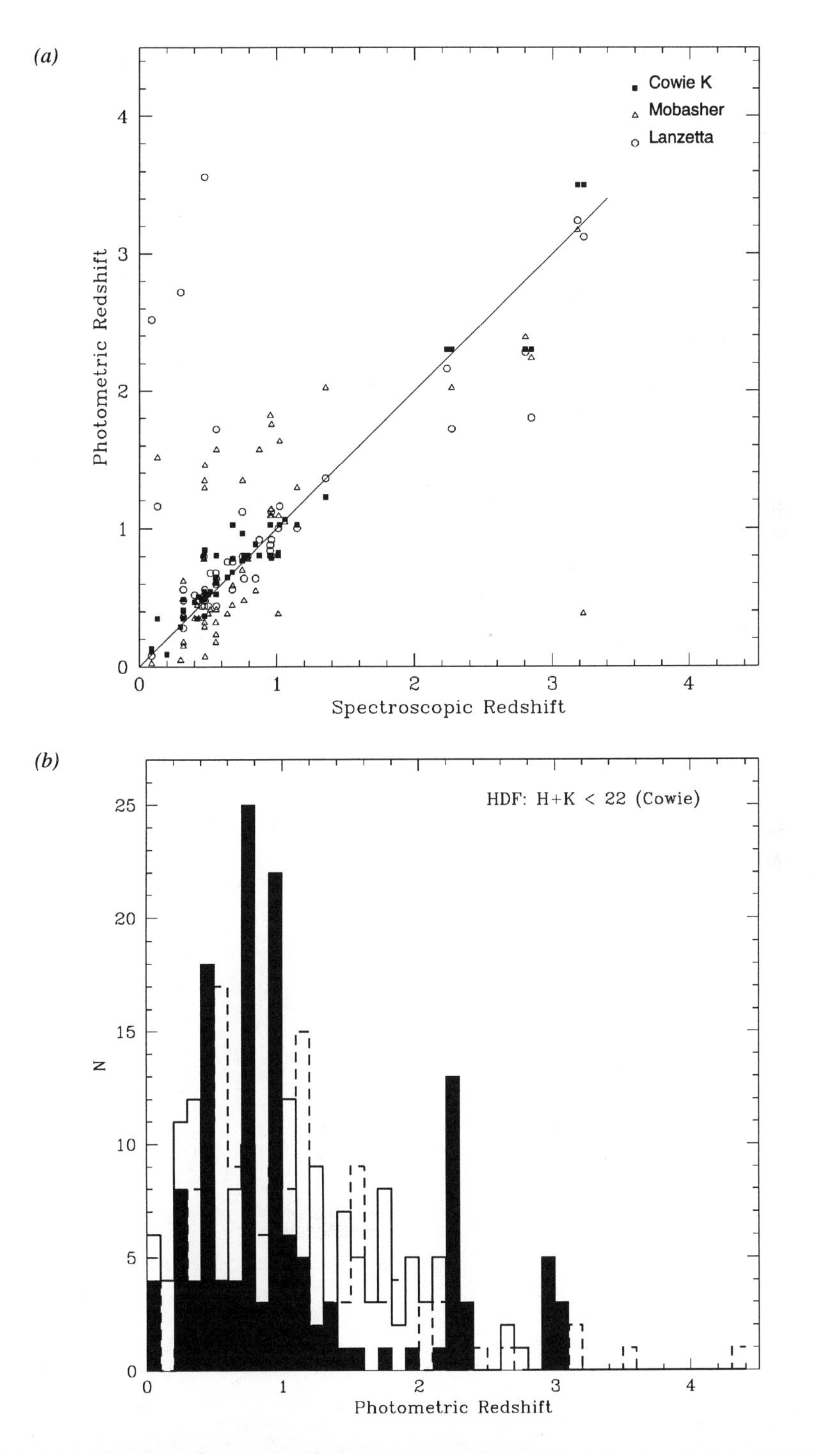
(a)
Cowie K
Mobasher
Lanzetta
Photometric Redshift
Spectroscopic Redshift
0
1
2
3
4
(b)
HDF: H+K < 22 (Cowie)
N
Photometric Redshift
0
5
10
15
20
25

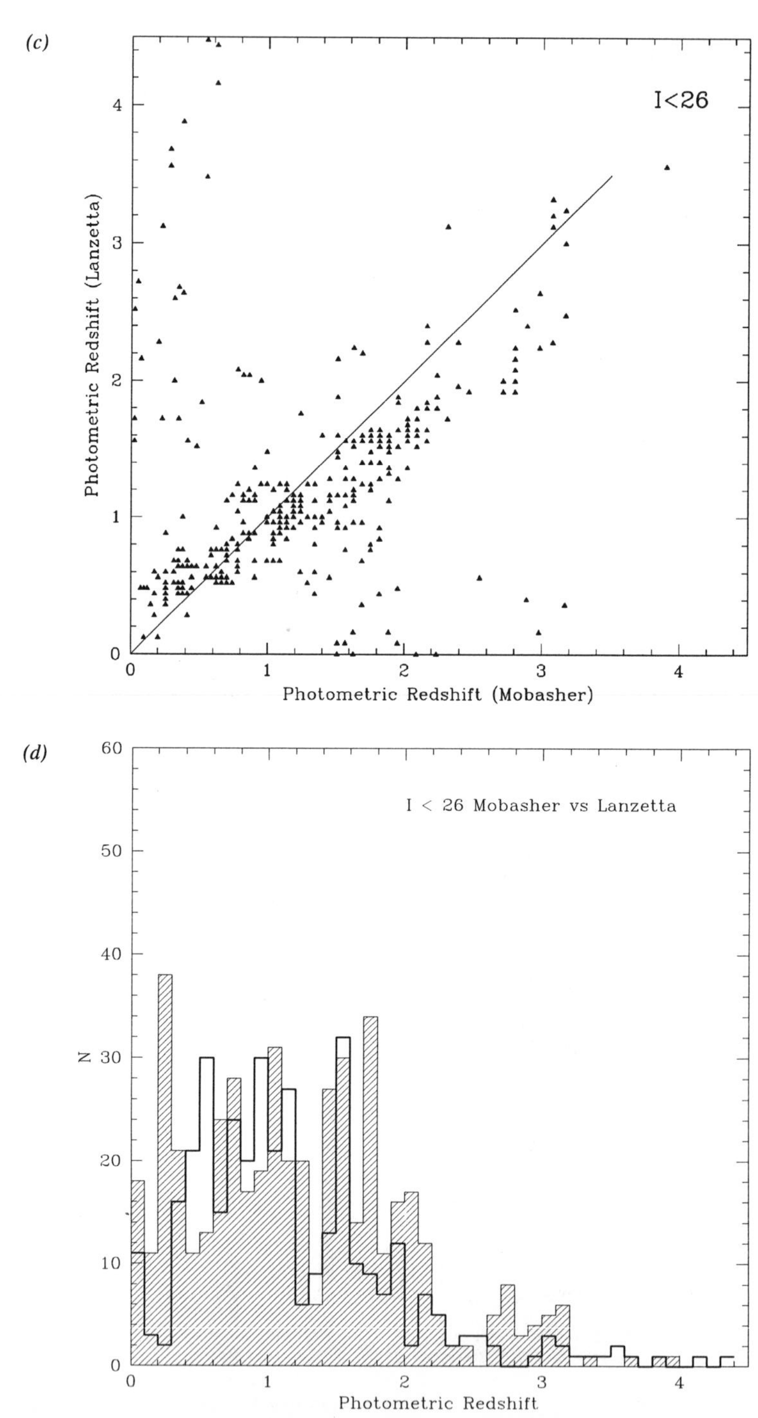
(c)
I<26
Photometric Redshift (Lanzetta)
Photometric Redshift (Mobasher)
(d)
I < 26 Mobasher vs Lanzetta
N
Photometric Redshift

$1 < z < 2$, with what seems like an artificial peak coincident with the location of the Lyman limit at the F300W filter.

$I = 26$ is a convenient deeper limit, as it represents that faintest I magnitude at which the signal-to-noise ratios for virtually all sources exceed $3\sigma$ in F300W. Figure 9 compares the redshift estimates of Mobasher et al (1996) and Lanzetta et al (1996) to this limit. Systematic differences are clearly present with effects similar to those seen in Figure 9. Nonetheless, again the overall distributions agree remarkably well (Figure 9). Of course, agreement between the various workers does not necessarily imply that the results are correct. Systematic errors could be introduced by the effects of dust, inaccurate UV SED slopes, or very strong emission lines. In this context, that the mean slope of the rest-frame UV SED for the Steidel et al (1996b) galaxies with $z > 2.3$ is less steep than the models predict is worrisome. Dust appears to be an unlikely explanation given the weak HI content. However, in delineating the overall fraction of faint galaxies with $z < 1$, 2, and 3, the photometric redshift techniques appear to give sensible results.

Several key points have emerged from these early studies. First, notwithstanding the uncertainties, both the lensing and HDF photometric redshift data point to a low mean redshift ($z \approx 1$–2) for the $I \approx 26$ population. The addition of near-infrared photometry to the HDF data appears to make the agreement with the lensing results even more convincing (Connolly et al 1997). Second, there is the small physical size (2–4 $h^{-1}$ kpc) of the faint star-forming population. Given that the bulk of the integrated light in the number counts arises from brighter systems with $z < 1$ (Section 3.1), the faint population beyond the number count break appears to represent an era of initial star formation at modest redshift $z < 2$. The complex morphology of many of the higher z luminous systems (Abraham et al 1996a, Giavilisco et al 1996), and the large ages inferred for spheroidal populations, at least in clusters (Bower et al 1992, Ellis et al 1996b), supports the suggestion that bulges form early and disk galaxies assemble by gradual infall and accretion (Cowie 1988, Baugh et al 1996). Specific cases where galaxies may be assembling in this way at high redshift have been proposed by Pascarelle et al (1996).

However, as with the $z < 1$ data, to make the connection with local systems, some reliable stable physical property such as mass is required. In fact, little is presently known about the dynamics of the most distant systems. Attempts to estimate masses from absorption line profiles are limited by a poor understanding of whether line broadening is due to shocks or turbulent motions. Examination of gravitationally magnified samples (cf Yee et al 1996, Ebbels et al 1996, Williams & Lewis 1996) will be particularly profitable as the boost of 2–3 mag will enable very detailed line studies.

# 6. THE STAR-FORMATION HISTORY OF FIELD GALAXIES

In the past year, there has been a remarkable synthesis of the star-formation history of field galaxies (as delineated by the observations reviewed here) and the gaseous and chemical evolution of the intergalactic gas (as delineated by the studies of QSO absorption lines). Lanzetta et al (1995), Wolfe et al (1995), and more recently Storrie-Lombardi et al (1996) analyzed the evolution of the cosmic density of neutral hydrogen with redshift via various QSO absorber samples. These data locate a redshift of z = 2–3, where the bulk of the present star-formation density is seen in neutral hydrogen clouds. Likewise, the chemical evolution of metallic clouds has been studied with redshift by Pettini et al (1994). Pei & Fall (1995) and Fall et al (1996) have shown, via a remarkably simple model, how these various data can be reconciled with the history of the volume-averaged star formation. Madau et al (1996) and Madau (1997) have updated this analysis with the most recent estimates of the high redshift SFR discussed in Section 5 (Figure 10). The emerging picture points to a redshift range of 1–2, in which the bulk of the present-day stellar population was assembled and perhaps most of the present-day metals produced (Cowie 1988, Songaila et al 1990, Ellis 1996b).

Appealing as this picture is, not least because it has much theoretical support (Baugh et al 1996, Kauffmann et al 1996, White 1996), it rests on preliminary observations and relies on the connection of disparate data sets, each using a different indicator of star formation and each selecting only some detectable subset of the overall population. At low z, is the star-formation density derived from the H$\alpha$ surveys (Gallego et al 1995) affected by fair sample problems that have been invoked to address the apparent low normalization of the local LF (Section 4.1)? At modest redshift, the rapid increase in the integrated star-formation density (Lilly et al 1996, Cowie et al 1996) relies on the conversion of blue light or [O II] line strength to the SFR and, quantitatively, via the extrapolation of the contribution to sources whose luminosities are fainter than the redshift survey limit. These uncertainties may affect the rate of increase with redshift beyond z = 0.8 (Lilly et al 1996). It will therefore be important to extend the redshift surveys fainter even at z < 1 to verify the extrapolation, as well as to use alternative diagnostics at moderate redshift such as H$\alpha$ fluxes secured via infrared spectroscopy. There are also promising new approaches to constrain the intermediate z SFR based entirely on emission line searching (Meisenheimer et al 1996). Similarly, at high z, the LF and spatial distribution of Lyman limit–selected samples needs to be explored via panoramic surveys based on the highly successful pilot studies (Steidel et al 1996c). The star formation densities derived from these data must be regarded as lower limits until the effects of dust have been properly explored.

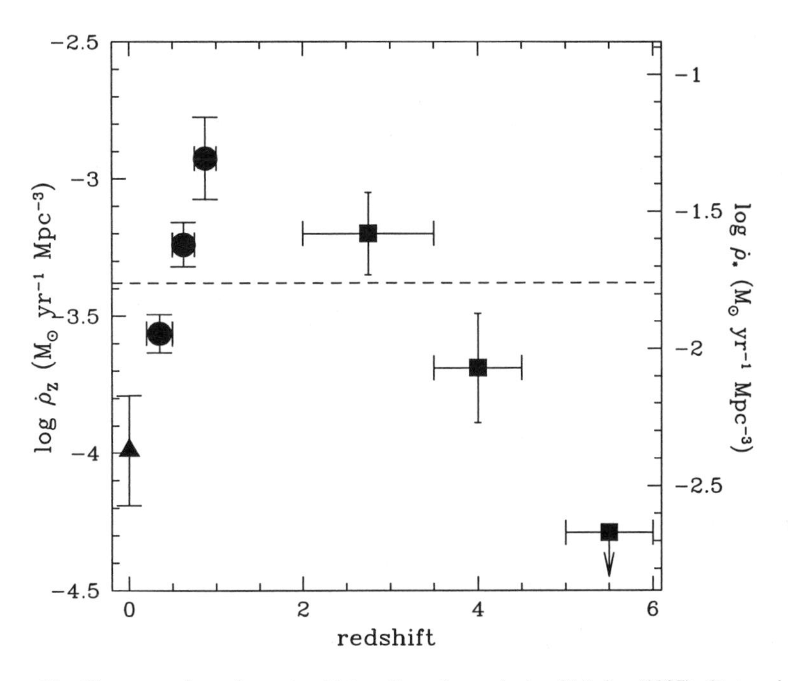


*Figure 10* Element and star-formation history from the analysis of Madau (1997). Data points provide a measurement or a lower limit to the universal metal ejection rate (*left ordinate*) and total star-formation density (*right ordinate*). *Triangle*: local H$\alpha$ survey of Gallego et al (1995). *Filled dots*: CFRS redshift survey Lilly et al (1996). *Filled squares*: Lyman limit galaxies observed in the HDF. The *dashed line* depicts the fiducial rate equivalent to the mass density of local metals divided by the present age of the universe ($\Omega = 1$, $H_0 = 50$ kms s$^{-1}$ Mpc$^{-1}$).

Despite these caveats, the trend is highly encouraging. A clear gap emerges, however, in our knowledge in the redshift range between that reviewed here at $z < 1$ and that revealed at $z > 2.3$ in the HST work. This has motivated the construction of a new generation of ground-based infrared spectrographs free from OH background light (Iwamuro et al 1994, LeFevre et al 1996c, Taylor & Colless 1996, Piché et al 1997), which aim to survey this difficult region in conjunction with optical-UV imaging. Together with NICMOS on HST, which should have powerful background-limited capabilities in both deep imaging and low-resolution grism spectroscopic modes, the intermediate redshift range can be systematically explored in a way that was highly successful for the $z < 1$ population.

The nonzero metallicity of the highest redshift absorbing clouds (Cowie 1996) also points to an earlier era of modest star formation possibly associated with the small but convincing population of high z galaxies already known to contain old stars (Dunlop et al 1996, Stockton et al 1995). The detection of high redshift sources at far infrared and submillimeter wavelengths (Omont et al 1996) together with the successful deployment of the Infrared Space Observatory and the SCUBA submillimeter array detector (Gear & Cunningham 1995) augurs well for surveying the $z > 4$ universe for earlier eras of star formation. This will remain an important observational challenge even if the bulk of the star-formation activity is convincingly demonstrated to occur at $z = 1$–$2$. To physically understand the processes that lead to galaxy formation, exploration of the high z tail is surely necessary.

Tremendous progress has been made in observational cosmology. The subject of galaxy formation and evolution has moved firmly from the realm of theoretical speculation into that of systematic observation. Some lessons are, however, being learned. Perhaps the most important of these is the dangers of relying purely on morphology and star-formation diagnostics to connect what may, in fact, be very different populations observed via different techniques at high and low z. Clearly we seek more representative physical parameters to subclassify the data sets over a range of look-back times in order to test detailed hypotheses. A further hindrance is the absence of well-defined local data of the kind needed for detailed comparisons with the high z samples (Koo & Kron 1992). However, notwithstanding the formidable challenges of studying the distant universe, the combination of deeper redshift surveys and morphologies from HST has demonstrated quantitatively the presence of rapid evolution in a subset of the population to $z = 1$. The absence of a dominant population of star-forming galaxies at $z = 3$ and the small physical sizes of the faint HDF images delineate a simple picture that is consistent with hierarchical galaxy formation and knowledge of the properties of intervening gas clouds as studied in QSO absorption lines. Galactic history seems to have been remarkably recent, which can only be our good fortune, given the power of our new facilities to observe these eras in considerable detail. The observational picture is already emerging very rapidly, but much work and ingenuity will be needed to identify the physical processes that drive the evolutionary trends now revealed.

## ACKNOWLEDGMENTS

I thank my collaborators, colleagues, and visitors at Cambridge, particularly Roberto Abraham, Matthew Colless, George Efstathiou, Masataka Fukugita, Simon Lilly, and Max Pettini for their critical and helpful comments on this review. Many workers sent detailed accounts of their views on the sensitive and complex issues discussed in Section 4. Special thanks are due to Emmanuel Bertin, Ray Carlberg, Stephane Charlot, Len Cowie, Julianne Dalcanton, Simon

Driver, Harry Ferguson, Luigi Guzzo, Karl Glazebrook, David Hartwick, David Koo, Huan Lin, Stacy McGaugh, Nigel Metcalfe, John Peacock, Tom Shanks, and Elena Zucca. I apologize to these and others if I have failed to represent their particular viewpoint fairly. I thank Bernard Sadoulet, Joe Silk, and Tom Broadhurst for their hospitality and support in Berkeley during the summer of 1996, when much of this review was written. Finally, I thank Allan Sandage for his numerous suggestions and encouragement.



*Literature Cited*


Abraham RG, Tanvir NR, Santiago B, Ellis RS, Glazebrook K, van den Bergh S. 1996a. *MNRAS* 279:L47–L52

Abraham RG, van den Bergh S, Glazebrook K, Ellis RS, Santiago B, et al. 1996b. *Ap. J. Suppl.* 101:1–17

Arimoto N, Yoshii Y, Takahara F. 1992. *Astron. Astrophys. Suppl. Ser.* 253:21–34

Baade W. 1957. In *Stellar Populations*, ed. DJK O'Connell, pp. 3–24. Rome: Vatican Obs.

Babul A, Ferguson H. 1996. *Ap. J.* 458:100–19

Babul A, Rees MJ. 1992. *MNRAS* 255:346–50

Barnes J, Hernquist L. 1995. *Annu. Rev. Astron. Astrophys.* 30:705–42

Baugh C, Cole S, Frenk CS. 1996. *MNRAS* 282:L27–L32

Baum WA. 1962. In *Problems of Extragalactic Research*, ed. G McVittie, *IAU Symp.* 15:390–400

Bernstein GM, Nichol RC, Tyson JA, Ulmer MP, Wittman D. 1995. *Astron. J.* 110:1507–25

Bernstein GM, Tyson JA, Brown WR, Jarvis JF. 1994. *Ap. J.* 426:516–23

Bershady M. 1995. *Astron. J.* 109:87–120

Bertelli G, Bressan A, Chiosi C, Fagotto F, Nasi E. 1994. *Astron. Astrophys. Suppl. Ser.* 106:275–302

Bertin E, Dennefeld M. 1997. *Astron. Astrophys.* 317:43–53

Bezecourt J, Soucail G. 1997. *Astron. Astrophys.* 317:661–69

Binggeli B, Sandage A, Tammann GA. 1988. *Annu. Rev. Astron. Astrophys.* 26:509–60

Blandford RD, Narayan R. 1992. *Annu. Rev. Astron. Astrophys.* 30:311–58

Bower RG, Lucey JR, Ellis RS. 1992. *MNRAS* 254:601–13

Brainerd TG, Smail IR, Mould JR. 1995. *MNRAS* 275:781–89

Broadhurst TJ. 1997. *Ap. J. Lett.* In press

Broadhurst TJ, Ellis RS, Glazebrook K. 1990. *Nature* 355:55–58

Broadhurst TJ, Ellis RS, Shanks T. 1988. *MNRAS* 235:827–56

Broadhurst TJ, Taylor A, Peacock JA. 1995. *Ap. J.* 438:49–61

Bruzual G. 1983. *Ap. J.* 273:105–27

Bruzual G, Charlot S. 1993. *Ap. J.* 405:538–53

Burkey JD, Keel WC, Windhorst RA, Franklin BE. 1994. *Ap. J.* 429:L13–17

Butcher H, Oemler A. 1978. *Ap. J.* 219:18–30

Carlberg R. 1992. *Ap. J. Lett.* 399:L31–L34

Carlberg R. 1996. In *Galaxies in the Young Universe*, ed. H Hippelein, K Meisenheimer, H-J Roeser, pp. 206–14. Berlin: Springer Verlag

Carlberg R, Charlot S. 1992. *Ap. J.* 397:5–13

Carlberg RG, Cowie LL, Songaila A, Hu EM. 1997. *Ap. J.* In press

Carlberg RG, Yee HKC, Ellingson E, Abraham R, Gravel P, et al. 1996. *Ap. J.* 462:32–49

Carroll SM, Press WH, Turner EL. 1992. *Annu. Rev. Astron. Astrophys.* 30:499–542

Charlot S. 1996. In *From Stars to Galaxies*, ed. C Leitherer, U Fritze, *Astron. Soc. Pac. Conf. Ser.* pp. 275–86

Charlot S, Worthey G, Bressan A. 1996. *Ap. J.* 457:625–44

Charlton J, Churchill CW. 1996. *Ap. J.* 465:631–45

Churchill CW, Steidel CC, Vogt S. 1996. *Ap. J.* 471:164–72

Clements D, Couch WJ. 1996. *MNRAS* 280: L43–L48

Cohen JG, Cowie LL, Hogg DW, Songaila A, Blandford R, et al. 1996. *Ap. J. Lett.* 471:L5–L9

Cole S, Aragon-Salamanca A, Frenk CS, Navarro JF, Zepf SE. 1994a. *MNRAS* 271: 781–806

Cole S, Ellis RS, Broadhurst TJ, Colless M. 1994b. *MNRAS* 267:541–47

Cole S, Treyer M-A, Silk J. 1992. *Ap. J.* 385:9–25
Coleman GD, Wu C-C, Weedman DW. 1980. *Ap. J. Suppl.* 43:393–416
Colless M. 1997. In *Wide Field Spectroscopy*, ed. M Kontizas, E Kontizas. Dordrecht: Kluwer. In press
Colless M, Ellis RS, Taylor K, Broadhurst TJ, Peterson BA. 1993. *MNRAS* 261:19–38
Colless M, Ellis RS, Taylor K, Hook RN. 1990. *MNRAS* 244:408–23
Colless M, Schade D, Broadhurst TJ, Ellis RS. 1994. *MNRAS* 267:1108–20
Colley WN, Rhoads JE, Ostriker JP, Spergel DN. 1996. *Ap. J. Lett.* 473:L63–L66
Connolly AJ, Csabai I, Szalay AS, Koo DC, Kron RG, Munn JE. 1995. *Astron. J.* 110:2655–664
Couch WJ, Ellis RS, Sharples RM, Smail I. 1994. *Ap. J.* 430:121–38
Cowie LL. 1988. In *The Post-Recombination Universe*, ed. N Kaiser, A Lasenby, pp. 1–18. Dordrecht: Kluwer
Cowie LL. 1997. See Tanvir et al pp. 67–74 (see also http://www.ifa.hawai.edu/cowie/hdf.html)
Cowie LL, Hu EM, Songaila A. 1995a. *Nature* 377:603–5
Cowie LL, Hu EM, Songaila A. 1995b. *Astron. J.* 110:1576–83
Cowie LL, Hu EM, Songaila A, Egami E. 1997. *Ap. J. Lett.* 481:L9–L13
Cowie LL, Lilly SJ, Gardner J, McLean IS. 1989. *Ap. J.* 332:L29–L32
Cowie LL, Songaila A, Hu EM. 1996. *Nature* 354:460–61
Cowie LL, Songaila A, Hu EM, Cohen JD. 1996. *Astron. J.* 112:839–64
da Costa L, Geller MJ, Pellegrini PS, Latham DW, Fairall AP, et al. 1994. *Ap. J. Lett.* 424:L1–L4
Dalcanton J. 1993. *Ap. J. Lett.* 415:L87–L90
Dalcanton J, Schectman S. 1996. *Ap. J. Lett.* 465:L9–L13
Dalcanton J, Spergel DN, Gunn JE, Schmidt M, Schneider DP. 1997. *Astron. J.* In press
Davis M, Geller MJ. 1976. *Ap. J.* 208:13–19
de Blok WJG, McGaugh SS, van der Hulst JM. 1996. *MNRAS* 283:18–54
Donas J, Milliard B, Laget M. 1995. *Astron. Astrophys.* 303:661–72
Djorgovski S, Soifer BT, Pahre MA, Larkin JE, Smith JD, et al. 1995. *Ap. J.* 438:L13–L16
Djorgovski S, Thompson D. 1992. In *Stellar Populations of Galaxies*, ed. B Barbuy, A Renzini, pp. 337–48. Dordrecht: Kluwer
Djorgovski S, Pahre MA, Bechtold J, Elston R. 1996. *Nature* 382:234–36
Dressler A, Oemler A, Butcher H, Gunn JE. 1994. *Ap. J.* 430:107–20
Driver SP, Phillipps S. 1996. *Ap. J.* 469:529–34
Driver SP, Windhorst RA, Griffiths RE. 1995a. *Ap. J.* 453:48–64
Driver SP, Windhorst RA, Ostrander EJ, Keel WC, Griffiths RE, Ratnatunga KU. 1995b. *Ap. J. Lett.* 449:L23–L27
Dunlop J, Peacock JA, Spinrad H, Dey A, Jiminez R, et al. 1996. *Nature* 381:581–84
Eales SA. 1993. *Ap. J.* 404:51–62
Ebbels T. 1997. See Tanvir et al, pp. 257–60
Ebbels T, LeBorgne J-F, Pello R, Ellis RS, Kneib J-P, et al. 1996. *MNRAS* 281:L75–L81
Efstathiou G. 1995. *MNRAS* 272:L25–L30
Efstathiou G, Bernstein G, Katz N, Tyson JA, Guhathakurta P. 1991. *Ap. J. Lett.* 380:L47–L50
Efstathiou G, Ellis RS, Peterson BA. 1988. *MNRAS* 223:431–61
Ellis RS. 1995. In *Stellar Populations*, ed. P van der Kruit, G Gilmore, IAU Symp. 164, pp. 291–300. Dordrecht: Kluwer
Ellis RS. 1996a. In *Unsolved Problems in Astrophysics*, ed. JN Bahcall, JP Ostriker, pp. 159–74. Princeton, NJ: Princeton Univ. Press
Ellis RS. 1996b. In *The Early Universe with the VLT*, ed. J Bergeron, pp. 65–78. Berlin: Springer
Ellis RS, Colless M, Broadhurst TJ, Heyl JS, Glazebrook K. 1996a. *MNRAS* 280:235–251
Ellis RS, Gondhalekar P, Efstathiou G. 1982. *MNRAS* 201:223–51
Ellis RS, Smail IR, Couch WJ, Oemler A, Dressler A, et al. 1996b. *Ap. J.* In press
Fall SM, Pei YC, Charlot S. 1996. *Ap. J.* 464:L43–L46
Ferguson H, McGaugh S. 1995. *Ap. J.* 440:470–84
Fort B, Mellier Y. 1994. *Astron. Astrophys. Rev.* 5:239–92
Fort B, Mellier Y, Dantel-Fort M. 1997. *Astron. Astrophys.* In press
Freeman KC. 1970. *Ap. J.* 160:811–30
Fukugita M, Takahara F, Yamashita K, Yoshii Y. 1990. *Ap. J. Lett.* 361:L1–L4
Gallagher J, Bushouse H, Hunter D. 1989. *Astron. J.* 97:700–7
Gallego J, Zamorano J, Aragon-Salamanca A, Rego M. 1995. *Ap. J. Lett.* 445:L1–L4
Gardner J. 1995. *Ap. J.* 452:538–48
Gardner J, Cowie LL, Wainscoat R. 1993. *Ap. J. Lett.* 415:L9–L12
Gardner J, Sharples RM, Carrasco BE, Frenk CS. 1996. *MNRAS* 282:L1–L6
Gear WK, Cunningham CR. 1995. In *Multi-Feed Systems for Radio Telescopes*, ed. DT Emerson, JM Payne, pp. 215–21. San Francisco: Astron. Soc. Pac.
Giallongo E, Charlot S, Cristiani S, D'Odorico S, Fontana A. 1996. In *Early Universe with the VLT*, ed. J Bergeron, pp. 208–18. Berlin: Springer

Giavilisco M, Steidel CC, Macchetto FD. 1996. *Ap J.* 470:189–94
Giraud E. 1992. *Astron. Astrophys.* 257:501–4
Glazebrook K, Ellis RS, Colless M, Broadhurst TJ, Allington-Smith JR, Tanvir NR. 1995a. *MNRAS* 275:157–68
Glazebrook K, Ellis RS, Santiago B, Griffiths RE. 1995b. *MNRAS* 275:L19–L22
Glazebrook K, Peacock JA, Miller LA, Collins C. 1995c. *MNRAS* 275:169–84
Griffiths R, Ratnaturga KU, Neuschaefer LW, Casertano S, Im M, et al. 1994. *Ap. J.* 437:67–82
Gronwall C, Koo DC. 1995. *Ap. J. Lett.* 440:L1–L4
Guhathakurta P, Tyson JA, Majewski SR. 1990. *Ap. J. Lett.* 357:L9–L12
Guiderdoni B, Rocca-Volmerange B. 1987. *Astron. Astrophys.* 186:1–21
Guzman R, Koo DC, Faber SM, Illingworth GD, Takamiya M, et al. 1996. *Ap. J. Lett.* 460:L5–L9
Gwyn SDJ, Hartwick FDA. 1996. *Ap. J. Lett.* 468:L77–L80
Hammer F, Flores H, Lilly SJ, Crampton D, LeFevre O, et al. 1997. *Ap. J.* 481:49–82
Heyl JS, Colless M, Ellis RS, Broadhurst TJ. 1997. *MNRAS* 285:613–34
Hu EM, McMahon RG. 1996. *Nature* 382:231–33
Huang J-S, Cowie LL, Gardner JP, Hu EM, Songaila A, Wainscoat RJ. 1997. *Ap. J.* 476:12–21
Hubble E. 1926. *Ap. J.* 64:321–69
Hubble E. 1936. *Ap. J.* 84:517–54
Hubble E, Tolman RC. 1935. *Ap. J.* 82:302–37
Huchra JP. 1977. *Ap. J. Suppl.* 35:171–95
Hudon JD, Lilly SJ. 1996. *Ap. J.* 469:519–28
Humason ML, Mayall NU, Sandage AR. 1956. *Astron. J.* 61:97–162
Illingworth G, Gallego J, Guzman R, Lowenthal JD, Phillips AC, et al. 1997. See Tanvir et al 1997. In press
Impey C, Bothun G, Malin DF. 1988. *Ap. J.* 330:634–60
Impey C, Sprayberry D, Irwin M, Bothun G. 1996. *Ap. J. Suppl.* 105:209–68
Infante L, Pritchet CJ. 1995. *Ap. J.* 439:565–83
Iwamura F, Maihara T, Oya S, Tsukamoto H, Hall D, et al. 1994. *Pac. Astron. Soc. J.* 46:515–21
Kauffmann G, Charlot S, White SDM. 1996. *MNRAS* 283:L117–22
Kauffmann G, Guiderdoni B, White SDM. 1994. *MNRAS* 267:981–99
Kennicutt RC. 1992. *Ap. J. Suppl.* 79:255–84
King CR, Ellis RS. 1985. *Ap. J.* 288:456–64
Kinney AL, Calzetti D, Bohlin RC, McQuade K, Storchi-Bergman T, Schmitt HR. 1996. *Ap. J.* 467:38–60
Kneib J-P, Mathez G, Fort B, Mellier Y, Soucail G, Longaretti P-Y. 1994. *Astron. Astrophys.* 286:701–17
Kneib J-P, Ellis RS, Smail I, Couch WJ, Sharples RM. 1996. *Ap. J.* 471:643–56
Koo D. 1981. *Multi-color analysis of galaxy evolution and cosmology.* PhD thesis. Univ. Calif. at Berkeley
Koo D. 1985. *Astron. J.* 90:418–40
Koo D, Gronwall C, Bruzual G. 1993. *Ap. J. Lett.* 415:L21–L24
Koo D, Guzman R, Faber SM, Illingworth GD, Berschady M, Kron RG, Takamiya M. 1995. *Ap. J. Lett.* 440:L49–L52
Koo D, Kron R. 1992. *Annu. Rev. Astron. Astrophys.* 30:613–52
Koo D, Szalay A. 1984. *Ap. J.* 282:390–97
Koo DC, Vogt NP, Phillips A, Guzman R, Wu KL, et al. 1996. *Ap. J.* 469:535–41
Kron R. 1978. *Photometry of a complete sample of faint galaxies.* PhD thesis. Univ. Calif. Berkeley
Lanzetta KM, Wolfe AM, Turnshek DA. 1995. *Ap. J.* 440:435–57
Lanzetta KM, Yahil A, Fernandez-Soto A. 1996. *Nature* 381:759–63
Larson R, Tinsley BM. 1977. *Evolution of Stellar Populations.* New Haven, CT: Yale Univ. Press
LeFevre O, Ellis RS, Lilly SJ, Abraham RG, Brinchmann J, et al. 1997. See Tanvir et al, pp. 81–90
LeFevre O, Hudon JD, Lilly SJ, Crampton D, Hammer F, Tresse L. 1996b. *Ap. J.* 461:534–45
LeFevre O, Vettolani P, Cuby JG, Maccagni D, Mancini D, et al. 1996c. In *Early Universe with the VLT*, ed. J Bergeron, pp. 143–50. Berlin: Springer
Leitherer C. 1996. In *From Stars to Galaxies*, ed. C Leitherer, U Fritze, *ASP Conf. Ser.* San Francisco: Astron. Soc. Pac.
Leitherer C, Alloin D, Alvensleben UF-V, Gallagher JS, Huchra JP, et al. 1996. *Publ. Astron. Soc. Pac.* 108:996–1017
Lilly SJ, Cowie LL, Gardner JP. 1991. *Ap. J.* 369:79–105
Lilly SJ, LeFevre O, Hammer F, Crampton D. 1996. *Ap. J. Lett.* 460:L1–L4
Lilly SJ, Tresse L, Hammer F, Crampton D, LeFevre O. 1995. *Ap. J.* 455:108–24
Limber DN. 1953. *Ap. J.* 117:134–44
Lin H, Kirshner RP, Schectman S, Landy SD, Oemler A, et al. 1996. *Ap. J.* 464:60–78
Lin H, Yee HC, Carlberg RG, Ellingson E. 1997. *Ap. J.* 475:494–501
Loh ED, Spillar EJ. 1986. *Ap. J.* 303:154–61
Loveday J, Peterson BA, Efstathiou G, Maddox SJ. 1992. *Ap. J.* 390:338–44
Lowenthal JD, Koo DC, Guzman R, Gallagher J, Phillips AC, et al. 1997. *Ap. J.* 481:673–88
Luppino G, Kaiser N. 1997. *Ap. J.* 475:20–28

Madau P. 1997. In *Star Formation Near and Far.* In press
Madau P, Ferguson HC, Dickinson ME, Giavalisco M, Steidel CC, Fruchter A. 1996. *MNRAS* 283:1388–404
Maddox SJ, Sutherland WJ, Efstathiou G, Loveday J, Peterson BA. 1990. *MNRAS* 247:1P–5P
Marzke RO, Huchra JP, Geller MJ, Huchra JP, Corwin HG. 1994. *Astron. J.* 108:437–45
Mattila K, Leinert C, Schnur G. 1991. In *The Early Universe from Diffuse Backgrounds*, ed. B Rocca-Volmerange, JM Deharveng, J Tran Thanh Van, et al, pp. 133–48. Gif-sur-Yvette: Ed. Front.
Mazzei P, Xu C, De Zotti G. 1992. *Astron. Astrophys.* 256:45–55
McGaugh S. 1994. *Nature* 367:538–41
McGaugh S. 1996. *MNRAS* 280:337–54
McGaugh S, Bothun GD. 1994. *Astron. J.* 107:530–42
Meisenheimer K, Beckwith S, Fockenbrock R, Fried J, Hippelein H, Hopp U, et al. 1996. In *Early Universe with the VLT*, ed. J Bergeron, pp. 165–72. Berlin: Springer
Mellier Y. 1997. See Tanvir et al, pp. 237–44
Metcalfe N, Fong R, Shanks T. 1995a. *MNRAS* 274:769–84
Metcalfe N, Shanks T, Campos A, Fong R, Gardner JP. 1996. *Nature* 383:236–39
Metcalfe N, Shanks T, Fong R. 1991. *MNRAS* 249:498–522
Metcalfe N, Shanks T, Fong R, Roche N. 1995b. *MNRAS* 273:257–76
Mihos JC. 1995. *Ap. J. Lett.* 438:L75–L78
Miralda-Escude J, Cen R, Ostriker JP, Rauch M. 1996. *Ap. J.* 471:582–616
Mobasher B, Rowan-Robinson M, Georgakakis A, Eaton N. 1996. *MNRAS* 282:L7–L14
Mobasher B, Sharples RM, Ellis RS. 1993. *MNRAS* 263:560–74
Moustakas LA, Davis M, Graham JR, Silk J, Peterson BA, Yoshii Y. 1997. *Ap. J.* 475:445–56
Munn JA, Koo DC, Kron RG, Majewski SR, Bershady MA, Smetanka JJ. 1997. *Ap. J. Suppl.* 109:45–77
Mutz S, Windhorst RA, Schmidtke PC, Pascarelle SM, Griffiths RE, et al. 1994. *Ap. J. Lett.* 434:L55–L58
Neuschaefer LW, Windhorst RA. 1995. *Ap. J.* 439:14–28
Neuschaefer LW, Windhorst RA, Dressler A. 1991. *Ap. J.* 382:32–43
Neuschaefer LW, Im M, Ratnatunga KU, Griffiths RE, Casertano S. 1997. *Ap. J.* 480:59–71
O'Connell RW, Marcum P. 1997. See Tanvir et al, pp. 63–66
Odewahn SC, Windhorst RA, Driver SP, Keel WC. 1996. *Ap. J. Lett.* 472:L13–L16
Omont A, Petitjean P, Guilloteau S, McMahon RG, Solomon P, Pecontal E. 1996. *Nature* 382:428–31
Pascarelle S, Windhorst RA, Keel WC, Odewahn SC. 1996. *Nature* 383:45–50
Patton DR, Pritchet CJ, Yee HKC, Ellingson E, Carlberg RG. 1997. *Ap. J.* 475:29–42
Peacock JA. 1997. *MNRAS* 284:885–98
Peebles PJE. 1971. *Physical Cosmology.* Princeton, NJ: Princeton Univ. Press
Peebles PJE. 1980. *The Large Scale Structure of the Universe.* Princeton, NJ: Princeton Univ. Press
Peebles PJE. 1994. *Principles of Physical Cosmology.* Princeton, NJ: Princeton Univ. Press
Pei YC, Fall SM. 1995. *Ap. J.* 454:69–76
Pence WJ. 1976. *Ap. J.* 203:39–51
Peterson BA, Ellis RS, Efstathiou G, Shanks T, Bean AJ, et al. 1986. *MNRAS* 221:233–55
Peterson BA, Ellis RS, Kibblewhite EJ, Bridgeland MT, Hooley T, Horne D. 1979. *Ap. J. Lett.* 233:L109–13
Pettini M, Smith LJ, Hunstead RW, King DL. 1994. *Ap. J.* 426:79–96
Phillipps S, Driver S. 1995. *MNRAS* 274:832–44
Phillipps S, Fong R, Ellis RS, Fall SM, MacGillivray HT. 1978. *MNRAS* 182:673–85
Piché F, Parry IR, Ennico F, Ellis RS, Pritchard J, et al. 1997. In *Optical Telescopes of Today & Tomorrow*, ed. A Ardeberg. SPIE 2871, pp. 1332–41 Bellingham, WA: Int. Soc. Opt. Eng.
Poggianti B. 1997. *Astron. Astrophys. Suppl. Ser.* 122:399–407
Pozzetti L, Bruzual G, Zamorani G. 1996. *MNRAS* 281:953–69
Rix H-W, Guhathakurta P, Colless M, Ing K. 1997. *MNRAS* 285:779–92
Rocca-Volmerange B, Guiderdoni B. 1990. *MNRAS* 247:166–72
Roche N, Ratnatunga K, Griffiths RE, Im M, Neuschaefer L. 1996. *MNRAS* 282:1247–73
Roche N, Shanks T, Metcalfe N, Fong R. 1993. *MNRAS* 263:360–68
Rousseau J, Di Nella H, Paturel G, Petit C. 1996. *MNRAS* 282:144–48
Sandage AR. 1961. *Ap. J.* 134:916
Sandage AR. 1995. In *The Deep Universe*, ed. B Binggeli, R Buser, pp. 1–229. Berlin: Springer-Verlag
Sandage A, Visvanathan N. 1978. *Ap. J.* 223: 707–29
Sawicki M, Lin H, Yee HKC. 1997. *Astron. J.* 113:1–12
Schade D, Carlberg RG, Yee HKC, Lopez-Cruz O, Ellingson E. 1996. *Ap. J. Lett.* 465:L103–6
Schade D, Lilly SJ, Crampton D, Hammer F, LeFevre O, Tresse L. 1995. *Ap. J. Lett.* 451:L1–L4
Schechter P. 1976. *Ap. J.* 203:297–306

Schombert JM, Bothun GD, Schneider SE, McGaugh SS. 1992. *Astron. J.* 103:1107–33

Shanks T. 1990. In *Galactic and Extragalactic Background Radiation*, ed. S Bowyer, C Leinert, pp. 269–81. Dordrecht: Kluwer

Simard L, Pritchet C. 1997. *Ap. J.* In press

Smail IR, Ellis RS, Aragon-Salamanca A, Soucail G, Mellier Y, Giraud E. 1993. *MNRAS* 263:628–40

Smail IR, Ellis RS, Fitchett MJ. 1994. *MNRAS* 270:245–70

Smail IR, Ellis RS, Fitchett MJ, Edge AC. 1995. *MNRAS* 273:277–94

Smail IR, Dressler A, Kneib J-P, Ellis RS, Couch WJ, et al. 1996. *Ap. J.* 469:508–18

Songaila A, Cowie LL, Hu E, Gardner JP. 1994. *Ap. J. Suppl.* 94:461–515

Songaila A, Cowie LL, Lilly SJ. 1990. *Ap. J.* 348:371–77

Sprayberry D, Impey C, Irwin M. 1996. *Ap. J.* 463:535–42

Steidel CC, Dickinson ME, Bowen DV. 1993. *Ap. L. Lett.* 413:L77–L80

Steidel CC, Dickinson M, Meyer DM, Adelberger K, Sembach KR. 1997. *Ap. J.* 480:568–88

Steidel CC, Dickinson ME, Persson SE. 1994. *Ap. J. Lett.* 437:L75–L78

Steidel CC, Giavalisco M, Pettini M, Dickinson ME, Adelberger K. 1996a. *Ap. J. Lett.* 462:L17–L21

Steidel CC, Giavalisco M, Dickinson ME, Adelberger K. 1996b. *Astron. J.* 112:352–58

Steidel CC, Hamilton D. 1992. *Astron. J.* 104:941–49

Stockton A, Kellogg M, Ridgeway SE. 1995. *Ap. J. Lett.* 443:L69–L72

Storrie-Lombardi L, McMahon RG, Irwin M, Hazard C. 1996. *Ap. J.* 468:121–38

Struck-Marcell C, Tinsley BM. 1978. *Ap. J.* 221:562–66

Tammann G. 1984. In *Clusters of Galaxies*, ed. F Mardirossian, G Giuricin, M Mezzetti, pp. 529–52. Dordrecht: Reidel

Tanvir NR, Aragon-Salamanca A, Wall JU, eds. 1997. *HST and the High Redshift Universe.* Singapore: World Sci.

Taylor K, Colless M. 1996. In *The Early Universe with the VLT*, ed. J Bergeron, pp. 151–55. Berlin: Springer

Telles E, Melnick J, Terlevich RJ. 1997. *MNRAS.* In press

Tinsley BM. 1972. *Ap. J.* 178:319–36

Tinsley BM. 1980. *Ap. J.* 241:41–53

Tresse L, Rola C, Hammer F, Stasinska G, LeFevre O, et al. 1996. *MNRAS* 281:847–70

Treyer MA, Lahav O. 1996. *MNRAS* 280:469–80

Tyson JA. 1988. *Astron. J.* 96:1–23

Tyson JA, Jarvis JF. 1979. *Ap. J. Lett.* 230:L153–56

Tyson JA, Valdes F, Jarvis JF, Mills AP. 1984. *Ap. J. Lett.* 281:L59–L62

Tyson JA, Valdes F, Wenk RA. 1990. *Ap. J. Lett.* 349:L1–L4

Vaisanen P. 1996. *Astron. Astrophys.* 315:21–32

Villumsen J, Freudling W, da Costa LN. 1997. *Ap. J.* 481:578–86

Vogt N, Forbes DA, Phillips AC, Gronwall C, Faber SM, Illingworth GD, Koo DC. 1996. *Ap. J.* 465:L15–L18

Weinberg S. 1974. *Gravitation & Cosmology.* New York: Wiley

Weir N, Djorgovski S, Fayyad UM. 1995. *Astron. J.* 110:1–20

Wells D. 1978. PhD thesis. Univ. Texas

White SDM. 1996. In *Science with Large Millimetre Arrays*, ed. P Shaver, Eur. South. Obs. Astrophys. Symp., p. 33. Berlin: Springer

Williams LLR, Lewis G. 1996. *MNRAS* 281:L35–L40

Williams RE, Blacker B, Dickinson M, Van Dyke Dixon W, Ferguson HC, et al. 1996. *Astron. J.* 112:1335–89

Windhorst RA, Fomalont EB, Kellermann KI, Partridge RB, Richards E, et al. 1995. *Nature* 375:471–74

Windhorst RA, Driver SP, Ostrander EJ, Mutz SB, Schmidtke PC, et al. 1996. In *Galaxies in the Young Universe*, ed. H Hippelein, pp. 265–72. Berlin: Springer-Verlag

Wolfe AM, Lanzetta KM, Foltz CB, Chaffee FH. 1995. *Ap. J.* 434:698–725

Woods D, Fahlman GG, Richer HB. 1995. *Ap. J.* 454:32–43

Worthey G. 1994. *Ap. J. Suppl.* 95:107–49

Yee HKC, Ellingson E, Bechtold J, Carlberg RG, Cuillandre J-C. 1996. *Astron. J.* 111:1783–94

Yoshii Y, Peterson BA. 1991. *Ap. J.* 372:8–20

Yoshii Y, Takahara F. 1988. *Ap. J.* 326:1–18

Zepf S, Koo DC. 1989. *Ap. J.* 337:34–44

Zucca E, Zamorani G, Vettolani G, Cappi A, Merighi R, et al. 1997. *Astron. Astrophys.* In press

Annu. Rev. Astron. Astrophys. 1997. 35:445–502


# VARIABILITY OF ACTIVE GALACTIC NUCLEI

*Marie-Helene Ulrich*
European Southern Observatory, Karl-Schwarzschild Strasse 2, D-85748 Garching bei München, Germany

*Laura Maraschi*
Brera Astronomical Observatory, via Brera 28, 20121 Milan, Italy

*C. Megan Urry*
Space Telescope Science Institute, 3700 San Martin Drive, Baltimore, Maryland 21218, USA

KEY WORDS: Seyfert galaxies, quasars, blazars, black holes, emission lines, accretion disks, jets

ABSTRACT

A large collective effort to study the variability of active galactic nuclei (AGN) over the past decade has led to a number of fundamental results on radio-quiet AGN and blazars. In radio-quiet AGN, the ultraviolet (UV) bump in low-luminosity objects is thermal emission from a dense medium, very probably an accretion disk, irradiated by the variable X-ray source. The validity of this model for high-luminosity radio-quiet AGN is unclear because the relevant UV and X-ray observations are lacking. The broad-line gas kinematics appears to be dominated by virialized motions in the gravity field of a black hole, whose mass can be derived from the observed motions. The "accretion disk plus wind" model explains most of the variability (and other) data and appears to be the most appropriate model at present. Future investigations are outlined.

In blazars, rapid variability at the highest energies (gamma-rays) implies that the whole continuum is relativistically boosted along the line of sight. The general correlation found between variations in TeV gamma rays and in X rays for Mrk 421, and between variations in GeV gamma rays and in the IR–optical–UV bands for 3C 279, two prototype objects, supports models in which the same population of relativistic electrons radiates the low-frequency continuum via synchrotron and the high frequency continuum via inverse Compton scattering of soft photons.



0066-4146/97/0915-0445$08.00

Identifying the dominant source of soft photons, which is at present unclear, will strongly constrain the jet physics.

---

## 1. INTRODUCTION

Active Galactic Nuclei (AGN) produce enormous luminosities in extremely compact volumes. Large luminosity variations on time scales from years to hours are common. The combination of high luminosity and short variability time scale implies that the power of AGN is produced by phenomena more efficient in terms of energy release per unit mass than ordinary stellar processes (Fabian 1979). This basic argument leads to the hypothesis that massive black holes are present in the cores of AGN. Accretion of matter onto a black hole or extraction of its rotational energy can in fact yield high radiative efficiencies (Rees et al 1982, Rees 1984).

The basic AGN paradigm developed thus far consists of a central supermassive black hole, surrounded by an accretion disk, or more generally optically thick plasma, glowing brightly at ultraviolet (UV) and perhaps soft X-ray wavelengths. In the innermost region, hot optically thin plasma surrounding and/or mixed with the optically thick plasma gives rise to the medium and hard X-ray emission. Clouds of line-emitting gas move at high velocity around this complex core and are in turn surrounded by an obscuring torus or warped disk of gas and dust, with a sea of electrons permeating the volume within and above the torus.

In some systems, highly relativistic outflows of energetic particles along the poles of the rotating black hole, accretion disk, or torus form collimated radio-emitting jets that lead to extended radio sources. These AGN are called radio loud because their radio emission is comparatively strong; AGN without collimated jets, which therefore have weaker (but detectable) radio emission, are called radio quiet.

Variability studies have been essential in understanding the physics of the central regions of AGN, which in general cannot be resolved even with existing or planned optical/infrared (IR) interferometers. The time scales, the spectral changes, and the correlations and delays between variations in different continuum or line components provide crucial information on the nature and location of these components and on their interdependencies.

In recent years progress has been made on two fronts. First, for a handful of objects, large international collaborations have led to improved sampling, duration, and wavelength coverage in AGN monitoring campaigns. Second, the availability of uniform data sets like the International Ultraviolet Explorer (IUE) archive and various X-ray archives has made statistical comparisons possible among different classes of AGN and different wavelength bands. This article describes these recent advances, with particular emphasis on multiwavelength variability studies. Several excellent reviews have covered or touched on the subject in previous years in this series: intraday variability (Wagner & Witzel 1995), X-ray spectra and time variability of AGN (Mushotzky et al 1993), unified models for AGN (Antonucci 1993), and the earlier presentation of the black hole models for AGN (Rees 1984). We also note reviews elsewhere on AGN continuum and variability (Bregman 1990, 1994), on the properties of the gas in the inner regions of the AGN (Collin-Souffrin & Lasota 1988), on reverberation mapping of the emission line regions (Peterson 1993), and an overview of the AGN field (Blandford et al 1990).

The AGN that are the subject of this review are those in which the central optical, UV, and X-ray source and the broad emission line region (if present) are viewed directly. The word AGN is used here regardless of redshift and luminosity and therefore encompasses the words Seyfert 1 and QSO or Quasar, which are often used to designate low- and high-luminosity AGN separately. In radio-loud AGN seen at small angles to the axis of the jet, the highly nonthermal radiation produced in the jet is strongly amplified by relativistic beaming and dominates the observed continuum. In these sources, called blazars, variability is the most violent and affects the whole electromagnetic range from the radio to the gamma-ray band.

The fundamentally different character of the radiation emitted by radio-quiet AGN and by blazars dictates different observational goals and techniques. For radio-quiet AGN the focus is on (*a*) the emission mechanisms of the optical–UV–X-ray continuum and (*b*) the kinematics of the gas, with the ultimate aim of investigating the mass accretion and mass loss, and of deriving the mass of the central black hole if, indeed, it can be shown that the kinematics is dominated by virialized motions. For blazars, the goal is to understand the structure and physical state of the plasma in the jet, i.e. the geometry, acceleration, and radiation processes.

Accordingly, this review is organized into two main parts, radio-quiet AGN (Sections 2–5) and blazars (Sections 6–8), with general conclusions in Section 9.

## 2. VARIABILITY OF THE CONTINUUM IN LOW-REDSHIFT RADIO-QUIET AGN

The continuum of radio-quiet AGN varies on all observable time scales, with amplitude up to a factor of 50 or so. The variability character depends on the wavelength, and there are correlations among the variations in different energy bands, as well as astrophysically important upper limits to the time delays between these variations, as detailed below.

The electromagnetic spectrum of radio-quiet AGN, after subtraction of the stellar continuum, extends from ~1 mm to ~100 keV with a prominent broad peak in the UV–extreme UV (EUV) range (for $\nu f_\nu$ versus $\nu$; Sanders et al 1989). The spectrum from 1200–5000 Å strongly suggests that the broad peak is primarily thermal emission from a very dense medium, probably an accretion disk (Lynden-Bell 1969, Shields 1978, Malkan & Sargent 1982; but see Ferland et al 1990). The temperature of the disk, however, is not set primarily by viscous effects, as initially thought, but by irradiation from the central X-ray source, as demonstrated by the recent variability studies (see below). The spectrum is not known in the range 1000 Å to ~0.1 keV.

## 2.1 *Continuum Variability in the Optical, UV, EUV, and IR Ranges*

THE DATA BASE: RICH AND INHOMOGENEOUS Most results come from three AGN—NGC 4151, NGC 5548, and 3C 273—that have been observed extensively in numerous coordinated campaigns in the IR, optical, UV, and X-ray ranges. NGC 4151 is an exceptional AGN in terms of the richness of its phenomenology and the quantity and variety of the data available. Of the brightest AGN, it varies on the shortest time scale. NGC 5548, another nearby bright AGN, has been observed in several multiwavelength campaigns, including 7 years of nearly daily spectrophotometric monitoring in the optical (Peterson et al 1994).

The quasar 3C 273 is the only extensively studied very high-luminosity AGN, with $L_{bol} \sim 2.5 \times 10^{47}$ erg s$^{-1}$. It displays some blazar characteristics (jet and VLBI source) but also has a strong blue bump and emission lines that indicate that the optical–UV is dominated by the non-blazar component. Two dozen AGN have been observed in smaller campaigns, some with multiwavelength coverage. The results are in agreement with and extend those obtained from the three intensively observed objects.

IR–EUV TIME SCALES, AMPLITUDES, AND SPECTRAL SHAPE On time scales of many decades, data are available only for NGC 4151. Multifractal analysis of B-band photometry from 1911–1991 clearly suggests a nonlinear intermittent behavior that, if confirmed, would rule out processes based on shot noise or on the superposition of a very large number of independent events (Longo et al 1996). The data were also searched for evidence of periodicity—none was found, confirming earlier results (Lyutyi & Oknyanski 1987, and references therein).

Figures 1 and 2 show examples of variability on time scales of years to a few hours. Remarkable "low states" or "minimum states" characterized by an exceptionally weak continuum flux and the quasi-absence of the broad components of the emission lines have been observed in some AGN (e.g. NGC 1566: Alloin et al 1986; Fairall 9: Clavel et al 1989, Recondo-González et al 1997;

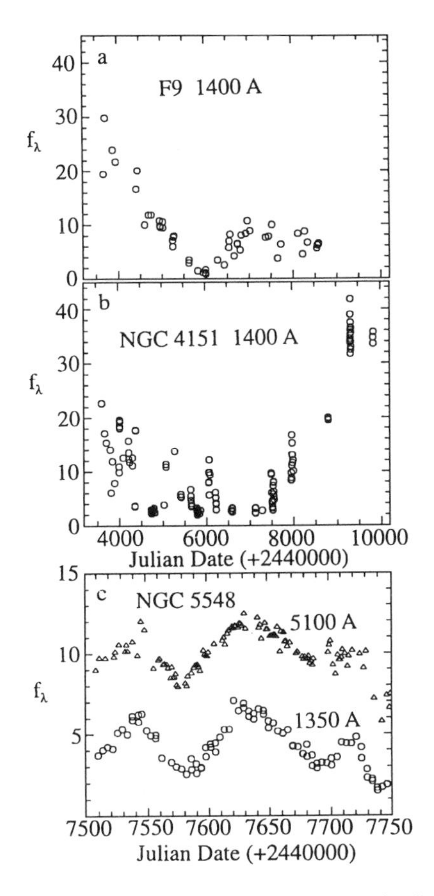


*Figure 1* Long- and short-term continuum variations in three low-luminosity AGN: (*a*) UV light curve of F9 over 14 years (Recondo-González et al 1997). (*b*) UV light curve of NGC 4151 over 17 years. The passage through the deep minimum was interrupted by short excursions to medium bright level. The vertical groups of points, unresolved on this scale, are IUE campaigns with an adequate sampling interval of typically three days. (*c*) Optical and UV light curves of NGC 5548 over 8 months (December 1988–July 1989; Clavel et al 1991, Peterson et al 1994). Ordinates in $10^{-14}$ erg s$^{-1}$ cm$^{-2}$ Å$^{-1}$.

NGC 4151) and, given time, could perhaps occur in all AGN. In NGC 4151 (Figure 1*b*) the prolonged minimum lasted from 1981–1987 with short spells at medium bright states (Perola et al 1986, Bochkarev et al 1991, Gill et al 1984, Ulrich et al 1985, 1991).

These extreme minima provide an opportunity to observe the central non-varying continuum (such as an underlying starburst or an extended scattered

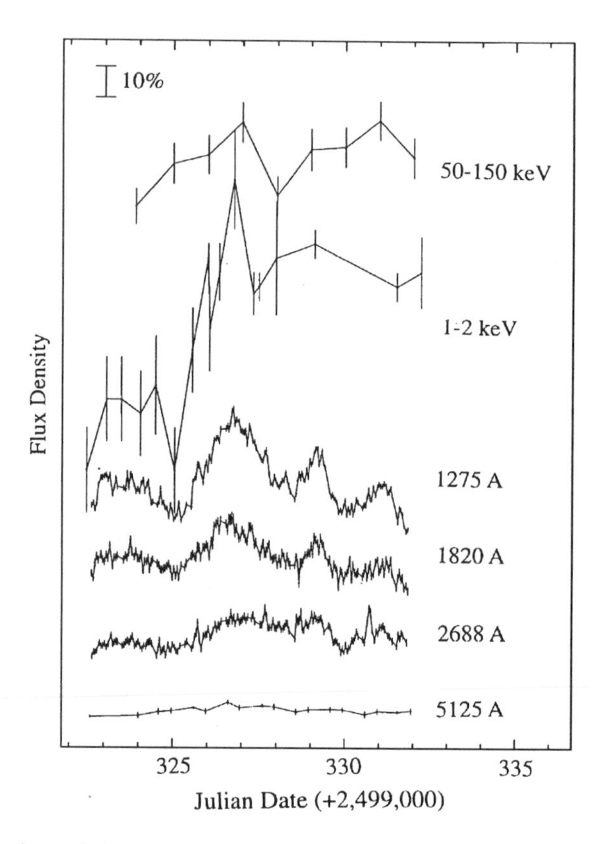


*Figure 2* Continuum light curves during the 10-day intensive period of multiwavelength observations of NGC 4151 in December 1993. The light curves are shifted vertically for convenience so a 10% change is indicated by the bar. The largest amplitude variations are in soft X rays. In the optical–UV, the amplitude decreases systematically with increasing wavelength. [From Edelson et al (1996).] Ordinates in erg $s^{-1}$ $cm^{-2}$ $Å^{-1}$.

continuum component). Hubble Space Telescope (HST) spectropolarimetric observations at the appropriate epochs (collected as Targets of Opportunity, if necessary) are highly desirable.

An example of optical–UV variability on time scales of months and weeks is shown in the light curves of NGC 5548 from December 1988 to July 1989 (Figure 1*c*). The power spectrum of the 1400-Å continuum variations is "red," with an exponent between −2 and −3 (Krolik et al 1991). A general feature of all radio-quiet AGN continuum variability is that the amplitude is inversely correlated with the time scale (see Table 1 for examples). In addition, the maxima and minima of the light curves appear to have statistically symmetrical shapes (see Figure 1*c*), but this property should be investigated more thoroughly.

**Table 1** Examples of continuum and emission line variations

| Object | Variations |
|---|---|
| Fairall 9 | Continuum variations at 1360 Å, July 1978–Oct 1984: |
| | $r_{max} = 24$ in $\Delta t \sim 5.5$ years[a] |
| | Line variations during the 1978–1987 IUE Campaign: |
| | Ly$\alpha$, C IV, and Mg II varied by factors 10, 7, and 3, respectively, with approximately the same $t_d$ of ~160 days (Clavel et al 1989, but see Recondo-Gonzalez et al 1997)[c] |
| 3C 273 | Continuum variations at 1400 Å: |
| | $r_{max} = 2$ in $\Delta t \sim 2$ years |
| | Variations of Ly$\alpha$ < 10% in 10 years (Ulrich et al 1993) |
| NGC 4151 | Short- and long-term continuum variations at 1400 Å: |
| | Factor of 1.3 in 1.5 days (Dec. 1993), 3.3 in 25 days (Nov.–Dec. 1991), 12 in 6 years (Nov. 1987–Dec. 1993) (Figures 1 and 2, and Ulrich et al 1991) |
| | Line variations during the Nov.–Dec. 1991 IUE Campaign: |
| | $t_d$[b] of C IV blue wing, red wing, and whole line are 2.6, 1.9, and 2.4 days, respectively (Ulrich & Horne 1996) |
| | During the Dec. 1987–July 1988 optical Campaign: |
| | $t_d$ of H$\beta$ was 9 days for variations by a factor 1.7 (Maoz et al 1991) |
| NGC 5548 | Continuum variations at 1400 Å, Dec. 1988–Aug. 1989: |
| | Factor 2.45 in 50 days |
| | Line variations during the same IUE Campaign: |
| | Ly$\alpha$, C IV, and He II lambda 1640 varied by a factor 1.8, 1.8, and 4 respectively with $t_d$ of about 12, 8, and 4 days (Krolik et al 1991). During the same period H$\beta$ varied by a factor 1.7 with $t_d$ of 19 days (Peterson et al 1994). |
| **Other time responses of H$\beta$** | |
| NGC 3227 | $t_d \sim 15$ days (Salamanca et al 1994, Winge et al 1995) |
| NGC 3516 | $t_d \sim 7$ days (Wanders et al 1993) |
| NGC 3783 | $t_d \sim 8$ days while $t_d$ of C IV was 5 days in the same period (Stripe et al 1994, Reichert et al 1994) |
| PG 0804+762 | $t_d \sim 93$ days (Kaspi et al 1996b) |
| PG 0953+414 | $t_d \sim 110$ days (Kaspi et al 1996b) |

[a] $\Delta t$: interval of time between successive maximum and minimum.

[b] $t_d$: time delay between the UV lines and UV continuum or between optical lines and optical continuum variations.

[c] Measurement errors and other uncertainties are detailed in references.

On time scales of days and hours, by far the best variability data were collected during the 10-day intensive optical, UV, and X-ray monitoring of NGC 4151 in December 1993 (Figure 2; Crenshaw et al 1996, Kaspi et al 1996a, Warwick et al 1996 Edelson et al 1996). The fastest variations disappear and the overall variability amplitude decreases toward longer wavelengths, as is seen in all AGN (e.g. Kinney et al 1991; Figures 1*c* and 2). The variability observed with the Extreme Ultraviolet Explorer (EUVE) satellite at 65–120 Å is consistent with this trend (Marshall et al 1997).

The shape of the optical–UV continuum always hardens when the nucleus brightens (Kinney et al 1991, Paltani & Courvoisier 1994), in contrast with BL Lac objects where the spectral shape changes little when the flux varies (Section 6.3). The spectral change occurs because two components with different spectra and variability time scales make up the continuum: the "small blue bump" (2300–4000 Å), which varies with the smallest amplitude and is a sum of permitted Fe II UV and optical multiplets plus the Balmer continuum (Wills et al 1985), and the more variable primary continuum component, whose spectral shape is difficult to determine because of various contaminations. In NGC 5548, this continuum appears to harden when the nucleus brightens (Wamsteker et al 1990, Maoz et al 1993), whereas in Fairall 9 (F9), the optical spectral slope was unchanged while the intensity increased by a factor of 20 (Lub & de Ruiter 1992).

Infrared variations, when detected at all, are of smaller amplitude and longer time scale than in the optical (Neugebauer et al 1989, Hunt et al 1994) and are consistent with being the delayed response of dust around the central source to large amplitude long-term variations of the UV flux, at least for nearby AGN (e.g. Clavel et al 1989, Nelson 1996). Neugebauer and Matthews (G Neugebauer & K Matthews, private communication) have monitored the continua of 25 quasars (5 radio loud) between 1 and 10 $\mu$ over a period of up to 25 years. If the 10-$\mu$ continuum is caused by thermal emission from dust grains that symmetrically surround the quasar and are in equilibrium with its radiation, the size of the emitting region is such that significant variations would have time scales longer than 100 years. Variations with time scales on the order of years are seen (not surprisingly) in all the radio-loud quasars. In at least one radio-quiet quasar, significant variability on the order of 5–10 years is seen, with correlated variations in all bands. These variations suggest either that the mid-IR emission in some radio-quiet quasars is nonthermal or that quite complicated structures are involved.

## 2.2 *Variability of the X-Ray Emission of Seyfert Galaxies*

Here we mention only the key points and most recent results because an excellent review is given by Mushotzky et al (1993). The X-ray emission of Seyfert galaxies consists of several components, including a power law in the medium energy X-ray range (1–10 keV; $\alpha \simeq 0.9$, where $f_\nu \propto \nu^{-\alpha}$), a soft excess usually below 1 keV, and a reflection hump in the 10- to 30-keV range. Superimposed on this continuum is a prominent Fe line and often (50%) absorption edges from highly ionized oxygen (Fabian 1996). In the hard X-ray range (>50 keV), the few OSSE data available indicate that the power law steepens, possibly with a cutoff around 100 keV (Johnson et al 1994).

Both the soft excess and the medium X-ray power-law component are variable, albeit differently; the soft excess is more strongly variable and is often but not always correlated with the medium energy X-rays. There are very few cases where the soft and medium X-ray variations appear uncorrelated. The best example is NGC 5548, observed 25 times at approximately daily intervals in December 1992 to January 1993. The soft excess component varied (factor 10) independently of the hard X-ray flux (factor 3), most noticeably in a soft X-ray flare lasting 8 days which had no medium X-ray counterpart (Done et al 1995).

The medium energy power-law component in general varies in intensity with very little change of the spectral index. In some well-observed AGN, the flux variations are accompanied by a softening of the spectrum with increasing intensity (e.g. Perola et al 1986, Mushotzky et al 1993 and references therein, Grandi et al 1992, Guainazzi et al 1994, Leighly et al 1996, Molendi et al 1993).

The most extreme soft X-ray variability occurs in Narrow-Line Seyfert 1 galaxies (NLS1), a subset of AGN with very steep soft X-ray spectra ($1 \lesssim \alpha \lesssim 4$ in the range 0.1–2.4 keV), narrow optical emission lines with full width at half maximum (FWHM) $\lesssim$2000 km s$^{-1}$, and prominent optical Fe II emission (Osterbrock & Pogge 1985, Boller & al 1996). To explain the absence of broad lines in NLS1, it has been proposed that the intense soft X-rays could blow away the inner broad-line region (BLR) or ionize it to states currently undetectable (Pounds et al 1995, Guilbert & Rees 1988). The steep soft X-ray spectrum of NLS1 may indicate a high accretion rate or a small black hole mass. In broad-line AGN, the medium–hard X-ray power law generally has luminosity significantly higher than the soft excess (Pounds & Brandt 1996), but in the NLS1 RE 1034+39, for example, the soft excess exceeds the luminosity of the medium–hard component, and moreover, the source has an exceptionally steep medium X-ray spectrum—characteristics shared with Galactic black hole candidates. This suggests that in NLS1, or at least in some of them, the accretion rate is close to the Eddington limit and the soft X-rays represent viscous heating of the accretion disk (Pounds et al 1995). Alternatively, a small black hole with an accretion rate of ~0.1 Eddington accretion rate could also emit a very hot spectrum with such an intense soft X-ray component.

Although the NLS1 display the most extreme soft X-ray variations, their range of variability merges with that of "classical" broad-line AGN. Here we summarize a few of the most spectacular examples of variability. The largest variation observed in one year was by a factor of 70 in RE J 1237+264; this object has remained weak and has the very same steep slope as in the high state (Brandt et al 1996). An optical spectrum taken a few months after the soft X-ray outburst shows emission lines of [FeX] and H$\alpha$ that are approximately 10 times brighter than those observed before or well after the burst (Pounds & Brandt 1996, Brandt et al 1996).

Strong variations are also seen in the Fe II strong AGN PHL 1092; if the radiation is isotropic, the rapid variability requires that mass be transformed into energy with an efficiency of at least 0.13, exceeding the theoretical maximum for a nonrotating black hole (Forster & Halpern 1996). Somewhat in contrast, a drastic spectral change from ultrasoft to typical soft X-ray Seyfert spectrum in RX J0134-42 occurred without change in count rate (Mannheim et al 1996; see also Pounds et al 1995).

An exceptional case is the persistent giant and rapid soft X-ray flux variability of the radio-quiet, ultrasoft, strong Fe II, narrow-line Seyfert 1 galaxy IRAS 13224-3809 (Boller, Brandt, Fabian & Fink 1997). In the first systematic monitoring of an ultrasoft NLS1, a 30-day Roentgen Satellite (ROSAT) High Resolution Imager observation revealed at least five giant amplitude variations, with the maximum observed amplitude of about a factor of 60. A variation by about a factor of 57 was detected in just two days. Variations by a factor of about 30 were also seen to occur during a 1994 observation with the Advanced Satellite for Cosmology and Astrophysics (ASCA; Otani et al 1996). Relativistic boosting effects provide the most plausible explanation of the X-ray data and may be relevant to understanding the strong X-ray variability of some steep spectrum Seyferts more generally. The variability is probably nonlinear in character, which suggests that flares and spots in the accretion disk interact nonlinearly or are affected by nonlinear flux amplification.

X-ray variability by a factor of about 50 has also been observed in one broad-line AGN, E1615+061 (Piro et al 1997). The high-state spectrum was very steep ($\alpha \sim 3$), whereas the low-state slope was near normal ($\alpha \sim 1$). There is only an upper limit to the variation time scale (16 years) and the soft X-ray activity of this broad-line AGN may not be related to the extremely fast soft X-ray variability of the NLS1 class. Finally, another extraordinary extragalactic X-ray transient is the NLS1 WPVS007, which decreased by a factor of more than 400 between the ROSAT All Sky Survey and the ROSAT pointed observations three years later (Grupe 1996).

## 2.3 *Simultaneity of the Flux Variations at Various Energies*

SIMULTANEITY OF THE UV AND OPTICAL FLUX VARIATIONS: CONSEQUENCES FOR THE NATURE OF THE OPTICAL–UV CONTINUUM The long viscous time scale for a standard optically thick, geometrically thin accretion disk (Shakura & Sunyaev 1973, Pringle 1981) is incompatible with models in which rapid flux variability is caused by variable fueling (Clarke 1987). That variable fueling does not cause rapid optical–UV–EUV variability is confirmed by the simultaneity of the variations in those wavelength bands; the current tightest limit on the time delay between the UV and optical flux variations is 0.2 days in NGC 4151 (Crenshaw et al 1996) and 0.25 days between the UV and EUV in NGC 5548 (Marshall et al 1997).

Consider the example of a disk around a black hole of mass $4 \times 10^7 M_\odot$. Because information cannot travel through the disk faster than the sound speed ($3 \times 10^6$ cm s$^{-1}$ for $T = 10^5$ °K), the delay between the rings in the disk emitting at 5400 Å and 1350 Å should be >3 years, in clear contradiction to the simultaneous flux variations in the UV and optical ranges (Krolik et al 1991, Collin-Souffrin 1991, Courvoisier & Clavel 1991).

SIMULTANEITY IN THE UV, SOFT-, AND HARD-X-RAY RANGES: REPROCESSING IS CURRENTLY THE BEST MODEL Given that viscous heating cannot be the emission process of the rapidly variable optical–UV continuum, the best alternative, suggested by the X-ray variations, is reprocessing. Simultaneity and proportionality on time scales of weeks to months between the medium energy X-ray flux and optical–UV flux have been observed in several AGN (NGC 4151: Perola et al 1986, Warwick et al 1996; NGC 5548: Clavel et al 1992, Tagliaferri et al 1996). This is readily explained by a model in which the optical–UV flux is emitted by an optically thick dense medium irradiated by a nearby variable central X-ray source (Collin-Souffrin 1991, Krolik et al 1991, Nandra et al 1991, Molendi et al 1992, Haardt & Maraschi 1993, Rokaki et al 1993, Petrucci & Henri 1997); this dense medium could be the accretion disk.

In one of the most actively investigated models, fueling occurs through an accretion disk, with angular momentum removed magnetically by field lines threading the disk surface. Explosive reconnections dissipate significant power via magnetic flares within the disk corona (Galeev et al 1979, Blandford & Payne 1982), and X-ray emission is produced via inverse Compton emission in the hot corona surrounding the cooler accretion disk (Haardt & Maraschi 1993).

These flares and variations in the optical depth of the corona (Haardt et al 1997) could be the primary cause of variability on time scales of days or less and could cause variations of the X-ray emission even with a constant accretion rate. This Comptonization model explains the average power-law slope at medium energy, the high-energy cutoff, the reflection hump, and the iron line emission. Reprocessing readily explains the correlation among power-law X rays, soft X rays, and UV emission on short time scales. In fact, it seems at present to be the only viable explanation.

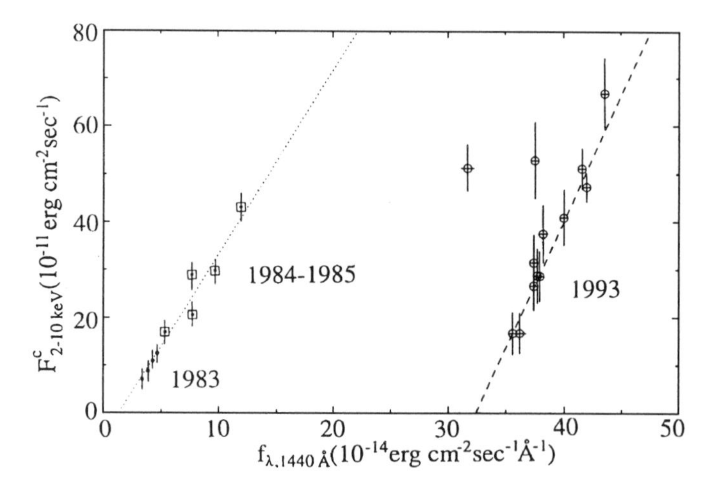


*Figure 3* X-ray flux (absorption corrected, 2–10 keV) versus UV flux (1440 Å) for NGC 4151. *Right*: ASCA observations in November 30–December 13, 1993, with best-fitting linear correlation. *Left*: EXOSAT observations in December 16, 1984, to January 28, 1985 (*large crosses*), and November 7–19, 1983 (*small crosses*), with best-fitting linear correlation. The good correlation of the UV and X-ray flux on a time scale of weeks and months breaks down at long (years) and short (days) time scales. [Adapted from Warwick et al (1996) and Perola et al (1986).]

If the corona is not uniform but is patchy (Haardt et al 1994, 1997), the reprocessed radiation is only a fraction of the UV emission and the rest is presumably accretion energy dissipated within the optically thick disk. The fraction of the UV emission due to reprocessed radiation can change "secularly." This is quite clear in NGC 4151, as can be seen in Figures 1*b* and 3. A major test of the reprocessing scenario is to verify the energy budget *among the variable components*, i.e. to check that the energy of the medium–hard X-ray component exceeds that of the the UV (and perhaps also the soft X-ray) component. This has to be checked with care and is at present very uncertain (cf Perola et al 1986, Ulrich 1994, Edelson et al 1996).

The origin of the soft X-ray emission in the reprocessing model is not definite. An attractive possibility is that the soft excess is due to reprocessing of the Comptonized emission into thermal radiation. This clearly has strong implications on the expected correlation of variability in different bands.

On time scales of years, the proportionality of the UV and medium X-ray fluxes breaks down, as is evident in NGC 4151 (Figure 3), NGC 5548, and F9, with the UV varying more than the X-ray flux (Morini et al 1986b, Perola et al 1986, Clavel et al 1992, Warwick et al 1996). In NGC 4151, a large slow-varying UV component has disappeared and reappeared during the lifetime of IUE, between 1978 and 1996 (Figures 1*b* and 3).

On time scales of days, the UV and X-ray variations are correlated, though not in a detailed way. In NGC 5548 and NGC 4151, the short time-scale variability amplitude is smaller in the UV than in the X-rays (Türler et al 1996 for NGC 5548; see Figure 2 for NGC 4151). The discrepancy from proportionality could be caused by variations of the soft X-ray emission due to viscous effects.

High-luminosity AGN tend to have larger UV/X-ray flux ratios when compared to low-luminosity AGN, which could pose an energy budget problem for the reprocessing scenario. This could be solved if the X-ray source is anisotropic and the disk receives more X-ray flux than can be inferred from the observed flux. Anisotropy of the Comptonized emission could account for a deficiency of only a factor of 2–3, but not more. In any case, at present, for high-luminosity AGN, the data are insufficient to establish whether or not there is simultaneity between the optical–UV and X-ray variations, and thus the relevance of the irradiation model for high-luminosity AGN is still uncertain.

VARIABILITY AND THE STARBURST MODEL A contrasting model views AGN as giant young stellar clusters (Terlevich et al 1992). In this case, variability results from the random superposition of "events"—supernova explosions generating rapidly evolving compact supernova remnants (cSNRs) due to the interaction of their ejecta with the high density circumstellar environment. This model is supported by the striking similarity between the optical spectra of AGN and of cSNRs (e.g. Filippenko 1989). The characteristics of an event (i.e. its light curve, amplitude, and time scale) result from the combination of complicated processes (Terlevich et al 1992). Still, the light curves of AGN of various absolute luminosities and redshifts can be predicted from this model and are found to be consistent with the observed dependence of the structure function (the curve of growth of variability with time) on luminosity and redshift (Cid Fernandes et al 1996, Cristiani et al 1996).

The light curves of cSNRs are still poorly known. At present, they appear to be consistent with the optical light curves of the low-luminosity AGN NGC 4151 and NGC 5548 (Aretxaga & Terlevich 1994). Much more detailed light curves of cSNRs are required to make a definitive check. The production of strong and rapidly variable X-rays is still difficult in this model.

## 3. VARIABILITY OF THE OPTICAL CONTINUUM IN HIGH REDSHIFT AGN

### 3.1 *Observed Characteristics of the Optical Variability*

The high surface density of high-z AGN, ~100 per square degree at $m_B$ brighter than 22, is such that a significant number of radio-quiet AGN can be recorded on a single image taken with a Schmidt telescope or a large telescope with a wide field. The technique of choice to study the optical variability of high-z radio-quiet quasars is thus the definition of large optically selected samples and their repeated broadband imaging at regular intervals (a few months to a year) over many years. This results in a light curve for each quasar in the sample, with the advantage that all the light curves have the same number of data points.

The very large effort in monitoring high-z quasars has recently come to fruition. There are now enough data in various samples to separate the effects of luminosity and redshift on the variability, avoiding the inherent correlation that exists in magnitude-limited samples. The observation that optical–UV spectra of both low- and high-luminosity AGN vary more at short than at long wavelengths (as found for high-luminosity AGN by Cutri et al 1985) accounts completely for the observed increase of variability with redshift (Giallongo et al 1991, Trevese et al 1994, Di Clemente et al 1996, Cristiani et al 1996). It also explains the long-puzzling absence of a time-dilation effect ($\Delta t_{rest\text{-}frame} = \Delta t_{observed}(1+z)^{-1}$), wherein quasars at higher redshift are sampled more frequently and for shorter time intervals in their rest frames, because this effect is compensated by the intrinsic increase of variability with decreasing rest-frame wavelength.

The recent developments come mostly from the four largest on-going programs (the first three based on Schmidt plates): (*a*) the South Galactic Pole sample of 300 radio-quiet quasars observed over 16 years (Hook et al 1994); (*b*) the monitoring program in field ESO/SERC 287 (Hawkins 1993, 1996), with more than 200 plates since 1975 (Hawkins & Véron 1993), which has produced the best quasar light curves to date; (*c*) the sample in SA 94 comprising 183 quasars observed in B over 10 years (Cristiani et al 1990); and (*d*) the sample in SA 57 based on prime focus plates at the Mayall 4-m telescope at Kitt Peak (Koo et al 1986, Trevese et al 1989).

The main result is that, in a given proper time interval and at a fixed rest-frame wavelength, more luminous AGN vary with a smaller fractional amplitude than less luminous AGN. Also, the maxima and minima of the light curves are symmetric, a result also suggested by the light curves of low-z AGN.

The analysis of the variability is normally done using the structure function, which is the curve of growth of variability with time. It is defined in slightly different ways in the literature but generally has the form $S(\Delta t_j) = < |m_{ik} - m_{i\ell}| >$, where $\Delta t_j = |t_k - t_\ell|$, $m_{ik}$ is the magnitude of the quasar $i$ at epoch $k$, and brackets signify the median of the ensemble (Hook et al 1994) or the average of the ensemble (Di Clemente et al 1996, Cristiani et al 1996).

Parametrizations of the structure function usually assume that the dependences on $\Delta t$ and luminosity are separable. In Hook et al (1994) for example (also in Trevese et al 1994 and references therein), the best-fit model has the form

$$|\Delta m| = [a + b(M_B + 25.7)]\Delta t_{rest}^{p}, \tag{1}$$

with $b$ about 0.022 and $p = 0.18 \pm 0.02$. This value of $p$ corresponds to a power spectrum of the light curve $P(\omega) = \omega^{-(1+\alpha)}$, with $\alpha = 0.36 \pm 0.04$. For comparison, in this description, $p = 0, 0.5$, and 1 correspond to uncorrelated measurements, random walk variations, and linear variations, respectively.

### 3.2 *Microlensing as a Possible Cause of Quasar Variability*

Variability of distant quasars could result from microlensing by compact bodies in intervening galaxies. The cosmological implications would be very important: If there were such a lens along all sight lines to high-z AGN, then $\Omega$ in compact objects would be close to one (Press & Gunn 1973, Blandford & Narayan 1992). (This potential cause of variability does not pertain to the nearby, rather faint AGN discussed in Section 2, as they have a low probability of having a lensing object in the line of sight.)

For the ESO/SERC 287 sample, the main predictions of microlensing for variability are borne out by the observations (Hawkins 1996). The smoothness of the light curves constrains the quasar emission region to be commensurate with the Einstein radius of the lenses. With a typical variability time scale of ~2 years and for a transverse velocity of 600 km $s^{-1}$, the Einstein radius is $\sim 8 \times 10^{-4}$ pc, corresponding to a Jupiter mass (Hawkins 1996).

Another view is that the quasar light curves give an upper limit to the lensing effect because some or all of the quasar variability can be intrinsic. Recently Schneider (1993), following an idea of Canizares (1982), made a theoretical analysis of the characteristics of the light curves in Hawkins' sample and calculated the properties of the lenses that can reproduce them. He can set an upper limit to the mass of dark matter in the form of compact objects in the specific mass range $\Delta$M 0.001–0.03 $M_\odot$. This limit corresponds to $\Omega(\Delta M) < 0.1$.

At present, the results of quasar variability studies appear to be consistent with microlensing. It is difficult to disprove that quasar variability is dominated by microlensing because the properties of the lenses—in particular, their space density and their Einstein radii—are free parameters.

## 4. EMISSION LINE VARIABILITY: RATIONALE AND METHODS

### 4.1 *Rationale for Emission Line Variability Study*

The exciting prospect of determining the mass of the central black hole is the main motivation for monitoring emission line variations in AGN. The mass is estimated in the following way. Variations in the emission line strengths of AGN are observed to echo the continuum variations with a time delay, which can be interpreted as the light travel time between the central source and the surrounding high velocity gas clouds (the BLR). Combining the radial distance to the line-emitting gas with its velocity (assuming virialized motions) allows determination of the mass of the central black hole.

It is clearly important to assess whether the gas is gravitationally bound, as well as to search for kinematic evidence of accretion and/or ordered gas motions such as infall or a rotating disk. Reverberation (or echo) mapping is a technique for inferring the structure and velocity field of the BLR from the time delays between continuum and line variations. The basic assumptions are that the BLR is ionized by a central continuum point source, the light travel time between continuum and gas clouds is much longer than the ionization or recombination times, and the line intensity is linearly correlated with the incident continuum flux (e.g. Peterson 1993).

## 4.2 *Inversion Methods, Cross-Correlations, and Modeling*

Methods of echo mapping are presented in numerous papers (e.g. Blandford & McKee 1982; Horne, Welsh & Peterson 1991; Krolik et al 1991) and in conference proceedings (Gondhalekar et al 1994). The light curve of a given line, $L(t)$, can be considered as a convolution of the continuum light curve, $C(t)$, and a transfer function (TF), $\Psi(\tau)$:

$$L(t) = \bar{L} + \int_{\tau_{min}}^{\tau_{max}} \Psi(\tau) \left[C(t-\tau) - \bar{C}\right] d\tau, \tag{2}$$

where $\bar{L}$ and $\bar{C}$ are the mean of the line and continuum intensities, respectively. The transfer function (the kernel) is a map showing where the line emission is produced in each interval of time delay by the gas along the paraboloid surfaces of constant delay. This can be generalized to give the time delay distribution at each velocity in the line profile.

A powerful and flexible inversion method is the Maximum Entropy Method, MEM (Horne 1994), especially in its velocity-resolved form, but less computationally expensive linear methods have also been proposed (Pijpers 1994, Krolik & Done 1995). The results of numerical techniques for inverting Equation 2 have been limited by the intrinsic indeterminacy of any inversion problem in the presence of noisy data collected at uneven time intervals (e.g. Vio et al 1994). Several authors have calculated theoretical velocity-delay maps for plausible BLR configurations and velocity fields (Perez et al 1992a,b, Goad et al 1993) for comparison with the result of the inversion of Equation 2.

A simpler estimate of the linear scale of the BLR comes from cross-correlating the line and continuum light curves. The peak of the cross-correlation function represents the material closest to the ionizing source, while the centroid gives an emissivity-weighted average time delay over the emission region (Edelson & Krolik 1988, Perez et al 1992a,b). These estimates are not unique, as they try to encapsulate in a single number the time-delay distribution of the emitting clouds. Note that the centroid of the transfer function can be obtained directly from the line-continuum cross-correlation function (Koratkar & Gaskell 1991, Robinson 1995).

## 5. EMISSION LINE VARIABILITY: RESULTS

Before reviewing the progress realized by the study of line variability, we briefly recall two critically important results obtained from spectroscopy alone. First, the high- and the low-ionization lines (HIL and LIL) are emitted by two different gas phases (Kwan & Krolik 1981, Collin-Souffrin et al 1986, Netzer 1987). The LIL come from a very dense medium with $N_e \geq 10^{11}$ cm$^{-3}$, an ionization parameter much less than 0.1, and a column density exceeding $10^{24}$ cm$^{-2}$. This medium must have a flat geometry, and there is the attractive possibility that the LIL are emitted by the accretion disk (Collin-Souffrin 1987). The HIL come from a more dilute medium, possibly a wind (van Groningen 1987, Collin-Souffrin & Lasota 1988) or an ensemble of clouds in a broad cone above and below the disk, with the gas density in the clouds not larger than a few $10^{10}$ cm$^{-3}$ and an ionization parameter of the order of 0.3.

Second, the gas velocity and its degree of ionization are correlated. Among the HIL, the broadest FWHM and the most extensive wings are those of the most highly ionized species (NGC 5548: Krolik et al 1991; NGC 4151: Antonucci & Cohen 1983, Ulrich et al 1984a). Similarly, in a few well observed AGN the H$\beta$ line has more extended wings than H$\alpha$. (We recall the nomenclature in this context: the main LIL are the Balmer lines, optical and UV Fe II multiplets, Mg II$\lambda$2796,2803, He I$\lambda$5876. Among the HIL, the strongest are C IV$\lambda$1548,1551, then C III]$\lambda$1909, Si IV$\lambda$1394,1403, He II$\lambda$4686, N V$\lambda$1239,1243, and Fe[X]$\lambda$6375. The Ly$\alpha$ line is special and has an intensity about twice that of C IV).

The crucial element contributed by variability studies is the linear scale of the LIL and HIL emission regions. For well-sampled high S/N data, the three-dimensional architecture and velocity field of the gas clouds can be reconstructed from the velocity-resolved transfer function. The results are at present dominated by the two bright, well-observed, strongly variable and intrinsically weak AGN, NGC 4151 and NGC 5548, and by 3C 273. Other AGN give consistent results. Most of the observational data on line variability come from the sources quoted for the continuum data in Section 2.

### 5.1 *Results of Variability: The Stratification of the Broad Line Region in Velocity and Degree of Ionization*

OBSERVED STRATIFICATION Among the HIL, the more highly ionized lines have shorter delays and larger amplitudes. The He II lines at $\lambda\lambda$1640,4686 and

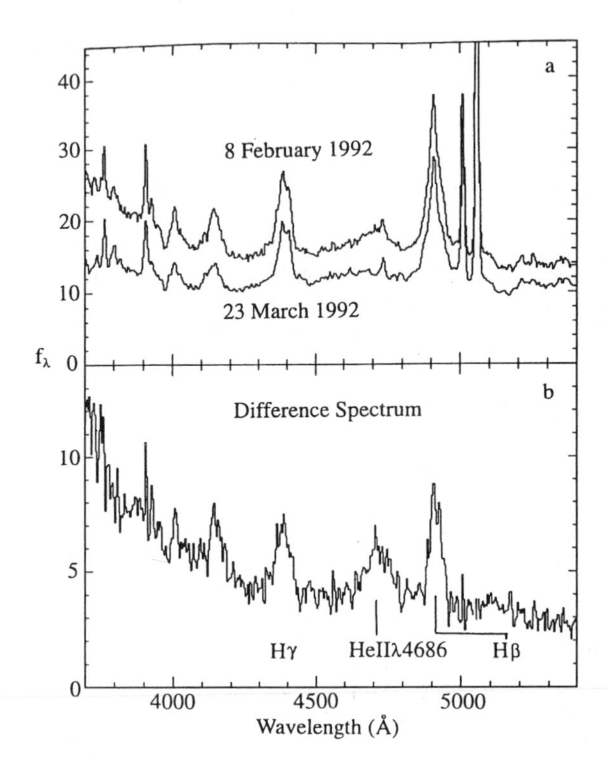


*Figure 4* *Top*: The optical spectrum of the nucleus of NGC 3783 on February 8 and March 23, 1992. *Bottom*: The difference spectrum showing the strong decrease and change of spectral shape of the continuum and the varying component of the main broad emission lines. Note the variations of the He IIλ4686. [From Stirpe et al (1994).] Ordinates in $10^{-15}$ erg $s^{-1}$ $cm^{-2}$ $A^{-1}$.

N Vλ1240 vary the fastest, then C IV and Ly$\alpha$ (Figure 4). Moreover, the wings vary faster than the core (NGC 5548: Krolik et al 1991, Dietrich et al 1993, Kollatschny & Dietrich 1996, Wamsteker et al 1990; NGC 3783: Stirpe et al 1994; NGC 4151: Ulrich et al 1984, Clavel et al 1987, Ulrich & Horne 1996; F9: Recondo-González et al 1997).

The picture that emerges is that of a stratified highly ionized BLR with the most highly ionized and fastest moving gas closest to the center, and the degree of ionization and velocity of the gas decreasing outwards. This is a "soft" stratification with ample overlap of the various ionization states. The absence of extended wings in the C III]λ1909 line implies an electron density exceeding $10^{10}$ $cm^{-3}$ in the innermost region of the HIL BLR.

The H$\beta$ and H$\alpha$ lines have, in general, a longer time response and smaller amplitude than the C IV line. In NGC 5548, we see some stratification in the

Balmer line emission region as well: The wings vary faster than the core, and the higher Balmer lines vary with larger amplitudes and shorter time delays than the lower ones (Dietrich et al 1993, Wamsteker et al 1990, Kollatschny & Dietrich 1996, van Groningen 1984). Representative values of the time delay and of $r_{max}$, the ratio of maximum to minimum emission during a given fluctuation are given in Table 1. The correlation coefficient is usually larger than 0.7.

PHYSICAL INTERPRETATION The variations of the HIL intensity ratios are generally consistent with the photoionization of ionization-bounded clouds, but there is also evidence for matter-bounded clouds: (*a*) a progressively weaker response of the C IV line to increases of the continuum flux above a certain level and (*b*) a flattening or decrease of the ratio C IV/Ly$\alpha$ at high-ionizing flux levels (NGC 3516: Ulrich & Boisson 1983; F9: Wamsteker & Colina 1986; NGC 5548: Dietrich & Kollatschny 1995; see also Binette et al 1989, Sparke 1993, Shields et al 1995). The BLR appears to be a mix of optically thin and optically thick gas clouds.

Some large variations of the Balmer decrement are associated with variations of the spectral shape of the optical continuum on time scales of 5–10 years and are entirely consistent with transient, strong, and variable dust extinction, possibly caused by clouds torn from the molecular torus (Goodrich 1989,1995, Tran 1995, Villar-Martin 1996).

A CAVEAT The BLR can respond only to continuum variations that last long enough to penetrate its volume significantly, and the amplitude of the continuum variations must also be large enough to alter the gas clouds' emissivity. That is, the BLR filters out continuum variations that are too fast or too small.

For example, in NGC 4151 the continuum variations occurring in $\sim$1 day (Figure 2, December 1993) did not result in any detectable variations of the C IV line intensity (Crenshaw et al 1996), although their amplitude, by a factor 1.3, was sufficient to produce line intensity variations in slower conditions.

Because the velocity and line emissivity vary with the radial distance ("stratification"), the line intensity and profile variations differ according to the duration and the amplitude of the continuum event (Netzer & Maoz 1990). Care should be exercised when comparing delays of line responses during different episodes or in different AGN. Only comparisons between events with similar continuum amplitudes are valid.

THE TREND WITH ABSOLUTE LUMINOSITY Although the data are few, spectroscopic observations of high-luminosity AGN show that time delays of lines with respect to the continuum appear to increase with intrinsic luminosity (Zheng et al 1987, Pérez et al 1989, Gondhalekar 1990, Hooimeyer et al 1992). The recent measurement of the H$\beta$ time delay in two high-luminosity PG quasars, combined with data for low-luminosity AGN (Table 1), suggests the BLR radius scales as $L^{-0.5}$ (Kaspi et al 1996b). This derived scaling is subject to the caveat above, but as the Kaspi et al (1996b) data have amplitudes in the range 1.4–4, it is probably valid.

3C 273 is the highest luminosity AGN (Table 1) for which long-term spectroscopic monitoring has been organized (regularly observed with IUE since 1978, and every other week since 1985). While the continuum flux has varied several times by a factor of 2 on time scales of ~2 years, no intensity variations of Ly$\alpha$ + N V above 10% have been detected (Ulrich et al 1993 and references therein). Therefore, only a very small fraction of the Ly$\alpha$ + N V emitting gas can be within 2 light-years of the continuum source. [The small amplitude Ly$\alpha$ + N V variations were judged by Ulrich et al (1993) not to be enough above the measurements errors to produce a reliable value for the time delay. With the same data set, time delays of 74 ± 33 days and 118 ± 57 days were found by Koratkar & Gaskell (1991) and O'Brien & Harries (1991). Whatever the robustness of these values of the delay, they apply to only a minute fraction of the Ly$\alpha$ + N V region. The most important result on 3C 273 is that the variations of the Ly$\alpha$ + N V line are, at most, of very small amplitude.]

This contrasts with lower luminosity AGN where a continuum flux variation by a factor of 2 always produces a response of the lines of comparable amplitude (e.g. Ulrich et al 1993, Figures 1 and 2). This implies the quasi-absence in 3C 273 of Ly$\alpha$-emitting gas at a distance less than $c\Delta t$ from the continuum source, $\Delta t$ being the characteristic time scale of the continuum variations, defined here as the time separating two maxima.

## 5.2 *Mapping the Velocity Field from the Emission Lines*

That the fastest moving gas is the closest to the center implies that radiative acceleration is less important than gravity and rotation. This dynamical information supports the connection of AGN line variability to black hole mass, providing that radial motions are not dominating the velocity field.

THE SEARCH FOR RADIAL MOTIONS FROM THE HIGH-IONIZATION LINES To first order, the HIL profile variations do not show the systematic change of one wing before the other that would be expected for purely radial flows (spherical winds, spherical accretion). This requires the main motions of the HIL clouds to be circulatory with only minor components of net infall or outflow. The data are consistent with the HIL clouds being in circular orbits in a disk, or having "chaotic" motions along randomly oriented orbits in the gravitational field of the central mass.

The velocity-resolved MEM inversion of Equation 2 applied to the best-sampled data of NGC 4151 and NGC 5548 shows that the time delays vary as function of velocity in a way that is roughly consistent with virialized motions. The central mass so derived is $\sim 10^7\ M_{\odot}$ for NGC 4151 (Ulrich & Horne 1996). In NGC 5548, it is between $2 \times 10^7\ M_{\odot}$ (for two-dimensional random motions) and $8 \times 10^7\ M_{\odot}$ (for three-dimensional random motions; Done & Krolik 1996).

In both NGC 4151 and NGC 5548 (and also in F9, Recondo-González et al 1997), however, the best-sampled data show small differences in the transfer functions of the blue and the red wing of C IV, with the response of the red wing being the stronger at small delays. This small asymmetry could, at first sight, be taken as a subtle indication of radial motions. On the other hand, optical depth effects in a rotating/outflowing wind produce a differential response across the C IV profile, with the red wing leading the blue wing (Bottorff et al 1997, Murray & Chiang 1997).

DIFFERENT EMITTING REGIONS FOR THE HIGH- AND LOW-IONIZATION LINES? As stated in the introduction, the LIL come from very dense gas with high column density, whereas the HIL come from a more diffuse photoionized medium. Can the variability data establish the existence of these two different emission regions? The answer is positive but provisional: In the few AGN with good variability data on C IV and H$\beta$, the inversion of Equation 2 gives a transfer function for H$\beta$ that is small near zero lag, indicating the near absence of matter along the line-of-sight, whereas in contrast, the transfer function of the C IV wings peaks near zero lag (Ulrich & Horne 1996, Done & Krolik 1996). This is what is expected if the Balmer lines come from a disk at small inclination (there is no matter close to the line of sight) and the HIL come from a broad cone or cylinder, in which case some matter lies along the line of sight producing a nonzero response near zero lag.

According to Wanders & Peterson (1996), however, the lack of response at zero lag for the transfer function of the H$\beta$ line is spurious and due to the combined effects of noise in the data and lack of resolution. On the other hand, Keith Horne, responding to a friendly challenge, has run his MEMECHO program on sets of simulated noisy data prepared by Dan Maoz and successfully identified which data corresponded to a transfer function peaking at zero and which ones did not (K Horne, private communication; Maoz 1997). This exercise gives weight to the results obtained from inversion of Equation 2, but, clearly, a final answer to this question awaits data of higher quality than presently available.

THE NATURE OF THE MOTIONS OF THE HIL GAS—IS THE DISK OPAQUE? If indeed a disk is present in AGN (and emits the Balmer lines and the Fe II multiplets), then two important points about the disk need to be specified before interpreting the velocity information on the HIL. First, is the disk opaque or transparent (can we see what is happening below the disk)? Second, could HIL clouds survive if they cross the disk in their chaotic orbits?

Calculations of accretion disk structure (Huré et al 1994) suggest that the disk becomes self-gravitating at 100 $r_S$ ($r_S = 2GM_{BH}/c^2$ is the Schwarzschild radius), much smaller than the radius of the region emitting the Balmer lines, although plausible mechanisms could stabilize the disk at a much larger radius (magnetic fields or internal dissipation; e.g. Sincell & Krolik 1997). It is possible, and we believe likely, that the disk (whether in a continuous structure or broken up in clouds in its outer parts) joins to the molecular torus, and nowhere is it sufficiently transparent that we can see the far side. This has profound consequences for the interpretation of the observed velocity field. In a biconal flow, only the near half would be visible, and if there were truly chaotic motions, only one half of each orbit would be seen. In addition to being opaque, the disk could have a column density such that free-flying clouds would be destroyed when crossing the disk. In this case, only stars can have chaotic motions, and if the HIL BLR clouds are the (modified) atmospheres of stars (Penston 1988, Alexander & Netzer 1994), they could collectively partake in pure gravitational motions, but we would still see only those clouds on the near side of the disk.

A solution to this puzzle is offered by the fact that in radio-quiet AGN, the HIL are blueshifted (by 0 to $\sim$1500 km s$^{-1}$) with respect to the low HIL, which themselves are at the host galaxy redshift (Gaskell 1982, Wilkes 1984, Corbin 1995, Sulentic et al 1995a, Marziani et al 1996). This indicates that the gas emitting the highly ionized lines is moving towards us, probably emanating from the disk, and still retaining a large part of the angular momentum it had in the disk (thus allowing a derivation of the central mass). The origin of the observed range of the HIL blueshifts is unclear—orientation can explain only part of it. Additional evidence for the presence of outflowing material in AGN includes (*a*) the blueshifted absorption lines in the emission line profile of a significant number of AGN, (*b*) the blueshift of the coronal lines (Wagner 1997) and (*c*) the shift of the reflected broad lines in Seyfert 2.

Magnetically accelerated outflows from accretion disks and radiatively driven winds are promising models for the formation and evolution of the highly ionized gas clouds (Blandford & Payne 1982, Emmering et al 1992, Königl & Kartje 1994, Murray & Chiang 1995, Bottorff et al 1997). Clouds, probably filaments, are pulled from the originally dense low-ionized material of the disk and subjected to the intense ionizing field, forming a more diffuse highly ionized outflowing medium. The densest, coolest inhomogeneities form the BLR clouds emitting the prominent lines (and if one such cloud happens to be crossing our line of sight to the AGN center, it should produce a blueshifted absorption line as, in fact, often observed in AGN). The hottest phase is detected as the fully ionized component of the warm absorber, which could also produce the blueshifted absorption lines in the wings of the HIL. Many features of these promising models remain unspecified and can be adjusted to accommodate the observations.

Two other models are compatible with an opaque accretion disk and the simultaneous response of the blue and red wings of the HIL. Both solve the confinement problem and include elements important for the formation and evolution of the BLR, but they do not specifically include magnetic forces. In the bouncing gas clouds model (Mathews 1993), the clouds congregate at a preferred radius where radiation forces and gravity are balanced. Fluctuations in the gas pressure move the clouds radially in and out, but most of the clouds tend to come back to the preferred radius set by the radiation level. In the bloated stars scenario, the BLR clouds are the modified atmospheres of some of the stars of the central stellar core and thus move along Keplerian orbits around the black hole (Penston 1988, Alexander & Netzer 1994).

## 5.3 *Other Issues Concerning the BLR*

PARTIAL REDISTRIBUTION OF THE GAS IN THE BLR Comparison of line profile variability during month-long campaigns separated by one or more years reveals that the line response is not stationary. This phenomenon, observed in the HIL and the LIL, is probably caused by changes in the distribution of the BLR gas in a few years (NGC 4151: Ulrich et al 1991, Perry et al 1994; NGC 3516: Wanders & Horne 1994; NGC 5548: Wanders & Peterson 1996). Although it is not possible at this time to offer a definitive interpretation of these changes, they suggest the presence of time variable inhomogeneities on the surface of the disk (possibly due to the magnetic field) which can (*a*) enhance the Balmer lines emission in some locations (e.g. bumps on the disk surface would intercept more continuum flux), thus altering the H$\beta$ profile and (*b*) strengthen the gas extraction from the disk surface (at local enhancements of the magnetic field), thus lifting additional HIL clouds above the disk and producing shoulders and other features in the C IV line (Figure 5).

That the line response evolves with time reduces the value of any single observation like the infall seen in NGC 4151 (November–December 1991) and in NGC 5548 (March–May 1993). Only if a certain behavior repeats itself over the years and for several AGN can it be considered to represent a standing feature of the velocity field or of the distribution of matter.

EMISSION LINES WITH DOUBLE PEAKS: ACCRETION DISKS, OUTFLOWS, OR BINARY BLACK HOLES? A small number of AGN show double-peaked Balmer lines, suggestive of emission from a rotating disk. (Double peaks are present only in the LIL.) This profile appears preferentially among broad-line radio galaxies and galaxies with a very compact central radio source (Eracleous & Halpern 1994, Gaskell 1996b). A relativistic disk would have an extended red

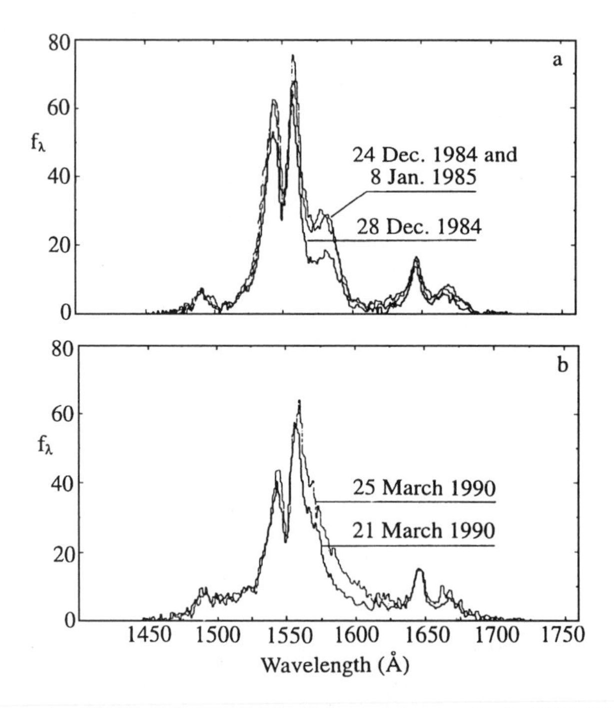


*Figure 5* Examples of long- and short-term variations of the C IV line in NGC 4151. Variations within days are shown on each panel. One can appreciate the variations on time scales of years by comparing the two panels and considering that between 1985 and 1990 the C IV line was observed to be perfectly symmetrical, e.g. November 1988–January 1989. [From Ulrich et al (1991).] Ordinates in $10^{-14}$ erg $s^{-1}$ $cm^{-2}$ $A^{-1}$.

wing because of gravitational redshift, a stronger blue peak because of beaming, and a range of peak intensity ratios, B/R, that is well constrained. Several double-peak profiles have been successfully fitted with an axisymmetric accretion disk model at some epochs (Halpern & Filippenko 1988, Rokaki et al 1992, Eracleous & Halpern 1994), but later observations revealed changes of B/R inconsistent with the disk model (Miller & Peterson 1990). Elliptical disks, warps and spiral shocks have been proposed but with no definitive conclusions (Chakrabarti & Wiita 1994, Eracleous et al 1995b, Bao et al 1996).

The model where the BLR is a biconal inhomogeneous flow illuminated by a variable double beam appears unlikely because of the difficulties in accelerating the very dense matter that produces the LIL, and the model is little constrained (3C 390.3; Zheng et al 1991). Double peaks have also been interpreted as the signature of two orbiting black holes, each with its own BLR (Gaskell 1996a), resulting from the merger of two galaxies with central black holes. In 3C 390.3, the period is estimated to be ~300 years, corresponding to a total mass of $7 \times 10^9\ M_\odot$. A difficulty of this model is that no other drift has been well observed (except possibly in OQ 208; Marziani et al 1993) and also that one would expect the double peaks to appear in the HIL as well as in the LIL, in contrast to the available observations (Arp 102B: Halpern et al 1996; 3C390.3: Wamsteker et al 1997).

Distinct double peak profiles form but one category of the complex profiles displayed by many AGN (Eracleous et al 1995a, Stirpe et al 1988). This suggests that the complex profiles may simply be caused by transient inhomogeneities and asymmetries in the emissivity and/or distribution of the emitting matter. This was also implied by the changes in the BLR gas distribution observed in NGC 4151 and NGC 5548 over several years (see Partial Redistribution of the Gas in the BLR). Symmetric inhomogeneities resulting in double peaks could be produced preferentially in the disk of radio sources through conditions related to radio jet formation.

EMISSION LINES DURING MINIMUM STATES; THE EFFECT OF THE LONG-TERM VARIATIONS The long minimum of NGC 4151 in 1981–1988 (interrupted by short episodes at medium bright states, Figure 1*b*), has caused the regular decrease of the central part of the C IV, C III, and Mg II lines between 1978 and 1991 (Ulrich et al 1991). A similar variation of a medium width component of Ly$\alpha$ has also been observed in 3C 390.3 (Clavel & Wamsteker 1987) and reveals, as in NGC 4151, the presence of an intermediate line region with a size of a few light-years and velocities of the order of 2000–3000 km s$^{-1}$.

During the deepest period of minimum NGC 4151 was the subject of two multi-month campaigns. The IUE spectra reveal the unexpected presence of two narrow emission lines, whose intensity varies by a factor of 2–3 in a few days apparently in phase with the small amplitude variations of the weak continuum (Ulrich et al 1985, Ulrich 1996). These lines, at $\lambda\lambda$(rest) 1518.5, 1594.4 Å and with FWHM less than 7 and 16 Å, respectively, are too narrow to be emitted by the entire BLR and must arise instead from two localized regions that have a special excitation mechanism.

These lines are best measured at minimum when the broad wings of C IV have faded but can also be seen, albeit with less contrast against the broad wings, at medium bright state (Clavel et al 1987, Kriss et al 1992). Their origin is unclear: That they are C IV components (at $-6100$ km and $+8500$ km s$^{-1}$) emitted by a two-sided flow is an attractive possibility, but it is based on the uncertain assumption that the accretion disk is transparent.

THE APPEARANCE AND DISAPPEARANCE OF BROAD EMISSION COMPONENTS Very large intensity changes in broad emission lines have been observed in different sets of circumstances, implying different origins for these large amplitude variations. First, some are in direct response to the large amplitude fluctuations of the ionizing continuum observed in a few AGN (see Sections 2.1 and 5.3).

The nonzero delay between continuum and line variations rules out, in these cases, obscuration by a dust cloud. Second, unexpected appearances of broad components have been reported in three AGN that, while displaying definite signs of activity, had before the event rather narrow emission lines or somewhat broad lines with triangular profiles. Interestingly, in these three cases (Pictor A, NGC 1097, and M81), the prominent new broad component is double-peaked. The circumstances that produce these lines and the cause of their double-peaked profiles remain unclear, although the accretion of a star followed by the formation of an elliptic disk has been proposed (Halpern & Eracleous 1994, Sulentic et al 1995b, Bower et al 1996, Storchi-Bergmann et al 1996).

The third case of unexpected appearance of a broad emission component occurred in the nucleus of NGC 4552, an apparently normal elliptical. Between two HST observations separated by two years, the continuum brightened by a luminosity of $10^6\ L_\odot$ (Renzini et al 1995). Spectroscopic observations carried out after the brightening revealed the presence of a broad Mg II line. Was it accretion of a passing star or of an interstellar cloud by a dormant central black hole? A systematic search for such events in apparently quiescent galaxies could be a new way to discover black holes in the general galaxy population. Finally, as mentioned in Section 5.1, dust clouds crossing our line of sight to the BLR cause large reddening and flux variations of the continuum with simultaneous characteristic Balmer decrement variations.

## 5.4 *Summary, Perspectives, and Emerging Fields in Emission Line Variability Studies*

THE MOST IMPORTANT RESULTS

1. The dimension of the line emitting region can be derived from the time delays between continuum and line variations. The delay depends on the line and on the velocity.

2. The velocity field of the broad line gas appears to be dominated by virialized motions in the gravity field of the black hole (the fastest moving gas is closest to the center; the blue and the red wings of the emission lines, to first order, vary simultaneously, as expected if radial motions are not important). The black hole mass can then be derived from the observed motions.

3. The variability observations can be best understood in the frame of the "accretion disk plus wind" model.[1] The results of variability studies strengthen

[1]This model is the simplest architectural sketch explaining the two basic results of spectroscopic studies: 1. The HIL and the LIL come from two media with distinctly different physical conditions, and 2. these media have different velocity fields because the HIL are blueshifted (a small shift when compared with the full width of the lines). This model attributes the emission of the LIL to the accretion disk, whereas the HIL are emitted by a hotter more diffuse outflowing medium; the disk is implicitly opaque in this model.

or are consistent with this model. Specifically, the first-order similarity between the responses of the blue and red wings of the lines supports the model. A second-order effect is observed in three AGN in the form of some asymmetry between the blue- and red-wing responses, with the red wing leading the blue; such a difference has a natural explanation in wind models. Finally, hints for different values of the transfer function at zero lag in HIL and LIL have been found in several AGN. If this effect is confirmed in those and other AGN, it will add strong support to this model.

In conclusion, the gas in the radio-quiet AGN central region has a roughly ordered velocity field: rotation in a disk for the low-ionization medium and rotation plus outflow for the high-ionization medium, which is pulled out from the disk by magnetic or radiative forces. The details of the disk plus wind structure are complex. For example, changes in the HIL and LIL transfer functions over several years are best understood as changes in the distribution of the HIL and LIL gas on time scales roughly commensurate with the dynamical time of the inner BLR in low-luminosity AGN. Together with the complexity of the line profile variations (Figure 5), they suggest the presence of inhomogeneities on the disk surface and in the outflowing medium.

THE LIMITS AND THE REMEDIES Derived transfer functions should in general be regarded with caution in spite of the successful challenge of Horne (Section 5.2). The physical situation is bound to be complicated by the gas inhomogeneities, local overlap in velocities (e.g. Done & Krolik 1996), and imperfect correlation between the measured continuum and the ionizing continuum. The calculation of the transfer function itself is limited not by the techniques but by the data. "Improving the Signal-to-Noise ratio and sampling by modest factors of 3 would greatly sharpen the velocity-delay maps" (Horne 1994).

Such an improvement represents a rather tall order but appears to be necessary in order to reach firm conclusions on the distribution and velocity field of the gas and the anisotropy of the line emission in each cloud (Horne 1994). In the future, such excellent data will allow full utilization of the information in the profile variations (something that has not yet been done systematically) and will motivate the development of inversion methods incorporating more specific geometry and physics into the models, such as photoionization codes. Meanwhile, the universality of results obtained predominantly from a few low-luminosity radio-quiet AGN selected as targets because they were the most active AGN about 10 years ago should be regarded with caution.

THE EMERGING FIELDS A number of avenues are ripe for future variability investigations. First, the entire parameter space defined by the black hole mass and the accretion rate should be explored. Several subsets of AGN populate the faint end of the optical/UV luminosity function. Intrinsically low-luminosity broad-line AGN (evidently with small accretion rates but with unknown black hole masses) have been discovered in 13% of nearby galaxies (Ho et al 1995). Their continuum and line variability is essentially unknown, and what little we know is puzzling, as evidenced by the fact that in M81, after about 15 years of no detectable change in the Balmer lines, a broad double-peaked component has recently appeared (Peimbert & Torres-Peimbert 1981, Ho et al 1996, Bower et al 1996).

From HST observations, LINERS (galaxies with low-ionization narrow emission-line regions whose spectra resemble neither an HII region nor the narrow-line spectrum of a broad-line AGN; Heckman 1980) have recently been found to harbor point-like nuclear UV sources in 25% of the cases (Maoz et al 1995). The absence of a nuclear UV source in 75% of the observed LINERS may reflect their duty cycle (Eracleous et al 1995a). The study of the continuum variability and the search for broad lines will greatly improve our understanding of LINERS.

Finally, for the NLS1—where the accretion rate may be exceptionally high (Section 2.2)—it is important to determine the central mass through systematic investigations of line variability.

All of the above AGN have in common that the broad lines are weak or only moderately strong. Their spectral variability studies require the use of HST or of large or very large optical telescopes. The question of the duty cycle of AGN, and the passage of a given AGN from one subset to another (from LINERS to "classical" broad-line Seyfert 1 to NLS1), requires good statistics on the number of AGN in each subset and long-term monitoring to witness the passage from one class to another, as occurred with Pictor A and NGC 1097 (Section 5.3, Appearance and Disappearance of Broad Emission Components). In fact, with the development of electronic archives, one could imagine light curves extending several centuries or more. (And we would not even be pioneers because records of supernovae can be found in Chinese and other archives that are a millennium or older).

At the bright end of the optical/UV luminosity function, the monitoring of line and continuum variability will allow us to follow the dependence of the line response on the AGN luminosity (cf the non-detection of significant Ly$\alpha$ variations in 3C 273, Section 5.1), and perhaps determine the central mass in the brightest objects in the universe. Powerful large field multi-object spectrographs (for example, installed on Schmidt telescopes) should be the instruments of choice.

Second, the recent observations of the profile and variability of the Fe K line (Tanaka et al 1995, Yaqoob et al 1996) and of the absorption lines originating in the hottest part of the warm absorber (Iwazawa et al 1997) are new ways to explore the innermost regions of AGN and will develop into a flourishing field with the Advanced X-Ray Astrophysics Facility, the X-Ray Multi-mirror Mission, and other X-ray missions.

The third promising avenue is that of theoretical investigations. This is hardly an emerging field, but new data will elicit new models and calculations. The accretion disk plus wind model appears particularly promising as it offers explanations for the variability of the broad lines, the two phases of the BLR gas, the blueshift of the emission and the absorption lines, and the emission of the UV continuum. This model and others should be pursued to the point where they can predict line intensity and profile variability patterns that can be compared with observations.

The effect of the certain presence of the central star cluster should also be investigated. The stars can be accreted (cf NGC 4552, Renzini et al 1995), collide with one another (Courvoisier et al 1996), or be trapped by the disk (Artymowicz et al 1993), and their atmospheres contribute to the BLR (Alexander & Netzer 1994, Armitage et al 1996). All these processes cause line and continuum variations.

# 6. VARIABILITY OF BLAZARS

## 6.1 *Overview and Relativistic Beaming*

Blazars exhibit the most rapid and the largest amplitude variations of all AGN (Stein et al 1976, Angel & Stockman 1980). The combination of extreme variability and relatively weak spectral features suggests the continuum is emitted by a relativistic jet close to the line of sight and hence that the observed radiation is strongly amplified by relativistic beaming (Blandford & Rees 1978). Here we take the point of view that all blazars, whether weak-lined like BL Lac objects or strong-lined like flat spectrum radio-loud quasars (FSRQ), contain essentially similar relativistic jets. They may differ in other aspects of nuclear activity; in particular, BL Lac objects may have less luminous accretion disks and BLRs than FSRQ.

Early multiwavelength studies provided the first global support for the idea of bulk relativistic motion in blazars: The observed radio emission was sufficiently luminous and rapidly variable that, assuming it was due to synchrotron radiation, high X-ray fluxes would be expected from Compton upscattering of the synchrotron photons (the so-called synchrotron self-Compton process), unless the radio emission was relativistically beamed (Hoyle et al 1966, Jones et al 1974a,b).

In addition, many blazars exhibited large-amplitude extremely fast X-ray variations (e.g. Morini et al 1986a, Feigelson et al 1986) that for isotropic emission would violate the limits on $\Delta L/\Delta t$ for Eddington-limited accretion (Fabian 1979), implying that relativistic effects are important (see also Bassani et al 1983). Finally, direct evidence for relativistic bulk motion has been obtained with VLBI observations of many blazars, which show that apparent superluminal motion is the rule (Mutel 1990, Vermeulen & Cohen 1994).

Here we summarize the data in wavebands that are particularly relevant to understanding the overall continuum (Sections 6.2–6.6) and discuss multi-wavelength studies, emphasizing the newest results (Section 7). Models are discussed in Section 8.

## 6.2 *Spectral Shape and Variability of the Blazar Continuum*

The continuum emission of blazars is remarkably smooth and steepens gradually towards shorter wavelengths from the radio to the UV range (Impey & Neugebauer 1988). The power emitted per decade exhibits a broad peak (where $\alpha \sim 1$) at IR through X-ray wavelengths. This long-wavelength component is almost certainly synchrotron emission, as evidenced by its high polarization from radio through optical wavelengths. The radio core is opaque, so that radio observations do not generally probe the region of the jet closest to the central engine; in general, the synchrotron continuum becomes optically thin at submillimeter or shorter wavelengths.

The wavelength of the peak of the synchrotron luminosity in blazars anti-correlates with the ratio of X-ray to radio flux. On this basis Padovani & Giommi (1995) divided BL Lac objects into high-frequency peaked (HBL) and low-frequency peaked (LBL) objects, according to whether $\alpha_{rx}$ (from 5 GHz to 1 keV) is $<0.75$ or $>0.75$, respectively (see schematic examples in Figure 6). The two subclasses, HBL and LBL, are sometimes referred to as XBL and RBL, respectively (see discussion in Urry & Padovani 1995).

Here we extend these definitions to all blazars. It is not yet known whether LBL and HBL represent two distinct populations or extremes of a continuous distribution of synchrotron peaks because X-ray selection favors HBL and radio selection, LBL. Arguments for continuity are given by Sambruna et al (1996a).

## 6.3 *Far IR–Optical–UV: The Thin Synchrotron Emission*

IRAS observations have shown that for LBL, a large fraction of the bolometric luminosity is radiated at mid-IR wavelengths (Impey & Neugebauer 1988) and that rapid (time scales of weeks) large-amplitude (factors of two) far-IR variability is common, whereas the nonblazar AGN do not vary significantly in the far-IR band (Edelson & Malkan 1986, 1987). Multiwavelength observations of 3C 345 from radio to UV wavelengths, close in time to the IRAS pointing, define clearly a peak in the power per decade between $10^{13}$ and $10^{14}$ Hz (Bregman et al 1986), one of the few LBL cases in which the peak is actually measured rather than inferred.

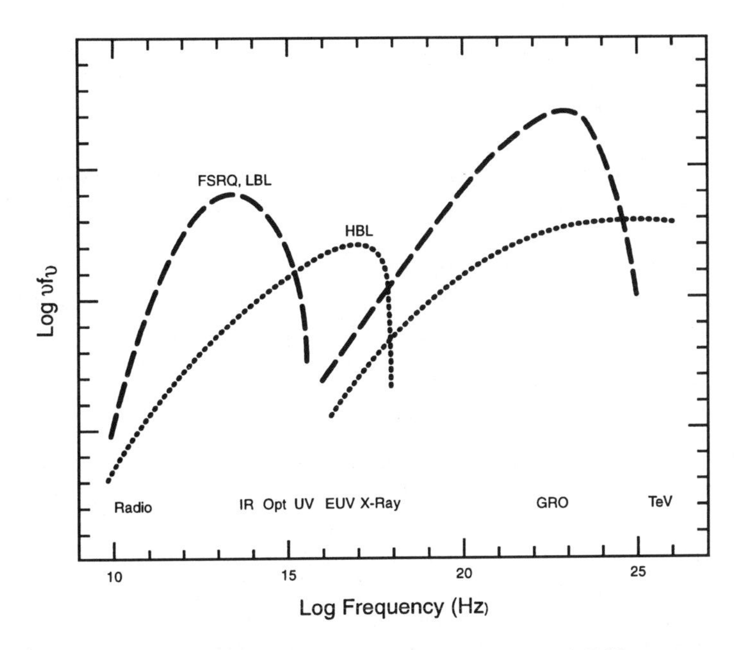


*Figure 6* Schematic broadband spectra of blazars from radio through TeV gamma rays. The low-energy component is probably due to synchrotron radiation and the high-energy component to Compton scattering of lower-energy seed photons, possibly the synchrotron photons or ambient UV/X-ray disk or line photons. Two different curves represent the average spectral shapes (Sambruna et al 1996) of HBL (high-frequency peaked BL Lac objects; *dotted line*) and LBL (low-frequency peaked BL Lac objects; *dashed line*) as defined by their ratios of X-ray to radio flux (Section 6.2). Strong emission-line blazars (i.e. flat-spectrum radio quasars, or FSRQ) have continua like those of LBL (Sambruna et al 1996).

The spectra of LBL fall off rapidly at wavelengths shorter than the synchrotron peak, i.e. in the optical–UV bands, whereas the spectra of HBL in the same bands are much flatter. Because of their relative brightness, LBL have been monitored extensively in the optical band, whereas at UV and X-ray wavelengths, mostly HBL have been observed.

The "historic" optical light curves of blazars on time scales of years show a variety of behaviors, from slow long-term trends to rapid repeated flares (Webb et al 1988). In general, long-term trends have typical time scales of 5 years in the rest frame, both for BL Lacs and FSRQ (Smith et al 1993, Smith & Nair 1995). Structure functions of the best data show slopes somewhat flatter than for radio light curves, indicating a transition from shot noise in the radio to flicker noise in the optical (Hufnagel & Bregman 1992), with relatively more power on short time scales. There are no obvious differences between BL Lac objects and FSRQ within the limited blazar samples available.

Optical variability extends to very short time scales, and intra-night small-amplitude variability has been observed in a number of blazars (Jang & Miller 1995, Heidt & Wagner 1996). The short time scale variations of radio-selected (LBL) are systematically larger in amplitude and have shorter duty cycles than those of X-ray selected (HBL) BL Lac objects (Heidt 1996). Within the radio-selected sample, there is a tendency for greater optical activity among higher luminosity sources (Heidt & Wagner 1996).

HBL are in general less polarized than LBL (Jannuzi et al 1993); however, the constancy of the polarized fraction and wavelength dependence over large flux variations observed in PKS 2155–304 argues against dilution by either starlight or an accretion disk continuum as the cause of this effect (Smith et al 1992).

High-quality UV light curves for more than a dozen blazars, obtained with IUE, show significant variability (amplitudes of 8–80%), which correlates with degree of optical polarization and with luminosity (Edelson 1992). This trend, which is opposite to the case of Seyfert galaxies, has been interpreted to mean that the most variable objects are the most beamed. Alternatively it could result from an inverse correlation between luminosity and peak frequency.

The densest IUE monitoring has revealed UV variations on extremely short time scales. In PKS 2155–304, the observed UV flux doubled in 1 h on one occasion (Pian et al 1997), comparable to the fastest X-ray variability observed in this object. A doubling time scale of 10 days is common (Urry et al 1993). At extreme ultraviolet wavelengths, rapid variations of slightly larger amplitude have been seen, although only two blazars have been monitored extensively with EUVE (HL Marshall et al, in preparation).

The spectral variability in the UV band is generally small, with only a weak tendency for larger amplitude variability at shorter UV wavelengths (Edelson 1992, Pian & Treves 1993, Paltani & Courvoisier 1994). The two blazars with the most UV observations, Mrk 421 and PKS 2155–304 (both HBL), show spectral hardening with increasing intensity only in a statistical sense (Ulrich et al 1984b, Maraschi et al 1986, George et al 1988a, Urry et al 1988). We note that LBL are faint in the UV and their slopes are difficult to measure with IUE.

## 6.4 *X Rays: The Crossing of Different Emission Components*

For HBL, the medium energy X-ray emission (2–6 keV) is typically steep, extrapolating smoothly from the UV and comprising part of the downward curving synchrotron spectrum. In LBL, the synchrotron component curves down well below the X-ray band, which is then dominated by a much flatter component (see Figure 6). Emission line blazars tend to fall in the LBL category. Borderline objects may have both components contributing in the X-ray range with the steep one prevailing in the soft X-ray range.

In HBL, rapid large-amplitude X-ray variability is the rule (flux doubling on time scales of hours). The spectra harden systematically with increasing intensity (Urry et al 1986, Treves et al 1989, George et al 1988b, Giommi et al 1990, Sembay et al 1993, Sambruna et al 1994). Comparing the UV and X-ray variability of HBL suggests that both spectral changes and variability amplitude are greater beyond the synchrotron peak, which is in the soft X rays for these objects.

ROSAT observations of a complete sample of radio-selected BL Lac objects (mostly LBL) show, in general, flatter spectra than for HBL and different variability behavior as well. In three cases there were "inverse" spectral changes, that is, a softening of the spectrum with increasing intensity (Cappi et al 1994, Urry et al 1996). This can be understood in terms of the relative variation of two spectral components that intersect each other in the X-ray range: a soft, highly variable one that swamps (high soft intensity) or uncovers (low soft intensity) a less-variable flatter component (see Figure 6). Such objects should be intermediate between LBL and HBL.

Less is known about the X-ray variability of emission line blazars because they are relatively weak X-ray sources. Einstein observations revealed extremely hard spectra in the 0.2–4 keV band (Worrall & Wilkes 1990), much flatter than the extrapolation of the optical–UV spectrum. Short-term variations were not detected in 3C 279 and NRAO 140, the only two blazars with sufficient intensity in the Ginga data base (2–20 keV), while long-term variations were seen in three out of four sources observed repeatedly (Tashiro 1995).

In summary, when a flat component is present in the X-ray band (as for LBL), it appears to vary on longer time scales and with lower amplitude than the steep X-ray component that is thought to be an extension of the longer wavelength synchrotron emission. However, this statement may be partly biased by the lower X-ray fluxes of LBL compared to those of the better-observed HBL.

## 6.5 *High Energy Gamma Rays: Where the Action Is*

Perhaps the most important progress in the last decade was the discovery with the Compton Gamma-Ray Observatory (CGRO) EGRET instrument that many blazars (presently 40–60, Thompson et al 1995; RC Hartman, private communication) emit enormous power in rapidly variable GeV gamma rays. This gamma-ray emission indicates a second peak in the overall spectral power distribution (Maraschi et al 1994, von Montigny et al 1995; Figure 6).

Practically all EGRET blazars with sufficient statistics and observations are strongly variable on time scales of months (Hartman 1996) and in several cases significant large amplitude flares have been observed on time scales of days, (3C 279: Kniffen et al 1993; PKS 0528+134: Hunter et al 1993). The most extreme example is PKS 1622–297, which brightened by at least a factor of 10

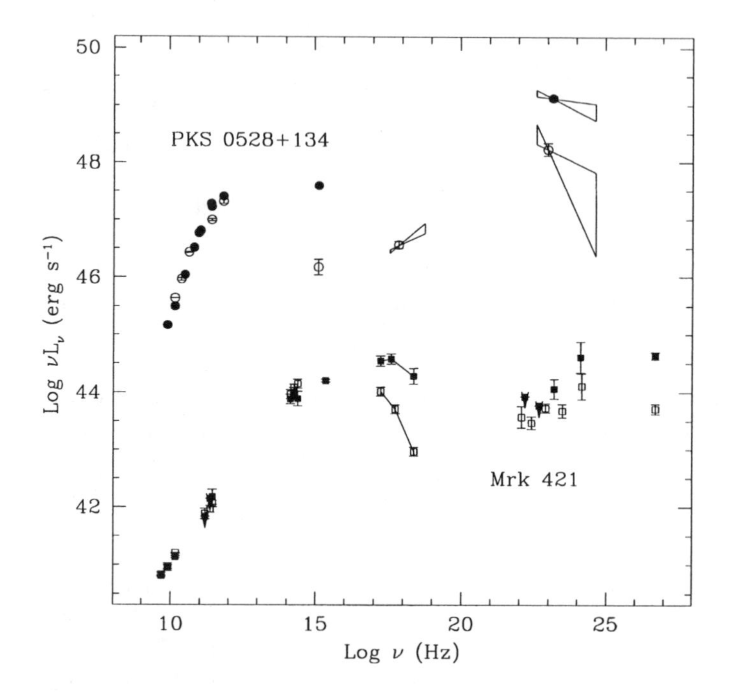


*Figure 7* Spectral energy distributions for two blazars in high and low states: Mrk 421 (*filled and open squares*) and PKS 0528+134 (*filled and open circles*). In both cases, the intensity is more highly variable above the peaks than below. [From Macomb et al (1995, 1996), Sambruna et al (1996).] The case of 3C 279 is illustrated in Maraschi et al (1994).

in two days, reaching a peak gamma-ray intensity 5 times that of any previously observed blazar (Mattox 1995, Mattox & Wagner 1996). The gamma-ray spectra of individual blazars appear to harden with increasing intensity (3C 279: Kniffen et al 1993; PKS 0528+134: Mukherjee et al 1996).

In some cases the gamma-ray emission extends to the TeV range (Figure 7). Mrk 421 was the first extragalactic source detected at such energies with the Whipple Observatory (Punch et al 1992), although it is only weakly detected at GeV energies with EGRET (Lin et al 1992). The TeV flux of Mrk 421 is variable by up to a factor of 10 on time scales of a day (Kerrick et al 1995) and by a factor of 5 on a time scale of 30 min (Gaidos et al 1996). Shorter time scales are not presently accessible owing to low event rates. Large-amplitude TeV variability must be frequent as such variations are commonly detected, in contrast with the quiet behavior of the same source in the GeV range.

The TeV power emitted by Mrk 421 (see Figure 7) dominates its bolometric luminosity, at least in the high state. Another nearby BL Lac object, Mrk 501 (also a modest EGRET source), has also been detected at TeV energies (Quinn et al 1996). Absorption by IR-optical photons locally or along the intergalactic path may prevent the detection of more distant blazars (Stecker et al 1996). Alternatively HBL, of which Mrk 421 is the brightest in the northern sky, may be intrinsically stronger TeV sources than LBL owing to the higher peak frequency of both spectral components (see Figures 6 and 7).

Some important conclusions can be drawn from the EGRET and Whipple discoveries. First, the gamma-ray emitting region must be transparent (i.e. the optical depth to pair production must be low), yet for minimal assumptions about ambient X-ray photon densities, the size limit imposed by the rapid gamma-ray variability implies very high optical depths. Therefore the gamma rays must be relativistically beamed (Maraschi et al 1992, Becker & Kafatos 1995, Dondi & Ghisellini 1995, Gaidos et al 1996).

Second, blazars emit a large fraction of their luminosity at very high energies. If the long-term gamma-ray light curve of 3C 279 is typical, even if the EGRET detections are biased toward exceptional states of gamma-ray activity, the "average" power output in gamma rays is comparable to that at all other wavelengths.

Third, the second peak of the spectral power distribution remarkably seems to fall at the highest energies for those objects whose first peak is also at high frequency: it probably lies near 0.1–1 GeV for LBL and 10–100 GeV for HBL (see Figure 7), although this is based on very incomplete data. A likely origin of the gamma-ray emission is Compton scattering of lower energy photons by the same relativistic electrons producing the low frequency component (Section 8.4). The correlation of variability at high and low frequencies is a crucial test for this class of models.

## 6.6 *Periodicity of OJ 287*

In one exceptional blazar, OJ 287, there is evidence for periodic flaring on a time scale of $\sim$12 years (Takalo 1994). Only three cycles have been seen during the epoch of dense monitoring, but the third of these was predicted to within six weeks from the light curve of the preceding 100 years (Sillanpaa et al 1988, Babadzhanyants et al 1992, Kidger et al 1992). Including the most recent data, the Fourier transform of the optical light curve shows six peaks, the two strongest of which correspond to 12.13 years, and its first harmonic, 6.07 years (Sillanpaa et al 1996). The period, if real, has a range of at least 1%; also, OJ 287 exhibits outbursts outside of the periodic oscillations, including some of comparable amplitude. A better assessment of the statistical significance of this period remains to be done.

If the light curve of OJ 287 does indeed contain a 12-year periodic component, its explanation in terms of precession of orbiting black holes is not straightforward: It requires two black holes with very high masses, $10^8$ and $1.7 \times 10^{10}$ $M_{\odot}$, and high eccentricity, $\epsilon = 0.7$, with a very short lifetime of $10^4$ years (Lehto & Valtonen 1996).

In no other blazar is there convincing evidence for either short- or long-term periodicities, so this is an extremely important precedent. Other periodicities—rotation of the position angle of the optical polarization with an apparent period of 27–35 days (Efimov & Shakhovskoy 1996), smaller fluctuations in intensity with periods of 10–20 min (Carrasco et al 1985, Komesaroff et al 1988, De Diego & Kidger 1990)—have also been reported for OJ 287 but have been (at best) of a transitory nature. We note that the observed rotation of the polarization position angle, even if not strictly periodic, is a remarkable phenomenon, which may be related to a lighthouse effect within a magnetized relativistic jet (Camenzind & Krockenberger 1992).

### 6.7 *Variability of Emission Lines*

Compared to quasars, blazars have generally weak emission lines (Miller et al 1978). This is especially true in BL Lac objects, where by definition the lines are practically absent ($W_{\lambda} < 5$ Å). Because the intrinsic line luminosities of BL Lac objects are systematically low compared to FSRQ (Padovani 1992), BL Lacs cannot be quasars with exceptionally beamed continua. Rather, an astrophysical explanation must be found for the intrinsic weakness of their line emission.

The detected lines in BL Lacs are usually of the narrow/forbidden type, which may suggest some fundamental difference between BL Lacs and strong emission line blazars, motivating a separate classification. However, broad $H\alpha$ has been seen in quite a few BL Lac objects (Miller et al 1978, Ulrich 1981, Sitko & Junkkarinen 1985, Moles et al 1987), including BL Lac itself (in which the broad line was clearly absent at earlier epochs; Vermeulen et al 1995, Corbett et al 1996); also, significant variations in the intensity of broad $H\alpha$ or $Ly\alpha$ have been seen (Ulrich 1981, Kidger et al 1996).

These observations demonstrate that the presence or absence of broad lines can be a transitory phenomenon, underscoring the continuity of properties between BL Lac objects and other blazars and supporting a unified view of the blazar phenomenon (Maraschi & Rovetti 1994, Bicknell 1994, Sambruna et al 1996a).

## 7. MULTIWAVELENGTH STUDIES OF BLAZARS

### 7.1 *Broadband Continuum Snapshots*

Spectral variability studies make it apparent that blazars are more variable at wavelengths shorter than the peak of the synchrotron emission, with amplitude increasing with decreasing wavelength (Bregman et al 1982, 1984, Maraschi et al 1983, Makino et al 1987, George et al 1988b, Falomo et al 1988, Treves et al 1989, Kawai et al 1991, Pian et al 1994). This is particularly clear for HBL in the X-ray range (Sambruna et al 1994, 1995), whereas for LBL this trend is observed at IR through UV wavelengths (Impey & Neugebauer 1988).

Recently, multiwavelength snapshots have been extended to gamma-ray energies. For example, the BL Lac object AO 0235+164 has a gamma-ray spectrum that matches well with an extrapolation from the flat X-ray spectrum (Madejski et al 1996), whereas nonsimultaneous ROSAT observations show that the soft X-ray flux is strongly variable and has a steep spectrum. The simplest interpretation is that in this object the synchrotron emission extends to the soft X-ray band, arguing for a classification intermediate between LBL and HBL.

In a few cases there exist simultaneous multiwavelength spectra of the same source at two epochs, one with high and one with low gamma-ray flux, notably 3C 279 (Maraschi et al 1994), Mrk 421 (Schubnell et al 1996; Figure 7), and PKS 0528+134 (Sambruna et al 1996b; Figure 7). The overall spectral variations of these three blazars show remarkable similarities. In all cases when the source was brighter in gamma rays, it was also brighter at longer wavelengths. The largest variations occurred at wavelengths shorter than the synchrotron and gamma-ray peaks. For Mrk 421 and PKS 0528+134, above gamma-ray peak frequency ($10^{25}$ Hz and $10^{22}$ Hz, respectively), the spectrum was harder in the brighter state. (For 3C 279, the gamma-ray spectrum in the faint state could not be determined owing to the low intensity.) Note that the spectral variation of Mrk 421 in the 0.5- to 10-keV range is analogous to that of 3C 279 in the IR to UV range. (PKS 0528+134 is highly reddened and therefore weak in the optical.)

In sum, the broadband variability behavior of different objects is similar when "normalized" to their respective peak frequencies. In each object, variations of the two spectral components are larger above the peak frequencies and appear to be correlated.

## 7.2 *Multiwavelength Light Curves and Correlations*

RADIO–OPTICAL On time scales of years, the optical emission from blazars is weakly correlated with the radio emission, with lead times of roughly one year (and with large uncertainties; Hufnagel & Bregman 1992). A stronger correlation appears between optical and high frequency (37 GHz) radio light curves, with lead times of months or less (Tornikoski et al 1994). Whereas the low frequency radio light curves are well sampled with respect to the variability time scales, at high radio and optical frequencies the flares are faster and variations may be simultaneous within the sampling ($\simeq$1–2 months). Nevertheless, not every optical flare has a radio counterpart even at high frequency, indicating that, although the particles radiating in the IR–optical and at submillimeter to centimeter wavelengths are physically related, most likely by a propagating shock, they occupy distinct spatial regions.

Simultaneous multiwavelength spectra of blazars from radio to IR bands usually show a self-absorption turnover, in some cases with multiple structure, evolving with time (Robson et al 1983, Gear et al 1986, Valtaoja et al 1988, Brown et al 1989). Systematic long-term monitoring of a number of objects from millimeter to IR bands shows that flares typically propagate from short to long wavelengths (Roellig et al 1986, Stevens et al 1994). A particularly well-studied example is 3C 279, in which the IR spectral index flattened as the intensity increased by a factor of $\sim$5 (Litchfield et al 1995). The delay of the submillimeter flare with respect to the IR appears to be $\sim$1 month, although the sampling is poor on such short time scales. Similar data for 3C 345 can be reconciled with a shock model only if the jet is nonadiabatic or curves away from the line of sight during the decline (Stevens et al 1996).

At millimeter wavelengths, blazar spectra generally flatten with increasing intensity, with BL Lacs (here, LBL) having flatter spectra in the submillimeter range than FSRQ (Gear et al 1994). This may depend on the different line strengths or could be related to the different luminosities, peak wavelengths, and redshifts of the two groups.

On very short (intraday) time scales, radio and optical variability seems to be correlated with no lag, at least for some blazars (Wagner & Witzel 1995).

MILLIMETER–X RAY Some well-monitored blazars, notably 3C 279, have shown a correspondence between millimeter and X-ray emission. The first suggestion of such a correlation was for 3C 279 (eight epochs over 1988–1991, showing two flares; Makino et al 1993). Subsequent observations in 1992–1996 showed another prolonged flare at 22–37 GHz, while the X-ray intensity remained low; however, the 1994 flare had a much lower self-absorption frequency than in 1991 or 1988 (Makino et al 1996). One could speculate that the lack of a corresponding X-ray flare is due to the latest millimeter flare originating in the outer regions of the jet, where the Compton process may be less important.

The X-ray flux of BL Lac also appears to correlate with the submillimeter flux (Kawai et al 1991). Both 3C 279 and BL Lac have similar continuum spectra, with synchrotron peaks in the far-IR and flat X-ray spectra that lie above the extrapolation of the UV spectrum.

A study of the historic light curves of 3C 279 at radio, millimeter, optical, UV, X-ray, and gamma-ray wavelengths (Grandi et al 1994) confirms that on time scales of months to years, the X-rays correlate with the high-frequency radio flux (particularly when a linear trend is subtracted from the radio light curves), whereas the optical–UV fluxes are correlated with the gamma-ray flux.

OPTICAL–GAMMA RAY Despite the obvious importance, it has proved very difficult to obtain simultaneous coverage of gamma-ray flares at optical wavelengths. In one case, PKS 1406–076, the optical flux rises by about 60% while the gamma rays increase by a factor of ~3, with the optical flare apparently leading the gamma-ray flare by about a day (Wagner et al 1995b). In another case, PKS 0420–014, gamma-ray high states correspond to optical flares, while gamma-ray nondetections coincide with optically faint states (Wagner et al 1995a).

Over long time scales, the optical and gamma-ray fluxes in 3C 279 are well correlated (Grandi et al 1994). During the rapid flare of 3C 279 in June 1991 a definite optical flare was observed, although only three optical measurements are simultaneous to the gamma-ray light curve. These show a maximum coincident with the gamma-ray maximum and a steep decline of 0.36 mag in a day, corresponding to a simultaneous decline of a factor ~3–5 in gamma rays (Hartman et al 1996).

GAMMA RAY–RADIO Comparison of gamma-ray data for blazars (detections and nondetections) with 37 GHz light curves shows that in a statistical sense, gamma-ray detections correspond to rising radio fluxes (Valtaoja & Teräsranta 1995). Large gamma-ray flares may be connected with the birth of new VLBI components, traditionally associated with the beginning of strong radio flares, as seems to occur in 3C 279 and PKS 0528+134 (Wehrle et al 1996, Zhang et al 1994).

## 7.3 *Intensive Multiwavelength Campaigns*

Most blazars vary substantially on very short time scales (hours to days), at least from the optical through the gamma-ray band. Therefore increasing efforts have been devoted to frequent, and when possible continuous, observations at many wavelengths. This has been possible only for very few objects.

The best candidates for UV–X-ray monitoring are the BL Lac objects PKS 2155–304 and Mrk 421 (HBL). Both can be sampled with ~1-h resolution due to their brightness at those wavelengths. Mrk 421 can also be monitored with daily sampling at TeV energies. In contrast, 3C 279, a highly luminous quasar with an LBL-like continuum, is too faint to monitor rapidly at UV wavelengths but is the brightest blazar at GeV energies and has a long history of multiwavelength observations. Finally, the LBL PKS 0716+71 is a well-monitored intraday variable source. These four blazars, the targets of the most extensive multiwavelength monitoring, are discussed in turn below.

PKS 2155–304 The first intensive campaign on PKS 2155–304 (November 1991), with ~5 days of quasicontinuous coverage with IUE and ~3.5 days with ROSAT, yielded a number of unprecedented results (Smith et al 1992,

Urry et al 1993, Brinkmann et al 1994, Courvoisier et al 1995, Edelson et al 1995). Variations at optical, UV, and X-ray wavelengths were extremely rapid. The fastest variations observed had ~10–30% changes over several hours, and the autocorrelation function of these fluctuations had a peak at ~0.7 days. Optical, UV, and X-ray light curves were closely correlated and the amplitude of variation was essentially independent of wavelength, ruling out an accretion disk origin of the UV continuum (as did also the polarization characteristics). The X-rays led the UV by a small but significant amount, roughly 2–3 h. Over the full month, the UV intensity changed by a factor of 2, as seen in the optical/IR light curves.

A second, longer campaign (May 1994)—with ~10 days of IUE, ~9 days of EUVE, 2 days of ASCA, and 3 short ROSAT observations—showed at least one prominent isolated flare rather than the "quasiperiodic" low amplitude variations seen in the first campaign. The light curves from this second campaign are shown in Figure 8 (Kii et al 1997, Pesce et al 1997, Pian et al 1997, Urry et al 1997; HL Marshall and collaborators, in preparation).

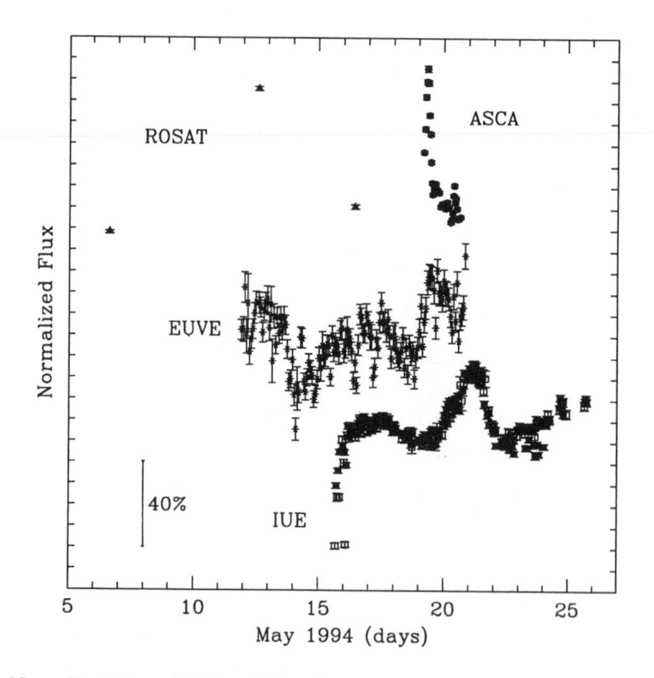


*Figure 8* Normalized X-ray, EUV, and UV light curves of PKS 2155–304 from the second intensive multiwavelength monitoring campaign, in May 1994. The ASCA data show a strong flare, echoed one day later by EUVE and two days later with IUE. The amplitude of the flares decreases and the duration increases with increasing wavelength. [From Urry et al (1997).]

The overall variability amplitude was much higher than in the first epoch, particularly in the X-ray band, where the flux rose and fell by a factor of $\sim$2.5 in about a day. The variability was strongly wavelength dependent: The central UV flare, which can most plausibly be identified with the strong X-ray flare, has an amplitude of $\sim$35%, a factor of $\sim$5 smaller than the X-ray flare. (Note however that the wavelength range covered by ROSAT is closer to that of EUVE than to ASCA, and the EUVE variations were only slightly larger than the IUE ones.) The X-ray flare appears to lead the EUV and UV fluxes by 1 and 2 days, respectively, an order of magnitude longer than the lag detected in the first epoch. Within the ASCA data, the 0.5- to 1-keV photons lagged the 2.2- to 10-keV photons by 1.5 h (Makino et al 1996).

Despite the differences, these two campaigns give the first direct evidence that the variations from 10 keV to 5 eV are correlated on short time scales and that high frequencies lead the lower ones.

MRK 421 The extraordinary high-energy spectrum of Mrk 421 motivated repeated campaigns of simultaneous observations with ASCA, CGRO, and the Whipple Telescope in May 1994 and in May 1995. In both cases, large variations of TeV and X-ray fluxes were observed. The 1995 campaign had better sampling in the TeV range and quasicontinuous coverage with EUVE. The spectral snapshots shown in Figure 7 derive from the 1994 data, and the 1995 light curves obtained with Whipple, ASCA, EUVE, and a ground-based optical telescope are shown in Figure 9 (Buckley et al 1996).

The X-ray and TeV light curves are highly correlated with no apparent lag, while the EUVE and optical light curves lag the TeV (and likely the X-ray) maximum by about a day. This behavior is very reminiscent of the PKS 2155–304 flare, at least in the spectral bands that were covered in both objects. (Unfortunately the southern declination of PKS 2155–304 has prevented observations in the TeV range up to now.) Moreover, as for PKS 2155–304, the 0.5- to 1-keV photons are found to lag the 2- to 7.5-keV photons by about 1 h (Takahashi et al 1996).

3C 279 3C 279 was the first blazar discovered to emit strong and variable GeV gamma rays, by EGRET in June 1991. A two-week multiwavelength monitoring campaign followed in December 1992–January 1993. Because 3C 279 was then at a very low intensity level, the scheduled daily sampling could not be performed with IUE and the gamma-ray observations yielded only an average flux; however, several simultaneous gamma-ray, X-ray, UV, optical (BVRI), millimeter, and radio observations were obtained, yielding a well-measured spectral energy distribution of 3C 279 in the low state. Compared to the high state in June 1991, the low-state spectrum decreased dramatically

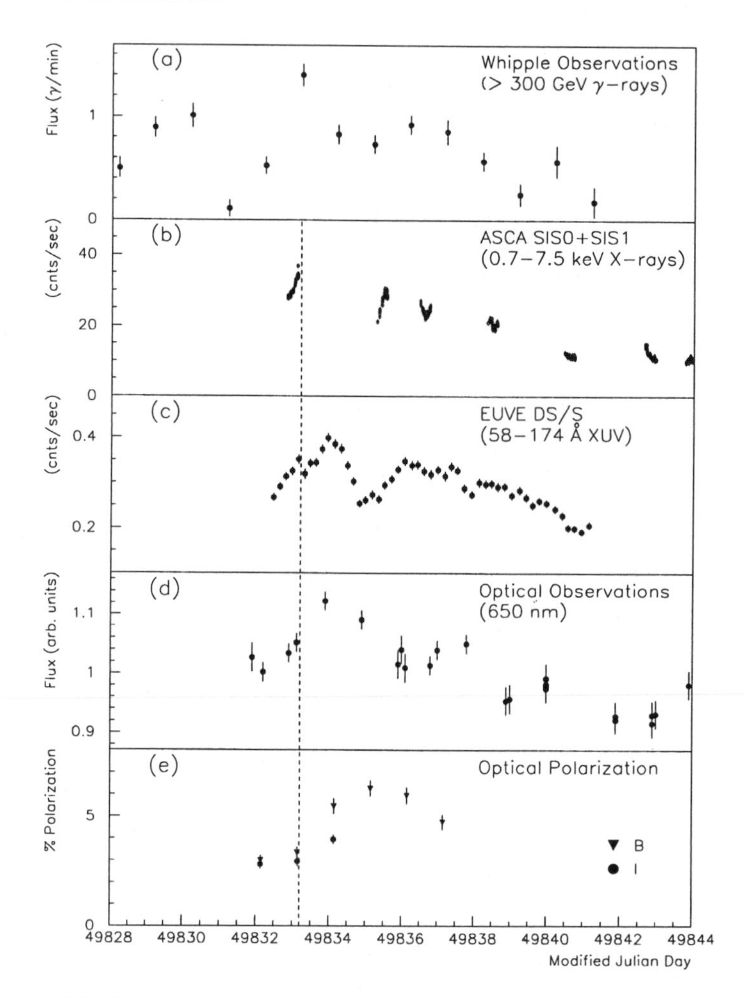


*Figure 9* TeV, X-ray, and optical light curves of Mrk 421 from April–May 1995. The largest amplitude variability is at TeV and X-ray energies; the optical amplitude is much smaller, but the optical light curve is well correlated with TeV light curve. [From Buckley et al (1996).]

and in a highly correlated fashion at frequencies above $10^{14}$ Hz (Maraschi et al 1994). The variability amplitude increased with frequency from the IR to the UV band, with a steeper spectral shape in the low state. Regarding the second spectral component, the variability amplitude was rather small at the lowest X-ray energies but increased with increasing frequency and was largest in gamma rays.

The optical monitoring during the 1992–1993 campaign evidenced large amplitude variability, almost a factor of 2 in 10 days (Grandi et al 1996), while at the same time the X-ray flux measured with ROSAT varied by less than 20%. This shows that the optical and X-ray emissions were not well correlated on short time scales.

In January 1996, intense gamma-ray flaring in 3C 279 was discovered and followed with multiwavelength coverage. These new data will yield information about the multiwavelength correlations on short time scales in this and, by extension, other LBL. Simultaneous observations with the Infrared Space Observatory (ISO) should determine the position of the synchrotron peak, which from ground-based data can only be constrained to lie between $10^{12}$–$10^{14}$ Hz.

S5 0716+71 After the discovery of intraday variability in S5 0716+71 (see Wagner & Witzel 1995), this BL Lac object was followed intensively at radio and optical wavelengths. IUE and X-ray variations were also seen but with sparser sampling (Wagner et al 1996). There is a definite trend for bluer colors during short flares, whereas the long-term variations occur at constant color (Ghisellini et al 1997). The spectrum in the radio band flattens when the optical flux brightens, indicating that the flaring component is self-absorbed. The UV and X-ray fluxes both seem to vary with the optical flux on short time scales (hours), but the optical to X-ray ratio is not constant from day to day.

The gamma-ray emission from S5 0716+71 is rather weak and shows a relatively flat spectrum, and it varies by factors of $\sim$2 over year-long time scales (Lin et al 1995). The broadband spectral properties of this object suggest that it is intermediate between LBL and HBL.

## 8. INTERPRETATION OF BLAZAR VARIABILITY

### 8.1 *Summary of Variability Results*

Recent observations of blazars have yielded six key points critical for understanding the continuum. First, blazar spectra are characterized by two broad spectral components (Section 6.2, Figures 6 and 7), one with peak power at low frequencies (IR to soft X rays) and one with peak power at very high energies (GeV to TeV gamma rays).

Second, variability in the low frequency component is much more pronounced above the peak frequency, and the variability amplitude increases with frequency so that the spectra harden with increasing intensity. Below the peak, both the amplitude of variability and spectral changes are much smaller.

Third, the high-energy component also seems to vary more and to harden with increasing intensity above its peak frequency. This inference is, however, based on very few observations.

Fourth, simultaneous snapshots of the full broadband spectra suggest that the intensities in the two spectral components are correlated, in the sense that when the short-wavelength (gamma-ray) component is bright, the long-wavelength emission is also bright. These snapshots do not constrain possible lags significantly.

Fifth, finite lags have been measured among flares in the optical through X-ray light curves of two HBL. For both PKS 2155–304 and Mrk 421, the soft X-ray photons lag the hard X-ray photons by $\sim$1 h (Makino et al 1996, Takahashi et al 1996). In addition, for PKS 2155–304 the EUV and UV light curves lag the X-ray curves by $\sim$1 and $\sim$2 days, respectively (Urry et al 1997; Figure 8). For Mrk 421, the EUV and optical light curves lag the TeV by $\sim$1 day (Schubnell et al 1996; Figure 9); a big gap in the X-ray light curve close to the flare peak prevents quantitative statements, but it is plausible that the behavior of Mrk 421 closely resembles that of PKS 2155–304.

Sixth, HBL and LBL exhibit completely analogous variability with respect to the (different) peak frequencies in their spectral distributions. For example, both vary more above their respective power peaks; it is just that this makes LBL highly variable in the optical and GeV gamma rays, whereas HBL are highly variable in X rays and in TeV gamma rays (and relatively quiescent in the optical and GeV ranges).

Below we discuss current theories of blazar spectra and variability (Sections 8.2–8.5) in the light of these six points. We start with the radio-emitting outer jet, which is reasonably well understood, and work progressively inward to smaller scales, higher energies, and lesser knowledge, with the goal of relating the observed variability to the physics of the jet. Mechanisms for variability involve essentially two possibilities, shock waves moving along the jet and rotation of the beaming cone across the line of sight; both could be relevant at different times or in different objects. In the first case, spectral variations are expected, whereas in the second achromatic variability can be produced (see Marscher 1996 and Dreissigacker & Camenzind 1996 for physical models).

## 8.2 *The Relativistic Jet: Synchrotron Radiation from the Outer Regions*

Self absorption is important in blazars even at high radio frequencies. The flat radio spectra likely derive from inhomogeneity along the jet—that is, from contiguous jet regions with different self-absorption cutoffs (Blandford & Königl 1979, Marscher 1980, Köonigl 1981). Often, discrete spectral components can be recognized, sometimes related to individual VLBI components or to the unresolved core. The spectral evolution of these components—in particular, the shift of the self-absorption cutoff to longer wavelengths and concurrent changes in polarization angle—can be well modeled as enhancements in the local radio emission due to shocks propagating along a relativistic jet whose particle density and magnetic field decrease outward (Marscher & Gear 1985, Hughes et al 1989, 1994, Litchfield et al 1995, Stevens et al 1996). The ability of such models to account at the same time for both total flux and polarization variability is encouraging, although the derived angle of view tends to be larger than for simple unification schemes (Kollgaard 1994, Urry & Padovani 1995).

The high polarization of the optical emission and the spectral continuity from the radio (albeit with increasing spectral index) imply that synchrotron radiation is the dominant emission mechanism in most blazars up to the UV range and in some cases up to the X rays. It is not yet clear, however, whether the relativistic electrons responsible for the high-frequency synchrotron emission are simply the high energy tail of those emitting at radio wavelengths or represent a distinct population in a different part of the jet, presumably closer to the central engine. Radio through optical light curves show some correlation, but the lags have not been quantified even for the best-observed sources (Stevens et al 1994, Tornikoski et al 1994). Uncorrelated flares are also present (Hufnagel & Bregman 1992). An understanding of the radio-optical connection is also hampered by the far-IR observational gap (now accessible with ISO). The self-absorption turnover of the optical synchrotron component cannot be determined unambiguously, and therefore its physical parameters are poorly constrained.

## 8.3 *High-Frequency Synchrotron Emission: Energy Stratification?*

At high frequencies (optical–UV for LBL, X-ray for HBL), the synchrotron spectrum steepens considerably and the largest spectral variability is observed. It is natural to associate the steepening with increasing radiative energy losses. Particle acceleration must also occur or the high-energy emission would vanish in a short time. Therefore the average spectrum probably represents an equilibrium among energy gains, losses, and escape. If the "injected" particle spectrum is a power law, its equilibrium spectrum (in a homogeneous region) is steeper by $\Delta\beta = 1$ above the energy $\gamma_b$ at which the radiative lifetime equals the escape time (Kardashev 1962). The peak in the synchrotron power would occur at $\nu_b \propto \gamma_b^2$, and the associated change in spectral slope would be $\Delta\alpha = 0.5$.

Spectral variability above $\nu_b$ (flattening with increasing intensity) could be due to fluctuations in the acceleration process on time scales shorter than those over which equilibrium can be established; the flatter injected spectrum would be visible briefly before energy losses set in. Another possibility is that increased power dissipation causes the injection of particles of higher energy. The second hypothesis is favored in model fits (assuming the simple homogeneous case) to the spectral variability of Mrk 421 (Mastichiadis & Kirk 1997). In both cases, the variations at the highest synchrotron and inverse Compton frequencies are quasisimultaneous, whereas at lower frequencies the time scales are longer and the peaks are delayed.

Acceleration and loss processes can depend on position in the jet, causing different electron energy distributions as a function of location (inhomogeneous model). The steeper spectrum above the peak can be reproduced (even $\Delta\alpha > 0.5$) if higher energy electrons occupy progressively smaller volumes (Ghisellini et al 1985). This occurs naturally behind a shock where the acceleration process is localized at the shock front (Marscher 1996). Thus, a disturbance propagating along an inhomogeneous jet or across a shock front would cause spectral variability, usually with shorter time scales and larger amplitudes at higher energies (Celotti et al 1991, Marscher & Travis 1991).

Indeed, the soft photons do lag the hard photons in both Mrk 421 and PKS 2155–304 (the two brightest HBL). In X rays, the frequency dependence of the measured lags is consistent with a $\nu^{-1/2}$ law, as expected from a radiative lifetime; the EUV and UV could also be consistent within the large uncertainties. At each frequency the decay times are longer than the lags, which is not easy to understand in a pure radiative case. A realistic model likely involves both radiative and travel time effects.

In any case, the data require a fast acceleration process injecting high-energy particles. This is not easy to reconcile with diffusive shock acceleration, whereby high energies build up through a cumulative process (see spectral evolution computed for time-dependent shock acceleration, Fritz & Webb 1990). A viable alternative could be particle acceleration through a large scale electric field (Kirk & Bednarek 1996). Note that the achromatic variability found in the first epoch of multiwavelength monitoring of PKS 2155–304 is not expected from a synchrotron flare in a jet and may be caused by a different physical process such as rotation or microlensing.

## 8.4 *Inverse Compton Models: The Gamma-Ray Jet*

Understanding the gamma-ray emission in blazars is particularly important because, at least during flare states, it dominates the bolometric luminosity. Here we discuss models that attribute the low-energy component to synchrotron emission and the high-energy component to inverse Compton emission from the same population of relativistic electrons.

For a homogeneous emission region, neglecting Klein-Nishina effects and assuming the seed photon spectrum is not too broad, the Compton emission closely mimics the synchrotron spectral shape—both depend primarily and in the same way on the electron spectrum. To first order, the peak of the Compton spectrum is due to the same electrons (of energy $\gamma_e$) as the synchrotron peak, independent of the seed photons. Pairs of wavelengths in the two spectral components in the same ratio as the peak wavelengths are also produced by the same electrons. Thus, if variations are due to changes in the electron spectrum, this class of models predicts a close correlation between flux variations for each wavelength pair. The coherent variability of the synchrotron and Compton components of LBL and HBL summarized at the beginning of the present section supports this simple scheme in broad terms. More realistic models should take into account the effects of inhomogeneity in the jet.

The seed photons could come from various sources. Synchrotron photons are produced copiously within the jet (Maraschi et al 1992) but we measure only their apparent density. The greater the beaming (bulk Lorentz factor $\Gamma$), the lower the intrinsic photon energy density in the jet frame ($\propto\Gamma^{-4}$). If an accretion disk is present, it is a strong source of photons (Dermer & Schlickeiser 1993), but at the distance implied by the gamma-ray transparency condition, their intensity as seen in the jet frame is greatly reduced. Alternatively, photons produced at the disk or nucleus can be reprocessed and/or scattered and therefore isotropized in a region of appropriate scale (Sikora et al 1994); their energy density is then amplified in the jet frame by a factor $\Gamma^2$.

The relative importance of the possible sources of soft photons—the jet itself, the accretion disk, or the broad emission line region, illuminated either by the disk or by the jet—must depend on the particular characteristics of the source. In different blazars, the origin of the seed photons may well be different: Perhaps the external-Compton (EC) case is more important in strong emission-line objects like 3C 279 and PKS 0528+134 (Sambruna et al 1996b), whereas the SSC model is applicable to weak-lined blazars like Mrk 421 and PKS 2155–304 (Zdziarski & Krolik 1993, Ghisellini & Madau 1996, Ghisellini & Maraschi 1996).

In a homogeneous SSC model, the ratio between the two peak wavelengths uniquely determines the energy of the radiating electrons: $\lambda_S/\lambda_C \sim \gamma_e^2$. For EC models, the relation is less straightforward and obviously leads to different physical parameters for the emitting region. For 3C 279, homogeneous SSC models yield lower magnetic fields than EC models and have difficulties accounting for the shortest variability time scales (Ghisellini & Maraschi 1996). Both models are consistent with the (presently) observed spectral variability; the larger amplitude in the Compton component compared to the synchrotron component is explained naturally by the SSC model (Maraschi et al 1994) but is also consistent with an EC scenario where nonlinearity can be caused by a variation of the bulk flow speed or by an external mirror effect for the synchrotron photons (Ghisellini & Madau 1996).

## 8.5 *The Invisible Jet Core*

The EGRET detections of blazars demonstrate that the observed power is in many cases dominated by high-energy gamma rays. Although the *degree* of dominance could result in part from selection effects—i.e. if there is a scatter, possibly due to variability, in the intrinsic gamma-ray to radio flux ratio of the population, objects with the highest ratio will be "selected" by gamma-ray observations (Impey 1996)—it is clear that gamma rays are a fundamental component of blazar power.

Paradoxically, rapid variability at gamma-ray energies implies that the observed radiation cannot be produced too near the center or the high-energy photons would never escape (Section 6.5). This raises the question of how power is transported from the central engine, and how and why it is radiated at a given distance. In FSRQ the gamma-ray variability time scale is $\sim$1 day, similar to that in the optical for some well-studied objects, so that the minimum radiative zone occurs near $\sim 10^{16}(\delta/10)$ cm. At roughly this scale, a magnetohydrodynamic wind from a rotating disk around a black hole would be collimated and accelerated to relativistic speed (Appl & Camenzind 1993); the energy transport would be magnetohydrodynamic and the radiative region would start "naturally" where a high bulk velocity is achieved.

Alternatively, the energy transport could be either purely electromagnetic, via a Poynting flux or through mildly relativistic ions with low internal entropy (Blandford 1993, Blandford & Levinson 1995). Assuming the emitting region is initially opaque to gamma rays, there will be a "gamma-ray photosphere" (due to interactions with lower energy photons) that is larger for higher energy gamma rays. The concept of a photosphere makes the emergent spectra independent of the details of the energy transport and conversion, and also allows a very strong prediction that gamma-ray flux variations are slower and/or later at higher energies. This prediction is violated for Mrk 421, where TeV gamma rays vary more rapidly and more frequently than GeV gamma rays. That the gamma-ray spectra of FSRQ are harder in brighter states also appears difficult to reconcile with this model.

Another way of transporting energy without dissipation is highly relativistic protons ($\gamma_p \sim 10^7$) (Mannheim & Biermann 1992, Mannheim 1993, 1996, Protheroe & Biermann 1997). The associated primary electrons, whose break energy is much smaller than that of the protons, produce the low frequency spectral component via the synchrotron mechanism. The very high-energy protons interact with these photons, initiating pair cascades that yield very flat spectra ($\alpha \simeq 1$) in the MeV–TeV range. This model predicts that high-energy variations should always follow low energy ones.

In an intermediate scenario, the jet starts out at high bulk Lorentz factor, $\Gamma_j \sim 10^4$, then decelerates through interactions with the local (disk) radiation field (Coppi et al 1993). This model has a difficulty in that the energy lost in the deceleration phase—i.e. the observed radiation—is much larger than that carried by the decelerated jet, which may therefore be insufficient to power the radio lobes.

All these models have difficulties, and in particular none of them yields a good fit to the overall spectral energy distribution. Nonetheless, they are of interest because they attempt to address crucial points about the origin of the radiating particles, which remain unexplained in the more phenomenological models.

## 9. CONCLUSIONS AND FUTURE INVESTIGATIONS

Fundamental results have been obtained from variability studies of AGN, but many important questions are still unanswered. Substantial progress is expected in the near future.

Regarding the radio-quiet AGN, the simultaneity of the continuum flux variations at optical and UV wavelengths and the very short doubling time scales (one week to several months) observed in low-luminosity AGN are incompatible with models wherein the rapidly variable component of the optical–UV continuum is emitted through viscous effects in an accretion disk. Because the amplitude of the variations on time scales of weeks is more than a factor of 2 at 1400 Å, at least half of the blue bump luminosity is involved. The best current model for the rapidly variable continuum emission is reprocessed emission from a dense medium irradiated by the variable X rays.

For high-luminosity AGN, the importance of this irradiation process is uncertain because the relevant observations have not yet been carried out. The most interesting question—whether and how the continuum fraction emitted through reprocessing varies with absolute luminosity (accretion rate) and/or with the mass of the central black hole—remains unanswered. The organization of a series of quasisimultaneous optical–UV–X-ray observations is therefore a priority in order to advance in this field.

The rather good knowledge of the variability characteristics that we now have, at least for low-luminosity AGN, should stimulate new theoretical investigations on the origin of the continuum variability, both time scales and amplitudes. While turbulence and magnetic reconnection may be a sufficient explanation for the short time-scale X-ray variability, the mechanisms for longer term variations are essentially unknown.

Results crucial for our understanding of the inner region of radio-quiet AGN have been obtained from variability studies of the broad-line gas emission, including (*a*) a measurement of the dimension of the emitting region (dimension that depends on the line and on the velocity); (*b*) indications that the gas has roughly virialized motions in the gravity field of the central black hole, whose mass can then be derived from the observed motions; and (*c*) support for the accretion disk plus wind working model wherein the LIL come from the accretion disk, and the HIL are emitted by a wind (which could be magnetically or radiatively driven from the disk).

Among future investigations, we see three priorities. The first is to confirm (or refute) the accretion disk plus wind model, which at present seems to be the most promising model, by collecting extensive data sets combining high S/N and large number of epochs. Second, and possibly even more important, is to verify if the models developed for low-luminosity AGN are valid for high-luminosity AGN, the brightest objects in the universe. This can be done only if we collect data on high-luminosity AGN that are of a quality similar to the low-luminosity AGN data. This requires more powerful instruments and a consistent effort over much longer periods than for low-luminosity AGN. The third priority is to explore the entire parameter space defined by mass and mass accretion rate. This requires monitoring the continuum and line variations in very low-luminosity AGN (low accretion but unknown mass), such as the LINERS and the galaxy centers displaying faint UV sources or very weak broad lines. Similar observations should be carried out on NLS1, which might be the AGN with the highest accretion rate.

Closely related to these questions are the recent observations of broad, variable Fe K lines (Tanaka et al 1995, Yaqoob et al 1996) and of absorption features originating in a warm absorber (Iwazawa et al 1997). All these observations will add impetus to the active field of investigations of the structure and stability of AGN accretion disks, the role of the magnetic fields, and the magnetically or radiatively accelerated outflows.

As for the strongly nonthermal activity of blazars, it is now reasonably well understood in phenomenological terms. The basic model consists of a relativistic jet closely aligned with the line of sight, filled with energetic particles that emit synchrotron and Compton-scattered radiation spanning radio through gamma-ray wavelengths. The radiating particles must have a bulk relativistic motion, beaming the radiation toward us and making the blazar appear more luminous and more rapidly variable. Beyond this, models differ, notably in the structure and physical parameters of the jet and in the nature of the seed photons to be upscattered to the gamma-ray range.

The detection of short lags in the UV to X-ray range in Mrk 421 and PKS 2155–304 indicates that, in HBL at least, energy injection in the jet occurs in a top down scenario. Because the spectral variability of LBL objects in the IR to UV range appears similar to that of Mrk 421 and PKS 2155–304 in X rays, it is important to look for similar lags in LBL at IR to UV wavelengths. In principle, this could be done with intensive ground-based monitoring but would be much more likely to succeed with space instrumentation (HST and ISO) owing to the more extended wavelength range accessible and to the quasicontinuous coverage and even sampling. Comparison of flare evolution in wavelength and time in these very different sources, HBL and LBL, will give important clues to the differences in their jets.

Determining the nature of the seed photons (possibly different in different objects) is important because it will allow better determination of the physical conditions in the jet and eventually understanding of whether and how the environment influences the jet properties. Because model predictions for correlated variability between the synchrotron and Compton emission differ, multiwavelength campaigns offer the best means to discriminate. Substantial progress is expected over the next few years. Repeated multiwavelength campaigns on the brightest sources should enable discrimination among different sources of seed photons, especially with studies of multiwavelength variability on short time scales, during which the external photon field should not change substantially.

It is unfortunate that the available time for these programs is limited by the residual amount of gas for the EGRET experiment. One of the critical difficulties is that ground-based observations in the optical (as IUE is not available any longer) cannot be arranged quickly enough after the discovery of inherently unpredictable gamma-ray flares, given their rapid intraday time scales. In order to measure cross correlations and possible lags between gamma-ray and optical flares, optical-UV monitors on board the next generation gamma-ray experiments are needed.

Jet models imply that energy is transported in a dissipationless way from very small scales, within a few Schwarzschild radii of the central black hole to relatively large distances along the jet, a light day or more. At present we have no clear picture of how this happens. We can infer from X-ray and gamma-ray observations that the innermost region is unlikely to be filled with photons, as the resulting pair cascades would redistribute energy from gamma-ray to X-ray energies. Further work in this area is critical, as it holds the key to understanding how energy is extracted from the central black hole, the fundamental issue in the study of AGN.

ACKNOWLEDGMENTS

It is a pleasure to thank our colleagues for comments and discussions at various stages of the preparation of this review, in particular R Blandford, S Collin-Souffrin, T Courvoisier, G Ghisellini, K Horne, B Hufnagel, L Lucy, G Madejski, S Phinney, E Pian, P Schneider, A Treves, R Terlevitch, and S Wagner.

*Literature Cited*


Alexander T, Netzer H. 1994. *MNRAS* 270:781–803

Alloin D, Pelat D, Phillips M, Fosbury R, Freemann K. 1986. *Ap. J.* 308:23–35

Angel JRP, Stockman HS. 1980. *Annu. Rev. Astron. Astrophys.* 18:321–61

Antonucci R. 1993. *Annu. Rev. Astron. Astrophys.* 31:473–521

Antonucci RRJ, Cohen RD. 1983. *Ap. J.* 271: 564–74

Appl S, Camenzind M. 1993. *Astron. Astrophys.* 270:71–82

Aretxaga I, Terlevich R. 1994. *MNRAS* 269:462–74

Armitage PJ, Zurek WH, Davies MB. 1996. *Ap. J.* 470:237–48

Artymowicz P, Lin DNC, Wampler EJ. 1993. *Ap. J.* 409:592–603

Babadzhanyants MK, Baryshev YuV, Belokon ET. 1992. In *Variability of Blazars*, ed. E Valtaoja, M Valtonen, pp. 45–51. Cambridge: Cambridge Univ. Press

Bao G, Hadrava P, Østgaard E. 1996. *Ap. J.* 464:684–89

Bassani L, Dean AJ, Sembay S. 1983. *Astron. Astrophys.* 125:52–58

Becker PA, Kafatos M. 1995. *Ap. J.* 453:83–94

Bicknell GV. 1994. *Austr. J. Phys.* 47:669–80

Binette L, Prieto A, Szuszkiewicz E, Zheng W. 1989. *Ap. J.* 343:135–41

Blandford RD. 1993. In *1st Compton Gamma-Ray Obs. Symp., Am. Inst. Phys. Conf. Proc. 280*, ed. M Friedlander, N Gehrels, DJ Macomb, pp. 533–40. New York: Am. Inst. Phys.

Blandford RD, Königl A. 1979 *Ap. J.* 232:34–41

Blandford RD, Levinson A. 1995. *Ap. J.* 441: 79–95

Blandford RD, McKee CF. 1982. *Ap. J.* 255: 419–39

Blandford RD, Narayan R. 1992. Annu. Rev. Astron. Astrophys. 30:311–58

Blandford RD, Netzer H, Woltjer L. 1990. In *Saas-Fee Advanced Course 20*, ed. TJL Courvoisier, M Mayor. Berlin/Heidelberg/New York: Springer-Verlag

Blandford RD, Payne DG. 1982. *MNRAS* 199:883–903

Blandford RD, Rees MJ. 1978. In *Pittsburgh Conference on BL Lac Objects*, ed. AM Wolfe, pp. 328–41. Pittsburgh: Univ. Pittsburgh Press

Bochkarev NG, Shapovalova AI, Zhekov SA. 1991. *Astron. J.* 102:1278–93

Boller T, Brandt WN, Fabian A, Fink H. 1997. *MNRAS.* In press

Boller T, Brandt WN, Fink H. 1996. *Astron. Astrophys.* 305:53–73

Bottorff M, Korista KT, Shlosman I, Blandford RD. 1997. *Ap. J.* 479:200–21

Bower GA, Wilson AS, Heckman TM, Richstone DO. 1996. *Astron. J.* 111:1901–7

Brandt WN, Pounds KA, Fink H, Fabian AC. 1996. *Max Planck Inst. Extraterr. Phys. Rep.* 263:429–30

Bregman JN. 1990. *Astron. Astrophys. Rev.* 2:125–66

Bregman JN. 1994. In *IAU Symp. 159*, ed. TJL Courvoisier, A Blecha, pp. 5–16

Bregman JN, Glassgold AE, Huggins PJ, Aller HD, Aller MF, et al. 1984. *Ap. J.* 276:454–65

Bregman JN, Glassgold AE, Huggins PJ, Kinney AL. 1986. *Ap. J.* 301:698–702

Bregman JN, Glassgold AE, Huggins PJ, Pollock JT, Pica AJ, et al. 1982. *Ap. J.* 253:19–27

Brinkmann W, Maraschi L, Treves A, Urry CM, Warwick R, et al. 1994. *Astron. Astrophys.* 288:433–47

Brown LMJ, Robson EI, Gear WK, Hughes DH, Griffin MH, et al. 1989. *Ap. J.* 340:129–49

Buckley JH, Akerlof CW, Biller S, Carter-Lewis DA, Catanese M, et al. 1996. *Ap. J.* 472:L9–L12

Camenzind M, Krockenberger M. 1992. *Astron. Astrophys.* 255:59–62

Canizares CR. 1982. *Ap. J.* 263:508–17

Cappi M, Comastri A, Molendi S, Palumbo GCC, Della Ceca R, Maccacaro T. 1994. *MNRAS* 271:438–48

Carrasco L, Dultzin-Hacyan D, Cruz-Gonzalez I. 1985. *Nature* 314:146–48

Celotti A, Maraschi L, Ghisellini G, Caccianiga A, Maccacaro T. 1993. *Ap. J.* 416:118–29

Celotti A, Maraschi L, Treves A. 1991. *Ap. J.* 377:403–16

Chakrabarti SK, Wiita PJ. 1994. *Ap. J.* 434:518–22

Cid Fernandes R Jr, Aretxaga I, Terlevich R. 1996. *MNRAS* 282:1191–202

Clarke CJ. 1987. *Bull. Am. Astron. Soc.* 19:732

Clavel J, Altamore A, Boksenberg A, Bromage GE, Elvius A, et al. 1987. *Ap. J.* 321:251–79

Clavel J, Nandra K, Makino F, Pounds KA, Reichert GA, et al. 1992. *Ap. J.* 393:113–25

Clavel J, Reichert GA, Alloin D, Crenshaw DM, Kriss G, et al. 1991. *Ap. J.* 366:64–81
Clavel J, Wamsteker W. 1987. *Ap. J.* 320:L9–L14
Clavel J, Wamsteker W, Glass IS. 1989. *Ap. J.* 337:236–50
Collin-Souffrin S. 1987. *Astron. Astrophys.* 179:60–70
Collin-Souffrin S. 1991. *Astron. Astrophys.* 249:344–50
Collin-Souffrin S, Dumont S, Joly M, Pequignot D. 1986. *Astron. Astrophys.* 166:27–35
Collin-Souffrin S, Lasota J-P. 1988. *Publ. Astron. Soc. Pac.* 100:1041–50
Coppi P, Kartje FJ, Königl A. 1993. In *1st Compton Gamma-Ray Obs. Symp., Am. Inst. Phys. Conf. Proc. 280*, ed. M Friedlander, N Gehrels, DJ Macomb, pp. 559–63. New York: Am. Inst. Phys.
Corbett EA, Robinson A, Axon DJ, Hough JA, Jeffries RD, et al. 1996. *MNRAS* 281:737–49
Corbin MR. 1995. *Ap. J.* 447:496–504
Courvoisier TJL, Blecha A, Bouchet P, Bratschi P, Carini MT, et al. 1995. *Ap. J.* 438:108–19
Courvoisier TJL, Clavel J. 1991. *Astron. Astrophys.* 248:389–94
Courvoisier TJL, Paltani S, Walter R. 1996. *Astron. Astrophys.* 308:17–20
Crenshaw DM, Rodríguez-Pascual PM, Penton SV, Edelson RA, Alloin D, et al. 1996. *Ap. J.* 470:322–35
Cristiani S, Trentini S, La Franca F, Aretxaga I, Andreani P, et al. 1996. *Astron. Astrophys.* 306:395–407
Cristiani S, Vio R, Andreani P. 1990. *Astron. J.* 100:56–59
Cutri RM, Wisniewski WZ, Rieke GH, Lebofsky MJ. 1985. *Ap. J.* 296:423–29
De Diego JA, Kidger M. 1990. *Astrophys. Space Sci.* 171:97–104
Dermer CD, Schlickeiser R. 1993. *Ap. J.* 416: 458–84
Di Clemente A, Giallongo E, Natali G, Trevese D, Vagnetti F. 1996. *Ap. J.* 463:466–72
Dietrich M, Kollatschny W. 1995. *Astron. Astrophys.* 303:405–19
Dietrich M, Kollatschny W, Peterson BM, Bechtold J, Bertram R, et al. 1993. *Ap. J.* 408:416–27
Dondi L, Ghisellini G. 1995. *MNRAS* 273:583–95
Done C, Krolik JH. 1996. *Ap. J.* 463:144–57
Done C, Pounds KA, Nandra K, Fabian AC. 1995. *MNRAS* 275:417–28
Dreissigacker O, Camenzind M. 1996. See Miller et al 1996, pp. 377–83
Edelson RA. 1992. *Ap. J.* 401:516–28
Edelson RA, Alexander T, Crenshaw DM, Kaspi S, Malkan M, et al. 1996. *Ap. J.* 470: 364–77
Edelson RA, Krolik JH. 1988. *Ap. J.* 333:646–59
Edelson RA, Krolik J, Madejski G, Maraschi L, Pike G, et al. 1995. *Ap. J.* 438:120–34
Edelson RA, Malkan M. 1986. *Ap. J.* 308:59–77
Edelson RA, Malkan M. 1987. *Ap. J.* 323:516–35
Efimov IS, Shakhovskoy NM. 1996. *Tuorla Obs. Rep.* 176, 32
Emmering RT, Blandford RD, Shlosman I. 1992. *Ap. J.* 385:460–77
Eracleous M, Halpern JP. 1994. *Ap. J. Suppl.* 90:1–30
Eracleous M, Livio M, Binette L. 1995a. *Ap. J.* 445:L1–L5
Eracleous M, Livio M, Halpern JP, Storchi-Bergmann T. 1995b. *Ap. J.* 438:610–22
Fabian AC. 1979. *Proc. R. Soc. London* 366:449–59
Fabian AC. 1996. *MPE Rep.* 263:403–8
Falomo R, Bouchet P, Maraschi L, Treves A, Tanzi EG. 1988. *Ap. J.* 335:122–25
Feigelson ED, Bradt H, McClintock J, Remillard R, Urry CM, et al. 1986. *Ap. J.* 302:337–51
Ferland GJ, Korista KT, Peterson BM. 1990. *Ap. J.* 363:L21–L25
Filippenko AV. 1989. *Astron. J.* 97:726–34
Forster K, Halpern JP. 1996. *Ap. J.* 468:565–70
Fritz KD, Webb GM. 1990. *Ap. J.* 360:387–95
Gaidos JA, Akerlof CW, Biller S, Boyle PJ, Breslin AC, et al. 1996. *Nature* 383:319–20
Galeev AA, Rosner R, Vaiana GS. 1979. *Ap. J.* 229:318–26
Gaskell CM. 1982. *Ap. J.* 263:79–86
Gaskell CM. 1996a. *Ap. J.* 464:L107–10
Gaskell CM. 1996b. In *Jets from Stars and Galactic Nuclei*, ed. W Kundt, pp. 165–95. Berlin: Springer-Verlag
Gear WK, Brown LMJ, Robson EI, Ade PAR, Griffin MJ, et al. 1986. *Ap. J.* 304:295–304
Gear WK, Stevens JA, Hughes DH, Litchfield SJ, Robson EI, et al. 1994. *MNRAS* 267:167–86
George IM, Warwick RS, Bromage GE. 1988a. *MNRAS* 232:793–808
George IM, Warwick RS, McHardy IM. 1988b. *MNRAS* 235:787–95
Ghisellini G, Madau P. 1996. *MNRAS* 280:67–76
Ghisellini G, Maraschi L. 1996. See Miller et al 1996, pp. 436–49
Ghisellini G, Maraschi L, Dondi L. 1996. *Astron. Astrophys. Suppl.* 120:503–6
Ghisellini G, Villata M, Raiteri C, Bosio S, De Francesco G, et al. 1997. *Astron. Astrophys.* Submitted
Ghisellini G, Maraschi L, Treves A. 1985. *Astron. Astrophys.* 146:204–12
Giallongo E, Trevese D, Vagnetti F. 1991. *Ap. J.* 377:345–48

Giannuzzo ME, Stirpe GM. 1996. *Astron. Astrophys.* 314:419–29

Gill TR, Lloyd C, Penston MV, Snijders MAJ. 1984. *MNRAS* 211:31–37

Giommi P, Barr P, Pollock AMT, Garilli B, Maccagni D. 1990. *Ap. J.* 365:432–55

Goad MR, O'Brien PT, Gondhalekar PM. 1993. *MNRAS* 263:149–67

Gondhalekar PM. 1990. *MNRAS* 243:443–58

Gondhalekar PM, Horne K, Peterson BM, eds. 1994. *Reverberation Mapping of the Broad-Line Region in Active Galactic Nuclei. Astron. Soc. Pac. Conf. Ser. 69.* San Francisco: Astron. Soc. Pac.

Goodrich RW. 1989. *Ap. J.* 340:190–202

Goodrich RW. 1995. *Ap. J.* 440:141–50

Grandi P, Tagliaferri G, Giommi P, Barr P, Palumbo GGC. 1992. *Ap. J. Suppl.* 82:93–116

Grandi P, Maraschi L, Urry CM, Wehrle AE, Aller MF, et al. 1994. *Adv. Space Res.* 15:23–26

Grandi P, Urry CM, Maraschi L, Wehrle AE, Madejski GM, et al. 1996. *Ap. J.* 459:73–81

Grupe D. 1996. *Properties of bright soft X-ray selected ROSAT AGN.* PhD thesis. Univ. Göttingen

Guainazzi M, Matsuoka M, Piro L, Mihara T, Yamauchi M. 1994. *Ap. J.* 436:L35–L39

Guilbert PW, Rees MJ. 1988. *MNRAS* 233:475–84

Haardt F, Maraschi L. 1991. *Ap. J.* 380:L51–L54

Haardt F, Maraschi L. 1993. *Ap. J.* 413:507–17

Haardt F, Maraschi L, Ghisellini G. 1994. *Ap. J.* 432:L95–L99

Haardt F, Maraschi L, Ghisellini G, 1997. *Ap. J.* 476:620–31

Halpern J, Eracleous M. 1994. *Ap. J.* 433:L17–L20

Halpern J, Eracleous M, Filippenko AV, Chen K. 1996. *Ap. J.* 464:704–14

Halpern J, Filippenko AV. 1988. *Nature* 331:46–48

Hartman RC. 1996. See Miller et al 1996, 110:333–39

Hartman RC, Webb JR, Marscher AP, Travis JP, Dermer CD, et al. 1996. *Ap. J.* 461:698–712

Hawkins MRS. 1993. *Nature* 366:242–45

Hawkins MRS. 1996. *MNRAS* 278:787–807

Hawkins MRS, Véron P. 1993. *MNRAS* 260:202–8

Heckman TM. 1980. *Astron. Astrophys.* 87:152–64

Heidt J. 1996. See Miller et al 1996, pp. 64–69

Heidt J, Wagner SL. 1996. *Astron. Astrophys.* 305:42–52

Ho LC, Filippenko AV, Sargent WLW. 1995. *Ap. J. Suppl. Ser.* 98:477–593

Ho LC, Filippenko AV, Sargent WLW. 1996. *Ap. J.* 462:183–202

Hooimeyer JRA, Miley GK, de Waard GJ, Schilizzi RT. 1992. *Astron. Astrophys.* 261:9–17

Hook IM, McMahon RG, Boyle BJ, Irwin MJ. 1994. *MNRAS* 268:305–20

Horne K. 1994. See Gondhalekar et al 1994, pp. 23–51

Horne K, Welsh WF, Peterson BM. 1991. *Ap. J.* 367:L5–L8

Hoyle F, Burbidge GR, Sargent WLW. 1966. *Nature* 209:751–53

Hufnagel BR, Bregman JN. 1992. *Ap. J.* 386: 473–84

Hughes PA, Aller HD, Aller MF. 1992. *Ap. J.* 396:469–86

Hughes PA, Aller HD, Aller MF. 1989. *Ap. J.* 341:54–79

Hughes PA, Aller MF, Aller HD. 1994. *IAU Symp. 159,* pp. 233–36. Dordrecht: Kluwer

Hunt LK, Zhekov S, Salvati M, Mannucci F, Stanga RM. 1994. *Astron. Astrophys.* 292:67–75

Hunter SD, Bertsch DL, Dingus BL, Fichtel CE, Hartman RC, et al. 1993. *Ap. J.* 409:134–38

Huré J-M, Collin-Souffrin S, Le Bourlot J, Pineau des Forêts G. 1994. *Astron. Astrophys.* 290:34–39

Impey C. 1996. *Astron. J.* 112:2667–83

Impey C, Neugebauer G. 1988. *Astron. J.* 95: 307–51

Iwazawa K, Fabian AC, Reynolds CS, Nandra K. 1997. *Ap. J.* In press

Jang MW, Miller HR. 1995. *Ap. J.* 452:582–87

Jannuzi BT, Smith PS, Elston R. 1993. *Ap. J. Suppl.* 85:265–91

Johnson WN, Grove JE, Kinzer RL, Kroeger RA, Kurfess JD, et al. 1994. In *2nd Compton Symp. Am. Inst. Phys. Conf. Proc. 304*, ed. CE Fichtel, N Gehrels, JP Norris, pp. 515–24. New York: Am. Inst. Phys.

Jones TW, O'Dell SL, Stein WA. 1974a. *Ap. J.* 188:353–68

Jones TW, O'Dell SL, Stein WA. 1974b. *Ap. J.* 192:261–78

Kardashev NS. 1962. *Sov. Astron. AJ.* 6:317–27

Kaspi S, Maoz D, Netzer H, Peterson BM, Alexander T. 1996a. *Ap. J.* 470:336–48

Kaspi S, Smith PS, Maoz D, Netzer H, Jannuzi BT. 1996b. *Ap. J. Lett.* 471:L75–L78

Kawai N, Matsuoka M, Bregman JN, Aller HD, Aller MF, et al. 1991. *Ap. J.* 382:508–14

Kerrick AD, Akerlof CW, Biller SD, Buckley JH, Cawley MF, et al. 1995. *Ap. J.* 438:L59–L62

Kidger M, Takalo L, Sillanpaa A. 1992. *Astron. Astrophys.* 264:32–36

Kidger M, Gonzales-Perez JN, De Diego JA, Mahoney TJ, Rodriguez–Espinosa et al. 1996. *Tuorla Obs. Rep.* 176:73–79

Kinney AL, Bohlin RC, Blades JC, York DG. 1991. *Ap. J. Suppl.* 75:645–717

Kniffen DA, Bertsch DL, Fichtel CE, Hartman RC, Hunter SD, et al. 1993. *Ap. J.* 411:133–36
Kollatschny W, Dietrich M. 1996. *Astron. Astrophys.* 314:43–50
Kollgaard RI. 1994. *Vistas Astron.* 38:29–75
Komesaroff MM, Roberts JA, Murray JD. 1988. *Observatory* 108:9–12
Königl A. 1981. *Ap. J.* 243:700–9
Königl A, Kartje JF. 1994. *Ap. J.* 434:446–67
Koo DC, Kron RG, Cudworth KM. 1986. *Publ. Astron. Soc. Pac.* 98:285–306
Koratkar AP, Gaskell CM. 1991. *Ap. J. Suppl.* 75:719–50
Kriss GA, Davidsen AF, Blair WP, Bowers CW, Dixon WV, et al. 1992. *Ap. J.* 392:485–91
Krolik JH, Done C. 1995. *Ap. J.* 440:166–80
Krolik JH, Horne K, Kallman TR, Malkan MA, Edelson RA, Kriss GA. 1991. *Ap. J.* 371:541–62
Kwan J, Krolik JH. 1981. *Ap. J.* 250:478–507
Lehto HJ, Valtonen MJ. 1996. *Ap. J.* 460:207–13
Leighly KM, Dietrich M, Waltmann E, Edelson R, George I, et al. 1996. *MPE Rep.* 263:467–68
Lin YC, Bertsch DL, Chiang J, Fichtel CE, Hartman RC, et al. 1992. *Ap. J.* 401:L61–L64
Lin YC, Bertsch DL, Dingus BL, Esposito JA, Fichtel CE, et al. 1995. *Ap. J.* 442:96–104
Litchfield SJ, Stevens JA, Robson EI, Gear WK. 1995. *MNRAS* 274:221–34
Longo G, Vio R, Paura P, Provenzale A, Rifatto A. 1996. *Astron. Astrophys.* 312:424–30
Lub J, de Ruiter HR. 1992. *Astron. Astrophys.* 256:33–44
Lynden-Bell D. 1969. *Nature* 223:690–94
Lyutyi VM, Oknyanski VL. 1987. *Sov. Astron.* 31(3):245–50
Macomb DJ, Akerloff CW, Aller HD, Aller MF, Bertsch DL, et al. 1995. *Ap. J.* 449:L99–L103
Macomb DJ, Akerloff CW, Aller HD, Aller MF, Bertsch DL, et al. 1996. *Ap. J.* 459:L111
Madejski GM, Takahashi T, Tashiro M, Kubo H, Hartman R, et al. 1996. *Ap. J.* 459:156–68
Makino F, Ebisuzaki T. 1996. *Ap. J.* 465:527–33
Makino F, Kii T, Fujimoto R, Otani C, Ohashi T, et al. 1993. In *Frontiers of Neutrino Astrophysics*, ed. Y Suzuki, K Nakamura, pp. 425–30. Tokyo: Univ. Acad.
Makino F, Edelson R, Fujimoto R, Kii T, Idesawa E, et al. 1996. *MPE Rep.* 263:413–16
Makino J, Tanaka Y, Matsuoka M, Koyama K, Inoue H, et al. 1987. *Ap. J.* 313:662–73
Malkan MA, Sargent WLW. 1982. *Ap. J.* 254: 22–37
Mannheim K. 1993. *Astron. Astrophys.* 269:67–76
Mannheim K. 1996. *Space Sci. Rev.* 75:331–40
Mannheim K, Biermann P. 1992. *Astron. Astrophys.* 253:L21–L24
Mannheim K, Grupe D, Beuermann K, Thomas HC, Fink HH. 1996. *MPE Rep.* 263:471–72
Maoz D. 1997. In *Emission Lines in Active Galaxies: New Methods and Techniques*, ed. BM Peterson, F-Z Cheng, AS Wilson, pp. 138–45. San Francisco: Astron. Soc. Pac.
Maoz D, Filippenko AV, Ho LC, Rix H-W, Bahcall JN, et al. 1995. *Ap. J.* 440:91–99
Maoz D, Netzer H, Mazeh T, Beck S, Almoznino E, et al. 1991. *Ap. J.* 367:493–506
Maoz D, Netzer H, Peterson BM, Bechtold J, Bertram R, et al. 1993. *Ap. J.* 404:576–83
Maraschi L, Ghisellini G, Celotti A. 1992. *Ap. J.* 397:L5–L9
Maraschi L, Ghisellini G, Celotti A. 1994. In *IAU Symp. 159*, ed. TJL Courvoisier, A Blecha, pp. 221–32
Maraschi L, Grandi P, Urry CM, Wehrle AE, Madejski GM, et al. 1994. *Ap. J.* 435:L91–L95
Maraschi L, Rovetti F. 1994. *Ap. J.* 436:79–88
Maraschi L, Tanzi EG, Maccagni D, Tarenghi M, Chiappetti L. 1983. *Ap. J.* 273:75–80
Maraschi L, Treves A, Tagliaferri G, Tanzi EG. 1986. *Ap. J.* 304:637–45
Marscher AP. 1980. *Ap. J.* 235:386–91
Marscher AP. 1996. See Miller et al 1996, pp. 248–61
Marscher AP, Gear W. 1985. *Ap. J.* 298:114–27
Marscher AP, Gear W, Travis J. 1992. In *Variability of Blazars* ed. E Valtaoja, M Valtonen, pp. 85–101. Cambridge: Cambridge Univ. Press
Marscher AP, Travis JP. 1991. In *Variability of Active Galactic Nuclei*, ed. HR Miller, PJ Wiita, pp. 153–56. Cambridge: Cambridge Univ. Press
Marshall HL, Carone TE, Peterson BM, Crenshaw DM, Kriss GA. 1997. *Ap. J.* 479:222–30
Marziani P, Sulentic JW, Calvani M, Perez E, Moles M, Penston MV. 1993. *Ap. J.* 410:56–67
Marziani P, Sulentic JW, Dultzin-Hacyan D, Calvani M, Moles M. 1996. *Ap. J. Suppl.* 104:37–70
Mastichiadis A, Kirk JG. 1997. *Astron. Astrophys.* 320:19–25
Mathews WG. 1993. *Ap. J.* 412:L17–L20
Mattox JR. 1995. *IAU Circ. No. 6181.*
Mattox JR, Wagner SJ. 1996. See Miller et al 1996, pp. 352–58
Miller HR, Webb JR, Noble JC, eds. 1996. *Blazar Continuum Variability, Astron. Soc. Pac. Conf. Ser.*, Vol. 110. San Francisco: Astron. Soc. Pac.
Miller JS, French HB, Hawley SA. 1978. *Pittsburgh Conference on BL Lac Objects*, ed. AM Wolfe, pp. 176–91. Pittsburgh:Univ. Pittsburgh Press
Miller JS, Peterson BM. 1990. *Ap. J.* 361:98–100

Molendi S, Maccacaro T, Schaeidt S. 1993. *Astron. Astrophys.* 271:18–24
Molendi S, Maraschi L, Stella L. 1992. *MNRAS* 255:27–31
Moles M, Masegosa J, del Olmo A. 1987. *Astron. J.* 94:1143–49
Morini M, Chiappetti L, Maccagni D, Maraschi L, Molteni D, et al. 1986a. *Ap. J.* 306:L71–L75
Morini M, Scarsi L, Molteni D, Salvati M, Perola GC, et al. 1986b. *Ap. J.* 307:486–96
Mukherjee R, Dingus BL, Gear WK, Hartman RC, Hunter SD, et al. 1996. *Ap. J.* 470:831–38
Murray N, Chiang J. 1997. *Ap. J.* 474:91–103
Mushotzky RF, Done C, Pounds KA. 1993. *Annu. Rev. Astron. Astrophys.* 31:717–61
Mutel RL. 1990. In *Parsec-Scale Radio Jets*, ed. AJ Zensus, TJ Pearson, pp. 98–109. Cambridge: Cambridge Univ. Press
Nandra K, Pounds KA, Stewart GC, George IM, Hayashida K. 1991. *MNRAS* 248:760–72
Nelson BO. 1996. *Ap. J.* 465:L87–L90
Netzer H. 1987. *MNRAS* 225:55–72
Netzer H, Maoz D. 1990. *Ap. J.* 365:L5–L7
Neugebauer G, Soifer BT, Matthews K, Elias JH. 1989. *Astron. J.* 97:957–77
O'Brien PT, Harries TJ. 1991. *MNRAS* 250:133–37
Osterbrock DE, Pogge R. 1985. *Ap. J.* 297:166–76
Otani C, Kii T, Miya K. 1996. *MPE Rep.* 263:491–92
Padovani P. 1992. *MNRAS* 257:404–14
Padovani P, Giommi P. 1995. *MNRAS* 277:1477–90
Paltani S, Courvoisier TJL. 1994. *Astron. Astrophys.* 291:74–88
Peimbert M, Torres-Peimbert S. 1981. *Ap. J.* 245:845–56
Penston MV. 1988. *MNRAS* 233:601–9
Pérez E, Penston MV, Moles M. 1989. *MNRAS* 239:75–90
Pérez E, Robinson A, de la Fuente L. 1992a. *MNRAS* 255:502–20
Pérez E, Robinson A, de la Fuente L. 1992b. *MNRAS* 256:103–10
Perola GC, Piro L, Altamore A, Fiore F, Boksenberg A, et al. 1986. *Ap. J.* 306:508–21
Perry JJ, van Groningen E, Wanders I. 1994. *MNRAS* 271:561–72
Pesce JE, Urry CM, Maraschi L, Treves A, Grandi P, et al. 1997. *Ap. J.* In press
Peterson BM. 1993. *Publ. Astron. Soc. Pac.* 105:247–68
Peterson BM, Berlind P, Bertram R, Bochkarev NG, Bond D, et al. 1994. *Ap. J.* 425:622–34
Petrucci PO, Henri G. 1997. In *The Transparent Universe*, ed. C Winkler, et al, *ESA-SP 382*. In press
Pian E, Falomo R, Scarpa R, Treves A. 1994. *Ap. J.* 432:547–53
Pian E, Treves A. 1993. *Ap. J.* 416:130–36
Pian E, Urry CM, Treves A, Maraschi L, Penton S, et al. 1997. *Ap. J.* In press
Pica AJ, Smith AG. 1983. *Ap. J.* 272:11–25
Pijpers FP. 1994. See Gondhalekar et al 1994, pp. 69–83
Piro L, Balucinska-Church M, Fink H, Fiore F, Matsuoka M, et al. 1997. *Astron. Astrophys.* 319:74–82
Pounds KA, Brandt WN. 1996. In *X-Ray Imaging and Spectroscopy of Cosmic Hot Plasmas*. Preprint
Pounds KA, Done C, Osborne JP. 1995. *MNRAS* 277:L5–L10
Press WH, Gunn JE. 1973. *Ap. J.* 185:397–412
Pringle J. 1981. *Annu. Rev. Astron. Astrophys.* 19:137–62
Protheroe RJ, Biermann PL. 1997. *Astropart. Phys.* In press
Punch M, Akerlof CW, Cawley MF, Chantell M, Fegan DJ, et al. 1992. *Nature* 358:477–78
Quinn J, Akerlof CW, Biller S, Buckley J, Carter-Lewis DA, et al. 1996. *Ap. J.* 456:L83–L86
Recondo-González MC, Wamsteker W, Clavel J, Rodríguez-Pascual PM, Vio R, et al. 1997. *Astron. Astrophys.* In press
Rees MJ. 1984. *Annu. Rev. Astron. Astrophys.* 22:471–506
Rees MJ, Begelman MC, Blandford RD, Phinney ES. 1982. *Nature* 295:17–21
Reichert GA, Rodríguez-Pascual PM, Alloin D, Clavel J, Crenshaw DM, et al. 1994. *Ap. J.* 425:582–608
Renzini A, Greggio L, Alighieri SD, Cappellari M, Burstein D, Bertola F. 1995. *Nature* 378:39–41
Robinson A. 1995. *MNRAS* 276:933–43
Robson EI, Gear WK, Clegg PE, Ade PAR, Smith MG, et al. 1983. *Nature* 305:194–96
Roellig TL, Werner MW, Becklin E, Impey CD. 1986. *Ap. J.* 304:646–50
Rokaki E, Boisson C, Collin-Souffrin S. 1992. *Astron. Astrophys.* 253:57–73
Rokaki E, Collin-Souffrin S, Magnan C. 1993. *Astron. Astrophys.* 272:8–24
Salamanca I, Alloin D, Baribaud T, Axon D, de Bruyn G, et al. 1994. *Astron. Astrophys.* 282:742–52
Sambruna RM, Barr P, Giommi P, Maraschi L, Tagliaferri G, Treves A. 1994. *Ap. J.* 434:468–78
Sambruna RM, Maraschi L, Urry CM. 1996. *Ap. J.* 463:444–65
Sambruna RM, Urry CM, Ghisellini G, Maraschi L. 1995. *Ap. J.* 449:567–75
Sambruna RM, Urry CM, Maraschi L,

Ghisellini G, Mukherjee R, et al. 1997. *Ap. J.* 474:639–49

Sanders D, Phinney ES, Neugebauer G, Soifer BT, Matthews K. 1989. *Ap. J.* 347:29–51

Schneider P. 1993. *Astron. Astrophys.* 279:1–20

Schubnell M, Akerlof CW, Biller S, Buckley J, Carter-Lewis DA, et al. 1996. *Ap. J.* 460:644–50

Sembay S, Warwick RS, Urry CM, Sokoloski J, George IM, et al. 1993. *Ap. J.* 404:112–23

Shakura NI, Sunyaev RA. 1973. *Astron. Astrophys.* 24:337–55

Shields GA. 1978. *Nature* 272:706–8

Shields JC, Ferland GJ, Peterson BM. 1995. *Ap. J.* 441:507–20

Sikora M, Begelman M, Rees MJ. 1994. *Ap. J.* 421:153–62

Sillanpaa A, Haarala S, Valtonen MJ, Sundelius B, Byrd GC. 1988. *Ap. J.* 325:628–34

Sillanpaa A, Takalo LO, Purisimo T, Lehto HJ, Nilsson K, et al. 1996. *Astron. Astrophys.* 305:L17–L21

Sincell MW, Krolik JH. 1997. *Ap. J.* 476:605–19

Sitko ML, Junkkarinen VT. 1985. *Publ. Astron. Soc. Pac.* 97:1158–62

Smith AG, Nair AD. 1995. *Publ. Astron. Soc. Pac.* 107:863–70

Smith AG, Nair AD, Leacock RJ, Clements S. 1993. *Astron. J.* 105:437–55

Smith PS, Hall PB, Allen RG, Sitko ML. 1992. *Ap. J.* 400:115–26

Sparke LS. 1993. *Ap. J.* 404:570–75

Stecker FW, De Jager OC, Salomon MM. 1996. *Ap. J.* 473:L75–L78

Stein WA, O'Dell SL, Strittmatter R. 1976. *Annu. Rev. Astron. Astrophys.* 14:173–95

Stevens JA, Litchfield SJ, Robson EI, Cawthorne TV, Aller MF, et al. 1996. *Ap. J.* 466:158–68

Stevens JA, Litchfield SJ, Robson EI, Hughes DH, Gear WK, et al. 1994. *Ap. J.* 437:91–107

Stirpe GM, de Bruyn AG, van Groningen E. 1988. *Astron. Astrophys.* 200:9–16

Stirpe GM, Winge C, Altieri B, Alloin D, Aguero EL, et al. 1994. *Ap. J.* 425:609–21

Storchi-Bergmann T, Eracleous M, Livio M, Wilson AS, Filippenko AV, Halpern JP. 1995. *Ap. J.* 443:617–24

Sulentic JW, Marziani P, Dultzin-Hacyan D, Calvani M, Moles M. 1995a. *Ap. J.* 445:L85–L89

Sulentic JW, Marziani P, Zwitter T, Calvani M. 1995b. *Ap. J.* 438:L1–L4

Tagliaferri G, Bao G, Israel GL, Stella L, Treves A. 1996. *Ap. J.* 465:181–90

Takahashi T, Tashiro M, Madejski G, Kubo H, Kamae T, et al. 1996. *Ap. J. (Letters)* 470:L89–L92

Takalo L. 1994. *Vistas Astron.* 38:77–109

Tanaka Y, Nandra K, Fabian AC, Inoue H, Otani C, et al. 1995. *Nature* 375:659–61

Tashiro M. 1995. *Study of X-ray emission mechanism from highly variable active galactic nuclei.* PhD thesis. Inst. Space Astronaut. Sci., Tokyo, Japan (ISAS Res. Note 549)

Terlevich R, Tenorio-Tagle G, Franco J, Różyczka M, Melnick J. 1992. *MNRAS* 255:713–28

Thompson DJ, Bertsch DL, Dingus BL, Esposito JA, Etienne A, et al. 1995. *Ap. J. Suppl.* 101:259–86

Tornikoski M, Valtaoja E, Terasranta H, Smith AG, Nair AD, et al. 1994. *Astron. Astrophys.* 289:673–710

Tran HD. 1995. *Ap. J.* 440:597–605

Treves A, Morini M, Chiappetti L, Fabian A, Falomo R, et al. 1989. *Ap. J.* 341:733–47

Trevese D, Kron RG, Majewski SR, Bershady MA, Koo DC. 1994. *Ap. J.* 433:494–509

Trevese D, Pitella G, Kron RG, Koo, DC, Bershady MA. 1989. *Astron. J.* 98:108–16

Türler M, Walter R, Schartel N. 1996. *MPE Rep.* 263:515–16

Ulrich MH. 1981. *Astron. Astrophys.* 103:L1–L2

Ulrich MH. 1994. In *Theory of Accretion Disks*, ed. WJ Duschl, J Frank, F Meyer, E Meyer-Hofmeister, WM Tscharnuter, pp. 253–59. Dordrecht: Kluwer Acad.

Ulrich MH. 1996. *MNRAS* 281:907–15

Ulrich MH, Altamore A, Boksenberg A, Bromage GE, Clavel J, et al. 1985. *Nature* 313:745–47 L1 L2

Ulrich MH, Boisson C. 1983. *Ap. J.* 267:515–27

Ulrich MH, Boksenberg A, Bromage GE, Clavel J, Elvius A, et al. 1984a. *MNRAS* 206:221–37

Ulrich MH, Boksenberg A, Bromage GE, Clavel J, Elvius A, et al. 1991. *Ap. J.* 382:483–500

Ulrich MH, Courvoisier TJL, Wamsteker W. 1993. *Ap. J.* 411:125–32

Ulrich MH, Hackney KRH, Hackney RL, Kondo Y. 1984b. *Ap. J.* 276:466–71

Ulrich MH, Horne K. 1996. *MNRAS* 283:748–58

Urry CM, Kondo Y, Hackney KRH, Hackney RL. 1988. *Ap. J.* 330:791–802

Urry CM, Maraschi L, Edelson R, Koratkar A, Krolik J, et al. 1993. *Ap. J.* 411:614–31

Urry CM, Mushotzky RF, Holt SS. 1986. *Ap. J.* 305:369–98

Urry CM, Padovani P. 1995. *Publ. Astron. Soc. Pac.* 107:803–45

Urry CM, Sambruna RM, Worrall DM, Kollgaard RI, Feigelson E, et al. 1996. *Ap. J.* 463:424–43

Urry CM, Treves A, Maraschi L, Marshall H, Kii T, et al. 1997. *Ap. J.* In press

Valtaoja E, Haarala S, Lehto H, Valtaoja L, Valtonen M. 1988. *Astron. Astrophys.* 203:1–20
Valtaoja E, Teräsranta H. 1995. *Astron. Astrophys.* 297:L13–L16
Van Groningen E. 1987. *Astron. Astrophys.* 186:103–13
Vermeulen RC, Cohen MH. 1994. *Ap. J.* 430:467–94
Vermeulen R, Ogle PM, Tran HD, Browne IWA, Cohen MH, et al. 1995. *Ap. J.* 452:L5–L8
Villar-Martin M. 1996. *Dust and Gas in Active Galaxies.* PhD thesis. Sheffield Univ., England
Vio R, Horne K, Wamsteker W. 1994. *Publ. Astron. Soc. Pac.* 106:1091–103
von Montigny C, Bertsch DL, Chiang J, Dingus BL, Esposito JA, et al. 1995. *Ap. J.* 440:525–53
Wagner SJ. 1997. In *Emission Lines in Active Galaxies: New Methods and Techniques*, ed. BM Peterson, F-Z Cheng, AS Wilson, pp. 298–301. San Francisco: Astr. Soc. Pac.
Wagner SJ, Camenzind M, Dreissagacker O, Borgeest U, Britzen S, et al. 1995a. *Astron. Astrophys.* 298:688–98
Wagner SJ, Mattox JR, Hopp U, Bock H, Heidt J, et al. 1995b. *Ap. J.* 454:L97–L100
Wagner SJ, Witzel A. 1995. *Annu. Rev. Astron. Astrophys.* 33:163–97
Wagner SJ, Witzel A, Heidt J, Krichbaum TP, Qian SJ, et al. 1996. *Astron. J.* 111:2187–211
Wamsteker W, Colina L. 1986. *Ap. J.* 311:617–22
Wamsteker W, Rodríguez-Pascual P, Wills BJ, Netzer H, Wills D, et al. 1990. *Ap. J.* 354:446–67
Wamsteker W, Wang T, Schartel N, Vio R. 1997. *MNRAS.* In press
Wanders I, Goad MR, Korista KT, Peterson BM, Horne K, et al. 1995. *Ap. J.* 453:L87–L90
Wanders I, Horne K. 1994. *Astron. Astrophys.* 289:76–82
Wanders I, Peterson BM. 1996. *Ap. J.* 466:174–90
Wanders I, van Groningen E, Alloin D, Aretxaga I, Axon D, et al. 1993. *Astron. Astrophys.* 269:39–53
Warwick RS, Smith DA, Yaqoob T, Edelson R, Johnson WN. 1996. *Ap. J.* 470:349–63
Webb J, Smith AG, Leacock RJ, Fitzgibbons GL, Gombola PP, Shepherd DW. 1988. *Astron. J.* 95:374–97
Wehrle A, Unwin SC, Zook AC, Urry CM, Marscher AP, Teräsranta H. 1996. See Miller et al 1996, pp. 430–45
Wilkes BJ. 1984. *MNRAS* 207:73–98
Wills BJ, Netzer H, Wills D. 1985. *Ap. J.* 288:94–116
Winge C, Peterson BM, Horne K, Pogge RW, Pastoriza MG, Storchi-Bergmann T. 1995. *Ap. J.* 445:680–90
Worrall DM, Wilkes BJ. 1990. *Ap. J.* 360:396–407
Yaqoob T, Serlemitsos PJ, Turner TJ, George IM, Nandra K. 1996. *Ap. J.* 470:L27–L30
Zdziarski A, Krolik JH. 1993. *Ap. J.* 409:L33–L36
Zhang YF, Marscher AP, Aller HD, Aller MF, Teräsranta H, Valtaoja E. 1994. *Ap. J.* 432:91–102
Zheng W, Burbidge EM, Smith HE, Cohen RD, Bradley SE. 1987. *Ap. J.* 322:164–73
Zheng W, Veilleux S, Grandi SA. 1991. *Ap. J.* 381:418–25

*Annu. Rev. Astron. Astrophys. 1997. 35:503–56*


# ABUNDANCE RATIOS AND GALACTIC CHEMICAL EVOLUTION

*Andrew McWilliam*
Carnegie Observatories, Room 33, 813 Santa Barbara Street, Pasadena, California 91101; e-mail: andy@marmite.ociw.edu

KEY WORDS: abundances, chemical composition, the Galaxy, nucleosynthesis, stars

ABSTRACT

The metallicity of stars in the Galaxy ranges from [Fe/H] = −4 to +0.5 dex, and the solar iron abundance is $\epsilon$(Fe) = 7.51 ± 0.01 dex. The average values of [Fe/H] in the solar neighborhood, the halo, and Galactic bulge are −0.2, −1.6, and −0.2 dex respectively.

Detailed abundance analysis reveals that the Galactic disk, halo, and bulge exhibit unique abundance patterns of O, Mg, Si, Ca, and Ti and neutron-capture elements. These signatures show that environment plays an important role in chemical evolution and that supernovae come in many flavors with a range of element yields.

The 300-fold dispersion in heavy element abundances of the most metal-poor stars suggests incomplete mixing of ejecta from individual supernova, with vastly different yields, in clouds of $\sim 10^6$ $M_\odot$.

The composition of Orion association stars indicates that star-forming regions are significantly self-enriched on time scales of 80 million years. The rapid self-enrichment and inhomogeneous chemical evolution models are required to match observed abundance trends and the dispersion in the age-metallicity relation.

## INTRODUCTION

Except for the lightest elements, the history of the chemical composition of the Galaxy is dominated by nucleosynthesis occurring in many generations of stars. Stars of low mass have long lifetimes, some comparable to the age of the Galaxy,



0066-4146/97/0915-0503$08.00

and their envelopes have preserved much of their original chemical composition. These stars are useful because they are fossils containing information about the history of the evolution of chemical abundances in the Galaxy. After the Big Bang, the story of nucleogenesis is concerned mostly with the physics of stellar evolution and nucleosynthesis in stars, with how the environment dictated the kinds of stars that formed to enrich the Galactic gas, and with how the enriched gas mixed with the interstellar medium to form subsequent stellar generations (Hoyle 1954). We can try to understand these processes and chemical evolution from theoretical models, but the best way to learn about the history of the elements in the Galaxy is to look at the fossils.

Time and space do not permit me to discuss in sufficient detail the many exciting developments that have occurred in the area of chemical evolution. Therefore, I restrict myself to areas most closely aligned with my research, which is usually concerned with high-resolution abundance analysis of stars in our Galaxy; in particular I do not discuss all the elements, or families of elements, and some elements may be conspicuous by their absence.

Despite its obvious flaws, a good starting point for developing a mental picture of chemical evolution is the Simple one-zone model (e.g. Schmidt 1963, Searle & Sargent 1972, Pagel & Patchett 1975). The model assumes evolution in a closed system, with generations of stars born out of the interstellar gas (ISM). In each generation, a fraction of the gas is transformed into metals and returned to the ISM; the gas locked up in long-lived low-mass stars and stellar remnants no longer takes part in chemical evolution. Newly synthesized metals from each stellar generation are assumed to be instantaneously recycled back into the ISM and instantaneously mixed throughout the region; thus, in this model, metallicity always increases with time, and the region is perfectly homogeneous at all times.

The ratio of mass of metals ejected to mass locked up, $y$, is a quantity commonly called the yield. The term yield has another meaning: Supernova (SN) nucleosynthesis theorists use it to refer to the mass of a particular element ejected in a SN model. The yield depends on the mass of metals ejected by stars (usually a function of mass) and the relative frequency of different mass stars born in a stellar generation (this is the initial mass function, or IMF). The mean IMF has been measured empirically (e.g. Scalo 1987) and over Galactic time appears to have been approximately constant; however, for individual molecular clouds, large deviations from the mean IMF occur.

Another chemical evolution parameter is the star formation rate (SFR), which has been postulated to be proportional to some power of the gas density and the total mass density. In the Simple model, the SFR affects the time evolution of the metallicity but does not affect the final metallicity function of the system after the gas has been exhausted.

Given the yield, the metallicity function of long-lived stars for the Simple model is as follows:

$$f(z) = y^{-1}\exp(-z/y)$$

If evolution continues to gas exhaustion, then the Simple model predicts that the average mass fraction of metals of long-lived stars is equal to the yield, $<z> = y$. In principle the mean metal content of a stellar system can tell us about the yield. Because the yield is the ratio of mass of metals produced to the mass in low-mass stars per generation, it is sensitive to the IMF: An IMF skewed to high-mass stars would have a higher yield because more stars are massive enough to produce metals as SN, and there are fewer low-mass stars to lock away the gas.

Abundance ratios can serve as a diagnostic of the IMF and SFR parameters and time scale for chemically evolving systems. Tinsley (1979) proposed that type Ia supernovae (SN Ia, resulting from mass accretion by a C-O white dwarf) are the major producers of iron in the Galaxy and that the SN Ia progenitors have longer lifetimes than the progenitors of type II supernovae (SN II, resulting from exploding massive stars), which are the source of Galactic oxygen; Tinsley argued that the time delay between SN II and SN Ia, of at least $10^8$ years, is responsible for the enhanced [O/Fe][1] ratios observed in halo stars. Theoretical predictions of SN II element yields show that [$\alpha$/Fe] (where $\alpha$ includes the elements O, Mg, Si, S, Ca, and Ti) increases with increasing progenitor mass (e.g. Woosley & Weaver 1995). In principle, the IMF of a stellar system could be inferred from the observed [$\alpha$/Fe] ratios. Note that if a stellar system is found to have a high average metallicity, and an IMF skewed to high-mass stars is responsible for increasing the yield, then the composition should reflect an increased [$\alpha$/Fe] ratio that is due to the increased [$\alpha$/Fe] from high mass SN II. In fact, this idea was used by Matteucci & Brocato (1990) to explain the putative high metallicity of the Galactic bulge, with the prediction that [$\alpha$/Fe] is enhanced in the bulge.

The [$\alpha$/Fe] ratio is also sensitive to the SFR in Tinsley's model: If the SFR is high, then the gas will reach higher [Fe/H] before the first SN Ia occur, and the position of the knee in the [$\alpha$/Fe] versus [Fe/H] diagram (Figure 1) will be at a higher [Fe/H]. Also, because the knee marks the time of the first SN Ia, then the formation time scale of a stellar system can be estimated by noting the fraction of stars with [Fe/H] below this point.

Another potentially useful diagnostic of the [O/Fe] ratio was pointed out by Wyse & Gilmore (1991): In a star-burst system, the O/Fe ratio of the gas is

[1][A/B] refers to an abundance ratio in $\log_{10}$ solar units, where A and B represent the number densities of two elements: [A/B] $= \log_{10}(A/B)_* - \log_{10}(A/B)_\odot$. Note that $\epsilon(M) = \log_{10}(M/H)$.

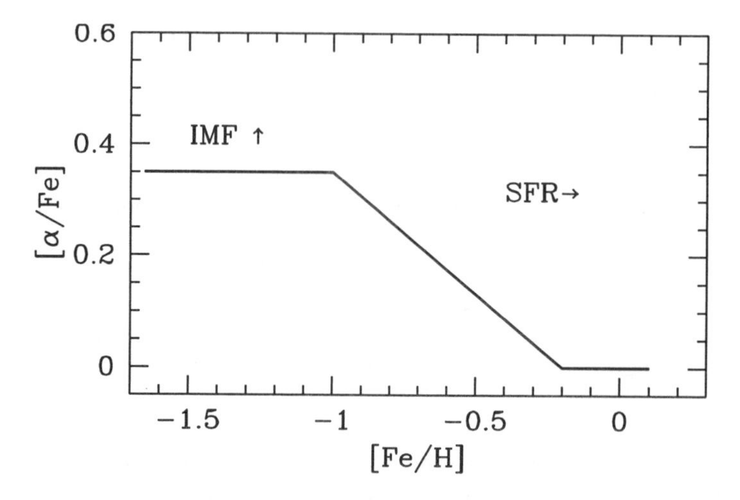


*Figure 1* A schematic diagram of the trend of $\alpha$-element abundance with metallicity. Increased initial mass function and star formation rate affect the trend in the directions indicated. The knee in the diagram is thought to be due to the onset of type Ia supernovae (SN Ia).

initially above solar owing to nucleosynthesis by SN II, but as time continues after the burst (with no new star formation) the SN II diminish, only SN Ia enrich the gas; ultimately subsolar [O/Fe] ratios occur. Wyse & Gilmore (1991) claimed that the composition of the LMC is fit by this model.

Elements like C, O, and those in the iron-peak, thought to be produced in stars from the original hydrogen, are sometimes labeled as "primary." The label "secondary" is reserved for elements thought to be produced from pre-existing seed nuclei, such as N and s-process heavy elements. The abundance of a primary element is expected to increase in proportion to the metallicity, thus [M/Fe] is approximately constant. For a secondary element, [M/Fe] is expected to increase linearly with [Fe/H] because the yield is proportional to the abundance of preexisting seed nuclei. One difficulty is that N and the s-process elements (both secondary) do not show the expected dependence on metallicity.

## THE SOLAR IRON ABUNDANCE

It is sobering, and somewhat embarrassing, that the solar iron abundance is in dispute at the level of 0.15 dex. This discrepancy comes in spite of the fact that more than 2000 solar iron lines, with reasonably accurate $gf$ values, are available for abundance analysis; that the solar spectrum is measured with much higher S/N and dispersion than for any other star; that LTE corrections to Fe I abundance are small, at only +0.03 dex (Holweger et al 1991); and that both theoretical and empirical solar model atmospheres are available, with parameters known more precisely than for any other star.

Anders & Grevesse (1989) reviewed published meteoritic and solar photospheric abundances for all available elements and found $\epsilon(\mathrm{Fe}) = 7.51 \pm 0.01$ for meteorites and $\epsilon(\mathrm{Fe}) = 7.67$ from the solar abundance analysis of Blackwell et al (1984, 1986), which was a notable increase from the earlier photospheric value of $7.50 \pm 0.08$ favored by Ross & Aller (1976). Blackwell et al's work utilized the Oxford group $gf$ values for Fe I lines, which are known to be of high accuracy.

Pauls et al (1990) found $\epsilon(\mathrm{Fe}) = 7.66$ from Fe II lines, but Holweger et al (1990), also using Fe II lines, found $\epsilon(\mathrm{Fe}) = 7.48$. Biémont et al (1991) measured the solar iron abundance of $7.54 \pm 0.03$ with a larger sample of Fe II lines. Holweger et al (1991) found $7.50 \pm 0.07$ based on gf values for Fe I lines measured by Bard et al (1991).

Two recent papers are characteristic of the conflicting solar iron abundance: those by Holweger et al (1995) and Blackwell et al (1995). Blackwell et al (1995) employed Oxford $gf$ values and the Holweger & Müller (1974) solar atmosphere and found $\epsilon(\mathrm{Fe}) = 7.64 \pm 0.03$ from the Fe I lines; although the Fe II line results indicated $\epsilon(\mathrm{Fe}) = 7.53$ dex.

When Blackwell et al (1995) computed iron abundances from the Kurucz (1992 unpublished) solar model, Fe I and Fe II lines gave better agreement, at 7.57 and 7.54 dex, respectively, but they claimed that the Kurucz model results are not valid because the solar limb darkening is not reproduced by the model. Blackwell et al (1995) concluded that neither the empirical Holweger-Müller model, nor the Kurucz theoretical model atmosphere, is adequate for measuring the solar iron abundance.

Holweger et al (1995) contested Blackwell et al's (1995) claim and argued that Fe I lines analyzed with the Holweger-Müller model give $\epsilon(\mathrm{Fe}) = 7.48 \pm 0.05$, or 7.51 with the 0.03-dex non–local thermodynamic equilibrium (non–LTE) correction. Holweger et al (1995) found the same low solar iron abundance from both Fe I and Fe II lines in their analysis.

Lambert et al (1995a) found that the $gf$ values of lines common to results of both Holweger et al (1995) and Blackwell et al (1995) had zero average difference, which suggests that gf values are not the source of the abundance difference. They attributed the difference mostly to variations in the measured equivalent widths and damping constants. Another low value of the solar iron abundance was found by Milford et al (1994), who found $\epsilon(\mathrm{Fe}) = 7.54 \pm 0.05$ with the Holweger-Müller solar model and new $gf$ values, from weak Fe I lines that are not sensitive to uncertainties in damping constants or microturbulent velocity.

Kostik et al (1996) attempted to resolve the differences between Blackwell et al (1995) and Holweger et al (1995). Kostik et al found that the Blackwell et al's equivalent widths are systematically higher than Holweger et al's values; remeasurement by Kostik et al favored the Holweger et al values. Kostik et al also found suspicious trends in the $gf$ values of the Holweger et al (1995) study, and they agree with Grevesse & Noels (1993) that the spread in iron abundance is dominated by uncertainties in the $gf$ values. They also noted that uncertainties in the microturbulent velocity and collisional damping constants are extremely important to the adopted value. Kostik et al provide a best estimate of the solar iron abundance of $7.62 \pm 0.04$, which favors the high solar iron abundance; although little weight was placed on the significance of this result.

Anstee et al (1997) measured the solar iron abundance from profile-matching 26 strong Fe I lines, using accurate laboratory collision-damping constants and $gf$ values. They found $\epsilon(\text{Fe}) = 7.51 \pm 0.01$ in complete agreement with the meteoritic iron abundance of Anders & Grevesse (1989), independent of non-thermal motions in the photosphere. Anstee et al traced the discrepancies between previous studies to the use of different atomic data, measured equivalent widths, and assumed microturbulent velocity.

It now seems that the weight of the evidence favors the low value of the solar iron abundance, and the issue may finally be settled; however, this statement has been made before . . . .

## SUPER METAL-RICH STARS

The existence of super metal-rich (SMR) stars was first claimed by Spinrad & Taylor (1969), based on low-resolution spectra. The term SMR is generally meant to signify that a star is more metal-rich than the sun by an amount that cannot be explained as simple measurement error. The existence of SMR stars is, historically, a controversial subject; the main question is whether SMR stars are really metal-rich or just appear so because of some kind of measurement dispersion or systematic error. Perhaps the notion of SMR stars became more acceptable with claims that the Galactic bulge red giant stars are on average more metal-rich than the sun (e.g. Whitford & Rich 1983, Frogel & Whitford 1987, Rich 1988). McWilliam & Rich (1994) showed that the average bulge [Fe/H] is the same as in the solar neighborhood, but that the most metal-rich bulge giant, BW IV-167, at $[\text{Fe/H}] = +0.44$ is almost identical to $\mu$ Leo, a metal-rich disk giant. Taylor (1996) has reviewed abundance estimates for SMR stars, including low- and high-resolution results, and concluded that true SMR stars do not exist.

Given the controversy and the potential significance for chemical evolution, it seems important to establish whether any firm cases of SMR stars exist at all. In the Galactic disk, the most well-studied SMR candidate is the K giant star $\mu$ Leo. High-resolution high-S/N model atmosphere abundance analyses of $\mu$

Leo have been performed by several groups: Gustafsson et al (1974), Branch et al (1978), Brown et al (1989), Gratton & Sneden (1990), McWilliam & Rich (1994), and Castro et al (1996) all found values near [Fe/H] = +0.45 for a solar scale of $\epsilon$(Fe) = 7.52; on the other hand, Lambert & Ries (1981), McWilliam (1990), and Luck & Challener (1995) found [Fe/H] from +0.1 to +0.2 dex.

Metal-rich stars in the McWilliam (1990) study (e.g. $\mu$ Leo) were affected by two systematic problems: CN blanketing depressed most of the $\mu$ Leo continuum regions in the two small 100-Å portions of the spectrum observed (found by McWilliam & Rich 1994), which resulted in smaller equivalent widths; second, McWilliam (1990) did not have access to metal-rich model atmospheres, which caused underestimation of the [Fe/H] for metal-rich stars (~0.1-dex underestimate for $\mu$ Leo). Both of these effects decreased the measured $\mu$ Leo [Fe/H] in the McWilliam (1990) work; accounting for the model atmosphere correction alone would increase [Fe/H] to +0.30 dex.

The Luck & Challener (1995) study concluded that their sample of strong-lined stars showed only small iron abundance enhancements at [Fe/H] ~ +0.1 dex; in the case of $\mu$ Leo they found [Fe/H] = +0.20 dex. Luck & Challener (1995) chose not to use a SMR model atmosphere for $\mu$ Leo, thus artificially lowering the computed [Fe/H] by ~0.08 dex (Castro et al 1996). Castro et al (1996) showed that the low Luck & Challener [Fe/H] must result from differences in analysis because of the good agreement between equivalent widths of lines in common. Furthermore, Luck & Challener confused the [A/H] = 0.0 of the Bell et al (1976) atmosphere grid with a solar iron abundance of $\epsilon$(Fe) = 7.67 (from Anders & Grevesse 1989), whereas the models were actually calculated with $\epsilon$(Fe) = 7.50. Castro et al noted that when these two problems are taken into account the Luck & Challener result for $\mu$ Leo becomes [Fe/H] = +0.43, assuming the solar $\epsilon$(Fe) = 7.52.

Thus the most recent high resolution abundance studies of $\mu$ Leo that are discordant with the notion of [Fe/H] = +0.45 can be readily resolved, and it appears that there is a convergence of the $\mu$ Leo iron abundance near [Fe/H] = +0.45 dex with the assumed low value for the solar iron abundance. I do not have an explanation for the Lambert & Ries (1981) low [Fe/H], although it seems possible that the heavy line blanketing and limited spectral coverage may have affected the continuum placement.

Studies with the highest S/N spectra, and the most detailed abundance analyses (e.g. Gratton & Sneden 1990, Branch et al 1978, Castro et al 1996, McWilliam & Rich 1994), consistently find [Fe/H] ~ +0.4 dex for $\mu$ Leo. In conclusion, the high dispersion abundance analyses confirm at least one case of super metallicity.

High-resolution abundance analyses of SMR stars have also been carried out by Edvardsson et al (1993), who found F dwarf stars up to [Fe/H] = +0.26

dex; Feltzing (1995), who extended the Edvardsson sample to find stars between [Fe/H] of −0.08 and +0.42 dex; and Castro et al (1997), who studied a subset of the sample identified by Grenon (1989) and found [Fe/H] ranging from +0.10 to +0.50 dex. McWilliam & Rich (1994) found two SMR Galactic bulge giants, BW IV-167 and BW IV-025, with [Fe/H] of +0.44 and +0.37 dex, respectively. It appears that high-resolution abundance studies do find SMR stars with [Fe/H] up to approximately 0.4–0.5 dex.

## OBSERVED METALLICITY DISTRIBUTION FUNCTION

In this section, I discuss some implications and uses of the most basic chemical composition information, namely metallicity. The word metallicity has more than one meaning: The precise definition is that metallicity is the mass fraction of all elements heavier than helium, denoted by the symbol $Z$; this is not always practical for observers because information usually does not exist for all elements. For observational stellar astronomy, metallicity is more often used to refer to the iron abundance. Unless explicitly stated the word metallicity used here refers to [Fe/H], the logarithmic iron abundance relative to the solar value.

### *The Disk*

Because the main-sequence lifetimes of G and F dwarfs are comparable to the age of the Galaxy, all the G dwarfs ever born are assumed to still exist (although see discussion of metallicity-dependent lifetimes by Bazan & Mathews 1990), and so these stars can provide a complete picture of Galactic chemical evolution. Early studies of the metallicity distribution of G dwarfs, within about 25 pc of the sun (vandenBergh 1962, Schmidt 1963, Pagel & Patchett 1975), showed that there is a deficit of metal-poor stars relative to the prediction of the Simple model; this is the well-known G-dwarf problem. The metallicities of these early studies were based on UV excesses (see Wallerstein & Carlson 1960, Sandage 1969), which are accurate to approximately $1\sigma = 0.25$ dex (Pagel & Patchett 1975); although Norris & Ryan (1989) claim uncertainties of $\pm 0.45$ dex. The observed metallicity distributions contain biases that must be taken into account in order to obtain the true metallicity function (e.g. see Sommer-Larsen 1991 and Pagel 1989).

Many possible explanations were presented to account for the G-dwarf problem (e.g. Audouze & Tinsley 1976), but infall of metal-poor gas onto the disk was the most favored solution. To fit the observed metallicity function by this scheme, the original disk was at most 5% of the present disk mass (Pagel 1989), with mass infall occurring over several billion years. Variants of the Simple model exist that include gas infall in various ways (e.g. Larson 1974, 1976,

Lynden-Bell 1975, Clayton 1985, 1988, Pagel 1989). All of these models predict a strict age-metallicity relation (AMR) with no abundance dispersion.

In these models, the halo could not have been responsible for the bulk of the gas infall because the present-day luminous halo mass is only a few percent of the disk (Sandage 1987, Pagel 1989); a metallicity function of the disk+halo still suffers a paucity of metal-poor stars relative to the simple model (e.g. Worthey 1996). Tosi (1988) showed that infall of gas with metallicity 0.1 $Z_\odot$ provides as effective an explanation of the observed disk metallicity distribution function as infall of zero metallicity gas; however, infalling gas with $Z = 0.4\ Z_\odot$ is excluded by observations.

A number of studies over the last decade and a half have combined star count and kinematic information with metallicities estimated from UV excesses (e.g. Sandage & Fouts 1987), $ubvy\beta$ photometry (e.g. Nissen & Schuster 1991), and low S/N spectra (e.g. Carney et al 1987, Jones et al 1995). The assembled databases have been used to imply the existence of various Galactic populations. For example, the thick disk of Gilmore & Reid (1983) is characterized by scale height of ~1.3 pc, mean [Fe/H] $\sim -0.6$ dex, and dispersion 0.3 dex (Gilmore & Reid 1983, Gilmore 1984, Gilmore & Wyse 1985, Wyse & Gilmore 1986), with no apparent metallicity gradient. Wyse & Gilmore (1995) conclude that the data are best fit by overlapping thick and thin disks; the thick disk has a mean metallicity of [Fe/H] $\sim -0.7$ dex, ranging from $-0.2$ to $-1.4$ dex. A low metallicity tail, extending down to [Fe/H] $\sim -2$ to $-3$, was claimed by Norris & Ryan (1991), Beers & Sommer-Larsen (1995), and Pagel & Tautvaisiene (1995). Typical star count models yield thick disk to thin disk ratios of a few percent (e.g. Majewski 1993). The thin disk metallicity peaks near [Fe/H] = $-0.25$ dex, ranging from $+0.2$ to $-0.8$ dex (Wyse & Gilmore 1995).

### *The Halo*

The Galactic halo does not appear to suffer from a severe G-dwarf problem (Laird et al 1988, Pagel 1989, Beers et al 1992). The halo metallicity ranges from $-4$ dex to just below the solar value, with a mean of $\sim -1.6$ (Laird 1988, Hartwick 1976); Hartwick (1976) noted that this low metallicity suggested that either the halo yield was much lower than in the disk or that gas was removed from halo star formation (e.g. Ostriker & Thuan 1975). The favored model is that the halo lost its gas before chemical evolution could go to completion. Carney et al (1990) and Wyse & Gilmore (1992) suggested that the missing spheroid mass fell to the center of the Galaxy and contributed most of the bulge mass, based on angular momentum considerations.

Whether or not there is a minimum metallicity level, below which stars do not exist, has been debated for at least 20 years. Hartquist & Cameron (1977) predicted that there was an era of "pregalactic nucleosynthesis" by very massive zero metallicity objects; as a result, the Galactic halo would have formed with a non–zero metal content.

Bond (1981) and Cayrel (1987) claimed that there is a paucity of stars below [Fe/H] $\sim -3$ relative to a Simple one-zone model of chemical evolution; this was attributed to a reduced efficiency of forming low-mass stars at low metallicity. Indeed several theoretical investigations (e.g. Kahn 1974, Wolfire & Cassinelli 1987, Yoshii & Saio 1986, and Uehara 1996) have predicted that at low metallicity the IMF is skewed to high-mass stars. Contrary to Bond's suggestion, the huge increase in the number of known metal-poor halo stars (e.g. Beers et al 1985, 1992) led to agreement between the observed metallicity function and predictions from modified Simple models (Beers et al 1985, 1992, Laird et al 1988, Ryan & Norris 1991) down to the lowest measurable abundance, consistent with no metallicity dependence of the IMF.

Audouze & Silk (1995) claimed that there is a lower limit to the metallicity that can form stars, based on predictions concerning the amount of material that can dilute and cool SN ejecta; they estimated the lower limit to be approximately [Fe/H] $\sim -4$.

The most metal-poor star presently known is CD $-38\,245$, at [Fe/H] $= -4.01$ (McWilliam et al 1995a,b), although it only narrowly beats CS 22949-037 for the record, at [Fe/H] $= -3.99$. This iron abundance for CD $-38\,245$ is supported by Gratton & Sneden (1988), who found [Fe/H] $= -3.97$, but it is higher than the metallicity of Bessell & Norris (1984), who found [Fe/H] $= -4.5$. Norris et al (1993) also analyzed stars from the list of Beers et al (1992), one of which was CS 22885-096, with a measured [Fe/H] $= -4.24$. McWilliam et al (1995a) found [Fe/H] $= -3.79$ for this star and explained the difference as due to systematic analysis effects of 0.4 dex; if applied to the Bessell & Norris (1984) result, the same zero point would bring all three analyses into agreement at [Fe/H] $= -4.0$ for CD $-38\,245$.

Thus, despite the heroic effort by George Preston of searching for metal-poor stars by visually inspecting over one million objective prism spectra (Beers et al 1985), the honor of the most metal-poor star known in the Galaxy still belongs to CD $-38\,245$.

### *The Bulge*

Measurement of the metallicity of Galactic bulge stars has been somewhat controversial in the last 15 years. Early bulge metallicity studies focused on stars in Baade's window, at Galactic latitude $-4°$. Initial low-resolution studies of 21 bulge giants by Whitford & Rich (1983) suggested that most of the bulge stars are super metal-rich.

Frogel & Whitford (1987) amassed photometric and spectral-type data for a large number of bulge giants. They found that bulge M giants have stronger TiO and CO bands than solar neighborhood M giants consistent with a metal-rich bulge.

Rich (1988) measured low-resolution indices of strong lines (Mg *b* and Fraunhofer Fe I lines) in 88 bulge giants in Baade's window and several bright standards. Calibration of the indices suggested a range of [Fe/H], from $-1.0$ to $+0.8$ dex for the bulge, with a mean value twice the solar value.

Terndrup et al (1991) found a mean bulge metallicity of $+0.3$ dex for M giant stars in Baade's Window, based on R=1000 spectrophotometry, which confirmed earlier results.

Geisler & Friel (1992) used Washington photometry to measure the metallicity of 314 red giants in the Galactic bulge, through Baade's window. They found the mean [Fe/H] $= +0.17 \pm 0.15$ dex, in good agreement with Rich (1988). They also found a high frequency of metal-poor stars, consistent with that expected from a simple closed box model, as found by Rich (1990).

Rich (1990) showed that the Galactic bulge contains a higher frequency of metal-poor stars than the solar neighborhood. In fact, the bulge metallicity function does not exhibit the G-dwarf problem. This is perhaps somewhat surprising because the bulge must be the final repository of infalling material (for example, the Sagittarius dwarf found by Ibata et al 1994). It may be that most of the infall occurred very rapidly, or that material that fell into the bulge, such as a dwarf galaxy, was stripped of its gas before reaching the bulge.

With the apparent convergence of different methods used to measure the bulge metallicity, it was a surprise that the first high-dispersion model atmosphere abundance analysis of bulge stars (McWilliam & Rich 1994) found that the bulge is slightly iron-poor relative to the solar neighborhood. McWilliam & Rich (1994) computed [Fe/H] for 11 bulge red giants, covering the full metallicity range, which had previously been measured by Rich (1988). A correlation of [Fe/H] values of McWilliam & Rich (1994) with those of Rich (1988) showed that Rich (1988) systematically overestimated the [Fe/H] of the most metal-rich stars. A regression relation between McWilliam & Rich's (1994) and Rich's (1988) [Fe/H] results was used to compute corrected [Fe/H] (Rich 1988) for the full sample of 88 stars. Rich's (1988) [Fe/H] values corrected in this way have a mean of $-0.25$ dex, slightly below the mean value of $-0.17$ dex for solar neighborhood red giants (McWilliam 1990). The corrected bulge metallicity function still shows the excess of metal-poor stars relative to the solar neighborhood noted by Rich (1990). McWilliam & Rich (1994) also found unusually high [Mg/Fe] and [Ti/Fe] ratios in the bulge stars, which might explain why previous investigators found high average metallicities.

Subsequent model atmosphere abundance analyses of two stars in the McWilliam & Rich (1994) sample (Castro et al 1996; A McWilliam, RC Peterson, DM Terndrup & RM Rich, in preparation) confirmed the McWilliam

& Rich (1994) [Fe/H] results. The low [Fe/H] of bulge stars in Baade's window found by McWilliam & Rich (1994) was supported by later low-resolution studies; for example, the analysis of low-resolution spectra of 400 bulge giants by Terndrup et al (1995) and Sadler et al (1996) found a low mean [Fe/H] $\sim -0.1$ dex.

## AGE-METALLICITY RELATION

The existence of an age-metallicity relation (AMR) in the disk is an important issue for developing chemical evolution models. There is currently some uncertainty whether an AMR exists: Studies of open cluster metallicities and ages (e.g. Arp 1962, Geisler 1987, Geisler et al 1992, Friel & Janes 1993) have resulted in the conclusion by some that there is no AMR in the Galactic disk (see the review by Friel 1995). The main factor in determining open cluster metallicity appears to be galactocentric radius (e.g. Geisler et al 1992). It is also clear that there is a large scatter in metallicity at any given age in the disk: The dispersion in the age-metallicity diagram is exemplified by the presence of very old open clusters with metallicities near or above the solar value. The open cluster NGC 188 has historically been used to illustrate this point (e.g. Eggen & Sandage 1969); but the most clear-cut modern case is NGC 6791, which is more metal-rich than the sun, with [Fe/H] $\sim +0.2$ to $+0.3$ dex (Peterson & Greene 1995, Montgomery et al 1994), but very old at $\sim 10 \times 10^9$ years (Montgomery et al 1994, Tripicco et al 1995).

The conclusion against an AMR is at odds with claims based on studies of field stars. For example, Twarog (1980), Meusinger et al (1991), and Jønch-Sørensen (1995) all employed $uvby\beta$ photometry and found a trend of decreasing metallicity with increasing stellar age. Edvardsson et al (1993) used spectroscopic abundance analysis to determine [Fe/H] and $uvby\beta$ photometry for the ages. They found an AMR consistent with the results of Twarog (1980) and Meusinger et al (1991) but with a considerable scatter about the mean trend (Figure 2). The Jønch-Sørensen data indicated a similar AMR slope and scatter as Edvardsson et al's data. The age-metallicity diagram from these studies (e.g. Figure 2) show a lower envelope to the observed metallicity of stars that increases with Galactic time; in particular, no young stars with [Fe/H] $\sim -1$ have been found in the solar neighborhood (although low metallicity stars at large galactocentric radii are known; e.g. Geisler 1987, Geisler et al 1992).

The large scatter in metallicity at all ages is the one consistent conclusion common to the age-metallicity diagrams for both the field stars and open clusters. François & Matteucci (1993) suggested that the scatter could be due to orbital diffusion; however, Edvardsson et al (1993) showed that this is not enough to reduce the observed scatter in the age-metallicity diagram.

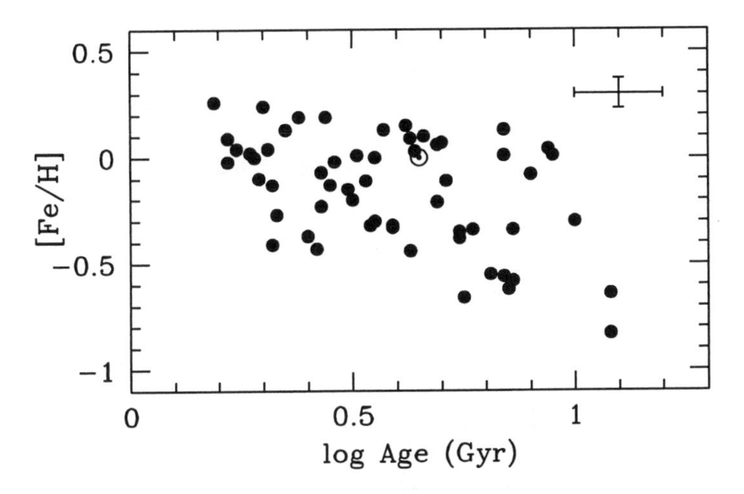


*Figure 2* The age-metallicity relation for the solar neighborhood, from the data of Edvardsson et al (1993). The sample is limited to galactocentric radius 7.7 $\leq R_m \leq 9.3$ kpc, maximum height above the plane $Z_{max} \leq 0.26$ kpc, and eccentricity $e \leq 0.16$. The position of the Sun is indicated.

It is clear that certain biases occur in samples of field stars that could conspire to create an apparent AMR, even if none exists (Knude 1990, Grenon 1987); indeed, Friel (1995) states that the age-metallicity trends seen by Twarog (1980) are the result of these selection effects. However, Twarog was aware of the selection biases and pointedly went to great effort to avoid them. Jønch-Sørensen (1995) estimated an upper limit to the number of metal-poor young stars and claimed that the selection bias against metal-poor young stars could not account for the apparent AMR. Edvardsson et al (1993) made a correction for a metallicity bias, but the AMR was still present. Obviously a definitive resolution to the existence or absence of a mean AMR in field stars would be extremely valuable. If age and metallicity data for the halo are added to Figure 2, as done by Eggen & Sandage (1969), a strong AMR would result; however, the validity of combining these two populations is not certain.

The large range of metallicities present for all ages suggests that chemical enrichment up to solar metallicity can occur on rapid time scales ($\sim 1 \times 10^9$ years) and that the disk has been chemically inhomogeneous throughout its development. The dispersion in the AMR at the solar circle (as seen in the Edvardsson et al 1993 study) shows that the composition of the Galactic disk did not evolve homogeneously. Traditional chemical evolution models, for example those of Lynden-Bell (1975), Larson (1976), Matteucci & François (1989), Pagel (1989), Sommer-Larsen (1991), and Pagel & Tautvaisiene (1995), cannot account for the observed AMR dispersion because they all assume instantaneous mixing of recycled gas and a homogeneous steady infall; as a consequence chemical homogeneity is preserved at all times.

Reeves (1972) suggested that significant spatial inhomogeneities in elemental abundances could occur as a result of self-enrichment of star-forming regions by SN events. However, Edmunds (1975) investigated this possibility and concluded that the Galactic disk is well mixed. White & Audouze (1983) developed analytical expressions that extended the standard chemical evolution model of Lynden-Bell (1975) to the case of inhomogeneous steady-state evolution. Two important mixing parameters dictated the inhomogeneity: (*a*) the mean mass of disk material mixed with a unit mass of enriched material from star formation events and (*b*) the mean mass of disk material mixed with a unit mass of infalling gas.

Recent models of Galactic evolution attempted to describe inhomogeneous chemical evolution: Pilyugin & Edmunds (1996a,b) and Raiteri et al (1996). Both studies adopt the Twarog (1980) AMR and the dispersion about this relation indicated by Edvardsson et al (1993).

Pilyugin & Edmunds (1996b) considered inhomogeneity by two mechanisms. In the first approach, self-enrichment of gas in star forming regions (H II regions) for $3 \times 10^7$ years is permitted, after which time the gas is instantaneously mixed with the ambient disk gas. This approximates a star-forming region in which SN ejecta enrich the region with metals until the energy input from SN is sufficient to disrupt the cloud in $3 \times 10^7$ years, followed by mixing with the disk in $\sim 10^8$ years. Justification for this assumption comes from Cunha & Lambert (1992, 1994), who showed that self enrichment in the Orion association has occurred in $\sim 80 \times 10^6$ years, based on enhancements in O and Si abundances as a function of age of the Orion subgroups.

Self enrichment of the H II regions gave a satisfactory fit to the dispersion in the oxygen abundance with time, but it was incapable of reproducing the observed dispersion in Fe abundance. The difficulty in reproducing the Fe dispersion was caused by the fact that Fe is produced mainly in SN Ia, whose progenitor lifetimes are thought to be $\sim 1 \times 10^9$ years, well in excess of the self-enrichment time scale. Pilyugin & Edmunds (1996b) suggest that self-enrichment of H II regions results in larger dispersion for oxygen abundances (SN II progenitors with short lifetimes) than iron abundances versus age. They concluded that the large observed dispersion for both O and Fe implicates another source of inhomogeneity.

Pilyugin & Edmunds (1996b) suggested that episodic gas infall could account for the large dispersions in the AMR for both Fe and O. If infalling gas fell onto the disk in a nonuniform fashion (both temporally and spatially), then disk gas could reach solar metallicity followed by substantial dilution to lower metallicities. Stars formed over such a cycle would exhibit equal Fe and O dispersion in the AMR because dilution affects all species equally. If this is the case, then the infalling gas cannot be pure hydrogen; otherwise the dilution would preserve solar abundance ratios even near [Fe/H] $= -1$, which is not observed. The gas would need to be of halo composition, with [Fe/H] $\sim -1$, to avoid the problem of solar ratios in low metallicity disk stars.

Raiteri et al (1996) have developed $N$-body/hydrodynamical simulations of Galactic chemical evolution. The method seems very promising and does produce an AMR similar to Twarog's (1980) with a large metallicity dispersion; it also predicts significant dispersion in the [O/Fe] ratio at all metallicities, which provides a basis for testing the model. There are some problems, however, such as a very high frequency of low metallicity stars.

## ABUNDANCE TRENDS WITH METALLICITY

### *Alpha Elements*

Enhancements of $\alpha$ elements in metal-poor stars were first identified by Aller & Greenstein (1960) and more firmly established by Wallerstein (1962), who found excesses of Mg, Si, Ca, and Ti relative to Fe. A corresponding enhancement for oxygen was first discovered by Conti et al (1967). The work of Clegg et al (1981) and François (1987, 1988) showed that S is also overabundant in metal-poor stars. These enhancements increase linearly with decreasing metallicity, reaching a factor of two above the solar [$\alpha$/Fe] ratios at [Fe/H] near $-1$; below [Fe/H] $= -1$ the enhancements are approximately constant. Figure 3*a* shows the general trend of [O/Fe] with [Fe/H]. It is important to emphasize that "$\alpha$ element" is simply a convenient phrase used to signify the observation that some even-Z elements (O, Mg, Si, S, Ca, and Ti) are overabundant relative to iron at low metallicity, and it does not signify that these are all products of a single nuclear reaction chain that occurs in the same astrophysical environment.

As mentioned in the introduction Tinsley (1979) suggested that the [$\alpha$/Fe] trend with [Fe/H] is due to the time delay between SN II, which produce $\alpha$ elements and iron-peak elements (e.g. Arnett 1978, Woosley & Weaver 1995), and SN Ia, which yield mostly iron-peak with little $\alpha$ element production (e.g. Nomoto et al 1984, Thielmann et al 1986). Thus, after the delay for the onset of SN Ia, the [$\alpha$/Fe] ratio declines from the SN II value. The SN Ia time scale is an important consideration for this model. Iben & Tutukov (1984) favor a mechanism with mass transfer during the merging of a CO+CO white dwarf binary system; time scales for SN Ia from this model range from $10^8$ to $10^{10}$ years, depending on progenitor masses and mass transfer parameters. Smecker-Hane & Wyse (1992) obtained estimates for the first SN Ia of $10^8$ years.

Other explanations for the $\alpha$-element trend have been put forward: Maeder (1991) suggested that exploding Wolf-Rayet stars (type Ib supernovae, SN Ib)

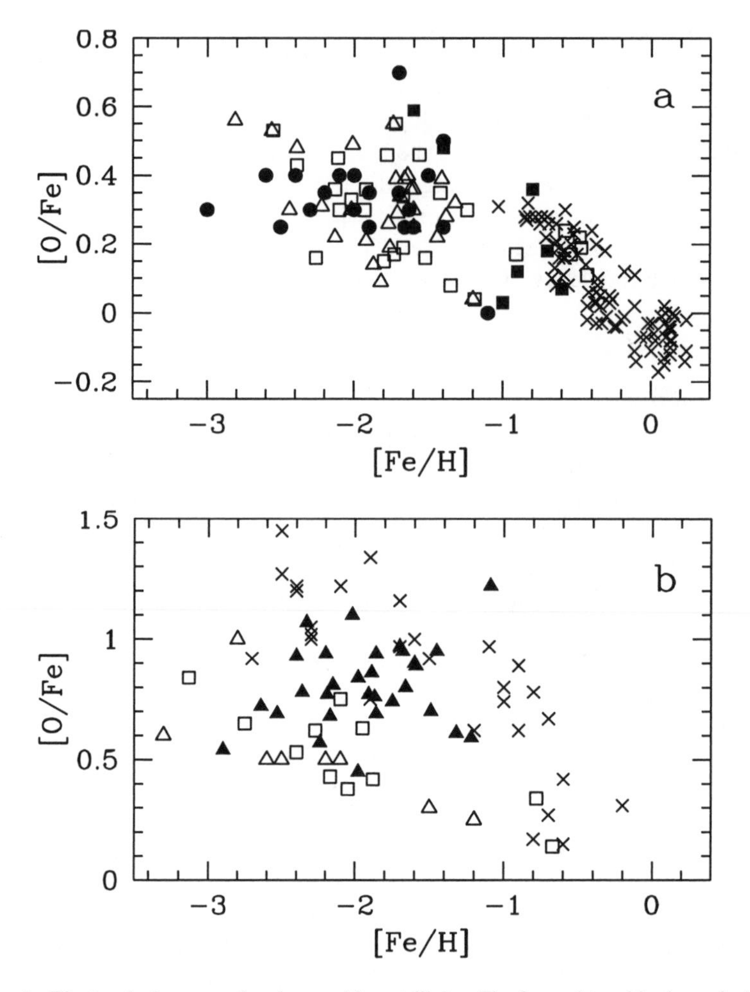


*Figure 3* The trend of oxygen abundance with metallicity. The favored trend is shown in (*a*), a compilation of [O I] results: *crosses* from disk data of Edvardsson et al (1993), *filled squares* from Spite & Spite (1991), *filled circles* from Barbuy (1988), *open triangles* from Kraft et al (1992) and Sneden et al (1991), *open squares* from Shetrone (1996a). (*b*) shows results from the O I triplet: *crosses* (Abia & Rebolo 1989) and *filled triangles* (Tomkin et al 1992); low S/N results from ultraviolet OH lines are indicated by *open squares* (Nissen et al 1994) and *open triangles* (Bessell et al 1991). Note the difference in the scale of the ordinate between (*a*) and (*b*).

might be responsible for the observed $\alpha$-element abundance trend. Wolf-Rayet stars are the bare cores of massive stars that have lost their outer envelopes through copious stellar winds. The radiatively driven winds are metallicity-dependent, producing significant numbers of Wolf-Rayet stars above [Fe/H] $\sim -1$. The chemical yields depend on the mass-loss rates: At high metallicity the strong winds remove much of the helium before it is further transformed into heavy elements.

Edmunds et al (1991) suggested that metallicity-dependent element yields could be the source of the $\alpha$-element abundance trend and predicted that SMR stars should possess subsolar [$\alpha$/Fe] ratios. The theoretical element yields from SN II (e.g. Woosley & Weaver 1995) do not show such a metallicity dependence; however, some star formation theories have predicted a metallicity-dependent IMF (e.g. Kahn 1974, Yoshii & Saio 1986), which might conceivably result in a steady increase of the SN Ia/SN II ratio with increasing metallicity and thereby account for the observed $\alpha$-element trend.

### *Disk Alpha Elements*

Studies of disk dwarf stars by several workers (e.g. Clegg et al 1981, Tomkin et al 1985, François 1986, Gratton & Sneden 1987, Edvardsson et al 1993) confirmed the trend of increasing [$\alpha$/Fe] with decreasing [Fe/H] in the Galactic disk, as established by the analysis of G dwarfs by Wallerstein (1962); typically [$\alpha$/Fe] $\sim +0.4$ at [Fe/H] $\sim -1.0$. The data of Tomkin et al (1986) and Edvardsson et al (1993) show that for Mg, Ca, and Si, there is a plateau at [$\alpha$/Fe]$= 0.0$ above [Fe/H] $\sim -0.2$ dex (see Figure 3*a*). This plateau suggests a transition from one kind of chemical evolution environment to another, which is consistent with the idea that above [Fe/H] $= -0.2$, the ratio of SN Ia/SN II had reached a constant value.

Edvardsson et al (1993) found that when the disk stars are separated into bins of mean galactocentric radius, $R_m$, the $\alpha$-element enhancements are seen to be maintained to higher [Fe/H] at small $R_m$ (see Figure 4). In Tinsley's picture of SN Ia and SN II this suggests that enrichment by SN II occurred to higher [Fe/H] in the inner disk than in the outer disk, before the first SN Ia occurred, in agreement with models of the disk that predict higher SFR in the inner disk than in the outer regions (e.g. Larson 1976, Matteucci & François 1989). Edvardsson et al's results also indicate that at the solar circle, old stars seem to show a distinctly different [$\alpha$/Fe] trend than young stars, and this suggests that the SFR increased with time in the disk. There is a hint that the inner disk stars of the Edvardsson et al sample show a bimodal [$\alpha$/Fe] ratio, rather than a slope with [Fe/H].

In order to reduce scatter in the trend with metallicity, $\alpha$-element abundances have often been averaged; Lambert's (1987) review popularized the mean relation between [$\alpha$/Fe] and [Fe/H]. The work of Edvardsson et al (1993) indicated

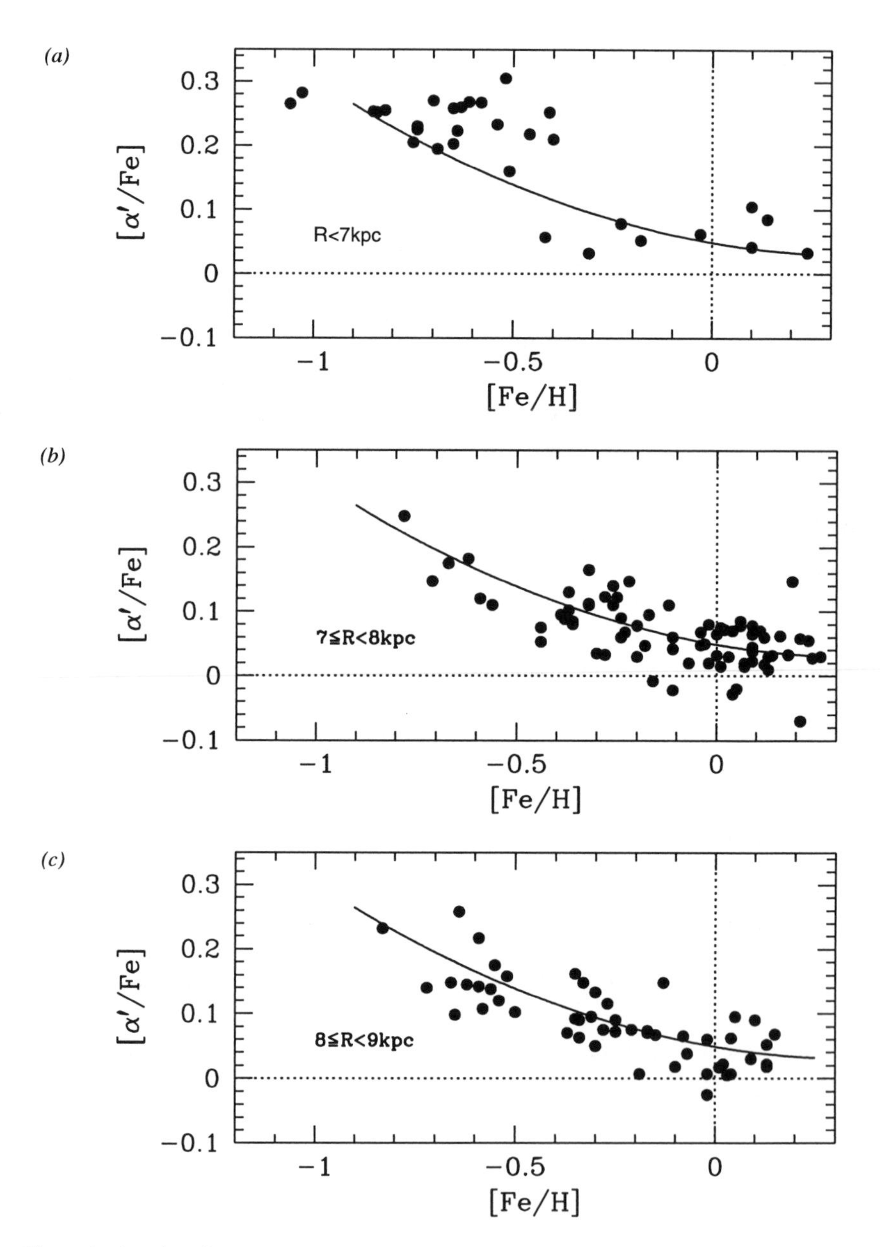


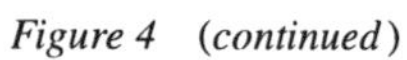
*Figure 4* (*continued*)

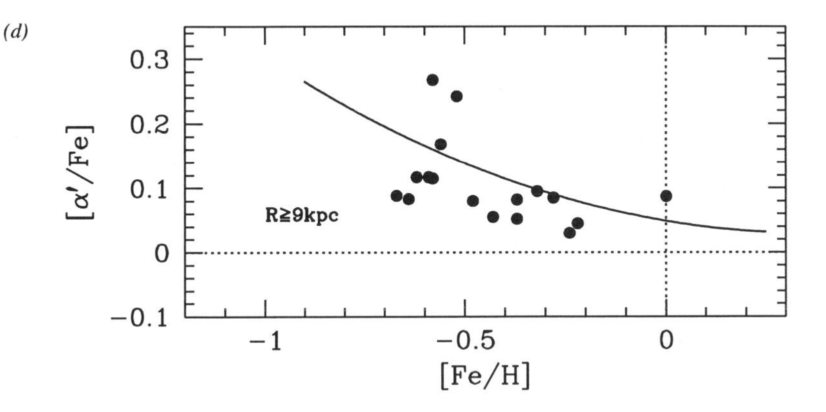


*Figure 4* The run of [$\alpha$/Fe], computed from [(Mg+Si+Ca+Ti)/Fe], versus iron abundance for four ranges in galactocentric radius, from the data of Edvardsson et al (1993). In the inner disk, $R_m \leq$ 7 kpc, [$\alpha$/Fe] is higher than the mean trend (indicated by the *solid line*), while the outer disk shows an [$\alpha$/Fe] deficiency. The trends at large and small $R_m$ seem to show a bimodal appearance, rather than the shift indicated in Figure 1 for different SFR.

that the trends are not the same for all $\alpha$ elements: Ca and Si abundances correlate very well, but both Mg and Ti are systematically over-enhanced relative to Ca and Si. These observations of subtle $\alpha$-element trends in the disk stars are similar to, but less extreme than, the enhanced Mg and Ti abundances found for Galactic bulge stars by McWilliam & Rich (1994). Nissen & Edvardsson (1992) found a somewhat steeper decline in [O/Fe] with [Fe/H] than other $\alpha$ elements from Edvardsson et al (1993). If these differences within the $\alpha$ element family withstand further scrutiny it shows that the $\alpha$ elements are not made in a single process but are produced in different amounts by different SN.

Cunha & Lambert (1992, 1994) studied the chemical composition of B stars in various subgroups of the Orion association ([Fe/H] $\sim -0.05$) and found evidence for self-contamination of the association by nucleosynthesis products from SN II. In particular the subgroups show an abundance spread of $\sim$0.3 dex for O, correlated with Si abundance, but no dispersion larger than the measurement uncertainties could be found for Fe, C, and N. This pattern of abundance enhancement is consistent with self-enrichment of the gas by SN II only. Additional support for this idea includes the spatial correlation of the O-Si–rich stars and the fact that the most O-Si–rich stars are found only in the youngest subgroup of the association. The time lag between the oldest and youngest subgroups is $\sim 11 \times 10^6$ years (Blaauw 1991), which is comparable to the lifetime of the massive stars. Thus, the massive stars had enough time to explode as SN and enrich the molecular cloud, but the time scale was too short

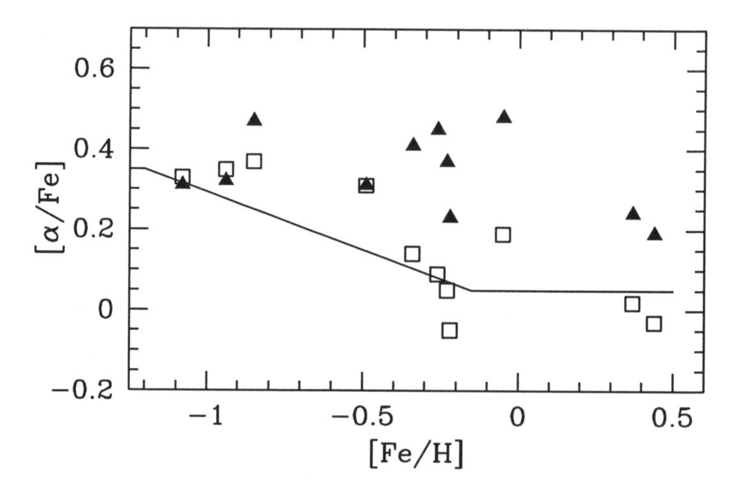


*Figure 5* Trends of $\alpha$-element abundances in the Galactic bulge, from McWilliam & Rich (1994). *Filled triangles* indicate the average [(Mg+Ti)/Fe] and *open boxes* indicate the average [(Si+Ca)/Fe]. For Si and Ca the trends follow the solar neighborhood relation (*solid line*), whereas the Mg and Ti abundances are enhanced by $\sim$0.4 dex for most stars, similar to the halo values.

to permit any pollution by SN Ia. If the same enrichment observed by Cunha & Lambert (1994) occurred in a similar cloud of zero-metal gas, the metallicity of the final generation would be approximately [Fe/H] $= -0.8$ dex.

As demonstrated by Cunha & Lambert (1992, 1994), chemical abundance studies of star-forming regions are a particularly useful way to study basic processes in chemical evolution and SN nucleosynthesis.

### *Bulge Alpha Elements*

To date, the only extant detailed abundance analyses of $\alpha$ elements for Galactic bulge stars are by McWilliam & Rich (1994) and A McWilliam, A Tomaney & RM Rich (in preparation). McWilliam & Rich (1994) found that Mg and Ti are enhanced by $\sim$+0.4 dex in almost all bulge stars, even at solar [Fe/H]; however, the abundances of Ca and Si appear to follow the normal trend of $\alpha$/Fe ratio with [Fe/H] (see Figure 5).

Some overlap exists between the chemical properties of the McWilliam & Rich (1994) bulge giant sample and the disk F dwarfs of Edvardsson et al (1993): In general, the disk results (Edvardsson et al 1993) show that Mg and Ti are slightly enhanced relative to Si and Ca, which is similar to, but less extreme than, the +0.4-dex enhancements of Mg and Ti in the bulge. Edvardsson et al (1993) identified a subgroup of stars with 0.1-dex enhancements of Na, Mg, and Al; these are conceivably related to the bulge giants, which have large Mg and

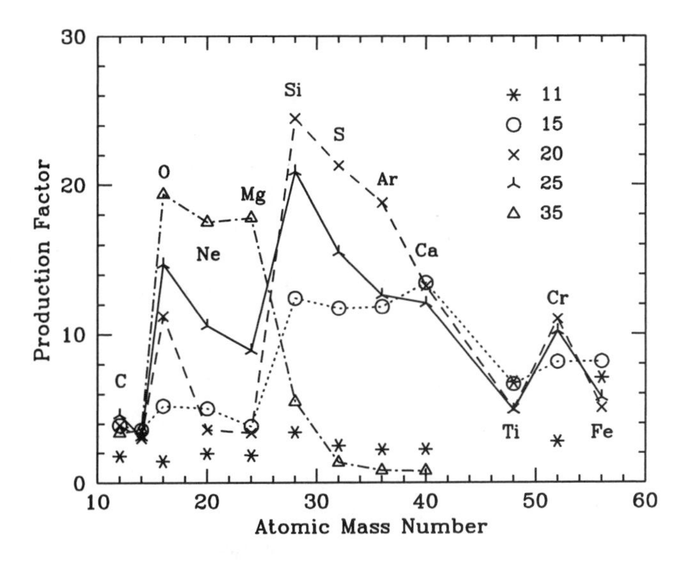


*Figure 6* Production factors from models of SN II by Woosley & Weaver (1995). Ejected element abundances for various progenitor masses are indicated by *connected symbols*; O and Mg are produced in large quantitiesat high mass ($\sim$35 $M_\odot$) but not in the lower mass (15–25 $M_\odot$) SN, which are responsible for most of the Si and Ca production. None of the models give significant enhancements of Ti relative to Fe, contrary to observations of stars in the Galactic bulge and halo. Note that production factor is defined as the ratio of the mass fraction of an isotope in the SN ejecta, divided by its corresponding mass fraction in the Sun. The mass of the progenitor making the indicated elements is given in the key in the upper right.

Al enhancements. The bulge [O/Fe] ratio is not well constrained: The extant data are insufficient to determine whether oxygen behaves like Mg and Ti or Si and Ca. However, any oxygen enhancement in the bulge must be less than +0.5 dex (A McWilliam, A Tomaney & RM Rich, in preparation).

The unusual mixture of $\alpha$-element abundances in the bulge is evidence that $\alpha$ elements are made in different proportions by different SN; i.e. there are different flavors of SN with different $\alpha$-element yields. This conclusion is borne out by predicted $\alpha$-element yields (e.g. Woosley & Weaver 1995), as shown in Figure 6. Figure 6 illustrates that enhanced Mg could occur with relatively more 35-$M_\odot$ SN progenitors than in the disk. The enhanced Ti is not explained by any SN nucleosynthesis predictions.

The Ti enhancements seen in bulge stars present a nice qualitative explanation for the well-known phenomenon that the spectral type of bulge M giants is later than disk M giants with the same temperature. Frogel & Whitford (1987) suggested that the later spectral types were due to overall super-metallicity of the bulge stars; McWilliam & Rich (1994) argued that the Ti enhancements are sufficient to create the stronger bulge M giant TiO bands, without affecting overall metallicity. The enhanced Mg abundances may also explain Rich's (1988) high [Fe/H] results, which were based on measurements of the Mg *b* lines and assumed that the bulge giants have the solar [Mg/Fe] ratio.

Unfortunately, the unusual mixture of $\alpha$-element abundances for the bulge makes it difficult to use these elements to estimate the bulge formation time scale; the simple picture of SN Ia and SN II implies a different time scale depending on which elements are considered. However, the observed Mg overabundances agree with the predictions of Matteucci & Brocato (1990) and a rapid formation time scale for the bulge.

Terndrup et al (1995) and Sadler et al (1996) analyzed low-resolution spectra of 400 bulge giants and found the average [Fe/H] $\sim -0.11$ dex, consistent with the result of McWilliam & Rich (1994). The [Mg/Fe] ratios +0.3 dex and +0.11 dex respectively.

Multi-population synthesis analysis of low-resolution integrated light spectra of the Galactic bulge by Idiart et al (1996a) indicated a mean bulge abundance ratio of [Mg/Fe] = +0.45 dex. Using the same technique for elliptical galaxies and bulges of external spirals, Idiart et al (1996b) showed a general Mg enhancement of $\sim$+0.5 dex. Worthey et al (1992), using single-population models, analyzed spectra of giant elliptical galaxies and found Mg enhancements relative to Fe between +0.2 to +0.3 dex. These results provide supporting evidence in favor of enhanced Mg in the bulge, as claimed by McWilliam & Rich (1994). An obvious question arising from the population synthesis results is whether Ti is enhanced in external bulges and elliptical galaxies.

The abundance results for $\alpha$ elements in the bulge show that chemical abundance ratios are a function of environmental parameters. In this regard, further study of the detailed chemical composition of Galactic components will lead to an understanding of how environment affects chemical evolution, which can be used to interpret low-resolution low-S/N spectra of distant galaxies. In particular, it is necessary to check the McWilliam & Rich (1994) results for O, Ca, and Si because results for these three elements are less reliable than for Mg and Ti.

### *Halo Alpha Elements*

The enhancement of $\alpha$ elements in the halo has been confirmed by numerous studies, both in the field and the globular cluster system (e.g. Clegg et al 1981, Barbuy et al 1985, Luck & Bond 1985, François 1987, 1988, Gratton & Sneden 1988, 1991, Zhao & Magain 1990, Nissen et al 1994, Fuhrmann et al 1995, McWilliam et al 1995a,b).

Because $\alpha$-element yields are predicted to increase with increasing SN II progenitor mass (e.g. Woosley & Weaver 1995), the [$\alpha$/Fe] ratio is sensitive to the IMF. Therefore it is interesting to know if the [$\alpha$/Fe] ratios in the halo are constant with changing [Fe/H], if there is a slope to the [$\alpha$/Fe] correlation with [Fe/H], or if there is a measurable dispersion at a given [Fe/H], which might indicate a change in the IMF.

Abundance studies of oxygen are frequently based on the weak [O I] forbidden lines at 6300 and 6363 Å for cool giants (e.g. Barbuy 1988) and the high excitation O I triplet lines at 7774 and 9263 Å for main-sequence stars (e.g. Tomkin et al 1992). Unfortunately, the O I lines have very high excitation potential, and the resulting abundances may be very sensitive to temperature uncertainties and non-LTE effects. Tomkin et al (1992) found [O/Fe] $\sim +0.8$ dex from the O I lines, with non-LTE calculations; but the strong temperature dependence suggests that oxygen abundances derived from the triplet lines are unreliable. On the other hand, the [O I] lines are very weak and frequently only the 6300-Å line can be measured; however, the [O I] results are considered more reliable than those from the O I lines because neither temperature or non-LTE effects are a problem. Recent oxygen abundances have been determined from OH lines in the UV by Bessell et al (1991) and Nissen et al (1994), whereas Balachandran & Carney (1996) used near-infrared OH lines. Both methods offer the advantage that many lines can be measured without severe non-LTE problems; but the reduced flux in the UV result in lower S/N and less reliable results for the UV OH lines than for the near-infrared OH lines.

All of these methods provide information on the free oxygen (uncombined into molecules) in the stellar atmospheres. However, for the total oxygen abundance, carbon abundances must also be known in order to account for the oxygen atoms locked up in the CO molecule.

The scatter in measured [O/Fe] values has been large: For example, Abia & Rebolo (1989) found [O/Fe] $= +1.0$ for stars near [Fe/H] $= -2.0$, based on the O I triplet at 7774 Å. This result is almost certainly too high, as shown by many investigations (e.g. Barbuy 1988, Bessell et al 1991, Spite & Spite 1991, Kraft et al 1992, Nissen et al 1994). King (1993) suggested that the Abia & Rebolo equivalent widths were too high by approximately 25%, which, when combined with a revision of the temperature scale by 200 K, resolves the differences between the abundance results for O I lines and other oxygen abundance indicators.

The low S/N OH line results of Bessell et al (1991) and Nissen et al (1994) suggest that [O/Fe] $= +0.5$ to $+0.6$ dex in the interval [Fe/H] $= -1$ to $-3.4$. Bessell et al (1991) claimed that the halo [O/Fe] ratios continue the slope of the [O/Fe] relation with [Fe/H] seen in the disk, down to [Fe/H] $= -1.7$; below this point the halo [O/Fe] ratio is constant.

From high S/N ($\sim$150) spectra, Barbuy (1988) measured a mean [O/Fe] $=$ $+0.35 \pm 0.15$ from the 6300-Å [O I] line in 20 halo giants with metallicities in the range $-2.5 \leq$ [Fe/H] $\leq -0.5$. Kraft et al (1992) and Sneden et al (1991b) measured [O/Fe] for many globular cluster giants and 27 field giants from S/N $\sim$150 spectra of the [O I] line. The field giants ranged in [Fe/H] from $-1.3$ to $-2.8$ with an average [O/Fe] = +0.34. For oxygen-rich giants in the globular clusters (those without envelope depletion of oxygen) measured by Kraft et al (1992) the mean [O/Fe] = 0.32.

Balachandran & Carney (1996) measured C and O abundances in a halo dwarf using high S/N spectra of near-infrared OH and CO lines; they also rederived abundances from published O I, [O I], and C I lines. In particular, the solar and stellar abundances were both computed from the same grid of model atmospheres with the same set of lines. Balachandran & Carney found [O/Fe] = +0.29 dex for this star and concluded that temperature corrections were not required to resolve differences between forbidden and high excitation O lines, as had been previously suggested by King (1993). The resolution of the high O abundance values of Abia & Rebolo (1989) was due to the use of a self-consistent solar and stellar model atmosphere grid.

The preferred results of Figure 3a indicate a trend of [O/Fe] with [Fe/H] that is flat between [Fe/H] $-1$ to $-3$, at [O/Fe] = +0.34 dex; the dispersion of 0.1 dex about this value is consistent with the measurement uncertainties. Thus, the oxygen abundances in the bulge are consistent with a constant IMF. In Figure 3*b* I show results from O I triplet lines and low S/N spectra of UV OH lines, which exhibit a large dispersion.

There have been many studies for Mg, Si, Ca, and Ti in halo field stars; some of the more recent examples include those of François (1986), Magain (1987, 1989), Gratton & Sneden (1987, 1988, 1991), Zhao & Magain (1990), Ryan et al (1991), Nissen et al (1994), Fuhrmann et al (1995), McWilliam et al (1995a), and Pilachowski et al (1996). Not surprisingly, the [$\alpha$/Fe] ratios from this list encompass a range of values; for calcium, the lowest measured mean ratio for halo stars is [Ca/Fe] = +0.18 (Gratton & Sneden 1987), and the highest is [Ca/Fe] = +0.47 (Magain 1989). Much of the scatter in the abundance ratios is probably due to systematic effects in the analysis of different researchers: For example Gratton & Sneden consistently find lower [$\alpha$/Fe] ratios than Magain Zhao & Magain; the usual differences are approximately 0.15 dex. Taking straight average abundance ratios for all the above studies gives the following results: [Mg/Fe] = +0.36, [Si/Fe] = +0.38, [Ca/Fe] = +0.38, and [Ti/Fe] = +0.29, with typical $1\sigma = 0.08$. The average of all four species gives [$\alpha$/Fe] = +0.35, with $\sigma = 0.05$ dex, which is very close to the adopted value for [O/Fe] of +0.34 dex. A conservative conclusion is that in the halo, the $\alpha$ elements O, Mg, Si, Ca, and Ti all show an enhancement, relative to Fe, of +0.35 dex; alternatively, the full range of measured [$\alpha$/Fe] ratios is well represented by $+0.37 \pm 0.08$.

François (1987, 1988) measured sulfur in halo stars from extremely weak, high excitation S I lines near 8694 Å and found [S/Fe] $= +0.6$; given the difficulty associated with abundance measurement of such weak lines, this result is approximately consistent with the general $\alpha$-element overabundances.

A small slope in the halo relation between [$\alpha$/Fe] abundance ratios and [Fe/H] may possibly be responsible for part of the dispersion in published abundance results. For example, the McWilliam et al (1995a,b) study has a lower mean metallicity than Gratton & Sneden's (1988, 1991) work, with some overlap; on average McWilliam et al's (1995a,b) [$\alpha$/Fe] ratios are $\sim$0.1 dex higher. However, the six stars common to both studies show a mean difference [McWilliam et al minus Gratton & Sneden] for [Fe/H], [Mg/Fe], [Si/Fe], [Ca/Fe], and [Ti/Fe] of only 0.00, 0.06, $-0.08$, $-0.03$, and 0.12 dex, respectively. Plots of [Mg/Fe] and [Ca/Fe] by McWilliam et al (1995a,b) showing the comparison with Gratton & Sneden (1988, 1991) could be interpreted as evidence for increases in both ratios with declining [Fe/H].

Gratton (1994) combined the $\alpha$-abundance results of several studies and found small increases in [O/Fe], [Ti/Fe], and [Mg/Fe] with declining [Fe/H] in the interval [Fe/H] $= -1$ to $-3$; this was claimed to be consistent with an increased production of O, Ti, and Mg at the lowest metallicity by high mass SN. The slopes were also consistent with chemical evolution model predictions of Matteucci & François (1992). Subtle slopes can also be seen in the [Ca/Mg] and [Ti/Mg] results of McWilliam et al (1995a,b), which may indicate a slight decrease in Mg, or an increase in Ca and Ti abundances, at the lowest metallicity. If true, these subtle trends indicate that the halo IMF was not constant with time. However, caution is warranted here because the small gradients could easily be the result of systematic measurement errors. Contrary to the above finding, Nissen et al (1994) concluded from analysis of halo field stars that the halo IMF was constant with time in the interval $-3.5 \leq$ [Fe/H] $\leq -1.8$.

Carney (1996) reviewed $\alpha$-element abundances in globular clusters and found no evidence for a decline in [O/Fe], [Si/Fe], or [Ti/Fe] from [Fe/H] $= -2.2$ to $-0.6$ dex. A hint of a decline in [Ca/Fe] with increasing metallicity was found, which might be real but could equally well signal an analysis problem. The conclusion was that there is a uniform enhancement of [$\alpha$/Fe] $\sim +0.3$ dex in the globular clusters, with no evidence of SN Ia nucleosynthesis products in the younger clusters. The $\sim 3 \times 10^9$ year dispersion in globular cluster ages implies that either the time scale for SN Ia is longer than $\sim 3 \times 10^9$ years or that the "old halo" and "disk" globular clusters do not share a common history; at least one of the classes presumably formed far from the Galaxy and was accreted at a later time. Also, no large changes in IMF occurred during the epoch of globular cluster formation.

In principle, it should be possible to estimate the mean mass of SN that occurred in the halo because SN nucleosynthesis predictions (e.g. Arnett 1991, Woosley & Weaver 1995) indicate that certain ratios (e.g. O/C and O/Mg) are sensitive to progenitor mass. Unfortunately the predictions of the two theoretical papers are not entirely consistent, which makes it difficult to constrain the IMF.

Establishing whether the all $\alpha$ elements exhibit the same level of enhancement in metal-poor stars is important; if so, this would favor a scenario in which the $\alpha$-element trend is due simply to the addition of iron-peak elements, as suggested by Tinsley (1979). In this regard, McWilliam et al (1995a,b) and Nissen et al (1994) measured [Ti/Fe] values $\sim$0.1 dex smaller than enhancements of other $\alpha$ elements (Mg, Si, Ca); McWilliam et al (1995a,b) claimed that this may be evidence that some Ti is produced in SN Ia.

There is increasing evidence for depletion of $\alpha$ elements abundances in some halo stars: Fuhrmann et al (1995) found [Mg/Fe] $= -0.28$ in BD + 3 740. McWilliam et al (1995a,b) found two stars (CS22968-014 and CS22952-015) near [Fe/H] $= -3.4$ with [Mg/Fe] $< 0.0$. These Mg depletions could be primordial, or they could be due to operation of the MgAl cycle in these stars (e.g. Shetrone 1996a,b), although the MgAl cycle would suggest large enhancements of Al, which are not seen in McWilliam et al's (1995a) stars.

Brown et al (1996) found low [$\alpha$/Fe] ratios in two young globular clusters (Rup 106 and Pal 12), with [Fe/H] of $-1.5$ and $-1.0$, respectively. In Pal 12, [Mg/Fe], [Ca/Fe], and [Ti/Fe] are approximately solar (i.e. below the halo value); in Rup 106, [O/Fe] and [Mg/Fe] are roughly solar, but [Ca/Fe] and [Ti/Fe] $\sim -0.2$ dex. Recently, Carney (in preparation) found a metal-poor field star with [Fe/H] $= -1.9$ and subsolar [$\alpha$/Fe] ratios. The low $\alpha$-element abundances, compared with the general halo, suggest that these two globular clusters and the field star formed from material with an unusually large fraction of SN Ia ejecta. One explanation is that star formation in the parent clouds proceeded over time scales longer than the time delay for SN Ia. Such an event could occur in low-mass clouds with relatively low star formation rates; because high-mass stars form much less frequently than low-mass stars, a fraction of clouds could be expected to escape SN II for long periods of time and thus permit enrichment by SN Ia. One is reminded of the Taurus molecular cloud, which is currently forming low-mass stars only. Another possibility is that the star and clusters were captured from a companion galaxy, like the LMC, which experienced chemical evolution over an interval of time longer than the characteristic SN Ia time scale.

Bazan et al (1996) found enhanced $\alpha$-element abundances for a number of "metal-rich" halo stars (compared to the mean halo [Fe/H] value of $-1.6$ dex); the sample ranges from [Fe/H] $\sim -1$ to 0.0, with a mean near $-0.5$ dex. If confirmed, this shows that the enhanced $\alpha$-element abundances are a characteristic of the halo as a population, regardless of metallicity. This is supported by the $\alpha$-element overabundances in Arcturus measured by Balachandran & Carney (1996). This underscores the fact that the knee in the [$\alpha$/Fe] versus [Fe/H] diagram (e.g. Figures 1 and 3) simply represents the intersection of the $\alpha$-element trends for the halo and disk and does not indicate an evolutionary connection.

Timmes et al (1995) made predictions of Galactic abundance ratios, from H to Zn, based on a Simple chemical evolution model and theoretical nucleosynthesis yields for SN II (from Woosley & Weaver 1995), SN Ia (from Nomoto et al 1984, Thielemann et al 1986), and 1- to 8-$M_\odot$ stars (Renzini & Voli 1981). The predicted trends with metallicity for O, Mg, Si, and Ca show reasonable agreement with observations; but the predictions for S lie below the observed [S/Fe] ratios and are just barely consistent with the 0.3-dex theoretical uncertainty. The predictions for Ti are by far the worst of all elements; the theoretical [Ti/Fe] ratio is almost 0.7 dex below the observed values near [Fe/H] $\sim -2$. It is clear that present SN nucleosynthesis calculations completely fail to account for the observed [Ti/Fe] ratios in the Galaxy; Ti is significantly enhanced in the bulge and halo, yet nucleosynthesis calculations suggest that it should scale with Fe. Thus, Ti provides an important constraint for SN nucleosynthesis theory.

## SOME LIGHT ELEMENTS

### *Carbon*

Carbon is one of those elements that can be greatly affected by late stages of stellar evolution. In the red giant stage, a star will dredge up material processed by the CNO cycle, which results in C depletions, increased $^{13}C$, and increased N abundances, and sometimes mild O depletions. In this review, I am mostly concerned with evolution of abundances in the Galaxy as a whole and not with the self-pollution of individual stars, unless this has a significant effect on the Galactic picture. For an excellent discussion of mixing in red giant branch stars, see Kraft (1994), Kraft et al (1997), and Shetrone (1996a,b).

Wheeler et al (1989) reviewed carbon abundances from Peterson & Sneden (1978), Clegg et al (1981), Laird (1985), Tomkin et al (1986), and Carbon et al (1987) and concluded that [C/Fe] $\sim 0.0$ independent of [Fe/H], but with a possible increase in [C/Fe] below [Fe/H] $\sim -1.5$.

Recent abundance studies of carbon indicate that [C/Fe] is enhanced with declining [Fe/H] in the Galactic disk (e.g. Friel & Boesgaard 1992, Andersson & Edvardsson 1994, Tomkin et al 1995), such that at [Fe/H] $= -0.8$, [C/Fe] $\sim +0.2$. Thus, in the disk [C/Fe] and [$\alpha$/Fe] show morphologically similar trends with [Fe/H].

Tomkin et al (1992) found that [C/O] $\sim -0.6$ for halo dwarfs in the interval $-1 \geq$ [Fe/H] $\geq -2.6$, based on high excitation C I and O I lines. Although the C I and O I results showed evidence of unaccounted non-LTE effects, Tomkin et al suggested that the errors cancel out for the [C/O] ratio. Tomkin et al also measured [C/Fe] from CH lines and found a trend of increasing [C/Fe] with decreasing [Fe/H]: Near [Fe/H] $= -1$, [C/Fe] $\sim -0.3$, with the trend suggesting that at [Fe/H] $= -2$, [C/Fe] $\sim 0.0$. Balachandran & Carney (1996) measured C and O from infrared CO and OH lines for one halo dwarf at [Fe/H] $= -1.2$, and they found [C/Fe] $= -0.32$.

Two puzzles arise from the Tomkin et al (1992) halo results: If [C/O] is constant, but [C/Fe] increases with declining [Fe/H], then [O/Fe] must also increase with declining [Fe/H]; yet Figure 3*a* rules out the required 0.3 dex change in [O/Fe] between [Fe/H] $= -1$ and $-2$. It is likely that the constant [C/O] ratio implied from the high excitation C I and O I lines may be suspect. The second difficulty, pointed out by Balachandran & Carney (1996), results from the large change in [C/Fe] between the halo and disk at similar metallicity; in the halo [C/Fe] $\sim -0.3$ near [Fe/H] $= -1.2$, while in the disk [C/Fe] $= +0.2$ at [Fe/H] $= -0.8$. According to Balachandran & Carney (1996), this would require a large contribution to Galactic carbon from intermediate-mass stars.

McWilliam et al (1995b) measured [C/Fe] values for 33 halo giants with $-4 \leq$ [Fe/H] $\leq -2$ and combined the results with the sample of Kraft et al (1982); no compelling evidence was found for a deviation of the mean [C/Fe] from the solar ratio. However, when compared with the results of Carbon et al (1987), a slight trend of increasing [C/Fe] could not be ruled out at the level of about 0.07 dex/dex in [Fe/H]. In either case, [C/Fe] is roughly constant over a range of 3.5 decades in [Fe/H]. McWilliam et al (1995b) found a large scatter in [C/Fe] for their giant sample, with a range of 1.6 dex, which is much larger than the measurement uncertainties. It seems possible that the scatter in [C/Fe] is due to an intrinsic dispersion in composition of the gas that formed the stars. If this is the case, then some of the halo carbon stars may not be the products of nucleosynthesis on the AGB, or mass transfer from an AGB star, but occurred because of stochastic enhancements in the carbon abundance of Galactic gas. This idea was supported by Kipper et al (1996), who claimed to have found at least three objects that formed as intrinsic carbon stars.

Thus, although the disk carbon abundances may resemble the $\alpha$ element pattern, in the halo the abundance trends are quite different and quite uncertain. It is clear that more work is required to properly understand [C/Fe] as a function of [Fe/H]; this might best be executed by taking advantage of new infrared spectrometers to measure C and O abundances from lines of CO and OH (as pointed out by Balachandran 1996).

For the bulge, the only carbon abundance measurement is that of A McWilliam, A Tomaney & RM Rich (in progress), who used the published values of the narrowband CO index for several bulge giants to estimate the average bulge [C/Fe] ratio, which was found to be ∼−0.2 dex. This value is consistent with typical [C/Fe] ratios seen in solar neighborhood red giants (Lambert & Ries 1981); the slight deficiency from the solar value is due to the normal red giant dredge-up of material processed though the CN cycle.

### *Aluminum and Sodium*

In the Galactic disk, the [Al/Fe] ratio increases with decreasing [Fe/H], reaching ∼+0.3 dex at [Fe/H] = −1 (Edvardsson et al 1993, Tomkin et al 1985), but this is only 0.2 dex above the mean [Al/Fe] ratio for solar metallicity stars. In the Edvardsson et al (1993) sample, these stars exhibit a 0.2-dex increase in [Na/Fe] from [Fe/H] = 0 to −1; however, Tomkin et al (1985) found [Na/Fe] ∼ 0.0 for their sample. Thus, from a phenomenological point of view, Al and perhaps Na could be classified as mild $\alpha$ elements, even though their nuclei have odd numbers of protons, which is consistent with a significant component of Al and Na synthesis from SN II.

The [Al/Fe] ratio in halo stars spans a range of approximately 2-dex (see Figure 7). The most extensive recent study of Al is that of Shetrone (1996a),

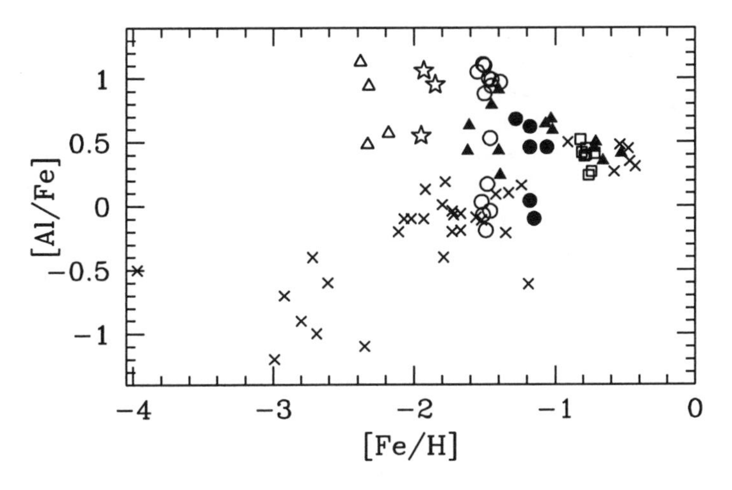


*Figure 7* The trend of [Al/Fe] with metallicity for field stars, indicated by *crosses* (from Shetrone 1996a and Gratton & Sneden 1988), and globular clusters from Shetrone (1996a); *open boxes* for M71, *open circles* for M13, *filled circles* for M5 and *open triangles* for M92; *filled triangles* for M22, M4, and 47 Tuc from Brown et al (1992), and *open stars* for NGC 2298 from McWilliam et al (1992). Note that the lower bound of the globular cluster values is consistent with the field star trend.

who showed that the globular cluster giants and the field halo giants have very different mean [Al/Fe] ratios, with little overlap. Shetrone (1996a) showed a plot of [Al/Fe] versus spectroscopic luminosity, indicating that the Al difference was not due to luminosity. However, luminosity is difficult to measure for field stars, so a more reliable estimate of the position of a star on the red giant branch for this comparison is the temperature; in this case, there is only a small region of overlap between field and globular cluster stars, near 4300 K. At 4300 K and above, for [Fe/H] $\sim -1$, no field giant has [Al/Fe] larger than +0.3, whereas many globular cluster giants with the same metallicity, luminosity, and temperature have [Al/Fe] larger than this value, up to a maximum value near +1 dex.

In Figure 7, the lowest [Al/Fe] ratios found in globular cluster giants appear similar to the field halo giants. This, at least, is consistent with the idea of self-pollution of Al from proton burning and deep mixing in evolved red giants (e.g. Denisenkov & Denisenkova 1990). In metal-poor stars, the declining [Al/Fe] ratio, or [Al/Mg], with decreasing metallicity is well known (e.g. Aller & Greenstein 1960, Arpigny & Magain 1980, Spite & Spite 1980) and has long been interpreted as consistent with the metallicity-dependent Al yields from explosive carbon burning, predicted by Arnett (1971). The Arnett predictions indicate that [Na/Fe] ratios should also decline with decreasing metallicity, but this is not observed; in fact, [Na/Fe] $\sim 0.0$ from [Fe/H] $= -1$ to $-4$ (e.g. McWilliam et al 1995b). Pilachowski et al (1996) analyzed a sample of 60 halo subgiants, giants, and horizontal branch stars in the interval $-3 \leq$ [Fe/H] $\leq -1$; they found a small [Na/Fe] deficiency of $-0.17$ dex in the mean and that bright field halo giants do not show the excess of sodium found in their globular cluster counterparts.

Globular cluster giants show large dispersions in Na and Al abundances. The Na and Al abundances are correlated, and they are correlated with N enhancements and O depletions. The abundance patterns have been interpreted either as evidence of internal nucleosynthesis and mixing operating in individual stars or, alternatively, as characteristic of a dispersion in the composition of the material out of which the stars formed (see Kraft 1994 and Shetrone 1996a,b, Kraft et al 1997, and references therein for details). Presently, the source of the Al and Na dispersion remains contested. Thus it is difficult to use Na and Al as probes of Galactic chemical evolution until the effects of individual stars can be quantified.

Studies of globular cluster main-sequence stars suggest that there may be a problem with the notion that globular clusters formed from chemically homogeneous material. Briley et al (1991) observed 10 main-sequence stars in 47 Tuc and found the same frequency of CN bimodality as present in the more evolved giant sequences; later, Briley et al (1996) found correlated CN and Na enhancements in main-sequence stars of the globular cluster 47 Tuc. These abundances, if found in giant branch stars, would have been attributed to dredge-up of internal nucleosynthesis products, i.e. self-pollution; however, this is not possible for main-sequence stars. Therefore, it is most likely that either Briley's CN and Na results are due to mass transfer from evolved companions, or the stars in 47 Tuc were formed from gas that was inhomogeneous in C, N, and Na. In this regard, it is interesting that the high frequency of CN-strong stars (near 50%) far exceeds the frequency in the field (at 5%). Spiesman (1992) found a nitrogen-rich metal-poor dwarf with Na and Al enhanced by +0.5 dex; citing the difficulty in producing Na and Al enhancements in main-sequence stars, Spiesman concluded that in this star the N, Na, and Al abundance anomalies are primordial. Suntzeff (1989) found an anticorrelation between CN and CH for main-sequence stars in NGC 6752. Pilachowski & Armandroff (1996) used a fiber spectrograph to acquire spectra of the [O I] region, at 6300 Å , for 40 stars at the base of the giant branch in M13. These stars are not expected to show oxygen depletions in the standard theory (e.g. Kraft 1994). The combined spectrum was of high S/N ($\sim$300), yet the undetected [O I] line indicated a limiting oxygen abundance below the solar [O/Fe] ratio; such deficiences have been interpreted in the past as due to red giant evolution.

Comprehending Al in globular cluster giants will have to await careful abundance analysis of larger samples of globular cluster main-sequence and red giant branch stars; only then will it be possible to know the scale of the Al production on the giant branch. This is an area in which the new large telescopes can make a significant impact. If the Al and Na abundance anomalies are due to primordial inhomogeneities in the cluster composition, there would be significant implications for ideas of globular cluster formation.

The sample of Galactic bulge giants studied by McWilliam & Rich (1994) also show marked Al enhancements, at approximately +0.7 dex, even at solar metallicity. This observation is consistent with self-pollution by evolved giants because the bulge stars observed were fairly luminous; however, this does not constitute proof of the self-pollution picture.

## HEAVY ELEMENTS

In this review I refer to heavy elements as those elements beyond the iron peak, with nuclear charge $Z \geq 31$. The reader is directed to Meyer (1994), Busso et al (1995), Lambert et al (1995b), and Käppeler et al (1989) for more detailed discussions of heavy element synthesis. Elements beyond the iron peak cannot be efficiently produced by charged-particle interactions owing to the large Coulomb repulsion between nuclei; temperatures high enough to overcome the Coulomb barrier tend to photodisintegrate even the iron-peak nuclei (e.g. Woosley & Weaver 1995). Burbidge et al (1957) showed that heavy elements can be synthesized by successive neutron captures onto iron-peak nuclei, followed by $\beta$ decays.

The neutron captures can occur on a time scale long enough for all $\beta$ decays to occur, which is called the s-process (for slow neutron capture), or on a time scale that is short compared to $\beta$ decay, called the r-process (for rapid neutron capture); these two processes lead to two characteristic abundance patterns.

In a steady flow of neutrons the abundance of each isotope is inversely proportional to its neutron capture cross section. The closed neutron shells with 50, 82, and 126 neutrons have small neutron capture cross sections, leading to abundance peaks for these nuclei. Similarly, even-numbered nuclei have smaller neutron capture cross sections than odd-numbered nuclei, resulting in higher abundances for the even nuclei; this is called the odd-even effect. The s-process abundance pattern is characterized by abundance peaks near mass numbers 87, 138, and 208 neutrons and a strong odd-even effect. The r-process abundance pattern is characterized by the abundance peaks shifted to mass numbers near 80, 130, and 195 with no odd-even effect.

Seeger et al (1965) showed that observed abundances of s-process–only isotopes can be represented by an exponential distribution of neutron exposures. For the Solar System material, the heavy element abundance pattern is best fit by a combination of two s-process exponentials (e.g. see Käppeler et al 1989): (*a*) the weak component, which corresponds to the light elements $A \leq 85$ (thought to occur in the cores of massive stars, $M \geq 10\,M_{\odot}$; see Raiteri et al 1991–1993), and (*b*) the main component, which fits the region approximately between Rb and Pb.

The s-process main component is thought to occur during the thermal pulse stage of low-mass (1–3 $M_{\odot}$) AGB stars at neutron densities of $10^7$–$10^9$ cm$^{-3}$. Quantitative calculations of AGB (asymptotic giant branch) nucleosynthesis were first performed by Iben (1975) and Truran & Iben (1977). Iben & Truran (1978) estimated that AGB nucleosynthesis in intermediate mass AGB stars could account for a significant fraction of the Galactic abundances of carbon and s-process elements. Since that time, much observational and theoretical work has converged on the idea that the s-process occurs during the AGB phase of low-mass stars (e.g. see Busso et al 1995, Lambert et al 1995b), between the H and He burning shells with neutrons liberated by the $^{13}C(\alpha, n)^{16}O$ reaction.

Smith & Lambert (1990) showed that the observed s-process abundances in M, MS, and S stars indicate a mean neutron exposure of $\tau_0 \sim 0.3$ at 30 keV. Because this is equal to the Solar System neutron exposure, it is consistent with AGB s-process nucleosynthesis as a major supply of the main component of the Solar System s-process elements. Recent observational information on the conditions of AGB s-process nucleosynthesis has come from neutron densities inferred from measurements of Rb and Zr isotopic abundances (Lambert et al 1995b).

Another exciting area of AGB nucleosynthesis research involves the study of presolar grains, embedded in meteorites (see Zinner 1996). Anomalous isotopic abundances of carbon and s-process elements in SiC grains indicate a carbon star origin (e.g. Anders & Zinner 1993). Boothroyd et al (1994) and Wasserberg et al (1995) infer the presence of deep circulation currents in AGB stars from the $^{12}C/^{13}C$ and $^{18}O/^{16}O$ ratios in these grains.

The site of the r-process is still in debate, although SN have been suspected from the beginning (Burbidge et al 1957). The most popular model is due to Meyer et al (1992), who suggested that the r-process occurs in the hot high-entropy bubble surrounding the nascent neutron star during the SN explosion. In this region, the high photon-to-baryon ratio favors photodissociation, thus keeping the number of free neutrons high and the number of nuclei low. Once the material cools, the nuclei are exposed to a sea of neutrons, at neutron densities of $\sim 10^{20}$ cm$^{-3}$, which drives the r-process.

### *Disk and Bulge Heavy Elements*

The notion that most heavy elements scale with [Fe/H] in the disk has been known for some time (e.g. Wallerstein 1962, Helfer & Wallerstein 1968, Pagel 1968, Huggins & Williams 1974, Butcher 1975). The most accurate and largest samples of disk dwarf abundances are the studies of Edvardsson et al (1993) and Woolf et al (1995); also McWilliam (1990) presented abundance results for a large number of disk giants.

The Edvardsson et al results demonstrate that the abundance of Y in the first s-process peak, and Ba and Nd in the second s-process peak, scale with metallicity down to [Fe/H] $= -1$. This observation is apparently at odds with the picture of primary and secondary elements: The s-process elements are made by the addition of neutrons to preexisting iron seed nuclei. Thus, the [s-process/Fe] abundance ratios are expected to behave like secondary elements, proportional to [Fe/H], rather than independent of the metallicity, as is observed. Clayton (1988) proposed that s-process abundances scale with metallicity if they were produced by the $^{13}C(\alpha,n)^{16}O$ neutron source in AGB stars; also, the increased neutron fluence in the model led to the prediction of increased [Ba/Y] ratios at low metallicity. This prediction was supported by the observed high [Ba/Y] abundance ratios found for CH stars (Vanture 1992, 1993) and $\omega$ Cen giants (e.g. Vanture et al 1994) and the abundance ratios in S and MS stars (see Busso et al 1995).

The metallicity dependence of the [Ba/Y] ratio produced in the s-process leads to a puzzle: If AGB stars are the source of the s-process elements in the disk, then why is the observed [Ba/Y] ratio approximately $\sim$0.0 dex over the full metallicity range of disk stars?

Part of the answer must be due to the transition from halo-like composition ([Ba/Y] $= +0.06$) gas near [Fe/H] $= -1$ to solar composition at [Fe/H] $= 0.0$;

both have similar [Ba/Y] values. Perhaps a more important factor is the presence of a large metallicity dispersion in the disk over most of Galactic history. In particular, the inhomogeneous chemical evolution models of White & Audouze (1983) showed that when there is a large dispersion in metallicity, the slopes for secondary elements can be erased; this is because at any given time the secondary elements were produced in sources with a large range of metallicity. Thus, the metallicity dispersion ensured that the [Ba/Y] ratio in the disk was always close to the average value.

Woolf et al (1995) measured [Eu/Fe] ratios in solar neighborhood F and G stars with $-0.9 \leq$ [Fe/H] $\leq +0.3$. Their results showed increasing [Eu/Fe] ratios with decreasing [Fe/H]; this trend was also reported by McWilliam & Rich (1994), although there was some considerable scatter about the mean relation. The [Eu/Fe] ratios match the trend of increasing $\alpha$-element abundances with [Fe/H]; from the [Eu/Fe] trends in the disk and the halo, one would classify Eu as an $\alpha$ element. Because Eu is a nearly pure r-process element, the observed trend with metallicity is consistent with the notion that the r-process and $\alpha$ elements are made in SN II (see Figure 8). For stars more metal-rich than the Sun, Woolf et al (1995) found subsolar [Eu/Fe] ratios, unlike the [$\alpha$/Fe] ratios that remain at the solar value above solar metallicity. The idea proposed by Maeder (1991) to explain the trend of [O/Fe] with [Fe/H] as due to the onset of Wolf-Rayet stars would have to affect the [Eu/Fe] ratio in the same way as

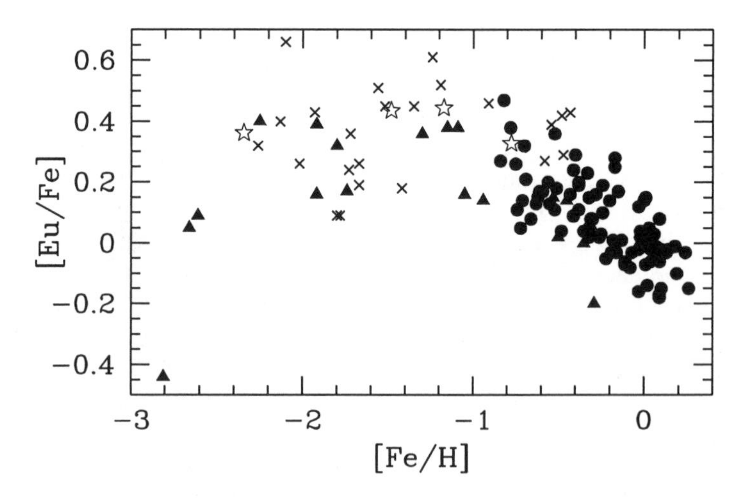


*Figure 8* The run of the (almost) pure r-process element europium with metallicity for field stars; indicated by *filled circles* for the disk (Woolf et al 1995) and in the halo by *crosses* (Shetrone 1996a) and *filled triangles* (Gratton & Sneden 1994). The *open stars* indicate mean values for the globular clusters M71, M13, M5, and M92 from Shetrone (1996a).

the [O/Fe] ratios. It is difficult to imagine how such a process might occur in current models of SN II, which holds that Eu is formed during the SN II event, deep inside the exploding star, whereas the ejected O is produced in higher regions of the star during the hydrostatic burning phase.

The Zr results from Edvardsson et al (1993) show a trend towards enhanced [Zr/Fe], reaching +0.2 dex in the lower metallicity disk stars; given the enhanced [Zr/Fe] ratios in the halo stars (Magain 1989, Gratton & Sneden 1994), this may suggest that SN II produce significant amounts of Zr.

Edvardsson et al (1993) showed that in the disk, [Ba/Fe] ratios increased with time and were very roughly independent of metallicity; from log age $\geq 0.9$ to $< 0.6$ Gyr, the mean [Ba/Fe] ratio increased from $-0.1$ to $+0.1$ dex. The halo fits into this picture, with a mean [Ba/Fe] ratio of $\sim -0.1$ dex (Gratton & Sneden 1994). This trend in [Ba/Fe] suggests that there is a source that produced Ba on a time scale longer than the time scale for Fe production. This conclusion is consistent with the idea that s-process nucleosynthesis is dominated by AGB stars over the mass range 1–3 $M_\odot$ (Meyer 1994) or 1–4 $M_\odot$ (Busso et al 1995). The steady increase in the disk [Ba/Fe] ratio with Galactic time is probably due to the delay in Ba production from the lower mass stars in this range, say 1–2 $M_\odot$, with main-sequence lifetimes of several billion years.

The results of Edvardsson et al (1993) show a steeper slope for [Ba/H] with age than [Fe/H] with age; also, despite the larger measurement uncertainties for Ba, the age–[Ba/H] relation shows less dispersion than the age–[Fe/H] relation. The steep slope of [Ba/H] with age must be due in part to the gradual increase in [Ba/Fe] with time due to the long-lived sources. If the intrinsic dispersion in the age–[Ba/H] relation is actually significantly smaller than for the age–[Fe/H] relation, then this must be understood in the context of disk chemical evolution models that describe the dispersion in the age-metallicity relation (e.g. Pilyugin & Edmunds 1996b). A characteristic of the Pilyugin & Edmunds (1996b) model is that it predicts roughly equal dispersion for *all* elements.

The fact that the [Ba/Fe] ratio is sensitive to age might have applications for other locations, such as the Galactic bulge. For the bulge, McWilliam & Rich (1994) found the mean [s-process/iron] ratio of $\sim$0.0 dex, from lines of Y, La, and Ba. In particular, a subsolar value of [s-process/Fe] near $-0.1$ dex, expected if the bulge formation time scale was rapid, is inconsistent with the data. Certainly more accurate and extensive measurements of s-process elements in the bulge are necessary to verify this point.

### *Halo Heavy Elements*

To summarize the principle result of this section: Heavy element abundances in the halo are characterized by a significant r-process component and a 300-fold dispersion in [heavy element/Fe] ratios below $[Fe/H] = -2.5$ at about a constant average value. The pattern of r-process nucleosynthesis in the halo heavy element abundances provides evidence that this dispersion reflects an inhomogeneous composition of the material from which the stars formed, and the observed [heavy element/Fe] values set the minimum [heavy element/Fe] range from SN II events. The decreasing dispersion with increasing metallicity is consistent with a gradual homogenization of low metallicity gas by the process of averaging yields from individual SN events.

Although present in the abundance results of Wallerstein et al (1963) Pagel (1968) was the first to recognize that very metal-poor halo stars show heavy element deficiencies. Evidence for a plateau with [heavy element/Fe]$\sim$0.0, followed by a systematic trend of decreasing [heavy element/Fe] below [Fe/H] $\sim -2.5$, was presented by Spite & Spite (1978, 1979), Luck & Bond (1981, 1985), Barbuy et al (1985), Gratton & Sneden (1988), Magain (1989), and Zhao & Magain (1990, 1991). In particular, the results of Gilroy et al (1988) and Magain (1989) showed that [Eu/Fe] and [Zr/Fe] are enhanced in the interval [Fe/H] $= -1.5$ to $-2.5$.

Perhaps the most accurate abundance results for the largest sample of heavy elements come from Gratton & Sneden (1994), with stars in the interval $-3 \leq$ [Fe/H] $\leq -0.5$. The abundance ratios [M/Fe] for Sr, Y, Ba, La, Ce, and Nd lie approximately between [M/Fe] $\sim$ 0.00 to $\sim$$-0.1$ dex, and Zr, Sm, Pr, Dy, and Eu are enhanced relative to the solar composition. The $\sim$0.3-dex enhancement in [Eu/Fe] is notable because Eu is an almost pure r-process element, thought to be produced only in SN II; its enhancement resembles the $\sim$0.3-dex enhancement observed for the $\alpha$ elements in the halo, which are also thought to be produced only in SN II.

The first evidence for a heavy element abundance dispersion in halo stars was due to Griffin et al (1982), who found heavy element enhancements in HD 115444. Luck & Bond (1985) found several halo stars with Ba and Sr enhancements. Both studies suggested that stars with heavy element enhancements were population II barium stars.[2] Gilroy et al (1988) claimed an r-process pattern and a large abundance dispersion for the halo heavy elements, which was consistent with heavy element abundance scatter in the material from which the stars formed. Ryan et al (1991) also claimed a dispersion in heavy element abundances larger than the measurement errors.

However, Baraffe & Takahashi (1993) suggested that the scatter was due entirely to measurement errors, based on the large scatter in published heavy element abundances for individual stars. The accurate abundance measurements of Gratton & Sneden (1994) indicated a heavy element dispersion of less than 0.1 dex for their sample, consistent with their measurement uncertainties.

[2]Barium stars are thought to arise from mass transfer from an AGB star that has polluted its envelope with s-process elements (e.g. McClure 1984).

McWilliam et al (1995a,b) analyzed a large sample of extremely metal-poor halo stars from the survey of Beers et al (1992), in the metallicity range $-4 \leq$ [Fe/H] $\leq -2$, and made reliable estimates of the measurement uncertainties. They found a decline in [heavy element/Fe] below [Fe/H] $= -2.5$, accompanied by a considerable scatter in [Sr/Fe] and [Ba/Fe] abundance ratios, with a range of 2.5 dex (see Figure 9); typical measurement uncertainties were $\pm 0.2$ dex. This scatter does not conflict with the small dispersion found by Gratton & Sneden (1994) because the Gratton & Sneden sample included only two stars below [Fe/H] $= -2.5$.

Ryan et al (1996) analyzed additional extremely metal-poor stars and found a large scatter in heavy element abundances for metal-poor halo dwarfs and giants, consistent with a primordial abundance scatter.

The McWilliam et al results show that for both [Sr/Fe] and [Ba/Fe] the mean ratios are the same above and below [Fe/H] $= -2.5$. Although there are many more stars deficient in heavy elements than with overabundances, the few heavy element–rich stars cause the average [Sr/Fe] and [Ba/Fe] ratios to be near the solar value. This explains why early studies found a trend of declining heavy element/Fe ratios as [Fe/H] declined; there is no trend, only a lower envelope of the dispersion that is skewed to low [Sr/Fe] and [Ba/Fe] ratios. Small samples preferentially picked the low [M/Fe] ratio stars because they are more frequent than the stars with heavy element enhancements.

Pagel (1968) and Truran (1981) noted that the near-solar value of the [heavy element/Fe] ratio in the halo implies that the formation time for these elements must be shorter than the lifetimes of stars that produce s-process elements. Truran concluded that the heavy elements in the halo were made in massive stars by the r-process.

Abundance studies of halo stars by Sneden & Parthasarathy (1983), Sneden & Pilachowski (1985), and Gilroy et al (1988) indicated heavy element abundance patterns consistent with nucleosynthesis dominated by the r-process.

The [Ba/Eu] ratio is particularly sensitive to whether nucleosynthesis of the heavy elements occurred by the s-process or r-process. The ubiquitous subsolar [Ba/Eu] ratios in halo stars (e.g. Magain 1989, François 1991, Gratton & Sneden 1994, McWilliam et al 1995b) show that the halo must contain a larger fraction of r-process material than the solar composition (e.g. Spite 1992). Figure 10 shows a compilation of [Ba/Eu] and [La/Eu] ratios for field halo stars, with pure r-process and s-process values indicated. The subsolar ratios indicate a larger fraction of r-process material than in the Solar System material; however, some s-process contribution may be required.

Cowan et al (1996) measured abundances of the r-process peak elements Os and Pt from UV lines in the metal-poor halo giant HD126238 ([Fe/H] $= -1.7$). When combined with abundances based on optical spectra, the best-fit heavy element pattern contains 80% r-process and 20% Solar System mixture. This

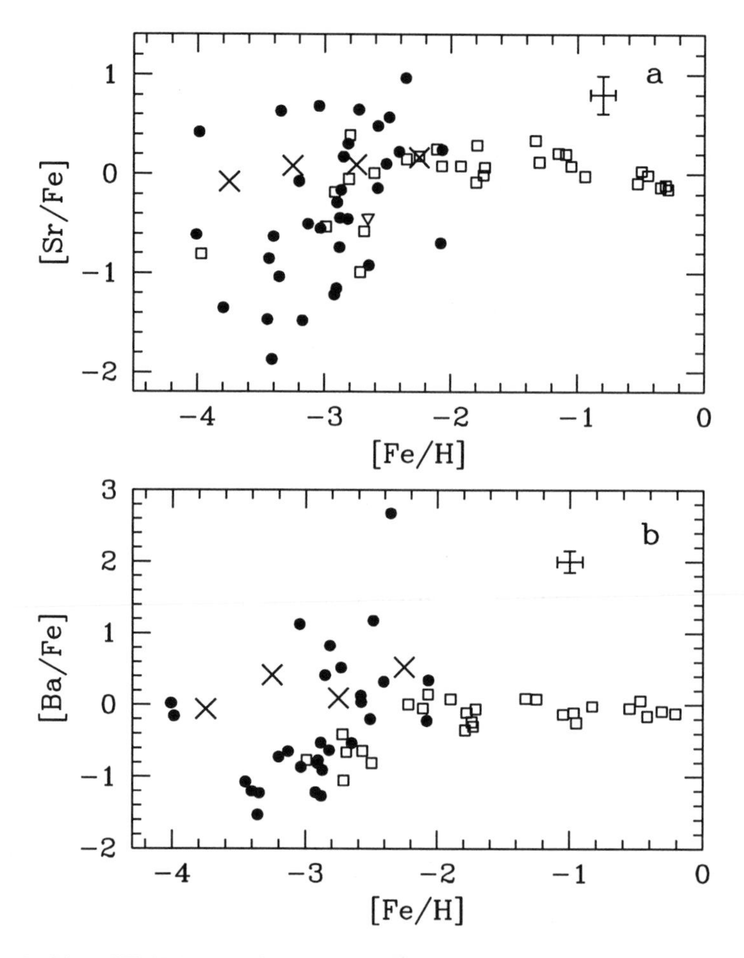


*Figure 9* Plots of [Sr/Fe] and [Ba/Fe] from McWilliam et al (1995; *filled circles*) and Gratton & Sneden (1994, 1988; *open squares*). The error bars indicate $1\sigma$ uncertainties on the McWilliam et al results. The general run of the [Sr/Fe] data indicates a downward trend below [Fe/H] = −2.5 with a dispersion of ~300-fold. The *large crosses* represent the average [Sr/Fe] and [Ba/Fe] ratios for the McWilliam et al sample taken for 0.5-dex bins. The [Ba/Fe] data are similar to the [Sr/Fe] trend, but there is a hint of a bifurcation. Note that the star at [Fe/H] = −2.36 and [Ba/Fe] = +2.67 represents CS 22898-027, a CH subgiant, which is contaminated by s-process material accreted from an evolved companion, and so was not included in the average.

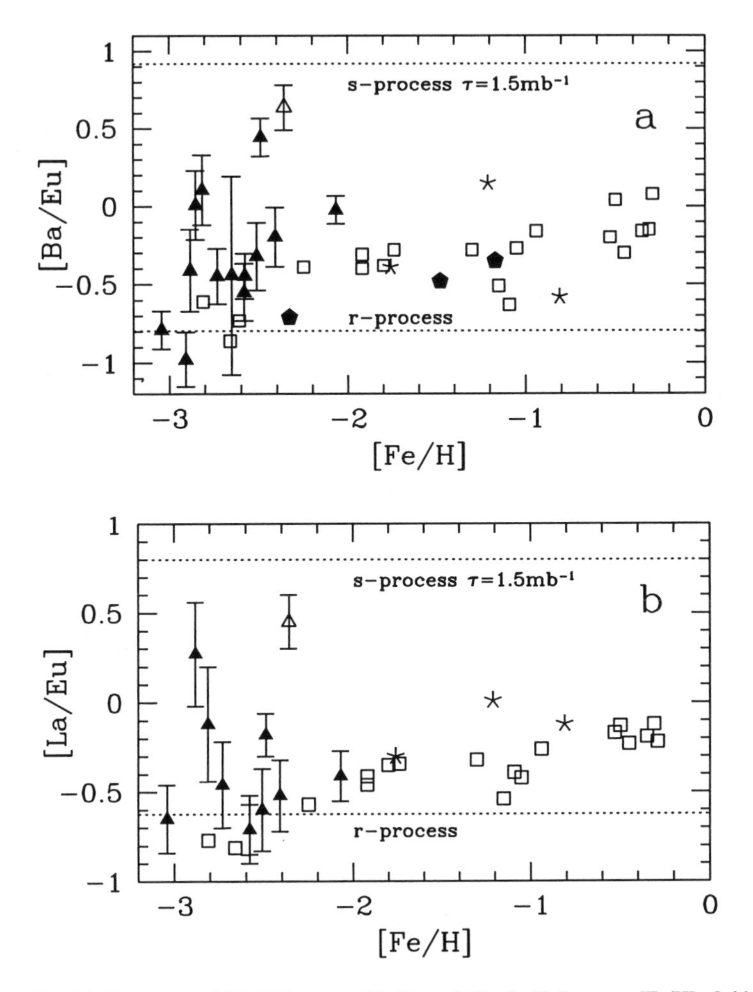


*Figure 10* (*a*) The trend of [Ba/Eu] versus [Fe/H] and (*b*) [La/Eu] versus [Fe/H]: field stars represented by *open boxes* (Gratton & Sneden 1994) and *triangles* (McWilliam et al 1995). *Star symbols* represent mean globular cluster values from Brown et al (1992), and *filled pentagons* indicate globular clusters from the data of Shetrone (1996a) and Armosky et al (1994). The *open triangle* indicates the CH subgiant, CS 22898-027, which is contaminated by s-process material. *Dashed lines* indicate the observed solar system r-process ratio (Käppeler et al 1989) and an extreme s-process value from Malaney (1987).

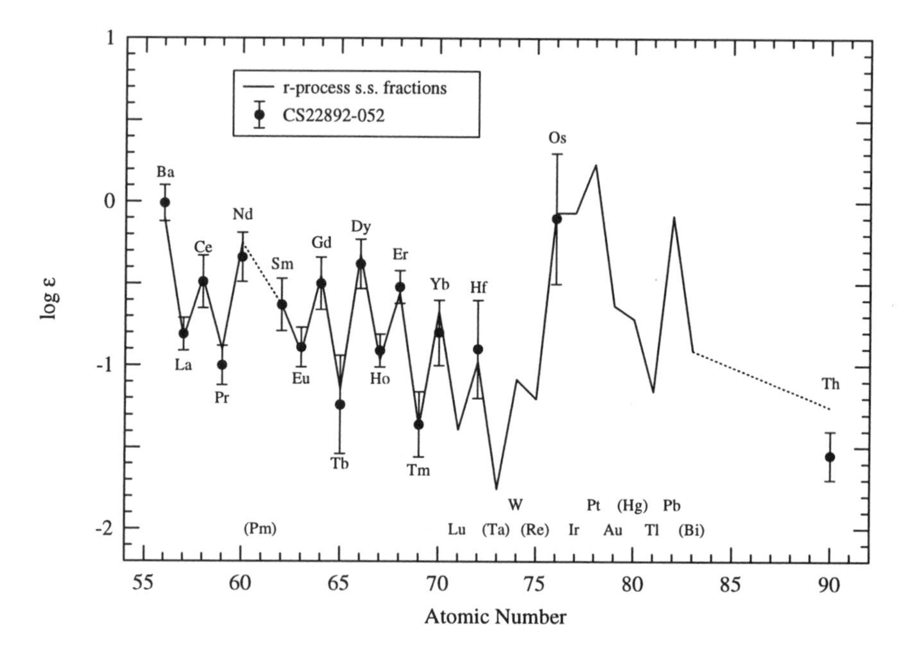


*Figure 11* The heavy element abundance pattern in star CS 22892-052, from Sneden et al (1996), scaled to the barium abundance. The *line* represents the observed Solar System r-process abundance pattern from Käppeler (1989). The excellent agreement suggests that nucleosynthesis was dominated by the r-process; the small scatter about the r-process line indicates that the *error bars* were overestimated.

is consistent with the value of [La/Eu] $= \sim -0.4$ for halo stars in the range $-2 \leq$ [Fe/H] $\leq -1$ seen in Figure 10.

Element abundances for the most heavy element rich star known, CS 22892-052, were measured by Sneden et al (1994), Cowan et al (1995), McWilliam et al (1995b), and Sneden et al (1996). Figure 11 shows the heavy element abundance pattern in this star, for elements heavier than Ba, which is identical to the Solar System r-process pattern (Käppeler et al 1989). Based on the large r-process overabundance and low [Fe/H], these authors concluded that the heavy elements in CS 22892-052 are dominated by the nucleosynthesis products of a single SN event. This does not mean that all the elements in this star are dominated by SN nucleosynthesis from a single event, only the heavy elements.

Cowan et al (1995) showed that some s-process contribution would help the fit to the observed Sr and Y abundances in CS 22892-052, but the Zr abundances cannot be explained by the s-process; thus the r-process probably at least contributes to the Zr abundance in the Sr-Y-Zr peak. McWilliam et al (1995b) showed that the [Sr/Ba] ratio is approximately constant in halo stars, despite the factor of 300 range of barium abundance due to the r-process. Therefore, if Sr has a significant contribution from the s-process, then the s-process to r-process ratio must be roughly constant in the halo; an alternative is that the r-process is a dominant source of Sr in the halo.

Magain (1995) measured the abundances of barium isotopes in one halo star in order to find the relative contribution of r- and s-process nucleosynthesis, from a profile fit to the Ba II line at 4554 Å. The best-fit profile indicated a Solar System mixture of barium isotopes, contrary to that expected from r-process nucleosynthesis. To resolve the discrepancy with element abundance ratios it would be very useful to have Ba isotopic compositions for a larger sample, especially for stars with both strong and weak Ba II lines.

François (1996) also disputed the claimed r-process source of halo heavy elements, based on a plot of [Eu/H] versus [Ba/H], arguing against the break in slope of the $\epsilon$(Ba) versus $\epsilon$(Eu) seen by Gilroy et al (1988). However, it is better to rely on diagnostic abundance ratios (like [Ba/Eu] and [La/Eu]) as a discriminant of the nuclear reactions involved, rather than on the presence or absence of a break in slope.

Sneden et al (1996) noted that total Ba abundances measured from strong Ba lines in halo stars depend upon the assumed r- and s-process fractions. This effect may result in a downward revision of many previously reported Ba abundances by ~0.1 to 0.2 dex and bring earlier measurements of the [Ba/Eu] ratio closer to the pure r-process value. Therefore, when considering published abundances, it is best to use Ba abundances based on weak lines or to substitute the abundance of the s-process element La in place of Ba, because lanthanum is dominated by a single isotope (99.9% $^{139}$La) with relatively weak lines.

Many astrophysical environments have been proposed as the main source of the r-process. Mathews & Cowan (1990) and Mathews et al (1992) list many of these and attempted to test the possibilities by comparing predicted abundances from a Simple model of Galactic chemical evolution to observed heavy element abundances. They claimed that low-mass SN II (7–8 $M_\odot$) were the most likely candidates. In their model the trend of increasing [heavy element/Fe] ratios with [Fe/H] was due to a time delay arising from the longer main-sequence lifetime of low-mass SN II progenitors relative to high-mass SN II progenitors. Thus at early times, when the lowest metallicity prevailed, only high mass SN II occurred with low [heavy element/Fe] yield ratios; at later times, and higher metallicity, the low mass SN II enriched the Galactic gas with high [heavy element/Fe] material. This model requires that in all situations the first low-mass SN II events were preceeded by high-mass SN II events, which probably would not occur in the case of chemical evolution in molecular cloud size masses or Searle-Zinn fragments (Searle & Zinn 1978). The model also requires that the mean [Sr/Fe] ratio increases with increasing [Fe/H] at low metallicity; however, the results of McWilliam et al (1995b) indicate a constant average [Sr/Fe] value. Thus, the time-delay mechanism cannot be used to explain the observed heavy element abundances in the halo, and no constraint can be placed on the mass of the SN chiefly responsible for heavy element synthesis.

McWilliam et al (1995a,b, 1996) and Sneden et al (1994) argued that the observed dispersion in heavy element abundances must reflect an intrinsic dispersion in the [heavy element/Fe] ratio of the gas from which the extremely metal-poor halo stars formed. In particular, the r-process abundance pattern and the high frequency of stars with heavy element enhancements rules out the possibility that these stars are population II barium stars.

The heavy element dispersions found by McWilliam et al (1995a,b) showed that the range in SN heavy element yields is at least a factor of 300. McWilliam et al (1996) argued that because the heavy element/Fe ratio for CS 22892-052 is ~15 times the asymptotic value, the progenitor SN must represent no more than 1/15 of all SN II. Because homogenization of the halo gas could only have occurred once the full range of SN yields was sampled, the metallicity of the homogenization point (at [Fe/H] = −2.5) corresponds to approximately 15 SN events. If 0.1 $M_\odot$ of iron is ejected per SN II event, then this metallicity requires mixing of the ejecta with $\sim 10^5$ to $10^6$ $M_\odot$ of hydrogen.

Searle & McWilliam (1997 in progress) have studied models of chemical enrichment by small numbers of SN II events, with [Sr/Fe] yields selected at random from the observed range in [Sr/Fe]. This stochastic model can reproduce the average, the dispersion, and the envelope of [Sr/Fe] values seen in metal-poor halo stars, with [Fe/H] $\leq -2.5$. The model is consistent with enrichment by single SN II events below [Fe/H] $\sim -3.3$, in regions of mass $\sim 10^6$ $M_\odot$, which is characteristic of present-day molecular clouds.

The large r-process enhancements in CS 22892-052 allowed Sneden et al (1996) to measure the abundance of thorium (Th, a pure r-process element) in this star. Owing to its $14 \times 10^9$ year half life, Th is potentially a useful Galactic chronometer (e.g. Butcher 1987). Based on the solar [Th/Eu] ratio, Sneden et al (1996) deduced a minimum age for CS 22892-052 of $15 \pm 4 \times 10^9$ years. Cowan et al (1997) employed r-process nucleosynthesis calculations and various Galactic chemical evolution models to predict the initial r-process [Th/Eu] ratio, which led to a minimum age of $15 \pm 4 \times 10^9$ years and a most likely age of $17 \pm 4 \times 10^9$ years.

The early work of Pilachowski et al (1983) indicated low, and even subsolar, [Eu/Fe] ratios for globular cluster stars; taken at face value these results suggest a difference between the composition of halo field stars and globular cluster stars. However, the more recent of François (1991), Brown & Wallerstein (1992), Shetrone (1996a), and McWilliam et al (1992) all indicate globular cluster

[Eu/Fe] ratios near +0.4 dex, which is similar to the results for halo field stars (e.g. Gratton & Sneden 1994, Shetrone 1996a, McWilliam et al 1995a,b, Magain 1989, recomputed here); the mean halo [Eu/Fe] value is +0.33 dex, which is the same enhancement as seen in the most metal-poor disk stars (Woolf et al 1995).

The chemical composition of the unusual globular cluster $\omega$ Cen differs from other globular clusters and field halo stars. Several recent abundance studies of $\omega$ Cen giants have been published: Vanture et al (1994), Norris & Da Costa (1995), Smith et al (1995), and Norris et al (1996). The cluster shows a metallicity spread from [Fe/H] = −1.9 to −0.6, with evidence for two star formation epochs. The [$\alpha$/Fe] ratios show the normal factor of 2 enhancement seen in halo stars, which implicates nucleosynthesis by SN II only. However, the heavy elements are enhanced well above the solar value and are consistent with significant contamination by s-process nucleosynthesis from AGB stars. This is evidence that the s-process occurs more rapidly than the time scale for enrichment by SN Ia. A puzzle noted by Smith et al (1995) is the subsolar [Eu/Fe] ratio, near −0.4 dex; if AGB stars produced s-process material, then SN II should have produced larger [Eu/Fe] ratios. It is as if the r-process SN II never occurred in $\omega$ Cen; perhaps this is an indication of a unique IMF that excluded r-process SN II events.

## IRON-PEAK ELEMENTS

The iron-peak elemental abundances showing conclusive evidence for deviations from the solar ratios are summarized in Figure 12. Wallerstein (1962) and Wallerstein et al (1963) were the first to find evidence of a nonsolar mixture of iron-peak elements: Deficiencies of Mn found by Wallerstein were confirmed by later studies (e.g. Gratton 1989). From [Fe/H] = 0.0 to −1.0, the [Mn/Fe] ratios are deficient in a manner that mirrors the $\alpha$-element overabundances, and in the interval [Fe/H] = −1.0 to −2.5 dex, [Mn/Fe] is constant at $\sim$−0.35 dex. Thus the [Mn/Fe] trend is similar, but in an opposite sense to the [$\alpha$/Fe] trend with [Fe/H]. A simple conclusion is that a significant source of Mn comes from SN Ia. McWilliam et al (1995a) discovered that below [Fe/H] $\sim$ −2.5, the [Mn/Fe] ratio decreases steadily with decreasing [Fe/H], like the trends exhibited by the heavy elements.

Wallerstein's (1962, 1963) observation of Mn deficiencies at low metallicity was claimed to be part of the neutron-excess–dependent yields of Arnett (1971). Even the recent Galactic nucleosynthesis predictions of Timmes et al (1995), based on the Woosley & Weaver (1995) calculations for SN II, predict deficiencies of Sc, V, Mn, and Co of $\sim$ −0.5 dex relative to Fe for metal-poor stars. Wallerstein claimed that V is also deficient in metal-poor stars, but this finding was not confirmed by subsequent analyses (e.g. Pagel 1968). The abundance

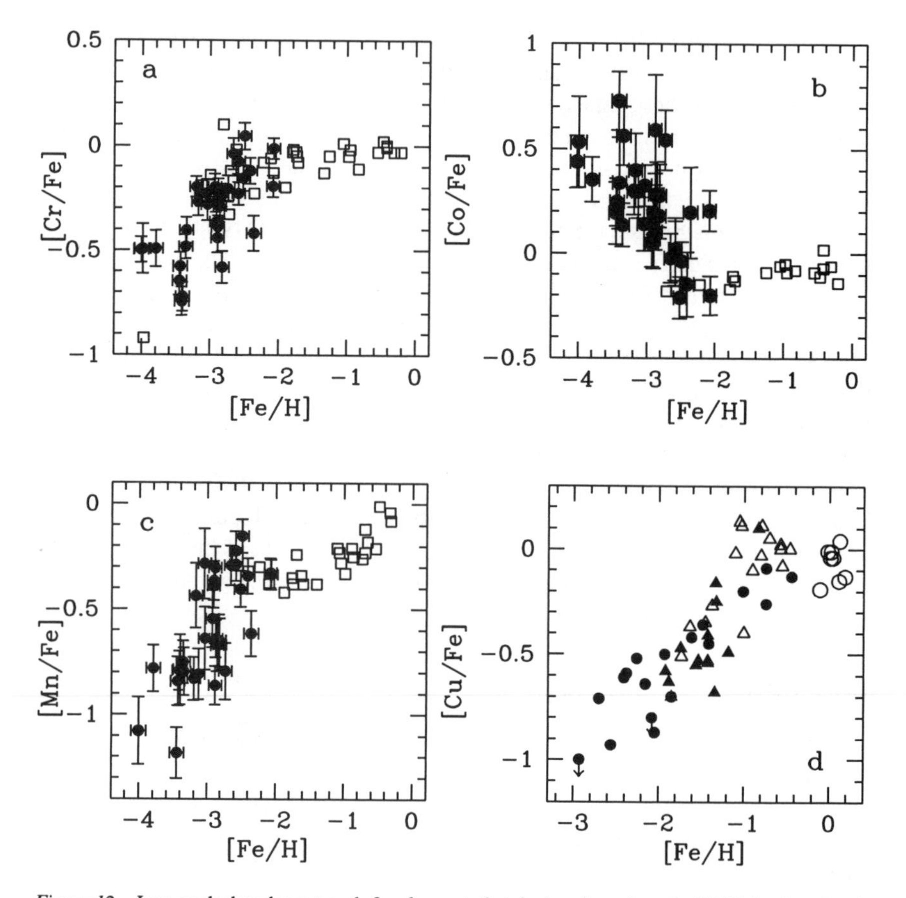


*Figure 12* Iron-peak abundance trends for elements that deviate from the solar [M/Fe] ratios. (*a–c*) Halo field stars: *filled circles* (McWilliam et al 1995), *open squares* (Gratton & Sneden 1988, 1991, Gratton 1988). Note that Cr, Co, and Mn each show a decline relative to Fe below [Fe/H] = −2.5; [Mn/Fe] also declines between [Fe/H] = 0.0 to −1. (*d*) [Cu/Fe] versus [Fe/H] from Sneden et al (1991): *filled circles* (field stars), *filled triangles* (globular clusters), *open triangles* (reanalyzed field star data), *open circles* (literature values for population I field stars).

of V has not been well studied. The most comprehensive analysis was done by Gratton & Sneden (1991), who found [V/Fe] ∼ 0.0 at all metallicities, which confirmed Pagel's conclusion. If the Mn deficiencies are due to a neutron-excess dependence, then V and Sc are also expected to follow the same trend, which is not observed.

Sneden & Crocker (1988) and Sneden et al (1991a) studied the abundances of Cu and Zn as a function of metallicity and discovered that [Cu/Fe] decreases linearly with declining metallicity, [Cu/Fe] $= 0.38$[Fe/H] $+ 0.15$ (curiously the trend resembles that of [Al/Fe] with [Fe/H]), while Zn is constant, at [Zn/Fe] $= 0.0$, for all metallicities. Sneden et al suggested that nucleosynthesis of Cu may occur mainly by the weak s-process in the cores of massive stars, with a small contribution from explosive burning in SN II. However, Matteucci et al (1993) suggested that the greatest production of Cu and Zn occurs in SN Ia. If this is true, then some SN Ia occurred for [Fe/H] $< -1$, which will have important consequences for chemical evolution models of the halo.

The constant [Zn/Fe] abundance ratio seen in the Galaxy is not universal: Abundance analyses of QSO absorption line systems show that Zn is enhanced relative to Fe (e.g. Pettini et al 1994, Lu et al 1996). Although the observed enhancements of the [Zn/Cr] ratios may indicate that the gas has been affected by dust depletion, if this were the case, then one would also expect to find large [S/Cr] ratios, which are not found. Thus the enhanced Zn abundance in QSO absorption line systems may have a nucleosynthesis origin.

Besides Zn, the abundance ratios of Sc, V, and Ni relative to Fe seem to scale with [Fe/H]. It should be noted that Zhao & Magain (1989, 1990) claimed a mean $+0.27$-dex enhancement of [Sc/Fe] in metal-poor dwarfs. However, high quality data of Gratton & Sneden (1991) and Peterson et al (1990), as well as the results of McWilliam et al (1995b) found no evidence for a deviation from [Sc/Fe] $= 0.0$ in metal-poor giant stars. The lower quality data of Gilroy et al (1988) actually indicated a deficiency of $\sim$0.2 dex.

Luck & Bond (1985) claimed enhanced [Ni/Fe] ratios in metal-poor stars, and Pilachowski et al (1996) found a mean [Ni/Fe] $= -0.27$ near [Fe/H] $= -2$; however, other studies tended to find solar [Ni/Fe] ratios everywhere. In particular, Peterson et al (1990) demonstrated that the Luck & Bond Ni overabundances were probably the result of selecting lines enhanced above the detection threshold by noise-spikes. The combined studies of Gratton & Sneden (1991), Peterson et al (1990), Edvardsson et al (1993), McWilliam et al (1995a), and Ryan et al (1996) are inconsistent with [Ni/Fe] more than $\pm 0.1$ dex from the solar ratio in the interval $-4 \leq$ [Fe/H] $\leq 0$; although some of the Gratton & Sneden points are subsolar near [Fe/H] $= -2.5$.

Until recently, [Co/Fe] and [Cr/Fe] ratios were commonly accepted to be independent of [Fe/H]; however, McWilliam et al (1995b) showed that Co and Cr deviate from a plateau at metallicities below [Fe/H] $\sim -2.5$ (see Figure 12). McWilliam et al (1995a) found evidence supporting their results in the data of Gratton & Sneden (1991), Ryan et al (1991), and Wallerstein et al (1963); these trends for Co, Cr, and Mn have subsequently been verified by Ryan et al (1996).

The divergence of [Co/Fe], [Mn/Fe], and [Cr/Fe] (and the heavy elements) from a plateau, below [Fe/H] $\sim -2.5$, suggests that chemical evolution was very different below the lowest globular cluster metallicities, perhaps indicating

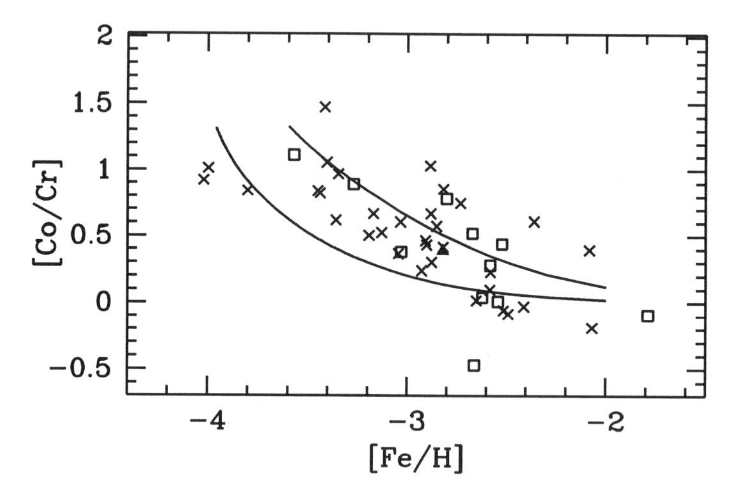


*Figure 13* [Co/Cr] versus [Fe/H] for extremely metal-poor halo stars: *crosses* from McWilliam et al (1995), *open boxes* from Ryan et al (1991). The *lines* trace the resultant [Co/Cr] ratios when solar-composition material is mixed with primordial material, characterized by a large [Co/Cr] value near +1.3 dex.

the existence of population III or early population II stars. If SN II were the dominant source of iron-peak elements at low metallicity, the observed range in [M/Fe] ratios indicates the minimum range of yield ratios for SN II.

McWilliam et al (1995b, 1996) argued that because the [Co/Fe], [Mn/Fe], [Cr/Fe], and heavy element/Fe ratios differ from the solar values, the metal-poor stars below [Fe/H] $\sim -2.5$ cannot be the products of simple dilution of higher metallicity gas (e.g. typical of globular cluster composition) by gas with zero metallicity. If metal-poor stars below −2.5 were products of dilution with pure hydrogen, then solar [Co/Cr] ratios would exist down to −4. In Figure 13, the tight correlation between [Co/Cr] and [Fe/H] sets tight constraints on the dispersion in dilution by zero metallicity gas that might have occurred, i.e. dilution with zero metal gas could not have differed from one star forming region to another by more than a factor of 2 because the [Fe/H] dispersion at fixed [Co/Cr] value is ~0.3 dex.

It would be interesting to understand why the [Co/Cr] ratios of extremely metal-poor stars are so tightly correlated with [Fe/H], whereas for the same stars, the heavy elements show such a large dispersion (Figure 9).

McWilliam et al (1995b) suggested that the heavy elements are produced in large amounts by a rare subclass of SN event and essentially not at all in most events. The tight correlation of [Co/Cr] with [Fe/H] might suggest that all or most SN produce iron-peak elements, like Cr, Mn, and Co, in similar proportions but with metallicity-dependent yields. One possibility suggested by McWilliam et al (1995b) is that the observed [Co/Cr] trend in Figure 13 could have occurred if the star-forming gas was in a process of steady chemical enrichment, with several generations of SN gradually enriching the parent cloud in the range $-4 \leq$ [Fe/H] $\leq -2.5$. In this model, different kinds of SN II existed with different [Co/Cr] yield ratios, and as metallicity increased, the average SN II changed in character until, when metallicity reached [Fe/H] $\sim -2.5$, the average SN produced solar [Co/Cr] yield ratios.

One way for the metallicity-modulated [Co/Cr] yields to occur may be by affecting the mean SN progenitor mass. For example, Wolfire & Cassinelli (1987), Yoshii & Saio (1986), and Kahn (1974) predicted that the IMF is weighted to higher mass stars at low metallicity. Another potential method of altering the SN progenitor mass range is through the effect of metallicity-dependent mass loss; for example, theoretical models by Bowen & Willson (1991) suggested that AGB stars with an initial mass of 2.4 $M_\odot$ reached the Chandrasekhar mass before the envelope could be ejected.

The observed [Co/Cr] versus [Fe/H] relation in Figure 13 cannot be explained by the time delay between SN II progenitors with different mass, for the following reason: Any time delay must be less than the maximim SN II progenitor lifetime, which, at $20 \times 10^6$ years, is much shorter than the dynamical time scale of the Galaxy; thus chemical evolution in the first $20 \times 10^6$ years must have taken place in isolated regions. The observed range of [Fe/H] at a fixed [Co/Cr] in Figure 13 excludes a range of SFR more than a factor of 2, which conflicts with the fact that SFR are known to vary by orders of magnitude.

Ryan et al (1996) suggested that the SN II [Co/Cr] yield is a function of the SN energy and that the SN energy dictates how much dilution of the ejecta occurs and, therefore, the metallicity of the next generation of stars. In this way the observed tight correlation between [Co/Cr] and [Fe/H] can occur.

An alternative mechanism (L Searle & A McWilliam, in preparation) is that there was a primordial composition, characterized by high [Co/Cr], and that this was later diluted with solar composition SN ejecta. In this model, the first generation SN, presumably zero-metallicity population III stars, produced high [Co/Cr] ratios, but all subsequent SN II produced solar [Co/Cr] yield ratios. Figure 13, shows two "dilution" curves that follow the combined composition of primordial plus solar mix material, with increasing amounts of solar composition material. It is clear that both dilution curves fit some but not all of the observed data points, which may suggest that the putative primordial material was characterized by a spread in metallicity as well as a high [Co/Cr] ratio, near +1 dex.

This mechanism produces a tight correlation of [Co/Cr] with [Fe/H], even with single SN events; thus it fits into the model of discrete chemical enrichment used by Searle & McWilliam to explain the observed heavy element dispersion. This model implies that stars with [Fe/H] $\leq -3.3$ may be the products of individual SN events.

The mechanisms proposed by Ryan et al (1996) and Searle & McWilliam are roughly consistent with the predictions of Audouze & Silk (1995), who considered the physics of mixing of SN ejecta and concluded that there must be a minimum possible metallicity, near [Fe/H] $= -4$.

An unusual kind of variation in iron-peak elements was found for the star CS22949-037 by McWilliam et al (1995b): The elements outside the range from Ti to Ni (i.e. C, Na, Mg, Al, Si, Ca, Sc, Sr, Ba) appear to be overabundant by typically $\sim$1 dex. The elements within the range Ti to Ni exhibit normal relative ratios. McWilliam et al (1995b) suggest that a simple explanation is that this star is actually deficient in the iron-peak elements. This observation suggests that SN exist that produce relatively small iron-peak yields. Ryan et al (1996) have found two more stars, CS 22876-032 and CS 22897-008, that show unusual chemical compositions, which are indicative of star-to-star scatter and a dispersion in element yields. These star-to-star variations indicate that at low metallicity, the intrinsic dispersion in SN yields can produce anomalous stellar compositions because the averaging process of combining yields from many SN is not yet complete.

## FINAL THOUGHTS

One of the themes of this article has been the idea that chemical composition is a function of Galactic environment. In order to learn how environmental parameters can affect chemical evolution, we must make accurate measurements of elemental abundances in all the accessible locations. The different components of our Galaxy are excellent places to make such detailed studies and will ultimately provide us with the means to interpret lower resolution, lower S/N spectra of external galaxies. Chemical analysis of stars in local group galaxies will also be of great use in this regard; in particular, the red giant branch stars permit us to sample the whole history of a stellar system.

The new large telescopes and efficient spectrographs will be of immense help for measuring the composition of the red giants at the distance of the local group galaxies. However, telescope aperture and spectrograph efficiency alone will not be enough to meet this task; we must also make routine the ability to measure chemical abundances from noisy spectra. For elements with many lines, one can derive abundances from noisy spectra by combining several line regions to produce an average line profile of high S/N. When abundances for elements with few lines are required, an average abundance can be measured by combining noisy spectra from many stars, which are acquired with a fiber spectrograph. Such methods will be commonplace in the future but have already been demonstrated in the works of Jones et al (1995), Carney et al (1987), and Pilachowski & Armandroff (1996).

I cannot stress enough the importance of accurate abundance measurement and reliable estimates of measurement uncertainty: If you don't have accurate measurements, or if you don't know how accurate your measurements are, you cannot draw reliable conclusions. In this regard, large surveys, as exemplified by Edvardsson et al (1993), are particularly useful; for large samples of homogeneous, high quality data with identical analysis, the zero-point uncertainties are reduced.

In the near future, the study of the composition of local group dwarf spheroidal galaxies (dSphs) will be particularly fruitful. These low-mass systems have experienced relatively low rates of chemical evolution, frequently with mean metallicities near [Fe/H] $= -2$; as such, these systems may contain a large fraction of stars from the first stellar generations. The low average metallicity of the dSphs and the large numbers of stars in a small area of sky offer the opportunity to make efficient searches for extremely low metallicity stars; we could learn whether a lower limit to metallicity really does exist, as predicted by Audouze & Silk (1995), and how the IMF is affected by metallicity. With more extremely low metallicity stars, we could accumulate additional evidence for composition dispersion at low metallicity and perhaps measure SN element yields, and we also might find more stars with super-enhanced r-process abundances, which are useful for measuring the age of the Galaxy. Furthermore, the observed populations in dSphs, which are indicative of star formation bursts (e.g. Smecker-Hane et al 1994), can be used to measure an approximate time scale for SN Ia.

In the bulge, the oxygen and carbon abundances are desperately needed and might best be measured with infrared spectra of OH and CO lines. Confirmation is required for the low abundances of Si and Ca relative to Mg and Ti in the bulge. The heavy element abundance pattern in the bulge offers a probe of the importance of SN II; thus, measurements of Eu and Ba abundance as a function of metallicity would be useful. The trend of carbon abundance with metalicity must be resolved for the halo. For the extant stars with metallicity in the range $-4 \leq$ [Fe/H] $\leq -2$, improved limits on the dispersion of Co, Cr, and Mn abundances with metallicity would test the primordial enrichment model suggested by Searle & McWilliam. Also in this metallicity range, heavy element abundances for a large sample of stars would test the idea that discrete enrichment events, by individual SN, are responsible for the observed dispersion. Abundances of C, N, O, Na, Mg, and Al for main-sequence stars in globular clusters would provide important evidence for or against the role of primordial abundance dispersion in globular clusters. In the Galactic disk, improved measurements of stellar metallicity towards the Galactic center and anti-center would be useful for understanding the radial metallicity gradient, both present and past. An extensive study of the composition of stars in star-forming regions, similar to the work of Cunha & Lambert (1994), will provide direct information on chemical enrichment and SN yields. It would also be nice to see the experts agree on whether there is a mean age-metallicity relation (AMR) in the solar neighborhood, what fraction of the halo heavy elements were made in the s-process, and the value of the solar iron abundance.

## ACKNOWLEDGMENTS

This work would not have been possible without the support of the Observatories of the Carnegie Institution of Washington. I would like to thank the many colleagues who have taught me a few things over the years and helped me on my way; they are Olin Eggen, David Lambert, George Preston, Mike Rich, Allan Sandage, and Leonard Searle. I'd also like to express my deep appreciation to my wife, Kim, who helped with figure preparations, kept me happy, and encouraged me to finish this manuscript.



*Literature Cited*

Abia C, Rebolo R. 1989. *Ap. J.* 347:186–94
Aller LH, Greenstein JL. 1960. *Ap. J. Suppl.* 5:139–86
Anders E, Grevesse N. 1989. *Geochim. Cosmochim. Acta* 53:197–214
Anders E, Zinner E. 1993. *Meteortics* 28:490–514
Andersson H, Edvardsson B. 1994. *Astron. Astrophys.* 290:590–98
Anstee SD, O'Mara JO, Ross JE. 1997. *MNRAS* 284:202–12
Armosky BJ, Sneden C, Langer GE, Kraft RP. 1994. *Astron. J.* 108:1364–74
Arnett WD. 1971. *Ap. J.* 166:153–73
Arnett WD. 1978. *Ap. J.* 219:1008–16
Arnett WD. 1991. In *Frontiers of Stellar Evolution,* ed. D Lambert, APS Conf. Ser. 20:389–401
Arp H. 1962. In *Problems in Extragalactic Research,* ed. GC McVittie, IAU Symp. 15:42–47. New York: MacMillan
Arpigny C, Magain P. 1983. *Astron. Astrophys.* 127:L7–L9
Audouze J, Silk J. 1995. *Ap. J.* 451:L49–L52
Audouze J, Tinsley BM. 1976. *Annu. Rev. Astron. Astrophys.* 14:43–79
Balachandran SC. 1996. In *Cosmic Abundances,* ASP Conf. Ser. 99:188–95
Balachandran SC, Carney BW. 1996. *Astron. J.* 111:946–61
Baraffe I, Takahashi K. 1993. *Astron. Astrophys.* 280:476–85
Barbuy B. 1988. *Astron. Astrophys.* 191:121–27
Barbuy B, Spite F, Spite M. 1985. *Astron. Astrophys.* 144:343–54
Bard A, Kock A, Kock M. 1991. *Astron. Astrophys.* 248:315–22
Bazan G, Mathews GJ. 1990 *Ap. J.* 354:644–48
Bazan G, Sneden C, Yoss K. 1996. In *Formation of the Galactic Halo... Inside and Out,* ed. H Morrison, A Sarajedini. ASP Conf. Ser. 92:351–54
Bell RA, Ericksson K, Gustafsson B, Nordlund A. 1976. *Astron. Astrophys. Suppl.* 23:37–95
Beers TC, Preston GW, Shectman SA. 1985. *Astron. J.* 90:2089–102
Beers TC, Preston GW, Shectman SA. 1992.

*Astron. J.* 103:1987–2034
Beers TC, Sommer-Larsen J. 1995. *Ap. J. Suppl.* 96:175–221
Bessell MS, Norris J. 1984. *Ap. J.* 285:622–36
Bessell MS, Sutherland RS, Ruan K. 1991. *Ap. J.* 383:L71–L74
Biémont E, Bardoux M, Kurucz RL, Ansbacher W, Pinnington EH. 1991. *Astron. Astrophys.* 249:539–44
Blaauw A. 1991. In *The Physics of Star Formation and Early Stellar Evolution,* ed. CJ Lada, ND Kylafis, pp. 125–54. NATO ASI Series. Dordrecht: Kluwer
Blackwell DE, Booth AJ, Haddock DJ, Petford AD, Leggett SK. 1986. *MNRAS* 220:549–53
Blackwell DE, Booth AJ, Petford AD. 1984. *Astron. Astrophys.* 132:236–39
Blackwell DE, Lynas-Gray AE, Smith G. 1995. *Astron. Astrophys.* 296:217–32
Bond HE. 1981. *Ap. J.* 248:606–11
Boothroyd AI, Sackmann IJ, Wasserburg GJ. 1994. *Ap. J.* 430:L77–L80
Bowen GH, Willson LA. 1991. *Ap. J.* 375:L53–L56
Branch D, Bonnell J, Tomkin J. 1978. *Ap. J.* 225:902–7
Briley MM, Hesser JE, Bell RA. 1991. *Ap. J.* 373:482–96
Briley MM, Smith VV, Suntzeff NB, Lambert DL, Bell RA, Hesser JE. 1996. *Nature* 383:604–6
Brown JA, Sneden C, Lambert DL, Dutchover E Jr. 1989. *Ap. J. Suppl.* 71:293–322
Brown JA, Wallerstein G. 1992. *Astron. J.* 104:1818–30
Brown JA, Wallerstein G, Zucker D. 1996. In *Formation of the Galactic Halo… Inside and Out,* ed. H Morrison, A Sarajedini, ASP Conf. Ser. 92:355–58
Burbidge EM, Burbidge GR, Fowler WA, Hoyle F. 1957. *Rev. Mod. Phys.* 29:547–650
Busso M, Lambert DL, Beglio L, Gallino R, Raiteri CM, Smith VV. 1995. *Ap. J.* 446:775–92
Butcher HR. 1975. *Ap. J.* 199:710–17
Butcher HR. 1987. *Nature* 328:127–31
Carbon DF, Barbuy B, Kraft RP, Friel ED, Suntzeff NB. 1987. *Publ. Astron. Soc. Pac.* 99:335–68
Carney BW, Latham DW, Laird JB. 1990. In *ESO/CTIO Workshop on Bulges of Galaxies,* ed. BJ Jarvis, DM Terndrup, 35:127–34. Garching: Eur. South. Obs.
Carney BW, Laird JB, Latham DW, Kurucz RL. 1987. *Astron. J.* 94:1066–76
Castro S, Rich RM, Grenon M, McCarthy JK. 1997. In press
Castro S, Rich RM, McWilliam A, Ho LC, Spinrad H, et al. 1996. *Astron. J.* 111:2439–52
Cayrel R. 1987. In *ESO Workshop on Stellar Evolution and Dynamics in the Outer Halo of the Galaxy,* pp. 627–35. Garching: Eur. South. Obs.
Clayton DD. 1985. *Ap. J.* 288:569–74
Clayton DD. 1988. *MNRAS* 234:1–36
Clegg RES, Lambert DL, Tomkin J. 1981. *Ap. J.* 250:262–75
Conti PS, Greenstein JL, Spinrad H, Wallerstein G, Vardya MS. 1967. *Ap. J.* 148:105–27
Corliss CH, Bozman WR. 1962. *NBS Monog.* No. 53
Cowan JJ, Burris DL, Sneden C, McWilliam A, Preston GW. 1995. *Ap. J.* 439:L51–L54
Cowan JJ, McWilliam A, Sneden C, Burris DL. 1997. *Ap. J.* In press
Cowan JJ, Sneden C, Truran JW, Burris DL. 1996. *Ap. J.* 460:L115–18
Cunha K, Lambert DL. 1992. *Ap. J.* 399:586–98
Cunha K, Lambert DL. 1994. *Ap. J.* 426:170–91
Denisenkov PA, Denisenkova SN. 1990. *Sov. Astron. Lett.* 16:642–51
Edmunds MG. 1975. *Astrophys. Space Sci.* 32:483–91
Edmunds MG, Greenhow RM, Johnson D, Klückers V, Vila BM. 1991. *MNRAS* 251: 33P–36P
Edvardsson B, Andersen J, Gustafsson B, Lambert DL, Nissen PE, Tomkin J. 1993. *Astron. Astrophys.* 275:101–52
Eggen OJ, Sandage A. 1969. *Ap. J.* 158:669–84
Feltzing S. 1995. PhD thesis. Univ. Uppsala, Sweden
François P. 1986. *Astron. Astrophys.* 160:264–76
François P. 1987. *Astron. Astrophys.* 176:294–98
François P. 1988. *Astron. Astrophys.* 195:226–29
François P. 1991. *Astron. Astrophys.* 247:56–63
François P. 1996. *Astron. Astrophys.* 313:229–33
François P, Matteucci F. 1993. *Astron. Astrophys.* 280:136–40
Friel ED. 1995. *Annu. Rev. Astron. Astrophys.* 33:381–414
Friel ED, Boesgard AM. 1992 *Ap. J.* 387:170–80
Friel ED, Janes KA. 1993. *Astron. Astrophys.* 267:75–91
Frogel JA, Whitford AE. 1987. *Ap. J.* 320:199–237
Fuhrmann K, Axer M, Gehren T. 1995. *Astron. Astrophys.* 301:492–500
Geisler D. 1987. *Astron. J.* 94:84–91
Geisler D, Claria JJ, Miniti D. 1992. *Astron. J.* 104:1892–905
Geisler D, Friel E. 1992. *Astron. J.* 104:128–43
Gilmore G. 1984. *MNRAS* 207:223–40
Gilmore G, Reid N. 1983. *MNRAS* 202:1025–47
Gilmore G, Wyse RFG. 1985 *Astron. J.* 90:2015–26

Gilmore G, Wyse RFG. 1986. *Nature* 322:806–7
Gilroy KK, Sneden C, Pilachowski CA, Cowan JJ. 1988. *Ap. J.* 327:298–320
Gorgas J, Faber SM, Burstein D, Gonzalez J, Courteau S, Prosser C. 1993. *Ap. J. Suppl.* 86:153–98
Gratton RG. 1989. *Astron. Astrophys.* 208:171–78
Gratton RG. 1994. In *Nuclei in the Cosmos,* 3:3–17. ed. M Busso, R Gallino, CM Raiteri. New York: Am. Inst. Phys.
Gratton RG, Sneden C. 1987. *Astron. Astrophys.* 178:179–93
Gratton RG, Sneden C. 1988. *Astron. Astrophys.* 204:193–218
Gratton RG, Sneden C. 1990. *Astron. Astrophys.* 234:366–86
Gratton RG, Sneden C. 1991. *Astron. Astrophys.* 241:501–25
Gratton RG, Sneden C. 1994. *Astron. Astrophys.* 287:927–46
Grenon M. 1987. *J. Astrophys. Astron.* 8:123–39
Grenon M. 1989. *Astrophys. Space Sci.* 156:29–37
Grevesse N, Noels A. 1993. *Phys. Scripta* T47:133–38
Griffin R, Griffin R, Gustafsson B, Vieira T. 1982. *MNRAS* 198:637–58
Gustafsson B, Kjaergaard P, Andersen S. 1974. *Astron. Astrophys.* 34:99–128
Hartquist TW, Cameron AGW. 1977. *Astrophys. Space Sci.* 48:145–58
Hartwick FDA. 1976. *Ap. J.* 209:418–23
Helfer HL, Wallerstein G. 1968. *Ap. J. Suppl.* 16:1–47
Holweger H, Muller EA. 1974. *Solar phys.* 39:19–30
Holweger H, Bard A, Kock A, Kock M. 1991. *Astron. Astrophys.* 249:545–49
Holweger H, Heise C, Kock M. 1990. *Astron. Astrophys.* 232:510–15
Holweger H, Kock M, Bard A. 1995. *Astron. Astrophys.* 296:233–40
Hoyle F. 1954. *Ap. J. Suppl.* 1:121–46
Huggins PJ, Williams PM. 1974. *MNRAS* 169:1P–5P
Ibata RA, Gilmore GF, Irwin MJ. 1994. *Nature* 370:194–96
Iben I Jr. 1975. *Ap. J.* 196:525–47
Iben I Jr, Truran JW. 1978. *Ap. J.* 220:980–95
Iben I Jr, Tutukov A. 1984. *Ap. J. Suppl.* 54:335–72
Idiart TP, de Freitas Pacheo JA, Costa RDD. 1996a. *Astron. J.* 111:1169–74
Idiart TP, de Freitas Pacheo JA, Costa RDD. 1996b. *Astron. J.* 112:2541–48
Jønch-Sørensen H. 1995. *Astron. Astrophys.* 298:799–817
Jones JB, Wyse RFG, Gilmore G. 1995. *Publ. Astron. Soc. Pac.* 107:632–47
Kahn FD. 1974. *Astron. Astrophys.* 37:149–62
Käppeler F, Beer H, Wisshak K. 1989. *Rep. Prog. Phys.* 52:945–1013
King JR. 1993. *Astron. J.* 106:1206–21
Kipper T, Jorgensen UG, Klochkova VG, Panchuk VE. 1996. *Astron. Astrophys.* 306:489–500
Knude J. 1990. *Astron. Astrophys.* 230:16–20
Kostik RI, Shchukina NG, Rutten RJ. 1996. *Astron. Astrophys.* 305:325–42
Kraft RP. 1994. *Publ. Astron. Soc. Pac.* 106:553–65
Kraft RP, Sneden C, Langer GE, Prosser CF. 1992. *Astron. J.* 104:645–68
Kraft RP, Suntzeff NB, Langer GE, Carbon DF, Trefzger CF, et al. 1982. *Publ. Astron. Soc. Pac.* 94:55–66
Kraft RP, et al. 1997. *Astron. J.* 113:279–95
Laird J. 1985. *Ap. J.* 289:556–69
Laird J, Rupen MP, Carney B, Latham D. 1988. *Astron. J.* 96:1908–17
Lambert DL. 1987. *J. Astrophys. Astron.* 8:103–22
Lambert DL, Heath JE, Lemke M, Drake J. 1995a. *Ap. J. Suppl.* 103:183–210
Lambert DL, Ries LM. 1981. *Ap. J.* 248:228–48
Lambert DL, Smith VV, Busso M, Gallino R, Straniero O. 1995b. *Ap. J.* 450:302–17
Larson RB. 1974. *MNRAS* 166:585–616
Larson RB. 1976. *MNRAS* 176:31–52
Lu L, Sargent WLW, Barlow TA, Churchill CW, Vogt SS. 1996. *Ap. J. Suppl.* 107:475–519
Luck RE, Bond HE. 1981. *Ap. J.* 244:919–37
Luck RE, Bond HE. 1985. *Ap. J.* 292:559–77
Luck RE, Challener SL. 1995. *Astron. J.* 110:2968–3009
Lynden-Bell D. 1975. *Vistas Astron.* 19:299–316
Maeder A. 1991. *Q. J. R. Astron. Soc.* 32:217–31
Magain P. 1987. *Astron. Astrophys.* 179:176–80
Magain P. 1989. *Astron. Astrophys.* 209:211–25
Magain P. 1995. *Astron. Astrophys.* 297:686–94
Majewski SR. 1993. *Annu. Rev. Astron. Astrophys.* 31:575–638
Mathews GJ, Bazan G, Cowan JJ. 1992. *Ap. J.* 391:719–35
Mathews GJ, Cowan JJ. 1990. *Nature* 345:491–94
Matteucci F, Brocato E. 1990. *Ap. J.* 365:539–43
Matteucci F, François P. 1989. *MNRAS* 239:885–904
Matteucci F, François P. 1992. *Astron. Astrophys.* 262:L1–L4
Matteucci F, Raiteri CM, Busso M, Gallino R, Gratton R. 1993. *Astron. Astrophys.* 272:421–29
McClure RD. 1984. *Publ. Astron. Soc. Pac.* 96:117–27

McWilliam A. 1990. *Ap. J. Suppl.* 1075–128
McWilliam A, Geisler D, Rich RM. 1992. *Publ. Astron. Soc. Pac.* 104:1193–204
McWilliam A, Preston GW, Sneden C, Searle L. 1995a. *Astron. J.* 109:2757–99
McWilliam A, Preston GW, Sneden C, Searle L, Shectman S. 1996. In *Formation of the Galactic Halo . . . Inside and Out,* ed. H Morrison, A Sarajedini, ASP Conf. Ser. 92:317–26
McWilliam A, Preston GW, Sneden C, Shectman S. 1995b. *Astron. J.* 109:2736–56
McWilliam A, Rich RM. 1994. *Ap. J. Suppl.* 91:749–91
Meusinger H, Reimann H-G, Stecklum B. 1991. *Astron. Astrophys.* 245:57–74
Meyer BS. 1994. *Annu. Rev. Astron. Astrophys.* 32:153–90
Meyer BS, Mathews GJ, Howard WM, Woosley SE, Hoffman RD. 1992. *Ap. J.* 399:656–64
Milford PN, O'Mara BJ, Ross JE. 1994. *Astron. Astrophys.* 292:276–80
Montgomery KA, Janes KA, Phelps RL. 1994. *Astron. J.* 108:585–93
Nissen PE, Edvardsson B. 1992. *Astron. Astrophys.* 261:255–62
Nissen PE, Gustafsson B, Edvardsson B, Gilmore G. 1994. *Astron. Astrophys.* 285: 440–50
Nissen PE, Schuster WJ. 1991. *Astron. Astrophys.* 251:457–68
Nomoto K, Thielemann FK, Yokio Y. 1984. *Ap. J.* 286:644–58
Norris J, Ryan S. 1989. *Ap. J.* 340:739–61
Norris J, Ryan S. 1991. *Ap. J.* 380:403–18
Norris JE, Da Costa GS. 1995. *Ap. J.* 447:680–705
Norris JE, Freeman KC, Mighell KJ. 1996. *Ap. J.* 462:241–254
Norris JE, Peterson RC, Beers TC. 1993. *Ap. J.* 415:797–810
Ostriker JP, Thuan TX. 1975. *Ap. J.* 202:353–64
Pagel BEJ. 1968. In *Origin and Distribution of the Elements,* ed. LH Ahrens, pp. 195–204. Oxford: Pergamon
Pagel BEJ. 1989. In *Evolutionary Phenomena in Galaxies,* ed. JE Beckman, BEJ Pagel, pp. 201–23. Cambridge: Cambridge Univ. Press
Pagel BEJ, Patchett BE. 1975. *MNRAS* 172:13–40
Pagel BEJ, Tautvaisiene G. 1995. *MNRAS* 276:505–14
Pauls U, Grevesse N, Huber MCE. 1990. *Astron. Astrophys.* 231:536–42
Peterson RC, Green EM. 1995. *BAAS* 187: #107.06
Peterson RC, Kurucz RL, Carney BW. 1990. *Ap. J.* 350:173–85
Peterson RC, Sneden C. 1978. *Ap. J.* 225:913–18
Pettini M, Smith LJ, Hunstead RW, King DL. 1994. *Ap. J.* 426:79–96
Pilachowski CA, Armandroff TE. 1996. *Astron. J.* 111:1175–83
Pilachowski CA, Sneden C, Kraft RP. 1996. *Astron. J.* 111:1689–704
Pilachowski CA, Sneden C, Wallerstein G. 1983. *Ap. J. Suppl.* 52:241–87
Pilyugin LS, Edmunds MG. 1996a. *Astron. Astrophys.* 313:783–91
Pilyugin LS, Edmunds MG. 1996b. *Astron. Astrophys.* 313:792–802
Raiteri CM, Busso M, Picchio G, Gallino R, Pulone L. 1991. *Ap. J.* 367:228–38
Raiteri CM, Gallino R, Busso M. 1992. *Astron. Astrophys.* 387:263–75
Raiteri CM, Gallino R, Busso M, Neuberger D, Käeppeler F. 1993. *Ap. J.* 419:207–23
Raiteri CM, Villata M, Navarro JF. 1996. *Astron. Astrophys.* 315:105–15
Reeves H. 1972. *Astron. Astrophys.* 19:215–23
Renzini A, Voli M. 1981. *Astron. Astrophys.* 94:175–93
Rich RM. 1988. *Astron. J.* 95:828–65
Rich RM. 1990. *Ap. J.* 362:604–19
Ross JE, Aller LH. 1976. *Science* 191:1223–29
Ryan SG, Norris JE. 1991. *Astron. J.* 101:1865–78
Ryan SG, Norris JE, Beers TC. 1996. *Ap. J.* 471:254–78
Sadler EM, Rich RM, Terndrup DM. 1996. *Astron. J.* 112:171–85
Sandage A. 1969. *Ap. J.* 158:1115–36
Sandage A. 1987. *Astron. J.* 93:610–15
Sandage A, Fouts G. 1987. *Astron. J.* 93:74–115
Scalo JM. 1987. In *Starbursts and Galaxy Evolution, Proc. Moriond Astrophys. Meet.,* ed. TX Thuan, T Montmerle, J Tran Thanh Van. 22:445–465
Schmidt M. 1963. *Ap. J.* 137:758–69
Searle L, Sargent WLW. 1972. *Ap. J.* 173:25–33
Searle L, Zinn R. 1978. *Ap. J.* 225:357–79
Seeger PA, Fowler WA, Clayton DD. 1965. *Ap. J. Suppl.* 11:121–66
Shetrone MD. 1996a. *Astron. J.* 112:1517–35
Shetrone MD. 1996b. *Astron. J.* 112:2639–49
Smecker-Hane T, Stetson PB, Hesser JE, Lehnert MD. 1994. *Astron. J.* 108:507–13
Smecker-Hane T, Wyse RFG. 1992. *Astron. J.* 103:1621–26
Smith VV, Cunha K, Lambert DL. 1995. *Astron. J.* 110:2827–43
Smith VV, Lambert DL. 1990. *Ap. J. Suppl.* 72:387–416
Sneden C, Crocker DA. 1988. *Ap. J.* 335:406–14
Sneden C, Gratton RG, Crocker DA. 1991a. *Astron. Astrophys.* 246:354–67
Sneden C, Kraft RP, Langer GE. 1991b. *Astron. J.* 102:2001–21
Sneden C, McWilliam A, Preston GW, Cowan JJ, Burris DL, Armosky BJ. 1996. *Ap. J.* 467:819–40

Sneden C, Parthasarathy M. 1983. *Ap. J.* 267:757–78
Sneden C, Pilachowski CA. 1985. *Ap. J.* 288:L55–L58
Sneden C, Preston GW, McWilliam A, Searle L. 1994. *Ap. J.* 431:L27–L30
Sommer-Larsen J. 1991. *MNRAS* 250:356–62
Spiesman WJ. 1992. *Ap. J.* 397:L103–5
Spinrad H, Taylor BJ. 1969. *Ap. J.* 157:1279–340
Spite M. 1992. In *The Stellar Populations of Galaxies,* ed. B Barbuy, A Renzini. IAU Symp. 149:123–32
Spite M, Spite F. 1978. *Astron. Astrophys.* 67:23–31
Spite M, Spite F. 1979. *Astron. Astrophys.* 76:150–57
Spite M, Spite F. 1980. *Astron. Astrophys.* 89:118–22
Spite M, Spite F. 1991. *Astron. Astrophys.* 252:689–92
Suntzeff NB. 1989. In *The Abundance Spread Within Globular Clusters,* ed. M Cayrel de Strobel, M Spite, T Lloyd Evans, pp. 71–81. Paris: Obs. Paris
Taylor BJ. 1996. *Ap. J. Suppl.* 102:105–28
Terndrup DM, Frogel JA, Whitford AE. 1991. *Ap. J.* 378:742–55
Terndrup DM, Sadler EM, Rich RM. 1995. *Astron. J.* 110:1774–92
Thielemann FK, Nomoto K, Yokio Y. 1986. *Astron. Astrophys.* 158:17–33
Timmes FX, Woosley SE, Weaver TA. 1995. *Ap. J. Suppl.* 98:617–58
Tinsley BM. 1979. *Ap. J.* 229:1046–56
Tomkin J, Lambert DL, Balachandran S. 1985. *Ap. J.* 290:289–95
Tomkin J, Lemke M, Lambert DL, Sneden C. 1992. *Astron. J.* 104:1568–84
Tomkin J, Sneden C, Lambert DL. 1986. *Ap. J.* 302:415–20
Tomkin J, Woolf VM, Lambert DL. 1995. *Astron. J.* 109:2204–17
Tosi M. 1988. *Astron. Astrophys.* 197:47–51
Tripicco MJ, Bell RA, Dorman B, Hufnagel B. 1995. *Astron. J.* 109:1697–705
Truran JW. 1981. *Astron. Astrophys.* 97:391–93
Truran JW, Cameron AGW. 1971. *Astrophys. Space. Sci.* 14:179–222
Truran JW, Iben I Jr. 1977. *Ap. J.* 216:797–810
Twarog BA. 1980. *Ap. J.* 242:242–59
Uehara H, Susa H, Nishi R, Yamada M, Nakamura T. 1996. *Ap. J. Lett.* 473:95–98
van den Bergh S. 1962. *Astron. J.* 67:486–90
Vanture A. 1992. *Astron. J.* 104:1997–2004
Vanture A. 1993. *Publ. Astron. Soc. Pac.* 105:445
Vanture AD, Wallerstein G, Brown JA. 1994. *Astron. J.* 106:835–42
Wasserburg GJ, Boothroyd AI, Sackmann IJ. 1995. ApJ 447:L37–L40
Wallerstein G, Carlson M. 1960. *Ap. J.* 132:276–77
Wallerstein G. 1962. *Ap. J. Suppl.* 6:407–43
Wallerstein G, Greenstein JL, Parker R, Helfer HL, Aller LH. 1963. *Ap. J.* 137:280–300
Wheeler JC, Sneden C, Truran JW. 1989. *Annu. Rev. Astron. Astrophys.* 27:279–349
White SDM, Audouze J. 1983. *MNRAS* 203:603–18
Whitford AE, Rich RM. 1983. *Ap. J.* 274:723–32
Wolfire MG, Cassinelli JP. 1987. *Ap. J.* 319:850–67
Woolf VM, Tomkin J, Lambert DL. 1995. *Ap. J.* 453:660–72
Woosley SE, Weaver TA. 1995. *Ap. J. Suppl.* 101:181–235
Worthey G, Dorman B, Jones LA. 1996. *Astron. J.* 112:948–53
Worthey G, Faber SM, Gonzalez JJ. 1992. *Ap. J.* 398:69–73
Wyse RFG, Gilmore G. 1986. *Astron. J.* 91:855–69
Wyse RFG, Gilmore G. 1991. *Ap. J.* 367:L55–L58
Wyse RFG, Gilmore G. 1992. *Astron. J.* 104:144–53
Wyse RFG, Gilmore G. 1995. *Astron. J.* 110:2771–87
Yoshii Y, Saio H. 1986. *Ap. J.* 301:587–600
Zhao G, Magain P. 1990. *Astron. Astrophys.* 238:242–48
Zhao G, Magain P. 1991. *Astron. Astrophys.* 244:425–32
Zinner E. 1996. In *Cosmic Abundances. ASP Conf. Ser.* 99:147–61

*Annu. Rev. Astron. Astrophys. 1997. 35:557–605*


# MIXING IN STARS

*M. Pinsonneault*
Astronomy Department, The Ohio State University, 174 W. 18th Ave., Columbus, Ohio 43210; e-mail: pinsono@tinsley.mps.ohio-state.edu

KEY WORDS: stellar evolution, stellar abundances, stellar rotation

ABSTRACT

Three cases for mixing not present in standard stellar models are presented: Light element depletion in low mass main-sequence stars, deep mixing in massive stars, and deep mixing in low mass giants. The review begins with the mixing indicators and the predictions of standard models. The observational evidence for anomalous mixing is then presented, followed by the physics of mixing outside the standard model. The status of theoretical models that include extra mixing is then examined.

## 1. INTRODUCTION

Stellar evolution is driven by changes in chemical composition caused by nuclear reactions. The extent to which the interiors of stars are mixed strongly influences their evolution and such properties as age of clusters, chemical evolution, and the relationship between the initial and surface abundances of stars. In standard stellar models, convection is the sole mixing agent, and these models make strong predictions about the surface abundances of stars as a function of mass, composition, and age.

To first order, the agreement between observation and standard stellar models is encouraging. For some time, however, there have been indications that the standard stellar model is incomplete. The long-standing problem of explaining the surface lithium abundances in low mass stars is one good example, and the evidence for $C \rightarrow N$ processing in the envelopes of red giants is another. In both cases, the data appear to require not just more mixing than expected in standard models, but mixing in mass ranges and phases of evolution where it is not expected. In the massive star regime, there is evidence for extensive mixing that could dramatically influence the evolution of at least the most rapidly rotating stars. Processes not included in standard models, such as mixing



0066-4146/97/0915-0557$08.00

driven by rotation, are required to understand the observed depletion pattern. By extension, some important properties inferred from studies of stars could potentially be affected, including globular cluster ages, the inferred primordial abundances of light elements, and chemical evolution. The purpose of this review is twofold: first, to assess the observational evidence for mixing not present in standard stellar models and, second, to examine the current state of the theoretical explanations for the data and the potential importance of effects not included in standard stellar models.

## 2. DIAGNOSTICS OF MIXING

The surface abundances of stars provide indicators of mixing to a wide range of temperatures. Figure 1 shows the normalized abundances of some of the

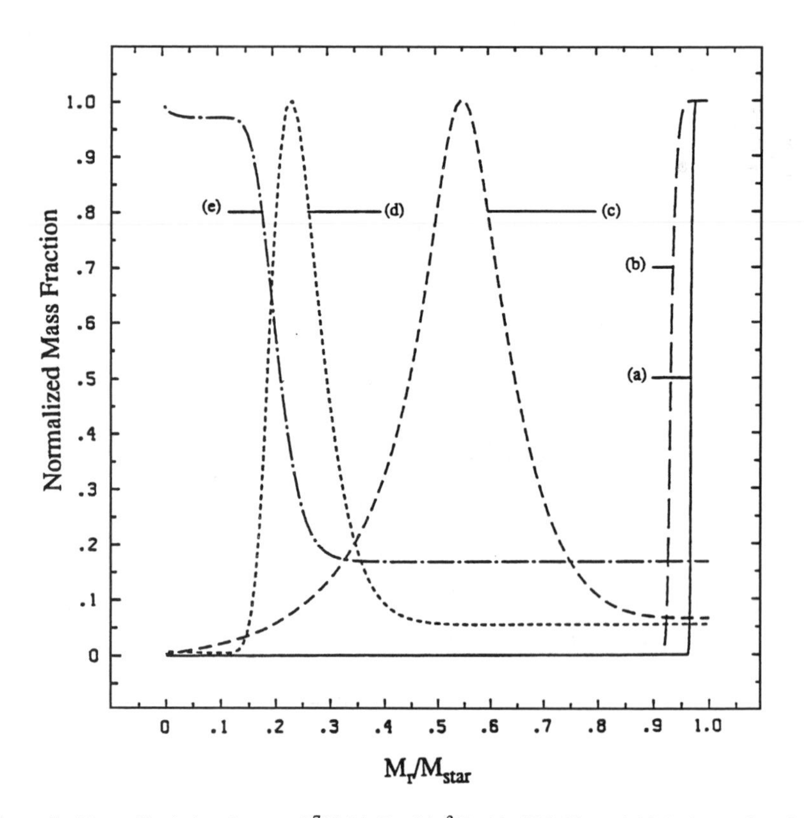


*Figure 1* Normalized abundances of $^7$Li (a), Be (b), $^3$He (c), 13C (d), and 14N (e) as a function of mass fraction within the Sun (from Pinsonneault 1988).

species of interest in the Sun. Some species, such as $^6$Li and $^7$Li, are destroyed at relatively low temperatures; others, like the CNO elements, are transformed at much higher temperatures.

## 2.1 *Outer Layers: Li, Be, B*

LITHIUM-7 $^7$Li is produced in the Big Bang, and the primordial $^7$Li abundance is of considerable interest for testing Big Bang nucleosynthesis (BBN) and $\Omega_{baryon}$ (Boesgaard & Steigman 1985, Alcock et al 1987, Malaney & Fowler 1988, Audouze & Silk 1989, Krauss & Romanelli 1990, Mathews et al 1990, Reeves et al 1990, Walker et al 1991, Wilson & Matteucci 1992, Copi et al 1995, Dearborn et al 1996). It is fragile in stellar interiors, surviving only to $\sim 2.5 \times 10^6$ K at typical envelope main-sequence (MS) densities. For stellar structure purposes, the extent of lithium depletion is the quantity of theoretical interest; measuring the amount of lithium depletion requires knowledge of both the initial and current abundance.

In metal-poor stars, the discovery of measurable $^7$Li (Spite & Spite 1982) at a level 10 times lower than the solar meteoritic value of $3.31 \pm 0.04$ (Anders & Grevesse 1989) (on the logarithmic scale where $H = 12$) raised the exciting possibility of a direct measurement of the primordial $^7$Li abundance. It remains controversial whether the observed abundances in metal-poor stars reflects their initial abundance or if they have been nearly uniformly depleted from a higher initial value. Little depletion is expected from standard stellar models (Deliyannis et al 1990), whereas models that include extra mixing can predict depletion factors as large as a factor of 10 (Pinsonneault et al 1992). The uncertainty in the initial abundance of halo stars makes it difficult to distinguish unambiguously between different stellar structure models. Disk stars provide a much cleaner test of theoretical models because the lithium depletion pattern can be mapped out as a function of age.

Abundances in the interstellar medium (ISM) are consistent with the solar meteoritic value (Lemoine et al 1993). Although higher abundances were initially reported in T Tauri stars, non–local thermodynamic equilibrium (non–LTE) effects and complications due to the presence of accretion disks appear to be responsible for anomalously high abundances (Magazzu et al 1992). The open cluster system also shows evidence for a nearly uniform initial abundance, on the order of 3.2–3.3 (Soderblom 1995).

LITHIUM-6, BERYLLIUM, AND BORON $^6$Li, Be, and B are produced primarily through cosmic-ray nucleosynthesis, with minimal primordial abundances expected from standard BBN (Reeves et al 1970, Vangioni-Flam et al 1990, Walker et al 1993), although a higher initial abundance may be possible for inhomogeneous BBN (Kajino & Boyd 1990). They differ dramatically in their sensitivity to destruction in stellar interiors: $^6$Li is destroyed at even lower temperatures than $^7$Li, whereas Be and B survive to much higher temperatures (~3.5 and 5 million K respectively). $^6$Li may have been detected in some hot halo stars, and if it is present, it can be used to place constraints on the degree of mixing in these stars (Section 4.1, Halo Stars).

There has been a fair amount of work on Be and B in metal-poor stars, with the goal of testing chemical evolution models (Duncan et al 1992, Ryan et al 1992, Boesgaard & King 1993, Thorburn & Hobbs 1996). The spectral features used to measure beryllium and boron abundances are in ultraviolet spectral regions that are difficult to model, and much of the recent observational work has focused on identifying the spectral features in the vicinity of the beryllium and boron features.

### 2.2 *Deep Mixing: $^3$He, CNO, and $^4$He*

HELIUM-3 AND HELIUM-4 $^3$He and $^4$He are important for BBN but difficult to measure directly in stars. The observational situation for measurements of $^3$He in HII regions has been reviewed by Wilson & Rood (1994); see also Balser et al (1994). $^4$He measurements are possible in hot stars.

CNO The solar surface CNO isotope ratios and the relative surface C, N, and O abundances differ dramatically from the nuclear equilibrium values. ISM measurements vary somewhat (e.g. Wilson & Mateucci 1992, Wilson & Rood 1994) but are also uniformly different from nuclear equilibrium. The mixing of CNO-processed material into the surface convection zone can therefore be measured by examining these ratios. C→N processing is completed at lower temperatures than O→N processing, but observational evidence exists for changes in the surface oxygen abundances of stars (Kraft 1994). CNO burning tends to occur at temperatures characteristic of the nuclear burning core rather than of the envelope. Observational evidence for surface CNO anomalies is seen in giants and massive stars; this indicates deep mixing.

## 3. PROPERTIES OF STANDARD MODELS

### 3.1 *Pre–Main-Sequence Light Element Depletion*

In a pioneering work, Iben (1965) explored the evolution of pre–main-sequence (pre-MS) stars. Low mass pre-MS stars are cool and luminous and have deep surface convection zones. They evolve to the MS over a Kelvin-Helmholtz time scale. During the course of the pre-MS, their surface convection zone can be hot and dense enough at the base to burn light elements such as Li and Be, even in stars that on the MS have relatively thin surface convection zones (Bodenheimer 1965, 1966). See Strom (1994) for a review of pre-MS lithium burning in standard models. More recent studies of pre-MS stars have emphasized the role of deuterium burning in the creation of a stellar birthline (Stahler 1988, 1994), and intermediate mass stars may emerge on the birthline close to the MS (Palla & Stahler 1993). This does not qualitatively change the light element depletion pattern because pre-MS light element depletion is only significant for low mass stars (below $\sim$1.2 $M_\odot$ at solar composition).

Recent calculations of pre-MS lithium depletion for standard stellar models have been presented by Proffitt & Michaud (1989), Pinsonneault et al (1990), D'Antona & Mazzitelli (1993), and Swenson et al (1994). The light element depletion pattern is a sensitive function of mass and composition, but there are some broad patterns that have been found by all investigators. A sample set of lithium depletion isochrones is shown in Figure 2*a* at 100 Myr for a range in [Fe/H] and in Figure 2*b* for different ages at two values of [Fe/H], 0.0 and +0.15 (relevant for the Hyades cluster). The major features of the theoretical models are as follows:

- The degree of light element depletion increases with decreased mass. Lower mass stars retain deep convective envelopes for a longer time and contract towards the MS more slowly. For the lowest mass stars, this trend reverses (Nelson et al 1993); the survival of lithium in old objects has been used both as a test for brown dwarfs (Rebolo et al 1996) and as an age indicator for open clusters (Basri et al 1996).

- Pre-MS lithium depletion decreases with lower metal abundance at a given mass. Even though the relevant masses for the onset of significant light element depletion are significantly lower for halo metal abundances, the drop in lithium abundance occurs at a comparable MS effective temperature (see Deliyannis et al 1990 for a detailed discussion of lithium depletion in low metal abundance standard models). As can be seen from Figure 2, even for near-solar metal abundance there is a strong sensitivity of pre-MS lithium depletion to metal abundance, a point emphasized by Swenson et al (1994) (see also Chaboyer et al 1995b).

- The magnitude of pre-MS depletion is a sensitive function of the input physics; this can be seen by comparing lithium depletion results from different investigators and by parameter variation studies (Swenson et al 1994).

- Both pre-MS and MS beryllium depletion are confined to the lowest mass stars (below 0.5 $M_\odot$ at solar composition).

We can therefore anticipate that there will be a mass- and composition-dependent pre-MS depletion of lithium apparent in open clusters; beryllium is not expected to be depleted except in the lowest mass stars.

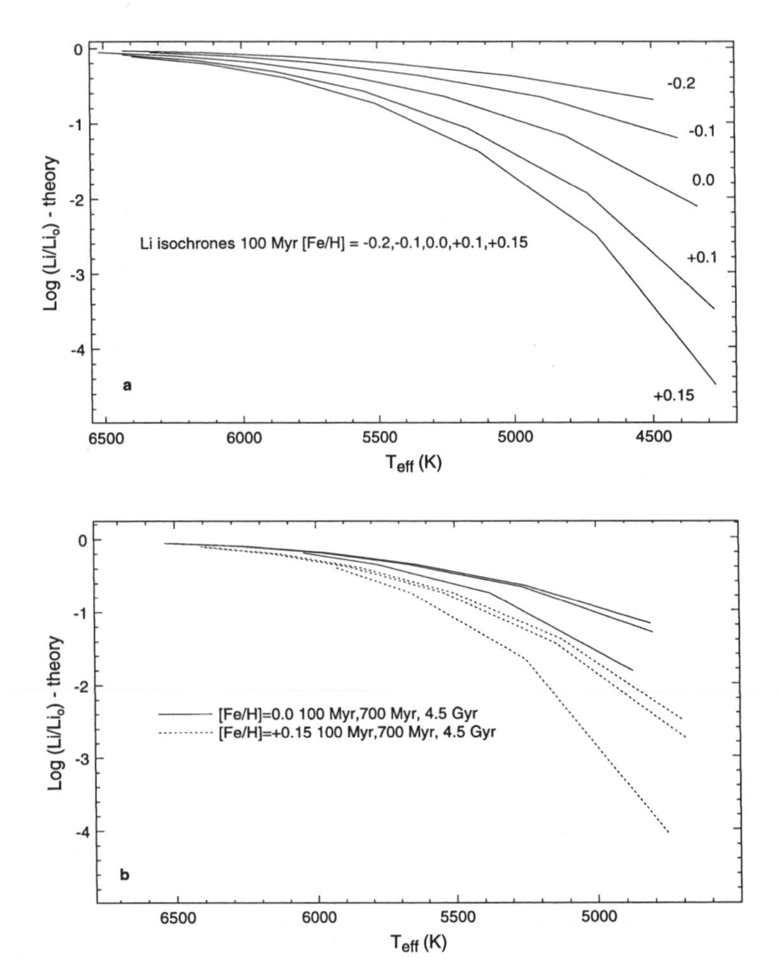


*Figure 2* Pre-MS (*top*) and MS (*bottom*) $^7$Li depletion isochrones from standard models. Theoretical lithium depletion factors as a function of $T_{eff}$ for [Fe/H] $= -0.2, -0.1, 0, +0.1$, and $+0.15$ at an age of 100 Myr are illustrated in the top panel. Lithium depletion for [Fe/H] $= 0$ (*solid lines*) and [Fe/H] $= +0.15$ (*dashed lines*) are shown for ages of 100 Myr, 700 Myr, and 4.5 Gyr in the *bottom panel.*

## 3.2 *Main Sequence Changes in Surface Abundance*

MAIN SEQUENCE: LIGHT ELEMENT SURVIVAL On the MS, the depth of the surface convection zone, and therefore the temperature at the base of the surface convection zone, is a strong function of mass (see Kippenhahn & Weigart 1990 for a discussion of the basis of this phenomenon in stellar structure). As a result, only the lowest mass stars (below 0.9 $M_\odot$ at solar composition) are expected to have surface convection zones deep and dense enough to burn significant amounts of lithium on the MS, as can be seen in Figure 2*b*. The top line corresponds to a typical age (100 Myr) for young systems; the second line (at 700 Myr) indicates the expected depletion pattern at intermediate ages where there is extensive open-cluster lithium data (Hyades, Praesepe, Ursa Major, Coma). The bottom line indicates the extent of depletion at 4.5 Gyr, which is comparable in age to the Sun and the open cluster M 67.

Little lithium depeletion is predicted for solar-mass models and above on the MS. The location of the base of the solar surface convection zone inferred from helioseismology is $0.713 \pm 0.003\ R_\odot$ (Christensen-Dalsgaard et al 1991), not hot enough to burn Li on the MS. This confirms the absence of MS lithium depletion in solar composition stars comparable to, and by extension more massive than, the Sun (Christensen-Dalsgaard et al 1992). Because of the steep dependence of the temperature at the base of the convection zone on mass, this result is insensitive to the assumed input physics. This is in direct contradiction to the observational data for open cluster stars and the Sun (Section 4.1); the interesting situation for halo stars is discussed in Section 4.1 (Halo Stars).

MASSIVE STARS There is an extensive literature on the evolution of massive stars; see Maeder & Conti (1994) for a review. Mass loss dominates stellar evolution on the upper MS, which produces objects in late phases of evolution with numerous and well-documented surface abundance anomalies (e.g. Wolf-Rayet stars). However, MS models, even with mass loss, are not expected to show surface processing of elements for $M < 50\ M_\odot$ (Maeder & Conti 1994). Evidence for mixing in massive MS stars not predicted by standard models is presented in Section 4.2.

## 3.3 *Post-Main Sequence: First Dredge-Up*

Post-Main Sequence (post-MS) stars develop deep surface convection zones as they travel from the MS to the first ascent giant branch (Iben 1967, 1974, Sweigart & Gross 1978, Kippenhahn & Weigart 1990). Once on the giant branch, the surface convection zone of low mass stars reaches a maximum depth and then gradually retreats in mass as material from the envelope falls onto the hydrogen-burning shell, is converted into helium, and is deposited in a degenerate core. A region with ~0.3 solar masses is not incorporated into the surface convection zone at solar composition, almost independent of the total stellar mass. In standard models, there will therefore be a first dredge-up between the MS and the giant branch where the surface abundances change to reflect the average abundances of the entire MS star outside the nuclear burning core. A second dredge-up can occur after helium core burning on the asymptotic giant branch (AGB) (Iben & Renzini 1983, Lattanzio et al 1991).

Recent calculations of the surface abundances predicted from the first dredge-up in standard models can be found in Dearborn (1992), Schaller et al (1992), Bressan et al (1993), El Eid (1994), and Wasserburg et al (1995). The mass and composition dependence of the first dredge-up constrains deep internal mixing on the MS, which could significantly alter the internal composition profile of species such as $^{13}C$ without altering their surface abundances (Section 4.3).

An important prediction of standard models is that changes in abundance on the upper giant branch are not expected for basic stellar structure reasons. The clear presence of such changes in low mass giants is an indication for deep mixing of their surface layers and is the third major piece of evidence for mixing outside the framework of the standard stellar model (Section 4.3). Oxygen isotope anomalies may also be an indicator of slow mixing in AGB stars.

## 3.4 *Mixing in Standard and Nonstandard Models*

In standard stellar models, convection is the sole mixing mechanism, and the time scale for convection is far less than the nuclear burning time scale. The strong temperature dependence of nuclear burning therefore implies that essentially complete burning of fragile species will precede significant processing of species that burn at higher temperatures. For example, complete destruction of $^{6}Li$ will occur before significant $^{7}Li$ destruction, which in turn would be completely destroyed before significant Be destruction. Similar considerations apply to species that burn at higher temperatures, such as B, $^{3}He$, and the CNO species. In the Sun, the surface $Li^{7}$ is depleted by a factor of $\sim$100, and Be is depleted by a factor of $\sim$2 with respect to meteorites. Such a depletion pattern is not consistent with short time scale mixing to an arbitrary depth.

In general, models with mixing over a time scale comparable to or greater than the nuclear burning time scale show a very different pattern. In the presence of slow mixing, a species such as lithium will be destroyed before it ever reaches very high temperatures, even if the mixing persists to depths sufficient to destroy more resilient species such as beryllium. As a result, the degree of lithium or beryllium depletion will depend, respectively, on the distance Li-poor or Be-poor material must travel to reach the surface and the time scale for mixing across this region. Simultaneous mixing of lithium and beryllium is therefore expected in solar models with mild envelope mixing on general grounds (e.g. Pinsonneault et al 1989).

The assumption of instantaneous mixing to an arbitrary depth, which is sometimes made when investigating the effects of mixing in stars (Vandenberg & Smith 1988, Sweigart 1997), is therefore severely limited in its physical application when applied to species that burn at different temperatures. Such models can indicate how deep mixing may be (the stated purpose of Vandenberg & Smith 1988, for example), but there is no a priori reason to expect the detailed properties to agree with full calculations. Even models that permit both the time scale and depth to vary (Wasserburg et al 1995) do not necessarily reflect the underlying physical picture; the time scale for rotational mixing, for example, depends strongly on the distance from the surface convection zone and the internal stellar rotation.

# 4. THE OBSERVATIONAL PATTERN

## 4.1 *Mixing on the Main Sequence: Low Mass Stars*

The challenge of explaining the light element abundances of low mass stars is an old one; Greenstein & Richardson (1951) first proposed extra mixing as the solution of the low solar photospheric lithium abundances relative to that on the Earth and in meteorites. A good review of early work on lithium can be found in Zappala (1972). The lithium problem in low mass stars was already recognized in much of its present form by the time of the Zappala article: The lithium abundances in the young Pleiades were higher than the Hyades, which in turn were higher than the Sun; this contradicted the expectations from standard models.

The advent of modern charge coupled device (CCD) detectors led to a large increase in both the accuracy and quantity of the observational data. There is an extensive literature on lithium observations in low mass stars (see the series of papers in *Memorie della Societa Astronomica Italiana*, 1995, Vol. 66). Observational reviews include Hobbs & Pilachowski (1988) for an overview of the open cluster data at the time; Strom (1994) for a review of pre-MS and young cluster lithium abundances; Balachandran (1994, 1995) for reviews of lithium on the MS in open clusters, with a special emphasis on the F star "lithium dip" in the latter; and Pallavicini (1994) for lithium in post-MS stars. Carlsson et al (1994) also summarizes the observational lithium literature. The most extensive data set for metal-poor stars is that of Thorburn (1994); see Molaro et al (1995) and Ryan et al (1996) for recent discussions of the observational situation in halo stars.

Beryllium and boron data are becoming available for some stars as well. Because these species burn at different depths than lithium, they provide powerful additional constraints on the allowed classes of theoretical models (Pinsonneault et al 1989, Boesgaard & King 1993, Deliyannis 1995, Primas 1996).

In this review, I concentrate on low mass MS stars in open clusters, which provide the best laboratory for testing the reliability of the light element depletion pattern inferred from stellar models. The observed depletion pattern is a complex function of mass, composition, and age; in addition, there is strong observational evidence for variations in abundance among cluster stars of the same mass, composition, and age. Clusters provide the cleanest means of quantifying these dependencies. In particular, there are numerous possible solutions for the abundances of any one cluster or the Sun, but the requirement that the overall pattern seen in low mass stars be reproduced is a far more challenging theoretical task. An approximately constant initial lithium abundance is also appropriate for the open cluster system.

By contrast, the initial abundance of metal-poor stars is the quantity of interest, with considerable cosmological impact. The validity of different models for relating the initial and current abundances in metal-poor stars must therefore be established in the open cluster system, where such models can be distinguished, rather than relying on subtle distinctions between different classes of models in their predictions for the halo stars themselves. Note that chemical evolution considerations (Steigman 1996), as well as observations of the survival of the isotope $Li^6$ in some metal-poor stars (Smith et al 1993, Hobbs & Thorburn 1994), can place interesting additional constraints on any possible depletion in metal-poor stars relative to their initial value.

The overall properties of the open cluster system can be summarized as follows. The mean trend in young open clusters is consistent with that expected from pre-MS depletion in standard models. There is a large and unexpected dispersion in lithium abundance for young cool stars, possibly correlated with rotation. For older clusters, there is anomalous MS depletion for late F, G, and K stars. The rate of excess depletion decreases with increased age in these stars. A range in lithium abundance at fixed $T_{eff}$ is seen in all of the older open clusters. This implies that mass, composition, and age do not uniquely determine the surface lithium abundance of a star. The majority of the halo star data shows relatively little scatter.

A population of highly Be-depleted stars exists; most of these stars lie in the mid-F star region. The Sun and $\alpha$ Cen A and B also appear to be depleted with respect to the solar meteoritic value. $^6Li$ detections have been claimed in some metal-poor stars. I quantify the observational picture in systems of progressively greater age below.

YOUNG CLUSTERS Young clusters provide a powerful test of pre-MS lithium depletion from standard models. The observational data also show some interesting phenomena not expected from standard models. Lithium data has been obtained for the young (30-Myr) cluster IC 2391 (Stauffer et al 1989) and the 50-Myr cluster $\alpha$ Per (Balachandran et al 1988, 1996, Zapatero et al 1996). The best-studied young cluster by far is the 70- to 110-Myr Pleiades. An extensive set of observations was performed by Soderblom et al (1993b) (this paper contains an extensive discussion of the history of the study of lithium in the cluster). Further observations were obtained by Jones et al (1996b). Russell (1996) studied the question of the dispersion in abundance for cool Pleiades stars (see also Stuik et al 1996).

The lithium abundances in the Pleiades and $\alpha$ Per are compared with standard models in the top two panels of Figure 3. The dashed line represents the effects of even a small amount of overshoot (0.05 pressure scale heights), which can have a dramatic effect on pre-MS depletion (e.g. Swenson et al 1994). The most important features are as follows:

- The mean depletion trend in both clusters is in good agreement with that predicted by standard models (Strom 1994, Jones et al 1996b). The revised $\alpha$ Per abundances (Balachandran et al 1996) are much more similar to the Pleiades than the pattern inferred from earlier work (Balachandran et al 1988). The consistency between the data and models strongly constrains increased pre-MS depletion as a solution for the abundance patterns seen in older systems (see for instance Pinsonneault et al 1990, Michaud & Charbonneau 1991, Chaboyer et al 1995b).
- A large dispersion is seen in lithium abundances for stars cooler than 5300 K. Most of the overabundant stars are rapid rotators, but not all, and most are chromospherically active (Soderblom et al 1993b). This phenomenon has proven challenging to explain theoretically (Section 6.1). Soderblom et al (1993b) examined a variety of means of explaining this range through observational error and concluded that the observed spread in abundance was real. This conclusion was recently challenged by Russell (1996), who examined the abundances for a subset of the Soderblom et al (1993b) sample derived from a weaker spectral feature (at 6104 Å) rather than the typical 6708-Å resonance line (see also Stuik et al 1996). It appears unlikely that the large observed range can be entirely removed, but the magnitude of the effect is controversial.
- A small depression exists in the lithium abundance pattern for mid-F stars. Although the effect appears small in comparison with the steep mass dependence for cool stars, it is highly statistically significant (as can be seen by noting the small abundance errors for these stars in Figure 3*a*). This is the earliest detection of the dramatic F star lithium dip first discovered by Boesgaard & Tripicco (1986).

INTERMEDIATE-AGE CLUSTERS There are five systems with ages ranging from 200–700 Myr with large lithium samples: M34 (200 Myr), Ursa Major (300 Myr), Coma (500 Myr), the Hyades, and Praesepe (both 500–700 Myr).

The Hyades has been the focus of much theoretical and observational work, owing to both its closeness and the wealth of other data available for the system (see Thorburn et al 1993 for a summary of earlier work). Thorburn et al (1993) performed a large survey in the Hyades designed to carefully quantify the

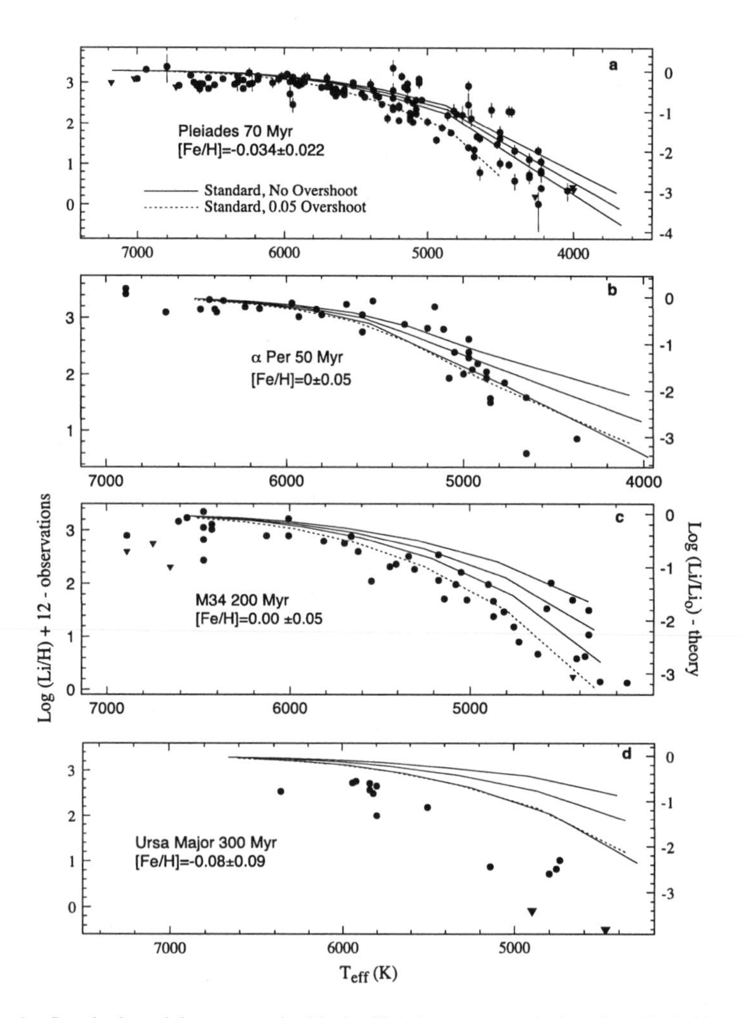


*Figure 3* Standard models compared with the Pleiades (*top panel*, data from Soderblom et al 1993b); $\alpha$ Per (*second panel from top*, data from Balachandran et al 1996); M34 (*third panel from top*, data from Jones et al 1996a,b), and Ursa Major (*bottom panel*, data from Soderblom et al 1993a). The *central solid line* indicates standard model depletion for the mean cluster [Fe/H] and the *solid lines above and below* represent 1-$\sigma$ errors in the [Fe/H] from Boesgaard & Friel (1990). The *dashed line* represents standard models with the addition of 0.05 pressure scale heights of overshoot.

dispersion in lithium abundance. Soderblom et al (1995) looked at lithium in K stars in the cluster, and Barrado y Navascues & Stauffer (1996) examined Hyades binary abundances.

Lithium in the younger (300 Myr) Ursa Major system was studied by Boesgaard et al (1988). Later work by Soderblom et al (1993c) found that most of the F stars in the Boesgaard et al sample were nonmembers, but it extended the sample down to much lower $T_{eff}$. Praesepe was examined by Boesgaard & Budge (1988) and Soderblom et al (1993a) (see also Balachandran 1995). Recent observations in M34 by Jones et al (1996a) provide valuable new insights into the process of lithium depletion in low mass stars. At 180–250 Myr, this cluster is young enough to retain a range of rotation rates at a given mass but old enough to have experienced MS depletion.

The data for M34 and Ursa Major are compared with standard models in the bottom two panels of Figure 3; Coma, the Hyades, and Praesepe are compared with standard models in the top two panels of Figure 4. Important features include the following:

- A dramatic lithium dip is seen in the mid-F stars. A hint of this phenomenon can be seen in the early photographic work of Wallerstein et al (1965). Jones et al (1996a) point out that this dip is already deep by the age of M34. It extends more than two orders of magnitude down in the Hyades, and is seen in all of the intermediate age systems that have data in the 6400–6800 K region. The existence of such a feature is an unambiguous contradiction of standard stellar models, as the convection zones in this mass range are far too thin to burn lithium.

- The mean abundance trend is significantly below that predicted by standard models in all of the systems, by an amount that increases with increased age. This is, in fact, the same conclusion reached by Zappala (1972) with more limited data. Even the models with overshoot sufficient to fit the lower bound of the Pleiades distribution fall above the mean cluster trends in Figures 3 and 4. Increasing depletion for clusters in this age range would result in inconsistencies for young clusters while not removing the observed discrepancy in the mean trend for older systems.

- A dispersion in abundance exists at a fixed $T_{eff}$ for stars near 6000 K, and anomalous abundances are found for at least some cooler stars, for Coma, Praesepe, the Hyades, and Ursa Major. M34 shows an intriguing pattern: a larger lithium dispersion for hotter stars than the Pleiades and a smaller dispersion for cooler stars than the Pleiades.

The claim is sometimes made in the literature that clusters such as the Hyades show no dispersion (for example, Swenson et al 1994, Russell 1996). Thorburn

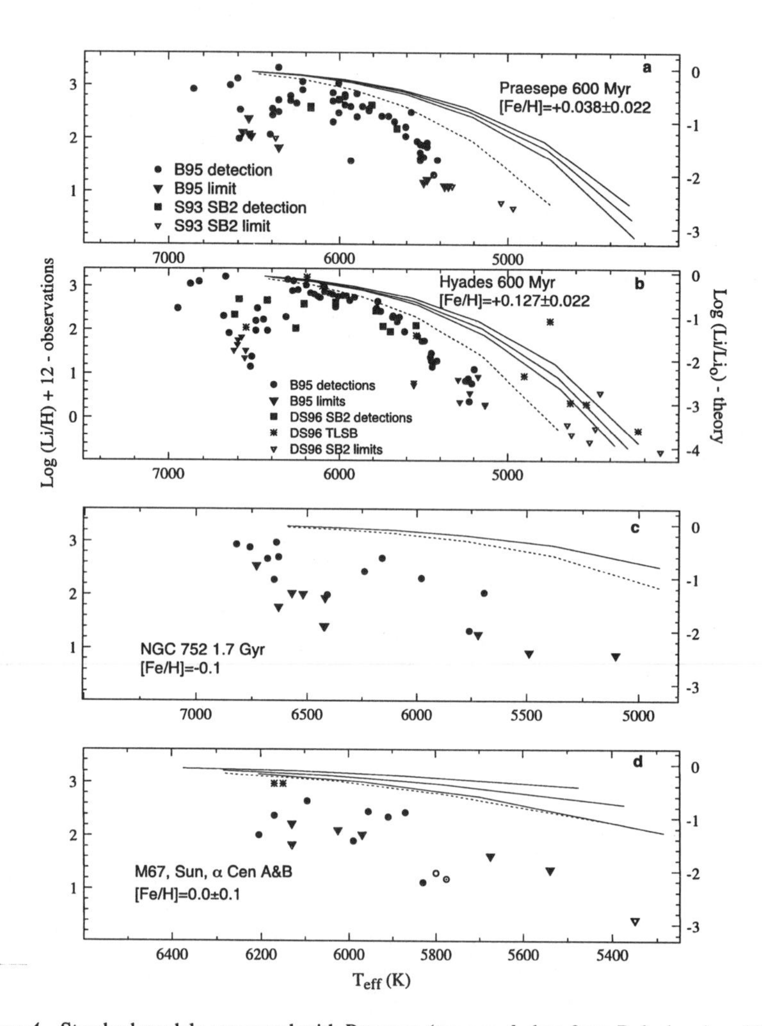


*Figure 4* Standard models compared with Praesepe (*top panel*, data from Balachandran 1995), the Hyades (*second panel from top*, data from Balachandran 1995, Barrado y Navascues & Stauffer 1996), and NGC 752 (*third panel from top*, data from Balachandran 1995). In the *bottom panel*, data for M67 (Balchandran 1995, Deliyannis et al 1994) are the *solid points*, with tidally locked binaries in M67 represented by *asterisks*. Lithium abundances for the Sun (Anders & Grevesse 1989) and $\alpha$ Cen A and B (Soderblom & Dravins 1984) are also shown; the *solar symbol* represents the Sun, $\alpha$ Cen A is the *open circle*, and $\alpha$ Cen B is an upper limit (*open triangle*). Additional cluster [Fe/H] data taken from Friel & Boesgaard (1992). Tidally locked binaries in the Hyades are represented by *asterisks*.

et al (1993) showed that the range in abundance for late F stars was greatly in excess of their errors, with *average* deviations from the mean trend at the 5-$\sigma$ level. A similar conclusion can be drawn from other clusters in this age range. In cooler stars, the strong dependence of lithium on mass makes detecting a modest dispersion difficult. Even so, there are clear examples of G and K stars that are normal in every other way except lithium (e.g. vB 9 in the Hyades) and of stars that differ in their lithium content but otherwise have similar spectra. It is true, and important, that there is a mean trend obeyed by most stars, i.e. that the range in depletion for most objects is less than the average depletion at a given $T_{eff}$.

- The cool star region where a large dispersion is seen in young systems is detected only in the younger M34 and Ursa Major systems. The Ursa Major sample shows scatter but with a small observational sample (Soderblom et al 1993c). The relative pattern in M34, namely cool faster spinners with higher abundances, is similar to that of the younger clusters, but the difference is less marked (Jones et al 1996a). Soderblom et al (1995) found that the mixture of detections and upper limits seen for the cool Hyades stars by Thorburn et al (1993) were all only upper limits when seen at higher resolution, and they concluded that there was no dispersion for cool Hyades stars. The absence of detections does imply a lack of distinctly underdepleted stars but does not permit a measurement of the intrinsic dispersion in their sample.

- Tidally locked binaries have higher abundances than single stars of the same $T_{eff}$ (Soderblom et al 1990, Thorburn et al 1993, Barrado y Navascues & Stauffer 1996 for the Hyades). The abundances of tidally locked binaries are normal in the Pleiades (Soderblom et al 1993b), so differing pre-MS depletion is not the cause of this phenomenon. The different behavior of tidally locked systems implies that the rotational history of stars influences their surface lithium abundances (Deliyannis et al 1990, Zahn 1994, Ryan & Deliyannis 1995). Significantly, the tidally locked binaries in older systems have abundances consistent with those predicted by standard models, as can be seen in Figure 5*c*. Barrado y Navascues & Stauffer (1996) reanalyzed the binary data in the Hyades and obtained some new measurements as well; they found that all of the cool short period binaries had abundances well above the mean trend from detached binaries and single stars. Two tidally locked binaries in Praesepe, KW 181 (Soderblom et al 1993a) and KW 367 (King & Hiltgen 1996), appear to have normal abundances. In both cases, however, the authors note that there is independent evidence that the systems may not be tidally synchronized, which is required in theoretical models for lithium underdepletion.

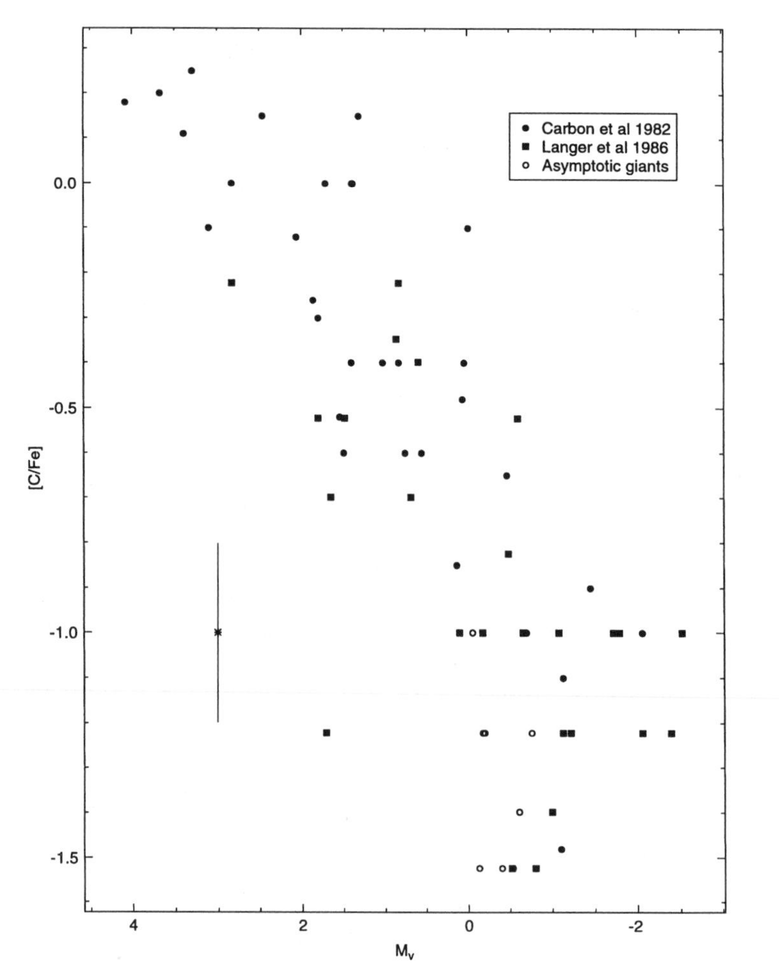


*Figure 5* Carbon abundance [C/Fe] as a function of absolute visual magnitude (Mv) in the globular cluster M92. Data from Carbon et al 1982, Langer et al 1986. *Typical error bars* from Carbon et al 1982 are illustrated at the *lower left*.

- Abundances in hot stars are generally compatible with the meteoritic value; there is some evidence for dispersion (Burkhart & Couprey 1989). The absence of deep mixing for spectral types earlier than the dip, along with the existence of chemically peculiar stars, places interesting constraints on the theory.

Beryllium abundances have been obtained for Hyades lithium dip stars by Boesgaard & Budge (1989), showing depletion of factors of 2–4 relative to the solar meteoritic value (Anders & Grevesse 1989). Cool Hyades stars have also been measured by Garcia Lopez et al (1995). There appears to be decreased surface beryllium with decreased $T_{eff}$, but this trend could also be caused by systematic errors (Garcia Lopez 1996).

OLD OPEN CLUSTERS, THE SUN, AND $\alpha$ CEN Old disk stars are an essential component of the study of lithium. The well-known discrepancy between the solar photospheric and meteoritic abundances (Anders & Grevesse 1989 and references therein) is one data point. Data also exist for field stars (Balachandran 1990, Lambert et al 1991, Favata et al 1996). Field star data is more difficult to interpret than cluster data because of the mixture of ages and compositions. There does exist, however, a population of highly overdepleted field objects cooler than the mid-F star region (Lambert et al 1991) and a large dispersion in abundance among post-turnoff F stars (Balachandran 1990).

Data for the old open cluster NGC 752 (1.7 Gyr) was originally obtained by Hobbs & Pilachowski (1986a) and Pilachowski & Hobbs (1988); it was reanalyzed by Balachandran (1995). Hobbs & Pilachowski (1988) obtained data for NGC 188, the oldest open cluster to be studied (6 Gyr). There is a more extensive and varied data set for M67, which at 4 Gyr is comparable in age to the Sun (Hobbs & Pilachowski 1986a, Spite et al 1987, Garcia Lopez et al 1988); abundances for tidally locked binaries in M67 were investigated by Ryan & Deliyannis (1995). The Sun and $\alpha$ Cen A and B also provide well-studied examples of both lithium and beryllium abundances in old disk stars. The data for the clusters NGC 752, M67, the Sun, and $\alpha$ Cen A and B are compared with standard models in the bottom two panels of Figure 4. Important features include the following:

- Significant depletion exists relative to younger systems, but an inferred rate of depletion decreases with increased age. Despite their differences in age, the old open clusters have abundance patterns much more similar to one another than to the intermediate-aged systems (Hobbs & Pilachowski 1988, Soderblom 1995).

- A large dispersion in abundance exists. The old open clusters are fainter than the younger systems, so the errors are significantly higher than for nearby systems such as the Pleiades (Hobbs & Pilachowski 1988). However, the observed magnitude of the dispersion is in excess of the errors as confirmed by recent data (Deliyannis et al 1997).

- The F star lithium dip is clearly seen in NGC 752. The stars in the dip region have evolved off the MS in older systems (see Section 4.3 for discussion of subgiants).

- The Sun relative to other stars. As an interesting aside, Pilachowski & Hobbs (1988) noted that the solar lithium abundance might be anomalously low relative to that of old open cluster stars. Because some classes of theoretical models are calibrated on the solar lithium, an unusual abundance for the Sun could affect theoretical predictions.

- Tidally locked binaries in M67 have abundances higher than other stars at the same $T_{eff}$ (Deliyannis et al 1994).

- Beryllium and boron. Boesgaard & King (1993) examined the observational database for beryllium in low mass stars, and concluded that extra mixing or a range in intrinsic abundance was needed to explain disk beryllium abundances. A population of beryllium-deficient field stars has long been known to exists in the same general temperature range as the lithium dip (Boesgaard 1976). Thorburn et al (1994) have reported that boron is depleted by a factor of 2 in 110 Her, a subgiant that came from the lithium dip (lithium-depleted by a factor of 100–200) region, and beryllium is depleted by a factor of 10 in this star. This combination of properties indicates mixing with a depth-dependent time scale.

Beryllium is depleted in the solar photosphere with respect to meteorites by 0.3 dex (Chmielewsky et al 1975, Anders & Grevesse 1989). In a recent analysis, King et al (1996) found that a range of solar depletion factors from 0.1–0.5 dex was consistent with the data, but that some beryllium depletion was needed to explain the difference between the photospheric and meteoritic abundances. $\alpha$ Cen A has a surface beryllium comparable to that of the solar photosphere, whereas $\alpha$ Cen B is depleted in beryllium with respect to both A and the meteoritic value. This could provide evidence for extra mixing in cool stars where lithium is unobservable, but standard models predict no beryllium depletion (Primas et al 1996, King et al 1996).

HALO STARS Spite & Spite (1982) discovered that hot metal-poor stars had a abundances of $\sim$2.1, on the order of 10 times greater than the solar photospheric value but 20 times lower than the meteoritic value. Furthermore, the abundances of hot stars appeared to show little dependence on $T_{eff}$. This lithium plateau, extending from 5600–6300 K, is now firmly established (Spite et al 1984, Hobbs & Duncan 1987, Rebolo et al 1988); the most comprehensive recent observational study has been performed by Thorburn (1994). Key features of the halo star lithium pattern include the following:

- Cool star depletion similar in morphology to that seen for the open clusters.

- A plateau with nearly uniform abundance across a wide range of metal abundances. A slight increase in lithium abundance with increased $T_{eff}$

and decreased lithium abundance with decreased metal abundance may be present in the data (Thorburn 1994, Norris et al 1994, Ryan et al 1996), on the order of 0.04 dex/100 K and 0.1 dex per order of magnitude in [Fe/H]. This claim has been disputed (Molaro et al 1995); the difference is largely caused by different temperature determination techniques. A decrease in abundance towards higher $T_{eff}$ is expected in some classes of theoretical models, and its absence is clear and significant.

- Relative to the old open clusters any dispersion at fixed $T_{eff}$ is small for the majority of stars, although some anomalous stars with only limits for Li are present (Thorburn 1994). For the overall halo system, a small intrinsic dispersion has been claimed by some authors, whereas others find the observed range to be consistent with observational errors (Deliyannis et al 1993, Thorburn 1994, Molaro et al 1995, Spite et al 1996). It is clear that any successful model must not predict a large scatter in abundance for the majority of halo stars. By contrast, subgiant stars in the globular cluster M92 (Deliyannis et al 1995) and NGC 6397 (Pasquini & Molaro 1996) appear to show a wide range in abundance at fixed $T_{eff}$.

- Detections of the isotope $^6$Li have been claimed in some halo stars (Smith et al 1993, Hobbs & Thorburn 1994). The detection of both isotopes of lithium is based on subtle changes in line profiles, which can be produced by other effects such as convection (Gray 1989); some caution in interpreting the data may therefore be in order. Because $^6$Li is fragile, its survival constrains the magnitude of deep mixing (Smith et al 1993, Steigman et al 1993); other possibilities such as flare production of $^6$Li have also been discussed (Deliyannis & Malaney 1995).

- Cool tidally locked binaries have a higher than normal abundance, which is consistent with the pattern seen in open clusters (Deliyannis et al 1994, Spite et al 1994). The situation for binaries in the plateau is more complex (Spite et al 1994, Deliyannis et al 1996); the abundances are similar to single stars, but it is unclear if the specific observed systems would be expected to have unusual abundances.

### 4.2 *Mixing on the Main Sequence: High Mass Stars*

Recent evidence exists for mixing in high mass stars that is not expected from standard stellar models. Lithium and beryllium are not observable in hot stars, so studies of mixing have focused on helium and CNO. Deep mixing produces enhanced surface helium and nitrogen, lowered surface carbon, and in the case of very deep mixing, lowered surface oxygen. A summary of the observational evidence for mixing can be found in Maeder & Conti (1994); see also Venn

(1995). The MS anomalies are not expected to be produced for masses less than about 40–50 solar masses, even in models that incorporate mass loss. The predictions of standard models for evolved stars depend on their evolutionary state. A and B stars leaving the MS are expected to have normal abundances (i.e. without extensive CNO processing), whereas stars on blue loops have undergone the first dredge-up. Either evolutionary state is possible in principle for A and B supergiants (Venn 1995).

The observational situation depends strongly on spectral type and thus on mass. Slowly rotating MS OB stars (Herrero et al 1992) and MS B stars (Gies & Lambert 1992) have essentially normal abundances. However, rapidly rotating OB stars show large surface helium abundances correlated with the rotation rate. Observational evidence is consistent with a lack of mixing in intermediate mass MS stars (see Section 4.3) from $\sim$1.8–7 $M_{\odot}$. The correlation between rotation and mixing found by Herrero et al (1992) is important, and it may indicate that such stars follow a completely different evolutionary path than traditional stellar evolution.

Evolved O stars, in general, show surface abundances indicating surface CN processing and helium enrichment (Walborn 1976, 1988, Howarth & Prinja 1989). Gies & Lambert (1992) found evidence for anomalous surface abundances in some B supergiants (see also Barbuy et al 1996); Venn (1995, 1996) examined A supergiants and, surprisingly, found N/C ratios intermediate between solar and that expected from the first dredge-up. The depletion pattern does not appear to be consistent with the pattern expected from the first dredge-up, which may force a revision of the usual evolutionary scenario for A supergiants. The observational analysis of supergiant abundances is challenging, so some caution regarding the interpretation of the data is in order (as stressed by the observers).

Recently, boron has been measured in B stars (Venn et al 1996), and boron may serve as a powerful discriminant between different physical scenarios. The measured boron abundance in the sample is well below solar, and boron depletion is correlated with nitrogen excess. Theoretical models of the mixing in massive stars are discussed in Section 6.2.

## 4.3 *Mixing in Evolved Low Mass Stars*

THE FIRST DREDGE-UP IN OPEN CLUSTERS The abundances of giants after the first dredge-up provide a test of the internal abundance profile of their MS precursors. Open clusters provide a sample with a range in mass, which is otherwise difficult to constrain for field giants. Lithium and $^{12}C/^{13}C$ data for open cluster giants with a range of masses were obtained by Gilroy (1989). The observed $^{12}C/^{13}C$ was in good agreement with theory for $M > 1.8\ M_{\odot}$ and was systematically lower than predicted in lower mass giants. $^{12}C/^{13}C$ and C/N abundances were consistent with the first dredge-up in low luminosity giants in the old open cluster M67, whereas lower $^{12}C/^{13}C$ was found for more luminous giants (Gilroy & Brown 1991). This suggests that mixing on the giant branch, and not internal MS mixing, is the cause of the deviation from standard models seen in the low mass giants. The absence of mixing can also be seen in the beryllium and boron abundances of the Hyades giants, which are also consistent with the predictions of standard models (Duncan et al 1994).

Lithium in giants presents a more complex picture. The maximum abundance seen in giants is less than predicted by standard models, which is probably caused by non-LTE effects (Carlsson et al 1994, Duncan et al 1994). There are also giants with anomalously low and high abundances (Gilroy 1989, Brown et al 1989, Fekel & Balachandran 1993). The active RS Cvn stars also tend to have higher than normal lithium abundances, similar to the MS tidally locked binaries (see Pallavichini 1994 for a review). The low lithium stars could have depleted lithium on the MS. The high lithium stars, however, present much more of a theoretical puzzle (even leaving aside the extremely high abundances seen in some asymptotic giants).

Subgiant lithium abundances can serve as a diagnostic of different MS models. Subgiants in the old open cluster M67 came from precursors in the lithium dip, and they show low abundances that indicate that lithium is destroyed and not hidden below the surface in the dip (Balachandran 1995, Deliyannis et al 1997). Intermediate mass stars in the Hertzsprung gap have also been studied for lithium, and they show a complex rotation-dependent abundance pattern (Balachandran 1990, Wallerstein et al 1994, Hiltgen 1996). There is a well-defined halo subgiant lithium pattern, followed by strong depletion in metal-poor giants (Pilachowski et al 1993).

ABUNDANCE ANOMALIES ON THE GIANT BRANCH A rich database exists of observations in metal-poor giants that require extra mixing (see Smith 1987, Suntzeff 1993, Kraft 1994 for reviews). The issue is usually raised in terms of either primordial abundance variations or mixing; this dichotomy is unnecessary, as evidence exists for both. Lithium abundances in nitrogen-rich halo dwarfs are normal (Spite & Spite 1986), which rules out internal MS mixing as the origin of the nitrogen anomalies. There is also evidence for abundance anomalies close to the MS (Bell et al 1983, Suntzeff 1989, Briley et al 1991, Briley et al 1996) and giants for which C + N is not the same in the same cluster (Carbon et al 1982, Trefzger et al 1983). This is difficult to understand from the point of view of the mixing hypothesis because deep mixing is not in general present in models that are consistent with the lithium data for metal-poor MS stars (Deliyannis et al 1989). $^{12}C/^{13}C$, C/N, and oxygen provide tests of mixing at increasingly high temperatures. In addition to the CNO isotopes, Na and Al variations may be produced by proton capture reactions on Ne and Mg, respectively, in the vicinity of the hydrogen-burning shell in giants. The trends inferred from the different indicators are summarized below.

*Low $^{12}C/^{13}C$ ratios* $^{12}$C/$^{13}$C ratios in globular cluster giants are low, close to the nuclear equilibrium value of 4 in some cases. The observed values are far from the values seen in dwarfs and the predictions of the first dredge-up in standard models (Sneden et al 1986, Brown & Wallerstein 1989, Suntzeff & Smith 1991, Briley et al 1994, Shetrone 1996, Briley et al 1997; see Charbonnel 1994 for a discussion). The more-or-less normal dilution of lithium in halo subgiants suggests that mixing on the red giant branch (RGB), rather than internal MS processing, is responsible for the low carbon isotope ratios (Pilachowski et al 1993; see also Sneden et al 1986). Modest but significant changes in $^{12}$C/$^{13}$C along the giant branch were also found in the old open cluster M67 (Gilroy & Brown 1991).

*C/N* Carbon is converted to nitrogen in the CN cycle; the globular clusters M92, M3, M4, NGC 6752, M13, and M15 all show evidence for decreased carbon with increased luminosity on the giant branch (Carbon et al 1982, Trefzger et al 1983, Briley et al 1990, Smith et al 1996). The carbon depletion as a function of luminosity on the giant branch is shown in Figure 5. There is a scatter in the mean C/Fe at a given luminosity that is real. C + N + O appears to be conserved in M3 and M13 (Smith et al 1996).

*Oxygen* Surface oxygen depletion requires that material be O $\rightarrow$ N processed, which requires higher temperatures than C $\rightarrow$ N processing. The first evidence for oxygen depletion in metal-poor cluster giants was found in M92 by Pilachowski (1988) and confirmed both in M92 and in M15 (Sneden et al 1991). Evidence for strong oxygen depletion exists for luminous giants in M13 (Brown et al 1991, Kraft et al 1992) and in $\omega$ Cen (Norris & DaCosta 1995). M3 and M10 show more normal oxygen abundances as a function of luminosity on the upper RGB (Kraft et al 1993, 1995); this implies that whatever governs the oxygen depletion varies from cluster. Because both M10 and M13 have blue horizontal branches, oxygen depletion is not a unique function of horizontal branch morphology. The observed depletion pattern is also metallicity dependent; the more metal-rich cluster M71 shows much-reduced evidence for oxygen depletion at the tip of the giant branch (Sneden et al 1994, Briley et al 1997). More normal oxygen abundances are seen in field giants (see Pilachowski et al 1996 for a discussion of this point). Kraft et al (1997) have traced O-depleted stars well down the giant branch in M13; the field and cluster data appear to be different.

*NaMgAl* A correlation between the strength of the CN band and the strength of the sodium and aluminum line strengths was initially found by Peterson (1980). Because changes in the abundance of Na and Al were thought to require neutron capture, this was taken as evidence for primordial variations in abundance. The observational evidence for Na and Al variations has accumulated (see Kraft 1994 for a discussion), but the interpretation has shifted. Proton capture on Ne and Mg can also occur under conditions typical for the hydrogen-burning shell in giants (Denissenkov & Denissenkova 1990, Langer et al 1993, Cavallo et al 1996). In an extensive study, Pilachowski et al (1996) found evidence for changes in the mean sodium abundance as a function of luminosity on the upper giant branch of M13. There was an abrupt change in behavior at log g = 1, possibly indicative of the onset of mixing at this point on the RGB. Simultaneous Mg and Mg isotope data (Shetrone 1996a,b) observations reveal a more complex pattern, and the abundance pattern among second ascent giants appears different from that in luminous first ascent giants.

There are significant trends with metal abundance in the data, in the sense that metal-poor stars have more marked composition anomalies (Shetrone 1996a,b, Kraft et al 1997). The open cluster data shows that the deviations from the standard model are strongly mass dependent, becoming pronounced only for the lowest-mass open cluster giants. By extension, mixing in metal-poor stars could also be mass (or equivalently age) dependent. Variations in abundance at a given luminosity are also present. The observed mixing pattern appears to depend on luminosity as well: Deeper mixing (as probed by O, Na, and Al) appears strongest in the brightest giants, whereas milder C $\rightarrow$ N mixing appears to be possible for less luminous giants. The status of theoretical models of RGB mixing is discussed in Section 6.3.

# 5. MIXING AND SEPARATION MECHANISMS

## 5.1 *Rotational Mixing*

Rotation has long been known to be capable of inducing mixing in the radiative interior of stars. Eddington (1925) found that meridional circulation must necessarily arise in rotating stars as a consequence of the von Zeipel paradox: namely the impossibility of maintaining hydrostatic equilibrium in a rotating star because the effective gravity on an equipotential varies with latitude. There is an extensive catalogue of other instabilities that arise in the presence of rotation (Zahn 1974, Tassoul 1978, Knobloch & Spruit 1982), especially in the presence of differential rotation as a function of depth. However, rotation has not in general been an ingredient of the standard stellar model. This can be attributed largely to the increase in both complexity and uncertainty of models that include rotation and to the availability of arguments that rotation is at most a perturbation to the structure and evolution of stars. In the present-day Sun, for example, the departure from spherical symmetry is on the order of $10^{-5}$, and the time scale for meridional circulation is on the order of $10^{12}$ years, which is far in excess of the nuclear burning time scale.

The observational data, however, requires a mixing mechanism not present in standard stellar models. For all of the problems of interest in this review, mixing driven by rotation is at least a potential candidate.

The problem of stellar evolution with rotation can be broken into a series of ingredients that must be included. The best overall approach remains that advocated by Endal & Sofia (1978), in their pioneering study of stellar models with rotation. The standard equations of stellar structure must be supplemented with appropriate corrections for the direct impact of rotation on the structure (Kippenhahn & Thomas 1970). The initial angular momentum distribution must be specified, a proper surface boundary condition is needed (including angular momentum and mass loss), and a prescription for the internal transport of angular momentum is required for a fully self-consistent model. This yields a solution of the angular momentum distribution within a stellar model as a function of time. A prescription for the mixing associated with a given angular momentum distribution is obviously necessary to compute rotational mixing; if vigorous enough, such mixing could also materially affect the structure and evolution of the star.

Broadly speaking, the study of rotational mixing therefore requires both an understanding of the angular momentum evolution of stars and an understanding of the degree of mixing for a given angular momentum distribution. The former subject is outside the scope of this review; the important features of angular momentum evolution are summarized at the beginning of the discussion of rotational mixing in Section 6.1. The remainder of the section is devoted to the degree of mixing and angular momentum transport for a given angular momentum distribution.

OVERALL TREATMENT OF ROTATIONAL MIXING AND ANGULAR MOMENTUM TRANSPORT Because internal angular momentum transport is the crucial ingredient for determining the extent of rotationally induced mixing, I describe the diffusion coefficient estimates that have been proposed. Following Endal & Sofia (1978), the transport of angular momentum and the associated material transport can be described by coupled diffusion equations

$$\frac{d\omega}{dt} = \frac{1}{4\pi\rho r^2 \tilde{I}} \frac{d}{dr}\left(4\pi\rho r^2 \tilde{I} D \frac{d\omega}{dr}\right), \tag{1}$$

and

$$\frac{dX_i}{dt} = \frac{1}{4\pi\rho r^2}\frac{d}{dr}\left(4\pi\rho r^2 f_c D\frac{dX_i}{dr}\right), \tag{2}$$

where $\tilde{I}$ is the moment of inertia per unit mass, $D$ is the diffusion coefficient, and $f_c$ is an efficiency factor relating the diffusion coefficient for material transport to that for mixing. The local angular velocity and its derivative allow us to determine the diffusion coefficient for angular momentum transport subject to considerations of stability. The efficiency factor $f_c$ can be calibrated empirically; in general, the diffusion coefficients for mixing must be about 30 times smaller than those for angular momentum transport (Pinsonneault et al 1989).

Chaboyer & Zahn (1992) derived a physical basis for this factor based upon the inhibition of vertical mixing by horizontal turbulence. Angular momentum transport is much less affected because it varies on a level surface if $\omega = \omega(r)$, and thus it will occur even in the presence of horizontal turbulence. Based on this reasoning, the same equation should be solved for material mixing, but the angular momentum transport equation is replaced by

$$\frac{d\omega}{dt} = \frac{1}{4\pi\rho r^2\tilde{I}}\left[\frac{d}{dr}(4\pi\rho r^2\tilde{I}v\omega) + \frac{d}{dr}\left(4\pi\rho r^2\tilde{I} f_c D\frac{d\omega}{dr}\right)\right], \tag{3}$$

where $v$ is the diffusion velocity, and for meridional circulation $D = vR$ ($R =$ radius). The efficiency factor $f_c$ can be computed from the velocity field (Zahn 1992; see below).

MERIDIONAL CIRCULATION The best-known hydrodynamic angular momentum transport and mixing mechanism is meridional circulation (Eddington 1925, Sweet 1950, Tassoul & Tassoul 1982, Zahn 1992). It is caused by the inability of a rotating star to maintain hydrostatic equilibrium in radiative regions because of variations in gravity between pole and equator on isobars. For rotation on cylinders, the potential is conservative and surfaces of constant thermodynamic variables ( P, T, $\rho$) coincide and lie on equipotential surfaces. In this case, the circulation velocity can be estimated reliably; Kippenhahn & Möllenhof (1974) give

$$v_{es} = \frac{-1}{g\delta}\frac{\nabla_{ad}}{\nabla_{ad} - \nabla}\left\{2\left(\frac{L}{M} - \varepsilon\right)\frac{\tilde{g}}{g} + \frac{L}{4\pi GM\rho r}\frac{d}{dr}(\omega^2 r^2)\right\}, \tag{4}$$

where $\delta$ is $(d \ln \rho)/(d \ln T)$, $\varepsilon$ is the energy generation rate, and $\tilde{g}/g$ is the departure from spherical symmetry.

In general, as stars evolve, differential rotation with depth is generated, both by structural evolution and angular momentum loss. Angular velocity gradients on surfaces of constant density are relatively easy to remove by shear instabilities (see Zahn 1992). The angular velocity is therefore likely to be constant on surfaces of constant pressure, but it can be a function of radius. In this case, the potential is nonconservative, and surfaces of constant ( P, T, $\rho$) do not coincide. This gives rise to additional terms in the circulation velocity estimate (Sakuri 1991, Zahn 1992). Zahn (1992) estimates the circulation velocity as

$$
\begin{aligned}
v_{es} = & \frac{-\nabla_{ad} r^2}{\frac{d\ln\rho}{d\ln T}(\nabla_{ad} - \nabla) GM} \\
& \times \left\{ 2\left[\frac{L}{M}\left(1 - \frac{\omega^2}{2\pi G\rho}\right) - \varepsilon\right]\frac{\tilde{g}}{g} \right. \\
& + \left[\left(\frac{H_T}{r} - \frac{1}{3}\right)\frac{L\omega r}{\pi GM\rho} - \frac{2\varepsilon\omega r^4}{3GM}(\varepsilon_T - \chi_T)\right]\frac{d\omega}{dr} \\
& \left. + \frac{L}{4\pi\rho r^2}\frac{d}{dr}\left(H_T\frac{d\theta}{dr} - \chi_T\theta\right) - \varepsilon H_T \frac{d\theta}{dr}\right\},
\end{aligned}
\tag{5}
$$

where $\chi = [(4ac)/(3\kappa\rho)]T^3$, $\varepsilon$ is the energy generation rate, $H_T$ is the temperature scale height, $\omega$ is the angular velocity. The departure from spherical symmetry $\tilde{g}/g$ is

$$
\frac{\tilde{g}}{g} = \frac{\omega^2}{3}\frac{d}{dr}\left(\frac{r^4}{GM}\right) - \frac{d}{dr}\left(\frac{\hat{\Phi}r^2}{GM}\right),
\tag{6}
$$

where $\hat{\Phi}$ is the quadrupole moment. $\theta$ is a function of the derivative of the angular velocity. The diffusion coefficient for meridional circulation is taken as $|v_{es}r|$. I note that the classical expression for the departure from symmetry, $(4\delta^2 r^3)/(3GM)$, is in significant error both in stellar cores (where $dM/dr$ is large) and for strong differential rotation (where the quadrupole is important). If the quadrupole dominates, the sign of $\tilde{g}/g$ will be reversed; this can also occur in some of the other terms in the circulation velocity estimate. The most important difference from the classic picture of meridional circulation is that a circulation-free state can be established in the Zahn (1992) framework, which may help greatly in explaining some of the observational data.

The Kippenhahn and Zahn prescriptions have been energetically challenged by Tassoul & Tassoul in a series of papers (e.g. Tassoul & Tassoul 1995 and references therein). The essential question is whether or not viscous stresses can prevent the large increase in circulation velocity implied by the above equations in the outer envelopes of stars with thin surface convection zones. There is an overall similarity between the philosophy of the latest works by Zahn and the Tassouls, namely the explicit recognition that the final meridional circulation velocity results from the balance of a series of effects. The different techniques reduce to similar results for stars with thick surface convection zones.

The above expression does not include the inhibition of meridional circulation by gradients in mean molecular weight $\mu$. Mean molecular weight gradients can inhibit circulation in two ways. A latitude-dependent $\mu$ distribution can choke off the circulation (Mestel 1953). Lifting material in the presence of a $\mu$ gradient also requires energy [see Roxburgh 1991 for a nice discussion of this point in the context of the Goldreich-Schubert-Fricke (GSF) instability]; this would dampen rather than suppress meridional circulation.

The generation of a latitude-dependent $\mu$ distribution is much more difficult to quantify. Because meridional circulation relies on small imbalances between pole and equator to drive large-scale circulation currents, even small latitude-dependent effects could stabilize a region with a $\mu$ gradient. Because horizontal turbulence is relatively easy to produce in stars (Zahn 1992), and it will tend to erase horizontal composition gradients, the net effect will be the balance between two (uncertain) processes, namely the rate at which latitude-dependent $\mu$ gradients can be produced (by vertical mixing in the presence of a $\mu$ barrier or by nuclear reactions occurring at slightly different T) and the rate at which they can be removed (by horizontal mixing). To further complicate matters, other instabilities [i.e. shear, GSF, axisymmetric baroclinic diffusive (ABCD)] can also be present, and such mechanisms will in general require a different latitude-dependent profile to suppress mixing; in fact, as discussed by Knobloch & Spruit (1983), variations in abundance on a level surface can themselves trigger instabilities and cause mixing (the "haloclinic" instabilities).

Empirical data, however, favors mixing in the presence of $\mu$ gradients in at least some cases. There is evidence in the Sun for both helium diffusion, which would create a $\mu$ gradient, and mixing sufficient to deplete the solar lithium and beryllium (Michaud & Proffitt 1993, Chaboyer et al 1995a). Data in massive stars and evolved low mass stars also requires that mixing be capable of penetrating $\mu$ gradients. The degree to which composition gradients can reduce mixing, however, is still important and cannot be regarded as a solved theoretical issue.

OTHER INSTABILITIES A variety of other rotational instabilities have been discussed in the literature. In most cases, the time scale estimates are significantly more uncertain than the time scale for meridional circulation. They can be divided into two general classes: dynamical instabilities, which operate on time scales much shorter than the evolutionary time scale, and secular instabilities, which operate on time scales comparable to or greater than the evolutionary time scale. See Tassoul (1978), Endal & Sofia (1978), and Knobloch & Spruit (1982) for a discussion of dynamical instabilities, which in general require larger angular velocity gradients than are expected in stars during most evolutionary phases. The major other instabilities that have been discussed in the astrophysical literature are as follows.

*Secular Shear* Differential rotation can survive in the presence of a density gradient in stars; the Richardson criterion expresses the condition for marginal stability (e.g. Shu 1992). Radiative heat losses can reduce the stabilizing effect of a density gradient (Townsend 1958, Zahn 1974); recent estimate of the stability conditions and vertical diffusion coefficient can be found in Zahn (1992), Maeder (1995), and Maeder & Meynet (1996). Gradients in $\mu$ can prevent the onset of shear instabilities, but recent work indicates that mixing by the shear instability may be important in semiconvective regions (Maeder 1995).

*GSF and ABCD instabilities* Stars are unstable against axisymmetric perturbation in the case where rotation is not on cylinders (Goldreich & Schubert 1967, Fricke 1968). Horizontal turbulence suppresses latitude dependence of the rotation profile, and evolutionay effects generate differential rotation as a function of radius. The net effect of the GSF instability will be to reduce radial angular velocity gradients. The time scale for the instability was initially thought to be dynamical, but more recent estimates indicate that it operates over a time scale comparable to meridional circulation (Kippenhahn et al 1980). The GSF instability is strongly damped by $\mu$ gradients, and even viscosity, but the ABCD instability is not inhibited by $\mu$ gradients (Knobloch & Spruit 1983, Spruit et al 1983).

## 5.2 *Internal Waves and Magnetic Fields*

Internal waves have recently received attention both as a potentially efficient mechanism for the internal transport of angular momentum and as a possible means of mixing in the envelopes of low mass stars. Magnetic fields can also be remarkably efficient at internal angular momentum transport. Because the internal angular momentum distribution is crucial for computing the degree of rotational mixing, magnetic fields may have an important indirect impact on mixing in stars. An excellent summary of the various mechanisms for angular momentum transport can be found in Charbonneau et al (1995). I begin with waves and then discuss magnetic fields.

WAVES AS ANGULAR MOMENTUM TRANSPORT AGENTS Gravity waves can be generated in stars, possibly by turbulence in convective regions. Waves have long been appreciated to be important for angular momentum transport in the Earth's atmosphere and ocean (see for example Lighthill 1978, Phillips 1977). Press (1981) demonstrated that significant internal gravity waves could exist in stars, and Goldreich & Nicholson (1989) investigated the behavior of gravity waves in stellar interiors in the context of tidal braking (see also Goldreich et al 1994). Two recent papers (Kumar & Quataert 1997, Zahn et al 1996) have raised the provocative prospect that gravity waves could enforce nearly rigid rotation in solar mass stars with a time scale on the order of $10^7$ years. If confirmed, this would reduce significantly the complexity of the problem of rotational mixing: After a brief period in the early MS, low mass MS stars could be treated as solid body rotators. An extension of the theory to higher mass stars (with convective cores) and evolved giants (with deep convective envelopes) would be highly desirable. Given the relatively short lifetimes of both massive stars and giants, it is possible that solid body rotation would not be enforced by waves in either case.

WAVES AND MIXING Gravity waves may theoretically produce mixing directly; in the astrophysical context, this issue was initially explored by Press & Rybicki (1981), who found only a small effect. Enhanced mixing may be produced with difficulty via waves in stellar interiors (Garcia Lopez & Spruit 1991, Schatzmann 1991, Montalban 1994, Montalban & Schatzmann 1996). The properties of wave-driven mixing are compared with the data in Section 6.1.

MAGNETIC FIELDS AND ANGULAR MOMENTUM TRANSPORT Even small internal stellar magnetic fields could potentially be remarkably effective agents for angular momentum transport (Spruit 1987). There have been vigorous debates in the literature about whether or not magnetic fields enforce rigid rotation in stars (for example, Mestel & Weiss 1987, Tassoul & Tassoul 1989). Charbonneau & MacGregor (1992, 1993) investigated the full magnetohydrodynamic problem for a variety of field strengths and configurations. They found a weak dependence on field strength but a strong dependence on the field morphology. Some configurations produced uniform rotation on a short time scale, whereas others permitted strong differential rotation with depth to survive to the age of the Sun. Some difficulties are present in reconciling the strong coupling case with the helioseismic data on the internal rotation of the Sun: The flatness of the solar rotation curve is naturally explained, but the latitude dependence present in the solar convection zone does not appear to be present in the radiative interior (Zahn et al 1996). The rich behavior of full-scale simulations, and the potential importance of magnetic instabilities (Balbus & Hawley 1994), is an indication that further study of magnetic angular momentum transport is needed. Given the difficulty in constraining the strength or the morphology of internal stellar magnetic fields, observational tests to distinguish the imprint of magnetic angular momentum transport on surface properties of stars would be highly desirable.

### 5.3 *Gravitational Settling, Thermal Diffusion, and Radiative Levitation*

Strictly speaking, the processes discussed in this section are element separation processes rather than mixing processes. They are germane to the discussion of mixing in stars because they are a competing explanation for some of the surface abundance anomalies I discuss above and because the interaction between atomic diffusion and mixing processes could be important. Heavier elements tend to sink relative to lighter elements because of both gravitational settling and thermal diffusion; radiative levitation can drive neutral or partially ionized species upwards (Burgers 1969, Chapman & Cowling 1970, Michaud et al 1976; see Vauclair 1983 for a review of the physics of microscopic diffusion). In general, radiative levitation is important only in stars with thin surface convection zones, particularly the chemically peculiar A stars. Recent calculations of the coefficients in a form convenient for astrophysical use can be found in Thoul et al (1994) and Proffitt (1994); see also Paquette et al (1986), Michaud & Proffitt (1993), and Richer & Michaud (1993).

Microscopic diffusion is a strong function of depth, with the time scale increasing rapidly with depth within a given model. The time scale for diffusion is also far shorter for stellar models with shallow convection zones than for stellar models with deep surface convection zones. Diffusion of helium will produce a $\mu$ gradient at the base of the surface convection zone, which provides an observational test of the ability of mixing to cross a barrier in mean molecular weight. Diffusion is expected to be a small effect for giants, MS stars with deep surface convection zones, and stars with strong mass loss. It could play a role in the origin of the mid-F lithium dip.

## 6. THEORETICAL RESULTS FROM NONSTANDARD MODELS

All of the phenomena discussed in Section 4 have been investigated with theoretical models in varying degrees of detail. Such models can be roughly divided into three classes. In initial studies, arbitrary mixing is induced to constrain the relevant depth and time scale needed to explain the data. Such models represent the state of the art for the treatment of mixing in giants. Mixing can also be tied to specific physical mechanisms, such as the time scale estimates for rotational mixing or wave-driven mixing discussed in Section 5. At this stage, a detailed comparison between the observational data and theoretical models is not generally carried out, but the observed systematics are tied to underlying physical causes. Models for extra mixing in massive stars fall into this category. In low mass stars, detailed models have been constructed that involve both mixing and angular momentum evolution, and a number of independent tests of the theory can be applied. After presenting the theoretical studies of mixing in low mass stars, I discuss mixing in massive and evolved stars.

## 6.1 *Mixing on the Main Sequence: Low Mass Stars*

Mild envelope mixing is the most likely explanation of the light element depletion pattern seen in disk MS stars. Both lithium and beryllium must be destroyed, lithium more efficiently than beryllium. The normal abundances seen on the RGB after the first dredge-up (Gilroy 1989) severely constrain deeper mixing, as does the excellent agreement between the thermal structure of unmixed solar models and the actual solar sound speed as a function of depth inferred from helioseismology (Bahcall et al 1997). Lithium must be preserved in intermediate mass MS stars and mixing cannot suppress the development of chemical peculiarities linked with microscopic diffusion and radiative levitation (Michaud & Charbonneau 1991, Charbonneau et al 1989). The absence of mixing in these cases places strong constraints on the theory. Several major classes of physical processes have been claimed as explanations for the light element abundance pattern seen in low mass stars:

- microscopic diffusion and radiative levitation,
- MS mass loss,
- mixing driven by waves, and
- mixing driven by rotation.

Some authors have explored models containing more than one of the above ingredients. Mass loss and microscopic diffusion have difficulty in reproducing the observed abundance pattern in open clusters (see below), so most theoretical work on nonstandard models has involved mixing. For the halo stars, standard models are competitive with nonstandard models. To establish that standard models can accurately predict the initial abundances of halo stars, however, simple demonstration of agreement between the halo star data and standard models is insufficient. The origin of anomalous abundances in open clusters must be identified, and a reason why it is not operating in the halo stars must be presented.

Many of the overall features will be important for the other cases for nonstandard mixing (massive and evolved stars) because the rotational properties of low mass MS stars are far better studied than the rotational properties of higher mass and evolved stars.

MICROSCOPIC DIFFUSION AND RADIATIVE LEVITATION Microscopic diffusion is an elegant explanation for the mid-F star lithium dip (Michaud 1986, Richer & Michaud 1993); the time scale is too long for it to contribute significantly to surface lithium depletion in cool stars. The time scale for downward diffusion decreases as the surface convection zone becomes shallower; this leads to a strong decrease in surface abundance with increased $T_{eff}$. Once lithium becomes partially ionized below the surface convection zone, radiative levitation can cause a rise in the surface abundance for hotter stars. A smaller beryllium dip, offset in temperature, is also expected from models with diffusion. The lithium is stored below the convection zone, and not destroyed, by diffusion. The surface lithium abundances of stars coming from the dip region would therefore be expected to increase dramatically as their surface convection zones become deeper and dredge up lithium-rich material. This is inconsistent with data for subgiants in M67 (Balachandran 1995). A dispersion in abundance is not expected for models with diffusion and is present in the data. Diffusion is a physical effect that should be included in models, but it is probably not the sole agent responsible for the observed light element depletion pattern, either in the dip or for cooler open cluster stars (Pinsonneault 1994, Balachandran 1995).

Another application of diffusion is in the lithium abundance pattern seen in halo stars (Proffitt & Michaud 1991, Chaboyer & Demarque 1994, Vauclair & Charbonnel 1995, Swenson 1995). Helium diffusion occurs over a comparable time scale and can have implications for the ages inferred for globular clusters (Stringfellow et al 1983, Proffitt & Vandenberg 1991, Chaboyer et al 1992). Models with diffusion predict little dispersion, which is compatible with the data, and a mean trend of decreased abundance with increased $T_{eff}$, which is in conflict with the observed flatness of the Li-$T_{eff}$ relationship for halo stars. Vauclair & Charbonnel (1995) and Swenson (1995) explored a combination of enhanced MS mass loss and diffusion as a possible solution; note that initial abundances higher than the observed value are needed in general. A combination of mixing and diffusion is more likely (Chaboyer & Demarque 1994) because the open cluster data is not consistent either with mass loss or diffusion alone. However, none of the models examined by Chaboyer & Demarque were entirely consistent with the halo star data, a rather troubling state of affairs for an important issue. The lithium data does constrain globular cluster age changes induced by helium diffusion (Chaboyer et al 1992).

MAIN SEQUENCE MASS LOSS Lithium survives only in the outer $\sim 0.02\ M_\odot$ of stars. MS mass loss could therefore potentially cause surface lithium depletion, provided the rate is far in excess of the solar wind rate on the order of $10^{-14}\ M_\odot$ year (Weymann & Sears 1965, Hobbs et al 1989, Swenson & Faulkner 1992). Swenson & Faulkner demonstrated that mass loss cannot produce the observed abundance pattern for cool stars with deep convective envelopes, but they left it as a possibility for the mid-F star lithium dip (see also Dearborn et al 1992). The measurement of normal lithium abundances in $\delta$ Scuti stars (Russell 1995), which has been postulated as the phase where extensive MS mass loss could be generated, and the lithium dilution pattern seen in subgiants (Pilachowski et al 1993, Balachandran 1995, Hiltgen 1996, Deliyannis et al 1997) argue against MS mass loss as an explanation for lithium depletion even in the mid-F star regime.

MIXING DRIVEN BY WAVES Garcia Lopez & Spruit (1991) investigated mixing driven by waves as a possible explanation of the F star lithium dip. Montalban (1994) developed a prescription for mixing following the method suggested by Schatzmann (1991) and explored wave-driven mixing in the Sun. These models were extended to low mass MS stars by Montalban & Schatzmann (1996). Wave-driven mixing occurs over a long time scale that can vary within the envelope and as a function of time. It therefore has some of the properties of rotational mixing including simultaneous beryllium and lithium depletion. The light element depletion expected from theoretical models can be adjusted to reproduce the mass dependence of open cluster lithium depletion in the Hyades and Praesepe clusters with a suitable choice of pre-MS depletion added to the MS models. A dispersion in abundance, however, would not be expected from mixing driven by waves. As the existence of a dispersion is one of the central features of the observed depletion pattern, this suggests that mixing driven by waves is not the sole mechanism for light element depletion. Qualitative suggestions have been made that mixing from waves could be modulated by other effects, such as rotation; these ideas should be quantified and compared with other mechanisms.

MIXING DRIVEN BY ROTATION Rotational mixing and light element depletion have been investigated by a series of authors; recent reviews of theoretical models of light element depletion can be found in Michaud & Charbonneau (1991) and Pinsonneault (1994, 1995).

Early theoretical work focused on using empirical constraints on turbulent diffusion coefficients, following the suggestion of Schatzmann (1969). Modest transport coefficients were found to have an appreciable impact on surface abundances, with some authors investigating the possibility of extensive deep mixing sufficient to explain the solar neutrino problem (Schatzmann 1977, Vauclair et al 1978, Schatzmann et al 1981, Baglin et al 1985, Lebreton & Maeder 1987, Schatzmann & Baglin 1991). Lack of data on the angular momentum evolution of stars, along with limited empirical data on stellar abundances, made it difficult to constrain theoretical models.

Since the late 1980s, the focus of theoretical work has changed, with mixing tied to specific physical mechanisms and extensive use of the observed constraints on angular momentum evolution. The method of Endal & Sofia (1978) has proved especially useful in this regard. Endal & Sofia advocated treating rotation as an initial value problem and solving simultaneously for angular momentum loss, the transport of angular momentum, and mixing. Full evolutionary calculations of rotation from the pre-MS have been computed for open cluster stars below the dip and halo stars, including angular momentum loss and assuming angular momentum transport from hydrodynamic mechanisms (Pinsonneault et al 1989, Pinsonneault et al 1990, Pinsonneault et al 1992, Chaboyer & Demarque 1994, Chaboyer et al 1995a,b). Zahn's (1983, 1992) prescriptions for mixing have also been investigated for open cluster and halo star models, which assume solid body rotation in the interior and include angular momentum loss (Vauclair 1988, Charbonnel & Vauclair 1992, Charbonnel et al 1992, Charbonnel et al 1994, Richard et al 1996). The similarity between the results of the two classes of models is indicative of the limited sensitivity of the results to the details of the treatment of angular momentum transport. Models of mixing from meridional circulation have also been computed for stars in the lithium dip and above (Charbonneau & Michaud 1988a,b, 1990, 1991). Zahn (1992, 1994) proposed a new model for meridional circulation and examined the behavior of tidally locked binaries; Martin & Claret (1996) looked at the structural impact of rotation on pre-MS lithium burning. Stars below the mid-F lithium dip suffer extensive angular momentum loss, which dominates their angular momentum evolution; stars in the dip and above have a different rotational history. This has a strong impact on the mixing properties of the models, which are examined separately below.

*Stars below the dip* A large and growing database of stellar rotation observations has been complemented by a series of theoretical studies of stellar angular momentum evolution. The improvement in our understanding of stellar rotation holds great promise in reducing the most significant source of uncertainty in rotational mixing, so I briefly summarize some of the most important features of stellar angular momentum evolution below (for recent papers with a discussion of angular momentum evolution, see Keppens et al 1996, Charbonneau et al 1997, Krishnamurthi et al 1997).

The natural starting point for stellar models with rotation is the pre-MS. The most widely accepted hypothesis for the origin of the range of rotation rates in low mass stars is that accretion disks enforce constant rotation periods on the order of 10 days, which is typical for classical T Tauri stars (Edwards et al 1993, Bouvier 1994). A range of accretion disk lifetimes will therefore produce a range of MS rotation rates, as stars that detach from their disks early will experience a larger change in their moment of inertia than stars that detach from their disks closer to the MS.

A wide range of rotation rates is seen in young stars. Angular momentum loss is more rapid in fast rotators, and the observed surface rotation rates converge by a few hundred million years for solar analogs (Stauffer 1994). At late ages the surface rotation velocities of stars decline as $v \sim t^{-1/2}$ (Skumanich 1972).

These overall properties lead to some general features of rotational mixing for models that are consistent with the observed stellar rotation data.

Rotational mixing is not effective in the pre-MS, so little dispersion in abundance is expected among young cluster stars (Pinsonneault et al 1990, Chaboyer et al 1995b). The observed dispersion for cool stars in young clusters is therefore not a manifestation of variations in early rotational mixing. The structural impact of rotation on pre-MS depletion has recently been suggested as a possible culprit (Martin & Claret 1996) and deserves further exploration.

Rotational mixing is tied to the absolute rotation rate and its derivative. On the MS, stars with higher rotation rates will therefore experience more rapid mixing than slow rotators at the same mass. The minimum dispersion is set by the time scale for the observed convergence of the surface rotation rates; the time scale for convergence of the internal rotation profile is model dependent but on the order of a few hundred million years in the most weakly coupled plausible models. The existence of a dispersion is thus a prediction of rotational mixing rather than a problem.

Theoretical models of rotational mixing in solar analogs produce a rate of depletion that decreases with age as stars spin down. The time scale for mixing also increases with distance from the surface convection zone; solar models with rotational mixing predict beryllium depletion of a factor of 2–3 for lithium depletion of a factor of 100–200 (Pinsonneault et al 1989, Richard et al 1996).

Tidally locked binaries have a different rotational history, and a period range exists where their rotation period evolves slowly. A circulation-free state is therefore possible for some tidally locked binaries (Zahn 1994, Ryan & Deliyannis 1995), which may explain their high abundances and better agreement with standard stellar models.

Rotational mixing therefore has the qualitative features needed to explain the depletion pattern seen in open cluster stars, with the possible exception of the dispersion in abundance among young cool stars. Detailed examination of the theoretical models, however, raises a series of questions: the validity of the underlying angular momentum and mixing models, the mass dependence of extra mixing, and the expected magnitude of the dispersion.

Models with internal angular momentum transport from hydrodynamic mechanisms in general predict the survival of a rapidly rotating core in the

Sun, which is in conflict with the helioseismic data (Tomczyk et al 1995). In general, the mixing expected from the theoretical models is relatively insensitive to the treatment of internal angular momentum transport in stars with angular momentum loss; the models adjust themselves to a rotation profile for which the diffusion coefficients for angular momentum transport balance the torque applied by the wind. Calibrating the models on the solar lithium depletion can further reduce the sensitivity of the results to errors in the physical model. The solar models of Richard et al (1996) and Pinsonneault et al (1989), for example, use very different treatments for angular momentum transport but have the same relative lithium and beryllium depletion. However, a proper matching of the surface rotation as a function of time can be important, and only recently have the theoretical models been able to reproduce the surface rotation rates of low mass stars as a function of mass and time (Keppens et al 1996, Krishnamurthi et al 1996). The uniqueness of the model is also an issue (Michaud & Charbonneau 1991, Schatzmann & Baglin 1991), although most of the degrees of freedom can be removed through the application of empirical constraints on the angular momentum evolution.

The mass dependence of depletion is an important test; the models of Chaboyer et al (1995b), for example, predict mixing that depends only weakly on mass for open cluster stars; the open cluster data appears to show increased mixing for lower mass stars (Jones et al 1996a). Models of halo stars with mixing show a plateau that is nearly flat, but they do not have the observed increase in lithium for the hottest and most metal-poor stars (Chaboyer & Demarque 1994). Because similar discrepancies exist in the mass dependence of the rotation rates for these models, this indicates the importance of an accurate treatment of the angular momentum evolution.

Rotational mixing predicts a dispersion in abundance, but the magnitude of the dispersion will depend on the distribution of initial conditions. Quantitative tests of the distribution of lithium depletion factors expected from the distribution of stellar rotation rates have not been performed. The majority of young stars are slow rotators, with a small population of fast spinning stars (Stauffer 1994, Allain et al 1996). This will tend to produce a well-defined mean trend with a subpopulation of overdepleted stars, rather than a large scatter in abundance. Consistency between the distribution of initial conditions needed to explain rotation and the distribution needed to explain lithium has been neither demonstrated nor disproven to date.

The rotational models predict a correlation between the rotation history of a star and lithium depletion, rather than a correlation between the current rotation of a star and its lithium abundance. This makes establishing a causal connection between rotation and mixing difficult; systems such as M34, which retain a range of surface rotation and have a dispersion in lithium, will be important in testing the detailed properties of the models. Future models will need to address the above issues, along with quantitative estimates of the degree of light element depletion in tidally locked binaries.

*The lithium dip and above* The effective temperature range where the lithium dip occurs is a transition zone from the point of view of stellar structure and angular momentum. Stars below the dip have deep surface convection zones and suffer angular momentum loss; stars above the dip have shallow surface convection zones and do not experience MS angular momentum loss from a magnetic wind. All of the explanations for the dip rely on these changes in behavior to explain the lithium dip. Important overall rotation properties in the dip are as follows.

The rotation velocities of stars increase across the lithium dip as the convection zones become shallower and angular momentum loss becomes ineffective. Stars earlier than spectral type F8 have rotation velocities uncorrelated with their age (Benz & Mayor 1984).

There is a wide range of surface rotation rates on the MS above the dip. In stars hotter than the dip, peculiar abundances indicative of gravitational settling, thermal diffusion, and radiative levitation are seen in slow rotators (Charbonneau & Michaud 1991), and cluster giants have normal CN abundances, with some anomalous lithium abundances (see Section 6.3).

All of the mixing explanations rely on the increase in rotation with increased $T_{eff}$ to produce the cool side of the lithium dip. There are varying suggestions for the increase in surface abundance on the hot side of the dip, which are described in the next few paragraphs.

For stars with a sufficiently thin surface convection zone, a separation into two distinct mixed zones is possible (Vauclair 1988, Charbonnel et al 1992); the region where such a separation would occur in cool stars is within their surface convection zone. The hot side of the lithium dip is therefore explained by the presence of a "quiet zone" that inhibits mixing to temperatures sufficient to destroy lithium. A complete separation into two distinct zones, however, is not found in more detailed models (Charbonneau & Michaud 1990, Charbonnel & Vauclair 1992, Zahn 1992, Tassoul & Tassoul 1995), although the efficiency of transport could be reduced.

The interaction between mixing and microscopic diffusion is another possible solution for such an increase in surface abundance (Charbonneau & Michaud 1988a,b, Michaud & Charbonneau 1991). Radiative levitation produces the rise on the hot side of the dip. Microscopic diffusion proceeds unimpeded up to some threshold rotation rate, above which it is damped. The threshold rotation rate for producing composition anomalies via diffusion was found to be lower ($v < 15$ km/s) than the observed rotation rates of stars in the middle of the lithium dip (~50 km/s), although the velocity cutoff for producing diffusion in higher mass stars was in good agreement with the data (Charbonneau & Michaud 1991).

A third possibility centers on the possibility of establishing a circulation-free state in stars without angular momentum loss (Zahn 1992). In this picture, stars adjust themselves to a state with zero net circulation in the absence of angular momentum loss (mass loss or rapid structural change could presumably also prevent the establishment of a circulation-free state). This explains the lack of mixing in stars above the dip. The picture advocated by Zahn is an intriguing suggestion that should be verified with full calculations.

## 6.2 *Mixing on the Main Sequence: High Mass Stars*

Nonstandard models of mixing in massive stars have now been investigated by a variety of authors, in response to the data indicating evidence for departures from the predictions of standard models. Rotational mixing may even prove an alternative to the various theories of convective overshoot and semiconvection that have traditionally been debated in the literature (Deng et al 1996a,b, Talon et al 1996). There are some important differences between the rotation and evolution of massive stars when compared with the low mass case.

- A wide range of rotation rates persists for the MS lifetime; a significant fraction of the stars rotate at or near the critical rotation rate (Fukuda 1982).

- Energy generation is in a convective core in massive stars. If rotational mixing is able to penetrate the core, it may permit fully mixed evolution; otherwise the development of a steep $\mu$ barrier during the lifetime will prevent deep mixing. Changes in the internal abundance profiles on the MS can influence subsequent evolution. Some extra mixing appears to be needed to explain the blue progenitor of SN 1987a (Langer 1991).

- The viscosity is far higher in massive stars than in low mass stars, which implies that the time scale for mixing may be comparable to, or less than, the MS lifetime (Maeder 1987).

- Mass loss is a strong effect on the structure and evolution of massive stars; the changes in surface abundance produced by mass loss need to be separated from those produced by mixing.

A successful model should also be able to reproduce the absence of mixing in intermediate mass MS stars and the increased mixing seen with increased mass and rotation. Maeder (1987) showed that a bifurcation in evolution is possible for the most massive stars, with ~15% of the stars experiencing nearly fully mixed evolution and the more traditional evolutionary path for the remainder. The inhibition of mixing by $\mu$ gradients was identified as the most important ingredient of the models. More recently, rotational models have been constructed for 9- to 20-$M_\odot$ stars (Langer 1992, Denissenkov 1994, Eryurt et al 1994, Fliegner et al 1996, Urpin et al 1996, Talon et al 1996), and different prescriptions for mixing have been explored. The major conclusions of these studies are summarized below.

Abundance anomalies in OBN stars can be explained within the framework of rotational mixing (Maeder 1987, Langer 1992, Eryurt et al 1994), and the effects of rotational mixing may be important for understanding the progenitor of SN 1987a. Full evolutionary models with rotation have not been computed for the most massive stars because of the complications of such effects as semiconvection, mass loss, and the possibility that a true zero-age MS may not exist for very high masses. The qualitative arguments of Maeder (1987) therefore need to be verified with direct calculations.

Full evolutionary models of a 9-$M_\odot$ model with rotation were computed by Talon et al (1996). Surface CN anomalies correlated with increased rotation were produced, but with little surface helium enrichment. Denissenkov (1994) found that a surface N enrichment that correlated with increased age was present in a 10-$M_\odot$ model. More recently, Fliegner et al (1996) examined boron and nitrogen abundances in 10- to 15-$M_\odot$ stars with rotation. They concluded that the N/B ratio was a diagnostic of deep mixing, with the measured B depletion not expected in models with mass loss but no extra mixing.

An important component in the absence of mixing for intermediate mass stars is the possibility of a circulation-free state. Urpin et al (1996) found that a circulation-free model can be produced within the framework of the Zahn (1992) prescription (see also Talon et al 1996). An extension of their static treatment to evolutionary models with mass loss would be desirable, in order to verify that mixing is indeed present when needed and suppressed when it is not.

The overall picture appears to be consistent with rotational mixing: Mixing is most rapid for the most massive stars and the fastest rotators. There are some important steps that need to be taken, however, to quantify the role of rotation in the evolution of massive stars. Most of the models to date have not evolved the angular momentum distribution from the pre-MS (see Eryurt et al 1994 for an exception); the origin of the range of rotation rates seen is almost certainly in the pre-MS, and the pre-MS evolution could influence the behavior at later ages. The effects of rotational mixing need to be clearly separated from other effects, such as the treatment of convection and mass loss. The distribution of surface abundance anomalies should also be consistent with the distribution of rotation rates, and the impact of mixing on chemical evolution models also needs to be addressed. Finally, it remains to be verified that the same physical model that produces mixing in massive stars does not produce excess mixing in intermediate mass stars.

## 6.3 *Mixing in Evolved Low Mass Stars*

The abundance pattern seen in giants is as rich as that for lithium in dwarfs and is equally compelling evidence for nonstandard mixing. Despite this, the status of the theory cannot be regarded as well developed, although most work implicitly assumes that mixing is driven by rotation. This can be traced to the greater computational difficulty in constructing giant branch models in part. However, there are also significantly greater physical uncertainties related to the paucity of rotational data for giants. The alternative possibilities of solid body rotation and constant specific angular momentum in the convective envelopes of giants, for example, give radically different prospects for mixing (e.g. Sweigart & Mengel 1979), and they cannot be distinguished observationally. Rotation data on the MS and in evolved stars is now available, and the constraints such data impose on the models are discussed below. Much of the recent interest in giant branch mixing stems from the realization that such mixing could have cosmological significance. $^3$He can be destroyed in low mass stars, which would affect Big Bang nucleosynthesis (BBN) yields depending on the chemical evolution model (Rood et al 1984, Deliyannis 1995, Hogan 1995, Wasserburg et al 1995, Weiss et al 1996). Mixing of $^4$He into the envelope could affect a variety of globular cluster properties, including age (Sweigart 1997).

LITHIUM IN INTERMEDIATE MASS GIANTS The presence of low lithium in some open cluster giants of intermediate mass (Gilroy 1989) could be explained if rapid rotators experienced mixing on the MS (Charbonneau et al 1989, Charbonnel & Vauclair 1992). There are strong observational selection effects against measuring lithium in hot rapid rotators, so MS surface depletion does not necessarily contradict the data. The measurement of high lithium in rapidly rotating subgiants (Wallerstein et al 1994) presents some difficulty for the Charbonneau et al (1989) model. Charbonnel & Vauclair (1992) proposed that surface lithium would be intact, but the size of the lithium preservation zone would vary with rotation, leading to differences in giant branch abundance. As noted in Section 6.1, there are some physical difficulties with the separation into two mixed zones that this model requires. The first ascent giants with high lithium abundance are a puzzle, with no compelling theoretical explanation (Fekel & Balachandran 1993); some speculative recent possibilities are discussed by de la Reza et al (1995).

DEEP MIXING IN LOW MASS GIANTS Giant branch mixing depends on both the structural evolution and the angular momentum evolution. The most important properties (see Sweigart & Mengel 1979 for a good discussion) are as follows.

A $\mu$ gradient exists at the base of the convection zone on the lower giant branch caused by the dredge-up of material that was partially burned by the proton-proton (pp) chain on the MS. This may prevent mixing on the lower giant branch (Sweigart & Mengel 1979, Charbonnel 1994). This $\mu$ gradient increases with increased metallicity.

Material falls into the hydrogen-burning shell with a time scale that decreases as the luminosity rises. At the same time, the outer radius expands. Giant branch mixing is therefore a threshold process. The time scale for mixing must also decrease with increased luminosity more rapidly than the infall time scale, given the observational evidence for enhanced mixing in luminous giants.

The ability of mixing to penetrate $\mu$ gradients determines how deep the mixing can go into the vicinity of the hydrogen-burning shell. Because some species ($^4$He, Na, O) are affected at higher temperatures than others (C, N), the sensitivity to $\mu$ gradients affects the observed mixing pattern. C $\rightarrow$ N processing occurs farther out in metal-poor stars than in more metal-rich stars, which makes mixing easier for lower Z (Sweigart & Mengel 1979).

Structural evolution on the giant branch produces strong differential rotation with depth. Constant angular momentum per unit mass (J/M) in the convection zone requires much lower rotation rates than solid body rotation in the convection zone, but the available angular momentum reservoir for either case is smaller than that required by Sweigart & Mengel (1979) to drive meridional circulation.

MS metal-poor stars rotate slowly, whereas rapid rotation is present in some evolved horizontal branch stars (Peterson 1983, Peterson 1985a,b, Peterson et al 1995). The degree of horizontal branch rotation differs from cluster to cluster; M13 has much higher typical rotation rates than M3 or NGC 288, for example. The combination of slow MS rotation and rapid horizontal branch rotation requires differential rotation with depth, either on the MS or in the convective envelope on the giant branch (Pinsonneault et al 1991). This is comforting, as either case makes it easier to sustain rotational mixing on the giant branch.

Sweigart & Mengel (1979) investigated meridional circulation as the possible agent for giant branch mixing. They assumed that mixing was inhibited by $\mu$ gradients, so neither $^4$He mixing nor mixing on the lower giant branch was expected in their model. They used the classical expression for meridional circulation and assumed no internal angular momentum transport. With these assumptions, they found that mixing could be driven, but with rotation rates that exceeded the surface v sin i limits (and which violate present constraints on the internal angular momentum content of such stars).

Charbonnel (1994) reviewed the open cluster $^{12}$C/$^{13}$C and C/N giant data, finding that the mass dependence could be explained if the $\mu$ gradient on the lower giant branch inhibited mixing. Note, however, that this gradient is smaller in more metal-poor stars, and the M92 carbon data (see Figure 7) seems to show mixing that begins earlier. Charbonnel (1995) then examined mixing by meridional circulation in giants. Her model assumes solid body rotation, inhibition of mixing by $\mu$ gradients, constant specific angular momentum in the convective envelope, and a constant surface rotation velocity of 1 km/s. Mixing can be generated in this model, although the angular momentum evolution it assumes should be verified with more detailed models. The onset of mixing in the data also appears to be more gradual than predicted in the model.

For further progress, self-consistent models that include both angular momentum evolution and mixing will be needed. Differential rotation can generate mixing, as discussed in Section 5, which may lower the rotation rates needed to drive mixing; angular momentum transport will reduce the rotation rates near the hydrogen-burning shell and will increase the rotation rates needed to drive mixing.

Other recent studies that include arbitrary mixing have been performed (Vandenberg & Smith 1988, Denissenkov & Denissenkova 1990, Langer et al 1993, Boothroyd et al 1995, Wasserburg et al 1995, Denissenkov & Weiss 1996, Weiss et al 1996, Sweigart 1997). These studies are consistent in requiring mixing to extend through the envelope essentially down to the hydrogen-burning shell. In particular, changes in surface O and Na require mixing in the vicinity of the hydrogen-burning shell, which implies some ability for the mixing to mix helium (Sweigart 1997). Explaining the details of the Mg, Na, and Al variations appears to present challenges for nuclear reaction rates, stellar evolution, or both (Zaidins & Langer 1997, Langer et al 1997).

Wasserburg et al (1995) also stress the importance of matching the $^{16}O/^{18}O$ ratios in asymptotic giants, and they argue for this as an additional indicator of mixing. In general, indicators that burn at lower temperatures exhibit changes at lower luminosity than indicators that burn at higher temperatures. $^{3}He$ destruction, rather than production, appears likely for the lowest mass stars.

The assumptions of instantaneous mixing to an arbitrary depth, a constant mixing time scale, or a constant diffusion coefficient are strong ones that have consequences for the physical behavior of the system. Model properties calibrated on one indicator can produce incorrect results for another that burns at a different temperature. For example, Wasserburg et al (1995) find that models that reproduce the $^{16}O/^{18}O$ ratios in asymptotic giants predict $^{12}C/^{13}C$ ratios that are too low. This probably indicates that the mixing is depth dependent (the $^{12}C$ in AGB stars arises from a deeper layer than the oxygen processing does). Chemical evolution models for $^{3}He$ depend sensitively on the properties ascribed to intermediate mass giants at low metal abundance (Dearborn et al 1996), and empirical checks on mixing in such stars are not available. A consistent physical model will probably be necessary for the problems of cosmological interest and for explaining all of the observed data.

## 7. SUMMARY: IMPLICATIONS OF MIXING

The need for mixing in stars rests on solid observational and theoretical ground. Three cases for mixing beyond that present in standard models have been reviewed: light element depletion in low mass stars, abundance anomalies in low mass giants, and deep mixing in massive stars. These departures from the predictions of standard stellar models will have implications for Big Bang nucleosynthesis (BBN) and possibly the ages of globular clusters. It is quite likely that an extensive reevaluation of the evolution of massive stars will also be needed.

Observational data on stellar abundances and stellar rotation have formed the basis for the study of mixing in stars. Nonstandard stellar models are more complicated than standard models, so empirical constraints are even more necessary for models that include such phenomena as rotational mixing than they are for standard models. With the advent of multiobject spectrographs, obtaining large abundance and rotation databases at high S/N is feasible. Lithium abundances in globular cluster MS stars and large samples of lithium abundances in old open clusters are two examples of projects that could place strong constraints on light element depletion in stars. In the case of mixing on the giant branch, establishing the luminosity at which mixing sets in would be an important step, as well as determining how the abundance pattern on the giant branch maps onto the horizontal branch and the AGB. The correlation between mixing on the giant branch, cluster age and composition, and rotation on the horizontal branch will also need to be carefully delineated. In the massive star regime, the existence of anomalies related to rotation has been demonstrated. Some such anomalies can be produced by models. The relationship between abundance anomalies, rotation, and evolutionary state in massive stars now needs to be clarified.

The angular momentum evolution of stars has been one of the greatest uncertainties in rotational mixing. Observational studies of stellar rotation, along with helioseismic studies of the internal rotation of the Sun, now place strong constraints on stellar models that include rotation. A particularly valuable contribution has been the realization of the connection between star formation and stellar rotation on the lower MS. A satisfactory theory of the origin of the observed range in rotation for massive stars does not yet exist. No true zero-age MS appears to exist for the most massive stars, for example. Further studies of rotation in stars will be essential for further progress.

Theoreticians need to take advantage of the work on the underlying physical mechanisms driving mixing. Recent revisions of the treatment of meridional circulation and the role of other hydrodynamic instabilities should be tested in stellar models. The long-standing questions of the roles of internal waves and magnetic fields on angular momentum transport have recently been addressed with quantitative work. The focus of theoretical work will likely shift from demonstrating that mixing is possible to quantifying how much mixing is expected for a given physical model and what the distribution of abundances should be. In addition, models should be checked in different phases of evolution and mass ranges. For examples, models that explain giant branch mixing should not overmix on the MS, and models that mix massive stars should not overmix intermediate mass stars. Hard and careful observational and theoretical work has established the limits of validity of the standard model of stellar structure and evolution. The implications of the deviations from the standard model now can, and should, be explored.

## ACKNOWLEDGMENTS

I would like to acknowledge R Kraft and C Deliyannis for their helpful comments on the manuscript.



*Literature Cited*


Alcock C, Fuller GM, Mathews GJ. 1987. *Ap. J.* 320:439–47

Allain S, Fernandez M, Martin EL, Bouvier J. 1996. *Astron. Astrophys.* 314:173–81

Anders E, Grevesse N. 1989. *Geochimica et Cosmochimic Acta* 53:197–214

Audouze J, Silk J. 1989. *Ap. J. Lett.* 342:L5–L9

Baglin A, Morel PJ, Schatzmann E. 1985. *Astron. Astrophys.* 149:309–14

Bahcall JN, Pinsonneault MH, Basu S, Christensen-Dalsgaard J. 1997. *Phys. Rev. Lett.* 78:171–74

Balachandran S. 1990. *Ap. J.* 354:310–32

Balachandran S. 1994. See Caillault 1994, pp. 234–43

Balachandran S. 1995. *Ap. J.* 446:203–27

Balachandran S, Lambert DL, Stauffer JR. 1988. *Ap. J.* 338:267–76

Balachandran S, Lambert DL, Stauffer JR. 1996. *Ap. J.* 470:1243–44

Balbus SA, Hawley JF. 1994. *MNRAS* 266:769–845

Barbuy B, de Meideros JR, Maeder A. 1996. *Astron. Astrophys.* 305:911–19

Barrado y Navascues D, Stauffer JR. 1996. *Astron. Astrophys.* 310:879–92

Basri G, Marcy GW, Graham JR. 1996. *Ap. J.* 458:600–9

Bell RA, Hesser JE, Cannon RD. 1983. *Ap. J.* 283:615–25

Benz W, Mayor M. 1984. *Astron. Astrophys.* 138:183–88

Bodenheimer P. 1965. *Ap. J.* 142:451–61

Bodenheimer P. 1966. *Ap. J.* 144:103–7

Boesgaard AM. 1976. *Ap. J.* 210:466–74

Boesgaard AM. 1987a. *Publ. Astron. Soc. Pac.* 99:1067–70

Boesgaard AM. 1987b. *Ap. J.* 321:967–74

Boesgaard AM, Budge KG. 1988. *Ap. J.* 332:410–20

Boesgaard AM, Budge KG, Burke EE. 1988. *Ap. J.* 325:749–58

Boesgaard AM, Budge KG. 1989. *Ap. J.* 338:875–87

Boesgaard AM, Friel ED. 1990. *Ap. J.* 351:467–91

Boesgaard AM, King 1993. *Astron. J.* 106:2309–23

Boesgaard AM, Steigman 1985. *Annu. Rev. Astron. Astrophys.* 23:319–78

Boesgaard AM, Tripicco MJ. 1986. *Ap. J. Lett.* 302:L49–L53

Boothroyd AI, Sackmann I-J, Wasserburg GJ. 1995. *Ap. J. Lett.* 442:L21–L24

Bouvier J. 1994. See Caillault 1994, pp. 151–62

Bressan A, Fagotto F, Bertelli G, Chiosi C. 1993. *Astron. Astrophys. Suppl.* 100:647–64

Briley MM, Bell RA, Hoban S, Dickens RJ. 1990. *Ap. J.* 359:307–18

Briley MM, Hesser JE, Bell RA. 1991. *Ap. J.* 373:482–96

Briley MM, Smith VV, King J, Lambert DL.

1997. *Astron. J.* 113:306–10
Briley MM, Smith VV, Lambert DL. 1994. *Ap. J. Lett.* 424:L119–22
Briley MM, Smith VV, Suntzeff NB, Lambert DL, Bell RA, Hesser JE. 1996. *Nature* 383: 604–6
Brown JA, Sneden C, Lambert DL, Dutchover E. 1989. *Ap. J. Suppl.* 71:293–322
Brown JA, Wallerstein G. 1989. *Astron. J.* 98:1643–47
Brown JA, Wallerstein G. 1992. *Astron. J.* 104: 1818–30
Brown JA, Wallerstein G, Oke JB. 1991. *Astron. J.* 101:1693–98
Burgers JM. 1969. *Flow Equations for Composite Gasses.* New York: Academic
Burkhart C, Couprey AA. 1989. *Astron. Astrophys.* 220:197–205
Caillault J-P. 1994. *Eighth Cambridge Workshop on Cool Stars, Stellar Systems, and the Sun.* San Francisco: Astron. Soc. Pac.
Carbon DF, Langer GE, Butler D, Kraft RP, Suntzeff NB, Kemper E, Trefzger CF, Romanishin W. 1982. *Ap. J. Suppl.* 49:207–58
Carlsson M, Rutten RJ, Bruls JH, Shchukina NG. 1994. *Astron. Astrophys.* 288:860–82
Cavallo RM, Sweigart AV, Bell RA. 1996. *Ap. J. Lett.* 464:L79–L82
Chaboyer BC, Deliyannis CP, Demarque P, Pinsonneault MH, Sarajedini A. 1992. *Ap. J.* 388:372–82
Chaboyer BC, Demarque P. 1994. *Ap. J.* 433: 510–19
Chaboyer BC, Demarque P, Pinsonneault MH. 1995a. *Ap. J.* 441:865–75
Chaboyer BC, Demarque P, Pinsonneault MH. 1995b. *Ap. J.* 441:876–85
Chaboyer BC, Zahn J-P. 1992. *Astron. Astrophys.* 253:173–77
Chapman S, Cowling TG. 1970. *The Mathematical Theory of Non-uniform Gasses.* 3rd edition. Cambridge: Cambridge Univ. Press.
Charbonneau P, MacGregor KB. 1992. *Ap. J. Lett.* 397:L63–L66
Charbonneau P, MacGregor KB. 1993. *Ap. J.* 417:762–80
Charbonneau P, Michaud G. 1988a. *Ap. J.* 327: 809–16
Charbonneau P, Michaud G. 1988b. *Ap. J.* 334: 746–60
Charbonneau P, Michaud G. 1990. *Ap. J.* 352: 681–88
Charbonneau P, Michaud G. 1991. *Ap. J.* 370: 693–708
Charbonneau P, Michaud G, Proffitt CR. 1989. *Ap. J.* 347:821–34
Charbonneau P, Schrijvner CJ, MacGregor KB. 1997. In *Cosmic Winds and the Heliosphere,* ed. JR Jokipii, CP Sonnett, MS Giampapa. Tucson, AZ: Univ. Arizona Press. In press
Charbonnel C, Vauclair S, Zahn J-P. 1992. *Astron. Astrophys.* 255:191–99
Charbonnel C, Vauclair S, Maeder A, Meynet G, Schaller G. 1994. *Astron. Astrophys.* 283:155–62
Charbonnel C. 1994. *Astron. Astrophys.* 282: 811–20
Charbonnel C. 1995. *Ap. J. Lett.* 453:L41–L44
Charbonnel C, Vauclair S. 1992. *Astron. Astrophys.* 265:55–64
Chmielewsky Y, Müller EA, Brault JW. 1975. *Astron. Astrophys.* 42:37–46
Christensen-Dalsgaard J, Gough DO, Thompson MJ. 1991. *Ap. J.* 378:413–437
Christensen-Dalsgaard J, Gough DO, Thompson MJ. 1992. *Astron. Astrophys.* 264:518–528
Copi CJ, Schramm DN, Turner MS. 1995. *Ap. J. Lett.* 455:L95–99
D'Antona F, Mazzitelli I. 1994. *Ap. J. Suppl.* 90:467–500
Dearborn DS. 1992. *Phys. Rep.* 210:367
Dearborn DS, Schramm DN, Hobbs LM. 1992. *Ap. J. Lett.* 394:L61–L64
Dearborn DS, Steigman G, Tosi M. 1996. *Ap. J.* 465:887–97
de la Reza R, Drake NA, da Silva L. 1995. *Ap. J. Lett.* 456:L115–18
Deliyannis CP. 1995. In *The Light Element Abundances,* ed. P Crane, pp. 395–410. Berlin: Springer-Verlag
Deliyannis CP, Boesgaard AM, King JR. 1996. *BAAS* 188:58.06
Deliyannis CP, Boesgaard AM, King JR. 1995. *Ap. J. Lett.* 452:L13–16
Deliyannis CP, Demarque P, Kawaler SD. 1990. *Ap. J. Suppl.* 73:21–65
Deliyannis CP, Demarque P, Pinsonneault M. 1989. *Ap. J. Lett.* 347:L73–L76
Deliyannis CP, King JR, Boesgaard AM. 1997. In *IAU Comm. 9 Int. Conf. Wide Field Spectroscopy,* ed. E Kontizas, et al. Dordrecht: Kluwer. In press
Deliyannis CP, King JR, Boesgaard AM, Ryan SG. 1994. *Ap. J. Lett.* 434:L71–L74
Deliyannis CP, Malaney RA. 1995. *Ap. J.* 453:810–18
Deliyannis CP, Pinsonneault MH, Duncan DK. 1993. *Ap. J.* 414:740–58
Deng L, Bressan A, Chiosi C. 1996a. *Astron. Astrophys.* 313:145–58
Deng L, Bressan A, Chiosi C. 1996b. *Astron. Astrophys.* 313:159–79
Denissenkov PA. 1994. *Astron. Astrophys.* 287:113–30
Denissenkov PA, Denissenkova SN. 1990. *Sov. Astron. Lett.* 16:275
Denissenkov PA, Weiss A. 1996. *Astron. Astrophys.* 308:773–84
Duncan DK, Peterson RC, Thorburn JA, Pinsonneault MH. 1994. *Bull. Am. Astron. Soc.* 184:07.02

Eddington AP. 1925. *Observatory* 48:73–75
Edwards S, Strom SE, Hartigan P, Strom KM, Hillenbrand LA, et al. 1993. *Astron. J.* 106:372–82
El Eid MF. 1994. *Astron. Astrophys.* 285:915–28
Endal AS, Sofia S. 1978. *Ap. J.* 210:184–98
Eryurt D, Kirbiyk H, Kizloglu N, Civelek R, Weiss A. 1994. *Astron. Astrophys.* 282:485–92
Favata F, Micela G, Sciortino S. 1996. *Astron. Astrophys.* 311:951–60
Fekel FC, Balachandran S. 1993. *Ap. J.* 403: 708–21
Fliegner J, Langer N, Venn KA. 1996. *Astron. Astrophys.* 308:L13–L16
Friel ED, Boesgaard AM. 1992. *Ap. J.* 387:170–80
Fricke KJ. 1968. *Zh. Astrophys.* 68:317–44
Fukuda I. 1982. *Publ. Astron. Soc. Pac.* 94:271–84
Garcia Lopez RJ. 1996. *Astron. Astrophys.* 313:909–12
Garcia Lopez RJ, Rebolo R, Beckman JE. 1988. *Publ. Astron. Soc. Pac.* 100:1489–96
Garcia Lopez RJ, Rebolo R, Perez de Tauro MR. 1995. *Astron. Astrophys.* 302:184–92
Garcia Lopez RJ, Spruit HC. 1991. *Ap. J.* 377:268–77
Gies DR, Lambert DL. 1992. *Ap. J.* 387:673–700
Gilroy KK. 1989. *Ap. J.* 347:835–48
Gilroy KK, Brown JA. 1991. *Ap. J.* 371:578–83
Goldreich P, Murray N, Kumar P. 1994. *Ap. J.* 424:466–79
Goldreich P, Nicholson PD. 1989. *Ap. J.* 342:1079–84
Goldreich P, Schubert G. 1967. *Ap. J.* 150:571–87
Gray DF. 1989. *Publ. Astron. Soc. Pac.* 101: 832–38
Greenstein JL, Richardson RS. 1951. *Ap. J.* 113:536–46
Herrero A, Kudritzki RP, Vilchez JM, Kunze D, Butler K, Haser S. 1992. *Astron. Astrophys.* 261:209–34
Hiltgen DD. 1996. *Lithium in evolved stars.* PhD thesis, Univ. Texas Austin.
Hobbs LM, Duncan DK. 1987. *Ap. J.* 317:796–809
Hobbs LM, Iben I, Pilachowski CA. 1989. *Ap. J.* 347:817–20
Hobbs LM, Pilachowski CA. 1986a. *Ap. J. Lett.* 309:L17–L21
Hobbs LM, Pilachowski CA. 1986b. *Ap. J. Lett.* 311:L37–L40
Hobbs LM, Pilachowski CA. 1988. *Ap. J.* 334:734–45
Hobbs LM, Thorburn JA. 1994. *Ap. J. Lett.* 428:L25–L28
Hogan CJ. 1995. *Ap. J. Lett.* 441:L17–L20
Howarth ID, Prinja RK. 1989. *Ap. J. Suppl.* 69:527–92
Iben I. 1965. *Ap. J.* 142:1447–67
Iben I. 1967. *Annu. Rev. Astron. Astrophys.* 5:571–626
Iben I. 1974. *Annu. Rev. Astron. Astrophys.* 12:215–56
Iben I, Renzini A. 1983. *Annu. Rev. Astron. Astrophys.* 21:271–342
Jones BF, Fischer D, Shetrone M, Soderblom DR. 1996a. *Astron. J.* In press
Jones BF, Shetrone M, Fischer D. 1996b. *Astron. J.* 112:186–91
Kajino T, Boyd RN. 1990. *Ap. J.* 359:267–76
Keppens R, MacGregor KB, Charbonneau P. 1996. *Astron. Astrophys.* 294:469–87
King JR, Deliyannis CP, Boesgaard AM. 1996. *Bull. Am. Astron. Soc.* 188:58.08
King JR, Hiltgen DD. 1996. *Publ. Astron. Soc. Pac.* 106:246–49
Kippenhahn R, Müllenhof C. 1974. *Astrophys. Space Sci.* 31:117–41
Kippenhahn R, Ruschenplatt G, Thomas H-C. 1980. *Astron. Astrophys.* 91:181–85
Kippenhahn R, Thomas H-C. 1970. In *Stellar Rotation,* ed. A Slettbak, pp. 20–29, Dordrecht: Reidel
Kippenhahn R, Weigart A. 1990. *Stellar Structure and Evolution.* Berlin: Springer-Verlag
Knobloch E, Spruit HC. 1982. *Astron. Astrophys.* 113:261–68
Knobloch E, Spruit HC. 1983. *Astron. Astrophys.* 125:59–68
Kraft RP. 1994. *Publ. Astron. Soc. Pac.* 106:553–65
Kraft RF, Sneden C, Langer GE, Prosser CF. 1992. *Astron. J.* 104:645–68
Kraft RF, Sneden C, Langer GE, Shetrone MD. 1993. *Astron. J.* 106:1490–507
Kraft RF, Sneden C, Langer GE, Shetrone MD, Bolte M. 1995. *Astron. J.* 109:2586–99
Kraft RF, Sneden C, Smith GH, Shetrone MD, Langer GE, Pilachowski CA. 1997. *Astron. J.* 113:279–95
Krauss LM, Romanelli P. 1990. *Ap. J.* 358:47–59
Krishnamurthi A, Pinsonneault MH, Barnes S, Sofia S. 1997. *Ap. J.* In press
Kumar P, Quataert EJ. 1996. *Ap. J. Lett.* 475:L143–46
Lambert DL, Heath JE, Edvardsson B. 1991. *MNRAS* 253:610–18
Lamers HJ, Maeder A, Schmutz W, Cassinelli JP. 1991. *Ap. J.* 368:538–44
Langer GE, Hoffman R, Sneden C. 1993. *Publ. Astron. Soc. Pac.* 105:301–7
Langer GE, Hoffman R, Zaidins CS. 1997. *Publ. Astron. Soc. Pac.* 109:244–51
Langer GE, Kraft RP, Carbon DF, Friel E, Oke JB. 1986. *Publ. Astron. Soc. Pac.* 98:473–

85
Langer N. 1991. *Astron. Astrophys.* 243:155–59
Langer N. 1992. *Astron. Astrophys. Lett.* 265:L17–L20
Lattanzio JC, Vallenari A, Bertelli G, Chiosi C. 1991. *Astron. Astrophys.* 250:340–50
Lebreton Y, Maeder A. 1987. *Astron. Astrophys.* 175:99–112
Lemoine M, Ferlet R, Vidal-Madjar A, Emerich C, Bertin P. 1993. *Astron. Astrophys.* 269:469–76
Lighthill J. 1978. *Waves in Fluids.* Cambridge: Cambridge Univ. Press.
Magazzu A, Rebolo R, Pavlenko IV. 1992. *Ap. J.* 392:159–71
Maeder A, Conti PS. 1994. *Annu. Rev. Astron. Astrophys.* 32:227–75
Maeder A, Meynet G. 1996. *Astron. Astrophys.* 313:140–44
Maeder A. 1987. *Astron. Astrophys.* 178:159–69
Maeder A. 1995. *Astron. Astrophys.* 299:84–88
Malaney RA, Fowler WA. 1988. *Ap. J.* 333:14–20
Martin EL, Claret A. 1996. *Astron. Astrophys.* 306:408–16
Mathews GJ, Alcock CR, Fuller GM. 1990. 349:449–57
Mestel L. 1953. *MNRAS* 113:716–45
Mestel L, Weiss NO. 1987. *MNRAS* 226:123–35
Michaud G. 1986. *Ap. J.* 302:650–55
Michaud G, Charbonneau P. 1991. *Space Sci. Rev.* 57:1–58
Michaud G, Charland Y, Vauclair S, Vauclair G. 1976. *Ap. J.* 210:447–65
Michaud G, Proffitt CR. 1993. In *Inside the Stars, IAU Colloq. 137*, ed. A Baglin, WW Weiss, pp. 246–59. San Francisco: Astron. Soc. Pac.
Molaro P, Primas F, Bonifacio P. 1995. *Astron. Astrophys. Lett.* 295:L47–L50
Montalban J. 1994. *Astron. Astrophys.* 281:421–32
Montalban J, Schatzmann E. 1996. *Astron. Astrophys.* 305:513–18
Nelson LA, Rappaport S, Chiang E. 1993. *Ap. J.* 413:364–67
Norris JE, Ryan SG, Stringfellow GS. 1994. *Ap. J.* 423:386–93
Palla F, Stahler SW. 1993. *Ap. J.* 418:414–25
Pallavicini R. 1994. See Caillault 1994, pp. 244–53
Paquette C, Pelletier C, Fontaine G, Michaud G. 1986. *Ap. J. Suppl.* 61:177–95
Pasquini L, Molaro P. 1996. *Astron. Astrophys.* 307:761–67
Peterson RC. 1980. *Ap. J. Lett.* 237:L87–L91
Peterson RC. 1983. *Ap. J.* 275:737–51
Peterson RC. 1985a. *Ap. J.* 289:320–25
Peterson RC. 1985b. *Ap. J. Lett.* 294:L35–L37
Peterson RC, Rood RT, Crocker DA. 1995. *Ap. J.* 453:214–28
Phillips OM. 1977. *The Dynamics of the Upper Ocean.* Cambridge: Cambridge Univ. Press. 2nd ed.
Pilachowski CA. 1988. *Ap. J. Lett.* 326:L57–L60
Pilachowski CA, Hobbs LM. 1988. *Publ. Astron. Soc. Pac.* 100:336–37
Pilachowski CA, Sneden C, Booth J. 1993. *Ap. J.* 407:699–713
Pilachowski CA, Sneden C, Kraft RF, Langer GE. 1996. *Astron. J.* 112:545–64
Pinsonneault MH. 1988. *Evolutionary models of the rotating sun and implications for other stars.* PhD thesis. Yale Univ.
Pinsonneault MH. 1994. See Caillault 1994, pp. 254–63
Pinsonneault MH. 1995. In *Proc. 32nd Liege Colloq.*, ed. A Noels, et al, pp. 65–74. Liege, Belgium: Univ. Liege Press
Pinsonneault MH, Kawaler SD, Sofia S, Demarque P. 1989. *Ap. J.* 338:424–52
Pinsonneault MH, Kawaler SD, Demarque P. 1990. *Ap. J. Suppl.* 74:501–50
Pinsonneault MH, Deliyannis CP, Demarque P. 1991. *Ap. J.* 367:239–52
Pinsonneault MH, Deliyannis CP, Demarque P. 1992. *Ap. J. Suppl.* 78:179–203
Press WH. 1981. *Ap. J.* 245:286–303
Press WH, Rybicki GB. 1981. *Ap. J.* 248:751–66
Primas F. 1996. *The Challenge of Beryllium Observations.* PhD thesis. Univ. Trieste
Primas F, Duncan DK, Pinsonneault MH, Deliyannis CP, Thorburn JA. 1997. *Ap. J.* In press
Proffitt CR. 1994. *Ap. J.* 425:849–55
Proffitt CR, Michaud G. 1989. *Ap. J.* 346:976–82
Proffitt CR, Michaud G. 1991. *Ap. J.* 371:584–601
Proffitt CR, Vandenberg DA. 1991. *Ap. J. Suppl.* 77:473–514
Rebolo R, Beckmann JE. 1988. *Astron. Astrophys.* 201:267–72
Rebolo R, Martin EL, Basri G, Marcy GW, Zapatero-Osorio MR. 1996. *Ap. J. Lett.* 469:L53–56
Rebolo R, Molaro P, Beckman JE. 1988. *Astron. Astrophys.* 192:192–205
Reeves H, Fowler WA, Hoyle F. 1970. *Nature* 226:727–29
Reeves H, Richer J, Sato K, Terasawa N. 1990. *Ap. J.* 355:18–28
Richard O, Vauclair S, Charbonnel C, Dziembowski WA. 1996. *Astron. Astrophys.* 312:1000–11
Richer J, Michaud G. 1993. *Ap. J.* 416:312–30
Rood RT, Bania TM, Wilson TL. 1984. *Ap. J.* 280:629–47
Roxburgh I. 1991. In *Angular Momentum Evo-*

*lution of Young Stars*, ed. S Catalano, J Stauffer, pp. 365–72. Dordrecht: Kluwer
Russell SC. 1995. *Ap. J.* 451:747–57
Russell SC. 1996. *Ap. J.* 463:593–601
Ryan SG, Beers TC, Deliyannis CP, Thorburn JA. 1996. *Ap. J.* 458:543–60
Ryan SG, Deliyannis CP. 1995. *Ap. J.* 453:819–36
Sakuri T. 1991. *MNRAS* 248:457–64
Schaller G, Schaerer D, Meynet G, Maeder A. 1992. *Astron. Astrophys. Suppl.* 96:269
Schaller G, Schaerer D, Meynet G, Maeder A. 1992. *Astron. Astrophys. Suppl.* 96:269–331
Schatzmann E. 1969. *Astron. Astrophys.* 3:331–46
Schatzmann E. 1977. *Astron. Astrophys.* 56: 211–18
Schatzmann E. 1991. *Soc. Astron. Mem. Ital.* 62:111–30
Schatzmann E, Baglin A. 1991. *Astron. Astrophys.* 249:125–33
Schatzmann E, Maeder A, Angrand F, Glowinski R. 1981. *Astron. Astrophys.* 96:1–16
Shetrone MD. 1996a. *Publ. Astron. Soc. Pac.* 112:1517–35
Shetrone MD. 1996b. *Publ. Astron. Soc. Pac.* 112:2639–49
Shu FH. 1992. *The Physics of Astrophysics*, vol. D., *Gas Dynamics*. Mill Valley, CA: Univ. Sci. Books. 104 pp.
Skumanich A. 1972. *Ap. J.* 171:565–67
Smith GH. 1987. *Publ. Astron. Soc. Pac.* 99:67–90
Smith GH, Shetrone MD, Bell RA, Churchill CW, Briley MM. 1996. *Astron. J.* 112:1511–16
Smith VV, Lambert DL, Nissen PE. 1993. *Ap. J.* 408:262–76
Sneden C, Kraft RP, Langer GE, Prosser CF, Shetrone MD. 1994. *Astron. J.* 107:1773–85
Sneden C, Kraft RP, Prosser CF, Langer GE. 1991. *Astron. J.* 102:2001–21
Sneden C, Pilachowski CA, Vandenberg DA. 1986. *Ap. J.* 311:826–42
Soderblom DR. 1995. *Mem. Soc. Astron. Ital.* 66:347–56
Soderblom DR, Dravins D. 1984. *Astron. Astrophys.* 140:427–30
Soderblom DR, Fedele SB, Jones BF, Stauffer JR, Prosser CF. 1993a. *Astron. J.* 106:1080–86
Soderblom DR, Jones BF, Balachandran S, Stauffer JR, Duncan DK, et al. 1993b. *Astron. J.* 106:1059–79
Soderblom DR, Jones BF, Stauffer JR, Chaboyer B. 1995. *Astron. J.* 110:729–32
Soderblom DR, Oey MS, Johnson DR, Stone RP. 1990. *Astron. J.* 99:595–607
Soderblom DR, Pilachowski CA, Fedele SB, Jones BF. 1993c. *Astron. J.* 105:2299–307
Spite F, Spite M. 1982. *Astron. Astrophys.* 115:357–66
Spite F, Spite M. 1986. *Astron. Astrophys.* 163:140–44
Spite F, Spite M, Peterson RC, Chaffee FH. 1987. *Astron. Astrophys. Lett.* 171:L8–L9
Spite M, Francois P, Nissen PE, Spite F. 1996 *Astron. Astrophys.* 307:172–83
Spite M, Maillard JP, Spite F. 1984. *Astron. Astrophys.* 141:56–60
Spite M, Pasquini L, Spite F. 1994. *Astron. Astrophys.* 290:217–27
Spruit HC. 1987. In *The Internal Solar Angular Velocity*, ed. BR Durney, S Sofia, pp. 185–200. Dordrecht: Reidel
Spruit HC, Knobloch E, Roxburgh IW. 1983. *Nature* 304:520–22
Stahler SW. 1988. *Ap. J.* 332:804–25
Stahler SW. 1994. *Publ. Astron. Soc. Pac.* 106:337–43
Stauffer JR. 1994. See Caillault 1994, pp. 163–73
Stauffer JR, Hartmann LW, Jones BF, McNarama BR. 1989. *Ap. J.* 342:295–294
Steigman G. 1993. *Ap. J. Lett.* 413:L73–L76
Steigman G, Fields BD, Olive KA, Schramm DN, Walker TP. 1993. *Ap. J. Lett.* 415:L35–L38
Steigman G. 1996. *Ap. J.* 457:737–42
Stringfellow GS, Bodenheimer P, Noerdlinger PD, Arigo RJ. 1983. *Ap. J.* 264:228–36
Strom S. 1994. See Caillault 1994, pp. 211–33
Stuik R, Bruls JH, Rutten RJ. 1996. *Astron. Astrophys.* Submitted
Suntzeff NB. 1989. In *The Abundance Spread Within Globular Clusters*, ed. G Cayrel de Strobel, M Spite, T Lloyd Evans, pp. 71. Paris: Obs. Paris
Suntzeff NB. 1993. In *ASP Conf. Series 48*, pp. 167. San Francisco: Astron. Soc. Pac.
Suntzeff NB, Smith VV. 1991. *Ap. J.* 381:160–72
Sweet P. 1950. *MNRAS* 110:548–58
Sweigart AV. 1997. *Ap. J. Lett.* 474:L23–L26
Sweigart AV, Gross PG. 1978. *Ap. J. Supp.* 36:405–37
Sweigart AV, Mengel JG. 1979. *Ap. J.* 229:624–41
Swenson FJ. 1995. *Ap. J.* 438:L87–L90
Swenson FJ, Faulkner J. 1992. *Ap. J.* 395:654–74
Swenson FJ, Faulkner J, Rogers F, Iglesias CA. 1994. *Ap. J.* 425:286–302
Talon S, Zahn J-P, Maeder A, Meynet G. 1997. *Astron. Astrophys.* Submitted
Tassoul J-L. 1978. *Theory of Rotating Stars.* Princeton: Princeton Univ. Press
Tassoul J-L, Tassoul M. 1982. *Ap. J. Suppl.* 49:317–50
Tassoul J-L, Tassoul M. 1989. *Ap. J.* 345:472–79
Tassoul J-L, Tassoul M. 1995. *Ap. J.* 440:789–

809
Thorburn JA. 1994. *Ap. J.* 421:318–43
Thorburn JA, Duncan DK, Peterson RC. 1994. *Bull. Am. Astron. Soc.* 184:07.03
Thorburn JA, Hobbs LM. 1996. *Astron. J.* 111:2106–14
Thorburn JA, Hobbs LM, Deliyannis CP, Pinsonneault MH. 1993. *Ap. J.* 415:150–73
Thoul AA, Bahcall JN, Loeb A. 1994. *Ap. J.* 421:828–42
Tomczyk S, Schou J, Thompson MJ. 1995. *Ap. J. Lett.* 448:L57–L60
Trefzger CF, Carbon DF, Langer GE, Suntzeff NB, Kraft RP. 1983. *Ap. J.* 266:144–59
Townsend AA. 1958. *J. Fluid Mech.* 4:361–75
Urpin VA, Shalybkov DA, Spruit HC. 1996. *Astron. Astrophys.* 306:455–63
Vandenberg DA, Smith GH. 1988. *Publ. Astron. Soc. Pac.* 100:314–35
Vangioni-Flam E, Cass M, Audouze J, Oberto Y. 1990. *Ap. J.* 364:568–72
Vauclair S. 1983. In *Astrophysical Processes in Upper Main Sequence Stars*, ed. B Hauck, A Maeder, pp. 167–252. Geneva: Geneva Obs. Press
Vauclair S. 1988. *Ap. J.* 335:971–75
Vauclair S, Charbonnel C. 1995. *Astron. Astrophys.* 295:715–24
Vauclair S, Vauclair G, Schatzmann E, Michaud G. 1978. *Ap. J.* 223:567–82
Venn KA. 1995. *Ap. J.* 449:839–62
Venn KA. 1996. *Publ. Astron. Soc. Pac.* 108:309
Venn KA, Lambert DL, Lemke M. *Astron. Astrophys.* 307:849–59
Walborn NR. 1976. *Ap. J.* 205:419–25
Walborn NR. In *Atmospheric Diagnostics of Stellar Evolution, IAU Colloq. 108*, ed. K Nomoto, pp. 70–78. Berlin: Springer-Verlag
Walker TP, Steigman G, Kang H-S, Schramm DM, Olive KA. 1991. *Ap. J.* 376:51–69
Walker TP, Steigman G, Schramm DM, Olive KA, Fields B. 1993. *Ap. J.* 413:562–70
Wallerstein G, Böhm-Vitense E, Vanture AD, Gonzalez G. 1994. *Astron. J.* 107:2211–21
Wallerstein G, Herbig GH, Conti PS. 1965. *Ap. J.* 141:610–16
Wasserburg GJ, Boothroyd AI, Sackman I-J. 1995. *Ap. J. Lett.* 447:L37–L40
Weiss A, Wagenhuber J, Dennissenkov PA. 1996. *Astron. Astrophys.* 313:581–90
Weymann R, Sears RL. 1965. *Ap. J.* 142:174–81
Wilson TL, Matteucci F. 1992. *Astron. Astrophys. Rev.* 4:1–33
Wilson TL, Rood RT. 1994. *Annu. Rev. Astron. Astrophys.* 32:191–226
Zahn J-P. 1974. In *Stellar Instability and Evolution, IAU Symp. 59*, ed. P Ledoux, A Noels, AW Rodgers, pp. 185–195. Dordrecht : Boston
Zahn J-P. 1983. In *Astrophysical Processes in Upper Main Sequence Stars*, ed. B Hauck, A Maeder, pp. 253–327
Zahn J-P. 1992. *Astron. Astrophys.* 265:115–32
Zahn J-P. 1994. *Astron. Astrophys.* 283:829–41
Zahn J-P, Talon S, Matias J. 1996. *Astron. Astrophys.* Submitted
Zaidins CS, Langer GE. 1997. *Pub. Astron. Soc. Pac.* 109:252–55
Zapatero OM, Rebolo R, Martin EL, Garcia Lopez RJ. 1996. *Astron. Astrophys.* 305:519–26
Zappala RR. 1972. *Ap. J.* 172:57–74

*Annu. Rev. Astron. Astrophys. 1997. 35:607–36*

# PARSEC-SCALE JETS IN EXTRAGALACTIC RADIO SOURCES[1]

*J. Anton Zensus*
National Radio Astronomy Observatory,[2] Charlottesville, Virginia 22903

KEY WORDS: compact radio sources, radio source morphology, radio jets, VLBI jets, superluminal motion, jet models

ABSTRACT

Observations of parsec-scale radio jets associated with active galactic nuclei are reviewed, with a particular emphasis on high-luminosity core-dominated sources where the most detailed information on individual objects exists. Extensive imaging surveys with very long baseline interferometry (VLBI) have made possible a morphologic classification of compact radio sources and systematic studies of the statistics of apparent (often faster-than-light) motions found in an increasing number of sources. VLBI monitoring studies at centimeter and millimeter wavelengths, enhanced by spectral and polarization imaging, of representative types of AGN can discriminate detailed physical models. The observations are especially discussed in light of the variant of the relativistic beaming model that explains kinematic, spectral, and polarization properties of parsec-scale jets through shocks in an underlying continuous jet flow.

## INTRODUCTION

The radio emission from radio-loud active galactic nuclei (AGN) ordinarily takes the form of collimated "jets" that connect a compact central region with kiloparsec-scale extended lobes and hot spots. For the past two decades, the structure and emission properties of these jets have been studied extensively

[1]The US Government has the right to retain a nonexclusive, royalty-free license in and to any copyright covering this paper.

[2]The National Radio Astronomy Observatory is a facility of the National Science Foundation, operated under cooperative agreement by Associated Universities, Inc.

with aperture synthesis radio imaging (see Bridle & Perley 1984, Bridle 1996). Very long baseline interferometry (VLBI) has become the primary tool to probe the most compact of these emission regions with angular resolution of 0.1 to 2 milliarcsec, corresponding to linear scales of 0.2 to 10 parsec (pc) (see Zensus & Pearson 1990, Pearson 1996).

The direct radio imaging of parsec-scale jets and the complementary study of activity in the associated AGN in all spectral regimes has broadly impacted our understanding of these objects; in particular, they have inspired and constrained the development of realistic physical models, helped to establish the relativistic jet paradigm, and influenced the thinking about general AGN unification, radio source evolution, and cosmology.

The relativistic jet model (e.g. Blandford & Königl 1979) has become the de facto generally current AGN paradigm; it postulates that AGN are fueled by a massive central black hole and accretion disk at the base of a relativistic flow in symmetric twin jets (Rees 1971, Scheuer 1974, Blandford & Rees 1974). This scenario is rooted in the ideas that the source characteristics are determined by relativistic beaming (Shklovskii 1963), by relativistic injection from galactic nuclei (Rees 1966), and by accretion of matter on a central black hole (see Begelman et al 1984); the inner jets themselves are thought to be formed magnetically from a rapidly rotating accretion disk around the black hole (Blandford & Payne 1982). The only direct evidence for the existence of bulk relativistic outflow along the radio jets comes from the detection of collimated, apparent faster-than-light ("superluminal"), outward-motion of parsec-scale jets (see Readhead 1993, Zensus & Pearson 1987). The discovery of a rapidly rotating molecular disk in NGC 4258 arguably is the most direct demonstration of the presence of a black hole in the nucleus of an active galaxy (Miyoshi et al 1995, Moran et al 1995, but see also Burbidge & Burbidge 1997). The intraday variability found in a number of compact sources hints at the presence of linear dimensions for ultra-compact jet components that are smaller than light hours (see Wagner & Witzel 1995).

The radio radiation from parsec-scale extragalactic radio sources is partially transparent synchrotron emission with characteristic spectral and polarization properties, as well as significant inverse-Compton emission (see Marscher 1990, Hughes & Miller 1991). The flat spectra of compact sources can be explained by superposition of the synchrotron spectra of the individual components that form the compact structure, the compact base of the jet, and the bright regions within the jet flow. The details of the formation of inner jets that connect the nucleus to the observed radio jet, their acceleration to near the speed of light, and the strong collimation remain poorly understood (e.g. Marscher 1995, Begelman 1995). On the other hand, considerable progress has been made in explaining the kinematic, spectral, and polarization properties of parsec-scale jets through shocks in an underlying continuous jet flow (Blandford & Königl 1979, Marscher & Gear 1985, Hughes et al 1989a). Simulations of hydro- and magneto-hydrodynamic relativistic jets have reached a quality where meaningful comparisons between numerical and observed jet properties and evolution are possible (see Wiita 1996, Norman 1996).

In the basic beaming model scenario, superluminal motion and morphologic differences in general between compact and extended radio sources are attributed to beaming and orientation effects; compact sources are thought to be basically extended lobe sources at a small inclination to the line of sight and with strongly Doppler boosted core flux density (Orr & Browne 1982). A unification of Seyfert galaxies of type 1 with type 2 is supported by optical studies of the polarization properties of broad- and narrow line regions (Antonucci 1993, Miller 1995). The original unification hypothesis for radio sources has also been extended and refined by adding opaque nuclear obscuring tori to the relativistic jets. Beamed, intermediate, and unbeamed populations of radio-moderate edge-darkened Fanaroff & Riley Class 1 galaxies and of radio-loud edge-brightened Class 2 galaxies have been identified (see Barthel 1989, 1995, Gopal-Krishna 1995, Urry & Padovani 1995). In particular, lobe-dominated, narrow-line radio galaxies such as Cygnus A are assumed to be the unbeamed counterparts to core-dominated, variable, and superluminal quasars like 3C 345; the FR I radio galaxies in turn are thought to be the unbeamed pendants to BL Lac type objects.

The observational study of parsec-scale jets has seen astonishing progress since early imagery of compact radio sources at milliarcsecond resolution was achieved with very long baseline interferometry (VLBI) (see Kellermann & Pauliny-Toth 1981). Extensive VLBI surveys (see Wilkinson 1995) have made possible a morphologic classification of compact radio sources (see Pearson 1996). The systematic measurement of motion in these large samples yields apparent velocity statistics and distribution of Lorentz factors that can be compared to other indicators of relativistic motion (see Ghisellini et al 1993, Vermeulen 1995). The surveys also provide clues for unification models and for cosmology (see Vermeulen 1995, Kellermann 1993, Gurvits 1994). Realistic physical models can be tested and constrained through in-depth studies of prototypical objects, in particular when VLBI observations are combined with information from other spectral regimes.

Advances in VLBI techniques have been especially successful in the areas of detailed long-term monitoring observations, multifrequency and polarization imaging, and sensitivity-enhancing phase referencing (see the reviews in Zensus et al 1995b). High-quality VLBI observations have become routinely possible, especially with the very long baseline array (VLBA) (Napier et al 1994). This versatile and dedicated instrument offers full polarization and spectral-line imaging capability, as well as spectral coverage from 300 MHz to 45 GHz. Independent regional VLBI networks in the northern and southern hemispheres, coordinated global VLBI campaigns between the VLBA and the European VLBI Network, and the "World Array" campaigns have made possible images of outstanding fidelity (see Schilizzi 1995). At millimeter wavelengths, ad hoc arrays have been operating that result in images with submilliarcsecond resolution (Kirchbaum et al 1994b, Schalinski et al 1994, Bååth 1994). The first true space VLBI mission, the Japanese VLBI Space Observatory Programme (VSOP), launched in early 1997, should provide high-fidelity images of strong sources (>1 Jansky) at even greater resolution (Hirabayashi 1996).

This review summarizes observational properties of parsec-scale radio jets in extragalactic sources, focusing in particular on the high-luminosity core-dominated sources where the most detailed information on individual jets has been gathered. Recent accounts can also be found in several conference proceedings (Zensus & Kellermann 1994, Cohen & Kellermann 1995, Hardee et al 1996, Ekers et al 1996). Complementary perspectives are offered by Readhead (1993), Pearson (1996), Cawthorne (1991), Marscher (1995). A Hubble constant of $H_0 = 100\, h$ km s$^{-1}$Mpc$^{-1}$, a deceleration parameter $q_0 = 0.5$, and a spectral index definition $S_\nu \propto \nu^\alpha$ are assumed throughout this paper.

## MORPHOLOGY

Flux-density limited radio sky surveys at short centimeter (e.g. 6 cm) wavelengths yield roughly equal numbers of steep-spectrum extended double-lobed sources and flat-spectrum objects that are unresolved on an arcsecond scale. Among these radio-loud objects, the flat-spectrum sources typically can be associated with blazars and quasars, whereas the steep-spectrum sources tend to be FR I or FR II radio galaxies. Parsec-scale jets are found in all types of powerful radio sources, although they occur predominantly in elliptical galaxies rather than in spirals (e.g. Seyferts, Liners). A classification based on overall radio size and morphology has become possible through VLBI surveys, which yield several major classes with distinct characteristics (see Wilkinson 1995, Pearson 1996): (*a*) luminous core-dominated flat-spectrum sources, (*b*) powerful double-lobed radio galaxies with hotspots (Fanaroff-Riley class II, FR II), (*c*) weaker double-lobed radio galaxies without leading hot spots (FR I types), (*d*) compact steep-spectrum sources (CSS sources), and (*e*) compact symmetric objects (CSOs). Figure 1 shows representative examples of compact source

⟶

*Figure 1* Examples of VLBI structures of compact sources at 15 GHz (from JA Zensus, KI Kellerman, RC Vermeulen & MH Cohen, in preparation).

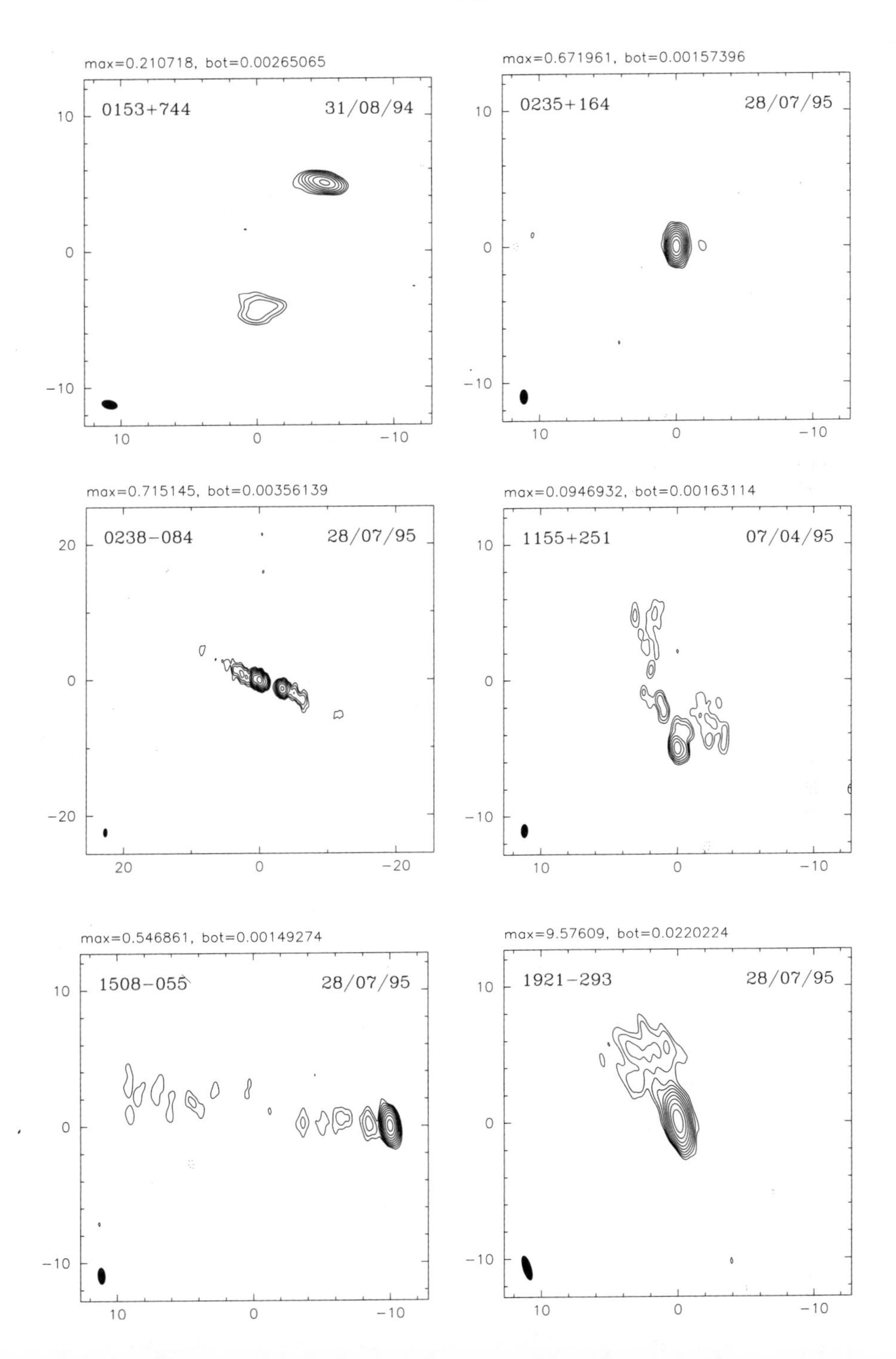
max=0.210718, bot=0.00265065
0153+744
31/08/94
max=0.671961, bot=0.00157396
0235+164
28/07/95
max=0.715145, bot=0.00356139
0238-084
28/07/95
max=0.0946932, bot=0.00163114
1155+251
07/04/95
max=0.546861, bot=0.00149274
1508-055
28/07/95
max=9.57609, bot=0.0220224
1921-293
28/07/95

structure typically found in high-frequency VLBI surveys (e.g. the 15 GHz survey of JA Zensus, KI Kellermann, RC Vermeulen & MH Cohen, in preparation). A range of astronomical phenomena has been attributed in part to gravitational lensing, and this effect has been suggested to be responsible for some of the distinguishing properties of AGN. Several of the known lens systems have VLBI jets, and their study reflects on the physics of AGN: For example, it can provide limits on source size (see Hewitt 1995).

### *High-Luminosity Core Dominated Objects*

The high-luminosity core dominated objects are typically found in systematic flux-density–limited surveys of compact sources (see Wilkinson 1995 for a comprehensive review, Pearson & Readhead 1988, Witzel et al 1988, Wehrle et al 1992, Polatidis et al 1995, Thakkar et al 1995, Xu et al 1995, Taylor et al 1994, Henstock et al 1995). Such surveys are biased towards selecting strongly beamed objects, and indeed a majority (about 80–90%) have "core-jet" type structures, i.e. they contain an unresolved flat-spectrum "core" and a steeper-spectrum one-sided "jet" that in turn may contain distinct structure "components."

The cores are identified based on their compactness, flat or inverted radio spectra, and sometimes their flux density variability. In several cases, phase referencing observations have demonstrated little or no discernible movement for the core component, including 3C 345 (Bartel et al 1986), 1038+528 (Marcaide et al 1994b) and 4C39.25 (Guirado et al 1995). Narrow collimated jets are typical in this class of source.

The degree of alignment between these parsec-scale and the kiloparsec-scale jets shows a bimodal distribution with peaks near $0^\circ$ and $90^\circ$, and BL Lac objects appear to show systematically stronger misalignments than quasars (e.g. Pearson & Readhead 1988, Wehrle et al 1992, Xu et al 1994, Appl et al 1996). This cannot be explained by a single population of slightly distorted jets, but it requires the presence of a second population of sources with small Lorentz factors and intrinsically large bending. Helical motion, perhaps arising in a binary black hole system, provides a likely mechanism to explain some cases that have been studied in detail (e.g. O'Dea et al 1988, Conway & Wrobel 1995). It is not yet clear if this is applicable for the population as a whole (Conway & Murphy 1993, Appl et al 1996), nor is it clear what causes the systematic differences found for BL Lac objects and quasars, of which the former have on average stronger misalignments (Conway & Murphy 1993, Conway 1994).

Structural variability and in particular apparent superluminal motion are frequently observed, and bulk relativistic motion is commonly inferred (Witzel et al 1988, Vermeulen 1995). Many of the observed jets are curved, and in some cases semioscillating trajectories or ridge lines have been observed (see below).

This effect is typically most pronounced near the core (Krichbaum & Witzel 1992).

A good example for a parsec-scale jet in a core-dominated source is the quasar 3C 345. Figure 2 shows an image at 5 GHz (Lobanov & Zensus 1997). This source has been regularly monitored with VLBI since 1979 (Biretta et al 1986, Zensus et al 1995a, Brown et al 1994, Wardle et al 1994, Rantakyrö et al 1995, Zensus et al 1995c). At 5 GHz, the jet appears continuous, but from comparison with images made at higher frequencies, it is known that the structure at a given epoch can be attributed primarily to a few distinct features (these are mostly

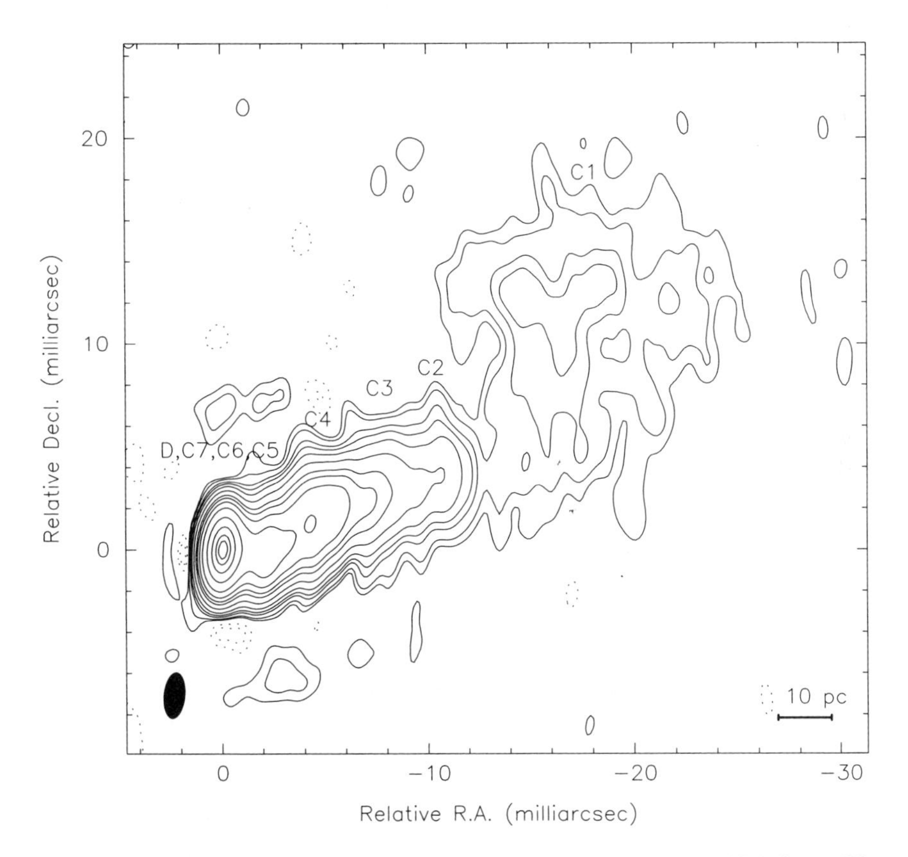


*Figure 2* The parsec-scale jet of 3C 345 at 5 GHz, June 1992 (from JA Zensus, AP Lobanov, KJ Leppänen, SC Unwin, and AE Wehrle, in preparation. Contours are −0.1, −0.06, 0.06, 0.1, 0.2, 0.3, 0.5, 1, 2, 3, 5, 10, 25, 50, 75, and 90% of the peak, 3.17 Jy per beam. The restoring beam is 2.22 × 1.0 milliarcsec, at PA −6°.

blended in the image shown). These features typically have been tracked for several years. In 3C 345, for example, high-dynamic range work has revealed a complex filamentary underlying jet flow (Unwin & Wehrle 1992).

An increasing number of core-dominated quasars like 3C 345 have been imaged in fascinating detail; two other examples are 3C 273 (Zensus et al 1990, Davis et al 1991, Unwin et al 1994a, Bahcall et al 1995) and 0836+710 (Hummel et al 1992a). The 18-cm image of 3C 273 has a dynamic range (defined as the ratio of image peak and lowest believable feature) exceeding 5000:1, and the image traces the jet to more than 150 milliarcsec from the core. An extended secondary feature detected as separate from the arcsecond jet has been speculated to be the elusive counterjet feature (Davis et al 1991). In general, however, no convincing counterjet has been found in any core-dominated, i.e. presumably strongly boosted, source. The 2000:1 dynamic range image of 0836+710 was made by combining VLBI data with VLBA and MERLIN observations and traces the complex continuous jet in this source from parsec to kiloparsec scales (Hummel et al 1992a).

These examples demonstrate a significant trend: Images of luminous sources typically show continuous jets with rich substructures, which are markedly different from the simple structures represented in maps from only a few years ago. In some cases, the jets are resolved in transverse directions, and filaments, limb-brightening, and edge-brightening have been reported. The technical advance in imaging capability has brought with it the need for more complicated models to explain the properties of the sources under study. The main features in these images still correspond to the "distinct components" seen in older maps, but state-of-the-art images tend to reveal weaker features and often directly reveal the underlying continuous jet emission. The evidence for the apparent superluminal motions in the best-studied sources has remained strong, but at the same time it is clear that infrequent sampling in time is bound to cause misidentification and confusion. Any motion is typically not uniform, and several values of motion have been observed in a given source. In particular, there coexist in some cases slow or stationary features and fast-moving regions. Little is known about the kinematic properties of the underlying continuous jet component. Although subluminal motions have been observed mostly in radio galaxies (e.g. Cygnus A, M 87, 3C 84, Centaurus A), there are some quasars with subluminal components as well (e.g. 0153+74).

There are a number of highly compact sources that show intraday variability (see Wagner & Witzel 1995). Although refractive interstellar scintillation can explain so far only the lower-frequency behavior in cases like 0917+624, it must have some effect in all very compact objects (Rickett et al 1995); the correlated variability between radio and optical and between optical and X rays, such as those found in 0716+714 (Wagner et al 1996), probably rules out a solely extrinsic origin (Qian et al 1995).

### *FR II Radio Galaxies and Quasars*

Relativistic beaming models attribute the diversity in properties of extragalactic radio sources to orientation effects (e.g. Scheuer & Readhead 1979, Blandford & Königl 1979, Orr & Browne 1982, Barthel 1989). The cores of extended radio galaxies and quasars tend to be much weaker and presumably are not strongly relativistically beamed towards us. The systematic study of morphology and parsec-scale jet velocities offers the opportunity to test the beaming hypothesis by measuring the distribution of parsec-scale jet speeds. This requires selection of samples with minimal orientation bias, e.g. samples of lobe-dominated quasars from low-frequency surveys. So far, there is an intriguing trend that the observed speeds (in eight cases to date) are in the range $v_{\rm app} \sim 1$–$5\ h^{-1}c$ (Hough et al 1996a, Hough 1994, Hough & Readhead 1989, Vermeulen et al 1993, Porcas 1987, Zensus & Porcas 1987). The quasar 3C 263 is fascinating; this source shows a one-sided core-jet morphology similar to that of core-dominated sources like 3C 345, with evidence for mild acceleration and nonradial motion of jet components (Hough et al 1996). So far, the relatively low superluminal speeds, together with these properties in 3C 263, tentatively support the unification of core- and lobe-dominated quasars.

Cygnus A (3C 405) is the closest luminous radio galaxy (see Carilli & Barthel 1996, Carilli & Harris 1996, and references therein). It is presumably oriented at a large angle ($>60°$) to the line of sight and presents the best-studied case of a parsec-scale jet in a FR II radio galaxy (Carilli et al 1994, 1996, Bartel et al 1995, Krichbaum et al 1996, 1997). The parsec structure contains a knotty jet and a weak counterjet. The jet extends to about 300 $h^{-1}$ pc from the core (at 1.6 GHz), it is well collimated, and it contains subluminal component motion, measured at 5 GHz in the range 0.35–0.55 $h^{-1}c$. There is no emission gap between jet and counterjet observed, in contrast to some inner-jet model predictions (Marscher 1996).

Close to the radio core of Cygnus A (Figure 3), the observed speed is even lower (0.1–0.2 $h^{-1}c$), which indicates the presence of acceleration or pattern speeds (Krichbaum et al 1997). Oscillations of the jet ridge line and jet width may reflect jet instabilities (e.g. Kelvin-Helmholtz), which in turn would predict a pattern speed slower than the fluid speed (Hardee et al 1995). Assuming intrinsic symmetry, the frequency-dependent jet-to-counterjet ratio can be explained by partial obscuration of the counterjet, perhaps by a disc or torus (see Carilli & Barthel 1996). Overall, the properties of Cygnus A jet are compatible with the simple relativistic beaming model; the speed of the jet fluid at larger

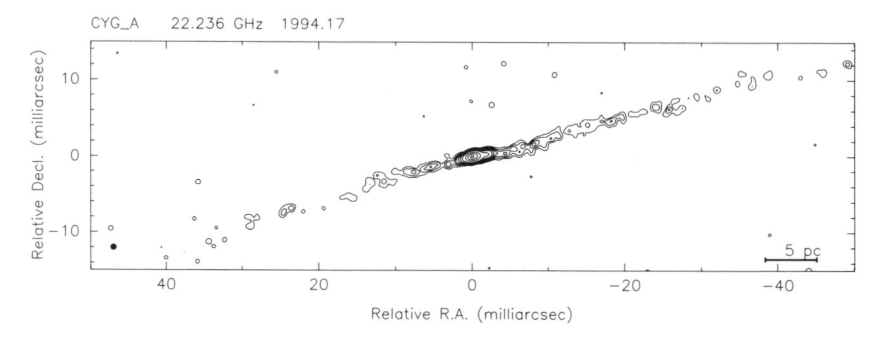


*Figure 3* The inner jet and counterjet of Cygnus A at 22 GHz, at epoch 1994.17 (from Krichbaum et al 1997). The field of view is ~100 × 30 milliarcsec. The size of the circular beam is 0.7-mas size and the contour levels are −0.05, 0.05, 0.15, 0.3, 0.5, 1, 2, 5, 20, 50, and 90% of the peak flux density of 0.73 Jy/beam. The total flux density of the map is 1.65 ± 0.05 Jy.

core-distances is $0.4 \lesssim \beta \lesssim 0.6$ and the inclination is $35° \lesssim \theta \lesssim 90°$ (Bartel et al 1995). However, given the slow apparent speed, the unification of Cygnus A type objects with blazars is difficult unless the intrinsic-jet Lorentz factor in this source is similar to that of quasars (Krichbaum et al 1997).

## *FR I Radio Galaxies and Quasars*

Systematic VLBI studies of FR I radio galaxy samples indicate that they are similar in morphology to FR II cores (Giovannini et al 1995, Pearson 1996): Asymmetric core-jet structures are typical (e.g. 3C 465, Venturi et al 1995) although in a few cases two-sided jets (e.g. 3C 338, Feretti et al 1993) or complex structures are found (e.g. 3C 272.1, Giovannini et al 1995). Jets tend to be well collimated close to the respective nucleus, they lie almost always on the same side as the larger-scale jets, and a higher degree of asymmetry seems to exist on small angular scales as compared with large angular scales. Stationary components are observed, as well as subluminal and slow superluminal motion (0.5–1.2 $h^{-1}$ $c$), which indicate that these jets are indeed relativistic. On the hypothesis that the radio structures are affected by Doppler favoritism, jet velocity and inclination to the line of sight can be derived from jet-to-counterjet ratios, from the ratio of core and total radio power, from Inverse-Compton arguments, and from imposing an upper limit to the deprojected size of a given source. Lorentz factors $\gamma \sim 3$ and viewing angles larger than 30° have been derived that are consistent with unification of low- and high-power radio galaxies (Giovannini et al 1994, 1995).

The best-studied objects in this category are the two nearby objects M 87 and Centaurus A. The radio galaxy M 87 (3C 274) contains the nearest powerful extragalactic jet in the northern hemisphere (Biretta 1996, Biretta & Junor 1995, and references therein) and offers unique opportunity for detailed study. This one-sided jet is characterized by filamentary features, limb-brightening, and side-to-side oscillations. The innermost part of the jet within $10^{16}$ cm from the center is well collimated and slightly curved (Junor & Biretta 1995). Prominent features on parsec and subparsec scales are stationary; at larger distances there is subluminal motion of $\beta_{\rm app} = 0.28 \pm 0.08$ (Reid et al 1989); and on a kiloparsec scale, apparent superluminal motion is evidenced up to $\beta_{\rm app} \approx 2.5$ (Biretta et al 1995). Together with the absence of a visible counterjet, these results are consistent with an underlying relativistic jet flow of Lorentz factor $\gamma$ of about 2 and initial collimation at distances less than 0.1 $h^{-1}$ pc from a suspected black hole/accretion disk at the center (Biretta & Junor 1995).

The giant radio galaxy Centaurus A (NGC 5128) is the closest radio galaxy and contains a straight jet of about 50 milliarcsec in extent in the same direction as the arcsecond and X-ray jet in the source (Tingay et al 1994, Jauncey et al 1995). The source varies rapidly on time scales shorter than four months and shows subluminal motion of about $0.15\,h^{-1}\,c$. The low speed and evidence for a subparsec-scale counterjet suggest that this jet is nonrelativistic or only mildly relativistic, that it is oriented at a large inclination, and that the innermost 0.4–0.8 $h^{-1}$ pc of the source is seen through a disk or torus of ionized gas that is opaque at lower frequencies owing to free-free absorption (Jones et al 1996).

The size of the compact radio nucleus of Centaurus A is $0.5 \pm 0.1$ milliarcsec (Kellermann et al 1997). The corresponding linear dimensions of 0.01 pc, about 10 light-days or $10^{16}$ cm, make this the smallest known extragalactic radio source. If the radio lobes are powered by a massive central engine, such as a black hole, their large total energy contents suggest that the central mass density may have exceeded $5 \times 10^{13}$ $M_\odot$ pc$^{-3}$, a value far larger than has been determined for any other AGN or quasar.

Two other prominent objects in this class are NGC 6251 (Jones & Wehrle 1994) and the core-dominated superluminal galaxy 3C 120, where the jet has been traced for about 500 milliarcsec from the core (Walker et al 1987, Benson et al 1988, RC Walker & JM Benson, personal communication).

A special case of a core-dominated source with FR I type radio morphology is 3C 84 (Figure 4), associated with the prominent Perseus cluster galaxy NGC 1275. On parsec scales, the source contains a broad complex jet with several components that exhibit subluminal motion of about 0.1 $h^{-1}c$ near the core and 0.5 $h^{-1}c$ at larger distances (see Romney et al 1995 and references therein; see also Venturi et al 1993, Krichbaum et al 1993b). A weak counter feature with strong low-frequency cutoff is found (Vermeulen et al 1994, Walker et al 1994); because of its low surface brightness, this cannot be explained by synchrotron self-absorption. Levinson et al (1995) have instead proposed obscuration through free-free absorption of the counterjet (but not the jet) by a

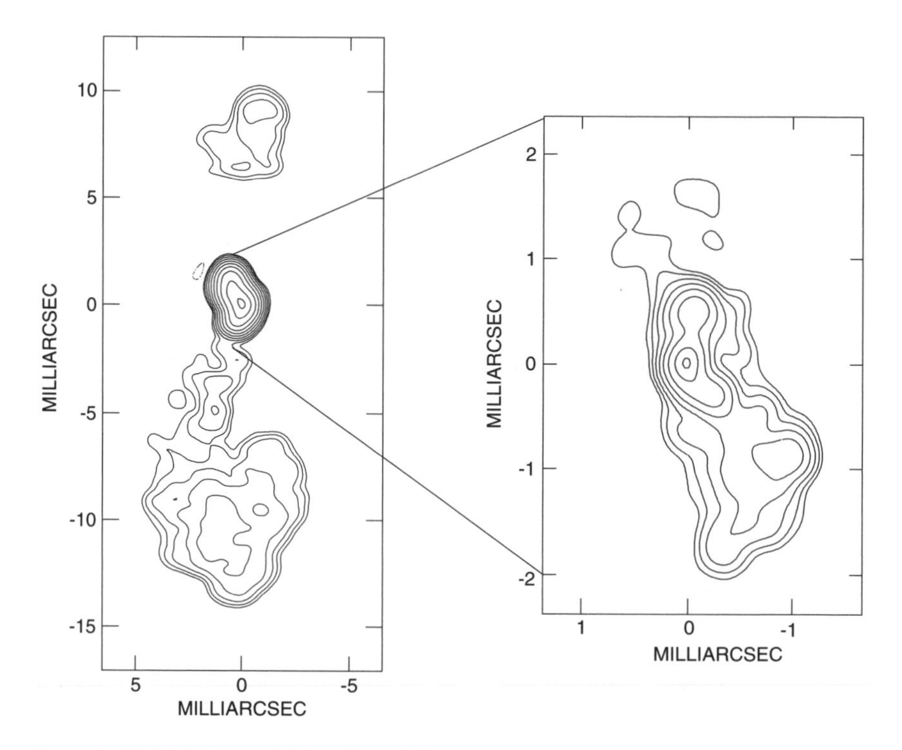


*Figure 4* VLBA images 3C 84 at 15 and 43 GHz (from Dhawan, Kellermann, & Romney, in preparation). (*left*): 15 GHz; contours are drawn at −0.25, 0.25, 0.5, 1, 2, 4, 8, 16, 32, 64, and 95% of the peak, 3.49 Jy/beam. (*right*): 43 GHz; contours are drawn at −1, 1, 2, 4, 8, 16, 32, 64, and 95% of the peak, 0.75 Jy/beam.

torus or disk region ionized by the central continuum source. This raises concern about the validity of jet-counterjet ratio arguments made in explanation of source properties by beaming, although Walker et al (1994) conclude that the obscuring region is probably irrelevant at high frequencies and that many properties including the motion in this source are consistent with a mildly relativistic symmetric twin jet at modest inclination. The nucleus in 3C 84 shows radio emission out to a total length of 0.1 arcsec (24 $h^{-1}$ pc), which indicates activity of the central engine for centuries preceding the last outburst observed in 1960 (Taylor & Vermeulen 1996).

## *Compact Steep Spectrum Sources and Compact Symmetric Objects*

On the order of 20–30% of compact sources in flux-density–limited surveys at lower frequencies (e.g. 3CR) belong to the compact steep spectrum (CSS) class

(Fanti et al 1990). These galaxies or quasars are typically small ($<2$ arcsec or linear sizes $\leq$10–15 $h^{-1}$ kpc) and exhibit a steep radio spectrum ($\alpha > -0.5$); few have been found to show evidence for the presence of relativistic components. The giga-Hertz–peaked spectrum sources (O'Dea et al 1991, O'Dea 1996), subkiloparsec-size objects with radio spectra that peak at giga-Hertz frequencies, are considered part of this class.

In a morphologic sense, three subgroups have been identified: compact symmetric objects (CSO, size $<0.5\ h^{-1}$ kpc), medium-size symmetric objects (MSO, size $\geq 0.5$ kpc), and complex sources (Readhead 1995, Fanti & Spencer 1996). The latter are mostly identified as quasars with core-jet structures where the jets dominate (Spencer 1994).

The distortions in the sources with complex structures have been attributed to beaming and/or "frustration" by interaction with a dense interstellar medium (Spencer 1994) that would inhibit the growth of the structures to large dimensions. Figure 5 shows VLBI polarization images of the superluminal CSS quasar 3C 138 (Cotton et al 1997). The magnetic field in this complex jet is aligned

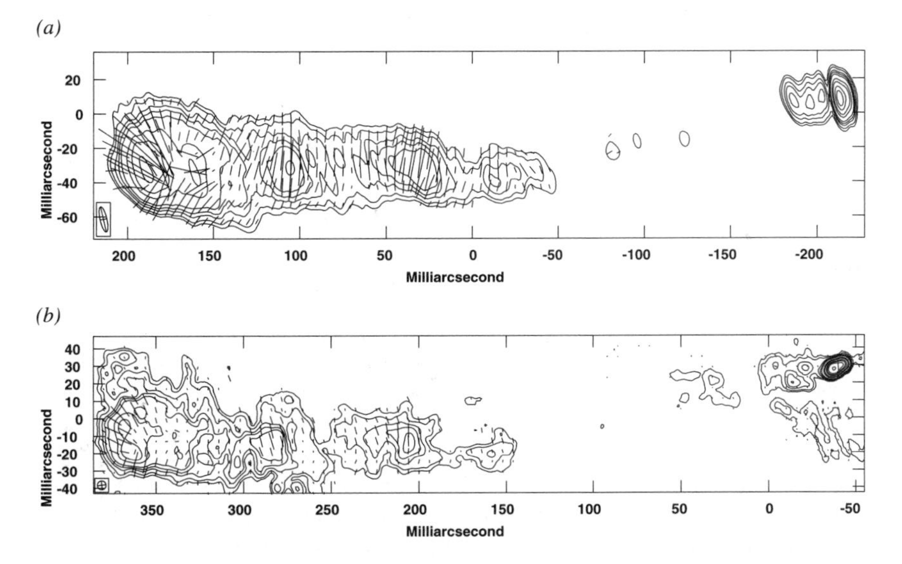


*Figure 5* The core and jet in the CSS quasar 3C 138 (from Cotton et al 1997). (*a*) Image at 18 cm, made with the EVN and the VLBA. The peak in the image is 241 mJy/beam and contours are drawn at $-5, 5, 8, 14, 19, 27, 54, 81, 135, 189$, and 270 mJy/beam. The lengths of the polarization vectors are proportional to the polarized intensity and have the orientation of the electric field vectors. (*b*) Image at 5 GHz, made with the VLBA. The resolution is 5 milliarcsec and the beam is shown in the lower left corner. The peak in the image is 215 mJy/beam and contours are drawn at $-2, 2, 3, 5, 7, 10, 20, 30, 50, 70, 100$, and 200 mJy/beam. Vectors as in *a*. Note that total intensity images of this source also show a weak feature on the opposite side from the jet.

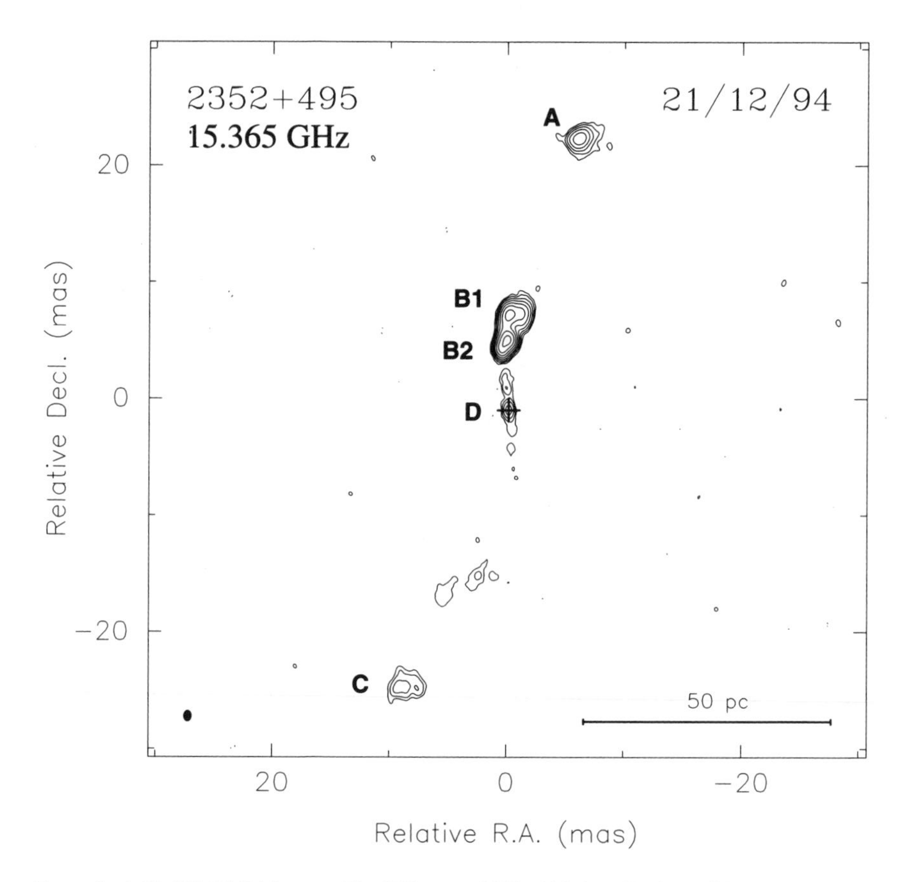


*Figure 6* A 15-GHz VLBA image of the CSS source 2352+495 (from Taylor et al 1996). Contours are drawn at −0.75, 0.75, 1.5, 3, 6, 12, 24, 48, 96 mJy/beam, and the beam is 0.91 × 0.67 milliarcsec. Component D is identified with the core, and A and C with hot spots.

along the jet axis and wraps around the head of the jet. The nuclear region is depolarized at 1.7 GHz, and there is no evidence for significant magnetized plasma in front of the end of the jet. At least in this case, the data suggest that this is not a jet that is frustrated by a dense interstellar medium.

The CSO sources comprise 5–10% of all high-luminosity AGN and are probably related to the "compact double" type (Phillips & Mutel 1982). Figure 6 shows an example, 2352+495 (Taylor et al 1996). The identification of the center of activity in this and other CSOs is often difficult due to low core fractions and complex structures. Most objects in this class are identified with galaxies (but see Perlman et al (1994) for a possible BL Lac object with CSO properties and a parsec-scale counterjet). Typically the CSOs show a very weak core, two symmetric lobes or hot spots, and resolved jet-like emission on one side. The ratio of jet length to counterjet length is close to unity, and higher pressure and smaller hot spots are seen on the jet side. It is still unclear if this asymmetry is intrinsic or instead caused by beaming and/or environmental effects. Within a simple unification model of radio sources, the CSOs and MSOs appear likely to be young and intermediate progenitors of large-scale FR II objects that show decreasing luminosity as they expand (Readhead et al 1996a,b, Fanti et al 1995).

# STRUCTURAL VARIATIONS

## *Apparent Superluminal Motion*

Statistical results from the study of motions in large samples of superluminal sources have been discussed by Ghisellini et al (1993), and Vermeulen & Cohen (1994), and Vermeulen (1995), for example. These studies aim at developing population models that test for the presence of a distribution in lieu of a constant value of Lorentz factor, pattern speeds that differ from fluid speeds, jet bending, and accelerations. For a sample of 81 flat-spectrum objects, Vermeulen (1995) found no evidence for intrinsically different populations of galaxies, BL Lac objects, and quasars, which is in contrast to other reports (see Gabuzda 1995). Apparent velocities $\beta_{\rm app}$ in the range 1–5 $h^{-1}c$ occur with roughly equal frequency (Figure 7), and in particular, higher values up to $\beta_{\rm app} = 10\,h^{-1}$ are rarer than assumed from earlier studies of superluminal sources (see Figure 7). This velocity distribution can be reproduced by assuming a wide range of Lorentz factors. Finally, a rising upper envelope to the $\beta_{\rm app}$ distribution appears to exist when plotted as a function of 5-GHz luminosity. This is consistent with the high-luminosity sources in the sample existing as strongly Doppler-beamed members of a parent population with much lower intrinsic luminosity.

In the majority of cases, the evidence for apparent superluminal motion is based on few observations, and accordingly, such statistical studies usually assume that the motion measured for a particular superluminal source component is along ballistic trajectories at constant speed. Detailed monitoring studies have shown that in a given source, not only can different components have different speeds, but accelerations and decelerations are present in several sources. Stationary features have been observed in a few instances. In addition, rarely are the reported motions along straight ballistic trajectories: Kinks and bends are frequently seen, and in a small number of cases, complicated curved trajectories have been determined. Note that such curvature sometimes is measured not directly for a given component, but indirectly by tracing several components, e.g. along a jet that might be characterized by a well-defined ridge line

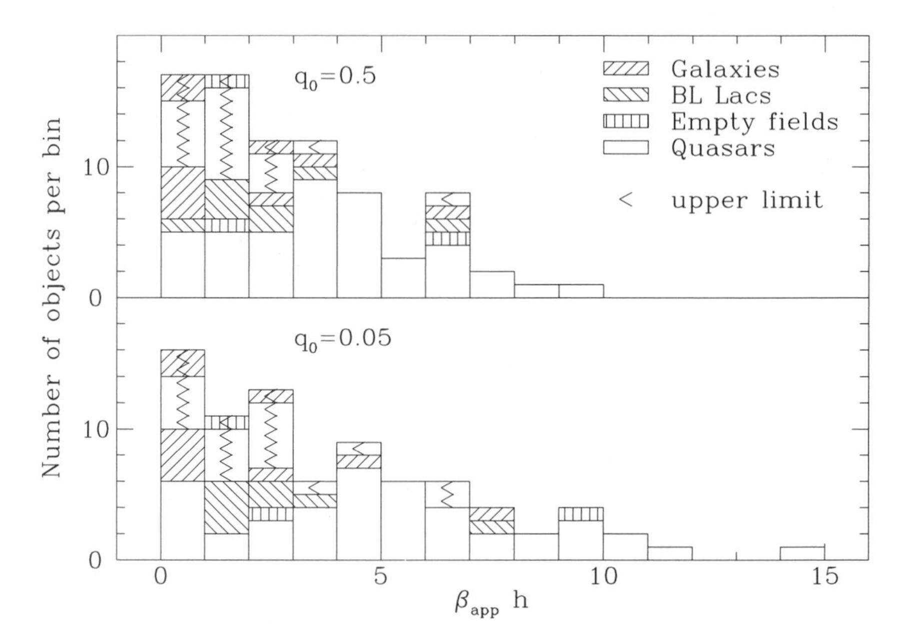


*Figure 7* The observed apparent velocity distribution for 81 objects in a homogeneous flat-spectrum sample is illustrated, showing that higher values are rare, compared with earlier work and predictions for beamed samples (from Vermeulen 1995).

or time-averaged mean jet axis. The curvature and nonlinear motion along bent trajectories that is seen so often is perhaps the strongest evidence against the existence of truly moving plasmons as the physical nature of superluminally moving components.

## *Curved Trajectories*

Figure 8 shows the different apparent trajectories (in the region 1 milliarcsec from the core) of the superluminal features of 3C 345 (JA Zensus, AP Lobanov, KJ Leppänen, SC Unwin, and AE Wehrle, in preparation), as measured with respect to the stationary core D (Bartel et al 1986). These tracks are substantially curved and show kinks and bends. Note that at larger distances from the core, the components appear to roughly follow the same curved trajectory in the northwest direction towards the arcsecond structure; near the core the trajectories differ significantly. Note that in this and other sources, the curvature appears most pronounced near the core—reflected in bends and wiggles with amplitudes of <0.2–0.5 milliarcsec. There should also be opacity effects causing frequency dependency of component shapes and positions and of the

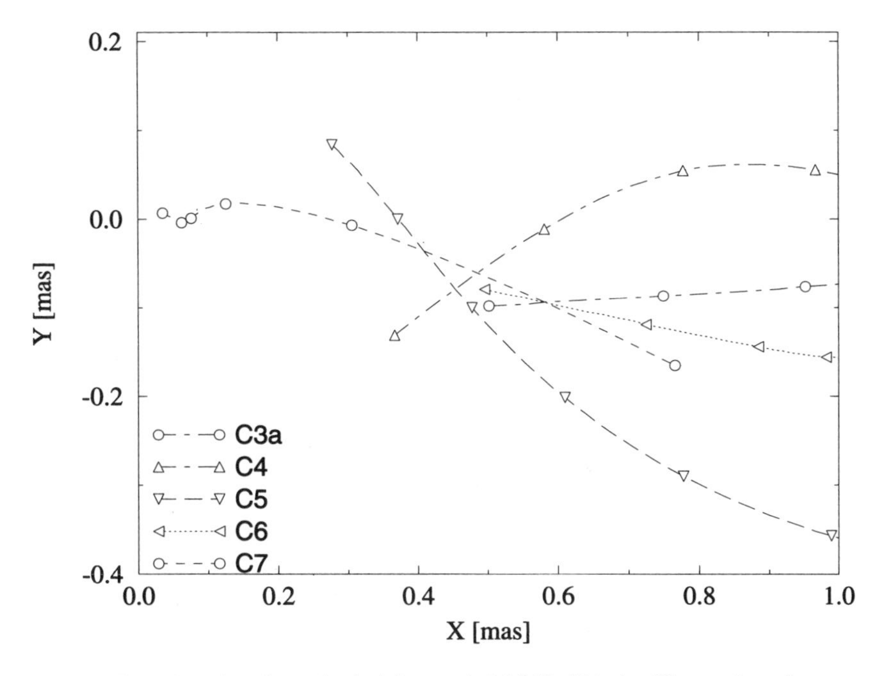


*Figure 8* The trajectories of superluminal features in 3C 345 within 1 milliarcsec from the core are shown (JA Zensus, AP Lobanov, KJ Leppänen, SC Unwin, and AE Wehrle, in preparation), measured with respect to the core D. Lines are from polynomial fits to $x$ and $y$ coordinates in time.

observed trajectories. Such behavior is seen, for example, in 3C 345 and also in 4C 39.25 (Alberdi et al 1997).

Similar curved trajectories or jet ridge lines have been reported in a growing number of cases, especially from high-resolution observations at millimeter wavelengths (Krichbaum et al 1994b, Bååth 1994, Krichbaum et al 1994a). The BL Lac object 1803+784 shows quasisinusoidal curvature in a parsec-scale jet (Krichbaum et al 1994b). The extrema in apparent velocity occur at or near the turning points in the curvature.

IIK Pauliny-Toth (personal communication) finds from observations during 1983–1990 that the jet in the quasar 3C 454.3 is curved, which is confirmed by observations at 3 mm (Krichbaum et al 1995). A correlation between curvature and velocity changes is also observed in 3C 84 (Krichbaum et al 1993a,c) at high frequencies, with components accelerating from $<0.1$ to $\sim 0.5\ h^{-1}\ c$ (see also Romney et al 1995, Dhawan et al 1997, and references therein). In 3C 273, it is not clear if all components follow the same path,

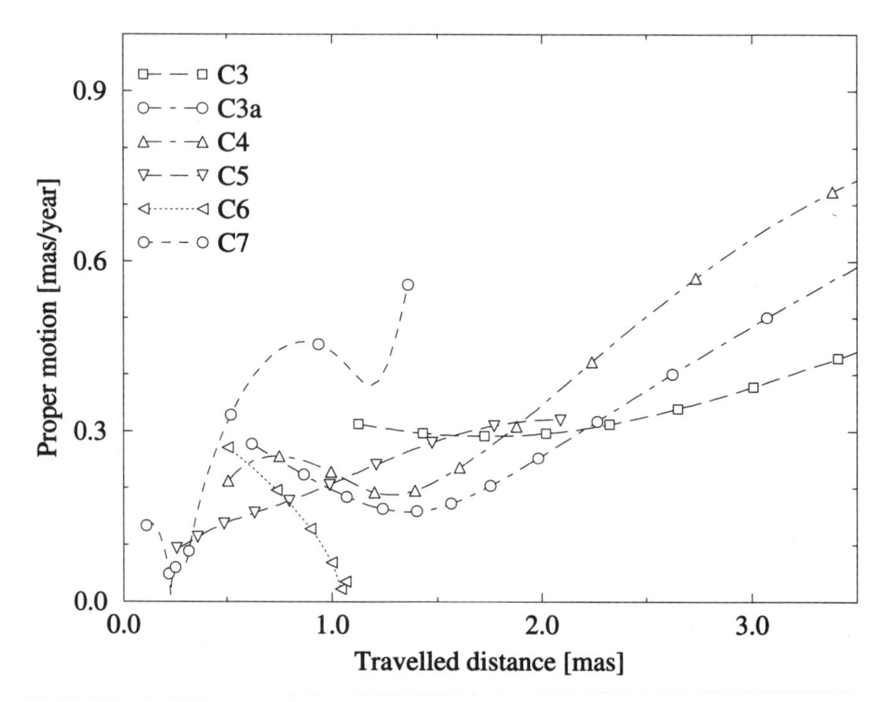


*Figure 9* Proper motions in 3C 345 (from JA Zensus, AP Lobanov, KJ Leppänen, SC Unwin, and AE Wehrle, in preparation) are shown, as derived from polynomial fits in relative coordinates $x$ and $y$ (see Figure 8).

although at least two component pairs have indistinguishable trajectories within the bent jet (Krichbaum et al 1993a, Abraham et al 1996).

## *Variable Component Speeds*

Figure 9 shows the apparent angular speed for components in 3C 345 versus distance traveled along the corresponding curved trajectory (JA Zensus, AP Lobanov, KJ Leppänen, SC Unwin, and AE Wehrle, in preparation). The plot highlights the complex nature of the velocity changes that occur in this source. To the first order, the component speeds increase with separation from the core, but there is evidence for a systematic minimum speed at 1–1.5 milliarcsec from the core, which perhaps indicates transition between physically different (the inner and outer) regions of the jet, although a purely geometric explanation is not ruled out. The kinematics and luminosity changes in 3C 345 suggest that the accelerations measured are intrinsic (Qian et al 1996). In the quasar 0836+71, three regions show different apparent speeds. There appears to be a correlation between superluminal speed and jet expansion in this source, i.e. in the narrow region of the jet there is no measured motion, and where the jet opens, speeds range between 5 and 10 $h^{-1}$ $c$, with the faster motion occurring close to the core (Krichbaum et al 1990b).

The quasar 4C 39.25—first thought to be a stationary double source and later thought to be a candidate for superluminal contraction (Shaffer & Marscher 1987)— shows a component moving between two stationary features (Alberdi et al 1993b), not unlike the case of 3C 395 (Simon et al 1988). Recent observations at 22 and 43 GHz demonstrate that the apparent speed slows down as the moving feature approaches the eastern stationary feature (see Alberdi et al 1993a). There is also some evidence that there are apparent differences in the jet curvature in the 22- and 43-GHz images (Alberdi et al 1997). This has been explained in terms of repeated bending of the jet; the stationary features correspond to the Doppler-enhanced underlying jet flow where it is oriented almost along the line of sight (Marcaide et al 1994a). Alternatively, the stationary features have also been explained by standing shocks (Gómez et al 1994, Hughes et al 1989a) and interactions with the narrow line region (Taylor et al 1989); e.g. see the extended component C1 in 3C 345 (Unwin & Wehrle 1992).

Acceleration (and superluminal brightening) have also been found in 3C 454.3 (Pauliny-Toth et al 1987; IIK Pauliny-Toth, personal communication); apparent speeds increase nonlinearly with separation from the core. For separations from the core of 1–2 milliarcsec, the speed is about 8.4 $h^{-1}c$; from 2- to 5-milliarcsec separation, the speed increases continuously to about 21 $h^{-1}$ $c$. Beyond that core distance, the speed is uncertain because the components become complex. All this is based on data between 1983 and 1990. The high-redshift blazar 0528+134 ($z = 2.07$) shows three moving features and at least one stationary component. One of these appears to be related to a $\gamma$-ray/millimeter flare (Krichbaum et al 1995). In the quasar 1928+738, there appears to be a transition zone at about 4–6 milliarcsec from the core, similar to the case of 3C 345, which separates two jet regions with different apparent component speeds: At separations from the core smaller than 4 milliarcsec, speeds of about 4$c$ were measured (Hummel et al 1992b), whereas between 6 and 18 milliarcsec from the core, the speeds are 6 to 7 $h^{-1}c$ (Schalinski et al 1992). Finally, in 3C 273 there are also different apparent component speeds and again a trend that the speeds are higher at larger distances from the core (Zensus et al 1990, Krichbaum & Witzel 1992, Abraham et al 1996).

In the sources observed with VLBI, there is often assumed to be no difference between pattern and fluid speed. The occurrence of different component speeds, as in M 87 (see Biretta & Junor 1995), 3C 120, or Cygnus A—if not interpreted by geometric effects—might be caused by variable pattern speed (Vermeulen & Cohen 1994). Some theoretical studies predict, for example, interacting instability modes that can produce variable observed pattern speeds (Hardee et al 1995).

### *Broadband Activity and Emergence of New Jet Components*

The study of multiwavelength total flux-density variability offers an important diagnostic in testing models for the "inner jet" in blazars, and it is complementary to direct observation of structural variability (see Marscher & Bloom 1994, Marscher 1995, Marscher 1996). In particular, nonthermal flaring activity from radio to $\gamma$-ray frequencies is expected to show frequency-dependent time delays that are model specific. The $\gamma$-ray emission detected in a large number of blazars (Von Montigny et al 1995) is most likely to be produced from inverse-Compton scattering from relativistic jet electrons, although the details of this process are not yet clear. Several such blazars are therefore being subjected to multiwavelength monitoring campaigns that are beginning to show model-discriminating results (see Marscher 1996). The superluminal quasar 3C 279 (e.g. Carrara et al 1993) has been intensively monitored over a large spectral range (Maraschi et al 1994, Grandi et al 1996), and there is VLBI component activity near coincident with $\gamma$-ray flaring (Wehrle et al 1994a,b). For the blazar PKS 0420−014, helical structure and new component ejection were predicted from simultaneous optical and $\gamma$-ray flaring (Wagner et al 1995), and so far VLBI imaging has confirmed curvature in the structure of this source. Several other $\gamma$-ray active blazars, including 0528+134, 3C 273, and 0836+710, are also being investigated with VLBI at millimeter wavelengths, and here new features have been observed near $\gamma$-ray flares and at the onset of millimeter-flux density activity, which perhaps suggests a causal connection (Krichbaum et al 1995, Britzen et al 1997). A correlation between $\gamma$-ray outbursts and superluminal motion might suggest the existence of relativistic $\gamma$-ray jets, i.e. within as close as a few light-days to the central engine in such objects. One of several alternative explanations is that the $\gamma$-ray bursts might be related to precessing relativistic $e^{\pm}$ beams (Roland et al 1994).

BL Lac is perhaps the best example for a correlation between variability in the centimeter regime and the appearance of superluminal features that can be modeled with shocks (Mutel et al 1990, Mutel et al 1994, Denn & Mutel 1996). In this source, the position angles of the axes of the curved trajectories of subsequent components shift, suggesting evidence for precession caused by a binary black hole system (Mutel et al 1994). Note that such a scenario was also proposed for the superluminal quasar 1928+738 (Hummel et al 1992b, Roos et al 1993).

In 3C 345, several components first appeared following a major flux density increase at centimeter wavelengths, but not all components in this source have been related to such pronounced flux density changes; for example, the recent feature C7 was first seen during a deep minimum in the total flux density. In this source X-ray variability is correlated to the radio variations supporting the synchrotron self-Compton model for this source (Unwin et al 1997). In 3C 273, a correlation appears to exist between the ejection of superluminal features with flares in the radio and infrared/optical regimes (Krichbaum et al 1990a, Abraham et al 1996). A possible correlation of the speed of a component in this source with the amplitude of a perhaps related optical outburst has been speculated to exist (Bhabadzhanyants & Belokon 1992). In 3C 454.3, a new component appeared around 1983, with an apparent mean motion of $0.81 \pm 0.04$ mas $yr^{-1}$ ($v/c = 20.4\,h^{-1} \pm 1.0$). Extrapolation gives a time of zero separation of 1982 $\pm$ 3, i.e. close to the time of a radio outburst (Pauliny-Toth et al, in preparation).

## PHYSICAL CONDITIONS IN THE PARSEC-SCALE JETS

Despite the observational and theoretical advances in the study of compact radio sources, the physical nature of the VLBI components observed in parsec-scale radio jets and their connection to the central engine remains unclear (e.g. Marscher 1996). Turbulent plasma inhomogeneities and shock waves have been proposed with considerable success in explaining specific source properties (Blandford & Königl 1979, Reynolds 1982). In the relativistic jet model, the optically thick cores in the VLBI images represent the base of the jet (located near the apex of the jet cone), and the superluminal features are regions of enhanced emission moving along the jet. Theoretical predictions of synchrotron self-Compton emission from the VLBI components can be compared with observed X-ray fluxes or limits, and an observed excess of predicted versus observed X rays is once again interpreted as evidence for bulk relativistic motion (see Marscher 1987). The corresponding Doppler factors constrain allowed ranges of model parameters. In a core-jet source, the inhomogeneous jet model for the core (Königl 1981) can be combined with (perhaps simplistic) homogeneous spheres for the moving components to estimate the basic properties of the parsec-scale emission regions and to derive Doppler factors for the relativistic motion (Marscher 1987, Zensus et al 1995).

Ghisellini et al (1993) derive Doppler factors for about 100 sources with known VLBI structures by comparing predicted and observed X-ray flux in the synchrotron self-Compton model. The main results agree with those from other beaming indicators (superluminal motion and core to extended-flux density ratio) and support a simple kinematic model of ballistic motion of knots in relativistic jets. The derived Doppler factors are largest for core-dominated quasars, intermediate for BL Lacertae objects, and smallest for lobe-dominated radio galaxies and quasars. For the subsample of 39 superluminal sources, apparent expansion speeds and Doppler factors correlate and have similar average numerical values. This is taken as evidence that the bulk motion causing the beaming also causes the superluminal expansion and that it does not require different pattern and bulk velocities. The corresponding Lorentz factors are about 10 on average, with no significant differences between core- and lobe-dominated quasars and BL Lacs. The derived viewing angles differ, with radio galaxies, lobe-dominated quasars, BL Lacertae objects, and core-dominated quasars representing increasingly aligned objects.

In the case of 3C 345, the synchrotron self-Compton model reconciles well the component radio spectra, including the flat radio spectrum of the core, with the observed flux density, spectrum, and correlated variability in the X rays (Zensus et al 1995, Unwin et al 1994, Unwin et al 1997). The best-fit model requires a very small opening angle ($\sim$0.5°) for the portion of the jet near the apex of the cone that represents the flat-spectrum radio core; the apparent opening angle is much larger owing to projection effects. The inverse Compton calculations suggest that typically the core and its nearest moving component dominate the X-ray emission. The model is consistent with a heavy proton/electron jet that would yield enough power to fuel the outer radio lobe, and it argues for a small angle to the line of sight of about 1°. This can be relaxed if a pattern speed smaller than the true fluid speed is allowed. Combination of the superluminal speed (from VLBI) and the Doppler factor deduced from the synchrotron self-Compton calculation for the more recently ejected component (C7) implies that the jet bends away from the line of sight (from $\theta \simeq 2°$ to $\simeq 10°$), and such combination also implies that it accelerates from $\gamma \simeq 5$ to $\gtrsim 10$ over the range of (deprojected) distance from the nucleus of $\simeq 3$ to $\simeq 20\ h^{-1}$ pc (Unwin et al 1997).

### *Helical Jets*

Superluminal motion requires that the jet must be relativistic and viewed at a narrow angle to the line of sight so that projection effects are likely to be significant. It is possible to reconstruct the three-dimensional trajectory of a moving feature if an assumption is made for the Lorentz factor, $\gamma$, of the motion. If, for example, a constant pattern speed is assumed, the motion of the outer components in 3C 345 can be reconciled with one fixed curved trajectory where the apparent acceleration is due to changes in angle to the line of sight (Wardle et al 1994). For component C4, the observed curvature and acceleration, and the modest decline in (Doppler-boosted) flux density, are consistent with a curved jet of constant Lorentz factor of $\gamma$ about 10 (Zensus et al 1995a). The derived intrinsic jet curvature is small, but it is greatly amplified by projection effects, and the angle of the jet to the line of sight increases smoothly with radius from the core from $\sim$1 to $\sim$4°.

The observed apparently bent jets with components moving on different ballistic trajectories found in a number of sources can be explained by precession in the region of the nucleus (Blandford 1987): for example, caused by gravitational interaction between galaxies, binary black holes, or black-hole/disk systems. However, the required periods are typically short and cannot easily be reconciled with realistic models (Linfield 1981, Roos 1988). On parsec scales, orbital motion may be a better explanation for the origin of the observed jet structure and kinematics (Roos et al 1993).

A number of models have also been proposed based on helical motion of some sort to explain the curved quasiperiodic trajectories seen in 3C 345, 1803+78, 4C39.25, and similar sources (Camenzind & Krockenberger 1992, Hardee 1987, Königl & Choudhuri 1985, Camenzind 1986, Qian et al 1991, 1996). Helical features in jets are also readily explained from three-dimensional Kelvin-Helmholtz jet simulations (Hardee et al 1995). For the case of 3C 345, the basic properties of component trajectories and apparent velocities for components C4 and C5 have been explained by a simple helical jet with a straight jet axis, which is based on the conservation of physical quantities (Steffen et al 1995). No clear case for a common helix for all components in this source could be made, perhaps because effects in addition to the helical motion are at work. Again, orbital motion at the base of the jet may be the natural explanation for the lack of a unique helix describing the kinematics of all components (Qian et al 1993).

## *Spectral Properties*

Until recently, most monitoring studies were confined to images of the total intensity of a given source in one or more particular observing bands. This suffices to determine basic morphology and component kinematics, but it lacks important physical information that can only be obtained from measuring the jet spectra from multifrequency work (e.g. Walker et al 1996) and the magnetic field distributions from polarization imaging (e.g. Roberts et al 1991, Cotton 1993, Leppänen et al 1995, Kemball et al 1996).

For 3C 345, by combining images taken at different frequencies at quasisimultaneous epochs (within about six months from each other), spectral information can be obtained and used to measure the basic parameters of the component synchrotron spectra: the turnover flux density and frequency and the integrated flux in the range 4–25 GHz (Lobanov & Zensus 1997). The observed luminosity variations suggest a variable pattern Lorentz factor in at least one component. The core turnover frequency and integrated flux show significantly weaker variations compared to the moving components, which suggests that the primary emission mechanism in the jet might differ from that of the core (although blending of the core with a new component can confuse the situation); at least two components, C4 and C5, show a peaked evolution of turnover frequency. Such spectral properties are important for testing the hypothesis that the jet components are caused by relativistic shocks (Rabaça & Zensus 1994, Marscher & Gear 1985, Hughes et al 1989b, Marscher et al 1992, Gómez et al 1993). However, the evidence in favor of the shock model is inconclusive so far. The total flux density variations in some sources have been adequately explained with shocks (Hughes et al 1989b). In 3C 345, strong shocks with a variable Doppler factor may well be the mechanism at work in the jet at least near the core, although at larger distances (>2 milliarcsec), the shock model alone does not suffice to explain the observed properties, e.g. the required intrinsic accelerations and long-component life times (Lobanov & Zensus 1997). Alternative models have been proposed that involve interaction with the ambient plasma (Rose et al 1987), e.g. the two-fluid model (Sol et al 1989, Pelletier & Roland 1990, Pelletier & Sol 1992) or nonsynchrotron emission mechanisms, i.e. bremsstrahlung (Weatherall & Benford 1991). For the scenario of an induced helical geometry of the jet, the two-fluid model predicts jet kinematics and flux density variability of a form not unlike those seen for at least one component (C7) of 3C 345 (Roland et al 1994).

The shock-in-jet hypothesis has been applied with success in a number of other sources. Marcaide et al (1994a) apply a detailed model of shock components moving on twisted trajectories to the quasar 4C39.25. BL Lac is perhaps the best-understood case for the interpretation of VLBI components by shocks (Mutel et al 1990).

### *Polarization Properties*

The initial motivation for the application of shock models to VLBI jets came from the modeling of total flux density variability and polarization behavior in sources with comparatively simple jet structure, e.g. BL Lac and 3C 279 (Hughes et al 1989b). Polarization imaging, especially at high frequencies, i.e. with adequate angular resolution, is a prerequisite to understanding the internal physics of sources with more complex VLBI jets (Roberts et al 1991, Wardle et al 1994, Aller et al 1994). Linear polarization–sensitive VLBI observations of compact sources yield direct information on the structure and order of the underlying magnetic fields, the presence and nature of thermal material, the energy of relativistic electrons, and the geometry of emission on submilliarcsec scales. There is evidence for a difference in the polarization properties of quasars and BL Lac objects, and the latter show significantly lower speeds, which argues for a physical and not merely an orientation difference between the two classes of sources (Roberts et al 1990, Gabuzda et al 1994a,b). Brown et al (1994) and Wardle et al (1994) have studied the polarization structure at 5 GHz of 3C 345 and interpreted their results in terms of a comprehensive shock model.

Leppänen et al (1995) obtained high-resolution polarization images of 3C 345 at 22 GHz with the VLBA. Assuming negligible depolarization in the core, they find the electric field predominantly oriented along the jet, i.e. the magnetic field orientation is perpendicular to the jet, in contrast to the lower frequency results that suggested a parallel magnetic field (Wardle et al 1994). Gabuzda (1995) found similar evidence for a perpendicular magnetic field from imaging 3C 345 at 10.7 GHz. An analysis applying the model of Wardle et al to the 22-GHz and subsequent 43-GHz observations suggests that the shock model remains applicable for the case of a moderately strong shock (KJ Leppänen & JA Zensus, in preparation). The 22-GHz polarized structure of 3C 345 shows two additional important features: a turn in the electric vector position angle and a steep decline to zero of the observed polarized intensity between C8 and C7. Both properties can, within the shock model, be explained by a simple helical motion of the moving feature (Steffen et al 1995). For the quasar 3C454.3, 7-mm VLBA imaging also reveals an electric field configuration that is aligned with the predominant jet orientation but orthogonal between the major components of the compact structure (Kemball et al 1996). In both cases, it is not yet clear if the observed changes depend on the strength of the shock, or if they result from changes in the local magnetic field configuration, a helical jet geometry, or from local differential Faraday rotation.

## CONCLUSION

Extensive VLBI monitoring studies at centimeter and millimeter wavelengths, enhanced by spectral and polarization imaging, of representative types of AGN can discriminate detailed physical models. When combined with broadband total flux density and polarization observations, and especially with X-ray and $\gamma$-ray data that measure the synchrotron self-Compton radiation component, they can be used to determine the overall physical conditions in parsec-scale radio jets. Here is an excellent opportunity to directly compare theoretical predictions with observational data, despite the complexity and the large number of free parameters of theoretical models. Some of the most intriguing areas where this may soon become feasible include the formation of the jets, possible jet acceleration mechanisms, the nature of any nonrelativistic matter involved (e.g. the ambient medium and thermal outflows), and the influence of the central region on the jet dynamics.

### ACKNOWLEDGMENTS

I thank Thomas Krichbaum for extensive discussions and helpful suggestions on the subject of this review. Thanks are also owed to K Kellermann, A Patnaik, and A Witzel for critical comments on the manuscript. I am grateful to B

Cotton, V Dhawan, K Kellermann, T Krichbaum, G Taylor, and R Vermeulen for permission to reproduce material, partly in advance of publication.



*Literature Cited*


Abraham Z, Cararra EA, Zensus JA, Unwin SC. 1996. *Astron. Astrophys. Suppl.* 115:543–49

Alberdi A, Krichbaum TP, Marcaide JM, Witzel A, Graham DA, et al. 1993a. *Astron. Astrophys.* 271:93–100

Alberdi A, Krichbaum TP, Graham DA, Greve A, Grewing M, et al. 1997. *Astron. Astrophys.* In press

Alberdi A, Marcaide JM, Marscher AP, Zhang YF, Elosegui P, et al. 1993b. *Ap. J.* 402:160–72

Aller HD, Hughes PA, Aller MF. 1994. See Zensus & Kellermann 1994, pp. 223–28

Antonucci RRJ. 1993. *Annu. Rev. Astron. Astrophys.* 31:473–521

Appl S, Sol H, Vicente L. 1996. *Astron. Astrophys.* 310:419–37

Bååth LB. 1994. In *VLBI Technology. Progress and Future Observational Possibilities*, ed. T Sasao, S Manabe, O Kameya, M Inoue, pp. 70–74. Tokyo: Terra Sci.

Bahcall JN, Kirhakos S, Schneider DP, Davis RJ, Muxlow TWB, et al. 1995. *Ap. J. Lett.* 452:L91–L93

Bartel N, Bietenholz MF, Rupen MP. 1995. *Proc. Natl. Acad. Sci. USA* 92:11374–76

Bartel N, Herring TA, Ratner MI, Shapiro II, Corey BE. 1986. *Nature* 319:733–38

Barthel PD. 1989. *Ap. J.* 336:606–11

Barthel PD. 1995. *Highlights Astron.* 10:551–57

Begelman MC. 1995. *Proc. Natl. Acad. Sci. USA* 92:11442–46

Begelman MC, Blandford RD, Rees MJ. 1984. *Rev. Mod. Phys.* 56:255–351

Benson JM, Walker RC, Unwin SC, Muxlow TWB, Wilkinson PN, et al. 1988. *Ap. J.* 334:560–72

Bhabadzhanyants MK, Belokon ET. 1992. See Valtaoja & Valtonen 1992, pp. 384–90

Biretta JA. 1996. See Hardee et al 1996, pp. 187–98

Biretta JA, Junor W. 1995. *Proc. Natl. Acad. Sci. USA* 92:11364–67

Biretta JA, Moore RL, Cohen MH. 1986. *Ap. J.* 308:93–109

Biretta JA, Zhou F, Owen FN. 1995. *Ap. J.* 447:582–96

Blandford RD. 1987. See Zensus & Pearson 1987, pp. 310–27

Blandford RD, Königl A. 1979. *Ap. J.* 232:34–38

Blandford RD, Payne DG. 1982. *MNRAS* 199: 883–903

Blandford RD, Rees MJ. 1974. *MNRAS* 169: 395–415

Bridle AH. 1996. See Hardee et al 1996, pp. 383–94

Bridle AH, Perley RA. 1984. *Annu. Rev. Astron. Astrophys.* 22:319–58

Britzen S, Witzel A, Krichbaum TP. 1997. In *Proceedings of the Heidelberg Workshop on Gamma-Ray Emitting AGNs*, ed. JG Kirk, M Camenzind, C von Montigny, S Wagner, pp. 109–12. Heidelberg: Max-Planck-Inst. Kernphys.

Brown LF, Roberts DH, Wardle JFC. 1994. *Ap. J.* 437:108–21

Burbidge EM, Burbidge G. 1997. *Ap. J.* 477: L13–L15

Camenzind M. 1986. *Astron. Astrophys.* 156:137–51

Camenzind M, Krockenberger M. 1992. *Astron. Astrophys.* 255:59–62

Carilli CL, Bartel N, Diamond P. 1994. *Astron. J.* 108:64–75

Carilli CL, Barthel PD. 1996. *Astron. Astrophys. Rev.* 7:1–54

Carilli CL, Harris DE, eds. 1996. *Cygnus A—Study of a Radio Galaxy.* Cambridge: Cambridge Univ. Press

Carilli CL, Perley RA, Bartel N, Sorathia B. 1996. See Hardee et al 1996, pp. 287–98

Carrara EA, Abraham Z, Unwin SU, Zensus JA. 1993. *Astron. Astrophys.* 279:83–89

Cawthorne TV. 1991. See Hughes 1991, pp. 187–231

Cohen MH, Kellermann KI, eds. 1995. *Quasars and Active Galactic Nuclei: High Resolution Radio Imaging, Proc. Natl. Acad. Sci. USA* 92:11339–450

Conway JE. 1994. See Zensus & Kellermann 1994, pp. 73–78

Conway JE, Murphy DW. 1993. *Ap. J.* 411:89–102

Conway JE, Wrobel JM. 1995. *Ap. J.* 439:98–112

Cotton WD. 1993. *Astron. J.* 106:1241–48

Cotton WD, Dallacasa D, Fanti C, Fanti R, Foley AR, et al. 1997. *Astron. Astrophys.* Submitted

Davis RJ, Booth RS, eds. 1993. *Sub-Arcsecond Radio Astronomy*. Cambridge: Cambridge Univ. Press

Davis RJ, Unwin SC, Muxlow TWB. 1991. *Nature* 354:374–76

Denn G, Mutel R. 1996. See Ekers et al 1996, pp. 41–42

Ekers R, Fanti C, Padrielli L, eds. 1996. *IAU Symp. 175: Extragalactic Radio Sources*. Dordrecht: Kluwer

Fanaroff BL, Riley JM. 1974. *MNRAS* 167:31P–35P

Fanti C, Fanti R, Dallacasa D, Schilizzi RT, Spencer RE, Stanghellini C. 1995. *Astron. Astrophys.* 302:317–26

Fanti R, Fanti C, Schilizzi RT, Spencer RE, Nan Rendong, Parma P, et al. 1990. *Astron. Astrophys.* 231:333–46

Fanti R, Spencer R. 1996. See Ekers et al 1996, pp. 63–66

Feretti L, Comoretto G, Giovannini G, Venturi T, Wehrle AE. 1993. *Ap. J.* 408:446–51

Gabuzda DC. 1995. *Proc. Natl. Acad. Sci. USA* 92:11393–98

Gabuzda DC, Cawthorne TV, Roberts DH, Wardle JFC. 1992. *Ap. J.* 388:40–54

Gabuzda DC, Mullan CM, Cawthorne TV, Wardle JFC, Roberts DH. 1994. *Ap. J.* 435:140–61

Ghisellini G, Padovani P, Celotti A, Maraschi L. 1993. *Ap. J.* 407:65–82

Giovannini G, Cotton WD, Feretti L, Lara L, Venturi T, Marcaide JM. 1995. *Proc. Natl. Acad. Sci. USA* 92:11356–59

Giovannini G, Feretti L, Venturi T, Lara L, Marcaide J, et al. 1994. *Ap. J.* 435:116–27

Gómez JL, Alberdi A, Marcaide JM. 1993. *Astron. Astrophys.* 275:55–68

Gómez JL, Alberdi A, Marcaide JM. 1994. *Astron. Astrophys.* 284:51–64

Gopal-Krishna. 1995. *Proc. Natl. Acad. Sci. USA* 92:11399–406

Grandi P, Urry CM, Maraschi L, Wehrle AE, Madejski GM, et al. 1996. *Ap. J.* 459:73–81

Guirado JC, Marcaide JM, Alberdi E, El'osegui P, Ratner MI, et al. 1995. *Astron. J.* 110:2586–96

Gurvits LI. 1994. *Ap. J.* 425:442–49

Hardee PE. 1987. *Ap. J.* 318:78–92

Hardee PE, Bridle AH, Zensus JA, eds. 1996. *Energy Transport in Radio Galaxies and Quasars*. San Francisco: Astronomical Society of the Pacific

Hardee PE, Clarke DA, Howell DA. 1995. *Ap. J.* 441:644–64

Henstock DR, Browne IWA, Wilkinson PN, Taylor GB, Vermeulen RC, et al. 1995. *Ap. J. Suppl.* 100:1–36

Hewitt JN. 1995. *Proc. Natl. Acad. Sci. USA* 92:11434–38

Hirabayashi H. 1996. See Ekers et al 1996, pp. 529–30

Hough DH. 1994. See Zensus & Kellermann 1994, pp. 169–74

Hough DH, Readhead ACS. 1989. *Astron. J.* 98:1208–25

Hough DH, Vermeulen RC, Wood DAJ, Standifird JD, Cross LL. 1996a. *Ap. J.* 459:64–72

Hough DH, Zensus JA, Porcas RW. 1996b. *Ap. J.* 464:715–23

Hughes PA, ed. 1991. *Beams and Jets in Astrophysics*. Cambridge: Cambridge Univ. Press

Hughes PA, Aller HD, Aller MF. 1989a. *Ap. J.* 341:54–67

Hughes PA, Aller HD, Aller MF. 1989b. *Ap. J.* 341:68–79

Hughes PA, Miller L. 1991. See Hughes 1991, pp. 1–51

Hummel CA, Muxlow TWB, Krichbaum TP, Quirrenbach A, Schalinski CJ, et al. 1992a. *Astron. Astrophys.* 266:93–100

Hummel CA, Schalinski CJ, Krichbaum TP, Rioja MJ, Quirrenbach A, et al. 1992b. *Astron. Astrophys.* 257:489–500

Jauncey DL, Tingay SJ, Preston RA, Reynolds JE, Lovell JEJ, et al. 1995. *Proc. Natl. Acad. Sci. USA* 92:11368–70

Jones DL, Tingay SJ, Murphy DW, Meier DL, Jauncey DL, et al. 1996. *Ap. J.* 466:L63–L65

Jones DL, Wehrle AE. 1994. *Ap. J.* 427:221–26

Junor W, Biretta JA. 1995. *Astron. J.* 109:500–6

Kellermann KI. 1993. *Nature* 361:134–36

Kellermann KI, Pauliny-Toth IIK. 1981. *Annu. Rev. Astron. Astrophys.* 19:373–410

Kellermann KI, Zensus JA, Cohen MH. 1997. *Ap. J. Lett.* 475:L93–L95

Kemball AJ, Diamond PJ, Pauliny-Toth IIK. 1996. *Ap. J.* 464:L55–L58

Königl A. 1981. *Ap. J.* 243:700–9

Königl A, Choudhuri AR. 1985. *Ap. J.* 289:173–87

Krichbaum TP, Alef W, Witzel A. 1996. See Carilli & Harris 1996, pp. 92–97

Krichbaum TP, Alef W, Witzel A, Zensus JA. 1997. *Astron. Astrophys.* Submitted

Krichbaum TP, Booth RS, Kus AJ, Rönnäng BO, Witzel A, et al. 1990a. *Astron. Astrophys.* 237:3–11

Krichbaum TP, Britzen S, Standke KJ, Witzel A, Schalinski CJ, Zensus JA. 1995. *Proc. Natl. Acad. Sci. USA* 92:11377–80

Krichbaum TP, Hummel CA, Quirrenbach A, Schalinski CJ, Witzel A, et al. 1990b. *Astron. Astrophys.* 230:271–83

Krichbaum TP, Standke KJ, Witzel A, Schalinski CJ, Grewing M, et al. 1994a. In *Proc. 2nd EVN/JIVE Symp.*, ed. AJ Kus et al, pp. 47–54. Torun: Torun Radio Astronomy Observatory

Krichbaum TP, Witzel A. 1992. See Valtaoja & Valtonen 1992, pp. 205–20

Krichbaum TP, Witzel A, Graham DA. 1993a.

In *Jets in Extragalactic Radio Sources*, ed. HJ Röser, K Meisenheimer, Number 421 in Lecture Notes in Physics, pp. 71–78. Berlin: Springer
Krichbaum TP, Witzel A, Graham DA, Schalinski CJ, Zensus JA. 1993b. See Davis & Booth 1993, pp. 181–83
Krichbaum TP, Witzel A, Graham DA, Standke K, Schwartz R, et al. 1993c. *Astron. Astrophys.* 275:375–89
Krichbaum TP, Witzel A, Standke KJ, Graham DA, Schalinski CJ, Zensus JA. 1994b. See Zensus & Kellermann 1994, pp. 39–44
Leppänen KJ, Zensus JA, Diamond PD. 1995. *Astron. J.* 110:2479–92
Levinson A, Laor A, Vermeulen RC. 1995. *Ap. J.* 448:589–99
Linfield R. 1981. *Ap. J.* 250:464–68
Lobanov AP, Zensus JA. 1997. *Ap. J.* Submitted
Maraschi L, Grandi P, Urry CM, Wehrle AE, Madejski GM, et al. 1994. *Ap. J. Lett.* 435:L91–L95
Marcaide JM, Alberdi A, Gómez JL, Guirado JC, Marscher AP, Zhang YF. 1994a. See Zensus & Kellermann 1994, pp. 141–48
Marcaide JM, Elosequi P, Shapiro II. 1994b. *Astron. J.* 108:368–73
Marscher A, Bloom SD. 1994. pp. 572–81. New York: Am. Inst. Phys.
Marscher AP. 1987. See Zensus & Pearson 1987, pp. 280–300
Marscher AP. 1990. See Zensus & Pearson 1990, pp. 236–49
Marscher AP. 1995. *Proc. Natl. Acad. Sci. USA* 92:11439–41
Marscher AP. 1996. See Hardee et al 1996, pp. 45–54
Marscher AP, Gear WK. 1985. *Ap. J.* 298:114–27
Marscher AP, Gear WK, Travis JP. 1992. See Valtaoja & Valtonen 1992, pp. 85–101
Miller JS. 1995. *Proc. Natl. Acad. Sci. USA* 92:11422–26
Miyoshi M, Moran J, Herrnstein J, Greenhill L, Nakai N, et al. 1995. *Nature* 373:127–29
Moran J, Greenhill L, Herrnstein J, Diamond P, Miyoshi M, et al. 1995. *Proc. Natl. Acad. Sci. USA* 92:11427–33
Mutel RL, Denn GR, Dryer MJ. 1994. See Zensus & Kellermann 1994, pp. 191–96
Mutel RL, Phillips RB, Su B, Bucciferro RR. 1990. *Ap. J.* 352:81–95
Napier PJ, Bagri DS, Clark BG, Rogers AEE, Romney JD, et al. 1994. *IEEE Proc.* 82:658–72
Norman ML. 1996. See Hardee et al 1996, pp. 319–26
O'Dea CP. 1996. See Hardee et al 1996, pp. 85–90
O'Dea CP, Barvainis R, Challis PM. 1988. *Astron. J.* 96:435–54
O'Dea CP, Baum SA, Stanghellini C. 1991. *Ap. J.* 380:66–77
Orr MJL, Browne IWA. 1982. *MNRAS* 200:1067–80
Pearson TJ. 1996. See Hardee et al 1996, pp. 97–108
Pearson TJ, Readhead ACS. 1988. *Ap. J.* 328:114–42
Pelletier G, Roland J. 1990. See Zensus & Pearson 1990, pp. 323–32
Pelletier G, Sol H. 1992. *MNRAS* 254:635–46
Perlman ES, Stocke JT, Shaffer DB, Carilli CL, Ma CP. 1994. *Ap. J. Lett.* 424:L69–L72
Phillips RB, Mutel RL. 1982. *Astron. Astrophys.* 106:21–24
Polatidis AG, Wilkinson PN, Xu W, Readhead ACS, Pearson TJ, et al. 1995. *Ap. J. Suppl.* 98:1–32
Porcas RW. 1987. See Zensus & Pearson 1987, pp. 12–25
Qian S, Witzel A, Krichbaum TP, Wagner SJ. 1995. *Acta Astron. Sinica* 36:34–46. *Transl. Chin. Astron. Astrophys.* 19:522
Qian SJ, Krichbaum TP, Witzel A, Quirrenbach A, Hummel CA, Zensus JA. 1991. *Acta Astro. Sinica* 32(4):369–79. *Transl. Chin. Astron. Astrophys.* 16:137–47
Qian SJ, Krichbaum TP, Zensus JA, Steffen W, Witzel A. 1996. *Astron. Astrophys.* 308:395–402
Qian SJ, Witzel A, Krichbaum TP, Quirrenbach A, Zensus JA. 1993. *Chin. Astron. Astrophys.* 17:150–60
Rabaça CR, Zensus JA. 1994. See Zensus & Kellermann 1994, pp. 163–68
Rantakyrö FT, Bååth LB, Matveenko L. 1995. *Astron. Astrophys.* 293:44–55
Readhead ACS. 1993. See Davis & Booth 1993, pp. 173–80
Readhead ACS. 1995. *Proc. Natl. Acad. Sci. USA* 92:11447–50
Readhead ACS, Taylor GB, Pearson TJ, Wilkinson PN. 1996a. *Ap. J.* 460:634–43
Readhead ACS, Taylor GB, Xu W, Pearson TJ, Wilkinson PN, Polatidis AG. 1996b. *Ap. J.* 460:612–33
Rees MJ. 1966. *Nature* 211:468
Rees MJ. 1971. *Nature* 229:312–17
Reid MJ, Biretta JA, Junor W, Muxlow TWB, Spencer RE. 1989. *Ap. J.* 336:112–20
Reynolds SP. 1982. *Ap. J.* 256:13–27
Rickett BJ, Quirrenbach A, Wegner R, Krichbaum TK, Witzel A. 1995. *Astron. Astrophys.* 293:479–92
Roberts DH, Brown LF, Wardle JFC. 1991. In *Radio Interferometry: Theory, Techniques, and Applications*, ed. TJ Cornwell, RA Perley, *ASP Conf. Ser.*, 19:281–88. San Francisco: Astron. Soc. Pac.
Roberts DH, Wardle JFC, Brown LF, Gabuzda DC, Cawthorne TV. 1990. See Zensus &

Pearson 1990, pp. 110–16
Roland J, Frossati G, Teyssier R. 1994. *Astron. Astrophys.* 290:364–70
Roland J, Teyssier R, Roos N. 1994. *Astron. Astrophys.* 290:357–63
Romney JD, Benson JM, Dhawan V, Kellermann KI, Vermeulen RC, Walker RC. 1995. *Proc. Natl. Acad. Sci. USA* 92:11360–63
Roos N. 1988. *Ap. J.* 334:95–103
Roos N, Kaastra JS, Hummel CA. 1993. *Ap. J.* 409:130–33
Rose WK, Beall JH, Guillory J, Kainer S. 1987. *Ap. J.* 314:95–102
Schalinski CJ, Witzel A, Hummel CA, Krichbaum TP, Quirrenbach A, Johnston KJ. 1992. In *Physics of Active Galactic Nuclei*, ed. WJ Duschl, SJ Wagner, pp. 589–91. Heidelberg: Springer-Verlag
Schalinski CJ, Witzel A, Krichbaum TP, Graham DA, Standke KJ, et al. 1994. See Zensus & Kellermann 1994, pp. 45–48
Scheuer PAG. 1974. *MNRAS* 166:513–28
Scheuer PAG, Readhead ACS. 1979. *Nature* 277:182–85
Schilizzi RT. 1995. See Zensus et al 1995b, pp. 397–408
Shaffer DB, Marscher AP. 1987. See Zensus & Pearson 1987, pp. 67–75
Shklovskii IS. 1963. *Sov. Astron.* 6:465–76
Simon RS, Hall J, Johnston KJ, Spencer JH, Waak JA, Mutel RL. 1988. *Ap. J.* 326:L5–L8
Sol H, Pelletier G, Asséo E. 1989. *MNRAS* 237:411–29
Spencer RE. 1994. See Zensus & Kellermann 1994, pp. 35–38
Steffen W, Zensus JA, Krichbaum TP, Witzel A, Qian SJ. 1995. *Astron. Astrophys.* 302:335–42
Taylor D, Dyson JE, Axon DJ, Pedlar A. 1989. *MNRAS* 240:487–99
Taylor GB, Readhead ACS, Pearson TJ. 1996. *Ap. J.* 463:95–104
Taylor GB, Vermeulen RC. 1996. *Ap. J. Lett.* 457:L69–L71
Taylor GB, Vermeulen RC, Pearson TJ, Readhead ACS, Henstock DR, et al. 1994. *Ap. J. Suppl.* 95:345–69
Thakkar DD, Xu W, Readhead ACS, Pearson TJ, Taylor GB, et al. 1995. *Ap. J. Suppl.* 98:33–40
Tingay SJ, Jauncey DL, Preston RA, Reynolds JE, Meier DL, et al. 1994. *Aust. J. Phys.* 47:619–24
Unwin SC, Davis RJ, Muxlow TWB. 1994a. See Zensus & Kellermann 1994, pp. 81–86
Unwin SC, Wehrle AE. 1992. *Ap. J.* 398:74–86
Unwin SC, Wehrle AE, Lobanov AP, Zensus JA, Madejski GM, et al. 1997. *Ap. J.* 480:596–606
Unwin SC, Wehrle AE, Urry CM, Gilmore DM, Barton EJ, et al. 1994b. *Ap. J.* 432:103–13
Urry CM, Padovani P. 1995. *PASP* 107:803–45
Valtaoja E, Valtonen M, eds. 1992. *Variability of Blazars*. Cambridge: Cambridge Univ. Press
Venturi T, Castaldini C, Cotton WD, Feretti L, Giovannini G, et al. 1995. *Ap. J.* 454:735–44
Venturi T, Readhead ACS, Marr JM, Backer DC. 1993. *Ap. J.* 411:552–64
Vermeulen RC. 1995. *Proc. Natl. Acad. Sci. USA* 92:11385–89
Vermeulen RC, Bernstein RA, Hough DH, Readhead ACS. 1993. *Ap. J.* 417:541–46
Vermeulen RC, Cohen MH. 1994. *Ap. J.* 430:467–94
Vermeulen RC, Readhead ACS, Backer DC. 1994. *Ap. J.* 430:L41–L44
Von Montigny C, Bertsch DL, Chiang J, Dingus BL, Esposito JA, et al. 1995. *Astron. Astrophys.* 299:680–88
Wagner SJ, Camenzind M, Dreissigacker O, Borgeest U, Britzen S, et al. 1995. *Astron. Astrophys.* 298:688–98
Wagner SJ, Witzel A. 1995. *Annu. Rev. Astron. Astrophys.* 33:163–97
Wagner SJ, Witzel A, Heidt J, Krichbaum TP, Quirrenbach A, et al. 1996. *Astron. J.* 111:2187–211
Walker RC, Benson JM, Unwin SC. 1987. *Ap. J.* 316:546–72
Walker RC, Romney JD, Benson JM. 1994. *Ap. J. Lett.* 430:L45–L48
Walker RC, Romney JD, Vermeulen RC, Dhawan V, Kellermann KI. 1996. See Ekers et al 1996, pp. 30–32
Wardle JFC, Cawthorne TV, Roberts DH, Brown LF. 1994. *Ap. J.* 437:122–35
Weatherall JC, Benford G. 1991. *Ap. J.* 378: 543–49
Wehrle AE, Cohen MH, Unwin SC, Aller HD, Aller MF, Nicolson G. 1992. *Ap. J.* 391:589–607
Wehrle AE, Unwin SC, Zook AC. 1994. See Zensus & Kellermann 1994, pp. 197–200
Wehrle AE, Unwin SC, Zook AC, Urry CM, Marscher AP, Teräsranta H. 1996. In *Blazar Continuum Variability*, ASP Conf. Ser. 110:430–35. San Francisco: Astron. Soc. Pac.
Wiita PJ. 1996. See Hardee et al 1996, pp. 395–404
Wilkinson PN. 1995. *Proc. Natl. Acad. Sci.* USA 92:11342–47
Witzel A, Schalinski CJ, Biermann PL, Krichbaum TP, Johnston KJ. 1988. *Astron. Astrophys.* 206:245–52
Xu W, Readhead ACS, Pearson TJ, Polatidis AG, Wilkinson PN. 1995. *Ap. J. Suppl.* 99:297–348
Xu W, Readhead ACS, Pearson TJ, Wilkinson PN, Polatidis AG. 1994. See Zensus & Kellermann 1994, pp. 7–10
Zensus JA, Biretta JA, Unwin SC, Cohen MH.

1990. *Astron. J.* 100:1777–84
Zensus JA, Cohen MH, Unwin SC. 1995a. *Ap. J.* 443:35–53
Zensus JA, Diamond PJ, Napier PJ, eds. 1995b. *Very Long Baseline Interferometry and the Very Long Baseline Array*, ASP Conf. Ser. 82. San Francisco: Astron. Soc. Pac.
Zensus JA, Kellermann KI, eds. 1994. *Compact Extragalactic Radio Sources.* Green Bank, WVa: Natl. Radio Astron. Obs.
Zensus JA, Krichbaum TP, Lobanov AP. 1995c. *Proc. Natl. Acad. Sci. USA* 92:11348–55
Zensus JA, Pearson TJ, eds. 1987. *Superluminal Radio Sources.* Cambridge: Cambridge Univ. Press
Zensus JA, Pearson TJ, eds. 1990. *Parsec-Scale Radio Jets.* Cambridge: Cambridge Univ. Press
Zensus JA, Porcas RW. 1987. See Zensus & Pearson 1987, pp. 126–28

*Annu. Rev. Astron. Astrophys. 1997. 35:637–75*


# GALACTIC BULGES

*Rosemary F. G. Wyse*
Department of Physics and Astronomy, The Johns Hopkins University, Baltimore, Maryland 21218, USA; e-mail: wyse@pha.jhu.edu

*Gerard Gilmore*
Institute of Astronomy, Madingley Road, Cambridge CB3 0HA, United Kingdom; Institut d'Astrophysique de Paris, 98bis boulevard Arago, 75014 Paris, France; e-mail: gil@ast.cam.ac.uk

*Marijn Franx*
Kapteyn Astronomical Institute, University of Groningen, PO Box 800, 9700AV Groningen, The Netherlands; e-mail: franx@astro.rug.nl

KEY WORDS: galaxy formation, the Galaxy, extragalactic astronomy, Local Group, dynamical astronomy

ABSTRACT

We discuss the present observational and theoretical understanding of the stellar populations of bulges and their implications for galaxy formation and evolution. The place of bulges as key to the Hubble Sequence remains secure, but some old paradigms are giving way to new ones as observations develop. Detailed studies of Local Group bulges and haloes provide a basis on which we consider higher redshift data. We present the evidence for and against the currently common preconceptions that bulges are old, above solar metallicity in the mean, and simply scaled-down versions of ellipticals. We conclude life is not so simple: Bulges are diverse and heterogeneous, and although their properties vary systematically, sometimes they are reminiscent of disks, sometimes of ellipticals. The extant observational data are, however, limited. New and future surveys will rectify this, and we discuss the questions those data will address.

## 1. MOTIVATION AND SCOPE OF REVIEW

### 1.1 *Introduction*

In his introduction to the report of IAU Symposium #1, *Coordination of Galactic Research*, held near Groningen, June 1953, Blaauw noted, "In the discussion



0066-4146/97/0915-0637$08.00

the terms 'halo', 'nucleus' and 'disk' are used to indicate different parts of the Galaxy. These general regions are not defined more precisely. Their introduction proved very useful, and one might rather say that their more exact description is one of the problems of galactic research." This statement provides an excellent example of the limitations of terminology and of the term galactic bulge in that this component continues to lack a clear definition (nucleus? halo?) of either its structure or its relationship to the other stellar components of the Galaxy. This is compounded by the difficulty of observing bulges even once one has decided which part of the galaxy that is.

The common usage of "bulge," for example in the term bulge-to-disk ratio, allocates all "non-disk" light in any galaxy that has a "disk" into the bulge. That is, the bulge contains any light that is in excess of an inward extrapolation of a constant scale-length exponential disk. Sandage (Sandage & Bedke 1994, *Carnegie Atlas of Galaxies*; panel S11 and p. 45) emphasizes that "one of the three classification criteria along the spiral sequence is the size of the central amorphous bulge, compared with the size of the disk. The bulge size, seen best in nearly edge-on galaxies, decreases progressively,while the current star formation rate and the geometrical entropy of the arm pattern increases, from early Sa to Sd, Sm and Im types." This is the clearest convenient description of a bulge, namely a centrally concentrated stellar distribution with an amorphous—smooth—appearance. Note that this implicitly excludes gas, dust, and continuing recent star formation by definition, ascribing all such phenomena in the central parts of a galaxy to the central disk, not to the bulge with which it cohabits. Furthermore, for a bulge to be identified at all it must, by selection, have a central stellar surface density that is at least comparable to that of the disk, and/or it must have a (vertical) scale height that is at least not very much smaller than that of the disk. The fact that this working definition can be applied successfully to the extensive classifications in the *Carnegie Atlas of Galaxies* illustrates some fundamental correctness. Bulges are also clearly very much a defining component whose properties underly the Hubble sequence, and hence the reason why we care—understanding how bulges form and evolve is integral to the questions of galaxy formation and evolution.

This review considers the current widespread beliefs and preconceptions about galaxian bulges—for example, that they are old, metal-rich, and related to elliptical galaxies—in the light of modern data. Our aim is to provide an overview of interesting and topical questions and to emphasize recent and future observations that pertain to the understanding of the formation and evolutionary status of bulges. We begin by considering some common preconceptions.

## 1.2 *Preconception Number 1: Bulges Are Old*

The expectation of "old age" arose, as far as we can ascertain, from the interpretation of the observed correlation between stellar kinematics and metallicity for local stars in the Milky Way by Eggen et al (1962). These authors proposed a model of Galaxy formation by collapse of a galaxy-sized density perturbation, generalized to models wherein the spheroidal components of galaxies—including the entire stellar mass of an elliptical galaxy—formed stars *prior* to the dissipational settling to a disk and so contained the oldest stars (e.g. review of Gott 1977). The high central surface brightnesses of bulges (and of ellipticals), assuming they correspond to high mass densities, also imply a higher redshift of formation, for a fixed collapse factor of the protogalaxy, because at higher redshift the background density was higher (Peebles 1989).

An older component in the central regions of the Milky Way Galaxy clearly exists. The first real work on the bulge (or "nucleus" as it was called at the time) used classical "halo" tracers, such as globular clusters, RR Lyrae, and planetary nebulae. Of course, one must remember that "older" is used here in the sense that the term was used until very recently, which meant much older than the local disk, which contains ongoing star formation. That is, "old" means "there is no obvious AF star population." The Baade-era concept of "old" meant a turnoff in the F-region, which is of course old only for a very metal-poor system (see Sandage 1986, and the *Carnegie Atlas* for thorough reviews of Baade's Population concept). Furthermore, the very idea of discriminating between ages of 10 Gyr and 15 Gyr is a recent concept, in spite of the large fractional difference between the two.

Constraints on the redshift of formation of bulges can be obtained by direct observations of high-redshift galaxies, for which morphological information may be obtained with the Hubble Space Telescope (HST) (see Section 4). In general, disentangling the effects of age and metallicity on stellar colors is difficult, even when the stars are resolved and color-magnitude diagrams may be examined. The state-of-the-art mean age determinations for lower redshift bulges and disks are discussed in Section 3, and the interpretations of color-magnitude diagrams are discussed in Section 2. Much ambiguity and uncertainty remains.

Implicit in the Eggen et al (1962) scenario was the hypothesis that the Galactic bulge was simply the central region of the stellar halo, traced at the solar neighborhood by the high-velocity subdwarfs. These stars are old by anyone's definition. Stellar haloes can be studied easily only in the Local Group, and we discuss the stellar populations in those galaxies in Section 2 below.

### 1.3 *Preconception Number 2: The Galactic Bulge Is Super-Solar Metallicity*

This belief was strongly supported by study of late M-giants in Baade's Window (cf Frogel 1988), motivated by the Whitford (1978) paper that compared the spectrum of the Milky Way bulge to that of the integrated light of the central regions of external bulges and giant elliptical galaxies (see Whitford 1986 for a personal interpretation of his research). Whitford's investigation aimed to determine whether or not the bulge of our galaxy was "normal," i.e. the same as others. Whitford was apparently influenced, as were most people at that time, by the interpretation of the color-magnitude relation of Faber (1973) to assume that bulges and ellipticals were differentiated only by luminosity, which determined the metallicity, and that ages were invariant and *old*, with a turnoff mass of $\sim 1\ M_{\odot}$ (Faber 1973), at least for the dominant population. In this case, the most metal-rich stars in a lower luminosity bulge, like that of the Milky Way, could be used as a template for the *typical* star in a giant elliptical.

Whitford (1978) concluded from his data that indeed "the strengths of the spectral features in the sampled areas of the nuclear bulge of the Galaxy are very close to those expected from measures on similar areas of comparable galaxies." However, Whitford's data were, by current standards, of low spectral resolution and were limited to the following: spectra, with a resolution of 32 Å in the blue and 64 Å in the red, for three regions in Baade's Window and for the central regions of five edge-on spirals of type Sa to Sb; lower spectral resolution data for the central regions of M 31; partial data—blue wavelengths only—for one elliptical (NGC3379, E1); and full wavelength coverage spectra for one other elliptical (NGC4976, E4), which he emphasized did not match the Milky Way and was anomalous. Furthermore, the data for Baade's Window in the blue wavelength region—where direct comparison with a "normal" elliptical galaxy was possible—were emphasized to be very uncertain, owing to the large corrections for reddening and foreground (disk) emission. Thus, while the Whitford paper was deservedly influential in motivating comparison between stars in the Milky Way bulge and the integrated population of external galaxies, its detailed conclusions rest on rather poor foundations.

The results of Rich (1988), based on his low-resolution spectra, that the mean metallicity of K/M giants in Baade's Window was twice the solar value, was very influential and widely accepted; however, it is now apparent that line-blending and elemental abundance variations contributed to a calibration error. We discuss below the current status of the metallicity-luminosity relation for bulges and for ellipticals and the detailed chemical abundance distribution for stars in the bulge of the Milky Way. Although super–metal-rich stars clearly exist in the bulge of the Milky Way, they are a minority, and their relationship to the majority population (are they the same age?) remains unknown.

### 1.4 *Preconception Number 3: Bulges Are Similar to Elliptical Galaxies*

Bulges and ellipticals have traditionally been fit by the same surface brightness profiles, the de Vaucouleurs $R^{1/4}$ law; for simplicity, one is tempted to assume that bulges are simply scaled-down ellipticals and that they formed the same way. N-body simulations (e.g. van Albada 1982), together with analytic considerations of "maximum entropy" end states (Tremaine et al 1986), suggested that this was through violent relaxation of a dissipationless, perhaps lumpy, system. These ideas incorporate the proposition (e.g. Toomre 1977, Barnes & Hernquist 1992) that equal-mass mergers destroy preexisting stellar disks and form bulges and ellipticals, of which these latter two are distinguished only by mass.

Furthermore, the stellar kinematics of ellipticals and bulges of the same luminosity are similar, in that each rotates approximately as rapidly as predicted by isotropic oblate models (Davies et al 1983). However, the two general categories of "bulges" and "ellipticals" are becoming clear to be somewhat heterogeneous and may cover systems that formed in a variety of ways.

The above preconceptions may be tested against modern data. We proceed with the systems for which the most detailed data may be obtained, the galaxies in the Local Group, and then outward in distance.

## 2. RESOLVED BULGES—LOCAL GROUP GALAXIES

The Local Group provides a sample of bulges in which one can determine the stellar distribution functions on a star-by-star basis, which allows a more detailed analysis than is possible based on the integrated properties of more distant bulges/haloes. In this comparison, one must be careful to isolate the essential features because there is much confusing detail, both observational and theoretical, specific to individual galaxies.

Obvious questions that can be addressed most efficiently locally include possible differences, similarities, or smooth(?) gradients in properties—kinematics, chemical abundance distribution, age distribution, scale-lengths, profiles, etc—from inner bulges to outer haloes, and from bulges to inner disks. Different tracers can be used that allow comparisons between, for example, globular clusters and field stars.

### 2.1 *Milky Way Galaxy*

Let us adopt for the moment the working definition of the bulge as the component constituting the amorphous stellar light in the central regions of the Milky Way. Although one might imagine that the Milky Way bulge can be studied in significantly more detail than is possible in other galaxies, our location in the disk restricts our view such that this is true only several kiloparsecs from the Galactic center. Most of the Galactic bulge is obscured by dust and stars associated with the foreground disk. We illustrate the situation in Figure 1 below.

CHEMICAL ABUNDANCES Chemical abundances of K and M giants in the central regions of the Galaxy have been determined by a variety of techniques, ranging from high-resolution spectra that allow elemental abundance analyses to intermediate-band photometry. Application to Baade's Window—approximately 500-pc projected distance from the Galactic center—determined that the metallicity distribution function (calibrated onto a [Fe/H] scale) of K/M giants is broad, with a maximum at $\sim -0.2$ dex (i.e. $\sim 0.6$ of the solar iron abundance) and extending down to at least $-1$ dex and up to at least $+0.5$ dex (e.g. McWilliam & Rich 1994, Sadler et al 1996). It remains unclear to what extent these upper and lower limits are a true representation of the underlying distribution function and to what extent they are observational bias, set by calibration difficulties and/or sensitivities of the techniques. Furthermore, the identification of foreground disk stars remains difficult.

At larger Galactocentric distances, Ibata & Gilmore (1995a,b) utilized fiber spectroscopy down many lines of sight to mimic "long-slit spectroscopy" of the Galactic bulge, in order to facilitate a direct comparison between the Milky Way bulge and those of external spiral galaxies. They obtained spectra of about 2000 stars; star count models, stellar luminosity classifications, and kinematics were used to isolate about 1500 K/M-giants from 700 pc to 3.5 kpc (projected distance) from the Galactic Center. These authors estimated metallicities from the Mg'b' index, calibrated against local field stars; thus there is a possible zero-point offset of up to $\sim 0.3$ dex, which is dependent on the element ratios of the Bulge stars compared to the local stars. Ibata & Gilmore truncated their distribution function above the solar value, owing to the great similarity

⟵

*Figure 1* An optical image of the central Galaxy, adapted from that published by Madsen & Laustsen (1986). The field covered is 70° × 50°. The Galactic plane is indicated by the horizontal line, and the Galactic center by the cross in the center of the image. Also shown is an outline of the COBE/DIRBE image of the Galactic center (*smooth solid curve*, from Arendt et al 1994), an approximate outline of the Sagittarius dSph galaxy (*complex curve*, from Ibata et al 1997), with the four Sgr dSph globular clusters identified as asterisks; Baade's Window (*heavy circle below the center*); the field of the DUO microlensing survey, which contains some of the other microlensing fields (*solid square*, overlapping the Sgr dSph rectangle; Alard 1996); the four fields for which deep HST color-magnitude data are available (*open squares*, near Baade's Window); and the six fields surveyed for kinematics and metallicity by Ibata & Gilmore (1995a,b: *black/white outline boxes*). The location of Kepler's supernova is indicated as a circle, north of the Galactic plane. Other features of relevance include the extreme extinction, which prevents optical/near-IR low-resolution observations of the bulge within a few degrees of the plane, and the pronounced asymmetry in the apparent bulge farther from the plane. The dust that generates the apparent peanut shape in the COBE/DIRBE image is apparent. The asymmetry at negative longitudes north of the plane, indicated by a large dotted circle, is the Ophiuchus star formation region, some 160 pc from the Sun. The Sagittarius spiral arm contributes significantly at positive longitudes in the plane.

in low-resolution spectra between foreground K dwarfs and such metal-rich K giants, which leads to an inability to identify contamination of the bulge sample by disk stars. They find that the outer bulge metallicity distribution function peaks at $\sim -0.3$ dex, and continues down beyond $-1$ dex (see Figure 2 below).

Minniti et al (1995) present the metallicity distribution function for $\sim 250$ K/M giants in two fields at projected Galactocentric distances of $R \sim 1.5$ kpc. Their results are calibrated only for stars more metal-poor than $\sim -0.5$ dex, and one of their fields was selected with a bias against high metallicities. Their data for their unbiased field again shows a broad distribution function, which is approximately flat from $-1$ to $+0.3$ dex. Minniti et al (1995) also summarize (and list the references to) results from extant photometric chemical abundance determinations (e.g. Morrison & Harding 1993); in general, these agree neither with each other nor with spectroscopic determinations. Further work is clearly needed.

The few large-scale kinematic surveys of the bulge (Ibata & Gilmore 1995a,b, Minniti et al 1995) find no convincing evidence for an abundance-kinematics correlation within the bulge itself, after corrections for halo stars and disk stars (see also Minniti 1996).

The most striking aspects of the bulge K/M giants' metallicity distribution function are its width and the fact that there is little if any radial gradient in its peak (modal) value when one considers only spectroscopic determinations. Further data are required to determine whether or not the wings of the distribution are also invariant. Certainly the very late spectral-type M giants have a significantly smaller scale height than do the K giants (Blanco & Terndrup 1989), a fact that could be a manifestation of either a metallicity gradient in the high-metallicity tail of the distribution function or of an age gradient, with a small scale height, metal-rich, younger population that is concentrated to the Galactic plane. Star formation clearly occurs in the very center of the Galaxy (e.g. Gredel 1996), so that a distinction between inner disk and bulge stellar populations remains problematic, and perhaps semantic, in the inner few hundred parsecs of the Galaxy. External disk galaxies do show color gradients in their bulge components, but the amplitude is luminosity dependent and expected to be small for bulges like that of the Milky Way (Balcells & Peletier 1994).

The little evidence there is concerning the stellar metallicity distribution of older stars in the inner disk is also somewhat confusing. An abundance gradient with the mean rising $\sim 0.1$ dex/kpc towards the inner Galaxy, but for data only relevant to Galactocentric distances of 4–11 kpc, has been plausibly established for F/G stars of ages up to $10^{10}$ years (Edvardsson 1993; their table 14—their few older stars show no evidence for a gradient). A similar amplitude of metallicity gradient is seen in open clusters older than 1 Gyr,

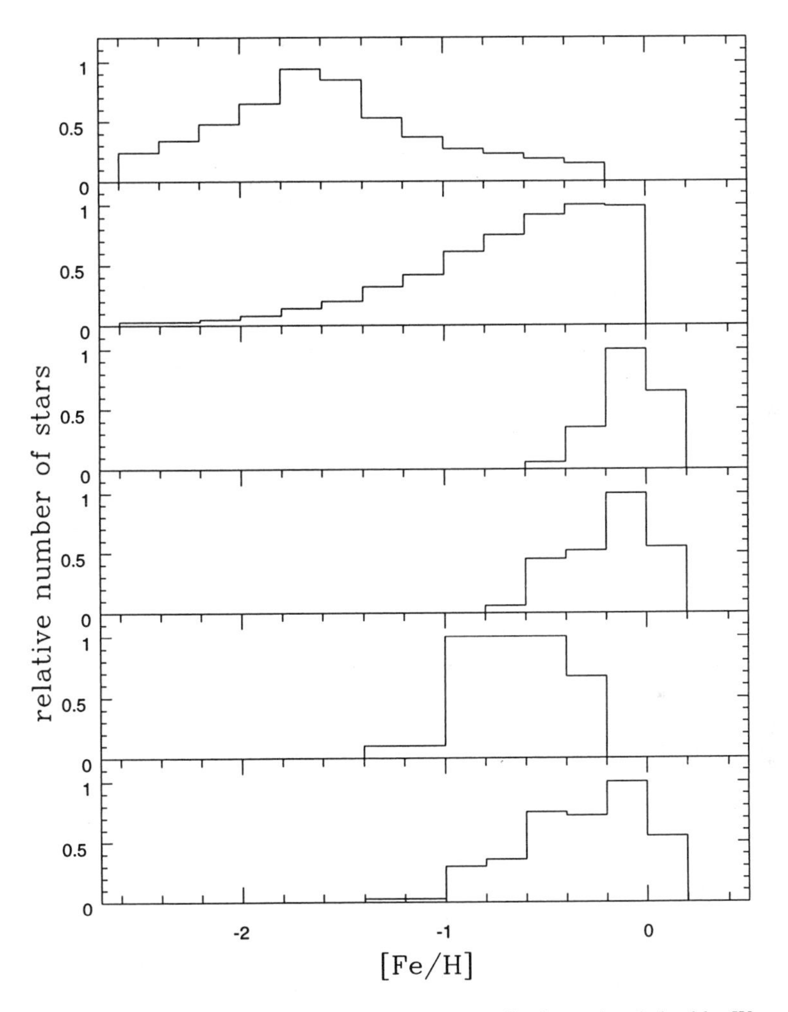


*Figure 2* Chemical abundance distribution functions, normalized to unity, derived by Wyse & Gilmore (1995), except where noted. The distributions are, from top to bottom, the solar neighborhood stellar halo (Laird et al 1988); the outer Galactic bulge (Ibata & Gilmore 1995b), truncated at solar metallicity; the younger stars of the solar neighborhood; a volume-complete sample of local long-lived stars; a volume-complete sample of local thick-disk stars; the column integral through the disk abundance distribution for the sum of the long-lived thin disk and the thick disk.

but for clusters that are exterior to the solar circle (e.g. Friel 1995). Earlier data for K giants, however, suggest no radial abundance gradient, with a mean [Fe/H] $\sim -0.3$ from exterior to the Sun to within 1 kpc of the center (Lewis & Freeman 1989), even though such stars should be no older than the F/G sample. Clearly, however, the abundance range that contains most of the bulge stars overlaps that of the disk, with probable disk gradients that are smaller than the range of the bulge metallicity distribution function. This is of particular interest given the correlations, discussed below, between the colors of bulges and inner disks in external galaxies (de Jong 1996, Peletier & Balcells 1996).

As discussed further below, the mean metallicity of field bulge stars is significantly above that of the globular cluster system of the Milky Way, even if only the inner, more metal-rich "disk" globular clusters with mean metallicity of $\sim -0.7$ dex (e.g. Armandroff 1989) are considered.

A characterization of the width of the metallicity distribution comes from the fact that the distributions for both Baade's Window (Rich 1990) and for the outer bulge (Ibata & Gilmore 1995b) are consistent with the predictions of the "Simple Closed Box" model of chemical evolution. This is in contrast to the disk of the Milky Way, at least in the solar neighborhood, which has a significantly narrower metallicity distribution and indeed a shortage of low-metallicity stars compared to this model (the "G-Dwarf problem"). This of course does not mean that any or all of the assumptions inherent in the simple closed box model were realized during bulge formation and evolution, but it is rather a way of quantifying the greater width of the observed metallicity distribution in the bulge compared to the disk at the solar neighborhood, two locations that have the same *mean* metallicity.

Elemental abundances provide significantly more information than does metallicity because different elements are synthesized by stars of different masses and hence on different time scales (e.g. Tinsley 1980, McWilliam 1997). Different scenarios for the formation of the bulge could in principle be distinguished by their signatures in the pattern of element ratios (Wyse & Gilmore 1992). The available data are somewhat difficult to interpret, in part owing to small number statistics (e.g. McWilliam & Rich 1994, Sadler et al 1996), but this can be rectified with the coming 8- to 10-m class telescopes.

AGE ESTIMATES RR Lyrae stars, the traditional tracers of an old metal-poor population, are found in significant numbers along bulge lines of sight, at characteristic distances that place them close to the Galactic center (Oort & Plaut 1975). This has been taken as supporting evidence for an old bulge. Indeed, Lee (1992) argued that, for a stellar population of high mean metallicity to produce significant numbers of RR Lyrae stars from the metal-poor tail of the chemical abundance distribution, the population must be older than a metal-poor population with the same RR Lyrae production rate. Lee hence concluded that the bulge contained the oldest stars in the Galaxy, older than the stars in the field halo. But are the observed RR Lyrae stars indeed part of the metal-rich bulge, or of the metal-poor stellar halo, whose density of course also peaks in the inner Galaxy?

The samples of RR Lyrae available for this experiment have been small. However, a side benefit of the recent interest in microlensing surveys of the Galactic bulge (e.g. OGLE, MACHO, DUO) has been well-defined catalogs of variable stars, including RR Lyraes. In an analysis of the projected spatial distribution of DUO RR Lyraes—which have been segregated statistically by metallicity based on periods and fit to density laws of halo, disk, and bulge—Alard (1996) has found that the great majority of RR Lyrae stars in his catalog are not associated with the bulge, but rather with the thick disk and halo. Nonetheless, a detectable fraction of the most metal-rich RR Lyrae variables of the 1400 discovered by DUO do indeed belong to a concentrated bulge population. These stars comprise only about 7% of the whole RR Lyrae sample. Thus, the microlensing surveys have in fact made the first discovery of true bulge RR Lyraes. The intermediate-abundance RR Lyraes are primarily thick disk, whereas the most metal poor are primarily halo, from this analysis.

Analysis of the variable stars detected by the IRAS satellite (mostly Mira variables) implied a significant intermediate-age population (e.g. Harmon & Gilmore 1988), perhaps that traced by the carbon stars (Azzopardi et al 1988, Westerlund 1991) and the strong red clump population (e.g. Pacynski et al 1994a,b).

Renzini (1994, 1995) has emphasized that the relative strength of the red clump and red giant branches is dependent on helium content as well as on age and argues that age is not an important parameter for stellar populations older than 1 Gyr. Thus, should the bulge stars be of high helium content—as expected if they had been found to be super–metal-rich—then the observed red clump would be consistent with an old age. However, the fact that the mean metallicity of the bulge is now established (from unbiased tracers) to be below the solar value, with a correspondingly much-reduced helium abundance, makes this unlikely.

Understanding the effects of dust along the line of sight to the central regions is crucial. The analysis of infrared (IR) data reduces some of the reddening problems of optical data, but again the interpretation in terms of stellar properties is far from unambiguous. A deep near-IR luminosity function for Baade's window was obtained by Tiede et al (1995). Houdashelt (1996), in a detailed analysis of the available IR photometry and spectroscopy for stars in Baade's Window, concluded that a typical age of perhaps 8 Gyr and mean metallicity of [Fe/H] $\sim -0.3$ are most consistent.

Optical/near-IR color-magnitude diagrams that extend well below the main sequence turnoff region may be used to make quantitative statements about mean age and age ranges of stellar populations: modulo uncertainties in this case that are due to large and highly variable extinction, to extreme crowding in the inner fields, and to the contribution of foreground stars. In spite of these complications, Ortolani et al (1995) concluded, from a comparison of HST color-magnitude data for the horizontal branch luminosity functions of an inner globular cluster with ground-based data towards Baade's Window, that the stellar population of the bulge is as old as is the globular cluster system and, furthermore, shows negligible age range. This contrasts with earlier conclusions based on prerefurbishment HST color-magnitude data for Baade's Window (Holtzman 1993), which suggests a dominant intermediate-age population. Future improved deep HST color-magnitude data are eagerly awaited.

An example of the information that can be obtained is given in Figure 3, which is a V–I, V color-magnitude diagram from WFPC2 data (planetary camera) obtained as part of the Medium Deep Survey (S Feltzing, private communication).

BULGE STRUCTURE The only single-parameter global fit to the surface brightness of the combined halo plus bulge of the Galaxy that implicitly assumes they are a single entity, is that by de Vaucouleurs & Pence (1978). From their rather limited data on the visual surface brightness profile of the bulge/halo interior to the solar Galactocentric distance, when assuming an $R^{1/4}$-law profile, they derived a projected effective radius of 2.75 kpc, which may be deprojected to a physical half-light radius of 3.75 kpc. As shown by Morrison (1993), the de Vaucouleurs & Pence density profile, extrapolated to the solar neighborhood, is brighter than the observed local surface brightness of the metal-poor halo, which was obtained from star counts, by 2.5 magnitudes. Because the density profile of the outer halo is well described by a power law in density, with index $\rho(r) \propto r^{-3.2}$, and oblate spheroidal axis ratio of about 0.6 (Kinman et al 1966, 1994, Wyse & Gilmore 1989, Larsen & Humphreys 1994), this result actually provides the first, though unappreciated, evidence that the central regions of the galaxy are predominantly bulge light and that the bulge light falls off faster than does the outer halo light. That is, the bulge and halo are not a single structural entity. More generally, because the spatial density distribution of the stellar metal-poor halo is well described by a power law, whereas the inner bulge (see below) is well described by another power law of much smaller scale length, the apparent fit of the single $R^{1/4}$-law profile must be spurious and misleading.

The limiting factors in all studies of the large-scale structure of the stellar Galactic bulge are the reddening, which is extreme and patchy, and severe crowding. The systematic difference between the best pre-HST photometry

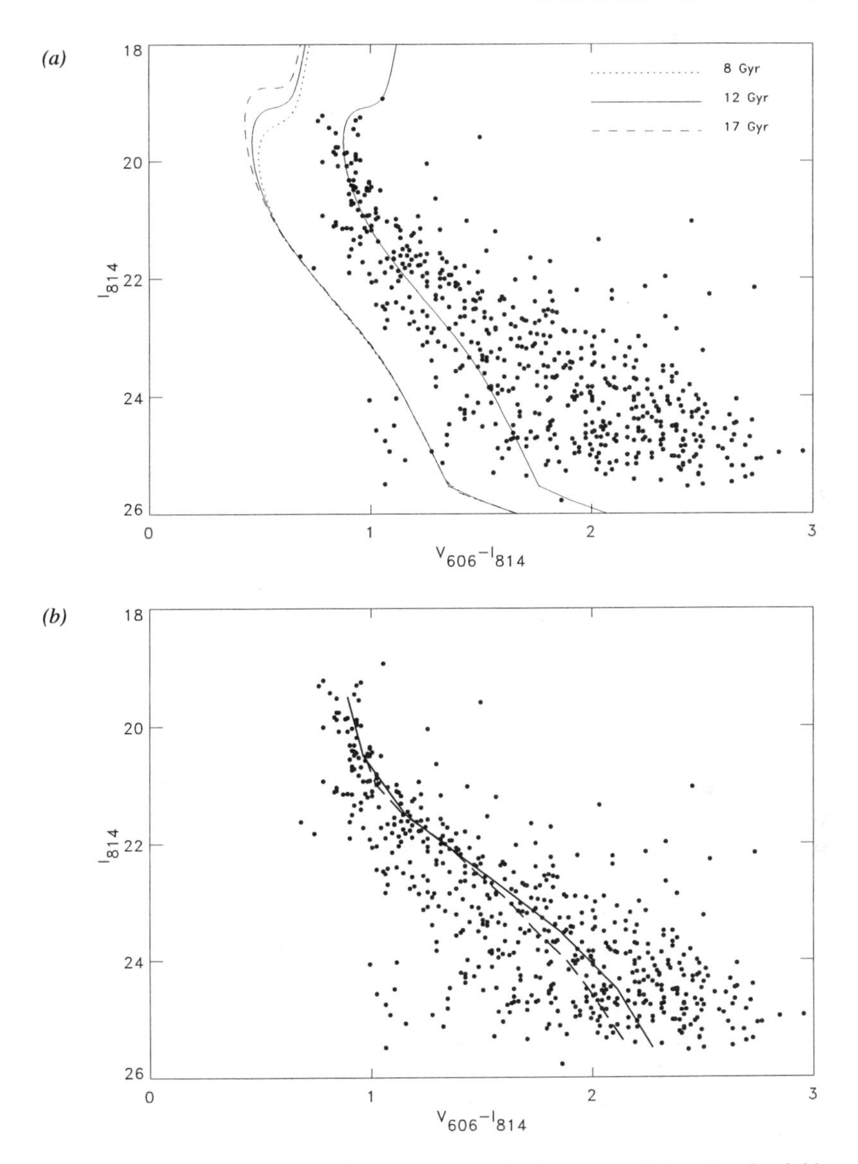


*Figure 3* The HST WF/PC2 color-magnitude data for the Galactic bulge, for the field at (l, b) = (3.6, −7) identified in Figure 1, from the Medium Deep Survey. (*a*) The panel shows the data. Overlaid, from a by-eye fit, is a 12-Gyr isochrone for metallicity [Fe/H] = −0.25, from Bertelli et al (1994), together with a range of other ages plotted to one side, to illustrate the precision required and the need for independent determinations of extinction at each point. (*b*) The panel shows the mean line through the data, excluding extreme points, together with the ridge line from similar HST data for the globular cluster 47 Tucanae (Santiago et al 1996), arbitrarily offset to match the mean line.

in crowded regions and the reality, as seen by HST, is now well appreciated after many studies of globular clusters. Near-IR studies within a few degrees of the Galactic plane show optical extinction that has a random variation, on angular scales down to a few arcseconds, of up to $A_V \sim 35$ mag (e.g. Catchpole et al 1990). At southern Galactic latitudes, however, more than a few degrees from the plane, extinction is both low (typically $E_{B-V} \sim 0.2$) and surprisingly uniform, as is evident in the optical bulge image in Figure 1 and as exploited by Baade. Nonetheless, detailed star-count modeling of the inner galaxy (Ibata & Gilmore 1995a,b; M Unavane, private communication) demonstrates that extinction variations are still larger in their photometric effects than are the photometric signatures of different plausible structural models. This sensitivity to extinction, together with the extreme crowding that bedevils ground-based photometry, is well illustrated by the recent history of structural analyses of the inner Milky Way disk by the OGLE microlensing group, based on low spatial-resolution optical data. Their initial analysis of their data suggested that there is no inner disk in the Galaxy, only prominent foreground spiral structure (Paczynski et al 1994a). After more careful consideration of crowding, and of alternative extinction models, this detection of a "hole" in the disk was retracted (Kiraga et al 1997). The true spatial density distribution of the inner disk remains obscure.

There are many analyses of the surface brightness structure of the bulge, which range from straightforward counts of late-type stars perpendicular to the plane along the minor axis (cf Frogel 1988 for references) through extensive two-dimensional analyses (Kent et al 1991), to detailed inversions of photometric maps (e.g. Blitz & Spergel 1991, Binney et al 1997). In all such cases, extreme reddening near the plane precludes reliable use of low spatial resolution data with $|b| < 2$, irrespective of the techniques used. The zero order properties of the photometric structure of the bulge are fairly consistently derived in all such studies and determine ~350 pc for the minor axis exponential scale height, as well as significant flattening, with minor:major axis ratio of ~0.5. Together with a disk scale length of around 3 kpc, this result places the Milky Way galaxy within the scatter of late-type disk galaxies on the correlation between disk and bulge scale lengths of Courteau et al (1996).

Considerable efforts have been expended in the last decade to determine the three-dimensional structure of the Galactic bulge. These efforts began at a serious level with analyses of the kinematics of gas in the inner Galaxy, following the prescient work of Liszt & Burton in the 1980s (see Liszt & Burton 1996 and Burton et al 1996 for recent reviews and introductions to the subject), by Gerhard & Vietri (1986) together with much other work reviewed by Combes (1991). A resurgence of interest in bar models has been motivated by (*a*) new dynamical analyses (e.g. Binney et al 1991, Blitz et al 1993), (*b*) the realization that near-IR data might reflect the pronounced molecular gas asymmetry (Blitz & Spergel 1991), (*c*) gravitational microlensing results (Paczynski et al 1994b), and (*d*) the new photometric COBE/DIRBE data (Weiland et al 1994).

It appears that all galaxies in their central regions have non-axisymmetric structures, often multiple structures such as bars within bars (e.g. Shaw et al 1995, Friedli et al 1996). The distinction between inner spiral arms, bars, lenses, local star formation, and the like is perhaps of semantic interest, except in cases where the distortions are of sufficiently large amplitude such as to affect the dynamical evolution. Is the Galaxy like that? The significant question is the existence of a substantial perturbation to the inner density distribution, and gravitational potential, associated with a bar. Secondary questions are the shape of that bar and its relationship to the disk or to the bulge. The extant three-dimensional models of the central regions of the Milky Way derived from the COBE surface photometry depend on systematic asymmetries of the derived "dust-free" surface brightness with longitude of less than 0.4 mag in amplitude, after statistical correction for extinction that is locally some orders of magnitude larger in amplitude (Binney et al 1997). Thus the models are crucially sensitive to reddening corrections made on a scale of 1.5 degrees (the COBE/DIRBE resolution), although reddening varies on much smaller scales (Figure 1).

The models also provide only a smooth description of most of the known foreground disk structure such as can be seen in Figure 1—the Ophiuchus star formation region, the Sagittarius (Sgr) spiral arm, etc—and do not work at low Galactic latitudes. A model of this disk must be subtracted before bulge parameters can be derived. The best available description of the stellar bulge derived this way suggests axis ratios $x$:$y$:$z$ $\sim$ 1.0:0.6:0.4 (Binney et al 1997).

It is worth noting that this model, although the best currently available, fails to explain either the high spatial frequency structure in the photometric data or the observed high rate of gravitational microlensing towards the inner Galaxy (Bissantz et al 1997), in addition to having remaining difficulties with the details of the gas kinematics in the inner Galaxy. Little evidence exists for non-axisymmetry in the potential from analyses of stellar kinematics—radial velocity surveys find consistency with an isotropic oblate rotator model (e.g. Ibata & Gilmore 1995, Minniti 1996), though with a mild bar allowed (Blum et al 1995). Although evidence for a bar is seen in proper-motion surveys (Zhao et al 1994, who analyzed proper motions from Spaenhauer et al 1992), this is very dependent on the distances assigned to the stars. Thus it must be emphasized that the best available models for the inner Galaxy remain poor descriptors of the very complex kinematics and spatial distribution of the gas (see Liszt & Burton 1996) and of the complex kinematics of some samples of stars (e.g. Izumiura et al 1995).

Analysis of the photometric structure of the inner galaxy is a very active field of research, which promises major progress in the next few years with the availability of the Infrared Space Observatory (ISO) imaging survey data of the inner galaxy (Perault et al 1996). ISO improves on the $\sim 1°$ spatial resolution of COBE, as it has typically 6-arcsec resolution in surveys. These data provide for the first time a detailed census of individual stars and the ISM in the inner Galaxy, with sufficient resolution and sensitivity to see single stars at the Galactic center, thereby allowing the first ever determination of the true three-dimensional spatial distribution of the inner Galaxy.

We consider the kinematics of the Galactic bulge, the halo, and the disk, and their implications for formation models, below (section 5).

## 2.2 *M 33 (NGC 598)*

The stellar population of M 33 was reviewed by van den Bergh (1991a), to which the reader is referred for details. We discuss the significant developments since then concerning the existence and nature of the stellar halo and bulge.

M 33 shows photometric evidence for nondisk light, in particular in the central regions. However, the nature of this light remains uncertain, as does whether or not there is a central bulge component that is distinct from the stellar halo.

Attempts to fit optical and IR data for the central regions with an $R^{1/4}$ law generally agree with a "bulge-to-disk" ratio of only $\sim 2\%$, or $M_{V,bulge}$ fainter than $\sim -15$ (Bothun 1992, Regan & Vogel 1994). Regan & Vogel emphasize that a single $R^{1/4}$ provided the best fit to their data. Some evidence is given from ground-based H-band imaging (Minnitti et al 1993) and from HST V–I/I CMD data (Mighell & Rich 1995) for asymptotic giant branch (AGB) stars in the central regions in excess of the number predicted by a simple extrapolation from the outer disk; these stars have been ascribed to a rather young centrally concentrated bulge. However, McLean & Liu (1996) contend that their JHK photometry, after removal of crowded regions, shows no resolved bulge population distinct from the smooth continuation of the inner disk.

Is the $R^{1/4}$ component metal-poor or metal-rich? The giant branch of the HST CMD data is consistent with a broad range of metallicity, ranging from M 15–like to 47 Tuc–like, some 1.5 dex in metallicity. The low end of this metallicity range is consistent with that estimated earlier from ground-based CMD data for fields in the outer "halo," [Fe/H] $\sim -2.2$ (Mould & Kristian 1986). These outer fields showed a narrow giant branch, which is consistent with a small dispersion in metallicity, and thus the two datasets together are suggestive of a gradient in the mean metallicity and metallicity dispersion. This may be interpreted as evidence for a centrally concentrated more metal-rich component, albeit following the same density profile as the metal-poor stars. Pritchet (1988) reported a preliminary detection of RR Lyrae stars in M 33, again evidence for old, probably metal-poor, stars.

The semistellar nucleus of M 33 has a luminosity similar to that of the brightest Galactic globular clusters, $M_V \sim -10$, and a diameter of ~6 pc. Analysis of its spectrum (Schmidt et al 1990) demonstrated that its blue color reflects the presence of young stars (age less than 1 Gyr) rather than extremely low metallicity; old and intermediate-age stars with metallicity greater than 0.1 of the solar value dominate. The relation of this nucleus to the "bulge," if any, is unclear.

The only kinematic data for nondisk tracers in M 33 are for a subset of its ~200 "large clusters of concentrated morphology" (Christian 1993), of which perhaps 10% have the colors of the classical old globular clusters of the Milky Way. Of these clusters, 14 have kinematics that are suggestive of halo objects, in that they define a system with little net rotation and with a "hot" velocity dispersion of order $1/\sqrt{2}$ times the amplitude of the HI rotation curve (Schommer et al 1991, Schommer 1993). Estimates of the metallicities and ages of the "populous" clusters, based on spectrophotometry, suggest a wide range of each, with even the "globular clusters" spanning perhaps ~−2 dex to just under solar metallicity (Christian 1993). Improved estimates from better data are possible and desirable. M 33 has a very large number of globular clusters per unit field halo light, but the meaning of this is unclear.

In summary, M 33 has a low luminosity halo, which is at least in part old and metal-poor. There is no convincing evidence for the existence of a bulge in addition to this halo.

## 2.3 *M 31 (NGC 224)*

The stellar population of M 31 was reviewed by van den Bergh (1991b), and again we restrict discussion to significant subsequent developments.

The field nondisk population has been studied by several groups, following Mould & Kristian (1986; see also Crotts 1986). These authors established, from V and I data that reach several magnitudes down the giant branch, that the bulge/halo of M 31, at 7 kpc from its center, has mean metallicity like the Galactic globular 47 Tuc, [Fe/H] $\sim -0.7$, and a significant dispersion in metallicity, when assuming an old population, down to ~−2 dex and up towards solar. Similar conclusions have been reached from HST data for the outer regions of M 31 (~10 kpc) by Holland et al (1996) and by Rich et al (1996) at ~30 kpc from the center, which limits the amplitude of any chemical abundance gradients, assuming always that one is dealing with an old stellar population.

These HST data also established firmly the scarcity of Blue Horizontal Branch (BHB) stars in the halo of M 31, which confirms the suggestion by Pritchet & van den Bergh (1987, 1988). A few BHB stars were found by Holland et al (1996), who suggest that the horizontal branch (HB) morphology is apparently too red for the derived broad metallicity distribution. If one assumes that the horizontal branch traces a population as old as the Galactic halo globular clusters, then the M 31 field population suffers a severe "second-parameter problem."

Assuming that the derived broad metallicity distribution is well-established, does this lack of a significant BHB population imply a young age for M31? Age can affect HB morphology in that younger populations are redder at a given metallicity, other things being equal (e.g. Lee 1993, who also demonstrates the effects of many other parameters), so that it is of interest to consider this possibility [while recalling that Richer et al (1996) argue quite convincingly, based on relative ages for those Galactic globular clusters with main sequence turn-off photometry, that age is not the dominant "second parameter" of HB morphology, at least in these systems]. Indeed, the presence of bright stars, identified as intermediate-age AGB stars, has been suggested from (prerefurbishment) WF/PC HST VI data at least within the inner 2 kpc of the bulge (Rich & Mighell 1995). Morris et al (1994) argued for a ubiquitous strong luminous AGB component, with a typical age of 5 Gyr, from their ground-based V and I data that reaches the bright giants in various fields of M 31, 16–35 kpc along the major axis of the disk and one probing the halo at 8 kpc down the minor axis (close to the field of Mould & Kristian 1986). Rich et al (1996), and also Holland et al (1996), find no evidence for an extended giant branch in their WF/PC2 HST data for fields in the outer halo, at 10–30 kpc from center, where again the RHB/clump is dominant, with essentially no trace of a BHB. Thus, the data describing possible metallicity/age effects remain unclear.

Large-scale surface photometry of the disk and of the bulge of M 31, in many broadband colors, was obtained and analyzed by Walterbos & Kennicutt (1988). They found that there was no color gradient in the bulge and that the inner disk and the bulge have essentially the same colors, i.e. those of "old, metal-rich stellar populations." This similarity of broadband colors has subsequently been found for a large sample of external disk galaxies, as discussed in Section 3, and clearly must be incorporated into models of the formation and evolution of bulges (see Section 5 below). Walterbos & Kennicutt also derived structural parameters for the disk and bulge that are consistent with the correlation between scale lengths found for the larger sample of more distant disk galaxies by Courteau et al (1996). In terms of total optical light, the bulge-to-disk ratio of M 31 is about 40%.

Pritchet & van den Bergh (1996) emphasize that a single $R^{1/4}$-law provides a good fit to their derived V-band surface photometry (from star counts), with no bulge/halo dichotomy. The $R^{1/4}$ component is significantly flattened, with axial ratio of 0.55, which is similar to the value for the metal-poor halo of the Milky Way (Larsen & Humphreys 1994, Wyse & Gilmore 1989).

In contrast to the metal-poor halo of the Milky Way, which is apparently flattened by anisotropic velocity dispersions, the bulge of M 31 has kinematics consistent with an isotropic oblate rotator, with mean rotational velocity of $\sim$65 km/s and velocity dispersion of $\sim$145 km/s (McElroy 1983), which are typical of external bulges (Kormendy & Illingworth 1982).

Thus, although Baade (1944a,b) identified the "bulge" of M 31 (which we may now define to be field nondisk stars at distances up to 35 kpc from the center of M 31) with Population II (similar to the Milky Way halo), the dominant tracers of the M 31 bulge do not share the characteristics of classical Galactic Population II, as they are neither of low mean metallicity nor have little net rotation (see Wyse & Gilmore 1988 for further development of this point, in the context of thick disks).

There are around 200 confirmed globular clusters associated with M 31 (e.g. Fusi Pecci et al 1993). The distribution of their metallicities has a mean of around $-1$ dex, which is more metal-poor than the field stars, with a range of perhaps 1 dex on either side (e.g. Huchra et al 1991, Ajhar et al 1996). The inner metal-rich clusters form a rapidly rotating system, whereas the outer metal-poor clusters have more classical "hot" halo kinematics (e.g. Huchra 1993; see also Ashman & Bird 1993 for further discussion of subsystems within the globular clusters). The overall globular cluster system has a projected number density profile that may be fit by a de Vaucouleurs profile (although the central regions fall off less steeply) with an effective radius of $\sim$4–5 kpc (Battistini et al 1993). This is more extended than the $R^{1/4}$ fit to the field stars. Thus, in terms of kinematics, metallicity, and structure, there may be evidence for a bulge/halo dichotomy in M 31 if the halo is traced by the globular clusters and the bulge by field stars. Note that, although there are exceptions, the spatial distributions of globular cluster systems and underlying galaxy light are similar to the first order (Harris 1991).

As seems to be the case for any system studied in sufficient detail, the morphology of the very central regions of M 31 is clearly complicated, with twisted isophotes (Stark 1977), gas kinematics that may trace a bar (e.g. Gerhard 1988), inner spiral arms (e.g. Sofue et al 1994), and two nuclei (Bacon et al 1993) that may indicate a tilted inner disk (Tremaine 1995). These phenomena have been modeled recently by Stark & Binney (1994) by a spherical mass distribution plus a weak prolate bar, with the bar containing one third of the mass within 4 kpc (the corotation radius). The association of the bulge with this bar, which one might be tempted to adopt by analogy with the Milky Way, is unclear.

## 2.4 *Large Magellanic Cloud*

The Large Magellanic Cloud (LMC) is the nearest barred galaxy, with the bar offset from the kinematic and isophotal center and embedded in an extensive disk. A minor metal-poor old component of the LMC is seen in deep HST color-magnitude data (Elson et al 1997), but its kinematics and spatial distribution are not yet well known. There is a significant amount of new information, from the several microlensing experiments, which will appear in the literature over the next few years concerning the variable star population of the LMC. Of particular relevance are data for the Long Period Variables (LPVs) and the RR Lyrae. The LPVs are believed to have low-mass progenitors and hence trace older stellar populations, while RR Lyrae variables are the traditional tracers of old metal-poor populations. However, most of the information has yet to be analyzed. There has been no kinematical analysis of the LPVs since that of Hughes et al (1991), who found tentative indications of classical hot halo kinematics. The old globular clusters of the LMC, despite prejudice, have kinematics consistent with being in a rotating disk (e.g. Freeman 1993). Thus, little evidence exists for a bulge or halo population in the LMC, except the observation that an old metal-poor stellar population exists.

### 2.5 *General Properties of the Local Group Disk Galaxies*

The diversity of properties of bulges, haloes, and disks evident in the four largest disk galaxies in the Local Group is striking. The essential properties seem to be the following. The two latest type galaxies (M 33, LMC) have no convincingly detected bulge, but both have at least some evidence for a small population of very old metal-poor stars. Both have old metal-poor globular clusters. The intermediate-type Milky Way galaxy contains what can be termed both a halo (metal-poor, old, extended, narrow abundance distribution, containing globular clusters) and a bulge (metal-rich, mostly, and perhaps exclusively, fairly old, with a very broad metallicity distribution function, and extremely compact in spatial scale). The earlier type M 31 has a prominent and extended bulge, which is both quite metal-rich and fairly old, and has a broad abundance distribution function. The only evidence for a metal-poor old halo in M 31 comes from its globular clusters and its—very few—RR Lyrae stars and BHB stars. In all cases, haloes are supported against gravitational gradients by their velocity dispersion (pressure-supported systems), very unlike disks, though this is perhaps as much a definition as an observation.

Thus, whereas the Local Group Spiral galaxies have a definable halo:disk ratio, which is apparently rather similar for all three, only the two earlier types have a definable bulge-to-disk ratio, which is greater for M 31 than for the Milky Way.

## 3. LOW-REDSHIFT UNRESOLVED BULGES

### 3.1 *Bulges and Ellipticals*

In the most simplified picture of galaxies, a galaxy consists of a bulge that follows an $R^{1/4}$ profile and an exponential disk, whereas elliptical galaxies are simply the extension of bulges in the limit of bulge-to-disk ratio tending to infinity.

The picture has been complicated by the discovery that most intermediate luminosity ellipticals (as classified from photographic plates) have significant disks (e.g. Bender et al 1988, Rix & White 1990). These disks can be very difficult to detect, especially when seen face-on. Kormendy & Bender (1996) have recently proposed that ellipticals with "disky" isophotes, which tend to be of lower luminosity than those with "boxy" isophotes, are the natural extension of the Hubble sequence of disk galaxies.

Futhermore, many ellipticals show nuclear disks, either from their kinematics or high-resolution imaging (e.g. review of de Zeeuw & Franx 1991). These disks are very concentrated towards the center and are therefore different from the extended disks in normal spiral galaxies. Sometimes these disks have an angular momentum vector opposite to that of the bulge (e.g. IC 1459, Franx & Illingworth 1988), implying that the gas that formed the disk did not have its genesis in the stars of the bulge but was accreted from elsewhere. Notice, however, that some spiral galaxies also show evidence for these "nuclear disks," including the Milky Way (Genzel et al 1996) and the Sombrero galaxy (Emsellem et al 1996).

HST observations confirm the similarity in some aspects of low-luminosity ellipticals and bulges. Most of these systems have power-law profiles in their inner parts, with steep profile indexes (e.g. Faber et al 1997). In contrast, most high-luminosity ellipticals show "breaks" in their surface brightness distribution within 1kpc or less from the center, i.e. relatively sudden changes where the intensity profiles flatten. It is not clear yet what formation processes have caused these variations, although it has been suggested that the dynamical effects of massive black holes may be responsible (Faber et al 1997). HST imaging of large samples of spirals is needed to determine better the structure of their bulges. Preliminary results (pre-refurbishment) indicate that a significant fraction of bulges in early-type spirals have power-law profiles in their inner parts, while late-type spirals have shallower inner profiles and often an unresolved nucleus (e.g. Phillips et al 1996).

These results suggest caution in the analysis of other data, as bulges are not necessarily the only important component near the center and as the formation histories of the centers of different galaxies may have been quite different from each other. Indeed, the central 1 kpc or so of most, if not all, galaxies clearly contain something unusual—even without the benefit of detailed HST images (e.g. note NGC 4314 in the Hubble Atlas, which is a barred galaxy that has spiral arms in the center of the bar).

Beyond the very central regions, a systematic variation of surface brightness profile with bulge luminosity has been established, in that bulges in late-type spiral galaxies are better fit by exponential profiles than by the de Vaucouleurs profile, which is appropriate for early-type spirals (e.g. Andredakis et al 1995, de Jong 1995, Courteau et al 1996). HST imaging of late-type spirals is needed to better determine the structure of their bulges. Preliminary results indicate that a significant fraction of bulges in late-type spirals have power-law profiles in their inner parts (e.g. Phillips et al 1996).

Much recent research into the properties of elliptical galaxies has demonstrated the existence of a "fundamental plane" that characterizes their dynamical state (e.g. review of Kormendy & Djorgovski 1989, Bender et al 1993). The bulges of disk galaxies in the range S0–Sc (T0–T5) have also recently been demonstrated to occupy the same general locus in this plane (Jablonka et al 1996). Furthermore, these bulges have a similar Mg2 line strength–velocity dispersion relationship to that of ellipticals, but the bulges are offset slightly to lower line strengths. This offset may be due to bulges having lower metallicity or lower age. Contamination by disk light can produce a similar effect. Jablonka et al argue in favor of a close connection between ellipticals and bulges. Balcells & Peletier (1994) find that bulges follow a color-magnitude relationship similar to that of ellipticals but that bulges have a larger scatter. Furthermore, they find that bulges and ellipticals of the same luminosity do not have the same colors and that bulges are bluer. The offset is similar to that seen by Jablonka et al in the strength of the magnesium index, but Balcells & Peletier interpret it as indicating a real, though complex, difference between bulges and ellipticals. In addition to the data noted above on the central parts of bulges, Balcells & Peletier (1994) find that the amplitude of radial color gradients also varies systematically with bulge luminosity. They interpret their results as consistent with bright bulges ($M_R < -20$) being similar to ellipticals (despite the color zero-point offset), whereas faint bulges are perhaps associated with disks.

The potential well of the outer regions of disk galaxies is clearly dominated by dark matter, whereas the properties of dark matter haloes around elliptical galaxies are less well known (e.g. de Zeeuw 1995). How do properties of bulges scale with dark haloes? Figure 4 shows the ratio of bulge dispersion divided by the circular velocity of the halo (derived from rotation of tracers in the disk) against bulge-to-disk ratios. The square on the right represents elliptical galaxies, derived from models by Franx (1993), which assume a flat rotation curve. The triangle on the left corresponds to the inner regions of pure disks, as derived for a sample of Sa–Sc galaxies by Bottema (1993) (it should be noted that the inner regions of disks are not cold, but warm). Bulges may be seen to lie on a rather smooth sequence between these two extreme points. This suggests that the bulges in galaxies with low bulge-to-disk ratios may have been formed at the same time as the disk, whereas bulges in galaxies with large bulge-to-disk ratios are so much hotter than the disk that it is more likely that they formed separately. More and better data would be valuable to improve the diagram.

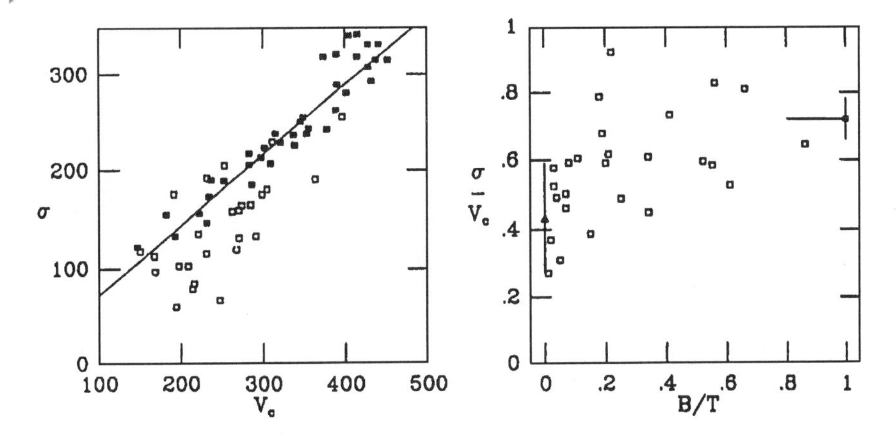


*Figure 4* (*a*) The central velocity dispersion of stellar tracers, $\sigma$, against dark halo circular velocity, $V_c$. Open symbols are bulges; closed symbols are ellipticals. Circular velocities for the ellipticals are derived from models, as described by Franx (1993). (*b*) The ratio of velocity dispersion in the bulge to dark halo circular velocity, $\sigma/V_c$, taken from Franx (1993), plotted as a function of bulge-to-total luminosity (B/T) ratio, for the entire range of Hubble Type. The triangle at left is valid for the inner regions of pure disks, the square at right for ellipticals. Note that systems with low B/T have kinematics almost equal to those of inner disks.

## 3.2 *Bulges and Disks*

Astronomical gospel declares that bulges are red and disks are blue. This is generally presumed to be derived from studies of nearby bulges. Unfortunately, there are very few data on which these rather strong statements are based. The observations were difficult to make before the advent of CCD cameras and have been lacking since then until very recently—perhaps because the problem was considered to be solved. Full two-dimensional imaging is needed for accurate bulge/disk decomposition and for exclusion of dusty areas, and large surveys with multicolor information are still rare. Notable exceptions are the recent studies of the colors of "normal" spiral galaxies by de Jong (1995, 1996) and by Balcells & Peletier (1994, Peletier & Balcells 1996).

A relationship between bulges and disks is seen clearly in their colors. We show in Figure 5 the correlation between bulge color and the color of the disk of the same galaxy, for the data of Peletier & Balcells (1996), taken from their table 1. The disk color is measured at two major axis scale lengths, and the bulge color at half an effective radius, or at 5 arcsec, whichever is the larger. Note that bulges are more like their disk than they are like each other, and the very wide range of colors evident. This sample consists of luminous ($M_R \lesssim -21$) nearby disk galaxies that span the range S0–Sbc.

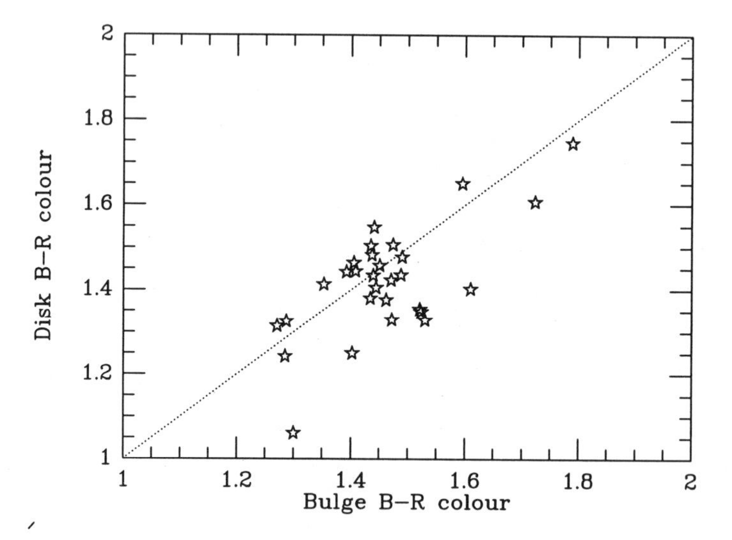


*Figure 5* The correlation between bulge color and the color of the disk of the same galaxy, for the data of Peletier & Balcells (1996), taken from their table 1. The disk color is measured at two major axis scale lengths, and the bulge color at half an effective radius, or at 5 arcsec, whichever is the larger. Note that bulges are more like their disk than they are like each other, and the very wide range of colors evident.

The color range for the bulges is noticeably large—almost as large as is the range of colors for the disks. Furthermore, although some bulges are quite red, blue bulges clearly exist, as do red disks. The sample of de Jong (1996) includes the later morphological types of disk galaxies (Sc and Sd) and shows a similar relationship between the colors of bulge and inner disks. These data show that there is little support for sweeping statements such as "bulges are red, and disks are blue." Color data for the "hidden" disks in elliptical galaxies would be very interesting.

Furthermore, the similarity in color between inner disk and bulge has been interpreted as implying similar ages and metallicities for these two components and an implicit evolutionary connection (de Jong 1996, Peletier & Balcells 1996). Given the difficulties of disentangling the effects of age and metallicity even with resolved bulges, any quantification of "similar" must be treated with caution (see Peletier & Balcells 1996, who derive an age difference of less than 30%, assuming old populations with identical metallicities). We notice in passing that the ages of ellipticals have not been determined yet to high accuracy. Measurements of various absorption line strengths have been interpreted to indicate a wide range of ages of the central regions of ellipticals, with no correlation between age and luminosity (Faber et al 1995), but this is far from rigorously established because of the coupling of age and metallicity in their effects on line strengths.

A close association between bulges and disks has been suggested by Courteau et al (1996), on the strength of a correlation between the scale lengths of the bulge and disk; they find that bulges have about one-tenth the scale length of disks. This correlation shows considerable scatter, especially for earlier galaxies of type Sa, and relies upon an ability to measure reliably bulge scale lengths that are a small fraction of the seeing. More and better data are anticipated.

### 3.3 *Bulges in Formation at $z < 0.1$?*

A few local exceptional systems are candidates for young bulges. Gravitational torques during interactions can act to drive gas to the central regions (e.g. Mihos & Hernquist 1994), where it may form stars, and which may, depending on the duration of star formation and of the interaction, be heated into a bulge. Schweizer (1990) discusses local disk galaxies with blue bulges, presenting them as evidence for recent bulge-building in this manner. These galaxies include (the dwarf) NGC 5102, an S0 galaxy with a bluer bulge than disk and strong Balmer absorption lines in its central regions. Classic merger remnants such as NGC 7252 are forming disks in their central parts, which may imply that these galaxies perhaps have evolved into S0s, or early type spirals (e.g. Whitmore et al 1993).

A more dramatic example of gas-rich mergers is Arp 230, which shows classical shells in the bulge component and a young disk rich in gas, as displayed in Figure 6 (D Schiminovich & J van Gorkom, private communication and in preparation).

## 4. HIGH-REDSHIFT BULGES

Direct searches for the progenitors of local bulges may be made by the combination of statistically complete redshift surveys of the field galaxy population, combined with photometric and especially with morphological data. As an example, the I-band–selected CFHT redshift survey contains galaxies out to redshifts of order unity, and these galaxies may be analyzed in terms of the evolution of the luminosity function of galaxies of different colors, presumed to correlate with morphological type (Lilly et al 1995). The data are consistent with very little evolution in the luminosity function of the red galaxies, over the entire redshift range $0 < z < 1$, and substantial evolution in the blue galaxies' luminosity function, with the color cut dividing the sample into blue and red taken as the rest-frame color of an unevolving Sbc galaxy. This lack of evolution for red galaxies may be interpreted as showing that the stars of

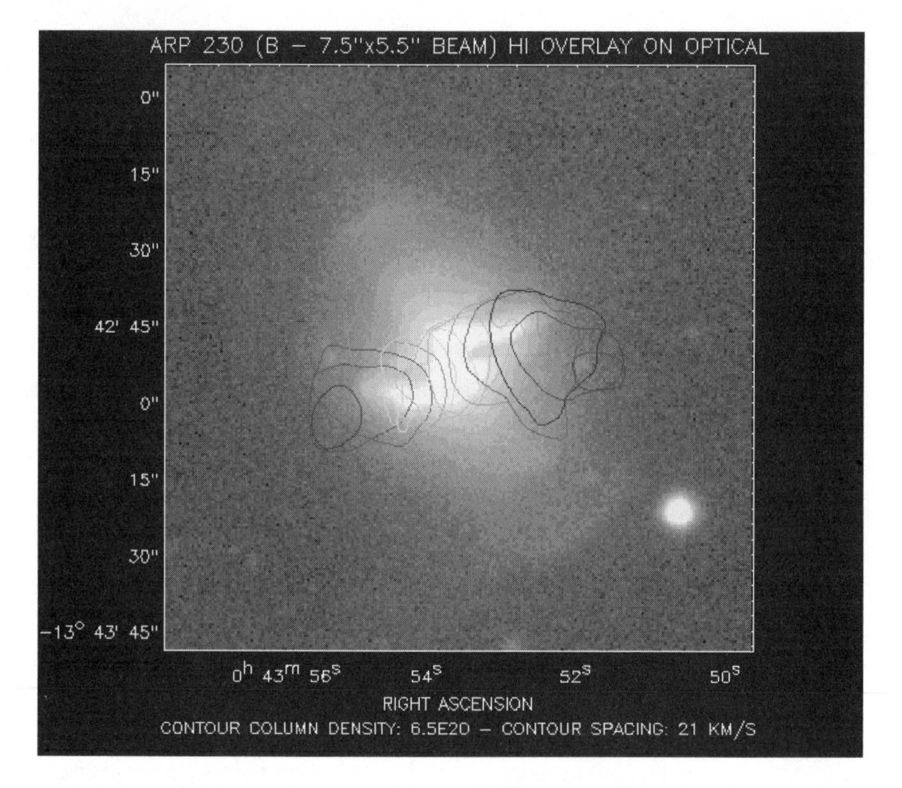


*Figure 6* An optical image of Arp 230, with overlaid HI contours. This galaxy shows evidence for shells in its outer bulge, which indicates a recent substantial accretion event, and also has a young gas-rich disk (D Schiminovich & J van Gorkom, private communication).

bulge-dominated systems—the red galaxies—were already formed at redshifts greater than unity, corresponding to a look-back time of greater than half of the age of the universe, or 5–10 Gyr (depending on cosmological parameters).

The high spatial resolution of the HST allows collection of morphological information. Schade et al (1995) obtained HST images for a subset (32 galaxies in total) of the CFHT redshift survey, mostly blue galaxies with $z > 0.5$. They found, in addition to the "normal" blue galaxies with exponential disks and spiral arms and red bulge-dominated galaxies, a significant population of high luminosity ($M_B < -20$) "blue nucleated galaxies" (BNG), with large bulge-to-disk ratio (B/T $\gtrsim 0.5$)—could these be bulges in formation, at look-back times of $\sim$5 Gyr? Small number statistics notwithstanding, most of the blue nucleated galaxies are asymmetric and show some suggestions of interactions. Schade et al (1996) found similar results for a larger sample, using just CFHT images for morphological classification, and confirmed that red galaxies tend to have high bulge-to-disk ratios.

Extending these results to even higher redshifts, and hence studies of progenitors of older present-day bulges, has been achieved by the identification of a sample of galaxies with $z \gtrsim 3$ based on a simple color criterion that selects systems with a Lyman-continuum break, superposed on an otherwise flat spectrum, redshifted into the optical (e.g. Steidel et al 1996a,b). Ground-based spectroscopy of 23 high-redshift candidates provided 16 galaxies at $z > 3$ (Steidel et al 1996b). The observed optical spectra probe the rest-frame 1400- to 1900-Å UV and provide a reasonable estimate of the reddening, and hence dust content, and of the star formation rate. The systems are inferred to be relatively dust-free, with the extinction at ~1600 Å typically ~1.7 mag, which corresponds to an optical reddening in the galaxies' rest-frame of $E(B-V) \sim 0.3$ mag. Whether the low dust content is a selection effect, perhaps due to fortuitous observational line of sight, or is a general feature of these high-redshift galaxies is not clear. The comoving space density of these systems is large—on the order of half that of bright ($L > L^*$, with $L^*$ the knee of the Schecter luminosity function) galaxies locally, which suggests that not too many of them can be hidden. The star formation rates, assuming a solar neighborhood IMF, are typically $\sim 10\ M_\odot$/year. There are interstellar absorption lines due to various chemical species; these lines may be interpreted as indicative of gas motions in a gravitational potential of characteristic velocity dispersion of ~200 km/s, which is typical of normal galaxies today.

Morphological information from optical HST images (Giavalisco et al 1996) for 19 Lyman-break candidates, of which 6 have confirmed redshifts, show that in the rest-frame UV (1400–1900 Å) these systems are mostly rather similar, in contrast to the wide range of morphological types seen at lower redshifts, $z \sim 1$, discussed above. Furthermore, the typical $z \sim 3$ galaxy selected this way is compact, at least in the UV, and has a half-light radius of ~2 kpc, which is reminiscent of present-day bulges in the optical. Some of these galaxies show faint surrounding emission that could be interpreted as "disks." The star formation rates inferred from the spectra build the equivalent of a bulge—say $10^{10}\ M_\odot$—over a few billion years, which spans the redshift range from $1 \sim z \sim 4$. Similar results are obtained from $z > 3$ samples derived from the HST Deep Field (Steidel et al 1996a) and for one galaxy at a redshift of $z = 3.43$, the central regions of which do, in fact, fit a de Vaucouleurs profile (Giavalisco et al 1995).

Thus, there is strong evidence that some (parts of some) bulges are formed at $z \gtrsim 3$. However, it is hard to draw definite conclusions about all bulges on the basis of these results because the observations at these redshifts can be biased. If, for example, half of all bulges form at $z \lesssim 0.5$, then we would simply not observe those at higher redshifts. At higher and higher redshifts, we would simply be selecting older and older bulges. Our conclusions would become strongly biased. This is very similar to the bias for early-type galaxies discussed by van Dokkum & Franx (1996).

## 5. FORMATION SCENARIOS

### 5.1 *Are Bulges Related to Their Haloes?*

Analyses of globular cluster systems in external galaxies conclude that they are more metal-poor in the mean than the underlying stellar light, at all radii in all galaxies (Harris 1991). It is worth noting that the Milky Way is sometimes considered an anomaly here, in that the metallicity distribution function for the (metal-poor, also known as halo) globular cluster system is not very different from that of field halo stars, with differences restricted to the wings of the distributions (e.g. Ryan & Norris 1991). It is important to note, however, that this comparison is done in the Milky Way at equivalent halo surface brightness levels well below those achievable in external galaxies. The higher surface brightness part of the Milky Way, that part which is appropriate to compare to similar studies in other galaxies, is the inner bulge. As discussed above, the metallicity there is well above that of the globular clusters. The Milky Way is typical. More importantly, this (single) test suggests the possibility that *all* spiral galaxies that have globular cluster systems have a corresponding field halo, which in turn is systematically more metal-poor and extended than is the more metal-rich observable bulge.

If this is true, the Local Group galaxies are typical, and the concept of "stellar halo" must be distinguished from that of "stellar bulge." In addition, although haloes seem ubiquitous, they are always of low luminosity and seem generally more extended than bulges. Bulges are not ubiquitous, as they are only found in earlier type galaxies, and cover a very wide range of luminosities. This is, in fact, clearly seen in the Hubble classification criteria from Sa to Sc types.

What is the evolutionary relationship, if any, between bulges and haloes? The Milky Way is an ideal case to study this because it has both bulge and halo. We noted above that the bulge is more metal-rich and possibly younger than the halo, contrary to the argument of Lee (1992). What of its dynamics?

In the Milky Way, the bulge stars do show significant net rotation (e.g. Ibata & Gilmore 1995b, Minniti et al 1995), but the very concentrated spatial distribution of these stars leads to low angular momentum orbits. Indeed, the angular momentum (per unit mass) distribution of the bulge is very similar to that of the stellar halo and very different from that of the disk (Wyse & Gilmore 1992, Ibata & Gilmore 1995b); see Figure 7. As discussed below, this is suggestive of the Eggen et al (1962) scenario, with the bulge as the central region of the

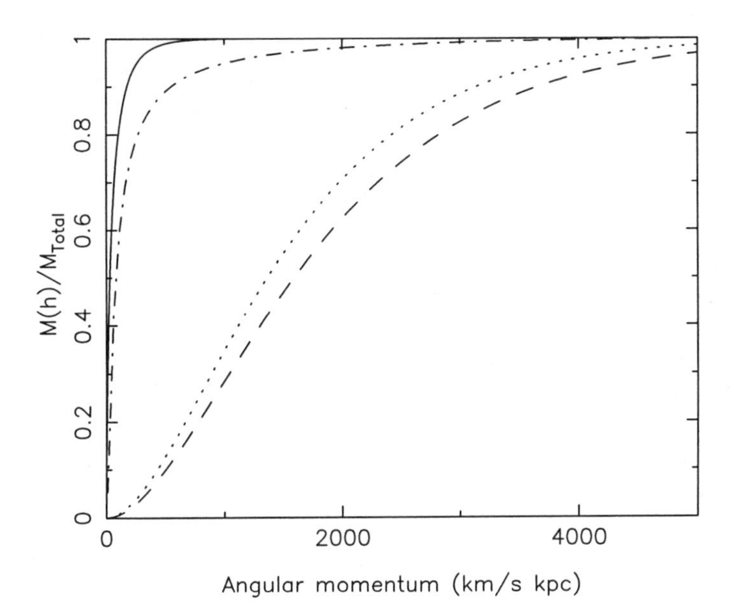


*Figure 7* Cumulative distribution functions of specific angular momentum for the four major Galactic stellar populations. The solid curve is the distribution for the bulge, from Ibata & Gilmore (1995b). The other curves are taken from Wyse & Gilmore (1992): The dashed-dotted curve represents the halo, the dotted curve represents the thick disk, and the dashed curve represents the thin disk. It is clear that the halo and bulge are more like each other than they are like the disk components.

halo but formed with significantly more dissipation. Furthermore, the available estimates of the masses of the stellar halo and bulge give a ratio of ~1:10, which is (coincidentally?) about the ratio predicted by models in which the bulge is built up by gas loss from star-forming regions in the halo (e.g. Carney et al 1990, Wyse 1995). The real test of this model is determination of the *rate* of formation and chemical enrichment of the stars in each of the halo and bulge. This is feasible and only requires good data on element ratios (e.g. Wyse & Gilmore 1992).

## 5.2 *Accretion/Merging*

DESTRUCTION OF DISKS BY MERGERS The current paradigm of structure formation in the universe is the hierarchical clustering of dominant dissipationless dark matter; galaxies as we see them form by the dissipation of gas into the potential wells of the dark matter, with subsequent star formation (e.g. Silk & Wyse 1993). The first objects to collapse under self-gravity are the highest density perturbations on scales which are characteristic of dwarf galaxies, and globular clusters, though globular clusters seem, on chemical evolution grounds, not to be the first objects to have formed. Large galaxies form by the merging of many smaller systems. The merging rate of the dissipationless dark haloes is reasonably straightforward to calculate (e.g. Lacey & Cole 1993). Unfortunately, many badly understood parameters are involved in the physics of gaseous heating/cooling and star formation, which determine how the baryonic components evolve. In the absence of understanding, the naive separation of different stellar components of galaxies is achieved by the following prescription (Baugh et al 1996, Kauffmann 1996): Star formation occurs in disks, which are destroyed during a merger with a significantly larger companion, with "significant" meaning a free parameter to be set by comparison with observations. In such a merger, all the extant "disk" stars are reassigned to the "bulge," the cold gas present is assumed to be driven to the center and fuel a burst of star formation, and a new disk is assumed to grow through accretion of intergalactic gas. Ellipticals are simply bare bulges, which are more likely in environments that prevent the subsequent reaccretion of a new disk—environments such as clusters of galaxies (e.g. Gunn & Gott 1972). One consequence (see Kauffmann 1996) of this prescription is that late-type spirals, which have a large disk-to-bulge ratio, should have older bulges than do early-type spirals, since to have a larger disk the galaxy must have been undisturbed and able to accrete gas for a longer time. This does not appear compatible with the observations discussed above. Bulge formation is highly likely to be more complex than this simple prescription.

ACCRETION OF DENSE STELLAR SATELLITES The central regions of galaxies are obvious repositories of accreted systems, as they are the bottom of the local potential well, provided that the accreted systems are sufficiently dense to survive tidal disruption while sinking to the center (e.g. Tremaine et al 1975). Should the accreted systems be predominately gaseous, then the situation is simply that described by Eggen et al (1962), with the chemical evolution modified to include late continuing infall. [It is worth noting that late infall of gas *narrows* resulting chemical abundance distribution functions (e.g. Edmunds 1990), and at least the Milky Way bulge has an observed very broad distribution.] We now consider models of bulge formation by accretion of small stellar systems.

As discussed above, the mean metallicity of the Galactic bulge is now reasonably well established at [Fe/H] $\sim -0.3$ dex (McWilliam & Rich 1994, Ibata & Gilmore 1995b, McWilliam 1997), with a significant spread below $-1$ dex and above solar. Thus, satellite galaxies that could have contributed significantly to the bulge are restricted to those of high metallicity. Given the fairly well-established correlation between mean metallicity and galaxy luminosity/velocity dispersion (e.g. Bender et al 1993, Lee et al 1993, Zaritsky et al

1994), only galaxies of luminosity comparable to the bulge can have been responsible. That is, one is immediately forced to a degenerate model, in which most of the stellar population of the bulge was accreted in one or a few mergers of objects like the Magellanic Clouds or the *most* luminous dwarf spheroidals (dSph). Because the metallicity distribution of the bulge is very broad, significantly broader than that of the solar neighborhood, a compromise model is viable, in which only the metal-poor tail of the bulge abundance distribution function has been augmented by accretion of lower luminosity satellite galaxies. Quantification of this statement awaits more robust measurement of the tails of the bulge metallicity distribution function and of appropriate element ratios.

Limits on the fraction of the bulge that has been accreted can be derived from stellar population analyses, following the approach utilized by Unavane et al (1996) concerning the merger history of the Galactic halo.

The Sagittarius dSph galaxy was discovered (Ibata et al 1994) through spectroscopy of a sample of stars selected purely on the basis of color and magnitude to contain predominantly K giants in the Galactic bulge. After rejection of foreground dwarf stars, the radial velocities isolated the Sagittarius dwarf galaxy member stars from the foreground bulge giants. The technique (serendipity) used to discover the Sagittarius dSph allows a real comparison between its stellar population and that of the bulge. Not only the radial velocities distinguish the dwarf galaxy, but also its stellar population—as seen in Figure 8 (taken from Ibata et al 1994), *all* giant stars redder than $B_J - R \gtrsim 2.25$ have kinematics that place them in the low velocity-dispersion component, i.e. in the Sagittarius dwarf. This is a real quantifiable difference between the *bulge* field population and this, the most metal-rich of the Galactic satellite dSph galaxies.

Furthermore, the carbon star population of the bulge can be compared with those of typical extant satellites. In this case, there is a clear discrepancy between the bulge and the Magellanic Clouds and dSph (Azzopardi & Lequeux 1992), in that the bulge has a significantly lower frequency of carbon stars.

Thus, although accretion may have played a role in the evolution of the bulge of the Milky Way, satellite galaxies like those we see around us now cannot have dominated. However, accretion is the best explanation for at least one external bulge—that of the apparently normal Sb galaxy NGC 7331, which is counter-rotating with respect to its disk (Prada et al 1996). It should also be noted that for S0 galaxies—those disk galaxies that at least in some models have suffered the most merging—Kuijken et al (1996) have completed a survey for counter-rotating components in the disks and found that only 1% of S0 galaxies contain a significant population of counter-rotating disk stars. This is a surprisingly low fraction and suggests some caution prior to adopting late merger models as a common origin of early-type systems.

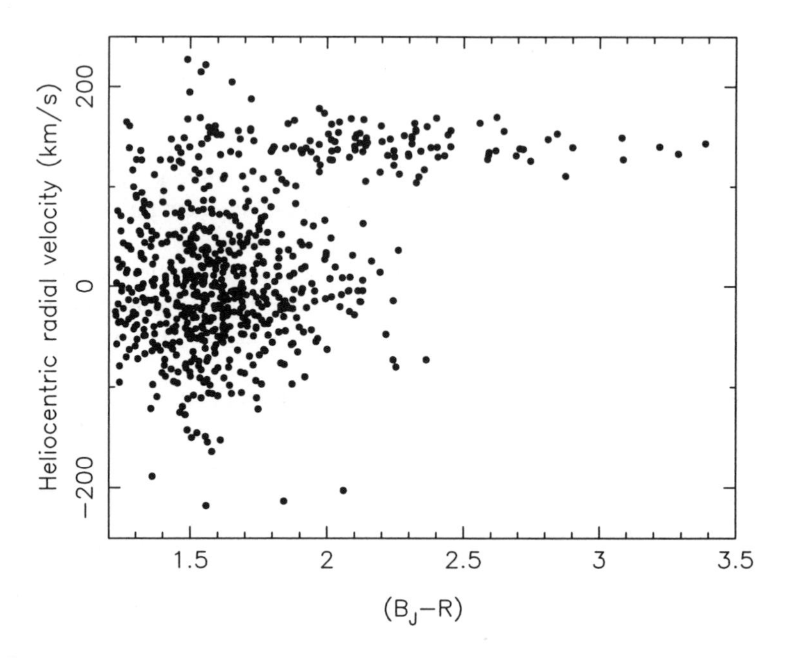


*Figure 8* Heliocentric radial velocities of the sample of stars observed by Ibata et al (1994), towards $\ell = -5°$, $b = -12°$, $-15°$, and $-20°$. The stars with velocities less than about 120 km/s are predominately bulge K giants. Those with velocities between about 120 and 180 km/s are members of the Sagittarius (Sgr) dSph galaxy, which was discovered from this figure. Note the real difference between the color distributions of bulge and Sgr members. Thus, the bulge cannot be built up by merger of several galaxies like the Sgr dwarf.

## 5.3 *Disk–Bars–Bulges, Etc*

Recall that the broadband color distributions of disk galaxies show smooth continuity across the transition between disk and bulge. In the mean, there is approximate equality between the colors of the inner disk and the bulge in any one galaxy (de Jong 1996, Peletier & Balcells 1996). These data may be interpreted as showing similar mean age and metallicity for inner disk and bulge (de Jong 1996, Peletier & Balcells 1996), but the degeneracy of age and of metallicity on the colors of stellar populations cause uncertainties (see, for example, Peletier & Balcells 1996). Courteau et al (1996) find further that the scale lengths of disk and bulge are correlated. They argue that this relationship implies that the bulge formed via secular evolution of the disk. In principle this is possible if disks are bar-unstable and bars are themselves unstable, and if significant angular-momentum transport is feasible.

The secular evolution of collisionless stellar disks has been studied in some detail recently, in particular through three-dimensional N-body simulations (Combes et al 1990, Raha et al 1991; see Combes 1994 and Pfenniger 1993 for interesting reviews). These simulations demonstrated that not only are thin disks often unstable to bar formation, but bars themselves can be unstable, in particular to deformations out of the plane of the disk, perhaps leading to peanut-shaped bulges. The kinematics of stars in "peanut bulges" lend some observational support for the association of peanut bulges with bars (Kuijken & Merrifield 1995). Thus stars initially in the inner disk end up in the bulge, which provides a natural explanation for the continuity observed in the properties of the stellar populations in disks and in bulges.

Merritt & Sellwood (1994; see also Merrifield 1996) provided a detailed description of the physics of instabilities of stellar disks. They demonstrated that the buckling instability of the stellar bar that produces a peanut bulge (Combes et al 1990, Raha et al 1991) is a collective phenomenon, similar to a forced harmonic oscillator. Thus the instability involves the bar in general, not only stars on special resonant orbits, as had been earlier proposed (e.g. Combes et al 1990). Not all instabilities form peanuts, which is just as well for this class of model for bulge formation, because, although box/peanut bulges are perhaps fairly common, comprising 20% of galaxies (Shaw 1987), the subset of these that rotate on cylinders is small (e.g. Shaw 1993 and references therein). Relevant photometric studies show that the light in a peanut bulge is additional to that in a smooth underlying disk, not subtracted from it (e.g. Shaw et al 1990, Shaw 1993), which rather weakens the case for these models.

The extant simulations of bar instabilities also find that a very small mass concentration at the center of the galaxy can destroy a bar. Such a mass concentration is very likely, since inflow, driven by gravitational torques, is probable after a bar is formed. Hasan & Norman (1990) suggested that a sufficiently large central mass concentration could eventually destroy the bar. Norman et al (1996) used three-dimensional N-body simulations to follow the evolution of a bar-unstable disk galaxy and attempted to incorporate the effects of gas inflow by allowing the growth of a very centrally concentrated component. Indeed, in time the fraction of material in this central component is sufficient to destroy the bar, fattening it into a "bulge-like" component. Bulges may be built up by successive cycles of disk instability–bar formation–bar dissolution (Hasan et al 1993). The time scales and duty cycles are not clear. Some simulations (e.g. Friedli 1994) find that as little as 1% of the mass in a central component is sufficient to dissolve a bar. This is a potential problem, as Miller (1996) points out, since the fact that one observes bars in around 50% of disk galaxies means that bars cannot be too fragile. A numerical example supporting Miller's important point is provided by Dehnen (1996), who finds that his bar is stable even with a cuspy density profile in the underlying disk. The simulations are clearly not yet mature.

A further potential problem with the general applicability of this scenario of bulge formation is the different light profiles of bars in galaxies of different bulge-to-disk ratio—early-type disk galaxies have bars with flat surface density profiles (e.g. Noguchi 1996, Elmegreen et al 1996), whereas late-type galaxies have bars with steeper surface brightness profiles than their disks. The Courteau et al correlation, that bulge scale lengths are around one-eighth that of disks, was found for a sample of late-type galaxies. In this scenario, the color of a bar should also be the same color as its surrounding disk, so that the subsequent bulge is the same color as the disk. While colors of bars are complicated by dust lanes and associated star formation, barred structures are often identified by means of color maps (e.g. Quillen et al 1996), suggesting problems for this class of model.

Specific counter-examples to models where the bulge forms through secular evolution of the inner disk are the high-luminosity but low surface brightness disk galaxies, such as Malin 1 (McGaugh et al 1995), which have apparently "normal" bulges (e.g. surface brightnesses and scale lengths typical of galaxies with high surface brightness disks) that clearly could not have formed by a disk instability.

Dissipationless formation of bulges from disks suffers yet a further problem, in that the phase space density of bulges is too high (Ostriker 1990, Wyse 1997). This also manifests itself in the fact that the spatial densities of bulges are higher than those of inner disks. Thus one must appeal to dissipational processes to form bulges, such as gas flows. The presence of color gradients in some external bulges would support a dissipative collapse with accompanying star formation (e.g. Balcells & Peletier 1994). Indeed, Kormendy (1993) has argued that many bulges are actually inner extensions of disks, formed through gas inflow from the disk, with later in situ star formation. This complicates the interpretation of the similarity between the colors of bulges and inner disks, which was a natural product of a stellar instability to form bulges from disk stars. One should note also that should bulges indeed not be formed at high redshift, then dissipation is also implicated in the production of the high spatial densities of their central regions.

It is also important to note that the term bar is used no less generically than is the term bulge. There is a fundamental, and rarely clarified, difference between a detectable perturbation to the luminosity distribution and a substantial $m = 2$ perturbation to the galactic gravitational potential. Inspection of the delightful pictures in the *Carnegie Atlas of Galaxies* (Sandage & Bedke 1994) suggests a continuum of structures, with all degrees of symmetry and asymmetry (i.e. $m =$

1, 2, ...) and relative amplitudes. When is a bar fundamentally more than the region where spiral arms meet the center? More important for the continuing debate about the center of the Milky Way, is it true that all these structures are seen in the cold disks only? Is there such a thing as a bar-bulge?

## 6. CONCLUSIONS

In the Local Group, all spiral galaxies, and probably all disk galaxies, have an old metal-poor spatially extended stellar population that we define to be a stellar halo. These seem to be the first stars formed in what would later become the galactic potential, though the possibility of later accretion of a *minor* fraction remains viable. The bulges of Local Group spiral galaxies are more diverse in properties, ranging from the very luminous, intermediate metallicity and very spatially extended bulge of M 31 through the intermediate luminosity, centrally concentrated bulge of the Milky Way, to no firm detection of a bulge in M 33.

In general, well-studied bulges are reasonably old, have a near-solar mean abundance, though with a very wide abundance distribution function, which is of importance, and are consistent with isotropic oblate rotator models for their kinematics, in which the basic support is provided by random motions and the flattening is consistent with additional rotational effects. Given these properties, bulges are most simply seen as the more dissipated descendents of their haloes.

However, diversity is apparent. All bulges of disk galaxies are not old, super–metal-rich, and simply small elliptical galaxies. This is not to say that such systems do not exist, but rather that bulges are heterogeneous. Higher luminosity bulges seem to have a closer affinity to ellipticals, whereas lower luminosity bulges prefer disks. But even this statement does not apply to all the properties of the stellar populations of bulges.

This diversity, together with the surprisingly limited database available concerning the photometric, structural, and kinematic properties of bulges, precludes firm conclusions. Much new and much needed data are about to become available, with the advent of 6- to 10-m class telescopes, with their exceptionally efficient spectrographs, and wide field array imaging systems on smaller telescopes. It will be interesting to see if the next review on bulges will be entitled "Disks and Ellipticals."

ACKNOWLEDGMENTS

RFGW and GG thank the North Atlantic Treaty Organization for a collaborative grant. RFGW acknowledges the support of the NASA Astrophysics Theory Program and the Seaver Institution, and she thanks the UC Berkeley Astronomy Department and Center for Particle Astrophysics, and the Institute of Astronomy, Cambridge, for hospitality during some of the writing of this review.



*Literature Cited*


Ajhar A, Grillmair C, Lauer T, Baum W, Faber S, et al. 1996. *Astron. J.* 111:1110–27

Alard C. 1996. PhD thesis. Univ. Paris VI

Andredakis YC, Peletier RF, Balcells M. 1995. *MNRAS* 275:874–88

Armandroff T. 1989. *Astron. J.* 97:375–89

Ashman K, Bird C. 1993. *Astron. J.* 106:2281–90

Azzopardi M, Lequeux J. 1992. In *The Stellar Populations of Galaxies, IAU Symp. 149*, ed. B Barbuy, A Renzini, pp. 201–6. Dordrecht: Kluwer

Azzopardi M, Lequeux J, Rebeirot E. 1988. *Astron. Astrophys.* 202:L27– L29

Baade W. 1944. *Ap. J.* 100:137–46

Bacon R, Emsellem E, Monnet G, Nieto J-L. 1994. *Astron. Astrophys.* 281:691–717

Balcells M, Peletier R. 1994. *Astron. J.* 107:135–52

Barnes J, Hernquist L. 1992. *Annu. Rev. Astron. Astrophys.* 30:705–42

Battistini P, Bonoli F, Casavecchia M, Ciotti L, Frederici L, Fusi-Pecci F. 1993. *Astron. Astrophys.* 272:77–97

Baugh CM, Cole S, Frenk CS. 1996. *MNRAS* 283:1361–78

Bender R, Burstein D, Faber S. 1993. *Ap. J.* 411:153–69

Bender R, Doebereiner S, Moellenhoff C. 1988. *Astron. Astrophys. Suppl.* 74:385–426

Bertelli G, Bressan N, Chiosi C, Fogatto F, Nasi E. 1994. *Astron. Astrophys. Suppl.* 106:275–302

Binney J, Gerhard O, Spergel D. 1997. *MNRAS* In press

Binney JJ, Gerhard O, Stark AA, Bally J, Uchida KI. 1991. *MNRAS* 252:210–18

Bissantz N, Englmaier P, Binney J, Gerhard O. 1997. *MNRAS* In press

Blanco V, Tendrup D. 1989. *Astron. J.* 98:843–52

Blitz L, Binney J, Lo KY, Bally J, Ho PTP. 1993. *Nature* 361:417–24

Blitz L, Spergel DN. 1991. *Ap. J.* 379:631–38

Blitz L, Teuben P, eds. 1996. In *Unsolved Problems of the Milky Way, IAU Symp. 169*. Dordrecht: Kluwer

Bothun G. 1992. *Astron. J.* 103:104–9

Bottema R. 1993. *Astron. Astrophys.* 275:16–36

Burton W, Hartmann D, West S. 1996. See Blitz & Teuben 1996, pp. 447–68

Buta R, Elmegreen BG, Crocker DA, eds. 1996. *Barred Galaxies, IAU Colloq. 157, ASP Conf. Ser. 91*. San Francisco: Astron. Soc. Pac.

Carney B, Latham D, Laird J. 1990. In *Bulges of Galaxies,* ed. B Jarvis, D Terndrup, pp. 127–35. Garching: Eur. Space Obs.

Catchpole R, Whitelock P, Glass I. 1990. *MNRAS* 247:479–90

Christian C. 1993. See Smith & Brodie 1993, pp. 448–57

Combes F. 1991. *Annu. Rev. Astron. Astrophys.* 29:195–238

Combes F. 1994. In *The Formation and Evolution of Galaxies,* ed. C Munoz-Turon, F Sanchez, pp. 317–98. Cambridge: Cambridge Univ. Press

Combes F, Debbasch F, Friedli D, Pfenniger D. 1990. *Astron. Astrophys.* 233:82–95

Courteau S, de Jong R, Broeils A. 1996. *Ap. J. Lett* . 457:L73–L76

Crotts A. 1986. *Astron. J.* 92:292–301

Davies RL, Efstathiou GP, Fall SM, Illingworth G, Schechter P. 1983. *Ap. J.* 266:41–57

Dehnen M. 1996. In *New Light on Galaxy Evolution*, ed. R Bender, RL Davies, p. 359. Dordrecht: Kluwer

de Jong R. 1995. PhD thesis. Univ. Groningen

de Jong R. 1996. *Astron. Astrophys.* 313:45–64

Dejonghe H, Habing H, eds. 1993. In *Galactic Bulges, IAU Symp. 153*. Dordrecht: Kluwer

de Vaucouleurs G, Pence W. 1978. *Astron. J.* 83:1163–74

de Zeeuw PT. 1995. See van der Kruit & Gilmore 1995, pp. 215–26

de Zeeuw PT, Franx M. 1991. *Annu. Rev. Astron. Astrophys.* 29:239–74

Edmunds M. 1990. *MNRAS* 246:678–87

Edvardsson B, Andersen J, Gustafsson B, Lambert DL, Nissen P, Tomkin J. 1993. *Astron. Astrophys.* 275:101–52

Eggen O, Lynden-Bell D, Sandage A. 1962. *Ap. J.* 136:748–66

Elmegreen B, Elmegreen D, Chromey F, Hasselbacher D, Bissell B. 1996. *Astron. J.* 111:2233–37

Elson R, Gilmore G, Santiago B. 1997. *MNRAS* In press

Emsellem E, Bacon R, Monnet G, Poulain P. 1996. *Astron. Astrophys.* 312:777–96

Faber SM. 1973. *Ap. J.* 179:731–54

Faber SM, Trager S, Gonzalez J, Worthey G. 1995. See van der Kruit & Gilmore 1995, pp. 249–58

Faber SM, Tremaine S, Ajhar EA, Byun Y-I, Burstein D, et al. 1997. *Ap. J.* In press

Franx M. 1993. See Dejonghe & Habing 1993, pp. 243–62

Franx M, Illingworth G. 1988. *Ap. J.* 327:L55–L59

Freeman KC. 1993. See Smith & Brodie 1993, pp. 27–38

Friedli D. 1994. In *Mass Transfer Induced Activity in Galaxies*, ed. I Shlosman, pp. 268–73. Cambridge: Cambridge Univ. Press

Friedli D, Wozniak H, Rieke M, Martinet L, Bratschi P. 1996. *Astron. Astrophys. Suppl.* 118:461–79

Friel E. 1995. *Annu. Rev. Astron. Astrophys.* 33:381–414

Frogel J. 1988. *Annu. Rev. Astron. Astrophys.* 26:51–92

Fusi-Pecci F, Cacciari C, Federici L, Pasquali A. 1993. See Smith & Brodie 1993, pp. 410–19

Genzel R, Thatte N, Krabbe A, Kroker H, Tacconi-Garman L. 1996. *Ap. J.* 472:153–72

Gerhard OE. 1988. *MNRAS* 232:P13–P20

Gerhard OE, Vietri M. 1986. *MNRAS* 223:377–89

Giavalisco M, Macchetto D, Madau P, Sparks B. 1995. *Ap. J. Lett.* 441:L13–L16

Giavalisco M, Steidel C, Macchetto D. 1996. *Ap. J.* 470:189–94

Gott JR. 1977. *Annu. Rev. Astron. Astrophys.* 15:235–66

Gredel R, ed. 1996. *The Galactic Center. ASP Conf. Ser.*, Vol. 102

Gunn J, Gott JR. 1972. *Ap. J.* 176:1–20

Harmon R, Gilmore G. 1988. *MNRAS* 235:1025–47

Harris W. 1991. *Annu. Rev. Astron. Astrophys.* 29:543–80

Hasan H, Norman C. 1990. *Ap. J.* 361:69–77

Hasan H, Pfenniger D, Norman C. 1993. *Ap. J.* 409:91–109

Holland S, Fahlman G, Richer HB. 1996. *Astron. J.* 112:1035–45

Holtzman JA, Light RM, Baum WA, Worthey G, Faber SM, et al. 1993. *Astron. J.* 106:1826–38

Houdashelt M. 1996. *PASP* 108:828

Huchra J. 1993. See Smith & Brodie 1993, pp. 420–31

Huchra J, Kent S, Brodie J. 1991. *Ap. J.* 370:495–504

Hughes SA, Wood P, Reid IN. 1991. *Astron. J.* 101:1304–23

Ibata R, Gilmore G. 1995a. *MNRAS* 275:591–604

Ibata R, Gilmore G. 1995b. *MNRAS* 275:605–27

Ibata R, Gilmore G, Irwin M. 1994. *Nature* 370:194–96

Izumiura H, Deguchi S, Hashimoto O, Nakada Y, Onaka T, et al. 1995. *Ap. J.* 453:837–63

Jablonka P, Martin P, Arimoto N. 1996. *Astron. J.* 112:1415–22

Kauffmann G. 1996. *MNRAS* 281:487–92

Kent S, Dame TM, Fazio G. 1991. *Ap. J.* 378:131–38

Kinman T, Suntzeff N, Kraft R. 1994. *Astron. J.* 108:1722–72

Kinman T, Wirtanen CA, Janes K. 1966. *Ap. J. Suppl.* 13:379–409

Kiraga M, Paczynski B, Stanek K. 1997. Preprint

Kormendy J. 1993. See Dejonghe & Habing 1993, pp. 209–30

Kormendy J, Bender R. 1996. *Ap. J. Lett* 464:L119–22

Kormendy J, Djorgovski S. 1989. *Annu. Rev. Astron. Astrophys.* 27:235–78

Kormendy J, Illingworth G. 1982. *Ap. J.* 256: 460–80

Kuijken K, Fisher D, Merrifield M. 1996. *MNRAS* 283:543–50

Kuijken K, Merrifield M. 1995. *Ap. J. Lett.* 443:L13–L16

Lacey C, Cole S. 1993. *MNRAS* 262:627–49

Larsen J, Humphreys R. 1994. *Ap. J. Lett.* 436:L149–52

Lee MG, Freedman W, Mateo M, Thompson I, Rath M, Ruiz M-T. 1993. *Astron. J.* 106:1420–62

Lee Y-W. 1992. *Astron. J.* 104:1780–89

Lee Y-W. 1993. See Smith & Brodie 1993, pp. 142–55

Lewis J, Freeman KC. 1989. *Astron. J.* 97:139–62

Lilly S, Tresse L, Hammer F, Crampton D, Le Fevre O. 1995. *Ap. J.* 455:108–24

Liszt H, Burton WB. 1996. See Blitz & Teuben 1996, pp. 297–310

Madsen C, Laustsen S. 1986. *ESO Messenger* 46:12

McElroy D. 1983. *Ap. J.* 270:485–506

McGaugh S, Schombert J, Bothun G. 1995. *Astron. J.* 109:2019–34

McLean I, Liu T. 1996. *Ap. J.* 456:499–503

McWilliam A. 1997. *Annu. Rev. Astron. Astrophys.* 35:503–56

McWilliam A, Rich M. 1994. *Ap. J. Suppl.* 91:749–91

Merrifield M. 1996. See Buta et al 1996, pp. 179–87

Merritt D, Sellwood J. 1994. *Ap. J.* 425:551–67

Mighell K, Rich RM. 1995. *Astron. J.* 110:1649–64

Mihos JC, Hernquist L. 1994. *Ap. J.* 425:L13–L16

Miller RH. 1996. See Buta et al 1996, pp. 569–74
Minniti D. 1996. *Ap. J.* 459:175–80
Minniti D, Olszewski E, Liebert J, White SDM, Hill JM, Irwin M. 1995. *MNRAS* 277:1293–311
Minniti D, Olszewski E, Rieke M. 1993. *Ap. J. Lett.* 410:L79–L82
Morris PW, Reid IN, Griffiths W, Penny AJ. 1994. *MNRAS* 271:852–74
Morrison H. 1993. *Astron. J.* 106:578–90
Morrison H, Harding P. 1993. See Dejonghe & Habing 1993, pp. 297–98
Mould JR, Kristian J. 1986. *Ap. J.* 305:591–99
Noguchi K. 1996. See Buta et al 1996, pp. 339–48
Norman C, Sellwood J, Hasan H. 1996. *Ap. J.* 462:114–24
Oort J, Plaut L. 1975. *Astron. Astrophys.* 41:71–86
Ortolani S, Renzini A, Gilmozzi R, Marconi G, Barbuy B, et al. 1995. *Nature* 377:701–4
Ostriker JP. 1990. In *Evolution of the Universe of Galaxies, ASP Conf. Ser.*, ed. R Kron, 10:25. San Francisco: Astron. Soc. Pac.
Paczynski B, Stanek KZ, Udalski A, Szymanski M, Kaluzny J, et al. 1994a. *Astron. J.* 107:2060–66
Paczynski B, Stanek KZ, Udalski A, Szymanski M, Kaluzny J, et al. 1994b. *Ap. J. Lett.* 435:L113–16
Peebles PJE. 1989. In *The Epoch of Galaxy Formation*, ed. CS Frenk, R Ellis, T Shanks, A Heavens, J Peacock, pp. 1–14. Dordrecht: Kluwer
Peletier R, Balcells M. 1996. *Astron. J.* 111:2238–42
Perault M, Omont A, Simon G, Seguin P, Ojha D, Blommaert J, et al. 1996. *Astron. Astrophys* . 315:L165–68
Pfenniger D. 1993. See Dejonghe & Habing 1993, pp. 387–90
Phillips A, Illingworth G, Mackenty J, Franx M. 1996. *Astron. J.* 111:1566–74
Prada F, Gutierrez CM, Peletier RF, McKeith CD. 1996. *Ap. J. Lett* . 463:L9–L12
Pritchet CJ. 1988. In *The Extragalactic Distance Scale, ASP Conf. Ser.*, ed. C Pritchet, S van den Bergh, 4:59–68. San Francisco: Astron. Soc. Pac.
Pritchet CJ, van den Bergh S. 1987. *Ap. J.* 316:517–29
Pritchet CJ, van den Bergh S. 1988. *Ap. J.* 331:135–44
Pritchet CJ, van den Bergh S. 1996. See Blitz & Teuben 1996, pp. 39–46
Quillen A, Ramirez S, Frogel J. 1996. *Ap. J.* 470:790–96
Raha A, Sellwood J, James R, Kahn FD. 1991. *Nature* 352:411–12
Regan M, Vogel S. 1994. *Ap. J.* 434:536–45
Renzini A. 1995. See van der Kruit & Gilmore 1995, pp. 325–36
Rich RM. 1988. *Astron. J.* 95:828–65
Rich RM. 1990. *Ap. J.* 362:604–19
Rich RM, Mighell K. 1995. *Ap. J.* 439:145–54
Rich RM, Mighell K, Freedman W, Neill J. 1996. *Astron. J.* 111:768–76
Richer HB, Harris WE, Fahlman GG, Bell RA, Bond HE, et al. 1996. *Ap. J.* 463:602–8
Rix H-W, White SDM. 1990. *Ap. J.* 362:52–58
Ryan S, Norris J. 1991. *Astron. J.* 101:1865–79
Sadler E, Rich RM, Terndrup D. 1996. *Astron. J.* 112:171–85
Sandage A. 1986. *Annu. Rev. Astron. Astrophys.* 24:421–58
Sandage A, Bedke J. 1994. *The Carnegie Atlas of Galaxies.* Carnegie Inst. Wash. Publ. 638. 750 pp.
Santiago B, Elson R, Gilmore G. 1996. *MNRAS* 281:1363–74
Schade D, Lilly S, Crampton D, Hammer F, Lefevre O, Tresse L. 1995. *Ap. J. Lett.* 451:L1–L4
Schade D, Lilly S, Lefevre O, Hammer F, Crampton D. 1996. *Ap. J.* 464:79–91
Schmidt A, Bica E, Alloin D. 1990. *MNRAS* 243:620–28
Schommer RA. 1993. See Smith & Brodie 1993, pp. 458–68
Schommer RA, Cristian C, Caldwell N, Bothun G, Huchra J. 1991. *Astron. J.* 101:873–83
Schweizer F. 1990. In *Dynamics and Interactions of Galaxies*, ed. R Wielen, pp. 60–71. Berlin: Springer-Verlag
Shaw M. 1987. *MNRAS* 229:691–706
Shaw M. 1993. *MNRAS* 261:718–52
Shaw M, Axon D, Probst R, Gately I. 1995. *MNRAS* 274:369–87
Shaw M, Dettmar R, Bartledress A. 1990. *Astron. Astrophys* . 240:36–51
Silk J, Wyse RFG. 1993. *Phys. Rep.* 231:293–67
Smith G, Brodie J, eds. 1993. In *The Globular Cluster – Galaxy Connection.* San Francisco: Astron. Soc. Pac.
Sofue Y, Yoshida S, Aoki T, Soyano T, Tarusawa K, et al. 1994. *PASJ* 46:1–7
Stark AA. 1977. *Ap. J.* 213:368–73
Stark AA, Binney J. 1994. *Ap. J. Lett.* 426:L31–L33
Steidel C, Giavalisco M, Dickinson M, Adelberger KL. 1996a. *Astron. J.* 112:352–58
Steidel C, Giavalisco M, Pettini M, Dickinson M, Adelberger KL. 1996b. *Ap. J. Lett.* 462:L17–L21
Tinsley BM. 1980. *Fundam. Cosmic Phys.* 5:287–388
Toomre A. 1977. In *The Evolution of Galaxies and Stellar Populations*, ed. RB Larson, B

Tinsley, pp. 401–416. New Haven: Yale Univ. Obs.
Tremaine S. 1995. *Astron. J.* 110:628–34
Tremaine S, Henon M, Lynden-Bell D. 1986. *MNRAS* 219:285–97
Tremaine SD, Ostriker JP, Spitzer L. 1975. *Ap. J.* 196:407–11
Unavane M, Wyse RFG, Gilmore G. 1996. *MNRAS* 278:727–36
van Albada T. 1982. *MNRAS* 201:939–55
van den Bergh S. 1991a. *PASP* 103:609–22
van den Bergh S. 1991b. *PASP* 103:1053–68
van der Kruit, Gilmore G, eds. 1995. *Stellar Populations, IAU Symp. 164.* Dordrecht: Kluwer
van Dokkum PG, Franx M. 1996. *MNRAS* 281:985–1000
Walterbos R, Kennicutt R. 1988. *Astron. Astrophys.* 98:61–86
Weiland JL, Arendt RG, Berriman GB, Dwek E, Freundenreich HT, et al. 1994. *Ap. J.* 425:L81–L84
Westerlund B, Lequeux J, Azzopardi M, Rebeirot E. 1991. *Astron. Astrophys.* 244:367–72
Whitford A. 1978. *Ap. J.* 226:777–89
Whitford A. 1986. *Annu. Rev. Astron. Astrophys.* 24:1–22
Whitmore BC, Schweizer F, Leitherer C, Borne K, Robert C. 1993. *Astron. J.* 106:1354–70
Wyse RFG. 1995. See van der Kruit & Gilmore 1995, pp. 133–50
Wyse RFG. 1997. Preprint
Wyse RFG, Gilmore G. 1988. *Astron. J.* 95:1404–14
Wyse RFG, Gilmore G. 1989. *Comments Astrophys.* 8:135–44
Wyse RFG, Gilmore G. 1992. *Astron. J.* 104:144–53
Wyse RFG, Gilmore G. 1995. *Astron. J.* 110:2771–87
Zaritsky D, Kennicutt R, Huchra J. 1994. *Ap. J.* 420:87–109
Zhao H, Spergel DN, Rich RM. 1994. *Astron. J.* 108:2154–63

# SUBJECT INDEX

## A

# B



# C

# D

# E



# F



# G

# H

# I

# N

# O

# S

# T

# CUMULATIVE INDEXES

## CONTRIBUTING AUTHORS, VOLUMES 25–35

# CHAPTER TITLES, VOLUMES 25–35